Lecture Notes in Networks and Systems

Volume 868

The series "Lecture Notes in Networks and Systems" publishes the latest developments in Networks and Systems—quickly, informally and with high quality. Original research reported in proceedings and post-proceedings represents the core of LNNS.

Volumes published in LNNS embrace all aspects and subfields of, as well as new challenges in, Networks and Systems.

The series contains proceedings and edited volumes in systems and networks, spanning the areas of Cyber-Physical Systems, Autonomous Systems, Sensor Networks, Control Systems, Energy Systems, Automotive Systems, Biological Systems, Vehicular Networking and Connected Vehicles, Aerospace Systems, Automation, Manufacturing, Smart Grids, Nonlinear Systems, Power Systems, Robotics, Social Systems, Economic Systems and other. Of particular value to both the contributors and the readership are the short publication timeframe and the world-wide distribution and exposure which enable both a wide and rapid dissemination of research output.

The series covers the theory, applications, and perspectives on the state of the art and future developments relevant to systems and networks, decision making, control, complex processes and related areas, as embedded in the fields of interdisciplinary and applied sciences, engineering, computer science, physics, economics, social, and life sciences, as well as the paradigms and methodologies behind them.

Indexed by SCOPUS, INSPEC, WTI Frankfurt eG, zbMATH, SCImago.

All books published in the series are submitted for consideration in Web of Science.

For proposals from Asia please contact Aninda Bose (aninda.bose@springer.com).

Sandeep Kumar · Balachandran K. ·
Joong Hoon Kim · Jagdish Chand Bansal
Editors

Fourth Congress on Intelligent Systems

CIS 2023, Volume 1

Springer

Editors
Sandeep Kumar
Department of Computer Science and Engineering
Christ (Deemed to be University)
Bengaluru, India

Balachandran K.
Department of Computer Science and Engineering
Christ (Deemed to be University)
Bengaluru, India

Joong Hoon Kim
School of Civil, Environmental and Architectural Engineering, Anam-ro
Anam-dong
Korea University
Seoul, Korea (Republic of)

Jagdish Chand Bansal
South Asian University
New Delhi, Delhi, India

ISSN 2367-3370 ISSN 2367-3389 (electronic)
Lecture Notes in Networks and Systems
ISBN 978-981-99-9036-8 ISBN 978-981-99-9037-5 (eBook)
https://doi.org/10.1007/978-981-99-9037-5

This Springer imprint is published by the registered company Springer Nature Singapore Pte Ltd.
The registered company address is: 152 Beach Road, #21-01/04 Gateway East, Singapore 189721, Singapore

Preface

This book contains outstanding research papers as the proceedings of the 4th Congress on Intelligent Systems (CIS 2023), held on September 4–5, 2023, at CHRIST (Deemed to be University), Bengaluru, India, under the technical sponsorship of the Soft Computing Research Society, India. The conference is conceived as a platform for disseminating and exchanging ideas, concepts, and results from academia and industry researchers to develop a comprehensive understanding of the challenges of intelligence advancements in computational viewpoints. This book will help in strengthening congenial networking between academia and industry. We have tried our best to enrich the quality of the CIS 2023 through the stringent and careful peer-review process. This book presents novel contributions to Intelligent Systems and is a reference material for advanced research in medical imaging and health informatics, agriculture, and engineering applications.

We have tried our best to enrich the quality of the CIS 2023 through a stringent and careful peer-review process. CIS 2023 received many technical contributed articles from distinguished participants from home and abroad. CIS 2023 received 870 research submissions from different countries. After a very stringent peer-reviewing process, only 104 high-quality papers were finally accepted for presentation and the final proceedings.

This book presents the first volume of 35 data science and applications research papers and serves as a reference material for advanced research.

Bengaluru, India — Sandeep Kumar
Bengaluru, India — Balachandran K.
Seoul, Korea (Republic of) — Joong Hoon Kim
New Delhi, India — Jagdish Chand Bansal

Contents

Editors and Contributors

About the Editors

Dr. Sandeep Kumar is a professor at CHRIST (Deemed to be University), Bengaluru. He recently completed his post-doctoral research at Imam Mohammad Ibn Saud Islamic University, Riyadh, Saudi Arabia, in sentiment analysis. He is an associate editor for Springer's *Human-Centric Computing and Information Sciences (HCIS)* journal. He has published over a hundred research papers in various international journals/conferences and attended several national and international conferences and workshops. He has authored/edited 12 books in the area of computer science. Also, he has been serving as General Chair of the Congress on Intelligent Systems (CIS), the International Conference on Communication and Computational Technologies (ICCCT), the International Conference on Sustainable and Innovative Solutions for Current Challenges in Engineering & Technology (ICSISCET) and the IEEE International Conference on Contemporary Computing and Communications (InC4). His research interests include nature-inspired algorithms, swarm intelligence, soft computing, and computational intelligence.

Dr. Balachandran K. is formerly Professor and Head of Computer Science and Engineering at Christ (Deemed to be University), Bengaluru, India. He has 38 years of experience in Research, Academia, and Industry. He served as Senior Scientific Officer in the Research and Development Unit of the Department of Atomic Energy for 20 years. His research interest includes Data Mining, Artificial Neural Networks, Soft Computing, and Artificial Intelligence. He has published more than 50 articles in well-known SCI/SCOPUS-indexed international journals and conferences and attended several national and international conferences and workshops. He has authored/edited four books in the area of computer science.

Prof. Joong Hoon Kim formerly Dean of Engineering College of Korea University, obtained his Ph.D. degree from the University of Texas at Austin in 1992 with the thesis title "Optimal replacement/rehabilitation model for water distribution systems". Professor Kim's major areas of interest include Smart Water Management Systems and Micro Water Grid Systems in Smart Cities, Design and Management of Water Distribution Systems, Maintenance and Management Techniques of Urban Drainage Systems, Real-time Management of Urban Water Systems Involving Smart Monitoring, Telemetered Systems, and Sensor Networks, Development of Alarm System against Urban Flood, Development of Optimal Decision Making Systems about Management of Water Supply and Sewage Networks. His publication includes "A New Heuristic Optimization Algorithm: Harmony Search", Simulation, February 2001, Vol. 76, pp. 60–68, cited over 7200 times by other journals of diverse research areas. He has been awarded the National Study Abroad Scholarship, Korean Ministry of Education, 1985.05; Quentin Mees Research Award, Arizona Water and Pollution Control Association 1994.03, Korea Water Resources Association Achievement Award 2005.12; National Emergency Management Agency Achievement Award 2007.11; Korea Minister of Government Administration and Home Affairs Award 2012.02, KSCE-Springer Award, 2013.12, Songsan Grand Prize (Academic) by KSCE, 2019.10, Order of Science and Technological merit, 2021.04, K-water Grand Academic Prize, 2022.12. He has been on the faculty of the School of Civil, Environmental, and Architectural Engineering at Korea University since 1993 and is now the professor emeritus. He has hosted international conferences, including APHW 2013, ICHSA 2014 and 2015, HIC 2016, HIC 2018, ICHSA 2019, ICHSA 2020, and ICHSA 2022, and has given keynote speeches at many international conferences, including AOGS 2013, GCIS 2013, SocPros 2014 & 2015, SWGIC 2017, RTORS 2017, ICHSA 2020, ICCIS 2020, and FSACE 2021. He is a member of the National Academy of Engineering of Korea since 2017.

Dr. Jagdish Chand Bansal is an Associate Professor (Senior Grade) at South Asian University New Delhi and a Visiting Faculty at Maths and Computer Science Liverpool Hope University UK. He also holds a visiting professorship at NIT Goa, India. Dr. Bansal obtained his Ph.D. in Mathematics from IIT Roorkee. Before joining SAU New Delhi, he worked as an Assistant Professor at ABV-Indian Institute of Information Technology and Management Gwalior and BITS Pilani. His Primary area of interest is Swarm Intelligence and Nature Inspired Optimization Techniques. Recently, he proposed a fission-fusion social structure-based optimization algorithm, Spider Monkey Optimization (SMO), which is being applied to various problems in the engineering domain. He has published over 70 research papers in various international journals/conferences. He is the Section Editor (editor-in-chief) of the journal *MethodsX* published by Elsevier. He is the series editor of the book series *Algorithms for Intelligent Systems (AIS)*, *Studies in Autonomic, Data-driven and Industrial Computing (SADIC)*, and *Innovations in Sustainable Technologies and Computing (ISTC)* published by Springer. He is also the Associate Editor of Engineering Applications of Artificial Intelligence (EAAI) and ARRAY, published by Elsevier. He is the general secretary of the Soft Computing Research Society (SCRS). He has also received Gold Medal at UG and PG levels.

Contributors

K. A. Abdul Nazeer National Institute of Technology, Calicut, India

M. Abinesh Department of Aeronautical Engineering, ACS College of Engineering, Bangalore, India

Shajal Afaq Centre for Advanced Studies, AKTU, Lucknow, Uttar Pradesh, India

P. S. Ambika Center for Computational Engineering and Networking (CEN), Amrita School of Artificial Intelligence, Amrita Vishwa Vidyapeetham, Coimbatore, India

T. Ambika Department of Aeronautical Engineering, ACS College of Engineering, Bangalore, India

Milos Antonijevic Singidunum University, Belgrade, Serbia

V. Aparna National Institute of Technology, Calicut, India

C. Arul Murugan Karpagam College of Engineering, Coimbatore, India

Senthil Kumar Arumugam Christ University, Bangalore, India

Shree Harsh Attri Department of Computer Science and Engineering, Sharda School of Engineering and Technology, Sharda University, Greater Noida, UP, India

Nebojsa Bacanin Singidunum University, Belgrade, Serbia

Abou Bakary Ballo LaMI, Université Felix Houphouët-Boigny, Abidjan, Côte d'Ivoire;
LMI Université Péléforo Gon Coulibaly, Korhogo, Côte d'Ivoire

Poonam Bansal Department of Artificial Intelligence and Data Sciences, IGDTUW, New Delhi, India

Daniel Bell Robotics Lab, School of Mathematics, Computer Science and Engineering, Liverpool Hope University, Liverpool, UK

Kesari Eswar Bhageerath Computer Science and Engineering, Gayatri Vidya Parishad College of Engineering (Autonomous), Madhurawada, Visakhapatnam, Andhra Pradesh, India

Rajesh Bhatia Punjab Engineering College (Deemed to be University), Chandigarh, India

Aleksandra Bozovic Technical Faculty "Mihajlo Pupin", University of Novi Sad, Zrenjanin, Serbia

Subhadeep Chakrarbarti Amity University Jharkhand, Ranchi, India

Lekshmy P. Chandran National Institute of Technology, Calicut, India

Kundan Kumar Chaubey School of Applied and Life Sciences, Uttaranchal University, Dehradun, Uttarakhand, India;
School of Applied and Life Sciences, Sanskriti University, Mathura, Uttar Pradesh, India

Ganesh Kumar Chellamani Department of Electronics and Communication Engineering, Amrita School of Engineering, Chennai, Amrita Vishwa Vidyapeetham, Chennai, India

Jyothsna Cherapanamjeri Jawaharlal Technological University, Anantapur, India

Adama Coulibaly LaMA Université Felix Houphouët-Boigny, Abidjan, Côte d'Ivoire

Sarat C. Dass School of Mathematical & Computer Sciences, Heriot-Watt University, Putrajaya, Malaysia

Subhojit Dawn Department of Electrical and Electronics Engineering, Velagapudi Ramakrishna Siddhartha Engineering College, Vijayawada, India

Eleonora Desnica Technical Faculty "Mihajlo Pupin", University of Novi Sad, Zrenjanin, Serbia

K. Dhananjay Rao Department of Electrical and Electronics Engineering, Velagapudi Ramakrishna Siddhartha Engineering College, Vijayawada, India

Ritika Dhaneshwar University Institute of Engineering and Technology, Panjab University, Chandigarh, India

R. Dharani Department of IT, Panimalar Engineering College, Chennai, India

Moustapha Diaby Lastic, Ecole Supérieure Africaine des Technologies de l'Information et de la Communication, Abidjan, Côte d'Ivoire

Lynette D'mello Dwarkadas J. Sanghvi College of Engineering, Mumbai, India

Gauravkumarsingh Gaharwar Navrachana University, Vadodara, Gujarat, India

Prakash Garia School of Management, Graphic Era Hill University, Bhimtal, Uttarakhand, India

Georgi Georgiev South-West University "Neofit Rilski", Blagoevgrad, Bulgaria

Urvashi Goyal Sanjay Gandhi Postgraduate Institute of Medical Sciences, Lucknow, Uttar Pradesh, India

D. N. D. Harini Computer Science and Engineering, Gayatri Vidya Parishad College of Engineering (Autonomous), Madhurawada, Visakhapatnam, Andhra Pradesh, India

S. Harsha Department of Artificial Intelligence and Machine Learning, RNS Institute of Technology, Visvesveraya Technological University, Belagavi, India

K. V. Hridika National Institute of Technology, Calicut, India

Navin K. Ipe Computer Science and Engineering, M. S. Ramaiah University of Applied Sciences, Bangalore, India

Anamika Jain Dr. Vishwanath Karad, MIT-World Peace University, Kothrud, Pune, Maharashtra, India

R. Jeena Department of CSE, Saveetha School of Engineering, Saveetha Institute of Medical and Technical Sciences, Chennai, India

Saju P. John Department of Computer Science and Engineering, Jyothi Engineering College, Cheruthuruthy, Kerala, India

Luka Jovanovic Singidunum University, Belgrade, Serbia

N. Juber Rahman Computer Science, PSG College of Arts and Science, Coimbatore, India

A. Kalavani Rajalakshmi Engineering College, Chennai, India

Soumen Kanrar Amity University Jharkhand, Ranchi, India

Mandeep Kaur University Institute of Engineering and Technology, Panjab University, Chandigarh, India

S. Kiruthiga Department of ECE, Saranathan College of Engineering, Trichy, India

R. Kottaimalai Kalasalingam Academy of Research and Education, Krishnankoil, India

Anjan K. Koundinya Department of Information Science and Engineering (Cyber Security), BMS Institute of Technology and Management, Visvesveraya Technological University, Belagavi, India

Raghavendra V. Kulkarni Computer Science and Engineering, Graphic Era Deemed to be University, Dehradun, India

K. R. Pruthvi Kumar Department of Computer Science and Engineering, BMS Institute of Technology and Management, Visvesveraya Technological University, Belagavi, India

Ritika Kumari Department of Artificial Intelligence and Data Sciences, IGDTUW, New Delhi, India;
USICT, Guru Gobind Singh Indraprastha University, New Delhi, India

K. Likitha Shree Department of Aeronautical Engineering, ACS College of Engineering, Bangalore, India

Diarra Mamadou LaMI, Université Felix Houphouët-Boigny, Abidjan, Côte d'Ivoire

Joseph P. Mani Modern College of Business and Science, Muscat, Sultanate of Oman

Ashapurna Marndi Council of Scientific and Industrial Research-Fourth Paradigm Institute, Bangalore, Karnataka, India;
Academy of Scientific and Innovative Research, Ghaziabad, Uttar Pradesh, India

S. Marulkar MATS School of IT, MATS University, Raipur, CG, India

R. Mente School of Computational Sciences, PAH Solapur University, Solapur, India

J. Midhunchakkaravarthy Lincoln University College, Petaling Jaya, Selangor, Malaysia

Amit Mittal Department of Allied Science, Graphic Era Hill University, Bhimtal, Uttarakhand, India

R. Mukesh Department of Aerospace Engineering, ACS College of Engineering, Bangalore, India

Lalasa Mukku CHRIST (Deemed to be University) Kengeri, Bangalore, India

S. Mythili Department of ECE, PSNA College of Engineering and Technology, Dindigul, India

G. S. Nagaraja Department of Computer Science and Engineering, RV College of Engineering, Visvesveraya Technological University, Belagavi, India

Pooja S. Nair National Institute of Technology, Calicut, India

Nikitha Reddy Nalla Department of Electronics and Communication Engineering, Amrita School of Engineering, Chennai, Amrita Vishwa Vidyapeetham, Chennai, India

B. Narain MATS School of IT, MATS University, Raipur, CG, India

B. Narendra Kumar Rao Sri Vidyanikethan Engineering College, Tirupathi, India

Ivan Nedyalkov South-West University "Neofit Rilski", Blagoevgrad, Bulgaria

P. Nithya Networking and Mobile Application, PSG College of Arts and Science, Coimbatore, India

P. S. Niveditha APJ Abdul Kalam Technological University, Thiruvananthapuram, Kerala, India

Ganesh Pai NMAM Institute of Technology-Affiliated to NITTE (Deemed to be University), Nitte, Karnataka, India

Sharnil Pandya Symbiosis International University, Pune, Maharashtra, India

Pooja Department of Aeronautical Engineering, ACS College of Engineering, Bangalore, India

Mukul Prabhu Dwarkadas J. Sanghvi College of Engineering, Mumbai, India

Tatiparti B. Prasad Reddy SENSE, Vellore Institute of Technology, Chennai, India

M. Prem Kumar Department of Electrical and Electronics Engineering, Velagapudi Ramakrishna Siddhartha Engineering College, Vijayawada, India

Shefali Raina Vasantdada Patil Pratisthan's College of Engineering and Visual Arts Sion, Mumbai, Maharashtra, India

Ranjana Department of Computer Science and Engineering, Guru Jambheshwar University of Science and Technology, Hisar, Haryana, India

Harish Rawat Department of Botany, Dhanauri P. G. College Dhanauri, Haridwar, Uttarakhand, India

Vidhi Ruparel Dwarkadas J. Sanghvi College of Engineering, Mumbai, India

Shruti Sahu Department of Artificial Intelligence and Data Sciences, IGDTUW, New Delhi, India

N. Sai Dhanush Center for Computational Engineering and Networking (CEN), Amrita School of Artificial Intelligence, Amrita Vishwa Vidyapeetham, Coimbatore, India

Aarti Sardhara Vishwakarma University Pune, Pune, Maharashtra, India

Madhvi Saxena Vishwakarma University Pune, Pune, Maharashtra, India

Emanuele Lindo Secco Robotics Lab, School of Mathematics, Computer Science and Engineering, Liverpool Hope University, Liverpool, UK

N. Senthamilarasi Department of CSE, Sathyabama Institute of Science and Technology, Chennai, India

Sonia Setia Department of Computer Science and Engineering, Sharda School of Engineering and Technology, Sharda University, Greater Noida, UP, India

Rushabh Shah Dwarkadas J. Sanghvi College of Engineering, Mumbai, India

Aarti Mehta Sharma Symbiosis Centre for Management Studies, Bengaluru Campus, Symbiosis International (Deemed University), Pune, India

Shweta Sharma College of Biotechnology, DUVASU, Mathura, Uttara Pradesh, India

M. Sharmila Kumari P. A. College of Engineering-Affiliated to VTU, Mangalore, Karnataka, India

Suman Shit Amity University Jharkhand, Ranchi, India

Chintala Rahulsai Shrinivas Punjab Engineering College (Deemed to be University), Chandigarh, India

Rishi Pal Singh Department of Computer Science and Engineering, Guru Jambheshwar University of Science and Technology, Hisar, Haryana, India

Aashna Sinha School of Applied and Life Sciences, Uttaranchal University, Dehradun, Uttarakhand, India

G. Soniya Priyatharsini Department of CSE, Dr.MGR Educational and Research Institute, Chennai, India

Gummadi Srinivasa Rao Department of Electrical and Electronics Engineering, Velagapudi Ramakrishna Siddhartha Engineering College, Vijayawada, India

Sasidhar Babu Suvanam Department of Computer Science and Engineering, Presidency University, Bangalore, India

Arnav Taya University Institute of Engineering and Technology, Panjab University, Chandigarh, India

M. Thilagaraj MVJ College of Engineering, Bengaluru, India

Jyothi Thomas CHRIST (Deemed to be University) Kengeri, Bangalore, India

Meena Ugale Lincoln University College, Petaling Jaya, Selangor, Malaysia

Naramula Venkatesh SR UNIVERSITY Warangal, Warangal, India

M. Vijay Department of Aerospace Engineering, ACS College of Engineering, Bangalore, India

Vydeki SENSE, Vellore Institute of Technology, Chennai, India

Shruti Wadhwa Punjab Engineering College (Deemed to be University), Chandigarh, India

Suhail Rashid Wani Department of Computer Science and Engineering, Sharda School of Engineering and Technology, Sharda University, Greater Noida, UP, India

Miodrag Zivkovic Singidunum University, Belgrade, Serbia

A Simple Way to Predict Heart Disease Using AI

Soumen Kanrar, Suman Shit, and Subhadeep Chakrarbarti

Abstract Early diagnosis of cardiovascular diseases in high-risk patients will help them make decisions about lifestyle changes and, in turn, minimise their complications. Due to asymptomatic illnesses like cardiovascular diseases, healthcare costs are exceeding the average national medical treatment cost and corporate budgets. The need for early identification and treatment of such diseases is critical. One of the developments in machine learning is the technology that has been used for disease prediction in many fields around the world, including the healthcare industry. Analysis has been attempted to classify the most influential heart disease causes and to reliably predict the overall risk using homogeneous techniques of data mining. This paper uses machine learning algorithms and selects the best one based on its classification report to find a simple way to predict heart disease. It helps to develop cost-effective software to predict heart disease for the betterment of mankind.

Keywords Heart disease · Chest pain · Cardiovascular · Machine learning · AdaBoost · Random forest

1 Introduction

Recent research has delved into amalgamating these techniques using methods such as hybrid data mining algorithms. To provide an approximate approach for the heart disease prediction, this paper has considered a rule-based model to compare the performance of applying regulations to the individual effects of classification and regression machine learning techniques. This work will follow these steps in order to develop a heart disease classifier. The well-known UCI heart disease dataset is used to develop the heart disease classifier. The purpose of this study is to classify the most important cardiac disease predictors and use classification and regression algorithms to predict the overall risks [1]. The classification algorithms in machine learning are thus used to classify the predictors [2]. Furthermore, to verify models,

S. Kanrar (✉) · S. Shit · S. Chakrarbarti
Amity University Jharkhand, Ranchi, India
e-mail: dr.soumen.kanrar@gmail.com

S. Kumar et al. (eds.), *Fourth Congress on Intelligent Systems*, Lecture Notes in Networks and Systems 868, https://doi.org/10.1007/978-981-99-9037-5_1

researchers conduct data analysis in Python using Jupyter Notebook [3]. Health policies and programs are frequently recommended based on the doctor's instructions rather than knowledge-rich data. Sometimes, this results in errors and high expenses, both of which have an impact on the quality of medical treatments. Clinical choices can be improved by using analytic tools and data modelling. As a result, the goal is to create a web application that will guide doctors in diagnosing cardiac disorders [4]. The main advantage is that if the doctors can make a proper diagnosis at the exact time, they can provide proper treatment at a considerable cost. This paper presents the exhibitor's analysis of the predictions of heart disease using algorithms for classification. These secret trends in medical data can be used for health diagnosis. Healthcare management may use the knowledge to get better treatment. For big data concerns and data science projects, the data science lifecycle has to be constructed first. Problem formulation, data collection, data preparation, data exploration, data modelling, model evaluation, and model deployment are the seven major steps normally used in data science. The human heart beats one hundred thousand times every day, pumping 7570.824 L of blood into the body. There are 96,560.64 km of blood veins inside the body. Men are at higher risk of heart attack than women, based on the basic symptoms. Generally, women may feel squeezing, pressure, fullness, or discomfort in the middle of their chest during a heart attack. It can also cause arm, back, neck, jaw, or stomach pain, as well as shortness of breath, nausea, and other symptoms. The basic symptoms observed for a heart attack in humans are chest pain, filling uneasiness, and mental stress. Sometimes, it is observed that humans have pain in their hands, arms, and shoulders, as well as in their neck, back, and jaw, as well as shortness of breath. To make the classification process more effective, the heart disease database is pre-processed. Classification and regression categories are the pre-processed data. In order to diagnose heart diseases early, it is important that health complications be reduced. In the current healthcare industry, machine learning has been commonly used to diagnose and predict the existence of diseases using data models. One of these relatively common machine learning algorithms for studies involving risk assessment is the classification technique for complex illnesses. Therefore, the research aims to classify the most relevant cardiovascular disease predictors and predict the overall risk using classification algorithms. On the Kaggle website, the dataset used for support vector machine, random forest, AdaBoost, and gradient boost analysis is available for various projects. To develop a classifier for the early stage of heart disease prediction, the researcher must collect and pre-process the data as prerequisites at the initial steps. Researchers used the UCI heart disease dataset for this purpose. The aim is to decide if the patient has a risk of future heart disease. The UCI dataset consists of 442 patient data records and 14 attributes. We have conducted the test through the Python program, which is run in a Jupyter Notebook. The Jupyter Notebook is equipped with a very powerful data science software package. For the last two decades, predicting heart attacks using data mining methods has gained importance. Much of the article has applied techniques to various patient datasets from around the world, such as SVM, neural networks, regression, decision trees, Naive Bayesian classifiers. The multiple regression model was also suggested in the research paper to forecast heart disease [5].

Maarten et al. (2022) addressed the issue of artificial intelligence needed to solve the targeted medical problem of cardiovascular disease [6]. It has been observed from the research papers that various linear regression models present the perfect conditions for predicting the chance of heart disease. The classification accuracy of the regression algorithm is good compared to other algorithms [7]. S. Das et al. have proposed the local discovery method of frequent disease prediction in the medical dataset [4]. S. Das et al. focus on examining information mining processes that are needed to classify for medical information mining, such as heart disease, lung malignancy, and bosom disease. The prediction algorithm used Naive Bayes analytic model, which is developed based on Baye's theorem [8]. Therefore, Naive Bayes could make independent assumptions [9]. Huang et al. have applied artificial intelligence to Wearable Sensor Data to identify and forecast cardiovascular disease [10]. V. Jackins et al. used a random forest classifier and Naive Bayes in the efficient prediction of clinical disease in the heart [11]. Big data tools such as Hadoop Distributed File System, MapReduce, and support vector machine are also considered for heart disease prediction [12]. Maarten et al. used the concept of data mining to forecast heart disease [6]. It is recommended to use HDFS on various nodes for storing a large volume of data and performing the prediction algorithm using SVM simultaneously. Concurrently, these results are obtained in faster computation time compared to other machine learning models. Another repository from which we can collect the sourced data is one of the leading institutes for diabetic research in Chennai, Tamil Nadu, India [13]. The dataset consists of more than 500 patients' data. In another dataset, the instrument implemented in Weka considers a seventy percent break for classification. Here, 86.419% of the data is the precision provided by Naive Bayes. Neurological computing is a little bit helpful for analysing the electric signal from the heart to determine the cardiovascular issue [14]. Ghrabat et al. [15] discussed the use of a nonlinear classification algorithm for the estimation of heart disease [16]. On the Kaggle website, the dataset uses support vector machine, random forest, AdaBoost, and gradient boost analysis, which are available from ongoing Kaggle projects. The aim of our study is to determine if the patient has a risk of future heart disease. We have conducted the study considering a training dataset consisting of 3000 cases with thirteen unique attributes. The dataset is separated into two clusters. One cluster contains seventy percent of the data that is used for training purposes, and another cluster contains thirty percent of the data that is considered for testing purposes.

2 Materials and Methods

In the body, the heart is like any other muscle. For the muscle to contract and pump blood to the rest of the body, it requires enough blood flow to deliver oxygen. The heart pumps blood into the coronary arteries. Arteries derive from the aorta (the largest artery of the body, which contains the heart's oxygenated blood) and then spread out along the heart wall. Table 1 presents the different types of heart disease with their common symptoms and causes. If humans change certain habits in their lives, they can reduce the chance of heart disease.

These include:

- Consume fibre-rich, nutritious diets.
- Regular exercise: These will help to improve the heart condition and circulatory system, decrease cholesterol, and maintain blood pressure.
- Control the body weight with respect to height.
- Control alcohol consumption: Per day may be consumed according to the prescription of the doctor.
- Managing underlying issues: Finding care for problems such as elevated blood pressure, obesity, and diabetes that impair heart well-being.
- Quitting or avoiding smoking: Smoke is an important risk factor for cardiovascular and heart disease.

Machine learning is commonly used in nearly all sectors of the healthcare industry around the world. ML technology helps to adopt new formations from experience. In comparison, the process of considering various methodologies to predict, forecast, and explore something in the universe based on machine learning is relatively simple [17]. Sometimes the complex heart disease problem can be predicted by machine learning, i.e. regression and classification. Regression algorithms are primarily used for numeric results and binary and multi-category problems. Algorithms for machine learning were further split into two groups, such as supervised learning and unsupervised learning. Supervised learning is basically carried out using prior knowledge of output values. Unsupervised learning does not possess fixed tags; consequently, it infers the spontaneous build-up in the dataset. Therefore, machine learning algorithm selection needs to be properly analysed. AdaBoost is an ensemble strategy in which a group of weak learners are joined to generate a powerful learner. The AdaBoost model is present in Fig. 1. Typically, each weak learner is created as a "decision stump" (a tree with only one major split). In the random forest procedure, different

Table 1 Heart disease

S. No.	Disease types		
	Heart disease types	Symptoms	Reason
1	Coronary artery disease	Chest pain or discomfort (angina)	Plaque build-up in the arteries
2	Congenital heart defects	Rapid heartbeat and breathing	Several genetic health conditions
3	Arrhythmia	A fluttering in the chest and a racing heartbeat	Low supply of oxygen and nutrients to the heart
4	Heart failure	Out of breath. Feel tired always. Swollen ankles and legs	High blood pressure and high cholesterol
5	Myocardial infarction	Chest pain, fatigue, and heartburn	Smoking and high cholesterol

weights are assigned to increase efficiency. When we need high accuracy, we need to assign more weight. Tree is created from the consecutive training set. AdaBoost can improve its efficiency by anticipating the updated weights. Gradient Boosting Machine (GBM), like AdaBoost, combines a number of weak learners to generate a powerful learner. The classifier's residual is used as the input for the next classifier, on which the trees are created, making this an additive model. The residuals are acquired by the classifiers in a step-by-step manner to bag major variation in the data, which is accomplished through massive learning. The Gradient Boosting Machine is present in Fig. 2. It is gradually moving in the correct direction towards improved prediction using this strategy. As a result, based on the number of classifiers, we arrive at a prediction value that is very close to the observed value. Initially, a single-node tree is constructed that predicts the aggregated value of Y in the case of regression or the log (odds) of Y in the case of classification issues, following which trees with greater depth are developed on the residuals of the preceding classifier. Olsen et al. [18] provide a good overview of machine learning procedures for the prediction of heart failure [18]. Joon-myoung et al. (July 2019) used a deep learning-based artificial intelligence algorithm to predict acute heart failure [19]. Akbilgic et al. [20] developed an electrocardiographic artificial intelligence model to predict heart failure [20]. They have used the electrocardiogram artificial intelligence model for the prediction of heart failure. Manimurugan, S., et al. used the Two-Stage Classification model for the identification of heart disease [21].

In GBM, each tree has 8–32 terminal nodes, and approximately, 100 trees are produced. For each tree, learning rates are set to a constant so that the model may take modest steps in the appropriate direction to capture variance and train the classifier. Unlike in AdaBoost, where all trees are given equal weights. The dataset used

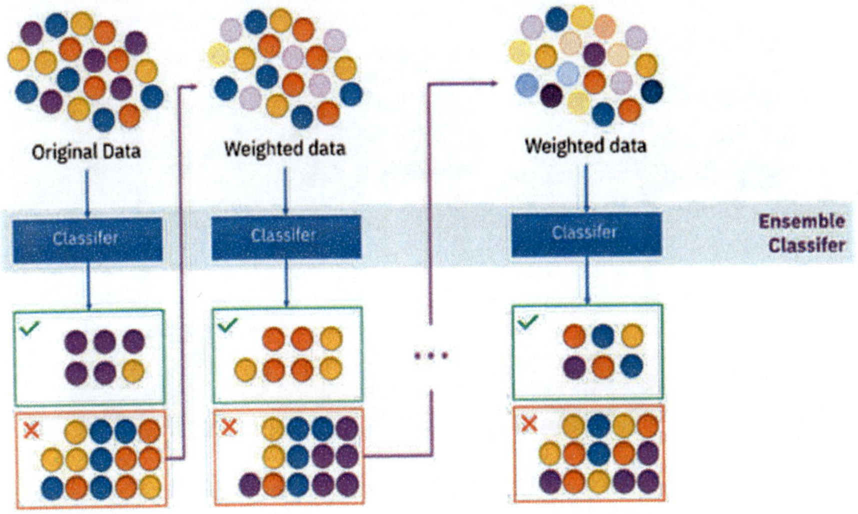

Fig. 1 AdaBoost model

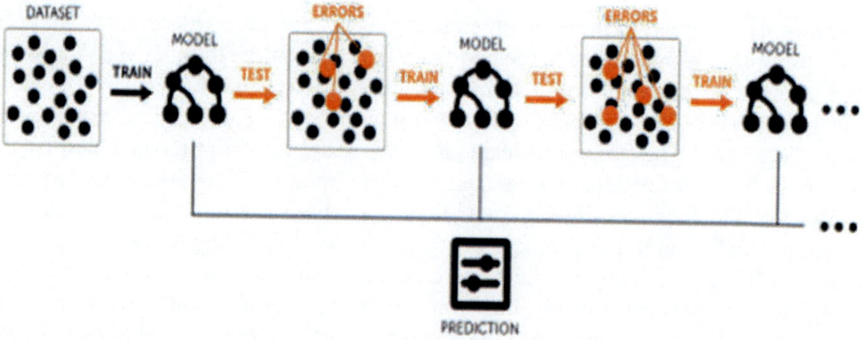

Fig. 2 Gradient boosting machine model

in the study of classification and regression algorithms is from an ongoing UCI cardiovascular study and is available on the Kaggle website.

The object of this analysis for classification is to predict that the patient has a chance of potential heart failure from the infections. With 442 pieces of information, the Framingham dataset consists of patient data and 14 attributes. Reviewing the Framingham data set and using Jupyter Notebook in Python programming, which is a more flexible and efficient technology tool for data science. Figure 3 indicates the steps followed to build the classification and regression models in machine learning.

In the data pre-processing stage, the initial process involves cleaning the data first to remove noise, identify inconsistent data, and obtain missing data. The initial dataset attributes are presented in Table 2. The regression model is created using the UCI Machine Learning Repository.

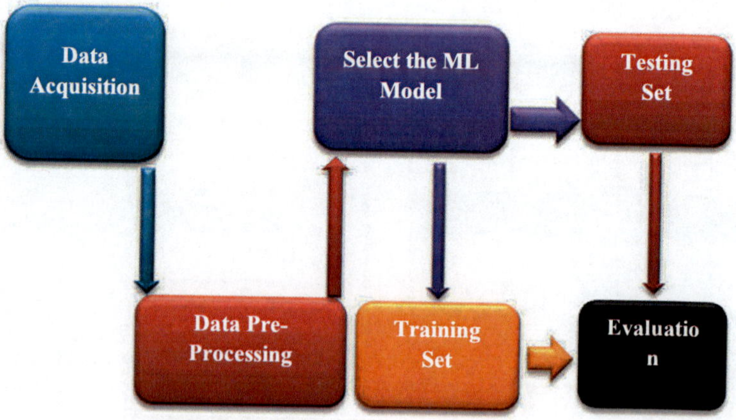

Fig. 3 Workflow of classification and regression model

Table 2 Dataset attributes

Attribute name	Attribute description, type, value			
	Description	Type/value		
Age	Age in years	Integer		
CP	Chest pain	Common angina		Value = 0
		Anomalous angina		Value = 1
		Non-angina pain		Value = 2
		No symptoms		Value = 3
trestbps	Resting blood pressure	In mm Hg		
chol	Serum cholesterol	In mg/dl		
fbs	Fasting blood sugar	More than 120 mg/dl	True	Value = 1
			False	Value = 0
restecg	Resting electrocardiographic	Normal		Value = 0
		ST-T wave abnormality		Value = 1
		Left ventricular hypertrophy		Value = 2
thalach	Heart rate	Heart rate/minimum		
Exang	Angina	Yes		Value = 1
		No		Value = 0
oldpeak	Depression	Exercise relative to rest		
slope	Slope of the peak exercise	Upsloping		Value = 0
		Flat		Value = 1
		Downsloping		Value = 2
ca	Number of major vessels	Coloured by fluoroscopy		Value range (0–3)
thal	Thalassemia	Normal		Value = 3
		Fixed defect		Value = 6
		Reversible defect		Value = 7
Gender	Male	Value = 1		
	Female	Value = 0		

3 Data Handle

The regression model is created using the UCI Machine Learning Repository. We have developed Python language code and tested it on the Jupiter notebook platform after removing some unnecessary fields from the collected data. In addition, the number of missed values for cleaning up current databases has been identified. A sample snapshot of the data header is presented in Table 3. The dataset is error-free and contains all of the data required for each variable. There are no issues regarding missing values or inconsistencies that are identified using the info (), describe (), and isnull () functions as defined in the Pandas Library. As a result, the dataset is fairly balanced. From the collected data, the proposed techniques approximate the (0.56) probability portion of the human population suffering from heart disease in the sample dataset presented in Table 3. Figure 4 presents the heart disease of the people with respect to the small sample dataset present in Table 3.

To build the attribute correlation, we consider the heatmap, which depicts the relationships between the dataset's attributes as well as how they interact. Attribute correlation is presented in Fig. 5.

Generally, from the observed prediction, we draw the inference that older people suffer from heart disease. The heart rate becomes lower at old age as he or she suffers from heart disease. The correlation between age and heart rate is presented in Fig. 6. The heatmap shows that the types of chest pain (CP), exercise-induced angina (Exang), exercise-induced ST depression, atypical, and non-anginal are all strongly linked to heart disease (target). We also see that cardiac illness and maximum heart rate have an inverse relationship (thalamus). It is presented in Fig. 7.

Table 3 Sample dataset

	Age	Sex	CP	trestbps	chol	fbs	restecg	thalach	Exang	oldpeak	slope	ca	thal	Target
0	70	1	4	130	322	0	2	109	0	2.4	2	3	3	2
1	67	0	3	115	564	0	2	160	0	1.6	2	0	7	1
2	57	1	2	124	261	0	0	141	0	0.3	1	0	7	2
3	64	1	4	128	263	0	0	105	1	0.2	2	1	7	1
4	74	0	2	120	269	0	2	121	1	0.2	1	1	3	1

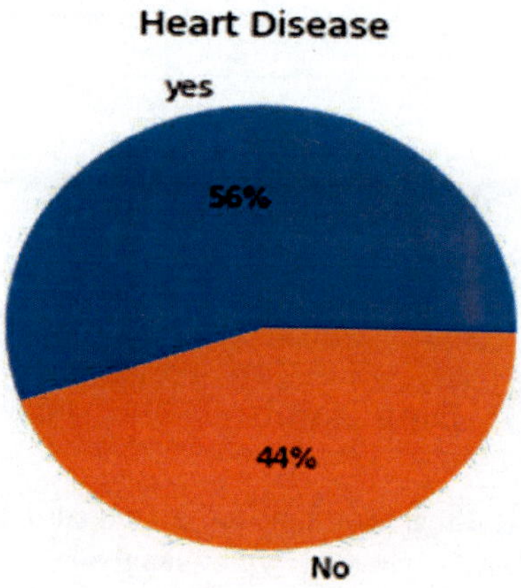

Fig. 4 Heart disease with respect to sample data

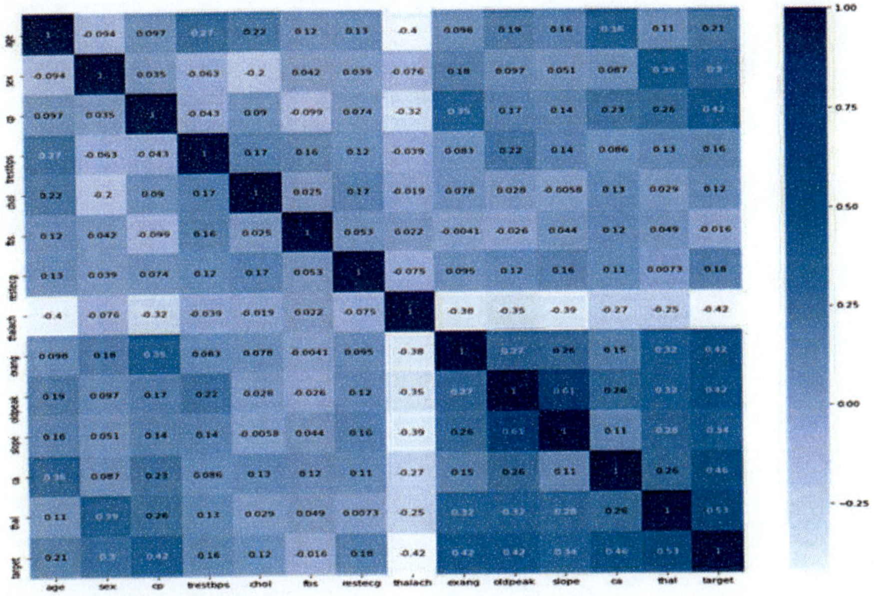

Fig. 5 Attributes correlation

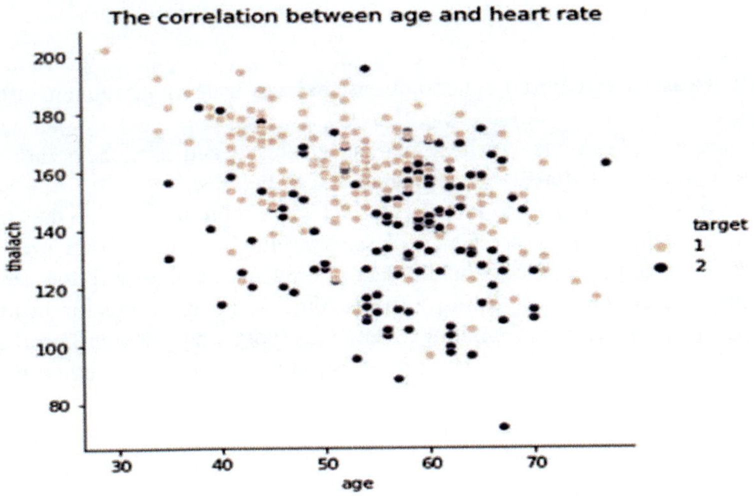

Fig. 6 Correlation between age and heart rate

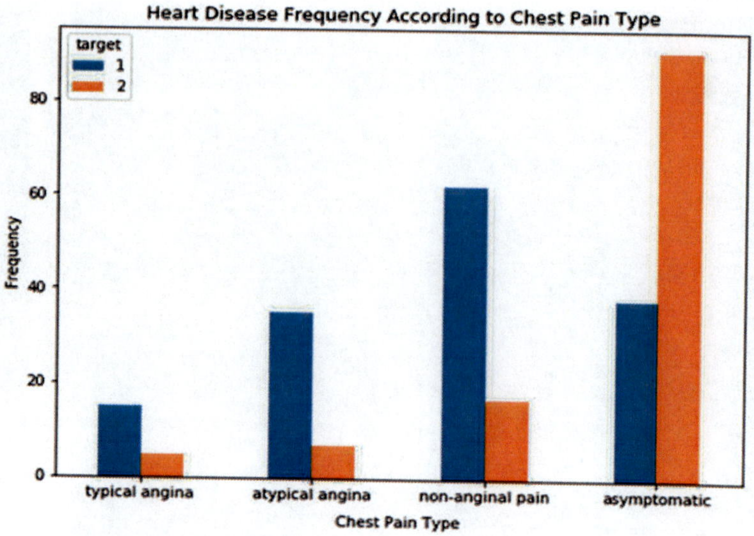

Fig. 7 Heart disease based on chest pain

4 Experiment and Results

From the result analysis, we get four different types of chest pain as the main cause of heart disease. The majority of heart disease patients have asymptomatic difficulty breathing.

The majority of patients with heart disease are elderly and have one or more main arteries coloured by Fluoroscopy. It is presented in Fig. 8.

We have logically subdivided the dataset. As some columns treat it as the "target" column, it is considered to serve as the class. A training set and a test set are created from the data. Seventy percent of the data is used for training purposes, and the remaining amount is used for testing purposes. Finally, the test results for the models for support vector machine, random forest, AdaBoost, and gradient boosting are exhibited in Figs. 9, 10, 11 and 12. Inference can be considered based on the best classification report.

Based on the aforementioned obtained results, with component-wise comparison based on precision, we can consider that gradient boosting is one of the best classifiers for heart disease prediction.

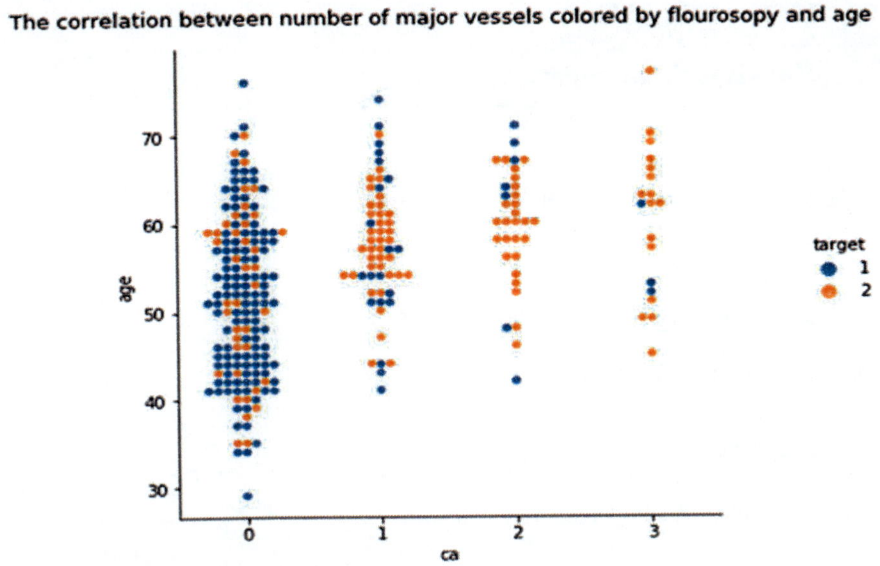

Fig. 8 Correlation between fluoroscopy and age

classification_report :

	precision	recall	f1-score	support
1	0.71	0.90	0.79	30
2	0.81	0.54	0.65	24
accuracy			0.74	54
macro avg	0.76	0.72	0.72	54
weighted avg	0.76	0.74	0.73	54

Fig. 9 Snap of the support vector machine used in the dataset

RandomForestClassifier

classification_report :

	precision	recall	f1-score	support
1	0.76	0.73	0.75	30
2	0.68	0.71	0.69	24
accuracy			0.72	54
macro avg	0.72	0.72	0.72	54
weighted avg	0.72	0.72	0.72	54

Fig. 10 Snap of the random forest classifier used in the dataset

```
AdaBoostClassifier

classification_report :
              precision    recall  f1-score   support

           1       0.75      0.80      0.77        30
           2       0.73      0.67      0.70        24

    accuracy                           0.74        54
   macro avg       0.74      0.73      0.73        54
weighted avg       0.74      0.74      0.74        54
```

Fig.11 Snap of the AdaBoost classifier used in the dataset

```
GradientBoostingClassifier

classification_report :
              precision    recall  f1-score   support

           1       0.79      0.73      0.76        30
           2       0.69      0.75      0.72        24

    accuracy                           0.74        54
   macro avg       0.74      0.74      0.74        54
weighted avg       0.74      0.74      0.74        54
```

Fig.12 Snap of the gradient boosting classifier used in the dataset

5 Conclusion

In conclusion, we learned how to build a suitable application for drawing predictions about heart diseases from medical datasets. We began with the definition of the problem and the moderate amount of accumulated data. Then we worked on data preparation, investigation, and building models. To create a classification model, we used a standardised dataset of heart disease patients. We began with data exploration and then moved on to data preparation. After that, we tested four heart disease classification models. Finally, we chose the finest model and saved it for later usage. The limitations of these works focus primarily on the use of classification strategies and algorithms for the prediction of heart disease. In this domain, researchers use different mechanisms for data cleaning and plan to construct a dataset that is suitable for their desired algorithms.

References

1. Hyeoun-Ae P (2013) An introduction to logistic regression: from basic concepts to interpretation with particular attention to nursing domain. J Korean Acad Nursing 43(2):154–164. https://doi.org/10.4040/jkan.2013.43.2.154
2. Strecht P, Cruz L, Soares C, Moreira MM, Abreu R (2015) A comparative study of classification and regression algorithms for modelling students' academic performance. In: Proceedings of the 8th international conference on educational data mining, 1–4
3. Mozaffarian D et al (2015) Heart disease and stroke statistics—2015 update. Circulation 131(4):29–322. https://doi.org/10.1161/CIR.0000000000000152
4. Das S, Dey A, Pal A, Roy N (2015) Applications of artificial intelligence in machine learning: review and prospect. Int J Comput Appl 115(9):31–41. https://doi.org/10.5120/20182-2402
5. Armin Z, Reiner K (2008) Vitamin D in the prevention and treatment of coronary heart disease. Curr Opin Clin Nutr Metab Care 11(6):752–757. https://doi.org/10.1097/MCO.0b013e328312c33f
6. Smeden VM et al (2022) Critical appraisal of artificial intelligence-based prediction models for cardiovascular disease. Eur Heart J 43(31):2921–2930. https://doi.org/10.1093/eurheartj/ehac238
7. Sathya R, Abraham A (2013) Comparison of supervised and unsupervised learning algorithms for pattern classification. Int J Adv Res Artif Intell 2(2):34–38
8. Ng YA, Jordan IM (2001) On discriminative vs. generative classifiers: a comparison of logistic regression and Naive Bayes. Adv Neural Inf Process Syst 14:841–848
9. Kanrar S (2016) Fast load balancing approach for growing clusters by bioinformatics. In: Proceedings international conference on signal processing, communication, power embedded system, SCOPES 2016, 382–385. https://doi.org/10.1109/SCOPES.2016.7955857
10. Huang J, Wang J, Ramsey E, Leavey G, Chico AT, Condell J (2022) Applying artificial intelligence to wearable sensor data to diagnose and predict cardiovascular disease: a review. Sensors 22(20):1–28. https://doi.org/10.3390/s22208002
11. Jackins V et al (2021) AI-based smart prediction of clinical disease using random forest classifier and Naive Bayes. J Supercomput 77:5198–5219. https://doi.org/10.1007/s11227-020-03481-x
12. Kanrar S, Mandal KP (2017) E-health monitoring system enhancement with Gaussian mixture model. Multimedia Tools Appl 76(8):10801–10823. https://doi.org/10.1007/s11042-016-3509-9
13. Amutha A et al (2021) Clinical profile and types of youth-onset diabetes in Chennai: the Indian council of medical research registry of YouthOnset diabetes in India–Chennai centres. J Diabetol 12(4):492–502. https://doi.org/10.4103/jod.jod_76_21
14. Kanrar S (2023) Machine learning model development using computational neurology. Smart Innov Syst Technol 313:149–158. https://doi.org/10.1007/978-981-19-8669-7_14
15. Ghrabat JJM, Ma G, Maolood YI, Alresheedi SS, Abduljabbar AZ (2019) An effective image retrieval based on optimized genetic algorithm utilized a novel SVM-based convolutional neural network classifier. Hum Cent Comput Inf Sci 9(31):1–29. https://doi.org/10.1186/s13673-019-0191-8
16. Abubaker M, Babayigit B (2022) Detection of cardiovascular diseases in ECG images using machine learning and deep learning methods. IEEE Trans Artif Intell 4(2):373–382. https://doi.org/10.1109/TAI.2022.3159505
17. Peng JC, Lee LK, Ingersoll MG (2002) An introduction to logistic regression analysis and reporting. J Educ Res 96(1):3–14. https://doi.org/10.1080/00220670209598786
18. Olsen RC et al (2021) Clinical applications of machine learning in the diagnosis, classification, and prediction of heart failure. Am Heart J 229:1–17. https://doi.org/10.1016/j.ahj.2020.07.009
19. Kwon J et al (2019) Artificial intelligence algorithm for predicting mortality of patients with acute heart failure. PLoS One 14(7):1–14. https://doi.org/10.1371/journal.pone.0219302J

20. Akbilgic O et al (2021) ECG-AI: electrocardiographic artificial intelligence model for prediction of heart failure. Euro Heart J Digital Health 2(4):626–634. https://doi.org/10.1093/ehjdh/ztab080
21. Manimurugan S et al (2022) Two-stage classification model for the prediction of heart disease using IoMT and artificial intelligence. Sensors 22(2):1–19. https://doi.org/10.3390/s22020476

Automating Dose Prediction in Radiation Treatment Planning Using Self-attention-Based Dense Generative Adversarial Network

V. Aparna, K. V. Hridika, Pooja S. Nair, Lekshmy P. Chandran, and K. A. Abdul Nazeer

Abstract Radiation treatment being a crucial step in cancer treatment, accurate radiation treatment planning is extremely important to reduce the effect of radiation on Organs-At-Risk (OAR). In order to determine the radiation treatment dose distributions received by the tumour and surrounding organs, this paper proposes a self-attention-based dense generative adversarial network (GAN). This is accomplished by adding a self-attention module to GAN which uses dense U-Net as generator. Using the DVH and dose scores, the performance is evaluated and compared to a few state-of-the-art models like simple GAN, self-attention-based GAN and dense GAN. The dataset was acquired from the OpenKBP 2020 AAPM Grand Challenge run by CodaLab. The DVH score and dose score of the self-attention-based dense GAN are 2.126 and 3.522, respectively. The proposed model performed better than other state-of-the-art models considered.

Keywords Organs-At-Risk (OAR) · U-Net · Dose volume histogram (DVH) · Convolutional neural network (CNN) · Generative adversarial network (GAN)

1 Introduction

Cancer is a disease that develops when cells in the body divide and grow at an uncontrollable rate. Nearly half of the population dies from cancer. The most common types are breast, lung, and prostate cancers. Radiation therapy (RT), chemotherapy, surgery, or a treatment combining these are used in the treatment of the cancer.

One of the main approaches of treating cancer is radiation treatment (RT). High energy X-ray beams are utilised in RT to target tumours with a controlled dose of radiation while sparing surrounding healthy tissue. A sophisticated design procedure including numerous medical specialists and numerous software systems leads to an RT treatment plan. Included in this is specific optimisation software that establishes the beam parameters necessary to produce the desired dose distribution. A number

V. Aparna · K. V. Hridika (✉) · P. S. Nair · L. P. Chandran · K. A. Abdul Nazeer
National Institute of Technology, Calicut, India
e-mail: hridikakv246@gmail.com

S. Kumar et al. (eds.), *Fourth Congress on Intelligent Systems*, Lecture Notes in Networks and Systems 868, https://doi.org/10.1007/978-981-99-9037-5_2

of computed tomography (CT) images of the patient, different dosimetric goals, and constraints, and other optimization-related parameters are input into the optimisation model. An oncologist then evaluates the model's recommended course of action. The treatment planner must resolve the optimisation model using new parameters when the oncologist typically suggests revisions to the original plan. Since the back-and-forth between the planner and oncologist is frequently repeated several times before the plan is finally approved, the entire process requires intensive labour, which is expensive and time-consuming [1].

In order to reduce treatment delay and error, there have been studies in the field of automation of radiation treatment planning. Various models have been evolved from knowledge-based planning to various deep learning models like U-Net, ResNet. Deep learning techniques are more effective and accurate at predicting dose and work well with unstructured data. Here, we have developed a self-attention-based dense GAN which can perform better than some of the other existing models.

2 Related Works

Automation of dose prediction in RTP is done using traditional knowledge-based planning and deep learning methods. Conventional KBP techniques involve investigations that call for features of anatomy in order to either discover a case that matches the most from a database of former treatment plans or to create dose prediction models. Deep learning methods such as U-Net ,CNN, and ResNet involve training neural networks to predict the dose distribution [2].

Kajikawa et al. [3] compared a 3D convolutional neural network (CNN) with the traditional machine learning approach for forecasting prostate cancer IMRT dose distribution using only contours. The CNN model's performance for dose distribution prediction was better or on par with that of the dose distribution produced by $RapidPlan^{TM}$. The work has given the motivation for using deep learning methods rather than knowledge-based planning for dose prediction.

Barragán-Montero et al. [4] investigated the use of a CNN architecture that combines U-net and DenseNet . In this model, variable beam configuration is used as input for predicting the 3D dose distributions. They investigated two models, one with only anatomial information as input, other with an extra channel for dosimetric information. The accuracy of the high dose region is similar in both the models but for medium to low dose region, the second model performed better. The paper suggests that for variable beam configuration, giving dosimetric information also as input to the network helps in building a reliable model for dose prediction.

An experimental study by Kearney et al. [5] examined the feasibility of a DoseNet in treatment planning for prostate stereotactic body radiotherapy (SBRT). In conclusion, DoseNet can accurately forecast volumetric dose for prostate patients undergoing non-coplanar SBRT (stereotactic body radiation treatment) while maintaining computing efficiency.

Jalaifar et al. [6] proposed a 3D residual network guided by self-attention to predict the outcome of local failure after radiotherapy. He compared the performance of simple 3D residual network and 3D residual network with convolutional block attention module(CBAM). The proposed model performs better than the vanilla residual network and residual network with CBAM attention. The paper has shown the potential for self-attention gate in improving the model.

Murakami et al. [7] developed a GAN for prostate cancer which learns the features of CT images and compared with that of a model provided with contoured images. The time required for treatment planning is reduced by this method.

Mahmood et al. [1] proposed a GAN approach for predicting ideal 3D dose distributions that minimise the exposure to healthy tissue. Studies using a dataset of oropharyngeal cancer patients reveal that this strategy performs much better than earlier ones on a number of clinical satisfaction and similarity parameters. This has motivated in choosing GAN for developing a better model.

Zhan et al. [8] proposed an approach based on multi-constraint GAN also called Mc-GAN. They gave importance to both global and local dose while extracting features. They also used a self-supervised perceptual loss and a dual attention module to further enhance the prediction result. In the cases of an internal dataset for cervical cancer and an internal dataset for rectal cancer, the technique was superior and feasible.

Babier et al. [9] proposed a knowledge-based automated planning (KBAP) pipeline consisted of a generative adversarial network (GAN) followed by two optimization models—one to learn objective function weights and other to generate fluence-based plans. They looked into three distinct GAN models. A three-channel colour map of dose was the initial model's prediction. The second model explicitly predicted the dose as a scalar value (one-channel). The third model simultaneously predicted the scalar doses for the entire 3D CT image. 3D GAN takes contoured CT image as input and predicts the full 3D dose distribution. This 3D GAN performs better compared to the other 2D GAN's.

Kearney et al. [10] proposed an attention-gated generative adversarial network (DoseGAN). Attention gates are incorporated into both generator and discriminator of DoseGAN. The proposed model predicted more realistic volumetric dosimetry, and it performed better than other state-of-the-art models.

Based on the literature survey conducted, we came to the conclusion that over the years different types of U-Net are studied in various papers. But only a few papers have explored the scope of GAN in medical image processing. In view of promising results from the available papers, we have decided to enhance the simple GAN by incorporating self-attention gates and dense U-Net.

3 Materials and Methods

3.1 Dataset

The dataset, which comprises 3D head and neck CT scans of patients with oropharyngeal cancer, is taken from CodaLab's OpenKBP 2020 AAPM Grand Challenge. 340 patients with oropharyngeal cancer's CT scans are included in the data set. Both the Planning Target Volumes (PTVs) and the Organs-At-Risk (OARs) contain segmentation masks. Dataset will undergo training, validation, and testing. In an 8:2 ratio, 240 patient records were used for training and validation. The data of 100 patients were used for testing. The performance of the selected model will be evaluated using the contoured CT images. Each patient image is a voxel tensor of $128 \times 128 \times 128$ pixels [11]. Figure 1 shows the input which is a contoured CT image when passed through the dose prediction model gives the output which is the predicted clinical dose.

3.2 Data Loading and Preprocessing

The information about the CT, dose, and contours for each patient is given in an encoded csv format. Using Python code, the csv files are transformed into 3D tensors.

CT, structural masks, and dose images are kept in distinct locations, and the data is loaded in batches. The mask images, dose images, and CT images are all reshaped to (128,128,128,1), (128,128,128,10), and (128,128,128,1), respectively. Hounsfield unit values are present in the CT images. The mask images are then concatenated with the normalised CT images to produce a 4D tensor, which is used as the models' input.

The validation and training sets are combined into a single folder with data of 240 patients. These 240 patient data are divided in the ratio of 8:2 for training and validation. The output dose distributions are the unnormalized 3D dose distribution images.

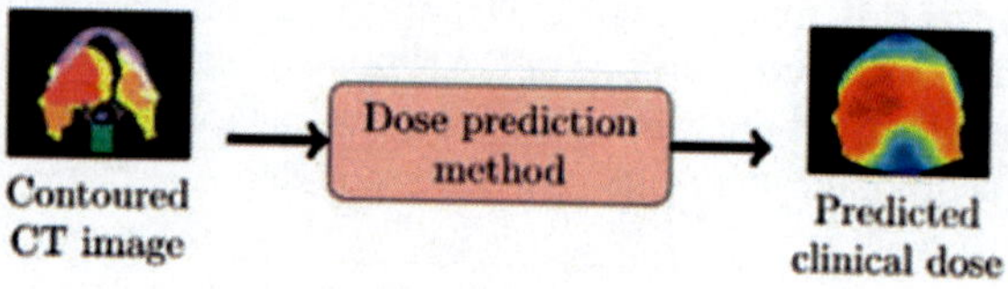

Fig. 1 Overview of dose prediction method [12]

3.3 GAN

GAN is a type of deep learning model that consists of two neural networks, a generator and a discriminator, that compete against each other in a zero-sum game to generate realistic-looking samples of data. Generator predicts the dose distribution using the CT image and contour information. Discriminator tries to identify the real dose distribution among the real dose distribution and predicted dose distribution. The predicted output is the generator output, when the discriminator can't distinguish between generator output and real dose distribution.

3.4 Self-attention Gate

Figure 2 shows a feature tensor when passed through a self-attention gate gives the self-attention feature tensor. Query, key, and value are the three feature maps obtained when the image is passed through three $1 \times 1 \times 1$ convolution filters. We transpose the query and matrix-multiply it by the key and take the softmax on all the rows to obtain the output attention map. The value is then matrix multiplied with the output attention map. The final output is multiplied by a learnable scale parameter (gamma), and the input is added back as a residual connection [13]. The attention gate directs the model's focus to important locations while inhibiting the activation of features in irrelevant regions. The self-attention model permits interactions between inputs.

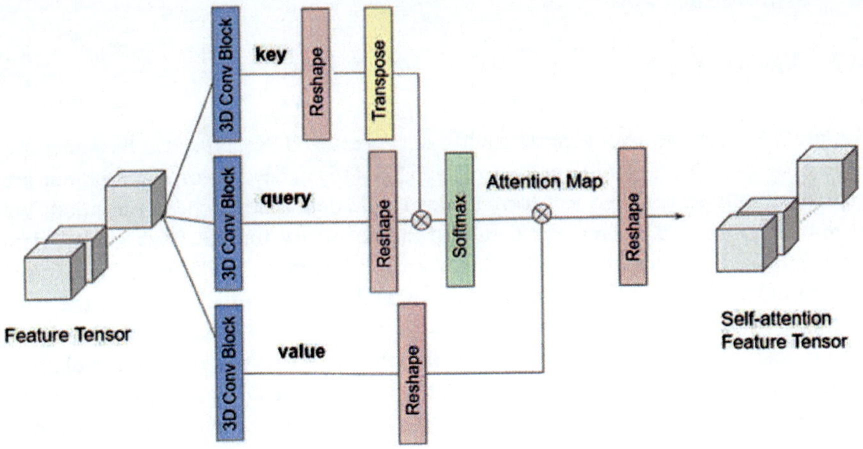

Fig. 2 Self-attention gate [13]

3.5 Dense U-Net

Dense U-Net [14] is a deep learning architecture that combines the U-Net and DenseNet architectures. The U-Net architecture is a good choice for medical image segmentation tasks because of its ability to capture contextual information and the fine details of the image. However, the U-Net has a drawback in that it can suffer from vanishing gradients during training, especially when the network is deep. A DenseNet is a type of CNN that utilises dense connections between layers, through dense blocks. The connection between layers in a dense block is accomplished by concatenating the output of each layer to the input of the next layer. By leveraging the features that were extracted in previous layers, the network is able to learn more efficiently. The dense U-Net architecture combines the strengths of these two architectures by introducing dense connections between the U-Net blocks. The dense connections allow for feature reuse and gradient flow through the network, while the U-Net blocks capture the contextual information and fine details of the image.

3.6 Hyperparameters

The model is trained for 200 epochs. The Adam optimiser parameters values $\beta 1 = 0.5$, The batch sizes for training and testing are 4 and 1, respectively.

4 Implementation

4.1 Simple GAN

Simple GAN is made up of generator and discriminator. U-Net is used as the generator of this GAN and discriminator is a patchGAN [15]. The data of each patient are kept separately in encoded csv format files in the data that has been provided. We have used classes to convert the data into the necessary format. For training, data is transformed into NumPy arrays. Concatenated CT, mask, and dose images serve as the model's input. The model was then trained, to predict the images of the dose distribution, for 200 epochs. There are four batches for the training set and one for the test set. The generator loss and discriminator loss are obtained. Relu and Sigmoid functions were the activation functions used. The DVH and dose scores are also obtained. Figure 3 shows the simple GAN architecture. Input image is given to generator which predicts a dose distribution. The predicted dose image and a true dose image is given to discriminator which discriminates between the two as real for true image and fake for predicted image. The training is done until the discriminator fails to discriminate between the two.

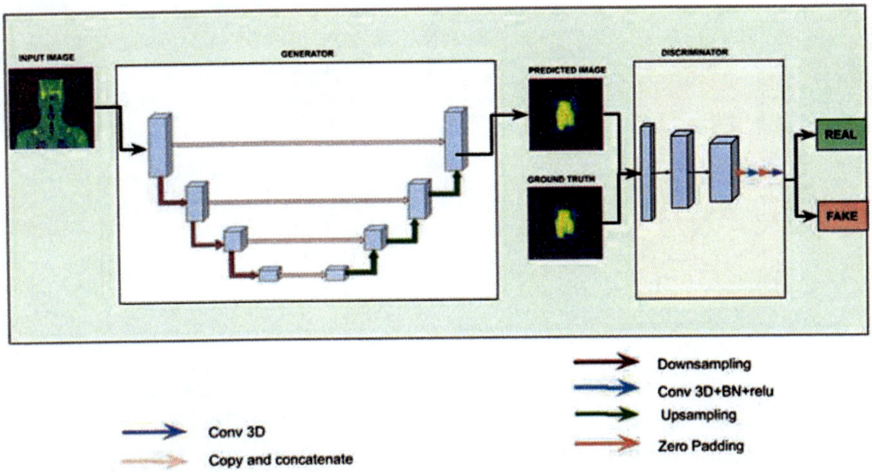

Fig. 3 GAN model architecture [16]

4.2 *Self-attention-Based GAN*

In self-attention-based GAN, self-attention gates [13] are incorporated into the generator of simple GAN as shown in Fig. 4. Input for this model is similar to that of the simple GAN. This model is also trained for same epochs with training batch size four and testing batch size one. The model is similar to the previous model except for the addition of self-attention gates.

4.3 *Dense GAN*

Dense GAN is a model where the generator of the simple GAN is replaced by dense U-Net. The discriminator is a patchGAN. Five dense blocks are used in upsampling and downsampling. Since this GAN utilises dense blocks, all features extracted are reused. Training procedure for this GAN is similar to the above mentioned GAN. Figure 5 represents a dense GAN whose generator is a dense U-Net and discriminator is a patchGAN.

4.4 *Self-attention-Based Dense GAN*

In this model, we merged the self-attention-based GAN and dense GAN. As shown in Fig. 6, the generator is made up of dense U-Net attached with self-attention gate and discriminator is made up of patchGAN. Self-attention gates are added at three layers in the generator. The training of model is similar to previous models.

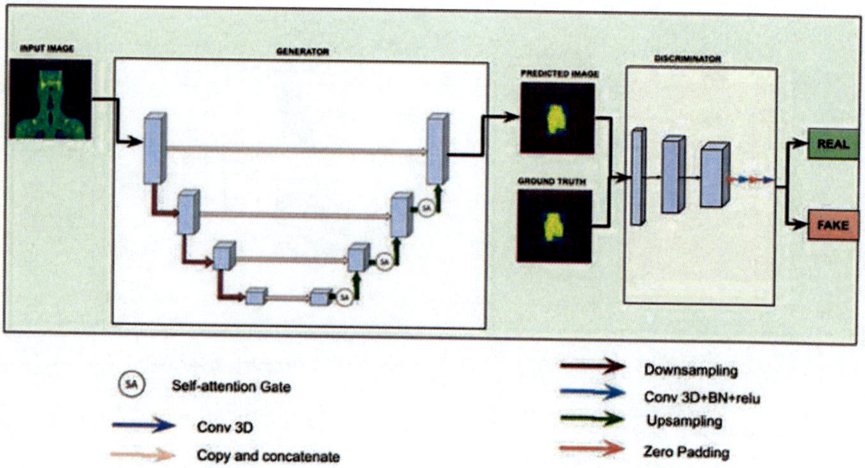

Fig. 4 Self-attention-based GAN model architecture

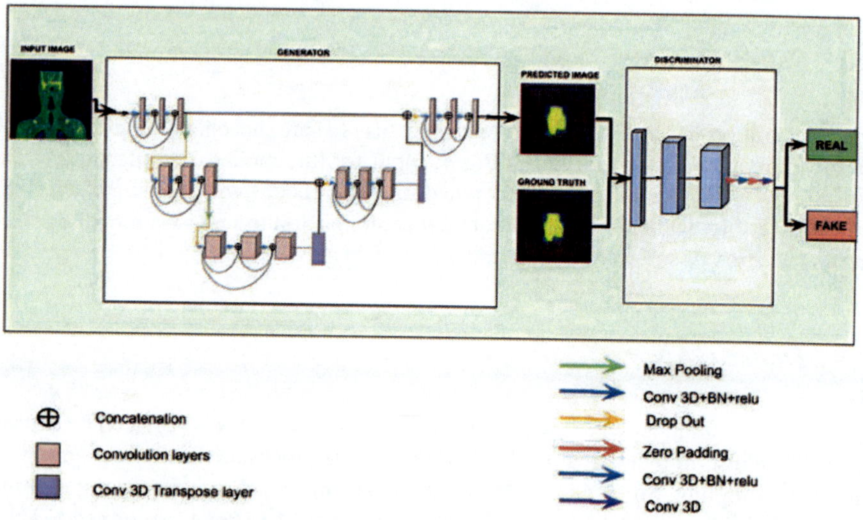

Fig. 5 Dense GAN model architecture

5 Results

In this paper, we used DVH score and dose score [11] as the evaluation metrics. Dose score is defined as the mean absolute difference between the ground truth dose distribution and predicted dose distributions. While dose score calculates the difference at all points, DVH score calculates the difference only at important DVH

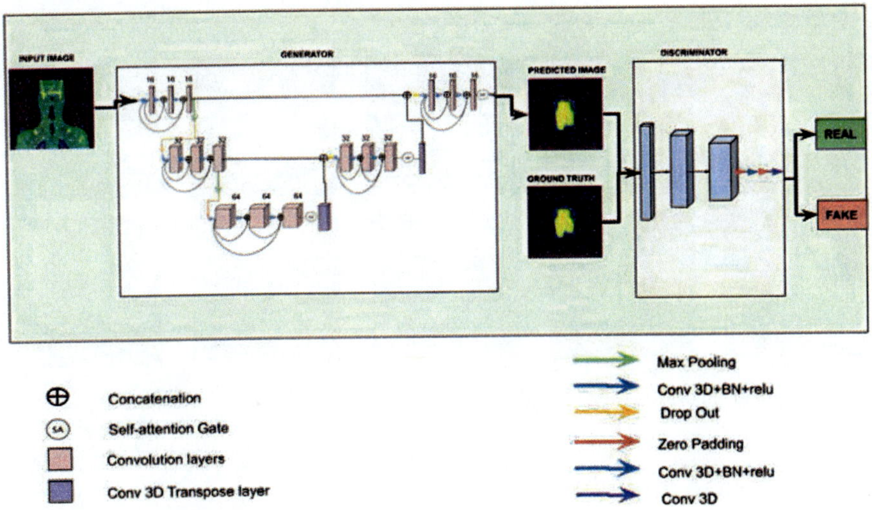

Fig. 6 Self-attention-based dense GAN model architecture

Table 1 Performance of models

Model	DVH score	Dose score
Simple GAN	3.520	3.915
Self-attention-based GAN	3.034	3.715
Dense GAN	2.447	3.689
Self-attention-based dense-GAN	2.126	3.553

metrics. Self-attention-based dense GAN with dense U-Net as generator is run for 200 epochs, and it's DVH score and dose score are noted.

The DVH score and dose score obtained in case of self-attention-based dense GAN are 2.126 and 3.553, respectively. With respect to simple GAN, there is an improvement of 1.394% in DVH score and 0.362% in dose score. We compared the performance of our model with simple GAN, self-attention-based GAN and dense GAN as depicted in Table 1 and Fig. 7.

The self-attention-based dense generative adversarial network uses a combination of dense GAN and self-attention gates. The dense U-Net architecture combines the strengths of the DenseNet and U-Net architectures by introducing dense connections between the U-Net blocks. The dense connections allow for feature reuse and gradient flow through the network, while the U-Net blocks capture the contextual information and fine details of the image. Self-attention gates are introduced into the GAN to enhance the feature selection. It weighs important features by considering dependencies at the global level. Thus self-attention-based dense generative adversarial network improved the results.

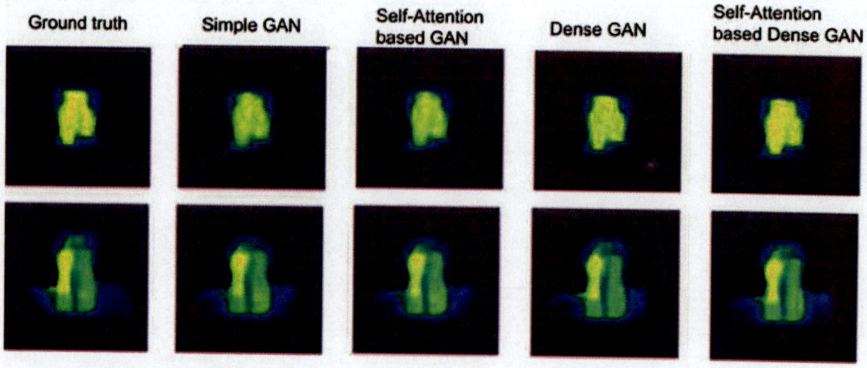

Fig. 7 Comparison of the models based on dose prediction

6 Conclusion and Future Work

Dose prediction in radiation therapy planning is one of the crucial steps in cancer treatment. It is important that this is to be done with accuracy and minimal time so as to avoid delay in treatment. So different models of automation are proposed over time which involves traditional knowledge-based planning methods and modern deep learning approach. Based on our literature survey, we concluded that though the GAN architecture is complex, introducing modifications can produce variants that can improve the accuracy of prediction. The proposed self-attention-based dense GAN is performing better than the considered state-of-the-art models in terms of DVH and dose scores.

Our model's architecture is limited by the availability of GPU. There is a possibility of enhancing the network architecture which can result in better performance. In this paper, we have only considered head and neck CT scan images, further body parts can also be included in the future study.

References

1. Mahmood R, Babier A, McNiven A, Diamant A, Chan TC (2018) Automated treatment planning in radiation therapy using generative adversarial networks. In: Machine learning for healthcare conference, PMLR, pp 484–499
2. Momin S, Fu Y, Lei Y, Roper J, Bradley JD, Curran WJ, Liu T, Yang X (2021) Knowledge-based radiation treatment planning: a data-driven method survey. J Appl Clin Med Phys
3. Kajikawa T, Kadoya N, Ito K et al (2019) A convolutional neural network approach for IMRT dose distribution prediction in prostate cancer patients. J Radiation Res 60:685–693
4. Barragan-Montero AM, Nguyen D, Lu W, Lin M, Geets X, Sterpin E, Jiang S (2019) Application of deep neural networks for automatic planning in radiation oncology treatments. In: 27th European symposium on artificial neural networks, computational intelligence and machine learning, ESANN 2019, ESANN (i6doc. com), pp 161–166

5. Kearney V, Chan JW, Haaf S, Descovich M, Solberg TD (2018) Dosenet: a volumetric dose prediction algorithm using 3D fully convolutional neural networks. Phys Med Biol 63(23):235022
6. Jalalifar SA, Soliman H, Sahgal A, Sadeghi-Naini A, IEEE (2022) A self-attention-guided 3D deep residual network with big transfer to predict local failure in Brain Metastasis after radiotherapy using multi-channel MRI
7. Murakami Y, Magome T, Matsumoto K, Sato T, Yoshioka Y, Oguchi M (2020) Fully automated dose prediction using generative adversarial networks in prostate cancer patients. PloS One 15(5):e0232697
8. Bo Z, Xiao J, Cao C, Zu XPC, Jiliu Z, Yan W (2022) Multi-constraint generative adversarial network for dose prediction in radiotherapy. Medical Image Anal 77:102339
9. Babier A, Mahmood R, McNiven A, Diamant A, Chan TCY (2018) Knowledge-based automated planning with three-dimensional generative adversarial networks. arXiv:1812.09309v1 [physics.med-ph], 21 Dec 2018
10. Kearney V, Chan JW, Wang T, Perry A, Descovich M, Morin O, Yom SS, Solberg TD (2020) DoseGAN: a generative adversarial network for synthetic dose prediction using attention-gated discrimination and generation. Sci Rep
11. Babier A, Zhang B, Mahmood R, Moore KL, Purdie TG, McNiven AL, Chan TC (2020) Openkbp: the open-access knowledge-based planning grand challenge and dataset. Med Phys
12. Hira S (2020) My 3rd place solution to the Openkbp challenge. https://medium.com/@sanchithira76/my-3rd-place-solution-to-the-openkbp-challenge-c0cbdd79de11. Accessed 13 June 2020
13. Zhang H, Goodfellow I, Metaxas D, Odena A (2018) Self-attention generative adversarial networks. arXiv:1805.08318
14. Cai S, Tian Y, Lui H, Zeng H, Wu Y, Chen G (2020) Dense-UNet: a novel multiphoton in vivo cellular image segmentation model based on a convolutional neural network. Quant Imaging Med Surg
15. Isola P, Zhu J-Y, Zhou T, Efros AA (2016) Image-to-image translation with conditional adversarial networks. arXiv:1611.07004 [cs.CV]
16. Jha S, Sajeev N, Marchetti AR, Chandran LP, Abdul Nazeer KA (2022) Performance evaluation of deep learning architectures for predicting 3D dose distributions in automatic radiotherapy treatment planning

Track Learning Agent Using Multi-objective Reinforcement Learning

Rushabh Shah, Vidhi Ruparel, Mukul Prabhu, and Lynette D'mello

Abstract Reinforcement learning (RL) enables agents to make decisions through interactions with their environment and feedback in the form of rewards or penalties. The distinction between single-objective reinforcement learning (SORL) and multi-objective reinforcement learning (MORL) is established, emphasizing the latter's ability to optimize multiple conflicting objectives simultaneously. The study explores various algorithms and approaches within the MORL framework, focusing on track navigation optimization. Key components of the implementation are detailed, including states, actions, rewards, and tracks used for training. The proposed algorithm, Pareto Q-learning, is highlighted as a powerful approach to simultaneously optimize multiple objectives. The architecture and methodology of the learning agent are presented, outlining the training process and the impact of hyper-parameters. Results from experimentation are discussed, revealing the agent's learning curve, crash avoidance, and successful achievement of multiple objectives. The study underlines the significance of MORL in enabling agents to manage complex decision-making scenarios, leading to more robust and optimal policies. The paper concludes by emphasizing the practical implications of the MORL approach in navigating challenging tracks with conflicting goals, such as minimizing steps, maximizing rewards, and avoiding collisions.

Keywords Reinforcement learning · Vector reward system · Multi-objective reinforcement learning · Pareto Q-learning

R. Shah (✉) · V. Ruparel · M. Prabhu · L. D'mello
Dwarkadas J. Sanghvi College of Engineering, Mumbai, India
e-mail: rushabhshah2001@gmail.com

V. Ruparel
e-mail: ruparelvd0919@gmail.com

M. Prabhu
e-mail: prabhumukul7@gmail.com

L. D'mello
e-mail: lynette.dmello@djsce.ac.in

S. Kumar et al. (eds.), *Fourth Congress on Intelligent Systems*, Lecture Notes in Networks and Systems 868, https://doi.org/10.1007/978-981-99-9037-5_3

1 Introduction

Reinforcement learning is one of the subfields of machine learning that is often used in situations where there is no labeled data available. It is especially useful for problems that require decision-making in complex and uncertain environments, where the optimal action may not be immediately obvious. Reinforcement learning is widely used in a variety of applications, and it is a fascinating field of study that might completely transform how we solve problems and determine the best course of action. Reinforcement learning differentiates itself from supervised learning in such a way that supervised learning uses labeled datasets where the desired outcome is already known. In contrast, reinforcement learning is the process to train an algorithm to make decisions through interaction with the surroundings and response received in terms of rewards or punishments.

This paper presents an implementation of a track learning agent whose goal is to navigate a track in the best viable way. In any given racetrack, there are multiple paths through which a car can go around the track from the starting line to the finishing line. The agent must learn by acting upon the environment to maximize its rewards. Sometimes, the rewards can be conflicting in nature. For instance, the agent is rewarded if it completes the track, i.e., reaching the finish line, but the time required to do so results in a penalty. Agent, thus, to maximize the reward, must reach the finish line in the minimum amount of time possible to have the lowest penalty. Apart from these attributes, other attributes can be added like a negative reward for collision in the wall or any other object, positive reward for collecting power-ups, negative reward for turning at an angle greater than a certain threshold (compromising comfort), positive reward for overtaking a competitor, etc.

Majority of research done on decision-theoretic planning and reinforcement learning either assumes that there is only one objective or that many objectives may be successfully addressed by a straightforward linear combination. These tactics could excessively simplify the underlying issue and lead to unsatisfactory outcomes. These many goals often compete with one another. Many reward functions are evaluated, each of which is a scalarization of the actual objectives. However, every reward function is not looked at. Therefore, even if the minimum standard for acceptable behavior is achieved, only a portion of all potential scalarizations is observed. To address this issue, multi-objective reinforcement learning can be employed.

MORL deals with learning of optimum strategies in situations having multiple goals. Identifying a set of policies that can optimize multiple objectives concurrently while also balancing trade-offs between them is the primary purpose of MORL. MORL algorithms use various techniques, such as Pareto optimality and multi-objective optimization, to achieve this goal.

2 Related Work

Numerous factors need to be considered in the majority of real-world issues, but most algorithms for agents that must interact with sequential decision problems concentrate on optimizing a single objective. Hayes et al. [1] gives us a full summary of the drawbacks of employing single-objective techniques to multi-objective problems, such as restricting our ability to quickly adjust to changing preferences by discovering various policies or handing over control of managing trade-offs from the problem stakeholders to the system developer. This could result in solutions that do not maximize user satisfaction. Their research also offers numerous inspiring instances of how MORL might be used. Hayes et al. [1] has written down the flow of how one should tackle a multi-objective decision-making problem. It includes factors affecting the design of MORL agents, relationship of the problem with other similar problems, all types of algorithms ranging from bandit algorithms to stateful single as well as multiple policy algorithms to interactive algorithms. It also includes using multiple agents in MORL for certain problems. They have provided certain metrics to measure the quality of our algorithm like axiomatic-based evaluation metrics, utility-based evaluation metrics, etc. It includes some benchmark problems and one example solved completely along with its results. It hints at the scarcity of high-level datasets which can be used as a benchmark to evaluate the agent. For MORL research, just a few benchmark issues have been put forth thus far, and many of them are straightforward.

For resolving multi-objective reinforcement learning (MORL) issues, [2] suggests a brand-new adaptable architecture entitled multi-objective deep reinforcement learning (MODRL). The Deep *Q*-Networks (DQN) framework allows both linear and nonlinear action selection methods as well as single and multiple policies. The mountain car and deep-sea treasure benchmark issues are used by the authors to assess the effectiveness of the MODRL framework in finding Pareto optimal solutions. The benefits of the MODRL architecture are also covered in the study, including its aptitude for handling complex issues as well as its modularity and adaptability to various problem domains. Overall, employing deep reinforcement learning approaches, the research proposes a promising method for MORL.

The study [3] introduces a model-based MORL algorithm, which uses dynamic programming to determine Pareto optimum policies after learning a multi-objective sequential decision-making model. Two exploration tactics are evaluated for effectiveness as the algorithm is put to the test on the deep-sea treasure challenge. The paper emphasizes the benefit of the model-based approach, which enables the dynamic programming method, after an accurate model has been built, to calculate all Pareto optimal policies. Near-optimal solutions to the issue are provided by the resulting policies. However, the agent's exploration strategy has a significant impact on how well the algorithm performs. Overall, the research shows how model-based reinforcement learning can be used to solve MORL challenges.

Gros et al. [4, 5] give an empirical study comparing the effectiveness of imitation learning (IL) and deep reinforcement learning (DRL) on a series of difficult continuous control objectives using a technique known as "race tracking." The research offers insightful analyses of the advantages and disadvantages of DRL and IL for continuous control jobs. In addition, [4, 5] compare the scores of *Q*-learning DRL algorithm with imitation learning (passive imitation learning and active imitation learning) in various settings. They have implemented normal start (NS) versus random start (RS), zero velocity (ZV) versus random velocity (RV), noisy (*N*) versus deterministic (*D*) with six different types of data set. Their results show that DRL performed significantly better than passive and active imitation learning, with a higher learning curve and fewer crashes. Passive IL was prone to taking more risky actions than DRL, leading to more crashes in imitation learning than DRL. Future study in the field of reinforcement learning may be influenced by these findings, which also offer information on the relative performance of DRL and IL.

To identify the best racing line around a racetrack and reduce the time it takes to get from the starting line to the finish line, Paper [6] investigates several reinforcement learning algorithms. Three distinct algorithms are compared in this study: value iteration, *Q*-learning, and a *Q*-learning variant. The algorithms consider variables such as the bot's position in x and y coordinates, acceleration, and maximum speed. The paper also considers two diverse ways of handling crashes and restarts: resetting the bot to the start line with zero initial velocity or spawning it at the closest point on the track from the location of the crash. Overall, the study provides insights into the performance of different reinforcement learning algorithms for solving racetrack problems.

Van Moffaert and Nowé [7] suggests a technique for multi-objective reinforcement learning (MORL) that locates and preserves a collection of rules that make up the Pareto frontier. Unlike conventional reinforcement learning techniques that concentrate on a single target, this methodology enables the simultaneous optimization of many objectives. The study introduces the Pareto *Q*-learning algorithm, which can discover and preserve the Pareto frontier set of policies. The authors demonstrate the effectiveness of their method on a few benchmark problems, including the mountain car problem and the inverted pendulum problem. The findings indicate that, in terms of accuracy and diversity of the resulting policies, Pareto *Q*-learning performs better than conventional reinforcement learning techniques. Overall, the study makes a significant contribution in reinforcement learning by outlining a unique method for MORL that enables agents to simultaneously optimize several goals.

In multi-objective Markov decision problems (MOMDPs), [8] proposes a policy-based method to learn continuous approximation of the Pareto frontier. The suggested method generates a more accurate continuous approximation of the Pareto frontier by doing a single gradient-ascent run on a function that builds a manifold in the policy parameter space. The purpose is to bring the objective space as close to the Pareto frontier as is physically possible. The goal is to move the objective space as near as feasible to the Pareto frontier. The authors also go into the computation and estimation of the gradient as well as the definition of a measure for judging the viability of potential Pareto boundaries. The efficacy of the suggested strategy is demonstrated by empirical assessments on two MOMDPs.

3 Reinforcement Learning

An agent is taught to make decisions by engaging with the environment and receiving feedback as rewards or penalties for its actions using a machine learning technique called reinforcement learning. The agent's goal is to identify a course of action that maximizes its long-term gain. Through trial and error, the agent learns the actions that will most probably lead to positive outcomes and updates its policy accordingly. RL provides a powerful framework for agents to learn from experience and make decisions in complex and dynamic environments shown in Fig. 1.

3.1 *Single-Objective Reinforcement Learning (SORL)*

Using the single-objective reinforcement learning method, an agent learns to optimize a single reward signal. The agent is trained to carry out a particular task, and its goal is to maximize a single-objective function. It learns by taking actions, receiving a reward signal, and adjusting its behavior to maximize the reward. Single-objective reinforcement learning is often used in tasks where there is a clear and well-defined objective, such as playing a game or navigating a maze.

Some of the agents explored using single-objective reinforcement learning algorithms such as *Q*-learning, value iteration, deep *Q*-learning, A* heuristic approach, Asynchronous Advantage Actor Critic (A3C), passive and active iterative learning had their own drawbacks.

For an agent, *Q*-learning can be a computationally expensive algorithm, especially in the beginning of learning. The algorithm's slow convergence speed and lag in data transmission can also negatively impact its performance. When there is a lot of training data available, the deep *Q*-learning algorithm performs well but it is computationally expensive to train using complex and large-scale data models, requiring significant computational resources. The value iteration algorithm suffers if there are many total states since each iteration of the process must touch every

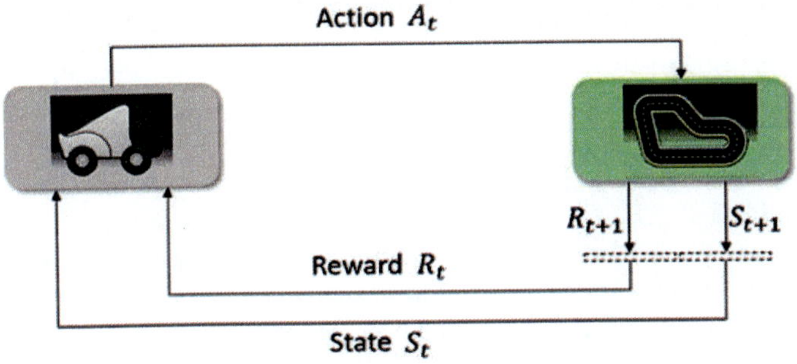

Fig. 1 Basic idea of how RL works in AWS DeepRacer. *Source* [9]

state. Passive imitation learning uses a data set and chooses the optimal action on each iteration. It does not seek a long-term goal and hence crashes more often than other algorithms.

To overcome these drawbacks and the inability to manage multiple conflicting objectives, MORL can be implemented.

3.2 *Multi-objective Reinforcement Learning (MORL)*

The goal of multi-objective RL is to concurrently optimize numerous objectives. In many real-world scenarios, there may be conflicting objectives, and the agent needs to find a balance between them.

For example, a robot may need to navigate through a cluttered environment quickly while avoiding obstacles and minimizing energy consumption. In deep-sea treasure problem, agent will have to collect treasures from under the water, while minimizing the amount of time spent underwater. Multi-objective reinforcement learning techniques enable the agent to optimize multiple objectives simultaneously, which can lead to more effective decision-making and better trade-offs.

MORL can deal with multiple conflicting objectives present in the problem. The scalar reward signal is multiplied by numerous feedback signals in multi-objective reinforcement learning (MORL), basically a single for every goal. The process of developing policies that simultaneously optimize many criteria is known as MORL. This helps eliminate the decision of choosing a scalar reward or scalarization of multiple rewards by the engineer.

MORL has the advantage of allowing agents to explore and learn the trade-offs between different objectives. The agent can learn which objectives are more important and which can be sacrificed to achieve better performance. This can lead to more optimal and robust decision-making, even when there are uncertainties in the environment or when the objectives change over time.

4 Methodology

4.1 *Pareto Q-Learning*

One of the most popular algorithms in the multi-objective reinforcement learning (MORL) field is Pareto Q-learning. Finding policies that concurrently optimize numerous objectives is what MORL is all about. An example of a value-based MORL algorithm is Pareto Q-learning, which aims to evaluate the value of every potential action in each potential state and use this knowledge to learn a policy.

Maintaining a set of Q-value functions, one for each objective being optimized, is how the method operates. Each Q-value function determines what reward is expected considering a specific action in a specific situation with respect to a specific goal. Employing a variation of the Q-learning update algorithm which takes into consideration all objectives being optimized, the collection of Q-value functions is updated iteratively.

A Pareto optimization strategy is used to identify the collection of policies that qualifies as Pareto optimal. This is the main concept behind Pareto Q-learning. If there is not another policy that is strictly superior in every aim, it is Pareto optimum. Pareto Q-learning maintains a set of non-dominated policies, which implies that no other policy from the set is strictly superior in all objectives, to discover this set of policies.

4.2 Architecture

Figure 2 shows the steps of our proposed method to make agent learn optimal policy. Firstly, all the hyper-parameters are initialized. These hyper-parameters are used to tweak the agent performance. Then the agent is trained for n training episodes. The agent tries to reach the finishing line in achieving maximum reward in minimum steps in each episode. On each step, an action is chosen based on the exploration rate and a random value. This action is used to compute the next state, based on which the agent receives a reward. To determine whether there has been a Pareto improvement, the reward is checked against the previous Q-values. Q-values are tuples of two values: primary Q-value and secondary Q-value. The primary Q-value corresponds to the expected reward for a given state-action pair, while the secondary Q-value corresponds to the expected time for that same state-action pair.

If there is Pareto improvement, the corresponding state-action pair's Q-values are updated to the new rewards, else to compute new Q-values Bellman's equation is used:

$$Q(s, a) = Q(s, a) + [\alpha * \left[R + \gamma m\left(Q\left(s', a'\right)\right) - Q(s, a)\right], \tag{1}$$

where $Q(s, a)$ stands for current estimate of action-value function for state s and action a. The variables α, γ_m and R stand for learning rate, maximum discount factor, and reward received. Next state and action are denoted by s' and a', respectively.

The agent repeats all the steps, starting from selecting the action until it reaches the terminal state, i.e., reaching the finish line. While the episodes are running, the agent continually updates the Q-values, learning a set of policies which slowly start converging. Once the Q-values have been converged, the agent seems to have learnt the optimal policy, maximizing the reward and minimizing the overall steps taken.

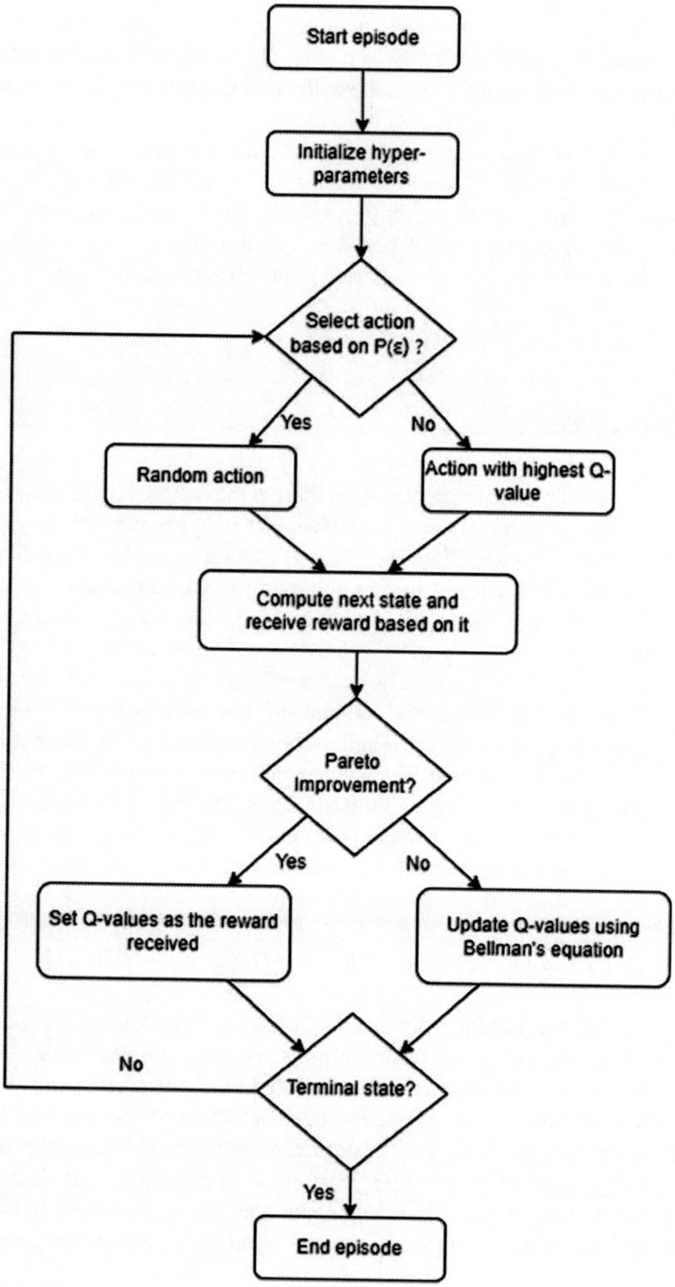

Fig. 2 Learning process of the agent

5 Implementation

5.1 States, Actions, and Rewards

The agent has two states: its position along the x-axis and y-axis. The agent can choose from eight different actions: move left, move right, move up, move down, or move diagonally on each corner. If there are any power-ups on the track, the agent's speed is doubled letting it take two steps instead of one. If the agent goes off the track, it is reset to the last valid position on the track and receives a negative penalty.

The MORL reward signal is a vector, not a scalar, and every component represents an objective. There are two rewards granted to the agent. The primary reward is the one that is given in accordance with the new state that the agent reaches after performing an action. It is positive if the agent crosses the finish line or lands on a power-up or negative if the agent leaves the track. As a secondary reward, the agent receives a penalty constant (negative reward) based on the number of steps it takes.

5.2 Algorithm

Algorithm 1: Track Learning Agent

```
Set the number of training episodes to n
for each training episode:
  Reset the environment
  while true:
    Choose an action using an epsilon-greedy policy
    if Pareto improvement:
      Update the Q-values to the new rewards received
    else:
      Compute new Q-values using Bellman's equation
    Current state = next state
    if terminal state:
      break;
```

5.3 Tracks Used

We have created two different sets of tracks that are used to teach the agent. Both sets contain tracks that differ in terms of size, difficulty, and features.

Set A: It has tracks with modest dimensions such as 10 * 10 and 20 * 20, making it fast enough to train the agent. The Q-table is likewise compact, making it easier to converge and learn optimal policies faster. Figure 3a displays track 1, which has a

(a) Track 1

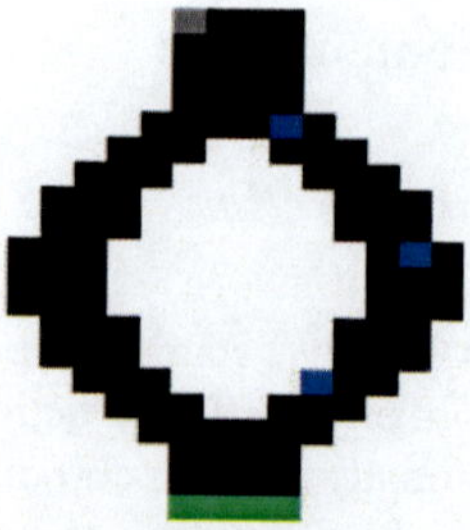

(b) Track 2

Fig. 3 Set A

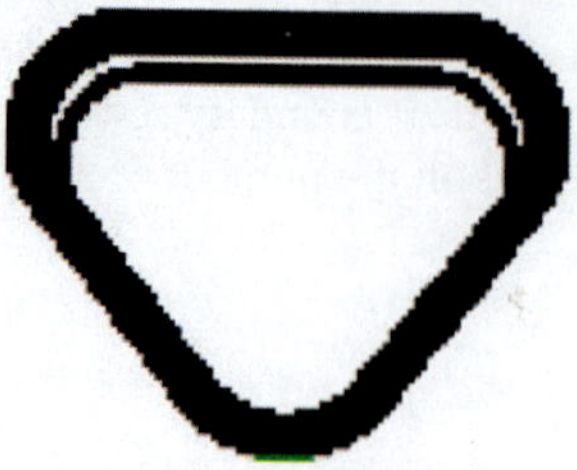

(a) Track 3

(b) Track 4

Fig. 4 Set B

right turn and two power-ups (colored blue). Track 2 is depicted in Fig. 3b, which has two paths to the finish line, one of which offers power-ups.

Set B: The tracks in this set are quite large. The start point and finish line are passed into a function, along with a track image. The agent is then trained using the matrix created from the image. Figure 4 shows a picture of tracks.

6 Results

The hyper-parameters of the algorithm have a significant impact on how well the training process goes. The learning rate, the exploration rate, and the discount factor are the three most important hyper-parameters in the current situation. Additional criteria include the reward system's penalty constant. The selected hyper-parameters are compiled in Table 1 after numerous iterations.

After running the agent on multiple different tracks, we can see that the agent has successfully attained the optimal policy, incorporating both objectives. We have used three different metrics to evaluate the agents' performance.

Table 1 Hyper-parameters

Parameter	Value
Learning rate (α)	0.1
Discount factor (γ)	0.9
Exploration rate (ε)	0.1
Penalty constant	− 1

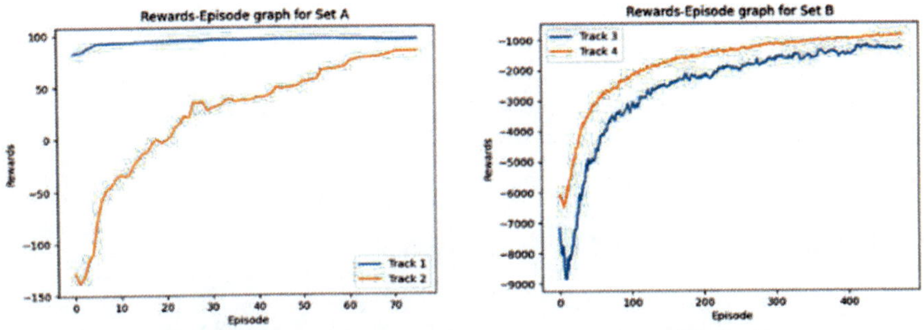

(a) Set A (b) Set B

Fig. 5 Primary reward-episode graph

6.1 Reward-Episode Graph

The reward-episode graph displays the convergence of the agent reaching the optimal policy. It shows the agent's learning process, starting from episode 1 until it starts converging. Figure 5 displays the graph of the agent on both sets of tracks. Since the tracks of Set B are large, the agent requires more episodes to learn the Pareto optimal policies.

6.2 Crash-Episode Graph

Figure 6 depicts the number of crashes encountered by the agent during the training process on the tracks. It illustrates the agent's learning progress and its ability to navigate the track successfully without crashing. As the number of training episodes increases, agent's performance improves which helps assess the effectiveness of the agent.

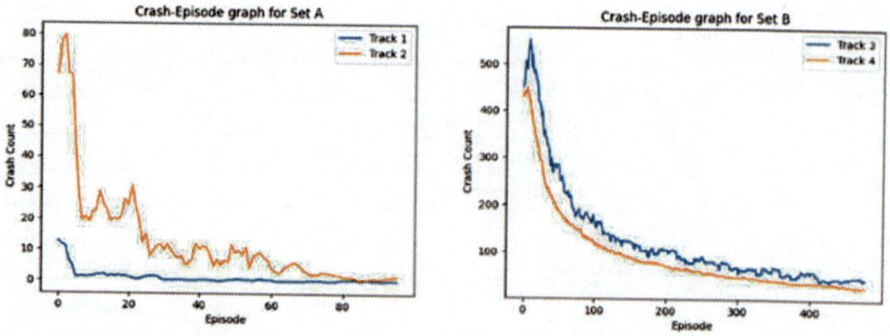

(a) Set A (b) Set B

Fig. 6 Crash-episode graph

6.3 Axiomatic Metric for the Objectives

Figure 7 illustrates that the agent learns the optimal policy keeping both the objectives into consideration. It maximizes the reward and minimizes the number of the steps. The dense part of the scatter plot depicts that the agent has converged.

The axiomatic metric for track 1 is shown in Fig. 7a. Most of the data points are in the bottom right corner, indicating that the agent has maximized the objective on the x-axis and minimized the objective on the y-axis. Figure 7d shows a straight line of data points that begin in the top left and become more numerous toward the bottom right. This demonstrates how the agent is learning from each episode and improving. Since track 1 is short, it was simpler for the agent to arrive at the best course of action than it was for track 4, which is more challenging and requires more time.

7 Conclusion and Future Work

This paper presents a multi-objective reinforcement learning-based track learning agent which accomplishes goals that are conflicting in nature, such as reaching the finish line (maximizing the reward) and minimizing the number of steps and collisions. The need for incorporating multiple conflicting goals for our agent led us to employ a novel approach of multi-objective reinforcement learning. Multiple MORL algorithms were researched, and Pareto Q-learning was found to be optimal to train the track learning agent. Our result concludes that the novel approach of MORL caters to the need for incorporating multiple objective goals and helps the agent find the optimal policy without the need of scalarizing the rewards or finding only a sub-optimal policy.

There exist many future improvements of this work. The features can be tweaked according to one's requirements. For example, in real-life driving scenarios, comfort

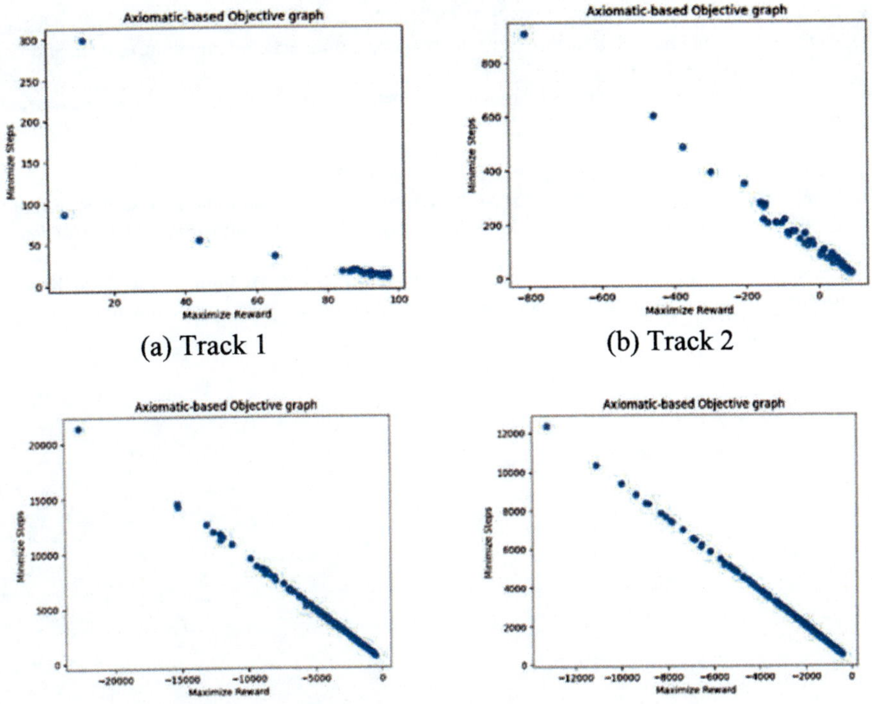

(a) Track 1 (b) Track 2

(c) Track 3 (d) Track 4

Fig. 7 Axiomatic graph of objectives on all tracks

is more important than speed so an objective comfort can be added incurring penalty if comfort is compromised. While in race car games and simulators, speed is given more priority than comfort. Steering angle and speed can also be incorporated as actions to the agent.

References

1. Hayes CF, Rădulescu R, Bargiacchi E et al (2022) A practical guide to multi-objective reinforcement learning and planning. Auton Agent Multi-Agent Syst 36:26. https://doi.org/10.1007/s10458-022-09552-y
2. Nguyen TT, Nguyen ND, Vamplew P, Nahavandi S, Dazeley R, Lim CP (2020) A multi-objective deep reinforcement learning framework. Eng Appl Artif Intell 96:103915
3. Wiering MA, Withagen M, Drugan MM (2014) Model-based multi-objective reinforcement learning. In: 2014 IEEE symposium on adaptive dynamic programming and reinforcement learning (ADPRL). IEEE, pp 1–6
4. Gros TP, Höller D, Hoffmann J, Wolf V (2020) Tracking the race between deep reinforcement learning and imitation learning—extended version. arXiv:2008.00766

5. Gros TP, Höller D, Hoffmann J, Wolf V (2020) Tracking the race between deep reinforcement learning and imitation learning. In: International conference on quantitative evaluation of systems. Springer, Cham, pp 11–17
6. Mohan V (2014) A comparison of various reinforcement learning algorithms to solve racetrack problems
7. Van Moffaert K, Nowé A (2014) Multi-objective reinforcement learning using sets of Pareto dominating policies. J Mach Learn Res 15(1):3483–3512
8. Pirotta M, Parisi S, Restelli M (2015) Multi-objective reinforcement learning with continuous Pareto frontier approximation. Proc AAAI Conf Artif Intell 29(1). https://doi.org/10.1609/aaai.v29i1.9617
9. Owen L (2020) Reinforcement learning in autonomous race car. Medium, 17 Oct 2020. https://towardsdatascience.com/reinforcement-learning-in-autonomous-race-car-c25822def9f8

Performance Assessment of Gaussian Filter-Based Image Fusion Algorithm

Kesari Eswar Bhageerath, Ashapurna Marndi, and D. N. D. Harini

Abstract Image fusion plays a vital role in many fields. Especially, fusion of infrared and visible images has high importance in every scenario from computer vision to medical sector. The objective of this work is to develop an effective method for producing clear objects with high spatial resolution along with background information by fusing infrared (IR) and visible (VIS) images. This integrated image can be efficiently utilized by humans or machines. To achieve this objective, we propose the use of Multi-Layer Bilateral Filtering (BF) and Gaussian Filtering (GF) techniques, which improvises the skewness and kurtosis of fused images. While the BF technique consistently produces higher quality images, the GF approach outperforms it by 86% in terms of statistical measures such as skewness and kurtosis. The findings demonstrate that the GF technique yields outputs with reduced noise and improved visual appeal. In this paper, we compare the assessment metrics of several outputs for both single images and a set of 100 images.

Keywords Infrared image · Visible image · Bilateral filter · Image fusion · Gaussian filter

K. E. Bhageerath (✉) · D. N. D. Harini
Computer Science and Engineering, Gayatri Vidya Parishad College of Engineering (Autonomous), Madhurawada, Visakhapatnam, Andhra Pradesh 530048, India
e-mail: eswarbhageerath2@gmail.com

D. N. D. Harini
e-mail: harinidhara@gvpce.ac.in

A. Marndi
Council of Scientific and Industrial Research-Fourth Paradigm Institute, Bangalore, Karnataka 560037, India

Academy of Scientific and Innovative Research, Ghaziabad, Uttar Pradesh 201002, India

A. Marndi
e-mail: asha@csir4pi.in

S. Kumar et al. (eds.), *Fourth Congress on Intelligent Systems*, Lecture Notes in Networks and Systems 868, https://doi.org/10.1007/978-981-99-9037-5_4

1 Introduction

Image fusion is a data fusion technique that represents the signal using individual pixels of an image. It involves combination of information from two or more images into a single image. Image fusion has five main subcategories, namely, multi-sensor fusion, multi-view fusion, multimodal fusion, multi-focus fusion, and multitemporal fusion. Among them, multi-sensor fusion plays a pivotal role in various domains including robotics, autonomous systems, surveillance, aerospace, healthcare, environmental monitoring, etc. We fuse the information from multiple sensors to get more accurate information and better performance. In this paper, the information is more toward multi-sensor fusion, which involves combining images captured for the same scene with different sensors. This is also observed more for satellite images, which are often taken with multiple color sensors and then processed to create a final image which has all the necessary information about the scene.

More information can be gained when a scene is captured using both visible (VIS) and infrared (IR) imaging systems, but more redundant information is also produced. Since the infrared sensor gathers data of an object's thermal radiation in a scene, it is possible to detect an object even in dim illumination [1]. Images captured by visible light sensors typically possess greater spectral data, sharper texture details, and higher spatial resolution [2]. Consequently, when infrared and visible images are fused, it leads to more clear description of the scene. Furthermore, fusion of the two types of images can enable observers to easily analyze the scenario and efficiently process the information present in the scene [3]. As a result of this, most of the unnecessary data can be deleted. Image fusion techniques are used in various applications such as video surveillance [4], object detection [5], and object tracking [6]. At present, most of the techniques available for fusing visible and infrared images are based on multi-scale transformation techniques [7, 8].

Commonly employed multi-scale transformation techniques include wavelet transformation, Non-Subsampled Shearlet Transform (NSST), and pyramid transformation [9].

Due to their user-friendly implementation and capability to maintain edge details in images, the utilization of edge-preserving filters in image processing has become increasingly popular in research. By employing multi-visual weight data and a multi-scale transformation model based on iterative guided filtering, image fusion can be effectively accomplished [10, 11]. Using this approach, the issue of spatial overlap is resolved and artifacts are prevented. Additionally, a multi-level GF image decomposition method was explored for the purpose of fusing infrared and visible images [12]. While the fusion technique may become more complex, the use of edge-preserving filtering remains advantageous due to its simplicity and effectiveness.

2 Methodology

Image fusion of IR and VIS is performed to generate a composite image that incorporates the maximum amount of information feasible. Both people and machines can easily understand the semantics of this fused image. Coarse-detail, fine-detail, and base layers are generated by fusing IR and VIS images. Because of the positional diversity of images collected by various sensors, image registration is necessary before image fusion [13].

2.1 Gaussian Filter

Gaussian Filter has many useful characteristics such as smoothing, preserving edges, separability. So, it is widely used in image processing and computer vision. It is basically a type of filter that can preserve important structural information in an image. For Gaussian Filter to work, input images are convolved with Gaussian kernel, which is a matrix of weights centered on a central pixel, where the weights of the kernel are calculated using the Gaussian function.

2.2 Bilateral Filter

The Bilateral Filter is a non-linear filter both spatial proximity and intensity similarity between pixels are taken into consideration. The process of performing the Bilateral Filtering is by applying a weighted average of neighboring pixels where the weights depend on both spatial distance and difference in intensities of the center pixel and its neighboring pixels. As additional intensity similarity calculations are required in Bilateral Filter, it is generally computationally intensive when compared to the Gaussian Filter.

2.3 Regional Variance Maximization (RVM)

Regional variance maximization (RVM) is a technique used for image and its contrast enhancement. By maximizing the local variance values, it enhances the important details in specific regions of the image.

2.4 Local Contrast (LC)-Based Saliency Detection

The goal of Local Contrast (LC)-based technique is to identify salient or visually distinctive regions in an image by considering contrast in different areas.

2.5 Image Fusion: Decomposition Model Based on Gaussian Filtering

We propose to decompose infrared and visible images into base layer and detail layer using Gaussian Filter. Figure 1 illustrates the schematic diagram of the fusion algorithm proposed in this paper. As depicted in Eq. (1), a three-layer Gaussian Filter is applied to the feature information of the original images (IR and VIS) for dividing the images into a base layer and a detail layer. Fusion of detail and base layer is carried out based on the technique of regional variance maximization and combining fusion weight map for the base layer respectively [14]. The fusion weight map for the base layer is generated using LC saliency detection method and employed to direct the fusion process. Finally, the fusion is done through linear transformation method.

The proposed image fusion algorithm has the following steps:

I. Three-layer Gaussian Filter

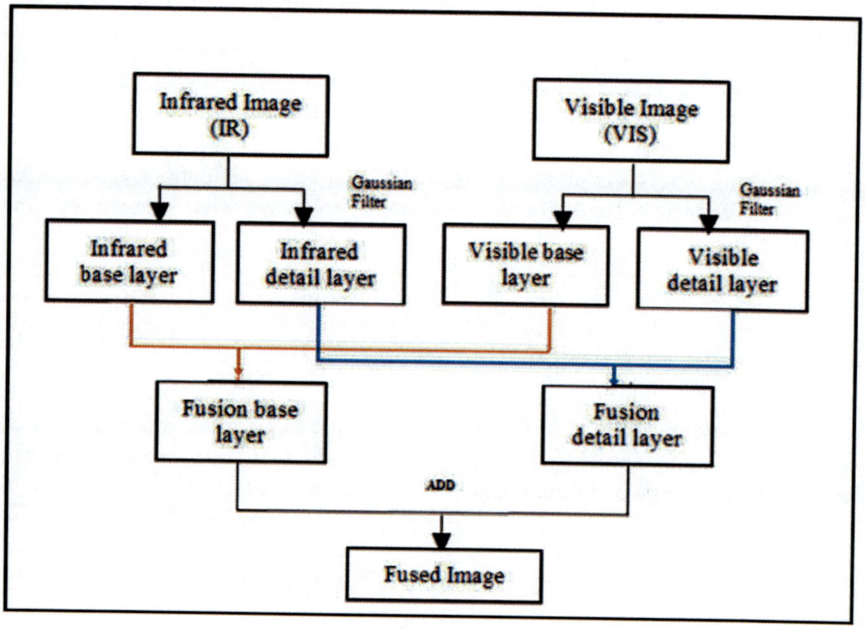

Fig. 1 Schematic diagram of the proposed algorithm

The original IR and VIS images are filtered using a three-layer Gaussian algorithm. With this filtering, the images are divided into two layers: the base layer and detail layer. Base layer has the overall structure, and the detail layer has more minute and specific features.

II. Fusion of Detail Layers

Through the application of regional variance maximization, the detail layers of both IR and VIS images are combined. Through this method, it is guaranteed that the fused detail layer retains the most important data, which enhances the quality of final image.

III. Fusion Weight Map for Base Layer

LC saliency detection method is used to create a fusion weight map for the base layer. Weight maps are used in the fusion process for the base layers. By including the saliency information, important regions are more prioritized which lead to better final results.

IV. Fusion Through Linear Transformation

Linear transformation technique is used to fuse base and detail layers, along with their respective fusion weight maps. This controlled fusion of IR and VIS images makes sure that important features from both images come together in the final combined image effectively.

The infrared and visible images are represented by I_{IR} and I_{VIS}, respectively. The detail layer and base layer are represented by D and B, respectively.

Equation (1) represents the result of performing Gaussian Filtering consecutively three times.

$$\begin{aligned} D1_{\text{IR}} &= \text{GF}(I_{\text{IR}}), D1_{\text{VIS}} = \text{GF}(I_{\text{VIS}}) \\ D2_{\text{IR}} &= \text{GF}(I_{\text{IR}} - D1_{\text{IR}}), D2_{\text{VIS}} = \text{GF}(I_{\text{VIS}} - D1_{\text{VIS}}) \\ D3_{\text{IR}} &= \text{GF}(I_{\text{IR}} - D1_{\text{IR}} - D2_{\text{IR}}), \\ D3_{\text{VIS}} &= \text{GF}(I_{\text{VIS}} - D1_{\text{VIS}} - D2_{\text{VIS}}) \end{aligned} \tag{1}$$

Equation (2) can be used to derive the decomposition model.

$$\begin{aligned} D_{\text{IR}} &= \{D1_{\text{IR}}, D2_{\text{IR}}, D3_{\text{IR}}\} \\ D_{\text{VIS}} &= \{D1_{\text{VIS}}, D2_{\text{VIS}}, D3_{\text{VIS}}\} \\ B_{\text{IR}} &= \{I_{\text{IR}} - D1_{\text{IR}} - D2_{\text{IR}} - D2_{\text{IR}}\} \\ B_{\text{VIS}} &= \{I_{\text{VIS}} - D1_{\text{VIS}} - D2_{\text{VIS}} - D2_{\text{VIS}}\} \end{aligned} \tag{2}$$

Some more research work related to image fusion using Gaussian Filtering could be found in [15].

2.6 Experimental Data

We have used images available in Maritime Imagery in the Visible and Infrared Spectrums [Dataset 12] in OTCBVS Benchmark dataset [http://vcipl-okstate.org/pbvs/bench/] [16] as input for testing various methods in this paper.

This image set contains simultaneously acquired unregistered thermal and visible images of unique 264 ships acquired from total 1088 piers. Total images lie into 6 basic categories. Visible and infrared images are captured from sensor 'ISVI IC-C25' and sensor 'Sofradir-EC Atom 1024', respectively.

3 Experimental Results and Analysis

This work has initial conditions of the experiment, including the experimental platform, comparison algorithm to ensure a best evaluation of algorithm's performance, the entire experiment was run on the experimental platform using a Windows 11 Operating System, Python 3.8, and the Google Colab editor.

As depicted in Fig. 2a, VIS images possess more detailed spatial information and Fig. 2b illustrates the distinct edges and structural features present in IR images. Consequently, before performing the fusion process, the key characteristics of the two distinct channels of the layers must be extracted.

The purpose of fusing infrared (IR) and visible (VIS) images is to integrate essential information from the IR image into the VIS image. The resulting fused images are shown in Fig. 2c, d. To demonstrate the effectiveness of the proposed technique, this paper compares the values of skewness, kurtosis, peak signal-to-noise ratio (PSNR), and image similarity ratio.

The PSNR block calculates the peak signal-to-noise ratio, measured in decibels, between two images. This ratio is utilized to assess the quality of the original and fused images. Prior to computing the PSNR, the block determines the mean-squared error using the following Eq. (3):

$$\text{MSE} = \frac{\sum |I_1(m,n) - I_2(m,n)|^2}{M * N} \tag{3}$$

The input images contain m rows and n columns. The block then uses Eq. (4) to get the PSNR:

$$\text{PSNR} = 10 \log_{10}\left(\frac{R^2}{\text{MSE}}\right), \tag{4}$$

where R refers to maximum fluctuation in the data type of input image.

Image Similarity Ratio: Image similarity is a measure of the degree of resemblance between two images which gauges the proximity of the intensity patterns of those

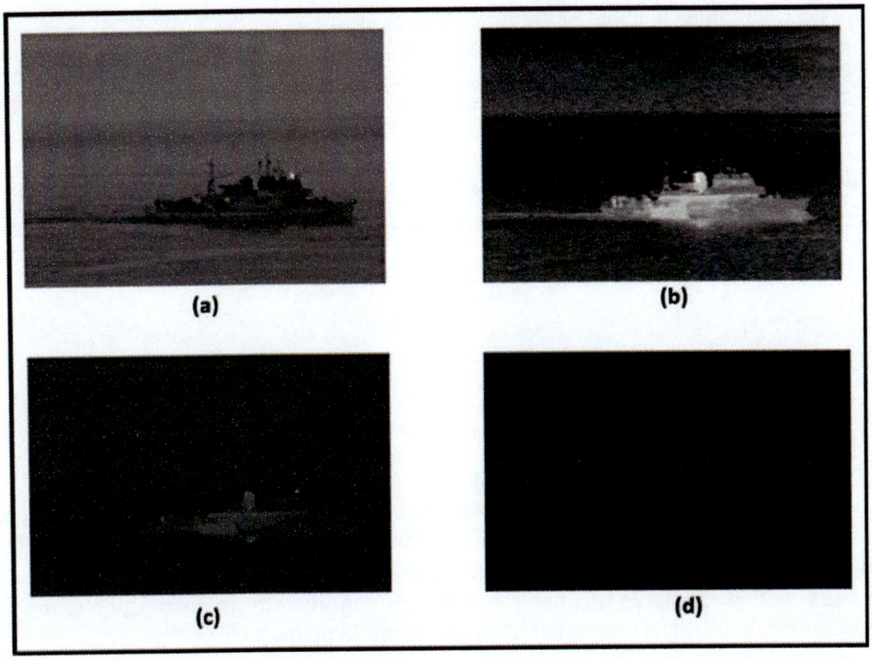

Fig. 2 The images of 'Ship' and the fused outcomes achieved by different algorithms. **a** VIS [16], **b** IR [16], **c** fused image using bilateral filtering method, **d** fused image using Gaussian filtering method

two images. Better image quality is associated with a higher image similarity ratio. For various windows inside an image, the Structural Similarity index (SSIM index) is calculated using Eq. (5). The distance between two windows, which are of the same size $n \times n$ is:

$$\mathrm{SSIM}(x, y) = \frac{(2\mu_x\mu_y + c_1)(2\sigma_{xy} + c_2)}{(\mu_x^2 + \mu_y^2 + c_1)(\sigma_x^2 + \sigma_y^2 + c_2)} \tag{5}$$

where

$\mu_x =$ pixel sample mean of x, $\mu_y =$ pixel sample mean of y,
$\sigma_x^2 =$ variance of x, $\sigma_y^2 =$ variance of y,
$\sigma_{xy} = x$ and y covariance,
$C_1 = (K_1L)^2$, $C_2 = (K_2L)^2$,
$L =$ pixel values dynamic range, $K_1 = 0.01$ and $K_2 = 0.03$.

Skewness: It is a measure of an image distribution's asymmetry. Distribution of pixel intensities can be evaluated using skewness within a picture. The intensities of pixels in an image indicate whether a certain area is bright or dark. The skewness of the pixel intensity distribution explains whether the distribution is symmetrical or it is toward the high or low end of the intensity range.

$$\text{Skewness} = \frac{\sum_i^N (X_i - \overline{X})^3}{(N-1) * \sigma^3}, \tag{6}$$

where

N is number of variables available in the pixel data distribution.
X_i is the value of current pixel

$$\overline{X} = \frac{\sum_i^N X_i}{N}$$

$$\sigma = \sqrt{\frac{\sum_i^N (X_i - \overline{X})^2}{N}}$$

Kurtosis: It is a way to measure how much the brightness of pixels in an image differs from what is considered normal. When the kurtosis value of image is high, there are many pixels with extremely high or low values. Images with low kurtosis values are better suitable for analyses that require more uniform distribution of pixel intensities. Kurtosis is calculated using the formula below.

$$k = \frac{n \sum_{i=1}^n (X_i - \overline{X})^4}{\left(\sum_{i=1}^n (X_i - \overline{X})^2\right)^2} - 3 \tag{7}$$

$$X = \frac{\sum_i^N X_i}{N} \tag{8}$$

Performance of Gaussian Filter and Bilateral Filter in terms of different performance metrics is presented in Table 1.

1. Image Similarity Ratio

The values of similarity ratios are higher for images processed using Bilateral Filter method (BF) when compared to the Gaussian Filter method (GF). For the VIS image, value of similarity ratio is 0.75 for GF method, whereas it is 1.50 for BF method. Similarly, for the IR image, the value of similarity ratio using BF method is 1.51, and similarity ratio using GF method is 0.68, which is considerably lesser.

2. Peak Signal-to-Noise Ratio (PSNR)

The PSNR values for both VIS and IR images are higher for the BF method compared to the GF method. When calculating for 100 images, for the VIS images, the value of PSNR using BF method is 7.83, while it is 5.62 when GF method is used. Similarly, for the IR images, the PSNR value using BF method is 7.95, while it is 6.88 when GF method is used.

Table 1 Evaluation metrics

Parameters	BF method	GF method	BF method (for 100 images)	GF method (for 100 images)
Image similarity ratio between VIS image and output image	**0.75**	0.31	**1.50***	0.15*
Image similarity ratio between IR image and output image	**0.68**	0.12	**1.51***	0.07*
PSNR between VIS image and output image	**28.17**	27.97	**7.83***	5.62*
PSNR between IR image and output image	**27.86**	27.69	**7.95***	6.88*
Skewness of output image	1062.739	**355.817**	1010.354	**342.742**
Kurtosis of the output image	201,344.225	**33,757.768**	200,571.104	**32,100.953**

*Average Value for 100 values, The numbers marked as bold signify the best results in each comparison

3. Skewness and Kurtosis

The GF method has better performance than BF method in terms of skewness and kurtosis. The GF method has significantly lower values of skewness and kurtosis, indicating that fused images generated through GF method are uniform.

4 Conclusion

This study compares two image fusion algorithms, (Decomposition of image into base layer and detailed layer using Bilateral Filtering and Gaussian Filter). The findings indicate that in terms of image quality and processing speed, the Bilateral Filter method gives better results than Gaussian Filter method in all cases. However, the performance of the Gaussian Filter method outperforms the Bilateral Filtering approach by 86% in terms of the statistical metrics skewness and kurtosis. Outputs produced by Gaussian Filter have reduced noise and enhanced visual quality. In this study, various outputs are compared for one single image and 100 images in terms of their evaluation metrics.

Acknowledgements The authors acknowledge sources of the datasets used for testing the image processing techniques in this research paper. First two authors (Kesari Eswar Bhageerath (KEB) and

Ashapurna Marandi) express their sincere thanks to Head CSIR Fourth Paradigm Institute (CSIR-4PI) to carry out this research work, also KEB wish to thank Dr Ashapurna Marandi to carry out internship under her guidance in CSIR-4PI. He also expresses his sincere gratitude and appreciation to the Head and the faculty in the Computer Science Dept of Gayatri Vidya Parishad College of Engineering for their encouragement and guidance.

References

1. Zhou Z, Wang B, Li S, Dong M (2016) Perceptual fusion of infrared and visible images through a hybrid multi-scale decomposition with Gaussian and bilateral filters. Inf Fusion 30:15–26
2. Ma J, Zhou Z, Wang B, Hua Z (2017) Infrared and visible image fusion based on visual saliency map and weighted least square optimization. Infrared Phys Technol 82:8–17
3. Li S, Kang X, Fang L, Hu J, Yin H (2017) Pixel-level image fusion: a survey of the state of the art. Inf Fusion 33:100–112
4. Kumar P, Mittal A, Kumar P (2006) Fusion of thermal infrared and visible spectrum video for robust surveillance. In: Indian conference on computer vision, graphics and image processing, pp 528–539
5. Guan D, Cao Y, Yang J, Cao Y, Tisse C (2018) Exploiting fusion architectures for multispectral pedestrian detection and segmentation. Appl Opt 57(18)(D):108–116
6. Qian X, Han L, Cheng Y (2014) An object tracking method based on local matting for night fusion image. Infrared Phys Technol 67:455–461
7. Li J, Peng Y, Jiang T (2021) Embedded real-time infrared and visible image fusion for UAV surveillance. J Real-Time Image Process 18:2331–2345
8. Li H, Wu X-J, Kittler J (2020) MDLatLRR: a novel decomposition method for infrared and visible image fusion. IEEE Trans Image Process 29:4733–4746
9. Liu X, Mei W, Du H (2017) Structure tensor and nonsubsampled shearlet transform based algorithm for CT and MRI image fusion. Neurocomputing 235:131–139
10. Zhu HR, Liu YQ, Zhang WY (2019) Infrared and visible image fusion based on iterative guided filtering and multi-visual weight information. Acta Photonica Sinica 48(3):0310002
11. Karim S, Tong G, Li J, Qadir A, Farooq U, Yiting Y (2023) Current advances and future perspectives of image fusion: a comprehensive review. Inf Fusion 90:185–217
12. Tan W, Zhou H et al (2019) Infrared and visible image perceptive fusion through multi-level Gaussian curvature filtering image decomposition. Appl Opt 58:3064–3073
13. Ma J, Qiu W, Zhao J, Ma Y, Yuille A, Tu Z (2015) Robust L2E estimation of transformation for non-rigid registration. IEEE Trans Signal Process 63:1115–1129
14. Wang C, Yang G, Sun D, Zuo J, Li Z, Ma X (2021) A novel lightweight infrared and visible image fusion algorithm. In: 2021 international conference of optical imaging and measurement (ICOIM), 978-1-6654-0354-2/21. https://doi.org/10.1109/ICOIM52180.2021.9524368
15. Hu Y, He J, Xu L (2021) Infrared and visible image fusion based on multiscale decomposition with Gaussian and co-occurrence filters. In: 4th international conference on pattern recognition and artificial intelligence (PRAI), Yibin, China, pp 46–50. https://doi.org/10.1109/PRAI53619.2021.9551089
16. Zhang MM, Choi J, Daniilidis K, Wolf MT, Kanan C (2015) VAIS: a dataset for recognizing maritime imagery in the visible and infrared spectrums [IEEE OTCBVS WS series bench]. In: Proceedings of the 11th IEEE workshop on perception beyond the visible spectrum (PBVS-2015)

A Cognitive Comparative Analysis of Geometric Shape-Based Cryptosystem

K. R. Pruthvi Kumar, Anjan K. Koundinya, S. Harsha, G. S. Nagaraja, and Sasidhar Babu Suvanam

Abstract Cryptography is one of the most important and widely utilized applications in our daily lives, particularly in the protection of user data and information numerous organizations such as banks, government institutions, and communication companies require the use of a cryptosystem to safeguard their data during Internet transmissions and ensure secure transfer from the sender to the receiver. Cognitive cities are regularly automating the day-to-day urban processes and constantly expanding the objective-driven communities' collection to share the personal data that must be stored securely. The cloud provides a desirable platform for cognitive smart cities to access user data, enabling them to adapt their current actions and learn from past experiences. Various algorithms are used in cryptosystems to secure user data and information by encryption and decryption in all fields. Symmetric and asymmetric are the two types of cryptographic algorithms, which are used to secure user communication. Cryptography assists users in achieving data confidentiality, integrity, availability, authentication, and non-repudiation. In this paper, various methodologies like the ElGamal algorithm, RSA algorithm, Ring algorithm, and Hermitian curve algorithm are used for geometric shape cryptosystems to secure the data effectively. Cryptosystem algorithms enable high-security performance,

K. R. P. Kumar (✉)
Department of Computer Science and Engineering, BMS Institute of Technology and Management, Visvesveraya Technological University, Belagavi 590018, India
e-mail: pruthvikumar31@gmail.com

A. K. Koundinya
Department of Information Science and Engineering (Cyber Security), BMS Institute of Technology and Management, Visvesveraya Technological University, Belagavi 590018, India

S. Harsha
Department of Artificial Intelligence and Machine Learning, RNS Institute of Technology, Visvesveraya Technological University, Belagavi 590018, India

G. S. Nagaraja
Department of Computer Science and Engineering, RV College of Engineering, Visvesveraya Technological University, Belagavi 590018, India

S. B. Suvanam
Department of Computer Science and Engineering, Presidency University, Bangalore, India

S. Kumar et al. (eds.), *Fourth Congress on Intelligent Systems*, Lecture Notes in Networks and Systems 868, https://doi.org/10.1007/978-981-99-9037-5_5

effectiveness, high encryption effect, low time complexity, and low computation complexity, as demonstrated in this survey.

Keywords Cryptography · Cryptosystem · Decryption · ElGamal · Encryption · Geometric shape · Hermitian curve · Ring · RSA

1 Introduction

In recent years, since the advancement of machine learning as a subfield of artificial intelligence, every field has seen an increase in the desire to use it for a variety of purposes to improve their capabilities or domains. This is also true of cryptography, on which numerous researchers have been working in the last ten years have made attempts to combine machine learning and cryptography. The majority of research in this field, however, has focused on integrating numerous cryptosystems in conjunction with a machine learning framework.

This field's applications are in high demand due to their volume and features. Machine learning will deliver the best solution for all applications. For example, the spam detection algorithm will not produce 100% correct results when looking for spam. Spam and non-spam data files are fed into machines on a huge scale to achieve the best solution. In general, more training is required for the machine to learn and produce the correct results. Machine learning techniques are utilized in various applications in business, commerce, and Walmart to discover the best forecast based on analysis.

The hallmark of cryptography in the realm of science is the ability to demonstrate that any data in any form (text, image, audio, video, numerical, etc.) in a data connection is safe. The conversion of data from plain to cipher and cipher to plain with or without the use of a key is the distinguishing feature of a cryptosystem. The key generation process is either symmetric (private key system) or asymmetric (public key system). The holy grail of this cryptosystem is creating a stable environment using a novel geometric method. With the help of multi-dimensional geometric shapes.

The following is the outline of the paper. The Literature Survey is covered in Sect. 2. The taxonomy of geometric shape-based cryptosystems is covered in Sect. 3. Section 4 compares and contrasts several models. The challenges faced during this research work are presented and elaborated in Sect. 5. In Sect. 6, concluding remarks and possible future research aspects are discussed.

2 Literature Survey

The term "Cryptography" is derived from the words of Greek such as "kryptos" represent hiding and "graphien" representing write, which describes the data hiding from unauthorized users while storing and transmitting [1]. The use of cryptography is crucial in maintaining the security of information shared between a sender and receiver during transmission over a network. Encryption is critical in protecting the user's identity, transactions, and other information from attackers, eavesdroppers, and unauthorized users. A large number of attacks have occurred between two authorized devices when unknown users attempt to intercept the communication. Security of data is very important to maintain, especially sensitive information that should be limited to certain users [2]. Cryptography used a wide variety of algorithms or ciphers and protocols to maintain effective data security. For data encryption, methods are primarily divided into two keys: symmetric key systems that use a single alphanumeric string/number for both encryption and decryption. This is also known as the private key. The other option is to use two keys where one is used for encryption and is public, the other key used for decryption is secret. This is an asymmetric key system. Some well-known examples of symmetric key or secret key encryption are AES, DES, RC2, RC4, RC5, and RC6. Similarly, some well-known asymmetric key cryptosystems are RSA, DSA, and ZKP [3].

The main objective of cryptosystem techniques is converting user data or information into a non-readable form. The data is only obtained by the authorized user and, in any case, the attacker never attempts to access the database server or the secret information of the user [4]. Cryptosystem is mainly divided into encryption and decryption. Encryption is the method of converting plaintext into a secret code that is unreadable by hackers and is known as ciphertext. Decryption is the process of reversing a process of encryption and it converts the ciphertext into plaintext [5]. Many cryptographic techniques have been developed to protect the data from cloud computing of different cloud service models and various deployment models such as private, public, and hybrid models [6]. Scientific research, tools, applications, scientists, computer systems, and software applications all progress with time and contribute to the breakdown of a secure cryptosystem [7]. Image encryption plays an important role and it provides some information, so that image is kept to be private and transmitted securely. The encryption process involved distorting the image pixel intensity for creating the cipher image as input, which is unrelated to the input image. By using the secret key, the receiver can decrypt the image into the original image [8].

This article analyze the issues of performance and security in the data considering different algorithm. The encryption algorithm employed key data, which is binary data of equal size to the original data. It also examines cryptography approaches for encrypting and decrypting digital data that use symmetric, asymmetric, fractal, and other methods. The public key carries the message in any operation, while the private key is kept secret, increasing data security.

3 Taxonomy of Geometric Shape Cryptosystem

The geometric shape-based cryptosystem is mainly divided into key-based and keyless algorithms. The key-based algorithm is divided into symmetric and asymmetric algorithms. The keyless contains fractal and other methods. Figure 1 shows the block diagram of geometric shape cryptosystem.

3.1 *Geometric Cryptosystem*

The cryptosystem is a structure that contains the set of algorithms used to convert the plaintext message into a ciphertext for encoding and ciphertext to plaintext for decoding the message securely. There is always a need of new advanced algorithms for strong cipher generation and there are many algorithms generated using mathematical fields. Geometric field is the one that needs to be explored more for long lasting security.

Any cryptosystem consists of three main algorithms such as key generation, encryption, and decryption. The geometric cryptosystem is a cryptology area that deals with different 2D and 3D geometric shapes and the ciphertexts of the messages are represented by using the geometric size as intervals and angles with the help of a compass and ruler. The geometric cryptosystem is mainly divided into key-based and keyless cryptosystems.

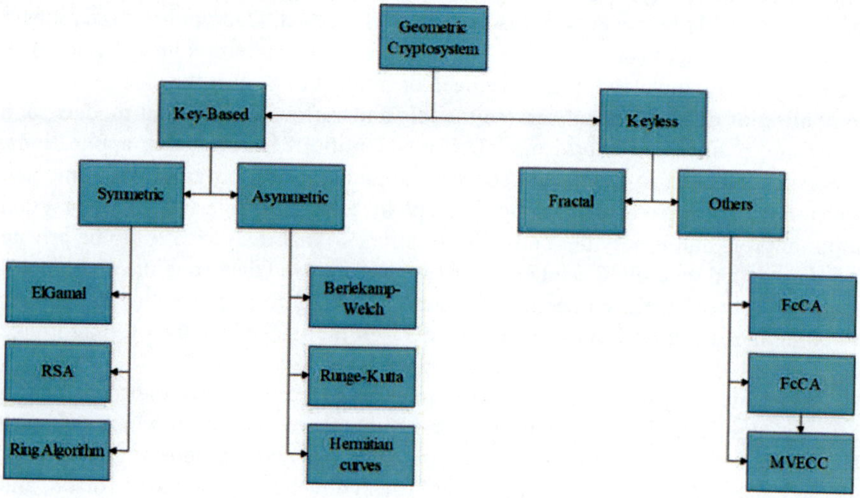

Fig. 1 Block diagram of geometric shape cryptosystem

Key-Based Cryptosystem

The Key-based cryptosystem is a piece of information that usually have a string of numbers, letters, and characters are stored in file or documents, and which is processed through the cryptographic algorithm. Two keys are used in cryptography: one for encryption and another for decryption. The key-based cryptosystem is mainly divided into symmetric and asymmetric keys.

Symmetric Key

The symmetric key algorithm, which is frequently referred to as secret key encryption, uses a single key for both the encryption of the plaintext and the decoding of the ciphertext. The different types of symmetric key algorithms are discussed as follows.

ElGamal Algorithm

Geometric shape-based features were used by Anitha Kumari et al. [9] to create a 3D-based ElGamal and Diffie Kellman (DH) technique using a two-server Password Authentication Key Exchange (PAKE) protocol. With the aid of geometrical qualities like circumcenter (ω) and the medians (θ), this method ensures a strong password for a strong encryption. There are certain limitations to the ElGamal method, such as reduced communication and computing complexity round, which results in poor computational efficiency.

Rivest-Shamir-Adleman (RSA)

Liang et al. [10] implemented a symmetric and optimization encryption method for protecting the 3-dimensional mesh models. A solution for numerous objective was used to detect a mapping range optimize for both encryption and decryption effectively. For establishing the encryption framework, the proposed method increases the encrypted model security and the secret key was no longer transmitted. In this result, the RSA model achieves shape loss balancing and low computation time, but the model requires several security measures for the 3D mesh model in the image extension security technology.

Ring Algorithm

In order to manage geographic data, Katambo et al. [11] designed a distributed protocol and a Parlier Homomorphic Encryption to mask the identities of the searcher and data provider involved in geographical data. The suggested system can encrypt both the searcher and the data supplier. The Parlier encryption system is less expensive and requires less computing than the other cryptosystems, and it can handle very big complex calculations. The issue is that both the user and the data provider are just interested in gathering information or providing data, and neither one is concerned with maintaining their anonymity.

Asymmetric Key

Asymmetric key or asymmetric key algorithm uses the different cryptography keys for encryption of plaintext and decryption of ciphertext, each key pair known as the public key and corresponding private key. The different types of asymmetric key algorithms are discussed as follows.

Berlekamp–Welch Algorithm

Yusof et al. [12] proposed a Bivariate Polynomial Reconstruction Problem (PRP), based on the Indistinguishable Chosen-Plaintext Attack (IND-CPA). This method provided used a strategy that used a modified Berlekamp–Welch algorithm. The IND-CPA model needed the cryptosystem to prevent a cryptographic protocol security attack. IND-CPA was a security idea for cryptosystems, in which a two-part session, such as the challenge phase and learning phase, were communicated through a Probabilistic Polynomial Time Adversary (PPTA) through a random oracle. The PRP appears difficult in this model, but it can solve the problem mathematically to improve the cryptosystem outcome; however, the proposed PRP approach is not secure against the IND-CPA.

Runge Kutta Algorithm

Masood et al. [13] implemented the chaotic fractal system hybrid technique, which includes a chaotic-based three-dimensional chaotic map and a fractional function with a shuffling mechanism. The 3D Lorenz Chaotic map was utilized to alter all image pixels throughout the diffusion process. To achieve greater security, the digital image was shuffled through a channel-wise priority for encryption. This suggested system has the advantages of strong initial security and sensitivity, minimal computational cost, and a low correlation coefficient.

Hermitian Curves

Alzub et al. [14] proposed a Hermitian-based cryptosystem (HBC) suited to a compact device with limited processing memory and power, which is more specific to IoT devices. The proposed system was built on algebraic-geometric (AG) curves superimposed atop a Hermitian curve. When compared to other AG-type curves, Hermitian curves can produce more points on curves, which is important for determining the encryption key. To strengthen its attack resistance, the binary representation was adopted. The suggested method has several advantages, including the use of random matrices to protect against mathematical assaults and the size of communication blocks.

Keyless Cryptosystem

The keyless cryptosystem is applied to the security of message transmission. This can be either key distribution directly in advance or protocol to key sharing between the communicating users, which is processed before the encryption or the decryption procedures. The keyless cryptosystem is mainly divided into fractal and other types.

Fractal Keys

A fractal is a curve or geometrical object with detailed non-regularity structure on all scales. Fractal cryptography is a technique for key generation, encryption, and decryption, and the fractal function is used to construct a variety of procedures.

A new cryptosystem approach based on the Dragon curve fractal was proposed by Bhowmik et al. [9]. For the level of security using multiple private keys, and it is important for effective data encryption and decryption. The fractal was a geometric shape and it requires an Iterative Function System (IFS). The private key components such as size, angle, and iterations are used to cause the fractal from their corresponding starting points. The advantage of the proposed method uses a smaller number of iterations and time complexity, but it is difficult to find the originated curve in the Dragon curve model.

Zadeh et al. [10] introduced a new color image encryption technique that makes use of a replacement box and a hyperchaotic dynamic with fractal keys. The chaotic approach was utilized to change the fractal image values, and the collection of images was evaluated to image encode in every pixel. The set of chosen fractal images and the original image's pixel values were then merged, with each pixel encrypted using the XOR method. The outcomes of this technology provide a larger safe key space, a highly sensitive secret key, and a strong encryption effect. For evaluation, the suggested system exclusively employs floating computations.

Bai et al. [15] proposed a photo encryption system based on fractals for Visually Meaningful Image Encryption (VMIE) approach. The original photos are encrypted into fractal images or scenes utilizing fractal graph creation parameters as keys in this suggested technique. By using those keys in reverse, the receiver obtains the original image from the fractal image. This technique delivers good security performance while being computationally simple. The VMIE approach has drawbacks, such as requiring a larger number of sensitive keys to create and a significant image size.

Others

A flexible Cellular Automata (FcCA)-based cryptosystem was proposed by Mizher et al. [16] to provide lossless encryption for three-dimensional (3D) objects and images. The FcCA's flexibility includes the option to adjust the encryption subkey sizes, enabling simple control of the processes' resilience and enabling the fitting of large objects and images of diverse shapes and sizes. This system is resistant to noise attacks and has a high level of encryption. The values do not change when the plain data is shuffled, and many generations were required to create the enormously vast key space.

Mizher et al. [17] proposed an improved flexible Cellular Automata-based cryptosystem (iFcCA) for lossless encryption and decryption of three-dimensional (3D) pictures and objects. The proposed approach encrypts all Meta Data, including the UV map texture, for 3D objects. This method gives a high level of scrambling, effective resistance to many attack types, and resilience to significant key. This proposed solution necessitates a greater image size and storage capacity for image sharing.

In order to increase the high-security efficiency of the Menezes-Vanstone Elliptic Curve Cryptosystem (MVECC), Ghadi and AL-Rammahi [18] proposed an improved version of the cryptosystem based on quadratic Bézier curve methods. In this paradigm, ASCII values were utilized to turn text into numbers by taking two letters in the message, separating them as a point, and ultimately converting to an ASCII value. The strategy improves high-level security while also being efficient in mathematical models. The proposed method's significant computing complexity was a constraint.

4 Comparative Analysis of Geometric Shape Cryptosystems

Table 1 compares geometric shape-based cryptosystems to improve the performance of the present model. Comparative analysis is a critical method for developing and improving the model's performance. The comparative analysis comprises the dataset, methodology, benefits, drawbacks, and performance criteria.

The compression of asymmetric and symmetric key methods with different keys and block sizes is shown in Table 2. It also describes the various sorts of keys (public and private) that are used for encryption and decryption.

The features of different algorithm suggests the pros and cons which drives to build a secure cryptosystem using multi-dimensional geometric shapes characteristics which last long during cryptanalysis.

5 Challenges

The problem in geometric shape cryptosystem is outlined and discussed in this section as follows:

- The encryption method should be able to decrypt encrypted data in order to recover the original data, which makes processing the model more complicated and time-consuming.
- In the geometric shape cryptosystem, using the Nondeterministic Polynomial time (NP) problem was hard to trace the triangle vertices in the circumcenter.
- The determinist system of mean for the encryption key and plaintext mean deterministic system causes some security results.
- 3-dimensional model causes difficulty to find a mapping range and has various demands for the users.
- Poor performance and slow execution occur when using the number of keys in an algorithm.
- The geometric shape cryptosystem used multiple key data for a large file sequential cryptosystem.

Table 1 Comparison of different cryptographic methods

Author	Methodology	Advantages	Limitations	Performance criteria
Anitha Kumari et al. [19]	3D ElGamal DH PAKE protocol	Reduced communication and computational complexity	This model provides the low computational efficiency	Attack resistance rate
Liang et al. [20]	Asymmetric and optimized encryption method	This approach balances the computation time and shape loss for achieving an optimal encryption result	Number of security measures needed for 3D mesh model for extensions of image security technology	MSE, entropy
Bhowmik et al. [9]	Koblitz encoding and decoding algorithm	The proposed method uses the less iterations, which reduces the time complexity	The proposed model was difficult to obtain the originated curve when using more iterations	Time calculation
Hasanzadeh et al. [10]	A novel color image encryption algorithm	The proposed method uses a larger secure key space, highly sensitive secret key, and highly effective encryption	The proposed system only uses floating calculations for evaluation	NPCR, UACI
Bai et al. [15]	Image cryptosystem based on fractal graph	The proposed method achieves high-security performance and low computation complexity	The proposed method requires a greater number of sensitive keys to develop the model	Redundant information capacity
Sulaiman et al. [16]	Flexible cryptosystem based on cellular automata (FcCA)	Effective in resisting noise attacks and it provides a high level of encryption	The values do not change for shuffling the plain data and many generations were required to make the very large key space	Entropy, correlation, PSNR, RMSE, MAE

(continued)

Table 1 (continued)

Author	Methodology	Advantages	Limitations	Performance criteria
Bulat et al. [21]	An infographic flow of basic algorithms	This approach helps in understanding the encryption individual digital info left in different forms in Social media	The proposed method takes longer time to collect privacy data of individuals	Min power computation
Levena et al. [17]	White box for elliptical curve cryptography	The proposed method increases the execution time as n increases to improve security	The proposed method fails to secure for longer time if its executed parallelly	Increased Time calculation
Bhardwaj et al. [22]	Securing QR code using Visual cryptography	The proposed method includes n different shares for higher security	The proposed system has limitation in number of keys size and shares	PSNR, SSIM

6 Conclusion and Future Scope

Information security is crucial in everyday life and is a key consideration in encryption techniques. Various techniques are used for geometric shape-based cryptosystems to secure user data effectively. The proposed work uses various cryptographic algorithms to frame a cryptosystem that can effectively adapt to the computing environment and take intelligent decisions for robust security. The techniques of the RSA algorithm, the ElGamal algorithm, the Ring algorithm, and the Hermitian curve algorithm were used for efficiency in security.

When compared to other cryptographic methods like RSA, AES, etc., the geometry for Adaptive Security computes encryption and decryption based on geometric properties, and the key size is huge. Future work will concentrate on increasing security features while also enhancing security performance.

Table 2 Comparison of different algorithms based on key features

Algorithm	Type of algorithm	Logic	Key size (in bits)	Block size (in bits)	Feature
RSA	Asymmetric key	Uses different keys for encryption and decryption	1024–4096	128	Good security with latency
Diffie Hillmen (DH)	Asymmetric key	Utilizes both a public key and a private key for encryption and decryption	Variable	–	Good security with low speed
DES	Symmetric key	Uses identical key of both encryption and decryption	64	64	Not flexible and not strong enough
AES	Symmetric key	Use same key is used to generate cipher text and back to plaintext	256	128	Uses 3 different key lengths (128, 192, 256), simple structure to break
3D-pytocrypt	–	Use different key for both encryption and decryption	683	–	It needs long time for cryptanalysis and more software resources to build

References

1. Raghunandan KR, Ganesh A, Surendra S, Bhavya K (2020) Key generation using generalized Pell's equation in public key cryptography based on the prime fake modulus principle to image encryption and its security analysis. Cybern Inf Technol 20(3):86–101
2. Logunleko KB, Adeniji OD, Logunleko AM (2020) A comparative study of symmetric cryptography mechanism on DES AES and EB64 for information security. Int J Sci Res Comput Sci Eng 8(1)
3. Chinnasamy P, Padmavathi S, Swathy R, Rakesh S (2020) Efficient data security using hybrid cryptography on cloud computing. In: Inventive communication and computational technologies: proceedings of ICICCT, pp 537–547
4. Kumar S, Gaur MS, Sharma PS, Munjal D (2021) A novel approach of symmetric key cryptography. In: 2021 2nd international conference on intelligent engineering and management (ICIEM), pp 593–598
5. Verma R, Dhiman J (2022) Implementation of an improved cryptography algorithm. Int J Inf Technol Comput Sci (IJITCS) 14(2):45–53
6. Abroshan H (2021) A hybrid encryption solution to improve cloud computing security using symmetric and asymmetric cryptography algorithms. Int J Adv Comput Sci Appl 12(6):31–37

7. Masood F, Ahmad J, Shah SA, Jamal SS, Hussain I (2020) A novel hybrid secure image encryption based on Julia set of fractals and 3D Lorenz chaotic map. Entropy 22(3):274
8. Chowdhary CL, Patel PV, Kathrotia KJ, Attique M, Perumal K, Ijaz MF (2020) Analytical study of hybrid techniques for image encryption and decryption. Sensors 20(18):5162
9. Bhowmik A, Menon U (2020) Dragon crypto—an innovative cryptosystem. arXiv preprint arXiv:2008.12645
10. Hasanzadeh E, Yaghoobi M (2020) A novel color image encryption algorithm based on substitution box and hyper-chaotic system with fractal keys. Multimedia Tools Appl 79:7279–7297
11. Katambo J, Nyirenda M, Zulu D (2022) Distributed spatial search using the Paillier cryptosystem and the distributed ring algorithm. Zambia ICT J 6(1):66–80
12. Yusof SN, Kamel Ariffin MR, Lau TSC, Salim NR, Yip SC, Yap TTV (2023) An IND-CPA analysis of a cryptosystem based on bivariate polynomial reconstruction problem. Axioms 12(3):304
13. Desoky A (2023) Keyless cryptosystem based on Latin square for blockchain and covert communications
14. Alzubi OA, Alzubi JA, Dorgham O, Alsayyed M (2020) Cryptosystem design based on Hermitian curves for IoT security. J Supercomput 76:8566–8589
15. Bai S, Zhou L, Yan M, Ji X, Tao X (2021) Image cryptosystem for visually meaningful encryption based on fractal graph generating. IETE Tech Rev 38(1):130–141
16. Mizher MAAJA, Sulaiman R, Abdalla AMA, Mizher MAA (2021) A simple flexible cryptosystem for meshed 3D objects and images. J King Saud Univ Comput Inf Sci 33(6):629–646
17. Levina A, Kamnev I, Zikratov I (2022) Implementation white-box cryptography for elliptic curve cryptography. In: 10th Mediterranean conference on embedded computing (MECO)
18. Ghadi DM, Adil AR (2020) Improvement of Menezes-Vanstone elliptic curve cryptosystem based on quadratic Bézier curve technique. J Comput Sci 16(5):715–722
19. Sudha Sadasivam G, Anitha Kumari K (2019) Two-server 3D ElGamal Diffie-Hellman password authenticated and key exchange protocol using geometrical properties. Mobile Netw Appl 24:1104–1119
20. Liang Y, He F, Li H (2019) An asymmetric and optimized encryption method to protect the confidentiality of 3D mesh model. Adv Eng Inform 42:100963
21. Bulat R, Ogiela MR (2022) Personalized cryptography algorithms—a comparison of classic and cognitive methods. In: IEEE/IFIP international conference on dependable systems and networks
22. Bhardwaj C, Garg H, Shekhar S. An approach for securing QR codes using cryptography and visual cryptography

A Lesion Feature Engineering Technique Based on Gaussian Mixture Model to Detect Cervical Cancer

Lalasa Mukku and Jyothi Thomas

Abstract Latest innovations in technology and computer science have opened up ample scope for tremendous advances in the healthcare field. Automated diagnosis of various medical problems has benefitted from advances in machine learning and deep learning models. Cancer diagnosis, prognosis prediction and classification have been the focus of an immense amount of research and development in intelligent systems. One of the major concerns of health and the reason for mortality in women is cervical cancer. It is the fourth most common cancer in women, as well as one of the top reasons of mortality in developing countries. Cervical cancer can be treated completely if it is diagnosed in its early stages. The acetowhite lesions are the critical informative features of the cervix. The current study proposes a novel feature engineering strategy called lesion feature extraction (LFE) followed by a lesion recognition algorithm (LRA) developed using a deep learning strategy embedded with a Gaussian mixture model with expectation maximum (EM) algorithm. The model performed with an accuracy of 0.943, sensitivity of 0.921 and specificity of 0.891. The proposed method will enable early, accurate diagnosis of cervical cancer.

Keywords Cervical cancer · Deep learning · Gaussian mixture modelling · Expectation maximization · Lesion feature extraction · Segmentation

1 Introduction

Cervical cancer is fourth on the list of malignancies that affect women across the globe [1]. The cancer is caused due to an infection of HPV left untreated for a long time [2]. Since there is no contributing genetic factor, cancer can be completely treated by surgically removing the infected tissue of the cervix. To that end, accurate and timely diagnosis is imperative to eliminate the burden of cervical cancer. Colposcopy is regarded to be the gold standard for cervical cancer diagnosis [3]. During the colposcopy procedure, prior to imaging, 5% acetic acid solution is applied

L. Mukku (✉) · J. Thomas
CHRIST (Deemed to be University) Kengeri, Bangalore 560074, India
e-mail: mlalasa2020@gmail.com

S. Kumar et al. (eds.), *Fourth Congress on Intelligent Systems*, Lecture Notes in Networks and Systems 868, https://doi.org/10.1007/978-981-99-9037-5_6

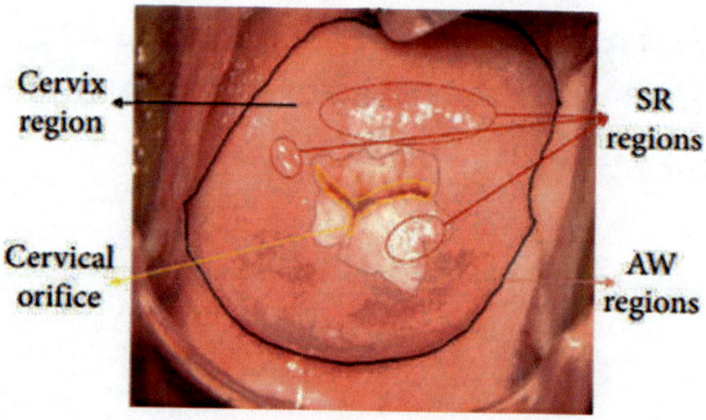

Fig. 1 Cervigram marked with **a** Cervix region of interest, **b** specular reflection (SR), **c** AW lesions, **d** orifice

to the cervix tissue, which turns the cancerous tissue into acetowhite lesions. The acetowhite lesions are the biomarkers of cervical intraepithelial neoplasia [4]. Experienced doctors and clinicians are reported to have 63.3% [5] sensitivity in identifying cancerous lesions leaving the scope of improving diagnosis with technological assistance [6]. In the early stages of cancer, the lesions are small and not obvious on the cervigram image. Hence, this study is inspired by the dire need for an automatic, robust and dependable lesion segmentation that can bridge the gap between diagnosis and biopsy. Figure 1 displays the cervix image with marked borders and features.

Latest innovations in technology and computer science have opened up ample scope for tremendous advances in the healthcare field. Automated diagnosis of various medical problems has benefitted from advances in deep learning and machine learning CAD models. Artificial intelligence-supported cancer diagnosis [7, 8], prognosis [9], prediction [9] and classification [10] have been the focus of an immense amount of research and development in intelligent systems [11]. Machine learning is a branch of artificial intelligence that can analyze and learn patterns from large datasets. It has proven to be a dependable support to many disease diagnoses as well as early prediction [12]. Deep learning, on the other hand, is a complex subarea of machine learning that majorly deals with computer vision problems, i.e., image processing [13]. It is being used extensively in cancer diagnosis, prognosis, prediction and personalized treatment [14].

In this paper, the crux of Gaussian mixture modelling (GMM) is used for efficient segmentation and detection of acetowhite lesions. The Gaussian mixture model acts on the principle that all instances of the dataset given are taken from a finite Gaussian distribution mixture. In addition to that, it has no use for parameters making it a nonparametric algorithm. GMM is the method that is popularly used in segmentation tasks of biomedical image processing [15]. GMM has displayed efficiency in the segmentation of medical images. The major contributions of this paper are:

(1) An optimized variant of multivariate Gaussian distribution is proposed to segment scope-based medical images, especially the colposcope image lesions.
(2) Expectation maximization algorithm is applied to the traditional Gaussian mixture model to enhance the segmentation.

(3) A fully connected deep learning framework is proposed for classifying the segmented colposcope images.

The paper is structured as follows: Sect. 2 presents the related literature, Sect. 3 describes the methodology of lesion detection algorithm (LDA) containing lesion feature extraction (LFE) and lesion recognition methods. Section 4 presents the results, and Sect. 5 is the conclusion of the study.

2 Related Works

Gaussian Mixture Model in Cervical Cancer

Ragothaman et al. [16] attempted to solve the whole slide image models of pap smear by using GMM as the segmentation method and reported satisfactory results. Phoulady et al. [17] attempted to segment clustered cervical cells through a Gaussian mixture model (GMM) algorithm on pixel levels. They declared a satisfactory dice coefficient of 0.861. Bouveyron [18] proposed a mixture modelling approach integrated with weakly supervised classification tasks on cell images. This would prove to be fruitful. Nevertheless, these works have primarily focused on cervical cell segmentation. However, this study targets colposcope images [19].

Colposcope images contain surrounding noise, speculum and vaginal walls around the cervix. So, extracting the cervix region before proceeding with segmenting AW region is imperative. Plenty of research works are aimed at segmenting and extracting the cervix region in a colposcope. The majority of the works have employed GMM, along with K means, Deeplab V3 and mask RCNN.

Ramaprabha and Ranganathan [20] defined a preprocessing method to remove the irrelevant information surrounding the cervix from a colposcope image by using mathematical morphological amalgamated with a Bayes classifier Gaussian mixture model. They have reported a drawback to the Bayes classifier's inability to properly distinguish the lesions from specular reflections. Meslouhi et al. [21] proposed a method to automatically extract and repaint the specular reflections based on the multi-resolution inpainting technique (MIT) and the dichromatic reflection model (DRM). Perperidis [22] further made use of the Gaussian mixture to perform two-level segmentations in the cervical lesions. The previous studies have focused segmenting cervigrams whereas the extraction of AW lesions presented to be more challenging. To overcome the limitations of the pervious works, a novel algorithm that uses EM is presented in this work. Table 1 lists the literature methods reviewed.

In this paper, a novel lesion detection algorithm to successfully extract and segment the AW lesions in a colposcope image is proposed. The model generates a new dataset after each step containing the marked and extracted lesion feature details. This dataset can be classified using a deep neural network model.

Table 1 Literature and methods

Paper	Method
[16]	Pap smear segmentation with GMM
[17]	Cluster cell separation using mixture modelling
[18]	Weakly supervised classifier
[20]	Bayes classifier
[21]	Multi-resolution inpainting dichromatic reflection
[22]	Traditional Gaussian mixture model

3 Methodology

3.1 Architecture

The schematic architecture of the current work is given in Fig. 2. The input is the cervix image taken post the acetic acid smearing. The key features are the presence and area of acetowhite regions, transition zone, cancerous build-up on the cervix, etc. These features selected from inside the region of interest of the cervix are fed into deep learning networks described in the methodology. The parameter threshold is set; thus, the area of the lesion can be determined. This study uses GMM with expectation–maximization and CNN to develop the model.

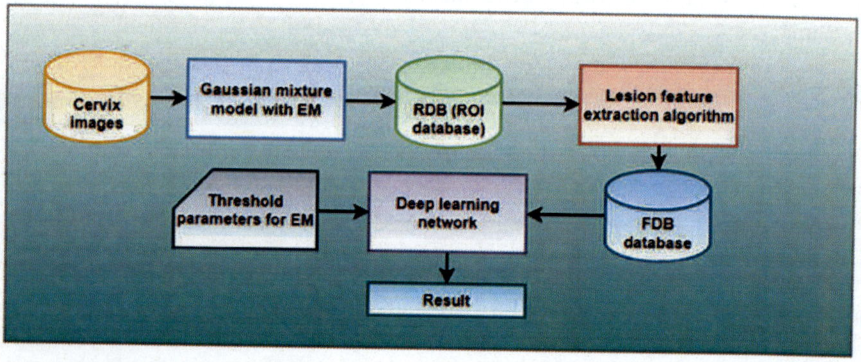

Fig. 2 Proposed model architecture

3.2 Gaussian Mixture Model with Expectation Maximization Algorithm

(a) Gaussian mixture modelling

Vaginal walls are speculum with medical devices that are often present in a colposcope image making the image noisy. Delineating the cervix region is imperative to precisely classify the same. A GMM model modified using EM has performed well in a work by Srinivasan et al. [23]. For $x \in \mathbb{R}d$, a Gaussian model can be expressed as considering K items' as Gaussian density components where the parameters μk and $\sum k$, each of which is a representation of multivariate Gaussian density. Parallelly, authors of [24] proposed a custom CaMeL-Net model for cervical cancer identification.

$$pk(x|\theta k) = \frac{1}{(2\pi)^{d/2}\left|\sum k\right|^{1/2}} e^{-1/2(x-\mu k)\sum(x-\mu k)\$}, \quad (1)$$

where $\theta k = \{\mu k, \sum k\}$. GMM is considered one of the few top choices for segmenting medical images. Hence, the paper adopted the same for ROI extraction.

(b) Expectation maximization of Gaussian mixture models

The expectation maximization (EM) algorithm works in iterations. The algorithm starts with ɵ, a random initial estimate and keeps updating the ɵ until the convergence is detected.

In every iteration, EM proceeds with two steps—*E* step and *M* step.

EM Algorithm

E step-ɵ, being a parameter value it must compute w_{ik} using the Eq. (1) and by allotting weights dynamically. The considered data points being xi while $1 \le i \le N$, considering the mixture items $1 \le k \le K$. The allotment of weights requires a calculation. It is given in (2).

$$\sum_{K=1}^{K} w_{ik} = 1 \quad (2)$$

The result of the above equation is a matrix of $N \times K$ matrix that contains the weights calculated by the summation of whose rows is 1.

M step: The weights calculated in *E* step are employed to generate parameter values. Consider *N* represents the count of weights and $\sum_i^k N\, w_{ik}$ is the summation of weights of "*k*"th item, the accumulated data instances allocated to *k* are calculated as

$$\alpha_k^{\text{new}} = \frac{N_k}{N}, 1 \le k \le K. \quad (3)$$

The weight representation of the mixture generated is given as

$$\propto_k^{\text{new}} = \left(\frac{1}{N_k}\right) \sum_{K=1}^{K} w_{ik} x_i 1 \le k \le K. \tag{4}$$

This results in updating the mean. The mean is calculated by standard statistical formula. Nevertheless, the vectors contain a fractional weight. For d-dimensional vectors "k" and x_i a fractional weight of w_{ik} and μ is allocated.

$$\propto_k^{\text{new}} = \left(\frac{1}{N_k}\right) \sum_{K=1}^{K} w_{ik}\left(x_i - \mu_k^{\text{new}}\right)\left(x_i - \mu_k^{\text{new}}\right)^t, 1 \le k \le K \tag{5}$$

The above equation denotes a $d \times d$ matrix on either side. The computation is carried out in a similar way as done in a covariance matrix empirically. However, the dynamic weight w_i is the weight of each data point's contribution.

The M step of EM is done. Subsequently, the E step needs to be updated by adapting the parameters derived from M step. One round of completing an E step and M step is termed as one complete iteration.

3.3 *Proposed Algorithm—Lesion Recognition Algorithm*

The region of interest in the cervix is extracted and is provided as input to the "lesion feature extraction" (LFE). The primary objective of the LFE is to optimize spatial redundancy. The means of accomplishing this objective is to allow the algorithm to only select lesion features from specific regions of interest and parallelly reduce the complexity in terms of implementation time and cost. The above algorithm is given by

Given input: Cervigram from database $D = \{L_1, L_2, L_3, L_4 \ldots L_n\} \forall \{L = 1, 2, 3, 4 \ldots n\}$
Attributes: $\{A_1, A_2, A_3, A_4 \ldots A_n\} \forall A_i = \{i = 1, 2, 3 \ldots n\}$
The representation of attributes is namely width, height, and structure of the lesion in the ROI
Parameters used: RoI Database (RID) $= \{R_1, R_2, R_3 \ldots Rn\} \forall R_i \{I = 1, 2, 3 \ldots n\}$
Output: Lesion Feature dataset FDB $= \{I_1, I_2 \ldots I_n\} \forall I_i \{i = 1, 2 \ldots n\}$
T = tuples collection
A = attributes set for lesion marking

Method

1. Given database is a collection of cervigrams (I)
2. Identify the cervix borders as ROI and eliminate surrounding noise. The resulting dataset is saved as a separate entity called Region of Interest Dataset (RIDs).
3. Perform the modelling to attain the lesion shape and area through a Gaussian mixture formula given by:

$$g\left(x|\mu,\sigma^2\right) = \frac{1}{\sqrt{2\pi\sigma_i}}\exp\left\{-\frac{(x-\mu_i)^2}{2\sigma_i^2}\right\} i - 1, 2, 3...n. \tag{6}$$

4. The output of (6) is the detected ROI shape. Taking it as the input to the EM algorithm gives the likelihood of lesion shape analyzed.
5. Thereafter the feature database (FDB) is constructed from the processed input of extracted cervix ROI. FDB has the attributes A_i and tuples T_i given by

$$\mathrm{T} = \bigcup T_i \forall i = 1, 2, 3..n, \tag{7}$$

$$\mathrm{A} = \bigcup A_i \forall i = 1, 2, 3..n, \tag{8}$$

$$\mathrm{FDB(Feature - Database)} = \sum A.T. \tag{9}$$

Lesion Recognition Algorithm

The focus of the "lesion recognition algorithm" (LR) is recognizing the spread of lesions within the ROI area provided in the earlier step for each cervix image. The objective of the algorithm is to locate the cancerous acetowhite lesions from the extracted features through LFE (which is taken as the training part). Thereafter parameters are bounded by a set threshold value specified for identifying lesion growth in the ROI. This provides support to doctors and clinicians in the early automated identification of cervical cancer. The lesion recognition algorithm is given below:

Input: Cervix image from database $D = \{I_1, I_2, I_3, I_4 \ldots I_n\} \forall \{I = 1, 2, 3, 4 \ldots n\}$
RID (Dataset of region of interests) = $\{R_1, R_2, R_3 \ldots R_n\} \forall R_i \{I = 1, 2, 3\ldots n\}$
Feature data FDB = $\{I_1, I_2 \ldots I_n\} \forall F_i \{i = 1, 2 \ldots n\}$

Parameters:
RID = Region of interest database
T = Tuples
FDB = Feature database
A = Attributes for lesion features

Method

1. Examine the RID dataset acquired from the region of interest from the group of cervigrams.
2. The feature level database, FDB, is extracted from region of interest RID, which is gathered and is used as training part of the dataset for deep learning neural network.
3. The algorithm's novelty appears in making the deep network's hidden layer count based on the features count, i.e., each feature extracted represents one layer in the network. Feature can be anything from shape specification, area of the circumference of the lesion, etc.

4. Thus, the number of hidden layers is determined by the number of feature inputs.
5. The final layer, also called the output layer, reads the threshold parameters and identifies the growth of the cancerous lesion with respect to those parameters. This results in an accurate diagnosis of the lesions.
6. The dataset of the retrieved images will be the lesion image database (LID) covering the specification like circumference, depth, size and weights attributed, etc.
7. The LID dataset is given by:

   LID = $\prec I, R, A, L \succ$,
   where
   R: Region of the image that is of interest (ROI)
   L: Lesion image
   I: Image
   A: Attributes/features.

The algorithm defined above performs the extraction followed by the identification of cancerous lesions in the cervix image. The features extracted through GMM from the cervigrams are the shape, size, and circumference of the lesion area. They are used by the deep learning network to train the model for the effectual diagnosis of cancerous lesions.

3.4 Deep Learning (DL) Strategies

Initially, using EfficientNet B7, which is an ImageNet pretrained model, images were trained, leading to an accelerated convergence rate for the network. The feature extraction and fusion networks were not trained separately. Instead, they were combined into a single network, so that the loss would not converge owing to the complexity of mastering the features. The feature encoding layers of EfficientNet-B3 were frozen, and the outputs of the feature maps were incorporated as input in a feature fusion system.

Frozen parameters would ensure that none of the layers are updated during training, thereby preserving the benefits of the features extracted of acetic acid images, ensuring there is no intercession between them, and avoiding any negative effect on results. Cross entropy loss is employed to minimize the loss shown in Fig. 3.

The main application of this method is to segment probing scope device medical images like colposcopy but not limited to the same. It can be extended to rectoscopy, endoscopy, colonoscopy, etc.

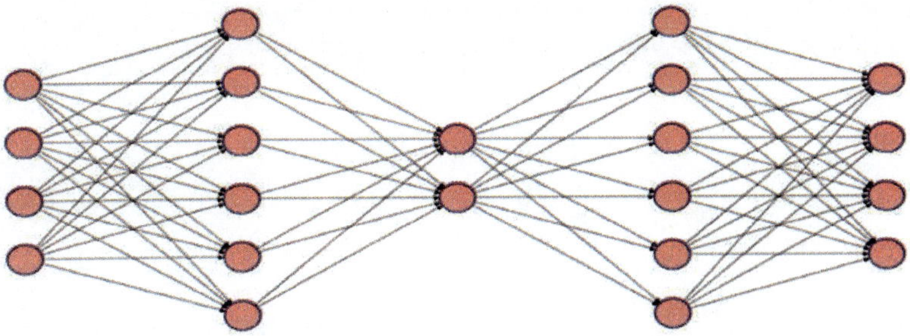

Fig. 3 Fully connected neural network using rectified linear units (RELU) in training

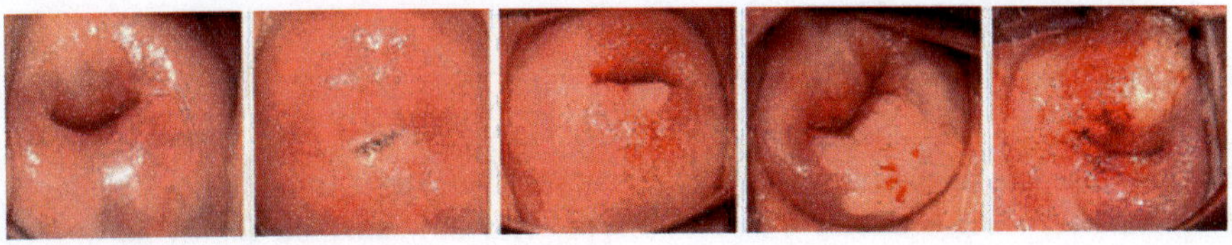

Fig. 4 Colposcope images of cervix

4 Results

4.1 Dataset

The colposcope images were collected from KIDWAI Memorial Institute of Oncology, Bangalore, India. A total of 872 images were collected from women between the ages of 24–55 who have tested positive for HPV infection. The images are labeled with corresponding biopsy result. Sample images from the dataset are shown in Fig. 4.

4.2 Evaluation Metrics

To evaluate the proposed approach, following metrics are used, namely sensitivity, accuracy, specificity, dice coefficient and Jaccard index.

$$\text{Jaccard Index} = \frac{|S_g^1 \cap S_t^1|}{|S_g^1 \cup S_t^1|} = \frac{\text{TP}}{\text{TP} + \text{FP} + \text{FN}}$$

$$\text{Accuracy} = \frac{\text{TP} + \text{TN}}{\text{TP} + \text{FP} + \text{TN} + \text{FN}} \times 100$$

$$\text{Dice Coefficient} = \frac{2|S_g^1 \cap S_t^1|}{|S_g^1| + |S_t^1|} = \frac{2\text{TP}}{2\text{TP} + \text{FP} + \text{FN}}$$

$$\text{Sensitivity} = \frac{\text{TP}}{\text{TP} + \text{FN}} \times 100$$

$$\text{Specificity} = \frac{\text{TN}}{\text{TN} + \text{FP}} \times 100$$

4.3 *Experimental Environment*

The proposed segmentation model is executed in Jupyter environment. The high-performance computer with 8 GB of RAM with 256 SSD on the Windows 10 operating system, an Intel Core i7 processor of 10th generation.

4.4 *Experimental Results*

The method introduced was tested on a dataset with 872 images gathered from a government hospital from India. The algorithm made with optimized Gaussian mixture model has worked with satisfactory results. The performance of the proposed model is compared with a traditional Gaussian mixture model and the outcome is shown in Table 2. The model performed with accuracy of 92.4%, sensitivity of 92.1% and specificity of 89.1%. The segmentation dice score is 0.901and the Jaccard index is 84.3. Figure 5 presents the ROC curve.

5 Conclusion

Recent advances in computational solutions for the accurate and timely diagnosis of medical cases are key factors in cancer identification. Artificial intelligence and deep learning have contributed to the critical advancements made in this field of study. Cervical cancer is fourth among the female cancers, across the globe. This study aimed to develop a framework to screen cervical cancer efficiently through

Table 2 Performance of the proposed model

Measures	GMM	*K* means	EM-GMM
Sensitivity	0.753	0.847	0.921
Specificity	0.61	0.331	0.891
Dice score	0.591	0.8319	0.901
Jaccard index	0.426	0.389	0.843
Loss	0.587	0.610	0.129
Accuracy	76.12	0.602	0.942

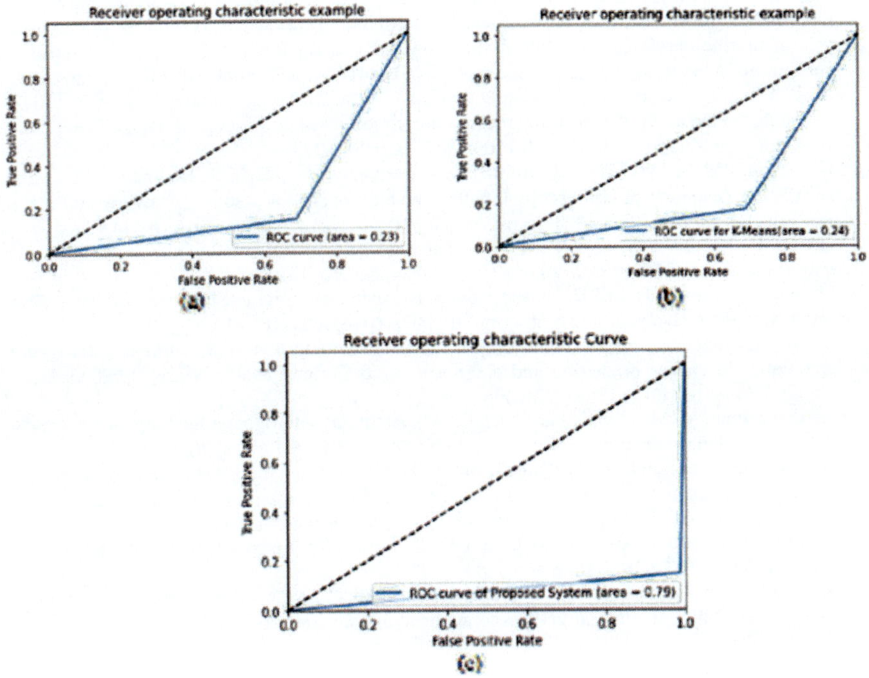

Fig. 5 ROC curve

colposcopy images. The lesion feature extraction algorithm (LFE) is built using Gaussian mixture modelling (GMM) with expectation maximization (EM) to enable the timely detection of cervical cancer and thus reduce the burden of disease and mortality. The model performed with accuracy of 92.4%, sensitivity of 92.1% and specificity of 89.1%. The segmentation dice score is 0.901and the Jaccard index is 84.3.The results and comparative performance analysis indicate that the current model is reliable as it extracts and segments the lesion features which are the key indicators of cancer. Hence, this method can be used for segmentation and automatic diagnosis of cervical cancer through colposcope images.

References

1. Sung H, Ferlay J, Siegel RL, Laversanne M, Soerjomataram I, Jemal A, Bray F (2021) Global cancer statistics 2020: GLOBOCAN estimates of incidence and mortality worldwide for 36 cancers in 185 countries. CA Cancer J Clin 71:209–249. https://doi.org/10.3322/caac.21660
2. Rodríguez AC, Schiffman M, Herrero R, Hildesheim A, Bratti C, Sherman ME, Solomon D, Guillén D, Alfaro M, Morales J, Hutchinson M, Katki H, Cheung L, Wacholder S, Burk RD (2010) Longitudinal study of human papillomavirus persistence and cervical intraepithelial neoplasia grade 2/3: critical role of duration of infection. J Natl Cancer Inst 102:315–324. https://doi.org/10.1093/jnci/djq001
3. Eheman CR, Leadbetter S, Benard VB, Blythe Ryerson A, Royalty JE, Blackman D, Pollack LA, Adams PW, Babcock F (2014) National breast and cervical cancer early detection program data validation project. Cancer 120(S16):2597–2603. https://doi.org/10.1002/cncr.28825
4. Gordon S, Zimmerman G, Long R, Antani S, Jeronimo J, Greenspan H (2006) Content analysis of uterine cervix images: initial steps towards content based indexing and retrieval of cervigrams. Med Imaging Image Process 6144:61444U. https://doi.org/10.1117/12.653025
5. Li Y, Chen J, Xue P, Tang C, Chang J, Chu C, Ma K, Li Q, Zheng Y, Qiao Y (2020) Computer-aided cervical cancer diagnosis using time-lapsed colposcopic images. IEEE Trans Med Imaging 39. https://doi.org/10.1109/TMI.2020.2994778
6. Underwood M, Arbyn M, Parry-Smith W, De Bellis-Ayres S, Todd R, Redman CWE, Moss EL (2012) Accuracy of colposcopy-directed punch biopsies: a systematic review and meta-analysis. BJOG 119:1293–1301. https://doi.org/10.1111/j.1471-0528.2012.03444.x
7. Hunter B, Hindocha S, Lee RW (2022) The role of artificial intelligence in early cancer diagnosis. Cancers 14. https://doi.org/10.3390/cancers14061524
8. Shastry KA, Sanjay HA (2022) Cancer diagnosis using artificial intelligence: a review. Artif Intell Rev 55:2641–2673. https://doi.org/10.1007/s10462-021-10074-4
9. Kumar Y, Gupta S, Singla R, Hu Y-C (2022) A systematic review of artificial intelligence techniques in cancer prediction and diagnosis. Arch Comput Methods Eng 29:2043–2070. https://doi.org/10.1007/s11831-021-09648-w
10. Al-shamasneh ARM, Obaidellah UHB (2017) Artificial intelligence techniques for cancer detection and classification: review study. https://doi.org/10.19044/esj.2016.v13n3p342
11. Hou X, Shen G, Zhou L, Li Y, Wang T, Ma X (2022) Artificial intelligence in cervical cancer screening and diagnosis. Front Oncol 12:851367. https://doi.org/10.3389/fonc.2022.851367
12. Uddin S, Khan A, Hossain ME, Moni MA (2019) Comparing different supervised machine learning algorithms for disease prediction. BMC Med Inform Decis Mak 19:281. https://doi.org/10.1186/s12911-019-1004-8
13. Shen D, Wu G, Suk H-I (2017) Deep learning in medical image analysis. Annu Rev Biomed Eng 19:221–248. https://doi.org/10.1146/annurev-bioeng-071516-044442
14. Zhu W, Xie L, Han J, Guo X (2020) The application of deep learning in cancer prognosis prediction.https://doi.org/10.3390/cancers12030603
15. Sekaran K, Chandana P, Krishna NM, Kadry S (2020) Deep learning convolutional neural network (CNN) with Gaussian mixture model for predicting pancreatic cancer. Multimed Tools Appl 79:10233–10247. https://doi.org/10.1007/s11042-019-7419-5
16. Ragothaman S, Narasimhan S, Basavaraj MG, Dewar R (2016) Unsupervised segmentation of cervical cell images using Gaussian mixture model. https://doi.org/10.1109/CVPRW.2016.173
17. Phoulady HA, Goldgof DB, Hall LO, Mouton PR (2016) A new approach to detect and segment overlapping cells in multi-layer cervical cell volume images. In: 2016 IEEE 13th international symposium on biomedical imaging (ISBI), pp 201–204. https://doi.org/10.1109/ISBI.2016.7493244
18. Bouveyron C (2009) Weakly-supervised classification with mixture models for cervical cancer detection.https://doi.org/10.1007/978-3-642-02478-8_128
19. Magaraja AD, Rajapackiyam E, Kanagaraj V, Kanagaraj SJ, Kotecha K, Vairavasundaram S, Mehta M, Palade V (2022) A hybrid linear iterative clustering and Bayes classification-based GrabCut segmentation scheme for dynamic detection of cervical cancer.https://doi.org/10.3390/app122010522

20. RamaPraba PS, Ranganathan H (2012) Automatic lesion detection in colposcopy cervix images based on statistical features. Commun Comput Inf Sci CCIS 270:424–430. https://doi.org/10.1007/978-3-642-29216-3_46
21. Meslouhi OE, Kardouchi M, Allali H, Gadi T, Benkaddour YA (2011) Automatic detection and inpainting of specular reflections for colposcopic images. https://doi.org/10.2478/s13537-011-0020-2
22. Perperidis A. Image segmentation of uterine cervix images
23. Srinivasan Y, Corona E, Nutter B, Mitra S, Bhattacharya S (2009) A unified model-based image analysis framework for automated detection of precancerous lesions in digitized uterine cervix images. IEEE J Sel Top Signal Process 3:101–111. https://doi.org/10.1109/JSTSP.2008.2011102
24. Lee J, Han C, Kim K, Park G-H, Kwak JT (2023) CaMeL-Net: centroid-aware metric learning for efficient multi-class cancer classification in pathology images. Comput Methods Progr Biomed: 107749

Energy Optimization of Electronic Vehicle Using Blockchain Method

Ranjana and Rishi Pal Singh

Abstract Numerous automakers are now developing a variety of electric car models, which bodes well for the future significant market penetration and dominance of electric vehicles. For instance, research predicted that by 2030, there will be about 15 million electric vehicles on the road. By giving a useful, autonomous, and adaptive mode of transportation, electric vehicles offer an environmentally benign alternative to existing transportation modes that are harmful to the environment. Electric vehicles use digital communication technologies to link to a network of other vehicles. In this research work, the energy consumption of the electric vehicle is optimized using blockchain model. Python is executed to simulate the presented approach, and the accuracy, precision, and recall are considered for analyzing the outcomes.

Keywords Electric vehicle · Blockchain · Energy optimization

1 Introduction

1.1 Background

The possibility of EVs to provide environment friendly and sustainable transportation alternatives has recently garnered increased attention. They can also act as a backup service for renewable energy systems or as an energy storage model during electrical power outages (such as from a car to a residence) [1]. Due to this potential, numerous automakers are now developing a variety of electric car models, which bodes well for the future significant market penetration and dominance of electric vehicles. For instance, research predicted that by 2030, there will be about 15 million

Ranjana (✉) · R. P. Singh
Department of Computer Science and Engineering, Guru Jambheshwar University of Science and Technology, Hisar, Haryana, India
e-mail: ranjana.arora27@gmail.com

R. P. Singh
e-mail: pal_rishi@yahoo.com

S. Kumar et al. (eds.), *Fourth Congress on Intelligent Systems*, Lecture Notes in Networks and Systems 868, https://doi.org/10.1007/978-981-99-9037-5_7

electric vehicles on the road. By giving a useful, autonomous, and adaptive mode of transportation, electric vehicles offer an environmentally benign alternative to existing transportation modes that are harmful to the environment. Electric vehicles use digital communication technologies to link to a network of other vehicles and exchange electricity with their peers at a price that is advantageous to both parties.

Information and communication technology (ICT) is currently being used by energy optimization methods for successfully handling power business tasks among EVs [2]. Additionally, electric vehicles can use information and communication technology (ICT)-based services to: (a) compute the battery power outage rate; (b) classify extremely reliable power charging banks; (c) predict the amount of energy needed after high demand duration; (d) discovering the smallest route from source to target power charging banks; (e) selling excess power to users; and (f) forecast the cost of power after high demand duration in a specific area. Most of the information and communication technology-based systems, however, are very disconnected, uncertain, opaque, and least reliable. Blockchain is a system which offers a transmission record and secures such records using cryptographic hashing algorithms [3]. It can make electric car energy optimization processes extremely effective, transparent, dependable, and detectable.

The operation of the distribution system is hampered by the expanding EV fleet, which is a power-demanding load. The amount of capacity needed at particular times and locations can be significantly impacted by the EVs' ability to charge from the power grid [4]. Charging frequently occurs at the same time as the current load demand, which strains the local grid and may have an impact on the supply's sufficiency and quality. As a result, it might necessitate more grid and power capacity investment. However, EVs can be seen as dispersed, adaptable loads and storage systems from the standpoint of the power grid. Because EV charging is independent of use, it can contribute to demand-side flexibility for improved grid stability and dependability [5]. Uncontrolled and controlled electric car charging are distinguished, and are frequently referred to as dumb and smart charging, respectively. The EV is plugged in and charged at maximum power until fully charged when using dumb charging.

Smart charging, on the other hand, is "when the charging cycle may be influenced by external events, allowing for adaptive charging habits, and giving the EV with the ability to integrate into the entire power system in a grid and user-friendly way," according to the definition [6]. The client benefits from smart charging since it lowers electricity bills while facilitating system operation. According to the control objective, the procedure to manage the charging can be divided into three main scenarios: peak shaving, renewables, and balancing. Thus, the controlled charging can provide a variety of ancillary services for the functioning of the power system. An EV can also enable bidirectional charging via vehicle-to-grid (V2G), which allows the EV to additionally pump power back into the grid and extend the operating range for distributed batteries used in demand response programs [7]. Direct and indirect control of the charge are further divisions of the control. Electric vehicle charging and battery changing are only two of the many services power grid stations offer.

To enable drivers in recognizing dependable and trustworthy facilities to charge and swap the battery, complete information on the vehicles as well as their trading priorities should be readily available. With the use of sensing or RF motes, Internet of Things technology enables practical interactions between automobiles [8]. By using centralized cloud-based resources, most of these mechanisms are saved and processed electric vehicle-related data for energy trading. However, the possibilities for cooperation among energy trading autos are constrained by unified data storage and processing. Unified systems are incredibly disjointed and unable to ensure data consistency and openness. A unified system's ledgers are vulnerable to changes, removals, and deceit by rivals [9]. The EVs are allowed to conduct ET functions in a way that is streamlined, effective, and trusted; BT is capable of dealing with such difficulties. A chronological record of data and transactions connected to the ET of EVs is kept on the blockchain, an openly distributed ledger [10]. By using a distributed system, blockchain is effective for removing the necessity of a centralized instruction structure which can resolve trade issues between ET participating devices like cars, power charging stations, etc. P2P approach is used to give landlords of EV direct access to each other so they can do business. Accusatory and untruthful drivers in the e-car network are unable to edit, remove, or revise the prior ledgers due to irrefutability of transmissions and data available in BT [11]. Additionally, blockchain offers the users of the energy trading network additional transparency because its data and transmissions can be easily accessed to all.

2 Literature Survey

Kakkar et al. [12] established a block chain-based platform of ET for EV in a smart grid platform [12]. The fundamental goal of this platform was to secure the power data required in electric vehicles. The 6G was adopted in this platform as a communicating way to ensure that that the system was reliable, scalable, and had lower latency in communicating the data. The energy demand data was analyzed and power consumed via EV was predicted using multiple linear regression (MLR). The smart grid was exploited for optimizing the energy for EVs, the power for energy vehicles so the benefit of provider and client of electric vehicle was increased. The results confirmed the reliability, security, and cost-efficacy of the established approach for EVs in comparison with the existing systems.

Li et al. [13] suggested a superconducting magnetic energy storage (SMES) unit [13]. Initially, this approach focused on implementing a model of the EV peer to peer ET, and planning the block chain model and SCs. Subsequently, transaction procedure aimed at analyzing the advantages of every user entity in the mechanism and optimizing the way to match the user. The pricing model ofP2P electricity transaction was implemented using the co-operative game algorithm. Eventually, the simulation outcomes demonstrated that the suggested approach was effectual to enhance adaptability and claim utilization rate of EV-energy trading.

Shekari et al. [14] presented the blockchain-based notions for DR programs in which electric vehicle (EV) and renewable energies (REs) were implemented in the electricity markets [14]. A number of attributes of blockchain technology, namely smart contracts were utilized to execute this procedure. The smart contracts were deployed in DR programs for improving the efficacy of these programs and mitigating their expenses as well as making of demand response more secure in communicating. Hence, these users attained real expenses of energy which provided more efficiency in consuming the energy and offered more suitability to the pricing signal of DR (demand response).

Chen et al. [15] developed a PR algorithm of electric vehicle drivers on the basis of their timings to drive and charge them at first [15]. A blockchain-based EVI framework was put forward with the objective of maximizing the deployment of RE. The developed algorithm offered more security, secrecy, and decentralization. The developed algorithm incorporated the values, drivers, CSPs for guiding the electric vehicle (EV) users so that the desired time frames were charged at superior efficacy to generate the RE. The simulation results proved that the developed algorithm was efficient. Moreover, the developed algorithm was useful for augmenting the usage of RE from the local microgrid.

Wang et al. [16] introduced a contract-based energy blockchain to charge the electric vehicle (EV) securely in smart community (SC) [16]. First of all, a permissioned energy blockchain system was adopted for deploying charging services securely for EVs after exploiting the smart contracts. After that, a reputation-based DBFT approach was introduced for attaining the consensus in the allowed blockchain in effective manner. The individual requirements of EVs were satisfied from the energy sources on the basis of the contract theory to increase the utility of operator. In the end, the outcomes depicted that the introduced approach was effective and reliable when compared with the conventional techniques.

Barnawi et al. [17] designed a blockchain-based technique to manage DR (demand response) that offered trading with energy-efficiency amid electric vehicles (EVs) and charging stations (CSs) [17]. The energy utilization and the energy processing were employed to select the miner nodes and block verifiers. These nodes were assisted in authenticating several transactions. A game theory (GT)-based solution was presented for aiding in managing the energy and controlling the load in high and lower demand circumstances. Several parameters were considered to quantify the designed technique. The simulations indicated the supremacy of the designed technique over the traditional approaches.

3 Research Methodology

The blockchain-based energy trading (ET) framework employs all the power generators and power load attributes as components which have connection with the retail markets. Every component has potential to publish and transmit the charging or discharging order to the smart grid public blockchain trading (SG-PBT) platform. A programmable charge installation is employed to charge and discharge the electric vehicles (EVs). The power transmission is switched on or off with this device. The energy providers (EPs) in the publicly available blockchain power exchange (BPE) platform are a kind of traditional power plants of enormous size, distributed micro renewable generators, the storage in which the EP side and electric vehicles are comprised. In addition, all the power loads as well as EVs are associated with BPE platform.

The information is exchanged in this platform after the time duration of 30-min. The components facilitate to decide the price for the energy whose generation is done for incentivized users. Hence, the supply and demand are balanced meanwhile for mitigating the peaks of energy generation and usage. The wholesale market employs the classic energy generators for trading with huge power demand offers. Furthermore, the traditional grid system (GS) utilizes the wholesale market to assist the transactional energy in providing a vision to coordinate with the retail customers. For this, an amount of frequent trenching transactions is deployed in automatic way on the basis of BPE platform. It leads to lessen the centralized attributes of GS of upcoming generation. An equal volume of information is exchanged among a huge generator, distributed energy resource, and renewable energy (RE) generators. The transactions are done from the retail to the wholesale markets for equalizing the opportunity for all elements. Besides, the transactions are also considered with regard to the broadcast and distribution limits and other physical restraints on the grid.

The electric vehicle status matrix X is expressed as:

$$X_{i,t} = \begin{cases} 1, & \text{if } \mathrm{EV}_i \text{ is connected at time } t \\ 0, & \text{otherwise} \end{cases} \tag{1}$$

The power demand of EV is based on the battery residual ($\mathrm{SOC_{ini}}$) in each electric vehicle and the expected $\mathrm{SOC}(\mathrm{SOC_{exp}})$ when they are charged. Thus, its formulation is done as:

Figure 1 illustrates a model of a transactional energy market system that encompasses both retail and wholesale markets within smart grids. In the diagram, arrows symbolize the movement of price offers and transactions.

$$P_{\mathrm{EV}}(t) = \sum_{i=1}^{I} \left(X_{i,t} \left(\mathrm{SOC_{exp}}(i) \pm \mathrm{SOC_{ini}}(i) \right) \right) \tag{2}$$

Therefore, the total residential load is defined as the addition of demand to charge or discharge the EV and load profile without EV for defining the EV charging issue.

$$P_{\mathrm{total}}(t) = P_{\mathrm{load}}(t) + P_{\mathrm{EV}}(t), t \in T \tag{3}$$

In this, a power load is illustrated with the P_{load} which the load leads to create and utilize within the microgrid network earlier. The entire utility function (UF) is

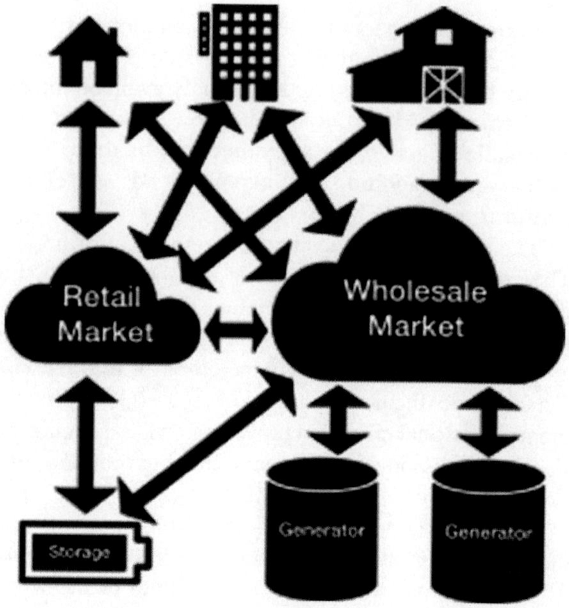

Fig. 1 Transactional energy market system model

exploited for the local region to implement the optimization methods for accomplishing the objective, such as to flatter the load, shave the peak, and preserve the privacy.

The demand is represented like an input to submit to the electricity trading stand book Std_{in}, that serves as a public order book for all stakeholders in the trading environment, as a vector that can be described in the following way:

$$\overrightarrow{O_i} = (\gamma, \text{Id}_i, \sigma_i, Q_i), \tag{4}$$

where Id_i be the specific identifier for the initiators of charging and discharging, which may be EVs or other elements. The price per unit that the customer is capable of paying for the electricity order is represented by σ_i. Q_i denotes the amount of electricity needed to fill this order's requirement, and γ is a matrix indicating whether the order is to purchase or sell electricity:

$$\gamma = \begin{cases} 1, \text{ buy order} \\ 0, \text{ sell order} \end{cases}. \tag{5}$$

The matching order should then be used to create the matched trades for each input order by applying it to the current stand book (Std_{in}). Additionally, every trade order that matches should have its non-error output sent to the (Std_{out}). The following is how the trade information format is demonstrated:

$$\overrightarrow{T_i} = \left(\mathrm{Id}_{\mathrm{sell}}, \mathrm{Id}_{\mathrm{buy}}, \sigma_m, Q_m\right). \tag{6}$$

If the matching electrical purchase and sell order identifiers are the $\mathrm{Id}_{\mathrm{sell}}$, $\mathrm{Id}_{\mathrm{buy}}$, respectively, the pence's matched price and the matched quantity for the order are represented by σ_m and Q_m, respectively. The implementations should show the current entire order book in the aforementioned manner after getting an order message, after getting any matches in the book, and after generating any created trade messages.

It is critical to supply the guiding price for this demand in to effectively handle transactional energy orders and guarantee stakeholder profit. Given that the energy trading market is price competitive, similar to the stock market, it is implied from a particular stock price model that the best charging guide price S_t should be constructed via a jump-diffusion process. $S_t < \overline{S}$, where S be the order's maximum price, can be written as follows:

$$S_t = S_0 \exp\left(\left(\mu - \frac{\sigma^2}{2}\right)t + \sigma W_t\right), \tag{7}$$

where μ and σ denote percentage drift and volatility, respectively. It is possible to adjust volatility and percentage drift to constants according to the limit range of this study, and the Wiener operation is W_t. The best price S_t can therefore be derived for a given greatest price value S_0 by applying derivatives to both sides, as illustrated in the given formulas.

$$\mathrm{d}S_t = \mu S_t \mathrm{d}t + \sigma S_t \mathrm{d}W_t \quad \text{with} \quad S_0 < \overline{S} \tag{8}$$

Additionally, it is possible to determine expectation and variance for S_t, with expectation serving as the trading operation's guide price.

$$E(S_t) = S_0 e^{\mu t} \tag{9}$$

The greatest and lowest prices are possible for the price function S_t, and it changes based on what users are willing to bid. It functions as a guide price for all participants, with the recommendation that no one surpasses it. Additionally, customers retain the freedom to select their trade price from a range between the lowest and the reference price. It is a closed double auction market with separate marketing closure time and price-time precedence. As a result, no central organization is required to carry out market trade.

4 Result and Discussion

4.1 Performance Analysis Parameters

Metrics for analyzing the performance of the proposed models are discussed below.

Accuracy. This metric can be expressed as the percentage of accurately classified points relative to the total number of points.

$$\text{Accuracy} = \frac{\text{Number of points corrrectly classified}}{\text{Total number of points}} * 100 \tag{10}$$

Precision. The quantity of data that may be gleaned from a number depends on its digits. A precise representation of the proximity of two or more measurements is made. It does not rely on accuracy. The precision is determined if the number of TPs is divided by the sum of the number of FPs and TPs (true positives). False positives are instances that the model wrongly classifies as positive but are actually negative.

$$\text{Precision} = \text{TruePositives}/(\text{TruePositives} + \text{FalsePositives}) \tag{11}$$

Recall. It is a measure determined by comparing the total relevant findings extracted by a model. This metric represents the proportion of correctly identified positive instances (true positives) relative to the sum of true positives and false positives (erroneous positive identifications). True positives are instances correctly classified as positive, while false negatives are instances mistakenly labeled as negative when they are actually positive.

$$\text{Recall} = \text{TruePositives}/(\text{TruePositives} + \text{FalseNegatives}) \tag{12}$$

As shown in Fig. 2, the plot shows the floor-wise zones. The plots show the temp on each day of the week.

As shown in Fig. 3, floor-wise plotting of each attribute. The attribute values are shown of each day.

Figure 4 illustrates a comparison of the proposed algorithm's performance with existent models, focusing on accuracy, precision, and recall. The results indicate that the proposed model outperforms the existing ones significantly, boasting superior levels of accuracy, precision, and recall.

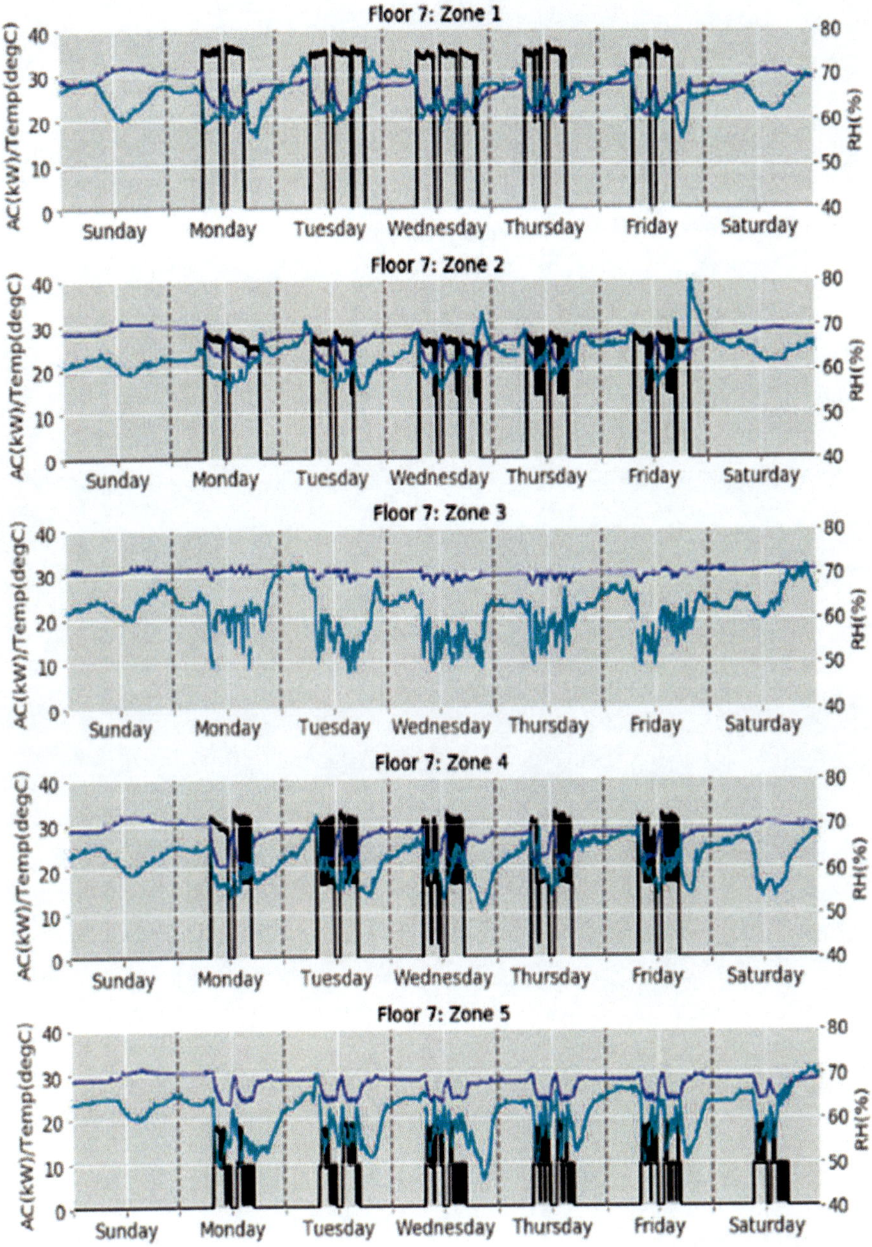

Fig. 2 Floor at each zone

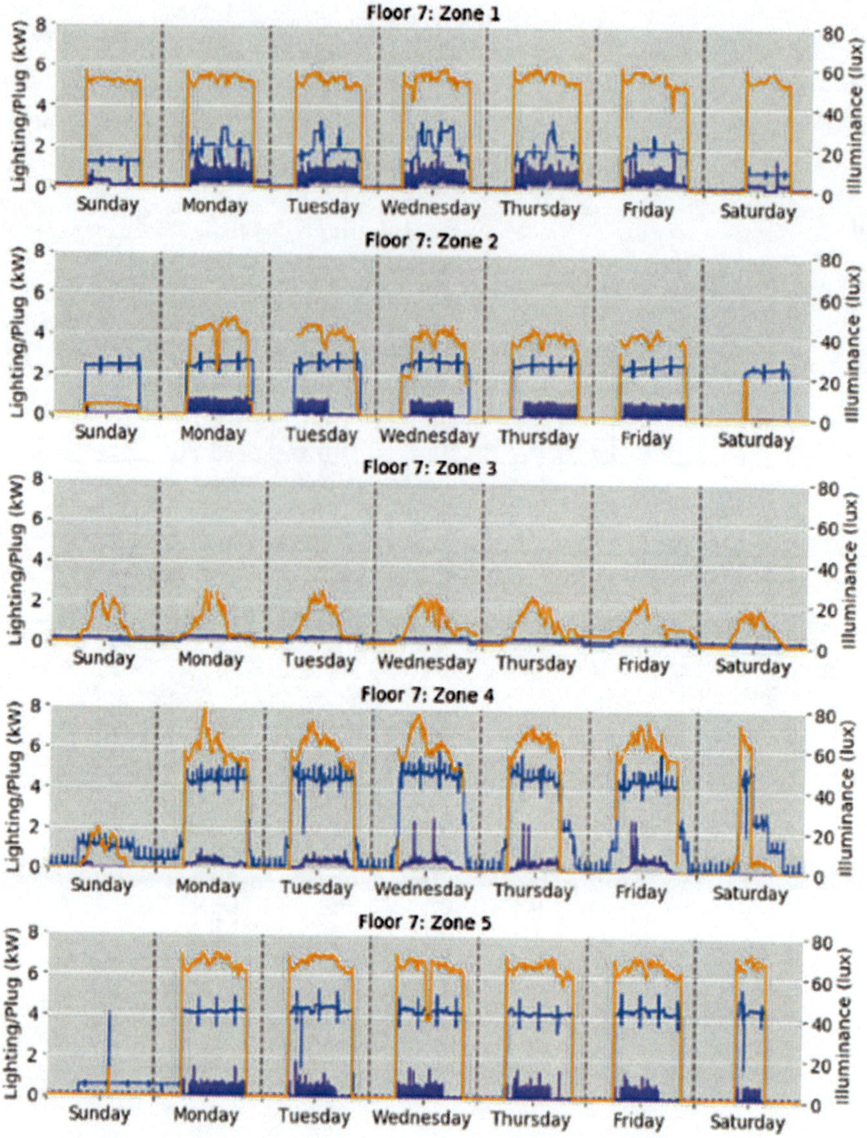

Fig. 3 Floor of each zone versus each attribute

5 Conclusion

Most IoT-based systems predominantly store and process data related to electric vehicles for the purpose of energy trading. An Internet of Electric Vehicles (IoEV) system incorporates blockchain-based smart contracts to record trade transactions. These

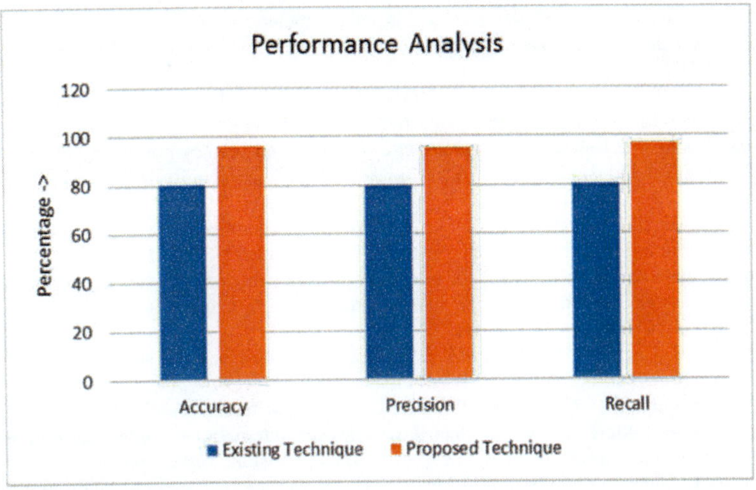

Fig. 4 Performance analysis

smart contracts facilitate the transfer of energy from energy producers (prosumers) to consumers, ensuring a secure and transparent process. The blockchain is traceable, immutable, and transparent; hence, the smart contracts offer potential to EVs for searching an EV which can sell its surplus charge. The proposed model leverages blockchain technology to optimize energy usage for electric vehicles. This model is implemented using Python, and its performance is assessed through an analysis of accuracy, precision, and recall. The results indicate a noteworthy improvement of up to 15% in comparison with existent algorithms.

References

1. Islam MM, Shahjalal M, Hasan MK (2020) Blockchain-based energy transaction model for electric vehicles in V2G network. In: International conference on artificial intelligence in information and communication (ICAIIC), Fukuoka, Japan, pp 628–630
2. Iqbal A, Rajasekaran AS, Nikhil GS (2021) A secure and decentralized blockchain based EV energy trading model using smart contract in V2G network. IEEE Access 9:75761–75777
3. Yang Y, Peng D, Wang W (2020) Block-chain based energy tracing method for electric vehicles charging. In: IEEE sustainable power and energy conference (iSPEC), Chengdu, China, pp 2622–2627
4. Priyanka, Raw RS (2020) The amalgamation of blockchain with smart and connected vehicles: requirements, attacks, and possible solution. In: 2nd international conference on advances in computing, communication control and networking (ICACCCN), Greater Noida, India, pp 896–902
5. Saha R (2021) The blockchain solution for the security of internet of energy and electric vehicle interface. IEEE Trans Veh Technol 70(8):7495–7508

6. Zhou Z, Wang B, Guo Y (2019) Blockchain and computational intelligence inspired incentive-compatible demand response in internet of electric vehicles. IEEE Trans Emerg Top Comput Intell 3(3):205–216
7. Yagmur A, Dedeturk BA, Soran A (2021) Blockchain-based energy applications: the DSO perspective. IEEE Access 9:145605–145625
8. Long Y, Chen Y, Ren W (2020) DePET: a decentralized privacy-preserving energy trading scheme for vehicular energy network via blockchain and K-anonymity. IEEE Access 8:192587–192596
9. Huang X, Xu C, Wang P (2018) LNSC: a security model for electric vehicle and charging pile management based on blockchain ecosystem. IEEE Access 6:13565–13574
10. Liu C, Chai KK, Zhang X (2019) Enhanced proof-of-benefit: a secure blockchain-enabled EV charging system. In: IEEE 90th vehicular technology conference (VTC2019-Fall), Honolulu, HI, USA, pp 1–6
11. Danish SM, Zhang K, Jacobsen HA (2021) BlockEV: efficient and secure charging station selection for electric vehicles. IEEE Tra Int Tra Sys 22(7):4194–4211
12. Kakkar R, Gupta R, Obaidiat MS, Tanwar S (2021) Blockchain and multiple linear regression-based energy trading scheme for electric vehicles. In: International conference on computer, information and telecommunication systems (CITS), Istanbul, Turkey, pp 1–5
13. Zugang L, Chen S, Zhou B (2021) Electric vehicle peer-to-peer energy trading model based on SMES and blockchain. IEEE Trans Appl Supercond 2(4):10–25
14. Shekari M, Moghaddam MP (2021) An introduction to blockchain-based concepts for demand response considering of electric vehicles and renewable energies. In: 28th Iranian conference on electrical engineering (ICEE), Tabriz, Iran, pp 13–20
15. Chen X, Zhang T, Ye W, Wang Z, Ho-Ching Iu H (2021) Blockchain-based electric vehicle incentive system for renewable energy consumption. IEEE Trans Circ Syst II Express Briefs 7(1):20–25
16. Wang Y, Su Z, Xu Q, Zhang N (2018) Contract based energy blockchain for secure electric vehicles charging in smart community. In: IEEE 16th international conference on dependable, autonomic and secure computing, 16th international conference on pervasive intelligence and computing, 4th international conference on big data intelligence and computing and cyber science and technology congress (DASC/PiCom/DataCom/CyberSciTech), Athens, Greece, pp 1019–1026
17. Barnawi A, Aggarwal S, Kumar N, Alghazzawi DM, Alzahrani B, Boulares M (2021) Path planning for energy management of smart maritime electric vehicles: a blockchain-based solution. IEEE Trans Intell Transp 9(9):637–642

Pattern Recognition: An Outline of Literature Review that Taps into Machine Learning to Achieve Sustainable Development Goals

Aarti Mehta Sharma and Senthil Kumar Arumugam

Abstract The sustainable development goals (SDGs) as specified by the United Nations are a blueprint to make the Earth to be more sustainable by the year 2030. It envisions member nations fighting climate change, achieving gender equality, quality education for all, and access to quality healthcare among the 17 goals laid out. To achieve these goals by the year 2030, member nations have put special schemes in place for citizens while experimenting with newer ways in which a measurable difference can be made. Countries are tapping into ancient wisdom and harnessing newer technologies that use artificial intelligence and machine learning to make the world more liveable. These newer methods would also lower the cost of implementation and hence would be very useful to governments across the world. Of much interest are the applications of machine learning in getting useful information and deploying solutions gained from such information to achieve the goals set by the United Nations for an imperishable future. One such machine learning technique that can be employed is pattern recognition which has applications in various areas that will help in making the environment sustainable, making technology sustainable, and thus, making the Earth a better place to live in. This paper conducts a review of various literature from journals, news articles, and books and examines the way pattern recognition can help in developing sustainably.

Keywords Sustainable development · Pattern recognition · Artificial intelligence · Machine learning · SDGs · Sustainability · Lower costs

A. M. Sharma (✉)
Symbiosis Centre for Management Studies, Bengaluru Campus, Symbiosis International (Deemed University), Pune, India
e-mail: director@scmsbengaluru.siu.edu.in

S. K. Arumugam
Christ University, Bangalore 560029, India
e-mail: senthilkumar.a@christuniversity.in

S. Kumar et al. (eds.), *Fourth Congress on Intelligent Systems*, Lecture Notes in Networks and Systems 868, https://doi.org/10.1007/978-981-99-9037-5_8

1 Introduction

The world is poised to become warmer by 1.5° within the next 20 years says a recent report by the United Nations [1]. The reason for this is that we have not respected the world's resources and this will lead to a catastrophe in the form of an unsustainable Earth and possible annihilation of the human species. To prevent this horror, the United Nations Organisation has set forth a series of goals to be achieved by member nations. The most recent form of this movement to save the Earth is known as the "Sustainable Development Goals" (SDGs). These goals are not restricted to only saving the environment but also to creating a just world with access to education, healthcare, and equality for all in the form of seventeen goals [2]. The 17 SDGs to transform our world are: (i) remove poverty, (ii) remove hunger, (iii) good health and well-being, (iv) everyone should get quality education, (v) there should be gender equality, (vi) everyone should have access to clean water and sanitation, (vii) energy should be clean and affordable, (viii) everyone should be able to get decent work and economic growth, (ix) industry should build resilient infrastructure, promote inclusive and sustainable industrialisation and foster innovation, (x) to reduce inequality, (xi) cities and communities should be sustainable, (xii) there should be responsible consumption and production, (xiii) take urgent action to combat climate change, (xiv) conserve and sustainably use all water bodies, (xv) protect, restore, and promote sustainable use of land, (xvi) promote peaceful, just, and inclusive societies for all, and (xvii) strengthen the means of implementation and revitalise the global partnership for sustainable development. These 17 goals are further broken down into 169 targets that have to be achieved so that the above goals can be reached by 2030.

2 Sustainable Development

Sustainability means "a capacity to maintain some entity, outcome or process over time" [3]. The term "Sustainable Development" was first brought into public consciousness by the Brundtland report of 1987 where it was defined as "development that meets the needs of the present without compromising the ability of future generations to meet their own needs" [4]. The concept is typically explained through the three pillars of society, the economy, and the environment [3]. These are depicted as building blocks through three intersecting circles as shown in Fig. 1. The diagram reinforces the fact that sustainable development will happen only when there will be sustainable development in each of these spheres.

The thrust towards sustainable development is pervasive throughout the world. The idea has caught the imagination of the world in a way other development ideas have not, and it is going to be the universal development paradigm for a long time. The movement towards the concept is largely due to the "2030 Agenda for Sustainable Development", adopted by all United Nations Member States in 2015. This agenda is a blueprint towards peace and prosperity for people and the planet, now and in

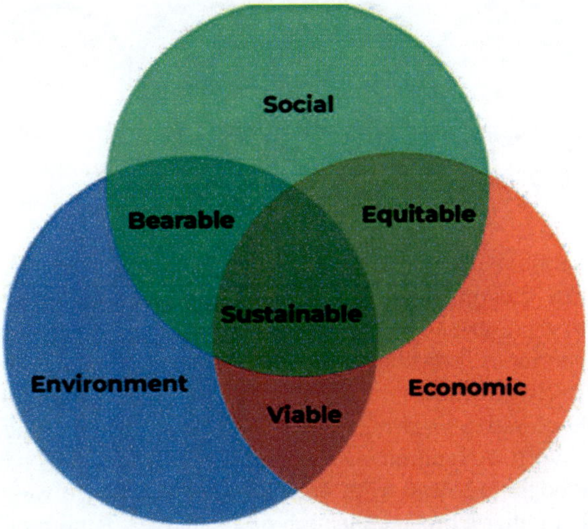

Fig. 1 Building blocks of sustainable development [3]

times to come. The agenda's goals, which are an urgent call for action by all countries—developed and developing—in a global partnership include ending poverty, improving health and education, reducing inequality, spurring economic growth, tackling climate change, and preserving our natural habits [2]. A major advantage of this call to action by the United Nations is that it has pushed the industry to examine ways in which they can help achieve these goals. A recent report by Boston Consulting Group (BCG) says that supply chains can reach a zero-carbon emission rate by making changes in their supply chains which will minimally affect the consumers in terms of cost but will have a great impact on the climate [5].

2.1 Pattern Recognition

As the world moves towards realising the ideals of the 2030 agenda, governments and societies are looking for means to achieve sustainable development. As we get more technologically advanced, there are massive changes in how we live and conduct business [6]. Pattern recognition is one of how technology is being harnessed and we define it as "the automated recognition of patterns and regularities in data". Pattern recognition analyses incoming data and tries to identify patterns. It has been defined by Bishop in his seminal work as "The field of pattern recognition concerned with the automatic discovery of regularities in data through the use of computer algorithms and with the use of these regularities to take actions such as classifying the data into different categories" [7]. The term "Pattern Recognition" is a familiar one which describes objects or classifies them using a collection of mathematical, statistical, heuristic, and inductive techniques [8]. This approach can also be understood as "the study of how machines can observe the environment, learn to distinguish

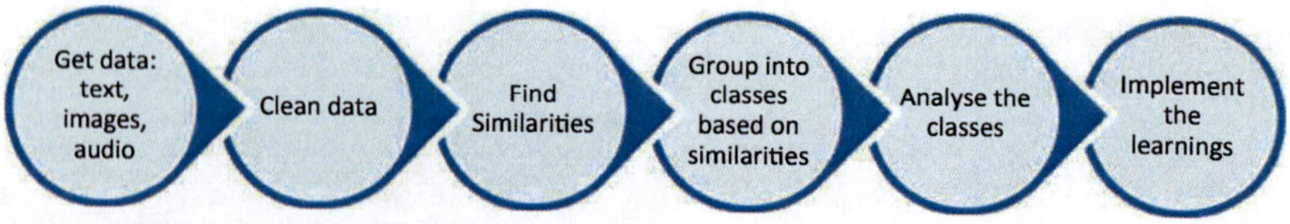

Fig. 2 Process of pattern recognition

various patterns of interest from its background, and make reasonable decisions about the categories of the patterns. During recognition, the given objects are assigned to a prescribed category" [7]. The factors responsible for its development are the enormous growth of computing power and the massive increase in data [9].

Broadly, there are two types of pattern recognition—explorative and descriptive. Explorative pattern recognition identifies overarching data patterns and descriptive pattern recognition starts categorising the detected patterns. In this manner, pattern recognition deals with both of these and hence we say that it is not a technique, but rather a broad collection of techniques. Pattern recognition capability is often a prerequisite for intelligent systems. Pattern can be recognised from words or texts, images, or audio files. Figure 2 depicts the process of pattern recognition.

Automatic and machine-based recognition, description, classification, and grouping of patterns are important problems in a variety of engineering and scientific disciplines like biology, medicine, marketing, analytics, artificial intelligence, etc. These pattern recognition algorithms include principal component analysis, partial least square regression, linear discriminant analysis, K-nearest neighbours, decision trees, random artificial neural networks, etc. [10]. These can be used to read to the blind, categorise documents sourced from the Internet, for biometric recognition, for forecasting crop yield, segment customers, speech recognition for monetary transactions or servicing complaints, etc. [11]. Data can be classified either through supervised or unsupervised classification techniques. In supervised classification, data is identified as a member of an existing category whereas in unsupervised classification data is assigned to an undefined class. Pattern recognition is constantly evolving, driven by emerging applications that are not only challenging but also more computationally intensive. Whereas earlier only numerical data were classified, now text, image, audio, and video data are also classified and analysed for patterns.

2.2 *Pattern Recognition and Artificial Intelligence*

Various machine learning algorithms like regression, clustering, etc., have been applied to measure the SDGs over time [9]. Artificial intelligence (AI) is the process where machines start thinking like human beings. This process might unconsciously also mimic human biases. These machines however can solve complex problems like recognising voice and faces fast. They do this by training on a large dataset and finding patterns which help them to identify and analyse new data. Accordingly, pattern recognition is a branch of artificial intelligence. Broadly, the following three approaches to pattern recognition can be applied [11].

(a) **Statistical Pattern Recognition (or Decision-Theoretic).** This method refers to the use of statistics to understand data and patterns and use this information to forecast. One of the prerequisites for this is working with clean and complete data. It can be used for both supervised and unsupervised learning and different kinds of data. The basic method is that in a dataset the significant features are selected and then they are used to understand, analyse, and interpret data. The algorithm learns and adapts as expected, and then uses the patterns for further processing and training. Data can be classified into clusters or categories and used for predictions. There are many applications for this method—it could be classifying email as spam or not, analysing transaction data at a store or in a bank, etc. Applications of statistical pattern recognition are for pattern recognition in road networks which can be used for map generalisations [12, 13].

(b) **Syntactic Pattern Recognition (or Structural).** In some cases, where there is a lot of data and hence a lot of features, the method mentioned earlier may not identify the correct patterns. Then a hierarchical approach would be preferred. Here complex patterns are first broken down into simpler patterns and then analysed. The simpler patterns are known as "primitive sub-patterns" (such as letters of the alphabet). The pattern is described depending on the way the primitives interact with each other. An everyday example of this interaction is for languages which algorithms study as a mix of words and letters. Each sentence has an underlying grammar to it and only when that is correctly used can a sentence be formed. While this is very useful in formulating sentences, correcting manuscripts, autocorrecting, etc., it also has other applications. Structural pattern recognition also provides a description of how the given pattern is constructed from the primitive sub-patterns and is used in areas where the patterns have a distinct structure that can be captured in terms of a rule set, such as electrocardiogram (ECG) waveforms or textured images. This also encapsulates "template matching" which discovers similarities between two similar types of data. It can be used for finding small images from larger pictures, for medical image processing, for quality control, for face recognition, etc. [14].

(c) **Neural Pattern Recognition.** AI pattern recognition and in particular deep learning methods like neural networks are very popular for pattern detection. These networks are based on data that is referred to as neurons that mimic human decision-making. There are different kinds of neural networks like artificial neural networks, convolutional neural networks, etc. Each can process data through backward propagation as well as forward propagation. They can be used for categorising data, forecasting time-series data, etc. It has been proven that their forecasting error is lesser than the conventional methods and hence is very popular among researchers. One application of a convolutional neural network, i.e. a type of neural network is to track poverty statistics. Using publicly available survey and satellite data, this technique can be trained to identify image features that can explain up to 75% of the variation in local-level economic outcomes and can lend a huge impetus in efforts to track poverty statistics [13]. Neural networks can also be used for map generalisations. Maps are no more static documents but by recognising data that we get with the help of satellites, they are living, breathing interactive comments of our times [12].

While the overarching utility of this deep learning method is to analyse data by identifying patterns, through the three approaches mentioned earlier, pattern recognition has been utilised for a variety of purposes. It is being used for character recognition, biometric recognition, human behaviour pattern analysis, medical image analysis, making maps, writing correctly, etc.

2.3 *Applications of Pattern Recognition*

The concept of the present study is represented in Fig. 3 which shows the main applications of pattern recognition and accomplishing SDGs.

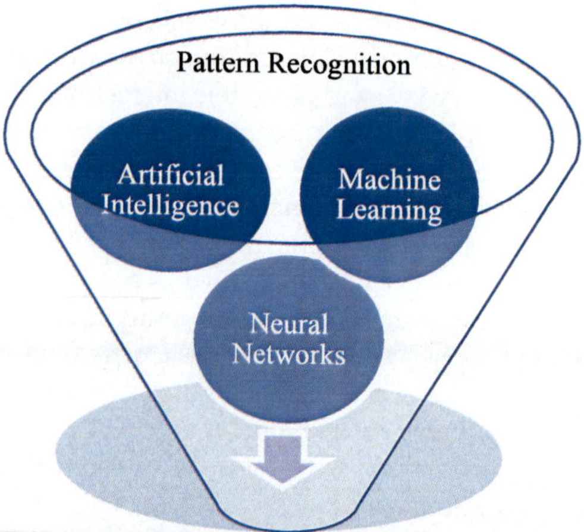

Applications			
Track Poverty	Web Mining	Traffic Networks	Power Load
Online Fraud Detection	Prevent Money Laundering	Agriculture	
Geography	Infrastructure	Character Recognition	Bioinformatics
Military Affairs	Internet of Things	Engineering	
Biometric Recognition	Voice Recognition	Medical Analysis	
Preserving Culture	Land Use and Land Cover		

Achieve SDGs

Fig. 3 Achieving SDGs through pattern recognition

Track Poverty. Governments across the world are applying various means to reduce poverty and help citizens achieve a certain standard of living. To this end, various schemes have been introduced to help citizens avail benefits accrued to them. These initiatives will be successful only if governments are aware of the number of people who are poor and where they live. This technique helps to track poverty by examining areas in which there are indicators of poverty like lack of nocturnal light or absence of roads and roofs [13].

Biometric Recognition. Useful for recognising people using eye scans, fingerprint scans or face scans, commonly used for attendance, etc. [6].

Voice Recognition. This is an interface for human and machine interaction where the machine understands simple voice commands and in certain cases can identify feelings of joy, sorrow, etc. [6, 15].

Medical Analysis. This technique is useful for medical imaging such as X-rays, etc., [6, 8] and to detect breast cancer and other diseases [15, 16].

Traffic Networks. Road, train and river networks can be identified to make large-scale and small-scale maps using pattern recognition. It makes it easier for computers to label roads correctly for example highways or service roads, etc. Urban development based on this kind of information pattern recognition makes it easier to design better communities keeping in mind the existing terrain and networks [14].

Online Fraud Detection. Online fraud can be in the form of sending false emails, stealing identities, making purchases using other people's credit cards, etc. Machine learning, and especially pattern recognition helps organisations to learn from previous frauds and hence prevent them from happening in the future [17].

Prevent Money Laundering. In most countries around the world, massive financial transactions take place on a daily system. These are monitored by rule-based systems and it becomes to identify behaviour which is different from normal customers. Some countries have already introduced intelligent algorithms into their systems to identify deviant transactions using pattern recognition. Adoption by more countries would lead to reducing money laundering [18].

Power Load. The world runs on power and the consumption and demand for power are continuously increasing. It is important to know the patterns of current consumption and foresee the load distribution for the future which should make the distribution of power safe and economical. The method of pattern recognition can improve the efficiency of power distribution, hence providing a great service in an electricity-starved nation like India [19].

Infrastructure. This technique can be used to design buildings and bridges. By working with historical data, this technique will help owners and governments to identify the condition buildings, bridges, etc., are in and predict when they will need to be repaired [20, 21]. This will alleviate suffering and expenditure by preventing the collapse of the same.

Character Recognition. Pattern recognition is very useful in sorting mail or processing cheques in banks, which saves both time and money [8].

Bioinformatics. Useful to conduct DNA sequence analysis which can help in finding the cure for diseases or research heredity [8].

Agriculture. Pattern recognition helps in evaluating the quality of soil, finding minerals in the soil, deciding on which crops to sow, etc. [8].

Geography. This technique is used to predict when earthquakes can happen and to analyse existing earthquake information. It is also helpful for classifying rocks into different types [8].

Engineering. In this field, pattern recognition is used to check for quality by finding faults if any and thus improving upon the safety of vehicles [8]. It can be applied to maintain and monitor expensive operations like aerospace structures [22]. It can be used to monitor the wear and tear of tools and hence maintain quality benchmarks while maintaining and monitoring processes [23, 24]. Pattern recognition along with other artificial intelligence tools helps in navigation for self-driving cars [25]. Image analysis is conducted to identify flaws at the nano or micro levels using pattern recognition [26] which is being applied to process industries to check for and correct against oscillations from set processes, thus maintaining quality [27]. Most manufacturing companies maintain strict quality control by using statistical process control charts which depend upon the data being normally distributed. However, in cases when the data is not normally distributed this method cannot be applied. In this scenario pattern recognition can be used to check for deviations from prescribed manufacturing standards and used to maintain quality [28].

Internet of Things. Pattern recognition can be successfully deployed to analyse data for smart homes so that better service is possible [27].

Web Mining. Makes analysing web logs faster thus helping websites optimise and provide customised service to customers, better e-services, and giving governments teeth to fight terrorism [28–31].

Military Affairs. It can be used to check for troublemakers by analysing land and aviation data [8].

Preserving Culture. Standardising local Indian languages like Devanagiri or Tamil which are handwritten by people of all ages. These handwritten documents are preserved and stored for posterity [32].

Land Use and Land Cover. As humans continue to grow, we convert a lot of land covered by forests and vegetation into agricultural land. This leads to an imbalance in the environment and sets the stage for global warming. Hence, it is important to have data on land occupation and land use so that the indiscriminate felling of trees and reduction of the earth's green cover does not take place. Current methods to do so have not been very accurate and the method of pattern recognition along with contextual information promises more accuracy [33].

Table 1 Mapping of applications of pattern recognition to the SDGs

Goal	SDG number	Application of pattern recognition
Remove poverty	1	1, 7
Remove hunger	2	7
Good health	3	2, 3, 4, 11
Quality education	4	
Gender equality	5	
Clean water and sanitation	6	7
Energy	7	7, 8
Decent work	8	2, 3, 6
Industry	9	6, 9, 10, 14, 16
Reduce inequality	10	4, 6
Sustainable cities	11	5, 8, 13, 14
Responsible consumption	12	12, 15
Climate action	13	19
Life below water	14	
Life on land	15	3, 5, 12, 13, 19
Peace	16	17
Partnerships	17	

2.4 *Sustainable Development Through Pattern Recognition*

The applications of pattern recognition are matched to the SDGs in Table 1 where they can aid in the achievement of the goals.

2.5 *The Way Forward*

In the next few years, the nations of the Earth have to work towards making the Earth sustainable. For this, they need data and techniques that can help them to gauge where they are placed currently concerning attaining sustainability and the amount they have to move forward. The techniques of pattern recognition could play an important part in gauging that and hence prove to be of immense help in attaining the sustainability goals of 2030.

3 Social Implications

The 17 SDGs are a war cry by the United Nations to make the Earth sustainable and fit for use for generations to come. If these goals are not realised, mankind will suffer due to increased temperatures, less food, and fewer resources. These goals cannot be achieved by a few countries acting in isolation. It is only when all member nations act actively on these goals will the Earth and society be sustainable. Governments across the world are putting in place mechanisms so that these goals can be achieved by 2030 which is not very far away. Technology in the form of artificial intelligence and machine learning is omnipresent and the application of technology to existing problems gives fast solutions. This paper attempts to give an impetus to the achievement of the SDGs by suggesting that the technique of pattern recognition, which falls under machine learning can be applied to existing data, to find patterns that can be applied in various areas like traffic monitoring, infrastructure maintenance, medical diagnosis, etc. Traditional methods can be time-consuming and expensive and this method of collecting and analysing data will be faster and less expensive. The machine learning method of pattern recognition will help in getting data in real-time and thus help in formulating policies and correcting courses where needed in the attainment of the SDGs.

4 Conclusion

The goals of sustainable development address values for both living and non-living organisms. Enhancing and identifying the right pattern integrate the overall development of the ecosystem. The current study was conducted based on secondary data. The literature discussion in the present study helps to understand the role of emerging technologies such as artificial intelligence, machine learning, and neural networks in pattern recognition and how it helps to attain the SDGs. The degrees of aliveness can also be observed in human interaction patterns, human-nature relationships, management of supply chain, and educational, political, and agricultural systems. Developing human skills benefits identifying and using interactional patterns that enhance the quality of life and may consequently pave the way for a more sustainable future. Applying concepts that result in interaction patterns are more functional and address various social, economic, and environmental issues [34]. Therefore, it is essential to search for emerging, co-evolving patterns of aliveness in the complexities of adaptive systems change that the SDGs seek to address. These interactional patterns can be recognised, used, and perhaps even further developed into more life-improving patterns. Such patterns appear when actors take part in several actions at various system levels, including local, regional, national, and international, utilising strategies that are somewhat similar to one another but not the same.

References

1. Global average temperature rise of 1.5 degree Celsius in next 20 years, Times of India. https://timesofindia.indiatimes.com/home/environment/global-warming/un-report-global-warming-is-likely-to-blow-past-paris-limit/articleshow/85174223.cms. Last accessed 10 Aug 2021
2. Department of Economic and Social Affairs. The 17 Goals Sustainable Development, United Nations, 2015. https://sdgs.un.org/goals. Last accessed 09 Aug 2021
3. Purvis B, Mao Y, Robinson D (2019) Three pillars of sustainability: in search of conceptual origins. Sustain Sci 14:681–695
4. World Commission on Environment and Development. Report of the World Commission on Environment and Development: Our Common Future, Oslo, Mar 1987. http://www.un-documents.net/our-common-future.pdf. Last accessed 22 Oct 2021
5. Mensah J (2019) Sustainable development: meaning, history, principles, pillars, and implications for human action: literature review. Cogent Soc Sci 5(1):1653531
6. Zapechnikov S (2021) Contemporary trends in privacy-preserving data pattern recognition. Procedia Comput Sci 190:838–844
7. Liu J, Sun J, Wang S (2006) Pattern recognition: an overview. Int J Comput Sci Netw Secur 6(6):57–61
8. Burchardt J, Fredeau M, Hadfield M, Herhold P, O'Brien C, Pieper C, Weise D (2021) Supply chains as a game-changer in the fight against climate change, BCG climate and sustainability. https://web-assets.bcg.com/b3/79/e18102e14739bb2101a49d8e63f0/bcg-supply-chains-as-a-game-changer-in-the-fight-against-climate-change-mar-2021.pdf. Last accessed 22 Oct 2021
9. Bishop CM (2006) Pattern recognition and machine learning. In: Jordan M, Kleinberg J, Scholkopf B (eds) Information science and statistics. Springer-Verlag, New York
10. Park I, Yoon B (2018) Identifying promising research frontiers of pattern recognition through bibliometric analysis. Sustainability 10(5):4055
11. Jain AK, Duin RPW, Mao J (2000) Statistical pattern recognition: a review. IEEE Trans Pattern Anal Mach Intell 22(1):4–37
12. Tian J, Song Z, Gao F, Zhao F (2016) Grid pattern recognition in road networks using the C4.5 algorithms. Cartogr Geogr Inf Sci 43(3):266–282
13. Kronenfeld BJ, Buttenfield BP, Stanislawski LV (2020) Map generalization for the future. Int J Geo-Inf 9(8):468
14. Boesch G (2021) What is pattern recognition? A gentle introduction. https://viso.ai/deep-learning/pattern-recognition/. Last accessed 09 Aug 2021
15. Jean N, Burke M, Xie M, Davis WM, Lobell DB, Ermon S (2016) Combining satellite imagery and machine learning to predict poverty. Science 353(6301):790–794
16. Djellali C, Adda M (2019) A new deep learning model for sequential pattern mining using ensemble learning and models selection taking mobile activity recognition as a case. Procedia Comput Sci 155:129–136
17. Bai L, Zheng W, Li W, Xu D, Chen N, Cui J (2020) Promising targets based on pattern recognition receptors for cancer immunotherapy. Pharmacol Res 159:105017
18. Kannagi A, Mohammed GJ, Murugan SG, Varsha M (2021) Intelligent mechanical systems and its applications on online fraud detection analysis using pattern recognition K-nearest neighbor algorithm for cloud security applications. Mater Today Proc 81(2):745–749
19. Tang J (2016) A survey of R&D of intelligent STR system based on behavior pattern recognition in China. J Money Laundering Control 19(2):109–121
20. Wan YY (2020) Power load pattern recognition algorithm based on characteristic index dimension reduction and improved entropy weight method. Energy Rep 6(9):797–806
21. Petrova E, Pauwels P, Svidt K, Jenson RL (2018) From patterns to evidence: enhancing sustainable building design with pattern recognition and information retrieval approaches. In: 12th European conference on product and process modelling, Copenhagen, Denmark

22. Alogdianakis F, Dimitriou L, Charmpis DC (2021) Pattern recognition in road bridges' deterioration mechanism: an artificial approach for analysing the US national bridge inventory. Transp Res Procedia 52:187–194
23. Perafán-López JC, Sierra-Pérez J (2021) An unsupervised pattern recognition methodology based on factor analysis and a genetic-DBSCAN algorithm to infer operational conditions from strain measurements in structural applications. Chin J Aeronaut 34(2):165–181
24. Hassan M, Damir A, Attia H, Thomson V (2018) Benchmarking of pattern recognition techniques for online tool wear detection. Procedia CIRP 72:1451–1456
25. Junior POC, Conte S, D'Addona DM, Aguiar PR, Baptista FG, Bianchi EC, Teti R (2019) Damage patterns recognition in dressing tools using PZT-based SHM and MLP networks. Procedia CIRP 79:303–307
26. Todorovic M, Simic M (2019) Clustering and pattern recognition in bioengineering and autonomous systems. Procedia Comput Sci 159:2364–2373
27. Belikov S, Su C, Enachescu M (2020) Image-based parametric pattern recognition for micro- and nano-defect detection. IFAC-Papers 53(2):8591–8598
28. Dambros JWV, Farenzena M, Trierweiler JO (2019) Oscillation detection and diagnosis in process industries by pattern recognition technique. IFAC-Papers 52(1):299–304
29. Guh RS (2002) Robustness of the neural network based control chart pattern recognition system to non-normality. Int J Quality Reliab Manage 19(1):97–112
30. Ezeife CI, Lu YI (2005) Mining web log sequential patterns with position coded pre-order linked WAP-tree. Data Min Knowl Disc 10:5–38
31. Talakokkula A (2015) A survey on web usage mining, applications and tools. Comput Eng Intell Syst 6(2):22–29
32. Prashanth DS, Mehta RVK, Sharma N (2020) Classification of handwritten Devanagari number—an analysis of pattern recognition tool using neural network and CNN. Procedia Comput Sci 167:2445–2457
33. Fonseca LMG, Körting TS, Bendini HDN, Girolamo-Neto CD, Neves AK, Soares AR, Taquary EC, Maretto RV (2021) Pattern recognition and remote sensing techniques applied to land use and land cover mapping in the Brazilian Savannah. Pattern Recogn Lett 148:54–60
34. Al Zamil MGH, Samarah SMJ, Rawashdeh M, Hossain MA (2017) An ODT-based abstraction for mining closed sequential temporal patterns in IoT-cloud smart homes. Cluster Comput 20:1815–1829

Novel Approach for Stock Prediction Using Technical Analysis and Sentiment Analysis

Gauravkumarsingh Gaharwar and Sharnil Pandya

Abstract Stock prediction is not new; people have tried to predict stock price or good quality stocks for ages. Machine learning has opened a new direction for the problem of stock prediction. The critical factor deciding the success or failure of the machine learning model depends on the quality of features computed. Moreover, the concept of ensemble learning can significantly enhance the quality of prediction. The proposed model gathers data price time series data and news articles from open access domain and do necessary pre-processing. The model also calculates technical and sentiment features and uses them for training the ensemble model. Performance is compared to definite matrices and with other similar research.

Keywords Stock prediction · Technical analysis · Sentiment analysis · Support vector machine · Ensemble learning

1 Introduction

1.1 Stock Market Analysis

Although equity (stock market) gives maximum return on investment for a long duration, very few investors do not participate in it due to inherent risk and uncertainty associated with investing in the stock market. Buying and selling equity at the right time can maximize the overall return on the investment. From the mathematical perspective, stock market can be modeled as a highly fluctuating nonlinear time series that can be future predicted using computational techniques.

G. Gaharwar (✉)
Navrachana University, Vadodara, Gujarat, India
e-mail: gauravsinghg@nuv.ac.in

S. Pandya
Symbiosis International University, Pune, Maharashtra, India
e-mail: sharnil.pandya@sitpune.edu.in

S. Kumar et al. (eds.), *Fourth Congress on Intelligent Systems*, Lecture Notes in Networks and Systems 868, https://doi.org/10.1007/978-981-99-9037-5_9

Not only stock market investment is ideal for long-duration investments, but stock trading is a good earning opportunity available. For stock price predictions, two types of analysis have been used for ages: fundamental and technical [1]. In this era of information, sentiment analysis is also gaining popularity in predicting the trend in stock price.

Fundamental Analysis

Usually, the price of shares changes with a company's financial performance, and fundamental analysis is the study of the company's financial health on various indicators. The company's marketcap, liquidity, debt, profitability, assets, liabilities, and shareholders' equity are widely used fundamental analysis indicators [2]. There are various reports that the company publishes on a quarterly, half-yearly, and annual basis. The most commonly referred reports for fundamental analysis are balance-sheet, income-expenditure, profit–loss, and cash flow statements. As fundamental analysis relies on published reports of the company, such analysis is often used for predicting stock price trends for a long duration.

2 Technical Analysis

Technical analysis relies on the most recent price and the movement in the stock for predicting the future price of the stock. In this school of thought, its basic notion is that if the stock has done well shortly, it is sure to do well shortly. In technical analysis, it is believed that most information about the stocks is reflected in recent prices, so if trends in the movements are observed, then prices can be easily predicted [3]. In technical analysis, several indicators are calculated from the stock price history to predict the trend. Also, for calculating a few indicators traded volume of the stock is used. The rationale behind technical analysis is that movements in the stock price are primarily due to an imbalance in demand–supply and which can be identified in the form of patterns from the noisy and historical stock price data.

In technical analysis, different charts are prepared from the stock price movements, and each chart suggests one of many technical indicators. The current values of these indicators help predict the future stock price trend. Such analysis is often used to predict short-term trend prediction of stock prices.

Sentiment Analysis

There are various ways in which news about the company dissipates information by the company. This news is dissipated through various news financial news portals. Financial news plays a significant role in predicting future trends in stock price. Twitter is becoming a trendy platform for sharing any type of news, including political, company-related, and financial news. These tweets can be categorized into news, which the company usually shares, and opinions from various experts. Sentiment analysis is a new technique for stock prediction, and very little work is done in this field.

2.1 *Technology Overview for Stock Prediction*

Various researchers for stock prediction employed various supervised techniques. Depending on the problem at hand, classification and regression techniques were used by different researchers. As the stock quality prediction falls under the classification category, ML algorithms applicable to classification will be utilized for this problem.

Support Vector Machine (SVM)

For the binary classification problem, like estimating the quality of stock for investments, the most recommended supervised machine learning algorithm is the support vector machine (SVM). In SVM, the basic idea is to identify $N - 1$ dimension hyperplanes in N-dimension space, where N is the total number of feature vectors computed from the data. Our objective is to find the one hyperplane which maximizes the margin between data points of both classes. Rosello et al. [4] took input parameters such as RSI and MACD of index data, applied them to the SVM algorithm, and predicted the up and down movements, i.e., bullish and bearish movements, for a week. Using SVM, only Dunis et al. [5] achieved a 100% hit ratio in predicting the market's direction using Madrid IBEX-35 stock index data. Grigoryan [6] recommended variable selection techniques before inputting data into the SVM model for better prediction.

K-Nearest Neighbors (kNN)

K-nearest neighbor (kNN) is a supervised machine learning algorithm used to solve both regression and classification types of problems. The basic concept behind the kNN algorithm is that similar things remain near each other. The correct value of k depends on the dataset, and one needs to run the kNN algorithm multiple times to select the correct value of k. Tanuwijaya and Hansun [7] used KNN to predict the LQ45 stock index and achieved an accuracy of 91.81 for a particular stock. Alkhatib et al. [8] applied the kNN algorithm with the $k = 5$ value for selected five companies of the Jordanian stock market and achieved a minimum error value in prediction. In the research work, Shi [9] applied the kNN algorithm to HANGSENG and DAX indices to label them with "winner" or "loser".

Artificial Neural Network

Artificial neural network (ANN) is inspired by human brain neural networks, in which the brain uses biologically inspired computing code with many simple, highly interconnected processing elements to process information model. ANN consists of three layers: the input, hidden, and output. Shastri et al. [10] applied sentiment analysis on data retrieved from news to predict stock price movement and achieved an accuracy of around 90% in most cases. Selvamuthu et al. [11] compared different ANN techniques on Indian stock market tick data and achieved an accuracy of about 99.90% for all the techniques. Karlsson and Nordberg [12] trained ANN models on the five most influential stock market indices, like S&P 500, DAX, Nikkei, OMX 30 Copenhagen, and OMX 30 Stockholm, and then cross-verified internally by grouping them into foreign and domestic models. They found that the ANNs perform similarly, regardless of the market data for which it is trained. Moghaddam et al. [13] developed the BPNN model, which used the NASDAQ Stock Exchange index value of the last four to nine working days and the day of the week as the input parameters and concluded that there is no distinction between the predicted values and actual values.

Ensemble Learning

It can be observed that different researchers have applied different algorithms like SVM, KNN, and ANN for predicting the performance of stock prediction models. We need to take the best approach from all the underlying ML algorithms, so we propose an ensemble approach for our model. Few researchers have already implemented an ensemble approach in stock prediction. Mehta et al. [14] combined SVR and LSTM in an ensemble approach and attained maximum accuracy with reduced variance. Wing et al. [15] developed a financial expert system based on historical stock prices, sentiment score of news articles, Google search trends, and unique visitors on Wikipedia pages of a stock using an ensemble approach to get better accuracy in the prediction. Nti et al. [16] developed an ensemble model by combining genetic algorithm and SVM to predict the stock price from Ghana Stock Exchange and identify that the newly developed model performs better the other machine learning algorithms like decision tree (DT), random forest (RF), and artificial neural network (ANN) for prediction of stock price ahead of 10 days.

3 Literature Review

Machine learning (ML)/deep learning (DL) comprises techniques that don't need an explicit algorithm for any predictions but learn with the help of the data provided. The process of this type of learning is also called model training. ML and DL are not new techniques for stock price predictions, and various researchers have done pioneering work in this area.

Researchers have done some research in using technical analysis for stock price predictions for short as well as long durations. In 2013, Dunis et al. [5] identified 35 stocks of the Madrid Stock Exchange for training a SVM-based model. They were able to predict the direction of market movement successfully. In the same year, Alkhatib et al. [8] successfully implemented the kNN algorithm to predict the stock prices of five selected Jordanian Stock Exchange stocks. In 2014, Rosillo et al. [4] implemented SVM to predict weekly change movement using S&P 500 index data from New York Stock Exchange. In 2015, Karlsson and Nordberg [12] worked on indices of countries like the USA, Denmark, Germany, Japan, and Sweden.

They successfully predicted the future price of these indices using ANN algorithm. Again in 2016, Moghaddam et al. [13] used ANN to predict the future price of the NASDAQ composite index. In 2016, Shi [9] successfully predicted the future price of HANGSENG and the DAX indices using the kNN algorithm. Boa et al. [17] have pioneered stock price prediction using DL. They have applied auto-encoders and Long Short-Term Memory (LSTM) to predict the future price of the CSI 300 Index, Nifty 50 Index, Hang Seng Index, Nikkei 225 Index, S&P 500 Index, and DJIA Index and predict the price for the duration up to six years. Chou and Nguyen [18] used the Linear Regression ML model to predict intraday stock prices for stocks listed on Taiwan Stock Exchange. In the same year, Grigoryan [6] successfully applied SVM with variable selection methods to predict the future price of the Dow Jones industrial average (DJIA) index. In 2018, Nava et al. [19] successfully predicted 25 min look ahead price for S&P 500 index, using Empirical Mode Decomposition (EMD) in SVR. In the same year, Zhang et al. [20] did a technical analysis of stocks of the Shanghai Stock Exchange (SSE50) index and predicted stock prices 30 days in advance using the ANN algorithm. In 2018 only, Guo et al. [21] used the SVR algorithm-based Particle Swarm Optimization (PSO) in the selected stocks of the Shanghai Stock Exchange. They could make predictions for different durations, including 5, 30 min, and one day. Tanuwijaya and Hansun [7] used the kNN algorithm successfully predict the Indonesian Stock Exchange-based index, namely LQ45. Selvamuthu et al. [11] have pioneeringly predicted stock prices for Indian Stock Exchanges. They have trained an ANN model to predict stock prices. In 2019, Mehta et al. [14] used the ensemble learning approach to combine various ML algorithms for stock price prediction listed on NASDAQ Stock Exchange. They were able to predict the 1-day future stock price for NASDAQ stocks. In 2020, Yuan et al. [22] successfully predicted stock prices for seven years duration. They integrated SVM, Random Forrest (RF), and ANN and predicted the price of Shanghai Stock Exchange stocks. In the same year, Nti et al. [16] demonstrated that the ensemble learning approach suits stock price prediction. They have selected two companies from the banking and petroleum sectors.

Any stock's price highly depends on news sentiments prevailing in the market. If appropriately analyzed, these sentiment values can be a significant factor in predicting stock price. Many researchers have contributed to predicting stock price from news sentiments. Ding et al. [23] used financial news from Reuters and Bloomberg and predicted price for a day, a week, and a month using deep neural network (DNN) approach. In 2016, Kirange and Deshmukh [24] studied financial news regarding two stocks listed on the Indian Stock Exchange, namely Wipro and Infosys. They used Naive Bayes, kNN, and SVM algorithms to predict the price for ten years. The same year, Pagolu et al. [25] predicted three days of price analyzing data on microblogging sites using RF and Logistic Regression algorithms. Ding et al. [26] use a knowledge-driven event embedding (KGEB) approach using news articles and predicted price of the S&P 500 index and individual stocks of the S&P 500 index. In 2017, Huynh et al. [27] predicted the price of the S&P 500 index and individual stock for one day, two days, five days, seven days, and ten days by applying DNN to financial news data. In the same year, Li et al. [28] applied Sentimental Transfer Learning to news articles impacting companies part of the HangSeng Index and predicting daily price in advance. In 2018, Bharathi et al. [29] applied the sentence level sentiment score (SSS) algorithm on RSS news feeds and predicted price for the duration of 5, 10, and 15 days for the stocks listed on the Amman Stock Exchange (ASE). In the same year, Batra and Daudpota successfully applied SVM on Twits related to Apple and successfully predicted price from 2010 to 2017. Chiong et al. [30] have also successfully demonstrated the application of the SVM algorithm on financial news for stock price prediction. Sohangir et al. analyzed the financial sentiment lexicon on Twitter data using Na¨ıve Bayes, SVM, and Logistic Regression algorithms. Shi et al. [31] proposed DNN-based Deep Stock Prediction using company-specific financial news for companies indexed in S&P 500 index. Weng et al. [15] applied an ensemble learning approach to news and Google search data and could predict stock price one day ahead. Shastri et al. [10] created a dataset using news articles from Apple, trained an ANN-based model, and successfully predicted future stock price.

Minimal research is done for combining technical and sentiment data and developing a comprehensive model for stock price prediction. Deng et al. [32] tried combining technical and sentimental analysis for three Japanese companies. They suggested a multi-kernel approach to the price data from the Japanese Stock Exchange and sentiment data from Engadget. Li et al. [33] successfully predicted intraday stock price by combining technical indicators and sentiment data from financial news for the companies listed on Hong Kong Stock Exchange. They have trained multi-kernel SVR for price prediction. Akita et al. [34] predicted stock prices for 50 companies in Tokyo Stock Exchange using the LSTM DL approach. The sentiment is computed by applying Bag of Words (BoW) to the new articles regarding companies. Ayman et al. [35] trained the model using Naïve Bayes and kNN algorithms, where they selected three companies, namely Yahoo, Microsoft, and Facebook, listed on NASDAQ for the prediction.

4 Novel Quality Stock Identification Model

As significantly less work is done for combining technical and sentiment analysis, we propose developing a ML model to combine technical and sentiment analysis features. Apart from this, we propose two novel approaches to the stock prediction problem. Firstly, investors are more concerned about identifying good quality stock than the exact price prediction. So, we propose to convert this classic regression-type problem of price prediction to the classification-type problem of good stock identification, which has the potential to give a CAGR of 15%. The second originality of this model is developing two new indicators, viz. Absolute Position Indicator (API) and Relative Position Indicatory (RPI), which are based on Current Market Price (CMP), 52 Week High (52WH), and 52 Week Low (52WL) of the stock.

$$\text{API} = \text{CMP}/52$$
$$\text{WHRPI} = (\text{CMP} - 52\text{WL})/(52\text{WH}-52\text{WL})$$

5 Abstract Model

Abstract model is described in Fig. 1. As shown in the Abstract model, financial data and sentiment data for individual stock are available in an open domain on the world wide web (WWW), where financial data is available in the form of stock price time series for the given duration, and sentiment data is available in the form of news articles on the company under investigation.

This data contains many relevant and irreverent fields, so data needs filtering along with extraction. Filtered data also needs additional processing to convert data into a usable format. Especially in the case of sentiment data, where data is only textual, much processing is to be done to extract numeric features from it. Processed feature data can be stored in the database for future computation. This data is used

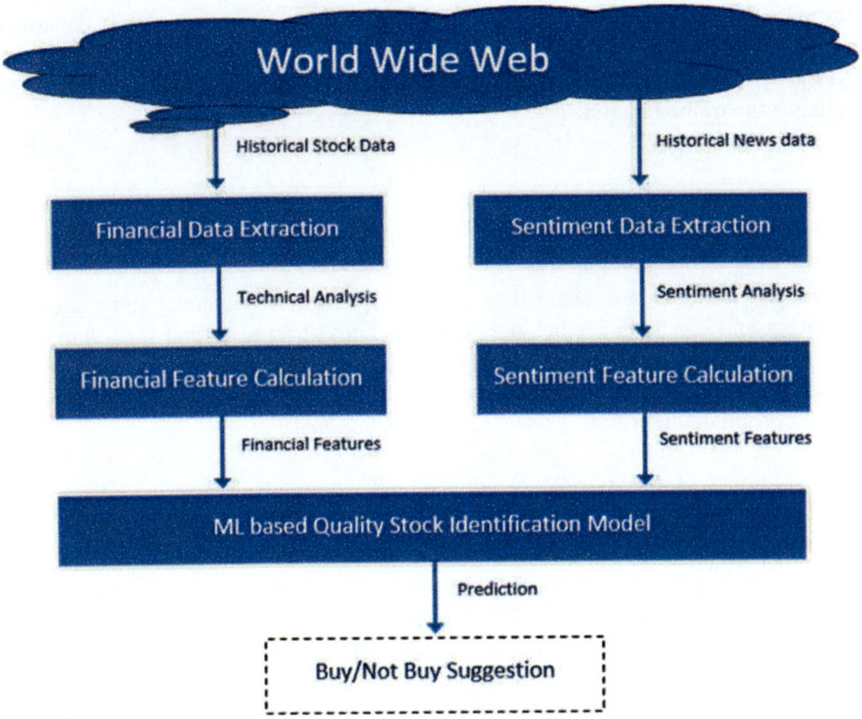

Fig. 1 Abstract model for the ensembled model combining technical and sentiment analysis

for training the selected ML model. The user can test the model by input, i.e., stock name. The system fetches all the relevant historical technical and sentiment data from the WWW and predicts whether the stock is a good buy for the future.

5.1 *System Architecture*

As described in Fig. 2, the system architecture consists of five different layers, viz. data gathering/extraction layer, data pre-processing layer, features calculation layer, features combining layer, model training layer, and performance evaluation layer. Each layer takes data from the previous layer, adds value to the data, and passes data to the next layer. The data gathering/extraction layer fetches stock-specific time series data and historical news data. The data pre-processing layer processes news data and converts time series data into various technical indicators. The feature calculation layer is crucial for the entire system, as any ML model is as good as the selected features. Two sets of features are calculated, one from financial (technical) data and the other from sentiment (news) data.

Each feature has an integer value; various computations will be done to get financial features, while different opinion-finding algorithms will be used to calculate sentiment features. Once different features are calculated, the next layer, the features combining layer, will combine financial and sentiment features. In model training layer, these features are used to train the selected ML model, and evaluation layer evaluates the trained model.

6 Conclusion and Future Work

The proposed model is developed to combine qualities of technical and sentiment indicators and can help predict the price for both short-term and long duration. Moreover, two new indicators, API and RPI, developed can be generalized and implemented for more detailed stock analysis. In the future, the model is to be implemented using appropriate algorithms and tested for predictions.

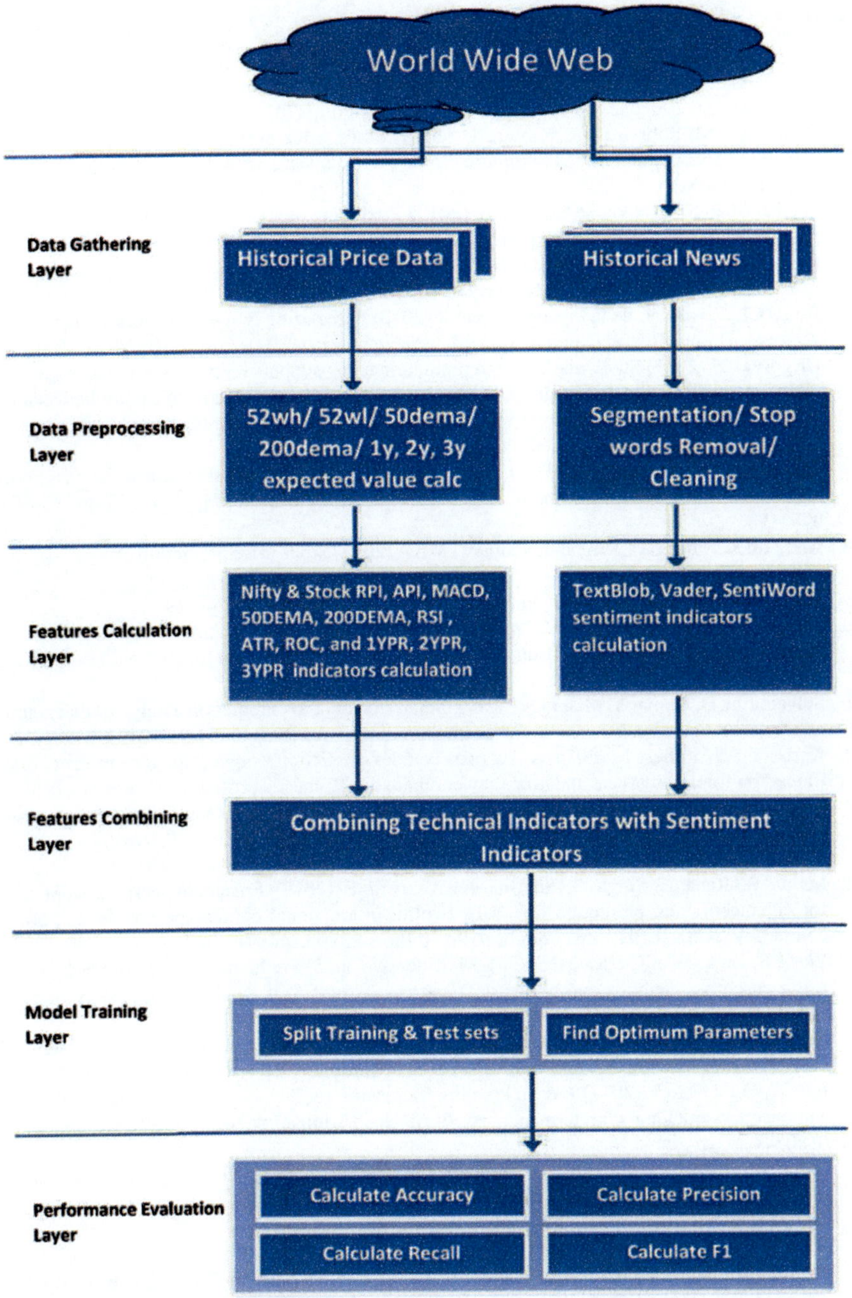

Fig. 2 Detailed system architecture of the ensembled model combining technical and sentiment analysis

References

1. Shah D, Isah H, Zulkernine F (2019) Stock market analysis: a review and taxonomy of prediction techniques. Int J Financ Stud 7. https://doi.org/10.3390/ijfs7020026
2. Henrique BM, Sobreiro VA, Kimura H (2018) Stock price prediction using support vector regression on daily and up to the minute prices. J Financ Data Sci 4. https://doi.org/10.1016/j.jfds.2018.04.003
3. Patel J, Shah S, Thakkar P, Kotecha K (2015) Predicting stock market index using fusion of machine learning techniques. Expert Syst Appl 42. https://doi.org/10.1016/j.eswa.2014.10.031
4. Rosillo R, Giner J, De la Fuente D (2014) Stock market simulation using support vector machines. J Forecast 33. https://doi.org/10.1002/for.2302
5. Dunis CL, Rosillo R, de la Fuente D, Pino R (2013) Forecasting IBEX-35 moves using support vector machines. Neural Comput Appl 23. https://doi.org/10.1007/s00521-012-0821-9
6. Grigoryan H (2017) Stock market trend prediction using support vector machines and variable selection methods. In: Proceedings of the 2017 international conference on applied mathematics, modelling and statistics application (AMMSA 2017). Atlantis Press, Paris, France. https://doi.org/10.2991/ammsa-17.2017.45
7. Tanuwijaya J, Hansun S (2019) LQ45 stock index prediction using k-nearest neighbors regression. Int J Recent Technol Eng 8:2388–2391. https://doi.org/10.35940/ijrte.C4663.098319
8. Alkhatib K, Najadat H, Hmeidi I, Shatnawi MKA (2013) Stock price prediction using K-nearest neighbor (kNN) algorithm
9. Shi Y (2016) kNN predictability analysis of stock and share closing prices
10. Shastri M, Roy S, Mittal M (2018) Stock price prediction using artificial neural model: an application of big data. ICST Trans Scalable Inf Syst. https://doi.org/10.4108/eai.19-12-2018.156085
11. Selvamuthu D, Kumar V, Mishra A (2019) Indian stock market prediction using artificial neural networks on tick data. Finance Innov 5. https://doi.org/10.1186/s40854-019-0131-7
12. Karlsson S, Nordberg M (2015) Stock market index prediction using artificial neural networks trained on foreign markets and how they compare to a domestic artificial neural network
13. Moghaddam AH, Moghaddam MH, Esfandyari M (2016) Stock market index prediction using artificial neural network. J Econ Finance Adm Sci 21. https://doi.org/10.1016/j.jefas.2016.07.002
14. Mehta S, Rana P, Singh S, Sharma A, Agarwal P (2019) Ensemble learning approach for enhanced stock prediction. In: 2019 twelfth international conference on contemporary computing (IC3). IEEE. https://doi.org/10.1109/IC3.2019.8844891
15. Weng B, Lu L, Wang X, Megahed FM, Martinez W (2018) Predicting short-term stock prices using ensemble methods and online data sources. Expert Syst Appl 112. https://doi.org/10.1016/j.eswa.2018.06.016
16. Nti IK, Adekoya AF, Weyori BA (2020) Efficient stock-market prediction using ensemble support vector machine. Open Comput Sci 10. https://doi.org/10.1515/comp-2020-0199
17. Bao W, Yue J, Rao Y (2017) A deep learning framework for financial time series using stacked autoencoders and long-short term memory. PLoS One 12. https://doi.org/10.1371/journal.pone.0180944
18. Chou J-S, Nguyen T-K (2018) Forward forecast of stock price using sliding-window metaheuristic-optimized machine-learning regression. IEEE Trans Ind Inform 14. https://doi.org/10.1109/TII.2018.2794389
19. Nava N, Matteo T, Aste T (2018) Financial time series forecasting using empirical mode decomposition and support vector regression. Risks 6. https://doi.org/10.3390/risks6010007
20. Zhang C, Ji Z, Zhang J, Wang Y, Zhao X, Yang Y (2018) Predicting Chinese stock market price trend using machine learning approach. In: Proceedings of the 2nd international conference on computer science and application engineering—CSAE'18. ACM Press, New York, USA. https://doi.org/10.1145/3207677.3277966

21. Guo Y, Han S, Shen C, Li Y, Yin X, Bai Y (2018) An adaptive SVR for high-frequency stock price forecasting. IEEE Access 6:11397–11404. https://doi.org/10.1109/ACCESS.2018.2806180
22. Yuan X, Yuan J, Jiang T, Ain QU (2020) Integrated long-term stock selection models based on feature selection and machine learning algorithms for China stock market. IEEE Access 8. https://doi.org/10.1109/ACCESS.2020.2969293
23. Ding X, Zhang Y, Liu T, Duan J (2014) Using structured events to predict stock price movement: an empirical investigation. In: Proceedings of the 2014 conference on empirical methods in natural language processing (EMNLP). Association for Computational Linguistics, Stroudsburg, PA, USA. https://doi.org/10.3115/v1/D14-1148
24. Kirange DK, Kirange MDK, Deshmukh RR (2016) Sentiment analysis of news headlines for stock price prediction image processing view project multispectral palmprint recognition view project sentiment analysis of news headlines for stock price prediction. Int J Adv Comput Technol 5. https://doi.org/10.13140/RG.2.1.4606.3765
25. Pagolu VS, Reddy KN, Panda G, Majhi B (2017) Sentiment analysis of Twitter data for predicting stock market movements. In: International conference on signal processing communication power embedded system SCOPES 2016—proceedings, 1345–1350. https://doi.org/10.1109/SCOPES.2016.7955659
26. Ding X, Zhang Y, Liu T, Duan J. Knowledge-driven event embedding for stock prediction
27. Huynh HD, Dang LM, Duong D (2017) A new model for stock price movements prediction using deep neural network. In: Proceedings of the eighth international symposium on information and communication technology. ACM, New York, NY, USA. https://doi.org/10.1145/3155133.3155202
28. Li X, Xie H, Wong T-L, Wang FL (2017) Market impact analysis via sentimental transfer learning. In: 2017 IEEE international conference on big data and smart computing (BigComp). IEEE. https://doi.org/10.1109/BIGCOMP.2017.7881754
29. Bharathi S, Geetha A (2017) Sentiment analysis for effective stock market prediction. Int J Intell Eng Syst 10. https://doi.org/10.22266/ijies2017.0630.16
30. Chiong R, Fan Z, Hu Z, Adam MTP, Lutz B, Neumann D (2018) A sentiment analysis-based machine learning approach for financial market prediction via news disclosures. In: Proceedings of the genetic and evolutionary computation conference companion. ACM, New York, NY, USA. https://doi.org/10.1145/3205651.3205682
31. Shi L, Teng Z, Wang L, Zhang Y, Binder A (2019) DeepClue: visual interpretation of text-based deep stock prediction. IEEE Trans Knowl Data Eng 31. https://doi.org/10.1109/TKDE.2018.2854193
32. Deng S, Mitsubuchi T, Shioda K, Shimada T, Sakurai A (2011) Combining technical analysis with sentiment analysis for stock price prediction. In: Proceedings of IEEE 9th international conference on dependable, autonomic and secure computing DASC 2011, 800–807. https://doi.org/10.1109/DASC.2011.138
33. Li X, Huang X, Deng X, Zhu S (2014) Enhancing quantitative intra-day stock return prediction by integrating both market news and stock prices information. Neurocomputing 142. https://doi.org/10.1016/j.neucom.2014.04.043
34. Akita R, Yoshihara A, Matsubara T, Uehara K (2016) Deep learning for stock prediction using numerical and textual information. In: 2016 IEEE/ACIS 15th international conference on computer and information science (ICIS). IEEE. https://doi.org/10.1109/ICIS.2016.7550882
35. Khedr AE, Salama SE, Yaseen N (2017) Predicting stock market behavior using data mining technique and news sentiment analysis. Int J Intell Syst Appl 9. https://doi.org/10.5815/ijisa.2017.07.03

Visualizing and Exploring the Dynamics of Optimization via Circular Swap Mutations in Constraint-Based Problem Spaces

Navin K. Ipe and Raghavendra V. Kulkarni

Abstract This paper presents an investigation into utilizing circular swap mutations and partial brute forcing in guiding a stochastic search toward an optimal solution. The findings have potential implications for computational intelligence approaches in massive search spaces with known constraints. The efficacy of the method is examined using Sudoku puzzles ranging from 17 to 37 clues. The study graphically depicts the magnitude of the problem space, thus revealing the spatial proximity of states and the nature in which intertwined constraints affect the scope for locating a solution. These insights potentially assist in comprehending the problem space when designing solutions for vast, multidimensional problems. Constraint-aware circular swap mutations can serve as a successful strategy in the design of computational intelligence algorithms that need to be made capable of escaping local optima under temporal constraints. Future directions for research are also suggested. These include mathematically examining paths to optimal solutions and reverse-generating fitness landscapes.

Keywords Computational intelligence · Combinatorics · Constraint satisfaction problems · Sudoku

1 Introduction

Sudoku is a number puzzle, originally consisting of a square grid of 9×9 cells, where each cluster of contiguous 3×3 cells (a sub-grid) contain unique numbers ranging from 1 to 9, such that each row and column of the 9×9 grid also contain unique numbers ranging from 1 to 9. The puzzle is pre-filled with numbers called

N. K. Ipe (✉)
Computer Science and Engineering, M. S. Ramaiah University of Applied Sciences, Bangalore, India
e-mail: navinipe@gmail.com

R. V. Kulkarni
Computer Science and Engineering, Graphic Era Deemed to be University, Dehradun, India
e-mail: arvie@ieee.org

S. Kumar et al. (eds.), *Fourth Congress on Intelligent Systems*, Lecture Notes in Networks and Systems 868, https://doi.org/10.1007/978-981-99-9037-5_10

"clues," where the number of clues given may range from 17 to 77 [1, 2]. Sudoku in its modern form was introduced in 1979 by Howard Garns as a game named "Number Place." Sudoku is an exact-cover type problem that can be modeled as a constraint satisfaction problem (CSP) and is NP-complete. Even the most difficult puzzles can be solved by exhaustive backtracking and pivot-based heuristics.

The purpose of this paper is not to provide an efficient method of definitively solving sudoku puzzles and being better than other algorithms presented in literature. Solving sudoku is easily done via recursive backtracking algorithms or the dancing links algorithm [3] which can solve even 17-clue puzzles in a fraction of a second. This paper does not claim to present an "evolutionary" algorithm either. Evolutionary algorithms are considered good enough if they can provide a near-optimal solution, but sudoku requires an optimal solution. This paper presents the realities of utilizing stochastic approaches in a vast domain. This is achieved by providing a fresh perspective that uses constraint-aware circular swaps as a mutation technique, visualizes the density of puzzle states, the proximity of non-optimal states to the solution and examining how a judicious use of techniques similar to minimum conflicts [4] CSP, can narrow down and channel the search toward a solution, even when complex heuristics are not used. These visualizations have the potential to assist future work on sudoku algorithms or CSP algorithms where properties of the search space are not well known.

Primary Contributions of This Paper

- Constraint-aware circular swaps have been proposed as a mutation technique that could improve the capability of evolutionary algorithms to escape local optima.
- Sudoku fitness landscape visualization technique has been proposed, which clearly depicts the vast, dense problem space and the proximity of sub-optimal solutions, to the optimal solution.
- Partial brute forcing is proposed and visualized as a method of narrowing down the problem space (Figs. 8, 7; Table 1).

The remainder of the paper is organized as follows: Evolutionary algorithms that attempt solving sudoku and concepts that utilize similar techniques for CSP are referenced in Sect. 2. Definitions used, concept introductions, and algorithm design are presented in Sect. 3. Visualization of the fitness landscape is presented in Sect. 3.2. Trial runs and results of solving the puzzle are presented in Sect. 4. The paper concludes with Sect. 5.

2 Related Work

The soft computing (SC) paradigm, having the human mind as its role-model has also been referred to as computational intelligence (CI) and although it does not have a precise definition, it has been defined in general by fuzzy logic, evolutionary computation (EC), machine learning, and probabilistic reasoning. These are a class of algorithms that operate on constraints by utilizing a soft stochastic approach that converges to a desired solution. Despite evolutionary algorithms not being appropriate for solving sudoku, various evolutionary and logical inference techniques have been attempted. These range from satisfiability problems (SAT) [5] and simulated annealing [6], to various evolutionary computing (EC) techniques like particle swarm optimization (PSO) [7–10], genetic algorithms [11, 12], and variants of harmony-search [13]. However, most algorithms do not attempt solving 17-clue puzzles due to the vastness of fitness landscapes containing approximately 3×10^{34} points (the number of permutations of a 17-clue puzzle). Among the most promising approaches that solved Sudoku via CI was the hybridization that used human heuristics and genetic algorithms (GA) [14]. Some authors also utilized logical deduction to augment GA and solved even 21-clue puzzles with 100% success in a matter of seconds [15]. However, it is important to note that even among tough 21-clue or 17-clue puzzles, some puzzles are easier to solve than others. Therefore, it is important to verify the algorithm for various categories of puzzles before making conclusions regarding its efficacy.

In the constraint satisfaction domain, the minimum conflicts heuristic which solved the eight queens problem, could solve even a one million queens problem in an average of less than fifty steps [4]. However, it was prone to getting stuck at a local optima and required execution of all possible iterations if all queens were initially positioned on the first row. Perhaps a more intelligent approach would be to exploit features of the problem domain, as encouraged by [6, 16], who also emphasized the importance of initialization, to reduce the search space. An important observation to note was the manner in which random perturbations [17] were capable of solving sudoku puzzles, sometimes offering quicker solutions than evolutionary approaches. The paper questioned the relevance of EC, when a purely random approach could solve even 21-clue puzzles. The utility of uniform randomness to locate optima in vast fitness landscapes was further validated in [18], for a fitness landscape that was similar in vastness to sudoku. Various authors [15] have already remarked on the futility of utilizing evolutionary approaches for solving sudoku, when simpler, quicker techniques like backtracking are sufficient. Despite this, efforts at utilizing hybrid techniques to improve on evolutionary approaches to solve sudoku continue, while tending to avoid difficult puzzles and offering minimal information on algorithm complexity or on how the solution was obtained.

Although this paper began as a challenge to find a new evolutionary method of solving sudoku, the vast literature we came across and our own observations while attempting to solve puzzles, exposed a research gap in the need to be able to visualize the magnitude of the problem space being tackled, the proximity of the optimal solution state to non-optimal states, puzzle densities, categories of solution difficulty and the hard truth that even puzzles that may appear to be simple (due to having a large number of given clues) may in-fact be difficult to solve. Preprints of this paper are available at [19, 20].

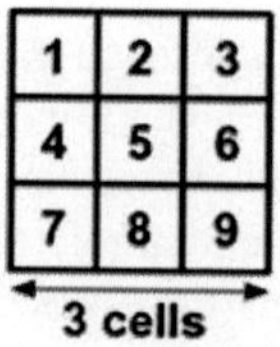

(a) Cells of one individual (3 × 3 sub-grid)

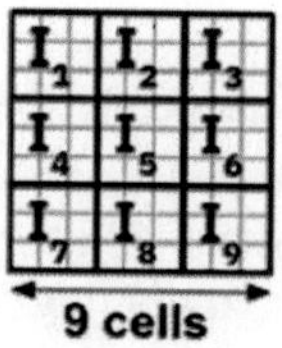

(b) Individuals of a puzzle in a 9 × 9 grid

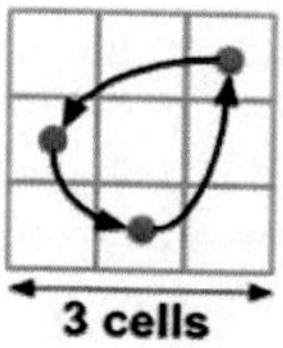

(c) Circular swap

Fig. 1 Some definitions

3 Definition of Concepts

A dataset was created, comprising of 9 × 9 puzzles categorized by the given number of clues. A few famous puzzles like the "easter monster," "2012s World's hardest puzzle by Arto Inkala," and "golden nugget" were manually added to the dataset. Puzzles were auto-selected in the order they were listed in the one million puzzle dataset [21] and the supersudokulib dataset [22]. A maximum of thirty puzzles were selected for each category. In a few categories, fewer than thirty puzzles were found, so the dataset comprised of a total of 438 puzzles. Figure 6 depicts number of puzzles per category. Results presented in this paper are obtained by attempting to solve puzzles from this assimilated dataset.

3.1 Definitions

Puzzle (S, S_E or S_F): The given "empty" Sudoku puzzle (S_E) in which some cells are pre-filled with numbers called "clues." When S_E is completely filled with nonzero numbers, it is denoted as S_F. In general, a filled puzzle state can be referred to as S.

Individual (I_i): Each non-intersecting 3 × 3 group of cells is an individual (Fig. 1a). The puzzle has $n = 9$ individuals, each containing unique numbers ranging from 1 to 9. Individuals I_1 to I_n are depicted in Fig. 1b.

Team (T_i): Although any "leader" individual I_i and the individuals toward the left, right, top, and bottom of I_i form a team, any reference to a team in this paper encompasses the individuals other than the leader. For example, in a team with I_1 as the leader, I_1's team members are $T_1 = \{I_2, I_3, I_4, I_7\}$. Similarly, $T_5 = \{I_2, I_8, I_4, I_6\}$ and $T_6 = \{I_3, I_9, I_5, I_4\}$.

Frozen cell (F_i): Any cell in I_i containing a value that is temporarily or permanently fixed in position. Cells with clues given at the start of the puzzle are always kept frozen. Algorithms can temporarily freeze any other cell when required. Frozen cells of a team are depicted with F_T.

Non-frozen cell ($\neg F_i$): Any cell in I_i that is not frozen.

No conflict cell ($\neg C$): A cell containing a number that does not match any frozen number in its row and column, spanning the entire puzzle.

Conflicting cell (C): A cell containing a number that matches the number in any frozen cell along the same row or column of the cell, spanning the entire puzzle (thus violating the rule of requiring unique numbers in every row and column).

Obvious cell (Ω): If any of the non-frozen numbers of an I_i can be inserted into only one $\neg F_i$ (due to being the only cell in which the number would not conflict with any F_T). An S_E matrix filled with Ω's but is not yet S_F will be considered as the new S_E.

Cardinality: The number of elements in a set is the cardinality of the set. For example, the phrase "the number of F_i cells is two" is denoted mathematically as: $|F_i| = 2$.

Fitness of an individual (f_i): The cardinality of C cells in I_i.

Total fitness ($\sum_{i=1}^{i=n} f_i$): The sum of fitness values of all individuals. A fully solved puzzle will have $\sum_{i=1}^{i=n} f_i = 0$.

Circular swap/mutation (${}_c\Delta_i$): If a number in $\neg F_i$ makes it a conflicting cell, the number is swapped with the number in another $\neg F_i$, but only if after the swap, the swapped cells would be $\neg C$. Figure 1c shows the pattern in which three cell values are swapped. The operation is represented as ${}_3\Delta_i$, where i is the individual id and $c = 3$ (the maximum number of cells undergoing swaps). When $c = 3$, either three cells can be circular swapped or the numbers in two cells can be swapped. If $c = 4$, the number of cells involved in a circular swap can be 2, 3, or 4. The range of c is $2 \leq c \leq 9$. Swaps can be referred to as "2-way swaps" or "3-way swaps."

Entropy (e): The number of times I_i attempts ${}_c\Delta_i$ when any individual is required to undergo swaps.

Permutations ($p \in {}^m P_r$): Permutations of a set of m numbers taken r at a time. Any one of those permutations can be referred to as p. Permutations are various ways in which numbers in $\neg F_i$ cells can be ordered.

Bruted puzzles (B, with β indicating the number of bruted individuals in any puzzle in B): Bruted matrices are created when permutations of $\neg F_i$ (generated as ${}^{|\neg F_i|}P_{|\neg F_i|}$) are inserted into β number of individuals. This is explained in Sect. 3.2. References to "n brutes" mean that n number of individuals were bruted.

Epochs (E): Each puzzle solution is attempted E times to check if a solution can be found reliably.

Generations (G): The number of times in each E, that all teams perform circular swaps, as described in Sect. 3.2.

3.2 *Method of Finding a Solution*

In some puzzles, the number of possible permutations of merely one individual may be as high as 362880, with the chances of locating the correct permutation being a mere $1/362880 = 0.0000027$. Considering the entire puzzle, the chances of finding a solution are lower. Therefore, it was necessary to reduce the search space with optimization.

Optimization 1: Once a few obvious cells were filled and frozen, a few more cells often became obvious and they could be filled too, until no more cells were obvious.

Optimization 2: Numbers were inserted into S_E only if it would not conflict with any frozen cells. If no such cells were found, the number was inserted into any empty cell, even if it conflicted with a frozen cell.

Puzzles with 29 or more clues would often get solved during these optimizations (Fig. 7). The puzzle was then run for $G = 2500$ generations in each epoch of $E = 100$ epochs. Circular swaps were performed on individuals that were not fully frozen, and individuals were processed in ascending order of the cardinality of non-frozen cells it possessed. Uniform pseudo-random numbers (Mersenne Twister [23]) were used to simulate stochasticity. During the selection of paths to traverse in the circular swap depth-first search (Fig. 2), a linear congruential algorithm (from glibc [24]) generated pseudo-random numbers. The random seed was initialized at the start of the program.

Examining What Makes Puzzles Difficult

Literature on evaluating the difficulty of puzzles [25] presents a "refutability" score and a "Richter-type" score [26]. Arto Inkala also presents a formula [27] that incorporates the techniques to make eliminations, the number of links needed to make eliminations and the difficulty of the next step, relative to the prior step. These measures were based on tactics that humans use to solve puzzles. However, evolutionary or random approaches do not depend on such tactics. Moreover, outliers observed (Fig. 8) revealed that locating a solution in some puzzles was significantly more difficult than others. An investigation was performed, to check if a correlation existed between the chances of obtaining a successful solution versus the given number of clues, the number of constraints of all team members on an individual, the proximity of each frozen cell to its nearest frozen cell and the number of permutations per individual. No strong correlation was observed. The difficulty therefore appears to be a function of how each individual influences its team members. This observation was the basis of designing the sub-grids of puzzles as individuals affected directly by the frozen cells of their team.

Designing Team Behavior

Figure 5 shows puzzle states that are much further from the solution (gray pixels), are spatially closer to the solved state (red pixel), than puzzle states which are closer to reaching the solution (green pixels). Therefore, standard evolutionary techniques of retaining the fittest individual and discarding other individuals were counterproductive, since often, it was necessary for the fitness to first worsen before it improved and reached the solution. This phenomenon was also observed by other authors [26] who remark on stochastic continuous time dynamics. Hence, it was decided that the algorithm would not have strategies that discarded individuals based on fitness. Three methods of approaching a solution were attempted:

1. Be the change: For each I_i, the $\neg F_i$ cells of T_i were temporarily frozen and circular swaps were performed on I_i. Temporarily frozen cells of T_i are then un-frozen. This method (Algorithm 3) checks if more constraints could help an individual converge to a solution faster.
2. Follow the leader (FTL): $\neg F_i$ cells in each leader I_i were temporarily frozen and circular swaps performed on T_i. Temporarily frozen cells were then un-frozen. Since at any given point of time there could be one or more leaders which were closer to the optimal fitness, this algorithm (Algorithm 2) was designed to allow team members to perform circular swaps to avoid conflicts with leaders, thus gradually bringing the entire puzzle closer to a solution. Starting the algorithm by processing individuals that possessed the least number of non-frozen cells reduced the exploration space since such individuals had fewer permutations of $\neg F_i$, thus being more likely to reach optimum fitness.
3. Follow the leaders (FLS): This algorithm differs from FTL, only by not freezing any $\neg F_i$ cells of the leader individuals. It examines whether mere stochasticity would help find a solution and is also designed to compare FTL with algorithms in literature that utilized pure randomness [17] or evolutionary algorithms. The difference being, that FLS utilized constraint-aware circular swaps, thus being less susceptible to a drastic worsening of fitness, compared to a purely random or conventional evolutionary approach.

Designing Circular Swaps

It was observed that simple "mutations" (swapping two numbers in $\neg F_i$ cells) were insufficient to converge difficult puzzles toward a solution. There often were situations where the only way a swap could be performed (without causing conflicts), was if swaps were performed in a "circular" manner, as explained in Sect. 3.1 and Fig. 1c. For efficiency, a pre-processing step was designed, where non-frozen numbers were identified for each non-frozen cell, based on whether the number could be inserted into the cell without making it a conflicting cell. The numbers formed a directed graph, as shown in Fig. 2, where {6, 9, 4, 8} are frozen cells, and {1, 5, 2, 7} are non-frozen. The number 1 could be moved to the cell that contains 5 or in the cell containing 2 (hence the arrows from node 1 of the graph, directed to nodes 2 and 5) without conflicting with any other frozen cell in the puzzle (the entire puzzle is not shown). The fact that 7 and 2 form a loop means that a 2-way circular swap is possible between them. Nodes 1, 2, and 7 also form a loop, revealing that a 3-way circular swap is possible (1 moved to 2's cell, 2 moved to 7's cell, and 7 moved to 1's cell). Ideally, some nodes of the graph should have been depicted with self-loops, but it is omitted to avoid cluttering the image. Graphs were implemented as adjacency lists, and circular swap possibilities were found using a depth-first search. Circular swaps

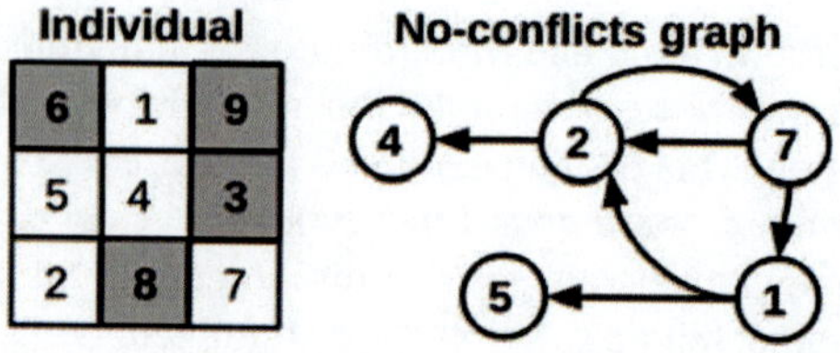

Fig. 2 No-conflicts graph of an individual, depicting which non-frozen numbers can be swapped into other non-frozen cells without conflicting with frozen cells of any other team member (here, 6, 9, 3, and 8 are frozen cells).

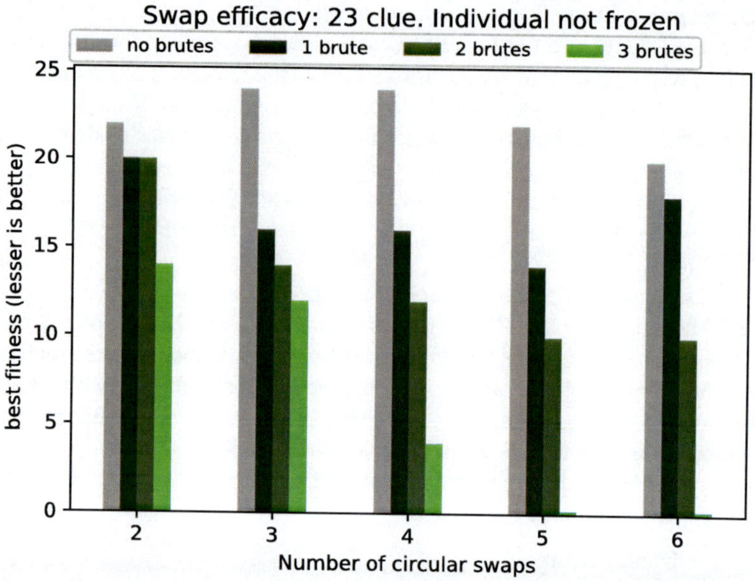

Fig. 3 More cells being involved in circular swaps, leads to faster convergence to a solution

were performed only on individuals belonging to teams having a fitness $f_i + f_{T_i} > 0$. If the algorithm uses $c = 4$, it means that circular swaps of $c = 2$, $c = 3$, or $c = 4$ can be performed. The improved fitness when using a larger c (the number of cells involved in the circular swap is c) is evident in Fig. 3. This is an important point to note when designing algorithms that require escaping local optima.

Proposed Fitness Landscape Visualization

When a puzzle undergoes circular swaps, it transitions between various states of existence. Such unique states were recorded. Puzzle states were also generated by filling multiple copies of S_E with permutations of non-frozen numbers of all individuals. This procedure was performed for three puzzles of 17, 28, and 29 clues. To keep the quantity of generated data within limits that could be graphically rendered comfortably, only approximately 15% of the permutations of each individual in the

Algorithm 1 Fitness Landscape Generation

```
function CALCULATE- POINT- XYZ(S)
  z = Σ_{i=1}^{i=n} f_i, x = 0, y = 0, p = 9 × 9
  θ = 0, d = 360/p
  for each I do
    for each number n in cells of I do
      θ = θ × π/180
      t_x = x + n, t_y = y
      a_1 = sin(θ), a_2 = cos(θ)
      x = a_2 × (t_x − x) − a_1 × (t_y − y) + x
      y = a_1 × (t_x − x) + a_2 × (t_y − y) + y
      θ = θ + d
    end for
  end for
  return x, y, z
end function
```

17-clue puzzle were considered (it generates almost 4×10^7 puzzle states). For the 28 and 29 clue puzzles, approximately half of 4×10^7 permutations were generated.

The $9 \times 9 = 81$ numbers in the puzzle's cells (considered in the order it is processed in Algorithm 1), needed to be represented in three dimensional (3D) space. Algorithm 1 received all numbers of the puzzle state as input and calculated for every unique puzzle state, a unique point in 3D space.

The z dimension of the point was the fitness value of the puzzle at that state. The x, y, and z points of each puzzle state were graphically rendered to create the fitness landscape visualization. When necessary, the x and y value was multiplied by a large constant number, to spread the dense cluster of points in space for easy viewing. Points with fitness $\sum_{i=1}^{i=n} f_i = 0$ (solved puzzle state) were red color, points with $1 \leq \sum_{i=1}^{i=n} f_i \leq 4$ (puzzle states close to being solved) were bright green, points with $5 \leq \sum_{i=1}^{i=n} f_i \leq 20$ were dark green, points with $\sum_{i=1}^{i=n} f_i \geq 21$ were shades of gray, where darker shades indicated lower fitness values. The Point Cloud Library [28] wrapped by the Cloud Compare viewer [29] software was used to visualize the generated points, and the size of each pixel was increased a little via the software, for improved visibility.

The fitness landscape generated is shown as a point cloud in Fig. 4b. The top view of the same points is shown in Fig. 4a. There are more than one red pixels (optimal solutions) in Figs. 4a, b since it is a puzzle which has more than one optimal solution. An important point to note from Fig. 4 is that the bright green points are not close to the red points. Also, the light gray and dark gray points are in very close proximity to each other as well as with the green and red points, (Fig. 5). The phenomenon of puzzle states appearing to be close to the solution when they were actually not, was also observed by other authors [26]. Fitness is a poor indicator of proximity to a solution. However, when the puzzle reaches a fitness of approximately 20 or lower, the landscape shows that it may help to explore within close proximity to the existing puzzle state, utilizing a lower value of c for circular swaps.

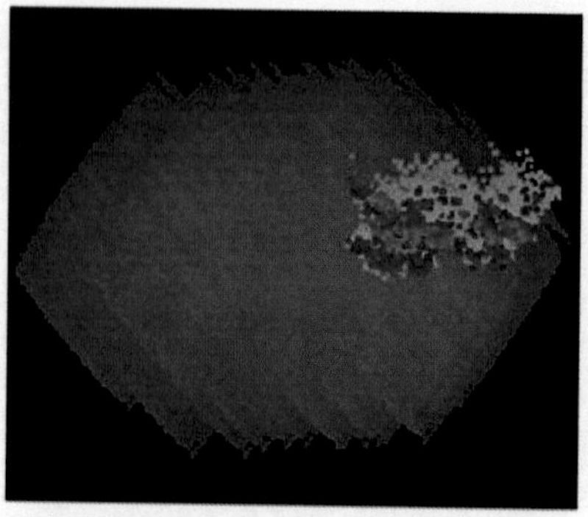

(a) 28 clue puzzle landscape top view

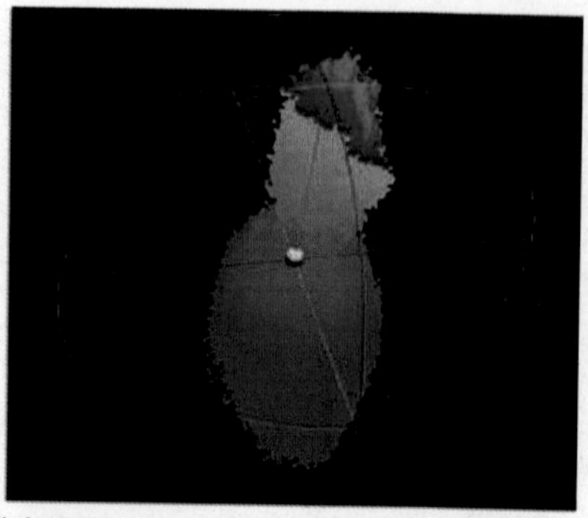

(b) 28 clue puzzle point cloud density side view

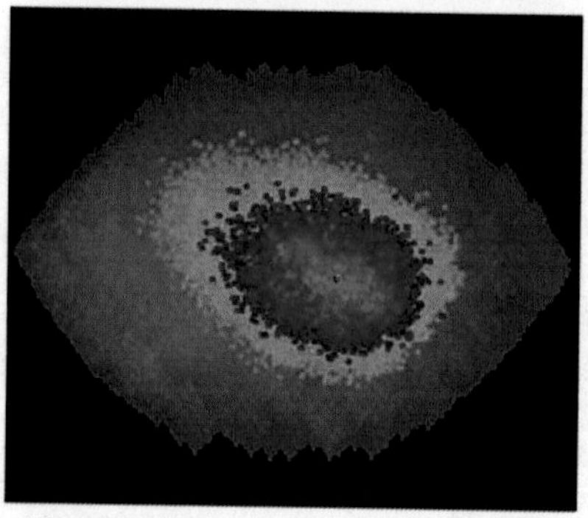

(c) 17 clue puzzle landscape top view

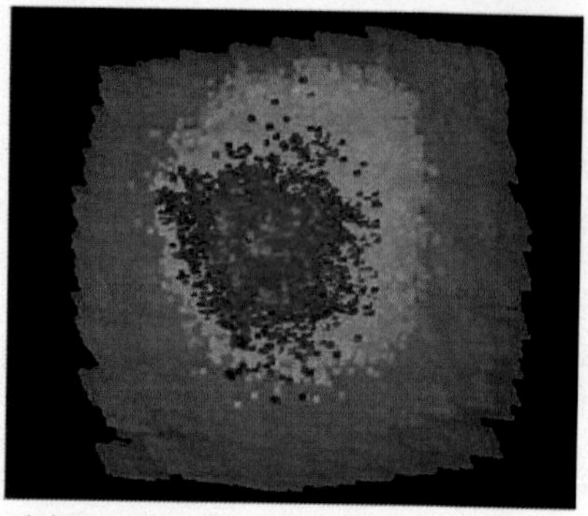

(d) 29 clue puzzle landscape top view

Fig. 4 Top and side views of fitness landscapes

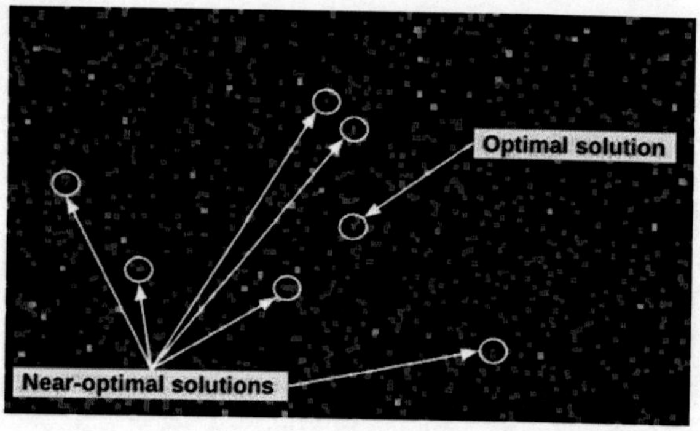

Fig. 5 Zoomed fitness landscape of puzzle from Fig. 4c

Table 1 Number of possible brutes versus valid brutes

Given clues	$\|\Omega\|$	β	Possible $\|B\|$	Valid $\|B\|$	% valid
30	6	1	6	3	50
30	6	2	36	12	33.33
30	6	3	216	24	11.11
17	2	1	120	20	16.66
17	2	2	14,400	400	2.77
17	2	3	72,576,000	320	0.0004
17	0	1	720	16	2.22
17	0	2	3,628,800	3840	0.10
17	0	3	18,289,152,000	31,104	0.0001

Partial Brute Forcing

In 21-clue puzzles, the sheer number of permutations of non-frozen numbers (calculated as per Eq. 1) is of the order 10^{31}. In 17-clue puzzles, it can reach 3×10^{34}. Therefore, it is insufficient to depend purely on constraint-based stochasticity of the given clues. Figure 7 showed that when the given clues were fewer than 23, the initial optimizations and circular swaps were also insufficient to find a solution. To guide the puzzle toward a solution, a strategy of partially brute forcing a few individuals was considered. Three individuals with the least number of non-frozen cells (such that $|\neg F_i| \neq 0$) were selected for brute forcing. A copy of S_E was created, and permutations of non-frozen cells were inserted into the first individual selected, as shown in Algorithm 4. During the insertion, if any inserted number conflicted with frozen cells, the permutation was discarded (it's not a "valid brute," since a puzzle cannot exist with conflicting cells) and the next permutation of numbers was attempted. All cells of the newly filled individual were frozen and the remaining empty cells of the puzzle were filled with non-frozen numbers. The puzzle was run for E epochs. Similarly, another copy of S_E was created and the next permutation of non-frozen cells of the individual was inserted and frozen. This was continued until all permutations were exhausted. Then, the same process was repeated, but using permutations of the first and second individual. The procedure is repeated again with permutations of all three individuals. Table 1 shows how only a small fraction of permutations form valid brutes, since most of the permutations would consist of conflicting cells. Among all valid brutes generated, at least one can lead to an optimal solution. Table 1 also shows how drastically the number of permutations reduce for puzzles with obvious cells filled.

$$\text{Total permutations} = \prod_{i=1}^{n} {}^{|\neg F_i|}P_{|\neg F_i|} \tag{1}$$

Algorithm 2 FollowTheLeader

```
function FOLLOW-THE-LEADER(i)
    Temporarily freeze ¬F_i
    for each I in T_i do
        for e iterations do
            Perform _cΔ_I
        end for
    end for
    Unfreeze temporarily frozen cells
end function
```

Algorithm 3 Be the change

```
function BE-THE-CHANGE(i)
    Temporarily freeze ¬F of T_i
    for e iterations do
        Perform _cΔ_i
    end for
    Unfreeze temporarily frozen cells
end function
```

Algorithm 4 Brute permutations

```
function BRUTE-PERMUTATIONS(β, S, B)
    if β == 0 then return B; end if
    Select I_i for β, based on |¬F_i|
    for each p ∈ ^{|¬F_i|}P_{|¬F_i|} do
        I_i of S is filled with p
        if filled cells are ¬C then
            Filled cells of S are frozen
            B = B ⋃ S
        else exit for loop
        end if
    end for
    β = β − 1
    return BrutePermutations(β, S, B)
end function
```

3.3 Algorithms

4 Numerical Simulation

Solutions were attempted for 438 puzzles (Algorithm 5), each having various numbers of given clues. The number of brutes β used is ranged from 0 to 3. The challenge was to check if even the most difficult puzzles could be solved within 500 generations, using constraints and stochasticity, while obtaining a realistic picture of how stochasticity within the puzzle manifests itself when subject to constraints. Solutions were considered under four categories, as shown in Fig. 7.

Algorithm 5 Main

```
Initialize S_E. Initialize β = 0.
while Ω newly found in S_E do
  S_E = S_E with newly found Ω cells frozen.
  if Σ_{i=1}^{i=n} f_i == 0 then exit program end if
end while
while β ≤ 3 do
  B = {∅} ; β_temp = β
  B = BrutePermutations(β_temp, S_E, B)
  if (B == {∅}) then B = {S_E} end if
  for each S in B do
    Insert ¬F in S
    if Σ_{i=1}^{i=n} f_i == 0 then exit B loop end if
    for each epoch of E do
      for each generation of G do
        for each I_i in order of least |¬F_i| do
          if f_{T_i} + f_i > 0 then
            Run Algorithm 2 or Algorithm 3 for i.
          end if
        end for
        if Σ_{i=1}^{i=n} f_i == 0 then exit G loop end if
      end for
    end for
  end for
  if Σ_{i=1}^{i=n} f_i == 0 for all E then exit while end if
  β = β + 1
end while
```

4.1 *Trial 1, with All Puzzles*

The first trial was run with $c = 6$, $e = 5$, $E = 2500$, and $G = 100$. Individuals in a puzzle were allowed to influence their teams or be influenced using frozen cells as constraints. Section 4.1 lists three such methods that attempted to influence $\neg F$ numbers to switch positions until they did not conflict with F cells. Any team with fitness equal to zero was not subjected to circular swaps. Among the algorithms attempted, "follow the leader" was most successful, as evident in Fig. 6.

Be The Change
With Algorithm 3, the team members of each of the 9 individuals chosen in a generation were fully frozen, and the individual attempted circular swaps within the entropy limit. This method was not very successful since the large number of frozen team members often left very few or no cells available for the individual to perform circular swaps. Additionally, the number of swaps performed was lesser than FTL and FLS, thus providing lesser opportunity to locate a solution.

Follow The Leaders (FLS)
In this method (similar to Algorithm 2, but without freezing any non-frozen cells), all team members including the leader performed circular swaps.

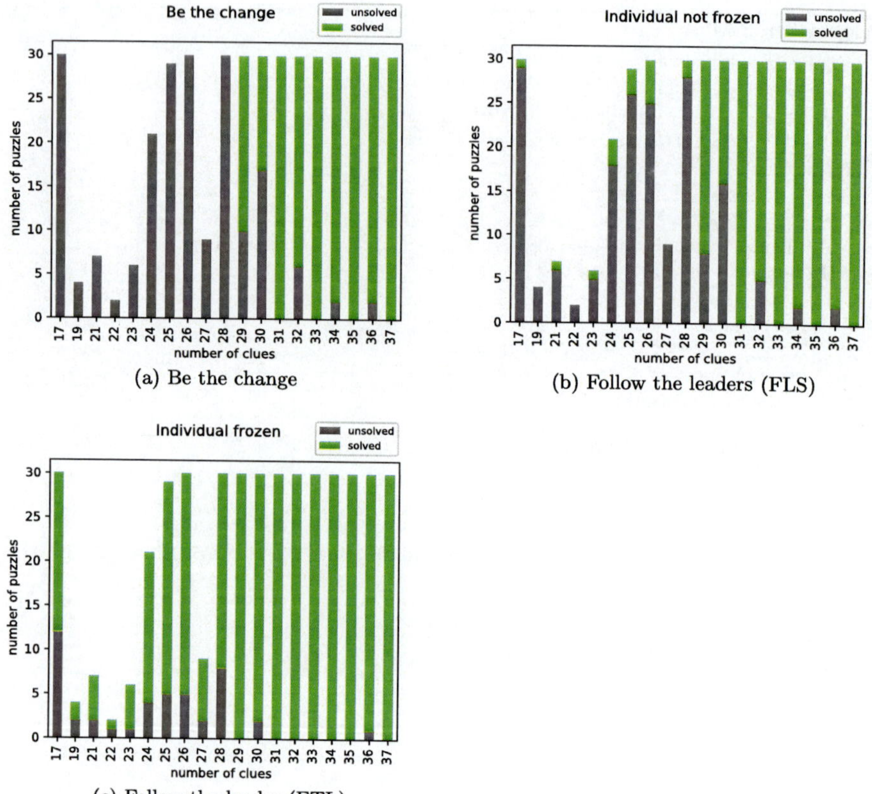

(a) Be the change

(b) Follow the leaders (FLS)

(c) Follow the leader (FTL)

Fig. 6 Successful solutions in trial 1, with $c = 6$

Follow The Leader (FTL)
Using Algorithm 2, for each of the 9 individuals chosen in an iteration, the chosen individual was considered the leader, and all its cells were temporarily frozen. As shown in Algorithm 5, the leader individuals were chosen in the order of the individual having the least number of non-frozen cells ($|\neg F_i|$). The team members then performed circular swaps.

Trial 1 Observations
Although FTL was capable of solving more puzzles, it was interesting to note that there were some 36-clue and 30-clue puzzles that remained unsolved. In Sect. 4.1, a large number of frozen members resulted in fewer cells being swappable. Similarly, in many FTL puzzles, the puzzle had a high success rate for 2 brutes and then an extremely small success rate at 3 brutes. It was also interesting to note that solved puzzles with given clues 17 to 23 were entirely dependent on brute forcing (shown in Fig. 7). Puzzles with given clues 24 to 28 were solved with or without brute forcing, and a majority of puzzles with given clues ranging from 29 to 37 could be solved

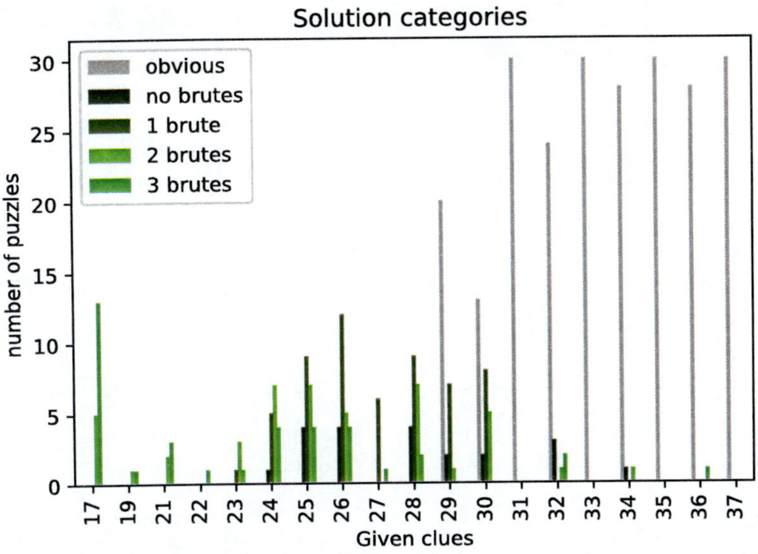

Fig. 7 Depiction of how many puzzles were solved at the obvious optimization stage, how many were solved without partial brute forcing and how many required various levels of partial brute forcing

as early as at the obvious optimization (Ω) stage. It was however surprising to note that even in the 36-clue category, there was a puzzle that required brute forcing. The 36-clue category also had a puzzle that remained unsolved until $G = 10{,}000$ was used (Sect. 4.2). On examining the number of generations it took for puzzles to reach a solution (Fig. 8), it was observed that utilizing 3 brutes, a majority of even the 17-clue puzzles could be solved within 500 generations. The figures also depict how the probability of finding a solution within fewer generations increases with a greater number of brutes (solutions obtained during Ω computation are considered as having being solved in generation zero, as depicted in Fig. 8a). The "no brutes" attempt (Fig. 8a) was run with all puzzles. Puzzles that could not be solved with "no brutes" were run using one brute (Fig. 8b). Puzzles that still remained unsolved were run with two brutes (Fig. 8c) and so on.

Among the 438 puzzles, 45 remained unsolved in trial 1. Based on the number of given clues, the unsolved puzzles were in the given clue categories: {17, 19, 21, 22, 23, 24, 25, 26, 27, 28, 30, 36}, and the cardinality of unsolved puzzles were {12, 2, 2, 1, 1, 4, 5, 5, 2, 8, 2, 1}, respectively. There is a lot of literature where authors tend to present only results that achieved a 100% success rate. However, in the true spirit of research, we believe that it is important to also present these results that were unsuccessful, since even such results can be compared and utilized to derive further insights.

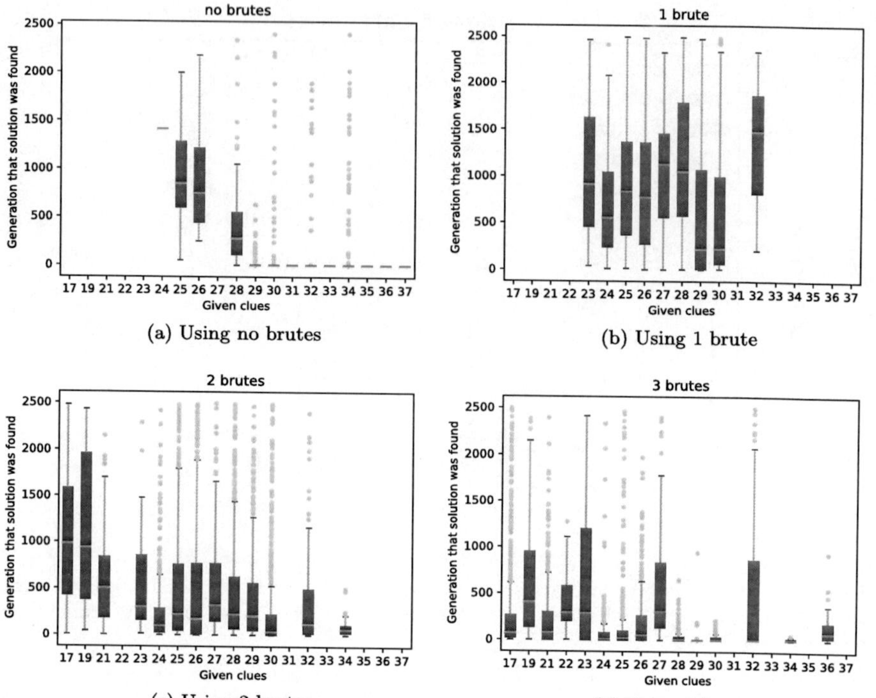

(a) Using no brutes

(b) Using 1 brute

(c) Using 2 brutes

(d) Using 3 brutes

Fig. 8 Trial 1: generations needed to find a solution during various epochs

4.2 *Trial 2, Attempting Unsolved Puzzles*

Unsolved puzzles of Sect. 4.1 were re-run with algorithms FLS and FTL. Entropy was $e = 20$ for both algorithms, where each was attempted with $c = 6$ and $c = 9$. It was observed that it was not necessary for a fitness value to be close to zero to reach a solution. Puzzles were reaching optimal fitness directly from fitness values of 16 or 12. As evidenced in Table 2, FTL consistently proved to be a better algorithm than FLS, and utilizing $c = 9$ did not provide an added benefit (as observed from the percentage of puzzles solved). However, the success of using a larger entropy value (compared to the value used in Sect. 4.1) demonstrated that the better opportunity available for each individual to adjust its values to avoid conflicts, benefits from such a stochastic search for solutions. Among the 45 puzzles run for FTL with $c = 6$, there were 24 puzzles that remained unsolved. They were in the given clue categories {17, 19, 24, 25, 26, 27, 28, 36}, with the number of unsolved puzzles being {9, 1, 2, 4, 3, 1, 3, 1}, respectively. Although $c = 9$ was used, it was observed that circular swaps with 8 or 9-way swaps almost never happened, since for such swaps to take place, it was necessary for individuals to have 8 or 9 non-frozen cells and required favorable constraints at team cells that allowed such swaps.

Table 2 Trial 2

Algorithm	c	e	% of puzzles solved
FLS	6	20	2.22
FLS	9	20	2.22
FTL	9	20	44.44
FTL	6	20	46.66

4.3 *Trial 3, Attempting Remaining Unsolved Puzzles*

An additional trial was run with the 24 unsolved puzzles using FTL, to check if a greater number of generations $G = 10{,}000$, a low entropy $e = 5$, and large circular swaps $c = 9$ would solve the puzzles. Ten puzzles were solved, with unsolved puzzles being in the given clue categories $\{17, 19, 25, 26, 27, 28\}$ with the number of unsolved puzzles being $\{5, 1, 2, 2, 1, 3\}$, respectively. The remaining 14 puzzles could have been solved if the algorithm was run a few more times. However, the objectives of the trials were to obtain a clear picture of the extent of success (under various puzzle categories) an algorithm can achieve, so more trials were not attempted.

5 Conclusion

The results of this study validate the fact that it is not merely the pattern of arrangement of clues in the puzzle that matters, but the pattern in which the clues place each team along a path on the fitness landscape, where the swapping of numbers finds a direct route toward the solution. There was a consistency with which some puzzles ended up with a solution while some did not. Additionally, the successful application of constraint-aware circular swaps of order greater than 2 underscores its capability as a mutation technique that could potentially benefit other evolutionary algorithms. Other important observations were the large number of outliers in Fig. 8, and the few puzzles that remained unsolved even after 10,000 generations were run (even with 28-clue puzzles). The need for partial brute forcing also revealed that a purely evolutionary approach is insufficient to solve hard sudoku puzzles. It is imperative to couple the algorithm with heuristics or improved logic to create a hybrid. While Sudoku rules and heuristics are well-established, this study illuminates their broader relevance in tackling vast constraint-based NP-complete problems whose properties remain unexplored. This lays the foundation for developing more effective algorithms, thus guiding researchers and practitioners in devising hybrid approaches that can advance the field of computational intelligence.

5.1 Future Work

This study can be extended in multiple directions. Here are some possibilities:

- Designing a gradient calculation or a path, based on an improved fitness landscape and given clues, could potentially assist in improving algorithms that could use the gradient or path to converge toward an optimal solution.
- Starting from an optimal puzzle state, a fitness landscape could be reverse-generated by varying puzzle states and observing the properties of the variations. It may very well be possible to also design an algebraic solution to solving the puzzle, by observing such patterns.
- Sudoku solving algorithms could additionally be compared based on the Big-O notation, rather than merely based on number of clues or fitness or time taken to solve the puzzle.
- A "tiredness" feature could allow the algorithm to decide when to stop using stochastic search and begin brute forcing. Ideally, such a program should be designed to understand the problem domain and be given sufficient time to formulate its own solution tactics. Similarly, the number of c-way swaps could initially be kept high, and as the puzzle approaches closer to a solution, the value could be decreased, to allow exploring within a limited range.
- Some 36-clue and 28-clue puzzles that were hard to solve could be compared with other puzzles of the same category to analyze what makes them different. The 17-clue puzzles that were simple to solve could also be compared to 17-clue puzzles that did not get solved.

References

1. McGuire G, Tugemann B, Civario G (2012) There is no 16-clue Sudoku: solving the sudoku minimum number of clues problem via hitting set enumeration. Exper Math 23
2. D.A.P. (2013) Maximum number of clues in a sudoku game that does not produce a unique solution. Mathematics Stack Exchange https://math.stackexchange.com/q/345255, https://math.stackexchange.com/users/31718/daniel-a-apelsmaeker (version: 2013-03-29)
3. Knuth DE (2000) Dancing links. Millennial Perspect Comput Sci 1:1–26
4. Johnston MD, Adorf HM (1989) Learning in stochastic neural networks for constraint satisfaction problems. In: Proceedings of the NASA conference on space telerobotics
5. Weber T (2005) A SAT-based Sudoku solver. In: Proceedings of the 12th international conference on logic for programming, artificial intelligence and reasoning, pp 11–15
6. Lewis R (2007) Metaheuristics can solve Sudoku puzzles. J Heuristics 13(4):387–401
7. Mantere T, Koljonen J (2008) Sudoku solving with cultural swarms. In: Proceedings of the 13th Finnish artificial intelligence conference (AI and machine consciousness), pp 60–67
8. Moraglio A, Togelius J (2007) Geometric particle swarm optimization for the Sudoku puzzle. In: Proceedings of the 9th annual conference on genetic and evolutionary computation. ACM, New York, NY, USA, pp 118–125
9. Hereford JM, Gerlach H (2008) Integer-valued particle swarm optimization applied to Sudoku puzzles. In: Proceedings of the IEEE swarm intelligence symposium, pp 1–7

10. McGerty S (2009) Solving sudoku puzzles with particle swarm optimisation. Final Report, Macquarie University
11. Wang C, Sun B, Du KJ, Li JY, Zhan ZH, Jeon SW, Wang H, Zhang J (2023) A novel evolutionary algorithm with column and sub-block local search for sudoku puzzles. IEEE Trans Games
12. Nicolau M, Ryan C (2006) Solving Sudoku with the GAuGE system. In: Collet P, Tomassini M, Ebner M, Gustafson S, Ekárt A (eds) Genetic programming. Springer, Berlin, Heidelberg, pp 213–224
13. Abdel-Raouf O, Abdel-Baset M, Henawy I (2014) A novel hybrid flower pollination algorithm with chaotic harmony search for solving Sudoku puzzles. Int J Eng Trends Technol 7:126–132
14. Pillay N (2012) Finding solutions to Sudoku puzzles using human intuitive heuristics. South African Comput J 49:25–34
15. Becker M, Balci S (2018) Improving an evolutionary approach to sudoku puzzles by intermediate optimization of the population. In: International conference on information science and applications. Springer, pp 369–375
16. Jones S, Roach P, Perkins S (2007) Construction of heuristics for a search-based approach to solving Sudoku. In: Proceedings of the 27th SGAI international conference on artificial intelligence. Cambridge, England, pp 37–49
17. Mcgerty S, Moisiadis F (2014) Are evolutionary algorithms required to solve Sudoku problems? Comput Sci Inf Technol 4:365–377
18. Ipe NK, Chatterjee S (2022) An in-memory physics environment as a world model for robot motion planning. In: Soft computing: theories and applications: proceedings of SoCTA 2020, vol 1. Springer, pp 559–569
19. Ipe NK (2021) The need to visualize sudoku, preprint at https://engrxiv.org/preprint/view/1649/
20. Ipe N, Kulkarni RV (2021) The need to visualize Sudoku, preprint at https://www.techrxiv.org/articles/preprint/The_Need_To_Visualize_Sudoku/14528928
21. Park K (2016) 1 million sudoku games. https://www.kaggle.com/bryanpark/sudoku
22. Cao Y (2008) Benchmarking Sudoku solvers. https://in.mathworks.com/matlabcentral/fileexchange/18921-benchmarking-sudoku-solvers
23. Matsumoto M, Nishimura T (1998) Mersenne twister: a 623-dimensionally equidistributed uniform pseudo-random number generator. ACM Trans Model Comput Simul (TOMACS) 8(1):3–30
24. McGrath R. The GNU C library (GLIBC). https://www.gnu.org/software/libc/
25. Pelanek R et al (2011) Difficulty rating of sudoku puzzles by a computational model. In: FLAIRS conference. Citeseer
26. Ercsey-Ravasz M, Toroczkai Z (2012) The chaos within Sudoku. Sci Rep 2:1–8
27. Inkala A. AI sudoku puzzle difficulty. http://www.aisudoku.com/index_en.html
28. Rusu RB, Cousins S (2011) 3D is here: point cloud library (PCL). In: 2011 IEEE international conference on robotics and automation. IEEE, pp 1–4
29. Girardeau-Montaut D. Cloud compare. https://www.danielgm.net/cc/

An Approach to Increase the Lifetime of Traditional LEACH Protocol Using CHME-LEACH and CHP-LEACH

Madhvi Saxena, Aarti Sardhara, and Shefali Raina

Abstract The underwater wireless sensor network (WSN) comprises multiple distributed sensing devices or nodes within the aquatic environment, collecting and transmitting data to a surface-based base station. The acquired data is utilized as needed. These battery-powered, compact-sized nodes necessitate energy-efficient routing protocols due to limited battery capacity. The network finds application in diverse fields, such as oil exploration and temperature monitoring. LEACH algorithm is mostly used algorithms for clustering, but the limited energy of nodes is a very critical challenge for WSN. This study enhances the original LEACH protocol through varied routing approaches, introducing novel algorithms like CHME-LEACH, which considers residual energy for cluster head selection, and CHP-LEACH, which employs probability for cluster head selection. Both algorithms outperform the traditional protocols in reduced energy consumption and lifetime of networks.

Keywords LEACH · Energy · CHME-LEACH · CHP-LEACH · Routing protocol

1 Introduction

WSN is a collection of sensor nodes that are randomly dispersed over any area and have limited battery backup. WSNs have a range of applications in different environments such as underwater surveillance, temperature, forest fire, battlefield, industrial areas, medical, home, etc. Maximum energy is drained in transferring the data between the nodes wirelessly [1]. Sensor nodes that are distributed across the environment are battery-powered devices and as the size of sensor nodes is small, the power of the battery is also compact. Since battery power is limited, it is essential to design a routing protocol that somehow reduces energy consumption.

M. Saxena (✉) · A. Sardhara
Vishwakarma University Pune, Pune, Maharashtra, India
e-mail: madhvisaxena12@gmail.com

S. Raina
Vasantdada Patil Pratisthan's College of Engineering and Visual Arts Sion, Mumbai, Maharashtra, India

S. Kumar et al. (eds.), *Fourth Congress on Intelligent Systems*, Lecture Notes in Networks and Systems 868, https://doi.org/10.1007/978-981-99-9037-5_11

The challenges that WSNs face are energy feasting, routing, load balancing, quality of service packet delivery, etc. [2]. Sensor nodes have a wide range of applications like oil exploration, temperature and pressure detection, etc. The sensor node is either static or dynamic as and when required. Consider an example; suppose we want to measure temperature and pressure in that case sensor nodes are installed at fixed bodies, whereas in dynamic if we want to track the movement of anybody, then we need to install the sensor node in moving creatures [3]. These tiny nodes have very limited battery backup and very limited memory space. The lifetime of sensors is a very active research area; the lifetime of sensors directly reflects the lifetime of networks. Many researchers have carried out various research to improve the lifetime of sensor nodes so that the overall lifetime of the network increases. LEACH is a famous clustering protocol for increasing the lifetime of the network [4]. This research has tried to improve the performance of original LEACH with effective CH selection algorithms. To increase the network lifetime, we need to design an energy-efficient protocol through various energy constraints are present.

1.1 Leach Protocol

The sensor nodes which are dispersed across an area restricted energy consumption and also limited storage for data storage. WSN is mostly used in critical fields where the replacement of batteries is not possible [5]. Many researchers contribute to various researches to improve the lifetime of networks. So, the LEACH protocol was discovered so that it could conserve energy anyhow. A wireless sensor network is judged by the amount of energy the network consumes. Routing protocol is an important component in WSN responsible for power consumption. All the sensor nodes are combined together and form a set of groups called clusters. Here, the immediate node forms a cluster and every cluster has a CH which dispatches the data packet to the base station/sink node upon receiving the data from cluster member sensor nodes [6]. The LEACH protocol process is divided into two phases. In the first phase, CHs are decided based on the criteria mentioned in the LEACH protocol. A node cannot become the cluster head for more than one cluster, and the second phase is called as steady phase in which there is communication between all the sensor node and CH take place. Once the cluster member sends the data to the cluster head, the CHs accumulate all the data and after that, it processes them and finally sends them to BS. LEACH protocol has two phases: the setup phase & steady phase.

Setup Phase

Every node has two choices whether or not to become cluster head in the current based on selecting a haphazard value in the range of 0 and 1. The threshold value $T(n)$ is compared with the chosen random value. If the number which we have chosen is lesser than $T(n)$, then the node is elected as CH (cluster head) for the current round.

$$T(n) = \begin{cases} \frac{P}{1 - P*(r \bmod P)} & \text{if } n \in G \\ 0 & \text{otherwise} \end{cases} \tag{1}$$

where p is the selection/probability rate of becoming a node as CH, the current round number is denoted by r, and the set of the node that have not been elected as CH in the recent $1/P$ round is termed as G, and range of n varies from 0 to 1.

Steady Phase

Every node except cluster head node grouped into cluster is decided upon this phase. When each of the sensor nodes makes a decision to which cluster it belongs, this information needs to be forwarded back to each CH. The CH receives the data from the entire sensor node that is included in the cluster and CH creates a TDMA schedule informing when each of the nodes can transmit. After cluster creation and TDMA scheduling, data transmission begin. The cluster head receives the entire message and combines this into a signal message and passes them to the base station.

1.2 Challenges in UWSN

WSN and its applications are very popular, but they are suffering with many challenges, and these challenges became a hurdle in the processing of WSN. Many types of research are already done and going on for the advancement of WSN [7]. In this section, a few challenges are discussed which create complexities in underwater WSN.

- **Limited battery power**: The most crucial hurdle in WSN is the limited battery power supply. Reduction and balancing the energy dissipation remains a challenge in improving the network lifetime [8].
- **Limited communication capability**: WSNs have a short range of communication, and thus, it hinders in transmission of data. So, designing an algorithm is complex.
- **Minimal computation capability**: The sensor node has fewer computing capabilities, memory, and also small processors [9]. Thus, it is challenging to design an efficient algorithm to minimize the complexity.
- **Node mobility**: The mobile sensor node moves frequently in the sensor network, and thus, designing a new route or path always is not feasible.
- **Single-hop communication**: As the sensor node sends the data only to neighboring nodes, the problem arises when a node needs to send the data to some other nodes [10].
- **Topology control**: Since WSNs have limited battery power and when some of the nodes die, then altering the network topology is tough [11].

1.3 Advantages and Disadvantages of Leach

In this section, some of the advantages and disadvantages related to LEACH are discussed [10].

- Aggregation of data by cluster head reduces the traffic of the network.
- LEACH follows a clustering approach so the scalability of the network is better here [12].
- Data aggregation in the LEACH protocol helps in eliminating redundant transmission [13].
- Lifetime of network increases [14].
- Global information is not required by the LEACH protocol [15].
- How many cluster heads need to be formed if not given?
- Once the CH dies, the data packet also gets lost and data never ever reaches the base station [16].

1.4 Network Lifetime Maximization Technique

It is very clear that sensor nodes are very small in size and have very compact architecture [17]. The power unit is also very small and has very limited energy backup. Energy efficiency is very important, and many researchers are working to maximize the energy utilization of each sensor node [18]. In this section, few techniques are discussed which improve the network lifetime:

- Energy Harvesting
- Routing/Clustering
- Data Correlation
- Data Gathering/Network Coding
- Coverage and Connectivity/Optimal Deployment
- Mobile Relays and Sinks
- Routing/Clustering
- Opportunistic Transmission Schemes/Sleep–Wake Scheduling
- Mobile Relays and Sinks.

2 Literature Survey and Related Work

Chit et al. in paper [1] propose the advancement of the underwater network lifetime with the help of the concept of remaining energy and routing between the nodes. The name of this protocol is RED_LEACH. Here, in RED_LEACH remaining energy of each of the sensor nodes and the distance of the sensor node to the base station are considered for energy efficiency. A node that becomes the cluster head advertises them, and then each of the sensor nodes joins with the nearest cluster head, and now the node is called a cluster member.

Here in paper Jambli et al. [6], have tried to find to what extent the LEACH protocol can be carried out better in parameters such as mean consumed energy and data losses at different rates. Here they have found the average energy consumption in a certain period of time. As the simulation duration increases, energy consumption also increases. They also analyze the average data packet lost. In another experiment, they have found that the data speed varies and that energy consumption also increases. But mean energy consumption doesn't vary so much. By increasing the speed, the average data packet loss also decreases.

In paper Tandel et al. [2], have discussed improving the cluster head selection criteria. They have a new method to calculate the threshold value. Data is transmitted between sensor nodes using inter-clustering and intra-clustering. Cluster heads in PR-LEACH are selected for various probabilities. In this paper, they have compared the PR-LEACH with the original LEACH and BEC. After the simulation, result is compared, and it is found that PR-LEACH overcomes both of the protocol in field of distributing energy among the entire sensor node uniformly. From simulation, it is observed that first node dies from the LRACH in this LEACH and BEC; there were many dead spots, i.e., some areas are not covered by cluster head due to non-uniform distribution of energy among all the sensor nodes.

In paper Priyadarshi et al. [10], have suggested a routing protocol for dissimilar network called as stable energy effectual network (SEEN). Here sensor nodes have to consider three different levels of energy. SEEN protocol is related to DEEC protocol. Cluster head in SEEN protocol is selected by using residuary energy concept and path from sensor node to the base station/sink node SEEN protocol uses three level of energy using three different kinds of energy. First type, second type, and third type sensor node are normal sensor node, advance sensor node, and super advance sensor node, respectively.

Tripathi et al. [12] proposed EELEACH-C, cluster-based protocol whose cluster head is randomly selected. Here a sorting algorithm is ran by base station and obtained a list of possible candidate's cluster head nodes sorted according to their residual energy in descending order. Then base station examined the cluster head, and node with max energy is selected.

In this paper Kim et al. [13], had recommended an improvised version of protocol LEACH-MOBILE where mobile nodes can declare themselves as member of cluster head as they move and also to confirm whether the mobile node is able to communicate within allocated TDMA schedule to establish a communication with the cluster head. LEACH-MOBILE is compared with original LEACH protocol.

In paper Bendjeddou et al. [19], proposed a new protocol which improvises the LEACH protocol termed as LEACH-S so that energy dissipation is reduced along with it reducing the network overhead. The simulation of new protocol shows that energy consumed by each of the node is reduced in new protocol and also the overall overhead in the network is reduced in our new protocol. There was more residual energy in case of EECDA in comparison with leach protocol.

Zhao et al. in paper [16] have considered the dynamic change of energy nodes for selecting the cluster head so that energy could be distributed evenly among all the nodes. Here they consider the dynamic change of energy nodes for selecting the cluster head so that energy could be distributed evenly among all the nodes. They also proposed a vice cluster method used in the communication for each of the CH, and selection of CH is to minimize the energy consumption in re-clustering as well as then prolonging the lifetime in steady phase. Since there is modification in steady phase, the time for choosing clusters head.

3 The Proposed Routing Protocol

(a) Energy Consumption Model

In this context, we have deliberated upon an energy consumption model that facilitates the calculation of energy usage after each successive round. In another section, the pseudocode of our algorithms is provided. In this work, the performance of original LEACH is improved. Proposed algorithm consumes less energy dissipation as well as for larger distances. A model was assumed with following assumptions [7]:

- All the sensor nodes are of same type (homogenous), fixed, fixed energy, and they have data available at their disposable every time.
- Energy dissipation used by the sensor node for receiving and transmitting data packet is same.
- Each of the sensor nodes can become cluster head.
- The sensor nodes are away from the base station/sink node which is fixed.
- The sensor node associated with cluster member transmits the data packet to CH in one hop, and CH transmitted the data packet to base station/sink node in multi-hop.

The proposed energy model is discussed below [2]. Energy consumption in sending the data of k bits is:

If distance $> d_o$.
$E_{tx} = E_{TX} * k + E_{mp} * k * \text{distance}^4$;
If distance $\leq d_o$.
$E_{tx} = E_{TX} * k + \text{Efs} * k * \text{distance}^2$.;
Energy consumption for receiving the data of k bits is:
$E_{rx} = E_{RX} * K + E_{DA} * k$;
E_{tx}: Energy consumption in sending the data.
E_{rx}: Energy consumption for receiving the data.
E_{fs}: It is termed as free space range model.
E_{mp}: This is termed as multipath model, which can be used only in case distance between source and target
d_o: Threshold value or distance calculated by taki ratio of square root of E_{fs} divided by E_{mp};

E_{DA}: Aggregation energy per bit.

From energy consumption model, it is to be noted how energy consumption increased from square of the distance to the quadruple power of the distance.

The threshold distance (d_o) as suggested in [2] is:

$$d_{o=}\sqrt{\varepsilon_{fs}/\varepsilon_{mp}} \tag{2}$$

(b) **Pseudocode Used**

Step 1: Assume N_i as a sensor node.
Step 2: N_1, N_2, N_3, and N_n considered as sensor nodes.
Step 3: The initial energy of each node is E.
Step 4: Each of the nodes is distributed randomly on a different set of X–Y coordinates on a given network.
Step 5: Then we compute the Euclidean distance of each of the sensor nodes to each of the other sensor nodes.
Step 6: Only those nodes are considered which follow the threshold distance (d_o) criteria.
Step 7: Find out the shortest path for data traveling between node and BS.
Step 8: The threshold value is computed using Eqs. 2 and 3.
Step 9: A random value-generating function for computing the RV and comparing it with the threshold value.
Step 10: Each of the sensor node which is nearest to cluster head forms a new cluster.
Step 11: Energy consumption model given in Sect. 3.(a) is used to calculate the amount of energy consumption.

All the steps from 5 are repeated till each of the sensor nodes in network dies.

(c) **Cluster Head Selection in CHME-LEACH Protocol**

However, energy is dissipated continuously while receiving or transmitting the data packet by cluster member sensor node or cluster head. But in the LEACH protocol, the CH is selected randomly so selecting hotspot node as CH is higher. Here in CHME-LEACH protocol, CH selection is done by modified energy by using random generator function but we are also specifying that select them as cluster head only if the energy of the sensor node is at least $E_o/2$. Here the possibility of selecting same node as cluster head is 0.1.

$$\text{tem}_{\text{rand}} \leq ((p/(1 - p * \text{mod}(r, \text{round}(1/p))) \text{ and } E \geq E_o/2) \tag{3}$$

where $\text{temp}_{\text{rand}}$ is random number b/w and $1, p = 0.1$ and $r =$ round number.

(d) **Cluster Head Selection in CHP-LEACH Protocol**

As it is mentioned in Eq. 1, cluster head in LEACH protocol is selected when probability is 0.1. But that is also a demerit point for the LEACH protocol because a node once become as cluster head has the probability to become as cluster head again after 10th round. So, there might be a situation arises when only a certain set of nodes become cluster head in rotation and rest of the sensor node remain idle. So, in that case several dead spots or energy hole would be generated and which would further deteriorate the energy consumption. And thus, our main motive to decrease the energy consumption doesn't fulfill. However, energy is dissipated continuously while receiving or transmitting the data packet by cluster member sensor node or cluster head. So here in CHP-LEACH protocol, the probability of selecting a node as cluster head depends on number of nodes.

$$p = 1/n$$

Thus, every node has an equal probability of becoming a cluster head. Now we have overcome the LEACH protocol demerits of the energy hole problem. Thus, the cluster head is selected by:

$$\boldsymbol{temp_{rand}} \leq \left(\frac{\frac{1}{n}}{1 - \frac{1}{n} \boldsymbol{*mod}\ (\boldsymbol{r}, \boldsymbol{n})} \right) \tag{4}$$

where $\text{temp}_{\text{rand}}$ is random number *b/w* 0 and 1 and $r =$ round number.

4 Simulation Result

CHME-LEACH and LEACH-CHP are analyzed on many factors like energy efficiency, lifetime of network and total packets delivery ratio. We have compared the performance both the new protocols with basic original LEACH protocol. The simulation is carried on MATLAB R2018b which spread the sensor node, receive the data packet, send the data packet, etc.

4.1 Simulation Parameter

Here in this simulation used total 100 sensor nodes dispersed randomly over an area of 100 m in length as well as 100 m in breadth. And the base station or sink node is situated on the center (50, 50). Initially, set the energy of each node as two joules. To calculate transmitted energy and reception energy dissipation, use energy consumption model. There were several parameters used in simulation where E_{fs} and E_{mp} is dependent on amplifying model and E_{fs} is amplification energy when distance between sensor node is less than threshold distance, and E_{mp} is amplification energy

Table 1 Simulation parameter

Parameter	Values
Total number of nodes	100
Size of network (x, y)	100, 100 m
Sink location	$X = 50$; $Y = 50$
Initial energy	2 J
E_{mp}	0.0013 pJ/Bit/m^2
E_{fs}	10 pJ/Bit/m^2
Data size (bits)	4000
Max rounds	5000
E_{lec}	50 nJ/bit
E_{da}	5 nJ/bit/message
E_{rx}	50 nJ/bit
E_{tx}	50 J/bit

when the two-sensor node is greater than threshold distance. E_{lec} (Electronic energy) is the energy used per bit to the transmitter, a receiver circuit which depends on factor such as channel coding, modulation, filtering, and spreading of signal. The parameter used in simulation is summarized in Table 1.

4.2 *Results*

Remaining Energy in the Network

We have simulated our protocol for different number of rounds. From Fig. 1, we can see that amount of remaining energy in new routing protocol. Energy consumption for LEACH protocol is more. Since we are considering residual energy of cluster head in designing CHME-LEACH protocol, we can see the effect of this how much energy is still left for the given network. From the figure, we can see with different rounds how the energy consumption changes. This result is plotted in Fig. 1.

Packet Delivery Ratio

In LEACH protocol, some of the situation arises when although the data packet is being sent to the CH; however, whole of the energy of the cluster head is dissipated in only receiving, transmitting, aggregating, and then it can't send to the sink node. So, all the information is lost. From Fig. 2, we can see that number of data packet sent to base station is best in CHP-LEACH then CHME-LEACH also send a greater number of data packet than the original LEACH protocol. Though for some round, LEACH protocol and CHME-LEACH sent almost the same number of data packet to the sink node.

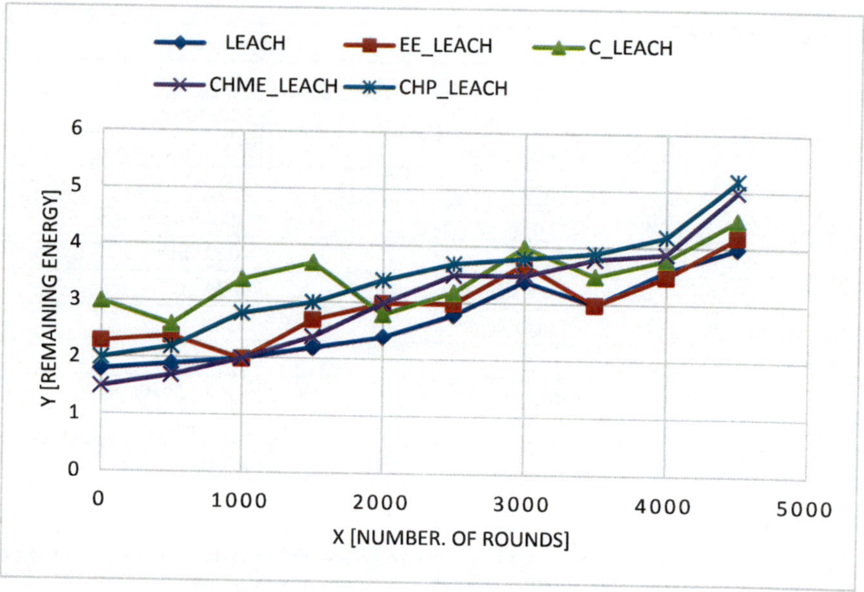

Fig. 1 Remaining energy in the network

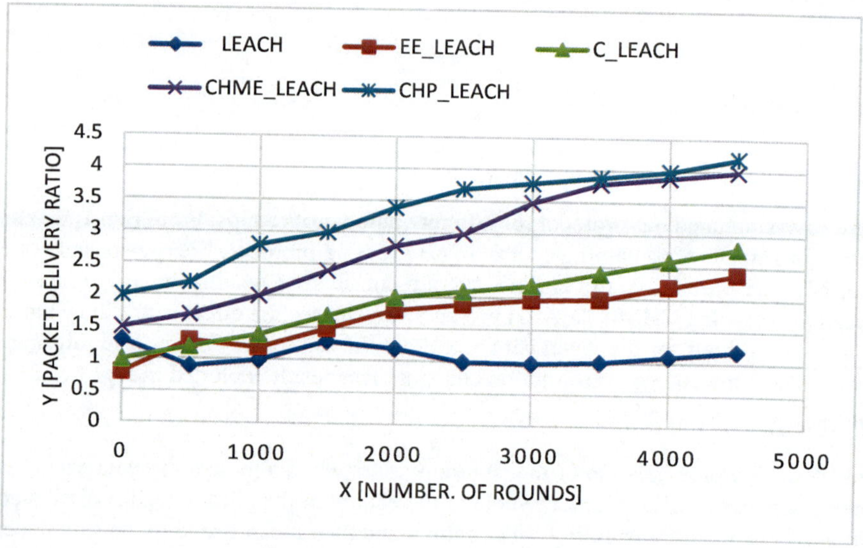

Fig. 2 Total number of packet sent to the base station

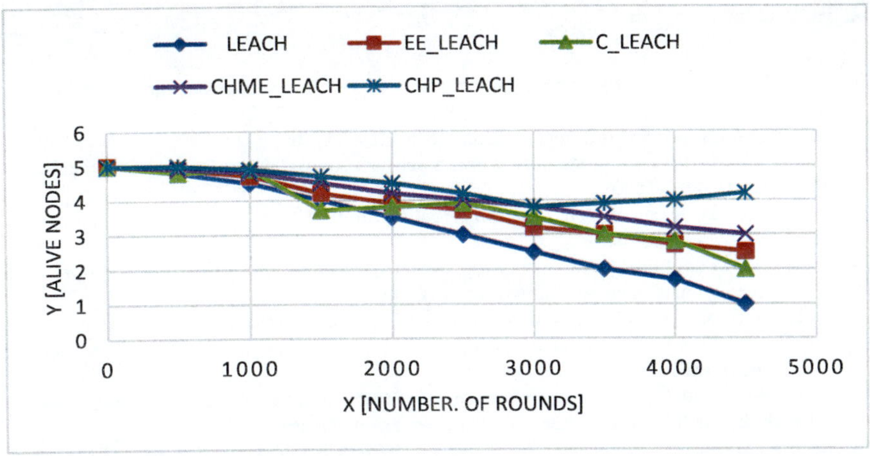

Fig. 3 Alive nodes after each round

Network Lifetime

From Fig. 3, the performance of CHME-LEACH and CHP-LEACH improved as compared to the original basic LEACH protocol. In leach protocol, each of the nodes almost died in 2500 rounds, but in CHME-LEACH and CHP-LEACH almost 60 nodes are still alive. So, we can easily say that our new algorithm outperforms in number of node alive. Thus, the main motive of our algorithm to increase the network lifetime is satisfied.

5 Conclusion and Future Scope

In this paper, an enhancement to the original LEACH protocol has been introduced, labeled as CHME-LEACH (Cluster Head Modified Energy LEACH) and CHP-LEACH (Cluster Head Probability LEACH). The primary objective of CHME-LEACH and CHP-LEACH is to enhance the network lifespan of the LEACH protocol. To achieve this, these proposed enhancements focus on reducing repetitive selection of the same CH based on the residual energy of sensor nodes. Simulation results demonstrate that both CHME-LEACH and CHP-LEACH not only decrease energy dissipation but also increase the successful transmission of data packets to the base station, mitigating losses observed in the original LEACH protocol. Future directions include incorporating distance parameters into the existing approach and potentially combining features from both protocols for improved results. Another avenue is to explore alternating execution of the algorithms to achieve enhanced outcomes.

References

1. Aye CT, Zar KT (2018) Lifetime improvement of wireless sensor network using residual energy and distance parameters on LEACH protocol. In: 2018 18th international symposium on communications and information technologies (ISCIT). IEEE
2. Tandel RI (2016) Leach protocol in wireless sensor network: a survey. Int J Comput Sci Inf Technol 7(4):1894–1896
3. Liu X (2012) A survey on clustering routing protocols in wireless sensor networks. Sensors 12(8):11113–11153
4. Al-Shalabi M et al (2018) Variants of the low-energy adaptive clustering hierarchy protocol: survey, issues and challenges. Electronics 7(8):136
5. Varshney S, Kuma R (2018) Variants of LEACH routing protocol in WSN: a comparative analysis. In: 2018 8th international conference on cloud computing, data science & engineering (confluence). IEEE
6. Jambli MN et al (2018) An analytical study of leach routing protocol for wireless sensor network. In: 2018 IEEE conference on wireless sensors (ICWiSe). IEEE
7. Behera TM, Nanda S, Mohapatra SK, Samal UC, Khan MS, Gandomi AH (2021) CH selection via adaptive threshold design aligned on network energy. IEEE Sensors J 21(6):8491–8500. https://doi.org/10.1109/JSEN.2021.3051451
8. Dhivya K, Premalatha G, Sindhuparkavi B (2019) Energy efficient power management scheme for WSN using network encoding protocol. In: 2019 international conference on intelligent computing and control systems (ICCS), Madurai, India, pp 78–82. https://doi.org/10.1109/ICCS45141.2019.9065334
9. Yetgin H et al (2017) A survey of network lifetime maximization techniques in wireless sensor networks. IEEE Commun Surv Tutor 19(2):828–854
10. Priyadarshi R et al (2018) SEEN: stable energy efficient network for wireless sensor network. In: 2018 5th international conference on signal processing and integrated networks (SPIN). IEEE
11. Saxena M, Joshi A, Dutta S et al (2021) Comparison of different multi-hop algorithms to improve the efficiency of LEACH protocol. Wireless Pers Commun 118:2505–2518
12. Tripathi M et al (2013) Energy efficient LEACH-C protocol for wireless sensor network: 402–405
13. Kim D-S, Chung Y-J (2006) Self-organization routing protocol supporting mobile nodes for wireless sensor network. In: First international multi-symposiums on computer and computational sciences (IMSCCS'06), vol 2. IEEE
14. Saxena M, Dutta S, Singh KB et al (2023) Multi-objective based route selection approach using AOMDV in MANET. SN Comput Sci 4:581. https://doi.org/10.1007/s42979-023-01982-z
15. Bharathi SA, Dandime GM, Nirmala GV, Baldania A, Sai Kumar CM, Fahlevi M (2022) Energy enhancement and optimization of WSN using firefly algorithm and deep learning. In: 2022 international conference on edge computing and applications (ICECAA), Tamil Nadu, India, pp 1432–1436. https://doi.org/10.1109/ICECAA55415.2022.9936146
16. John A, Rajput A, Babu KV (2017) Dynamic cluster head selection in wireless sensor network for Internet of Things applications. In: 2017 international conference on innovations in electrical, electronics, instrumentation and media technology (ICEEIMT), Coimbatore, India, pp 45–48. https://doi.org/10.1109/ICIEEIMT.2017.8116873
17. Saxena M, Dutta S (2020) Improved the efficiency of IoT in agriculture by introduction optimum energy harvesting in WSN. In: 2020 international conference on innovative trends in information technology (ICITIIT), Kottayam, India, pp 1–5. https://doi.org/10.1109/ICITIIT49094.2020.9071549
18. Kumar LKS, Gumudavally H, Sanku SK (2017) Comparative study of LEACH and EECDA protocols. In: 2017 IEEE international conference on power, control, signals and instrumentation engineering (ICPCSI). IEEE

19. Bendjeddou A, Laoufi H, Boudjit S (2018) LEACH-S: low energy adaptive clustering hierarchy for sensor network. In: 2018 international symposium on networks, computers and communications (ISNCC). IEEE

DenseMammoNet: An Approach for Breast Cancer Classification in Mammograms

Shajal Afaq and Anamika Jain

Abstract In women all around the world, cancer of the breast is a condition that is both prevalent and on the rise. Breast cancer can be formed due to the lumps in the mammary region in females. Early detection of breast cancer masses (BCM) can save the lives of many women. In this paper, we have proposed an automated method that it is feasible to spot cancer of the breast in its infancy from mammographic images. To validate the authenticity of the proposed work, we have experimented with the publicly available dataset, Mammography Image Analysis Society (MIAS), and achieved an accuracy of 99.4%.

Keywords Breast cancer mass · Transfer learning · DenseNet201 · Mammogram

1 Introduction

In a recent report by WHO, breast cancer has been listed as the most common cancer detected among adult women worldwide. According to the Indian Council of Medical Research (ICMR), breast cancer recorded about two lakhs of cases, i.e., 14.8% in the year 2020 [1]. According to ICMR, one out of every five women suffers from breast cancer at some point in their lives [1]. A breast cancer mass (BCM) is caused by the unusual development of mammary cells. In the female's body, there are two kinds of tumors in the breast area. One is the non-cancerous cells forming a benign or harmless mass, which grows just internally and does not spread across the organ. Cancer is a group of disorders in which the cell of the body create a cluster and form a mass that can be termed as malignant tumors [2]. On the other hand, malevolent tumors which are made up of cancerous cells will spread throughout the tissue and take a malignant tumor that causes cancer. Early breast cancer discovery

S. Afaq (✉)
Centre for Advanced Studies, AKTU, Lucknow, Uttar Pradesh 226021, India
e-mail: shajalafaq@gmail.com

A. Jain
Dr. Vishwanath Karad, MIT-World Peace University, Kothrud, Pune, Maharashtra 411038, India
e-mail: anamika06jain@gmail.com

S. Kumar et al. (eds.), *Fourth Congress on Intelligent Systems*, Lecture Notes in Networks and Systems 868, https://doi.org/10.1007/978-981-99-9037-5_12

can lead to the successful treatment of breast cancer. Therefore, the accessibility of appropriate screening techniques is necessary for identifying the early symptoms of breast cancer. Thermography, ultrasonography, and mammography imaging are used to screen breast cancer. Early detection of breast cancer includes mammography as one of the important imaging tool. It is a low-intensity X-ray method that allows us to see inside the anatomy of mammary [3]. A biopsy can also be used to diagnose BCM. The differences in shape, size, and position of cancerous cells are the most difficult aspects of examining BCM. Researchers have developed techniques to create and enhance image processing due to the challenges like noisy images, inter-class similarity, and intra-class variation. Healthcare sectors are using artificial intelligence (AI) and machine learning (ML) [4] to reduce the challenges of medical images.

Organization of the rest of the paper is divided into six sections. In Sect. 2, the related research work has been discussed. Section 3 has a discussion about the proposed method. Experimental setup and results have been elaborated in Sects. 4 and 5, respectively. In Sect. 6, the conclusion and future work have been discussed.

2 Related Work

In adult women, the most common sort of cancer appears to be breast cancer. Researchers have proposed methods to provide solutions to find breast cancer at its earliest stages. In this section, we have discussed some of the techniques proposed by the researcher. In [5], authors used several Machine Learning algorithms to detect and diagnose breast cancer with respect to confusion matrix. The authors have used SVM and achieved an accuracy of 97.2% on the WBCD database. Das et al. in [6] proposed a method using ensemble learning in order to achieve desired accuracy. RF, Nave Bayes, SVM with two separate kernels (RBF, polynomial), K-NN, and DT are among the five machine learning (ML) classifiers used by the ensemble voting method. They tested with the WBCD dataset. It involves data preprocessing and ensemble to combine the performance of all the used algorithms and got the accuracy of 99.28% [6]. In [7], authors used the MIAS mammography dataset. The authors developed the automated feature extraction and classification system, and the deep CNN model VGG16 was used to evaluate the performance of their work. They utilized the benefits of transfer learning to decrease training time and improve accuracy [7]. In [8], authors developed a framework based on transfer learning to focus the efforts on classifying histopathological and imbalanced images using the Imagenet dataset. They used the VGG19 as the base model and complemented it with several existing techniques to boost the system's overall performance [8]. In [9], authors utilized a deep Convolutional Neural Network-based method. There was 1 input, 28 hidden, and 1 outcome layers in this structure. The feature-wise data augmentation (FWDA) approach was used to mitigate over-fitting problem. For specificity, accuracy, and sensitivity, they obtained 99.71 %, 90.5 %, and 89.47 %, respectively [9]. Fatima et al. in [10] reviewed around 14 papers to highlight all the earlier researches of machine learning, deep learning, and ensemble approaches that were employed in the breast cancer detection. Authors listed out the problems regarding set of data having negative and positive disparity leading to bias forecasting, imbalanced BC images leading to improper diagnose, and a restricted set of data overcoming by data augmentation methods [10]. The possibilities and restrictions of breast cancer diagnosis using algorithmic learning are looked at in a research via Mohi ud din et al., in part by comparing earlier studies on the topic [11]. Support Vector Machine is mentioned as a typical method in regard to the categorization of mammograms. In earlier work, categorization on the basis of BI-RADS density level was performed using a light-weighted CNN, with a standard reliability of 83.6%. Reviewing prior developed algorithms reveals that even though a lot of them are efficient, they all have drawbacks that make it difficult for them to work consistently. The inadequate contrast of mammography is one difficulty, which underlines the value of picture preprocessing for reducing the number of falsely positives and negatives [11]. Finally, it is stated that more studies may focus on CNNs, which are underrepresented in research yet could improve processes like density categorization and asymmetry detection. There are many articles approaches to cancer screening utilizing CNNs, despite CNNs being underrepresented in the literature [12]. One such study by Tsochatzidis et al. analyzes the performance of eight well-known deep CNNs (D-CNNs). The models are trained entirely from scratch in one scenario, and pre-trained networks are fine-tuned in the other. The researchers talk about the challenges of good training because it needs a lot of data, which frequently is not available, especially for difficulties with medical imaging. The CBIS-DDSM dataset is one of the few publicly accessible large-scale datasets, that why used [12].

3 Proposed Work

In the proposed work, we have presented a method for the classification of the BCM. In the work, we have used modified DenseNet201. The complete illustration of the proposed work is shown in Fig. 1. The proposed methodology has been categorized into three parts: (1) preprocessing, (2) transfer learning, and (3) classification. Each module has been described in the following section.

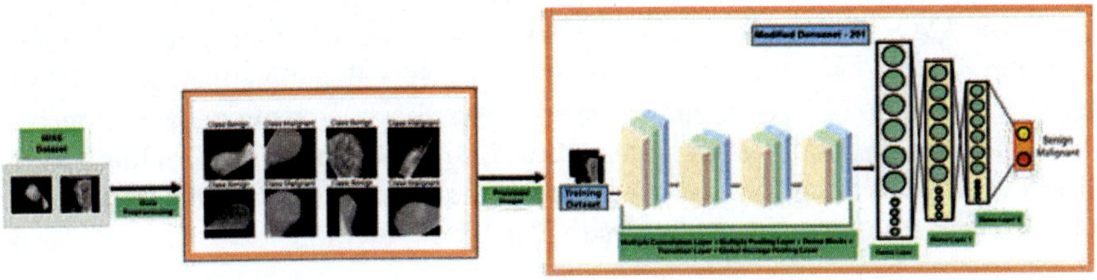

Fig. 1 Block diagram of the proposed mammographic detection framework

3.1 Preprocessing

MIAS dataset has different size of images, in order to use the transfer learning in our proposed work, the input images should be of size 224 × 224. We have resized the input images and augmented the dataset to eliminate the problem of class imbalance.

To reduce the problem of class imbalance in the dataset we have utilized the SMOTE oversampling (Synthetic Minority Oversampling Technique) algorithm [13]. The oversampling process makes a minor category (class malignant) has the same number of images as the other classes (benign).

3.2 Transfer Learning

In the field of deep learning, transfer learning plays a very vital role. It provides the power to use the pre-trained network in different domains. In the proposed work, we have used DenseNet201 architecture. Transfer learning is very useful in concept in deep learning field because most of the learning do not require to train complex algorithms using millions of labeled data points from everyday scenarios [14].

3.2.1 DenseNet201

DenseNet is a deep Convolutional Neural Network-based architecture, which uses dense connectivity among the layers via dense blocks. These dense blocks link every layer (with matched featuring-maps volumes) simultaneously. DenseNet201 is a variant of DenseNet, which comprises 201 layers. DenseNet201 has been trained over ImageNet dataset. Every layers of DenseNet receives extra information from the prior layers and then sends its unique feature-maps to the next layers. With the skip connection from the previous layers to the next layers, the network provides better operational efficacy and storage economy [15]. DenseNet has a better flow of gradients throughout the network, this makes the model training easier. This leads us to use the DenseNet201 in our proposed method.

In our work, we have slightly modified the architecture of the DenseNet201. Figure 2 depicts the modified DenseNet architecture. The input layer of the DenseNet has 224 × 224 × 3 size. The DenseNet201 architecture has four dense and three transition blocks. The DenseBlock consists of 4 layers-Batch Normalization (BN), ReLU, Convolutional Layer, and Dropout Layer. Each Transition Layer consists of Batch Normalization, ReLU, Convolutional Layer, Dropout, and Pooling Layer. Global average Pooling Layer is present after the DenseBlock4. The GAP layer is followed by 3 Fully Connected Layers (FCL).

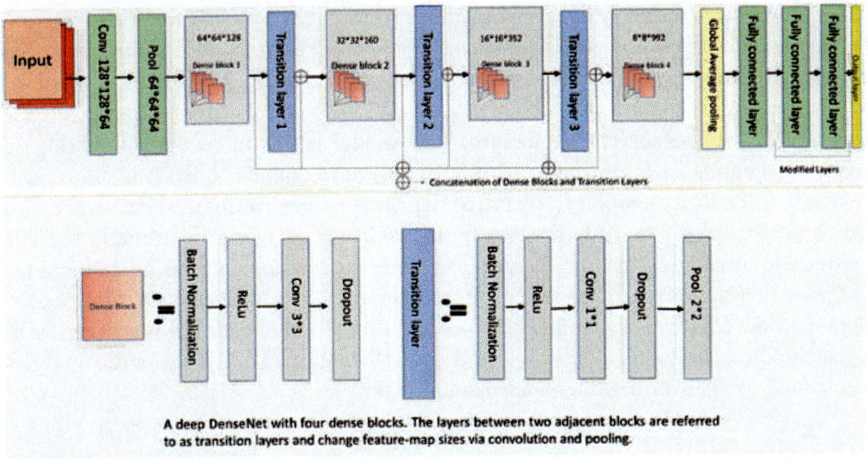

Fig. 2 Network architecture of modified DenseNet201

Table 1 Hyper-parameter of the modified DenseNet201

Name of the parameter	Value
Learning rate (lr)	0.0001
Optimizer	Adam
Batch-size	32
No. of epochs	20
Loss	Binary cross entropy

3.2.2 Model Training

The initial layers of the models are responsible for the generalized features like edges, so we have freezed starting 169 layers. We have added two fully connected layers in the pre-trained DenseNet201. The last fully connected layer has two neurons for the classification of benign and malignant tumors.

Table 1 describes the hyper-parameters utilized for training the modified DensNet201. With the learning rate, the optimizer will control the rate of convergence to reach global optima. With experimentation, we have selected a learning rate of 0.0001 for our work. In the proposed work, we have used Adam optimizer with 32 batch size with the binary cross entropy loss function. We have performed the training of the model for 20 epochs.

Prior to the study, DenseNet, which uses deep learning architecture, has not been thoroughly studied for use in identifying cancers of the breast in mammograms. By exploring and evaluating the application of the DenseNet design for malignancy identification in mammograms, this study seeks to fill this research gap. We conducted the architecture's performance in properly distinguishing between benign and malignant mammography images. The performance of the DenseNet model was validated and trained using MIAS dataset. To increase the robustness and generalizability of the model, the use of data preprocessing has been done.Intricate patterns in mammography pictures have been effectively captured by utilizing the distinctive property of the DenseNet architecture. This model learns to recognize subtle and complex features of breast cancer by utilizing these connections. This study successfully filled the research gap by proposing and demonstrating the effectiveness of the DenseNet architecture in breast cancer classification using mammograms. The uniqueness indicated that the DenseNet architecture outperformed other methods in terms of accuracy, demonstrating its potential as a robust tool for this cancer. This methodology was systematically contrasted with conventional machine learning techniques and other deeper learning models that are often used in the analysis of medical images during the experimental phase.

4 Experiment

The proposed work has been performed in Python language over a personal computer with Intel i7 processor with GTX 1650 NVIDIA graphic card with 4 GB memory and 8 GB RAM.

4.1 Dataset

The proposed approach makes use of the Mammography Image Analysis Society (MIAS) dataset, which is open to the general public. The Mini Mammographic Database (MIAS) is a scanned film that was created by UK research teams to better understand mammograms by creating a database of digital mammograms. 322 pictures were initially present in the MIAS collection. After oversampling, we get 3816 pictures with 50 micron resolution in Portable Gray Map (PGM) format and pertinent real data form, of which 2376 are benign and 1440 are malignant. Figure 3 shows the sample images of MIAS dataset.

5 Results and Discussion

We have achieved 99.4% accuracy with the proposed work. For a fair comparison of the proposed work with the state-of-the-art methods, we have calculated the other parameters like precision, recall, and $f1$-score which are given in Table 2. Training accuracy and loss curve has been depicted in Figs. 4 and 5, respectively.

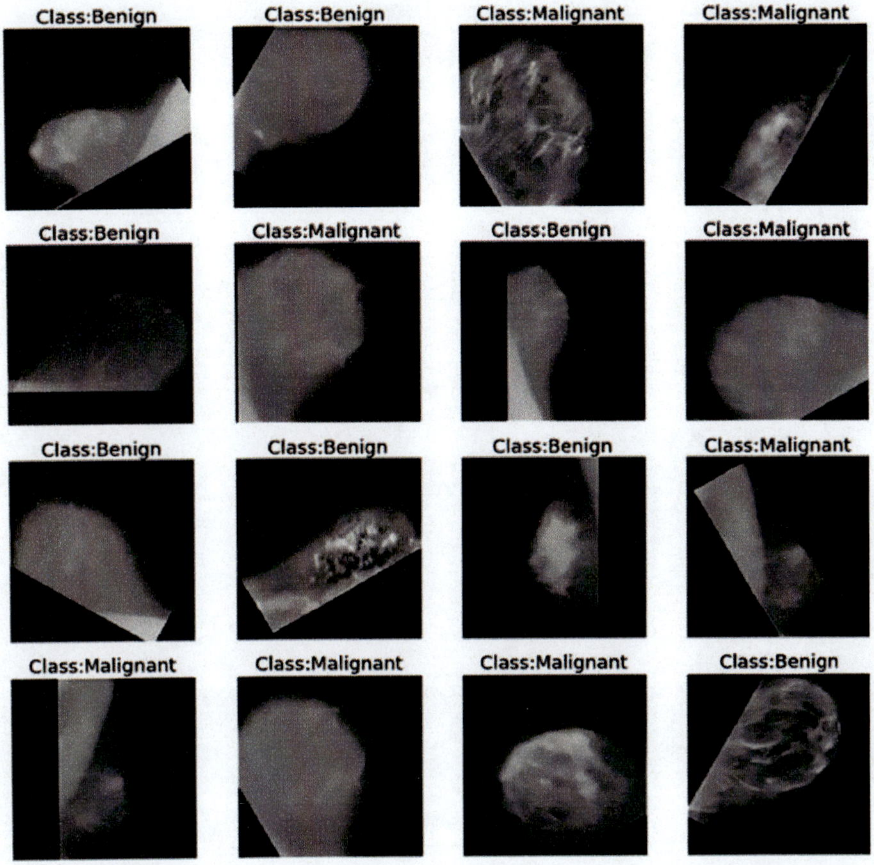

Fig. 3 Sample images of benign and malignant classes from MIAS dataset

Table 2 Proposed DenseNet model performance matrices

Classes	Precision	Recall	$F1$-score
Benign masses	0.97	0.99	0.98
Malignant masses	0.99	0.99	0.98

We have also performed experiments on other two pre-trained models, Inception V3 and VGG19, but the performance we received was not comparable to the state-of-the-art methods.

In Table 3, we have compared our method with the state-of-the-art methods. In Table 3, we can see that the proposed model gives a higher performance as compared to [16–21]. With Table 3 we state that the Modified DenseNet201 transfer learning model efficiently diagnoses breast cancer with low misclassification rate.

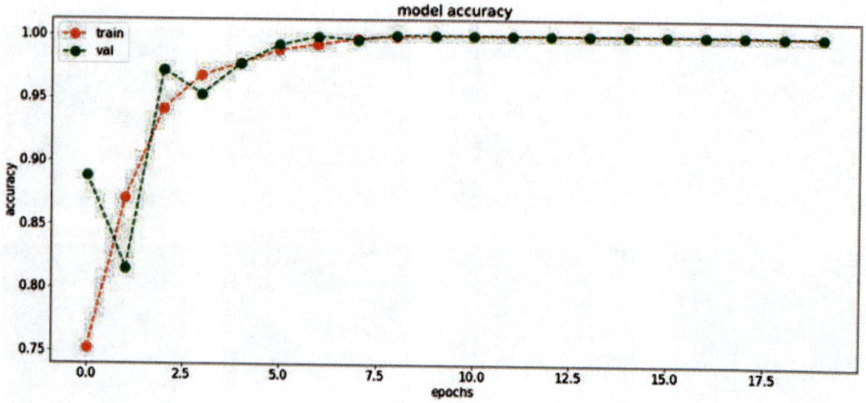

Fig. 4 Accuracy curve

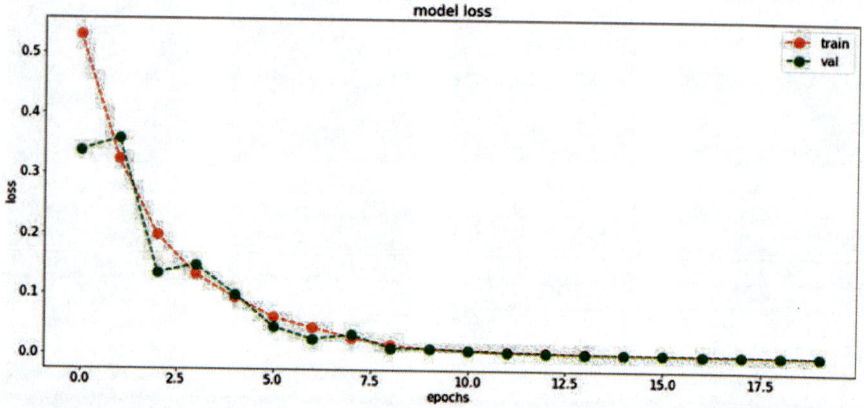

Fig. 5 Loss curve

Table 3 Comparison of the existing work with proposed method MIAS dataset

Features	Classifier	Accuracy (%)
Texture energy [17]	ANN	93.9
GLCM [18]	RBFNN	94.29
DTMBWT [19]	SVM	94.01
GLCM [20]	RBFNN	94.29
DTW [21]	K-NN	92.81
GLCM [16]	FFANN	96
Modified DenseNet (proposed)	DenseNet	99.4

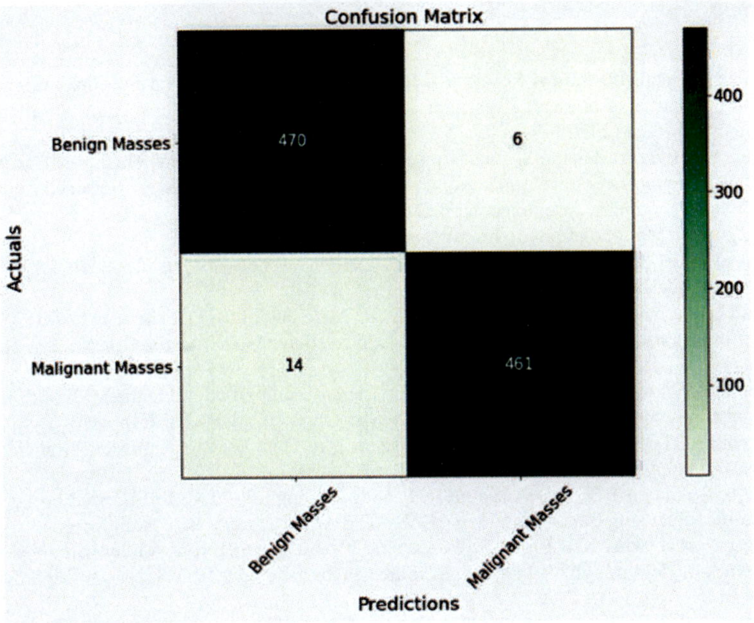

Fig. 6 Confusion matrix

Figure 6 depicts the confusion matrix over MIAS dataset. Confusion matrix analyzes the performance of a classification model on unseen test data.

6 Conclusion and Future Work

Breast cancer is a deadly disease in women if not treated on time. It is very crucial to predict breast cancer with limited resources. We have proposed a method to predict breast cancer using mammographic images. Modified DenseNet201 architecture has been used in the proposed work. The proposed work can incorporate the qualities of identity mappings, deep supervision, and diverse depth while obeying a simple connectivity rule. It enables feature reuse across networks and, as a result, can learn more compact and accurate details, according to our evaluations. The proposed model classifies the mammographic images with 99.4% accuracy on the publicly available MIAS dataset.

References

1. Indian Council of Medical Research Department of Health Research Press Note on Cancer, ICMR Department of Health Research, Research Ministry of Health & Family Welfare Government of India (2020)
2. Feng Y, Spezia M, Huang S, Yuan C, Zeng Z, Zhang L, Ji X, Liu W, Huang B, Luo W et al (2018) Breast cancer development and progression: risk factors, cancer stem cells, signaling pathways, genomics, and molecular pathogenesis. Genes Diseases
3. Dongola N. Mammography in breast cancer
4. Masud ARM, Hossain MS (2020) Convolutional neural network-based models for diagnosis of breast cancer. Neural Comput Appl 5
5. Filali S, Aarika K, Naji M, Benlahmar EH, Ait Abdelouahid R, Debauche O (2021) Machine learning algorithms for breast cancer prediction and diagnosis. Procedia Comput Sci 191:487–492
6. Das S, Biswas D (2019) Prediction of breast cancer using ensemble learning. In: 5th international conference on advances in electrical engineering (ICAEE). IEEE, pp 804–808
7. Krishna CRTH (2021) Mammography image breast cancer detection using deep transfer learning. Adv Appl Math Sci 20:1187–1196
8. Singh R, Ahmed T, Kumar A, Singh A, Pandey A, Singh S (2020) Imbalanced breast cancer classification using transfer learning. IEEE/ACM Trans Comput Biol Bioinform: 1
9. Toğaçar M, Özkurt KB, Ergen B, Cömert Z (2020) Breastnet: a novel convolutional neural network model through histopathological images for the diagnosis of breast cancer. Physica A Stat Mech Appl 545:123592
10. Fatima N, Liu L, Hong S, Ahmed H (2020) Prediction of breast cancer, comparative review of machine learning techniques, and their analysis. IEEE Access 8:150360–150376
11. Rayees Ahmad Dar AA, Rasool M (2022) Breast cancer detection using deep learning: datasets, methods, and challenges ahead. Comput Biol Med 149
12. Tsochatzidis L, Costaridou L, Pratikakis I (2019) Deep learning for breast cancer diagnosis from mammograms—a comparative study. J Imaging MDPI 5
13. Smote: https://www.jair.org/index.php/jair/article/view/11192
14. Torrey L, Shavlik J. Transfer learning
15. Huang G, Liu Z, Van Der Maaten L, Weinberger KQ (2017) Densely connected convolutional networks. In: Proceedings of the IEEE conference on computer vision and pattern recognition, pp 4700–4708
16. Sheba K, Gladston Raj S (2018) An approach for automatic lesion detection in mammograms. Cogent Eng 5(1):1444320
17. Setiawan AS, Elysia JW, Purnama Y (2015) Mammogram classification using law's texture energy measure and neural networks. Procedia Comput Sci Int Conf Comput Sci Comput Intell (ICCSCI) 59:92–97
18. Pratiwi M, Alexander JH, Nanda S (2015) Mammograms classification using gray-level co-occurrence matrix and radial basis function neural network. Procedia Comput Sci Int Conf Comput Sci Comput Intell (ICCSCI) 59:83–91
19. Suba C, Nirmala K (2015) An automated classification of microcalcification clusters in mammograms using dual tree m-band wavelet transform and support vector machine. Int J Comput Appl 115(20)
20. Candès EJ, Donoho DL (2005) Continuous curvelet transform: I. Resolution of the wavefront set. Appl Comput Harmonic Anal 19(2):162–197
21. Gardezi SJS, Faye I, Sanchez Bornot JM, Kamel N, Hussain M (2018) Mammogram classification using dynamic time warping. Multimedia Tools Appl 77(3):3941–3962

Aspect-Based Sentiment Classification Using Supervised Classifiers and Polarity Prediction Using Sentiment Analyzer for Mobile Phone Tweets

Naramula Venkatesh and A. Kalavani

Abstract Twitter platform finds an important part in social marketing, election campaigns, academia and news. Sentiment analysis came into existence for finding polarities on different domains based on customer tweets online such as to check the opinions of people on a particular topic. Aspect-based sentiment analysis (ABSA) is a text analysis which identifies an entity or feature present in a given sentence and provides an opinion according to the given aspect. In this paper, we proposed an automatic approach for aspect sentiment classification and polarity prediction using mobile phone tweets. So, research contribution is given in the modules of tweets collection and preprocessing, implicit aspect term extraction towards sentiment classification and polarity prediction for mobile phone tweets. In order to predict sentiment polarity for aspect category, we use Vander sentiment analyzer and predict overall polarity as a compound value.

Keywords ABSA · ATP · ATE · Mobile phone

1 Introduction

In current technology, words spoken by different customers in social media show much impact on purchase behavior. Therefore, for social media comments, reviews are much more useful to different organizations to compete with different competitors and to enhance their products in advance. Sentiment analysis also deals with human behaviour where it extracts emotion and opinion from given textual data. Sentiment analysis is a simple process that identifies textual data as positive value, negative value or neutral value. Sentiment Analysis may be complex when defining attitude

N. Venkatesh (✉)
SR UNIVERSITY Warangal, Warangal, India
e-mail: venkatesh.naramula@sru.edu.in

A. Kalavani
Rajalakshmi Engineering College, Chennai, India
e-mail: kalaivani.a@rajalakshmi.edu.in

S. Kumar et al. (eds.), *Fourth Congress on Intelligent Systems*, Lecture Notes in Networks and Systems 868, https://doi.org/10.1007/978-981-99-9037-5_13

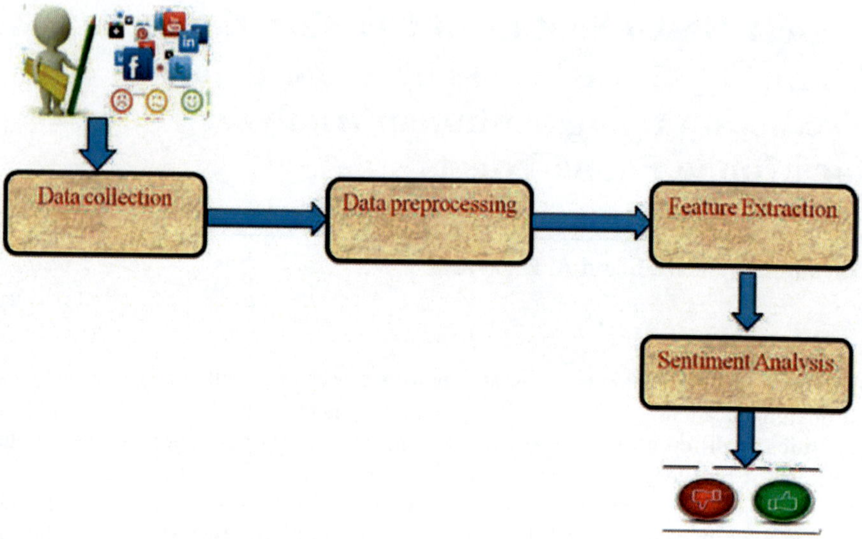

Fig. 1 General architecture of sentiment analysis

as ranking − 1 to + 1. Sentiment analysis is the process of extracting opinions, customers' feelings and users' attitudes on a particular topic. The general architecture of sentiment analysis is shown in Fig. 1 which starts with data collection in which raw tweets are extracted for opinion extraction followed by data preprocessing that converts unstructured to structured data. Feature extraction uses different methods to extract features from structured data and machine learning methods are applied for various sentiment analyses for required data for prediction of positive and negative opinion words.

Consider the below tweets given by online users:

These speakers are good for small areas like home but its sound quality is not good.

This example talks positive sentiment about speaker when coming to location but negative when it is about sound quality.

The most emerging social networking is Twitter, which allows the user for broadcasting updated personal shares, community shares, nationwide or international wide events by specifying through short messages called "tweets". Tweets are composed of textual, audio and visual formate or links to external websites. Twitter may be playing a key role in many fields such as social marketing teams, election campaigns persons, academics and news. Most of the Twitter posts are short form and constantly generated which makes the public use at a wider rate in providing public sentiment about an event.

2 Sentiment Analysis-Based Level Classification

Sentiment analysis can be categorized into three different types:

1. Document-based sentiment analysis which considers the whole documents in a single entity and provides sentiment value as positive and negative.
2. Sentence-based sentiment analysis makes use of a single sentence given in a document and provides an opinion as positive or negative.
3. Aspect-based sentiment analysis was text analysis for identifying an entity or feature present in a given sentence and providing an opinion based on the aspect. It is also termed feature-based sentiment analysis used in many applications for fine-grained sentiment analysis.

Sentiment analysis is considered as for classification problem and further subdivided into document-level sentiment analysis, sentence-based sentiment analysis and aspect-level sentiment analysis. In document-level, sentiment analysis considers whole documents as a single entity and provides sentiment value as positive and negative. It is possible to determine whether a text of writing was positive, negative or neutral using "computational" methods.

In sentence-based, sentiment analysis makes use of a single sentence given in a document and provides an opinion as positive or negative. The initial stage is to identify a given Sentences may have subjective or objective are not. This polarity of subjectivity sentences is determined, and objective sentences are eliminated. If the given sentence may be subjective, sentence-level sentiment analysis will be determining whether given sentence expressed has positive or negative opinions.

3 Aspect-Based Sentiment Analysis

Aspect-based sentiment analysis (ABSA) is text analysis that identifies an entity or feature present in a given sentence and provides an opinion according to the given aspect. It is also named feature-based sentiment analysis as it points to a single entity, it has been used in many applications for fine-grained opinions and can be identified.

Aspect-based feature term extraction and also sentiment classification will be two major tasks for ABSA. The primary step is for aspect term for the given feature to be identified in the task of the feature term extraction process. Based on different ABSA studies, feature term extraction will be extracting in considering all noun foams of groups proposes a hybrid approach using fusion as two approach's, machine learning approaches and lexicon-based approaches. These will use different lexicon similar to sentiment-Word2Net for detection of sentiment words and also assign POS tags for feature selection process.

Aspect-based sentiment analysis focused mainly on an important aspect of a product on which sentiment polarity is to find out. Aspect-level sentiment analysis first examines which are the important features of a product, extracts the features, analyzes the features and identifies polarities on them. General aspects which are considered and to be omitted are specified.

4 Applications of Aspect-Based Sentiment Analysis

Aspect-based sentiment analysis is implemented in many applications such as social media, medical and health purposes, industrial applications and so on. Social media applications for sentiment analysis may not be limited to predicting for violence impact and also for by analyzing in social media content for the polarity of violence, tracking students' opinion about the education system and election results prediction from different voters. Industry applications include stock market predicting, monitoring a different company's brands and measuring satisfaction regarding a product from different users. In Medical applications that will be used for Assigning ranks to different doctors based on doctors' experience in some aspects and patient's satisfaction oriented, also prediction of the depression rates. The application areas for sentiment analysis to predict what's happening on social media.

1. Business and Organization: For getting brand analysis, product and service benchmarking.
2. Social Media: Finding general opinions about trending topics.
3. Ads Placements: To get appraisal about the product and to compete with different competitors.
4. Question-answering type sentiment analysis.
5. Developing "hot mail filters" analogous to "spam mail detection".

5 Issues in Aspect-Based Sentiment Analysis

For achieving excellent performance in natural language processing (NLP) and aspect-based sentiment analysis (ABSA), lot of effort is required. Aspect term extraction and sentiment classification score determination may be the most important challenges for aspect-based sentiment analysis that may not be handled in a single common solution. Based on the challenges stated here for aspect-based sentiment analysis, an effective tool should be designed for extracting the aspects and will be increasing performance of sentiment at the aspect level. The issues that can be addressed by the researchers (Nazir et al. 2022) are listed below:

- Data Pruning and Cleansing
- Cross-Domain Transfer Learning
- Contextual Semantic Relationship
- Aspect Summarization
- Predicting Dynamicity of Sentiment.

6 Research Objective

In aspect-based, sentiment analysis system is used to identify polarity based on aspects retrieved from the user tweets. The fine-grained polarity sentiments are obtained based on the following subtasks:

- Tweets Collection: It is the process of extracting real-time tweets for mobile phones of multiple companies from Twitter API.
- Tweets Preprocessing: The extracted raw tweets should be preprocessed for sentiment identification.
- Aspect Term Extraction: The preprocessed tweets are analyzed and their aspects are extracted. The extracted aspects are grouped into aspect categories. The aspect categories are used to identify sentiments based on supervised machine learning.
- Sentiment Analyzer: The aspect categories identified are further fine-grained for polarity prediction. The overall polarity prediction for each mobile phone company is identified.

7 Literature Survey

Many researchers have done tremendous work on aspect-based sentiment analysis and different approaches. Previous work in tweets preprocessing deals with different techniques for converting structured text from given raw tweets. The data collected through Twitter data generated generally unprocessed raw format. Twitter API was a useful web service that extracts real-time tweets from the Twitter database and this collection of tweets is saved into a CSV file. The computer acts as the basic source for the dataset. For achieving high precision and accuracy, a large set of data is saved. For accuracy to be increased for the given reviews which are planned for extracting in a later stage, the dataset needs to be filtered. The tweets which are obtained from gibberish tweets have been created by BOTs rather than humans, they need to filter this kind of tweets which may produce noise and irrelevance for the results. Sharma et al. [4] applied tweets preprocessing methodology by removing useless words and words like I, am, and are, me. Punctuations marks removal, URLS removal and stop words are also removed using the NLTK library which are available in Python. The stemmer process will remove all the similar words like play, playing, played, plays, etc., and a root word is processed.

The parts of speech assigns the POS tag to the terms that appear most frequently in the training corpus Ghabayen and Ahmed [5]. A context-based method will perform extraction of implicit aspect extraction for Chinese product reviews proposed by Sun. The implicit feature extraction done by the first term frequency matrix will show relationship between product aspects and opinion for given words. The presence of an implicit feature is done and possible implicit features available are extracted. The scores of possible implicit features are calculated by checking opinion words and the correct implicit feature is detected.

Ganganwar et al. [6] specified the raw information showing to have polarity which was highly redundant, inconsistency which was suspected. Data preprocessing of tweets may involve removal of URLs (e.g. www.xyz.com), hashtags values (e.g. #topic), Internet language words (@username), correction of spellings, removing of repeated words, replacing all the emoticons with words and assigning with their sentiment, removing all punctuations, different symbols, removing of numbers, removing stop words, expanding acronyms and removing non-English tweets.

Tokenization means the process that converting different words in a tweet into tokens. Tokens are basic building blocks for document which give help to understand text meanings that are derived from document and also derive relationship with one another. The automatic assigning of a tag to words into a phrase is referred to as POS tagging. It takes a statement and turns it into a list of terms with their associated tags. The pair (tag, word) is generated which cannot be completed at the lexical level as it requires consideration of sentence structure. For example, *"I like to read newspaper"* is represented as: [("I", "Preposition"), ("like", "Verb"), ("to", "To"), ("read", "Verb") and ("newspaper", "Noun")].

Supervised learning approaches are used to identify aspect terms for product and restaurant review dataset. The implicit aspects list, list of unique lemmas and with respective frequencies is generated first and then score for individual implicit aspects is computed. Therefore aspects are identified based on the scores greater than the threshold defined. Dosoula et al. [7] extended the previous proposed implicit feature algorithm. A classifier is used to predict the occurrence of multiple implicit features and is built using a score function. The score function is computed depending on the numbers of nouns, adjectives, commas and numbers of 'and' words. Using logistic regression, the function parameters are estimated, and a threshold is set for performance improvement. The feature detection part of the algorithm then checks for one or more implicit features using the prediction of the above classifier. This method allocates POS tags depending on the likelihood that a specific tag sequence will occur. The probabilistic ways to assign a POS tag are conditional random fields (CRFs) and hidden Markov models (HMMs) Kumawat and Jain [15]. Chatterji et al. [8] stated that there is a close association between explicit and implicit aspects. Hence, instead of considering explicit and implicit aspects separately, all these aspects are being correlated with network-like structure. The proposing framework with aspect FrameNet is used for sentiment analysis. The system learns the aspect pattern in review text and the patterns of the aspects and their sentiments for these aspects are aggregated. Yiran and Srivastava [16] proposed an unsupervised method and supervised techniques. The association rule mining is applied on term-document frequency for the data obtained from a corpus to find aspect categories.

Mohammed et al. [9] proposed a hybrid approach that combines WordNet-dictionary based and Corpus-based methods for extraction of implicit aspect terms. In the first phase, for all adjectives from corpus, a list of WordNet words related are extracted and for each aspect from training data, all the relative adjectives are listed. In the second phase using implicit aspect representation from phase 1, the Naïve Bayes classifier is trained for detecting implicit aspects.

Khattak et al. [12] identifies different aspects based on opinion words. This paper presented a method to recognize multiple implicit aspects for a given sentence. Separate models are created for the identification of implicit and explicit aspects. These models used manually labelled training datasets of 1000 restaurant reviews. The first model uses the maximum entropy classification technique for explicit aspect identification. The second model finds opinion words and their associated implicit aspects. To improve the accuracy in presence of multiple interrelated aspects, entities and their aspects are modelled as a hierarchy which is used as the training dataset.

Bansal et al. [14] proposed a framework and algorithm for the identification of the explicit and implicit aspects in the tourism domain by identifying explicit aspects using noun phrases. The supervised machine learning algorithm is used for identifying implicit aspects based on noun aspects. The issue with the proposed system is that sometimes the same word may have different POS recognition, which may not pick the correct aspect.

Mitra [3] based on train and test datasets, machine learning algorithms are useful for identifying sentiment polarities (e.g. negative, positive and neutral). These methods are classified as supervised learning, unsupervised learning, semi-supervised learning and reinforcement learning. In supervised technique was used for classifying for particular classes, and also when determining this set for problematic due as a lack of a labelled set of data, as well as in unsupervised approach can be useful. Reinforcement learning algorithms employ trial-based processes to assist the agent in interacting with the environment to maximize cumulating rewards.

For a finite number of input data to classify with preset classes, the space for training data was divided into hierarchically on a condition on the attribute value for this approach Charbuty and Abdulazeez [1]. The more words presence or absence is a condition on attribute values. This technique has a flowchart-like structure, with each core node denoting an attribute test, each branch denoting a test outcome, with nodes of leaf denoting children nodes or distributions of class. Once the decision tree classifier was simple to comprehend to analyze, it will also handle data with noise. They are over-fitting for unstable and prone on the other hand. Because the decision tree approach may perform well for large datasets, this cannot be recommended for small datasets.

The main idea behind the ensemble learning approach Athar et al. [13] method is to merge multiple different classifiers to create a classifier that outperforms them all. To make a better decision, this technique uses all of the commonly used classifiers. In general, group of rules, in which most Majority of Vote procedure is used to get a conclusion. Because the classifiers collaborate, an ensemble system has a better generalization and accuracy, but the fundamental disadvantage is that required high processing and also training time for one method. As a result, it's a good idea to choose algorithms carefully.

Aspect-based sentiment analysis Dang et al. [10], Poria et al. [2] was fine-grained sentiment analysis activity that attempts to predict the sentiment polarity values on specified aspects based on target phrases present in a given text (e.g. products or resources). Attributes, traits or qualities of the goal might all be considered aspects. There are two stages to aspect-based sentiment classification: aspect term extraction and category-based sentiment classification. In the first stage, it seeks to extract aspects terms based on noun phrases for the same object and group-based similar words present in a given text, while the second stage main aim is to determine the emotions of words in each aspect. For this degree of sentiment analysis, several methodologies have been suggested.

Rebeen et al. [11] use cross-validation and partitioning techniques for the data across using supervised machine learning algorithms for classifying tweets into their required classes. This approach was also used to collect real-time Twitter microblogging data on mentioned topics as iPod and iPhone based on different locations. They have processed and analyzed that data according to emotions such as anger, anticipation, fear, joy, surprise and also classified total tweets in required polarity as positive and negative.

Literature surveyed on the associated topic by using various techniques, prior works and observations drawn on three different parts mentioned are tweets preprocessing, aspect term extraction and also a sentiment analyzer. It explored different open issues that are focusing on, related to aspect-based sentiment analysis on mobile phone tweets.

8 Materials and Methods

An efficient aspect-based sentiment analysis system is proposed with a methodology that includes tweets preprocessing, aspect term extraction and sentiment analyzer. The experimental results for different sets of mobile phone tweets are presented and demonstrated through different approaches to produce high-quality performance.

9 Tweets Preprocessing

Twitter, the popular micro blogs has abundant rich information in the form of short text posted by users. Processing these colloquial and informal sentences require specific preprocessing techniques to understand and analyze the tweets to identify the sentiments. A tweet preprocessing technique is needed to convert the raw unstructured tweets into clean structured tweets. A methodology for preprocessing raw tweets are projected for the issues tackled from the literature survey and will be proceeded further with aspect categories sentiment classification and sentiment analyzer for polarity prediction. The proposed system architecture for tweets collection and tweets preprocessing are shown in Fig. 2 to collect the real-time tweets from Twitter API for mobile phone products for five top companies. To collect the real-time tweets, a Twitter API account is created using Developer Twitter APPs. Once the account is created, secret key and access keys are generated and given to Twitter users by APP. Twitter users can log into their account and the user is authenticated

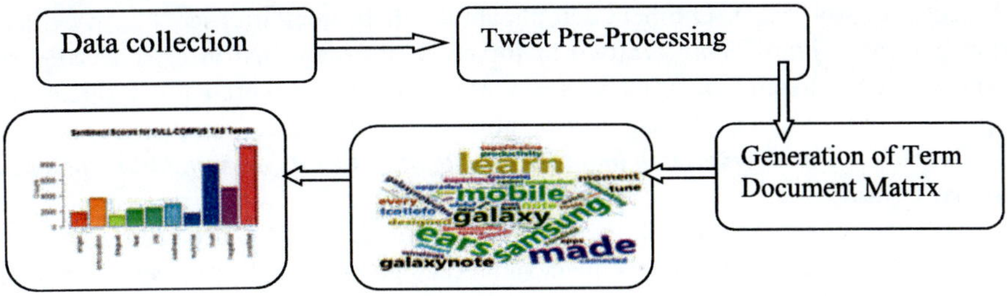

Fig. 2 Proposed system architecture for tweets collection and tweets preprocessing

using access keys and secret key. Twitter users can give the hash tag and time stamp to extract the raw tweets into a CSV file.

Tweets preprocessing takes raw mobile phone tweets as the input. Tweets are preprocessed for each tweet in a mobile phone category one after the other. The various steps for text preprocessing are removal of URL, lowercase conversion, removal of punctuations, removal of numbers and two-letter words, stop words elimination, POS tagging and generation of term frequency graph. The additional preprocessing steps are also included such as removal of unwanted words, replacing patterns and removal of whitespace characters. The preprocessed tweets are further processed to generate a terms document frequency matrix. Based on the term document frequency matrices, visualization of word clouds is formed. The sentiment score is based on a NRC dictionary for the presence of eight different emotions words present in each tweet. Finally, preprocessed tweets are grouped into positive and negative sentiments based on emotions sentiment score.

9.1 Aspect Categories Sentiment Classification

Aspect term extraction (ATE) methodology is used to identify opinionated aspects terms in textual data. Aspect term extraction includes sub-task in aspect-based opinion mining that focuses on extracting aspect terms from online user reviews. Aspect term extraction is applied to identify different phrases targeted by opinion indicators. Aspect terms are extracted and aspect terms are grouped into aspect categories. Aspect categories are further grouped into sentiments based on machine learning models. The various steps to be followed in extracting aspect category sentiments are processed using preprocessed tweets.

1. Analyzing given textual information and reviewing comments.
2. Analysis of aspect term and group different aspect category.
3. Analysis and sentiment classification.

Aspect-based sentiment analysis (ABSA) is used for the prediction of aspect categories present in textual data which are corresponding to overall sentiment polarities.

Aspect category-based sentiment classification is dependent on aspect term extraction present in data. There are two major processes identified in aspect category sentiment classification, i.e. first is aspect term extraction as a predefined aspect and second as performing sentiment classification based on aspect term present.

Example: "***I was tempted to buy this product as I like its design, but its price is not very good***".

In the above example tweet, the two aspects are identified which are design and price. Based on the above tweets the design is categorized as positive sentiment and price is categorized as negative sentiment.

Aspect term extraction identifies aspects from the given text and also expresses sentiment for individual aspects. Aspect categories (e.g. battery, camera and price) identify more common features other than aspect terms, and which are not needed to be aspect terms for a sentence.

Researchers identified different approaches to extract aspect terms and their value from the processed tweets. The different approaches found in the literature are discussed below:

1. Frequency-Based Approach: This approach involves finding frequently used noun phrases from a given review, tweet, blog, etc., and also identifies the type of aspects.
2. Relation-Based Approach: The main aim of this approach is to find relationships between aspects and identify opinion words as aspects. It also uses grammatically and syntax pattern relation between opinion and aspect for finding extraction rules.
3. Supervised Learning Approach: Supervised learning approach involves training data to infer a model for applying on unlabelled data. Therefore identifying aspects, opinion and their polarities can be labelling problems that are learned from labelled data and applied the same syntactic dependency and patterns on unlabelled data.
4. Topic-Based Approach: It is an unsupervised learning approach where different topics are discovered from text documents that contain many topics and finding the probability of distribution over different words present in a topic.

The issues to be addressed in the extraction of aspect terms are listed below:

- Identification of implicit and explicit aspect terms
- Handling of the ambiguous opinion words
- Removal of redundancy aspects
- Identification of multiple aspects in a sentence
- Identifying the rare words can contribute to aspects
- Long feature space problem.

The preprocessed tweets are considered as inputs for this module and generate sentiments based on aspect categories. The proposed architecture in aspects categories-based sentiment classification is shown in Fig. 3.

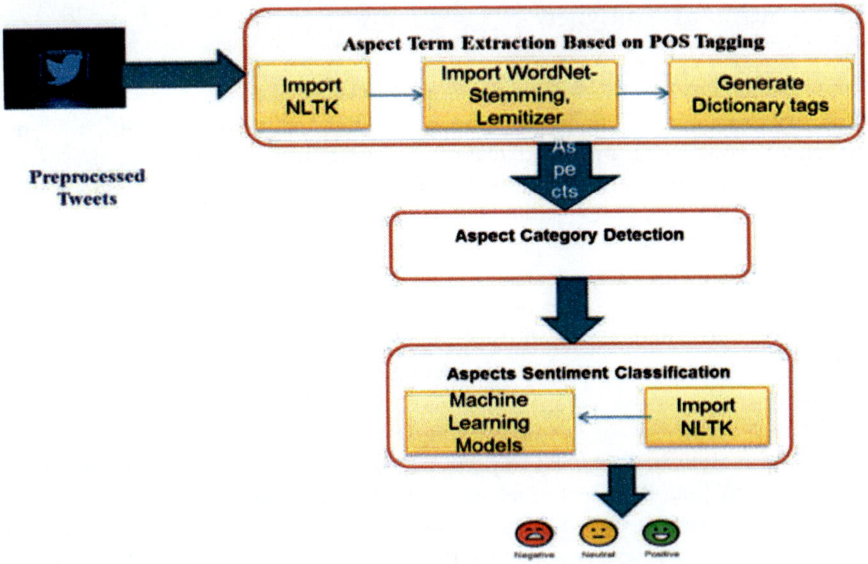

Fig. 3 Proposed architecture for aspect categories-based sentiment classification

The proposed architecture shown includes three modules such as aspect term extraction (ATE), aspect categories detection (ACD) and aspect sentiment classification (ASC). The proposed system takes preprocessed tweets generated in the previous module as input. The preprocessed tweets are processed individually, tweet by tweet and the aspects present in the tweets are identified. The aspects obtained from the tweets are large quantities, so they are grouped into an aspect category based on the similarity of features. Machine learning models are applied to classify the sentiments based on aspect categories.

Algorithm
IATE—SC: Implicit Aspect Term Extraction Towards Sentiment Classification (IATE-SC) Method

// Input: Processed tweets.

//Output: Aspects Categories tweets Importing NLTK library.

for each doc in tweets apply wordnet.

for each words in docs apply stemmer and lemmatize for each words in text generate POS tagging if docs = NLP(textdata).

aspects = [tokens.textdata for tokens in docs if tokens.pos = = "noun"] then return aspects.

Each aspects are grouped as Aspect category.

Aspect Category are classified into sentiments based on machine learning model.

end for end for end for.

10 Sentiment Analyzer

Sentiment analysis is used by various organizations to understand the sentiment of their customers through reviews, tweets and social media conversation. A simple model will automate the process of classifying tweets or reviews based on user sentiments. Sentiment classification helps business executives to save costs and promote business through accurate decisions. The polarity of the reviews helps business users to rate the product accurately and also helps new online consumers to filter the products based on user sentiments.

Sentiment analysis is considered as natural language processing technology that computes peoples' opinions as positive, negative and neutral within the unstructured text. The proposed architecture is for sentiment analyzer for prediction of overall polarity.

The preprocessed tweets are taken as input for sentiment analysis and predicting polarity for individual tweets. By importing the NLTK toolkit, the aspects are identified and overall polarity is identified by using a Vander sentiment intensity analyzer. The polarity score is predicted for given tweets and by summing all polarities we get compound values. These compound values will define the overall polarity present in the given tweets based on aspect terms.

The proposed architecture identifies the sentiments through VADER and Text Blob as a sentiment analyzer to identify overall polarity predictions of the given tweets.

VADER is capable of identifying sentiment for aspects that are based on valence-based lexicon and produces both intensity and polarity values as shown in Fig. 4. Different NLTK libraries are used with SentimentIntensity Library and a predefined method is used Vadersentimentintensity() which define polarities as positive, negative and neutral values based on the given range provided.

TextBlob is a Lexicon-based sentiment analyzer that can work on predefined rules or polarity scores. The polarity score is generated for the word with weights associated with the dictionary. So, only text blob is termed as both lexicon-based sentiment analyzer and rule-based sentiment analyzer.

11 Experimental Results and Discussion Tweets Preprocessing

Table 1 contains datasets of bench mark tweets and real-time tweets which are also used in the algorithms and also extracted raw tweets from Twitter API and is tested over tweets preprocessing on different mobile phone tweets, also bench mark tweets for comparison using a 10 TB Hard Disk, 8 GB RAM system and different metrics are evaluated from the tested results.

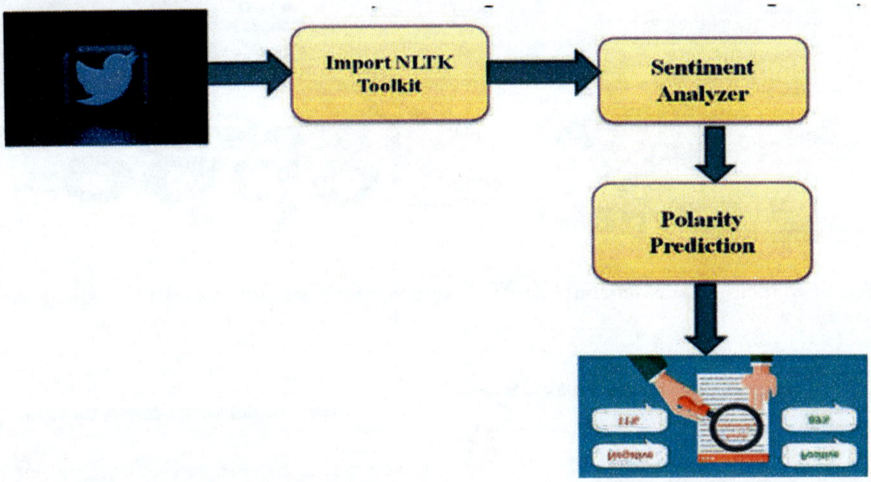

Fig. 4 Proposed architecture for polarity prediction

Table 1 Experimental set-up of considering for attributes

Input category	Mobile brands	Total tweets
Real-time tweets	SAMSUNG	1512
	OPPO	1500
	IPHONE	1510
	REDMIK 10	100
	ONE PLUS 3	650
Bench mark tweets	STC	1091
	TAS	7573
	FGD	4357
	ATC	1642

12 Results

Twitter preprocessing is a crucial step used to convert unstructured text to structured text. Twitter preprocessing is implemented for real-time and benchmark tweets. A sample of five real-time tweets are shown and tweet preprocessing is applied. The tweets are filtered by removing irrelevant punctuations such as hashtags, @, $ and !. The tweets are further processed to remove Internet language, alphanumeric characters, whitespace removal and expressions removal. After tweet preprocessing stage, words occurrence has been shown using word cloud2 in Fig. 5 and generated sentiment score using emotion in Fig. 6 for both real-time tweets and benchmark tweets.

Fig. 5 Word cloud2 based on term frequency—Samsung Mobile star pattern and OPPO—triangular pattern

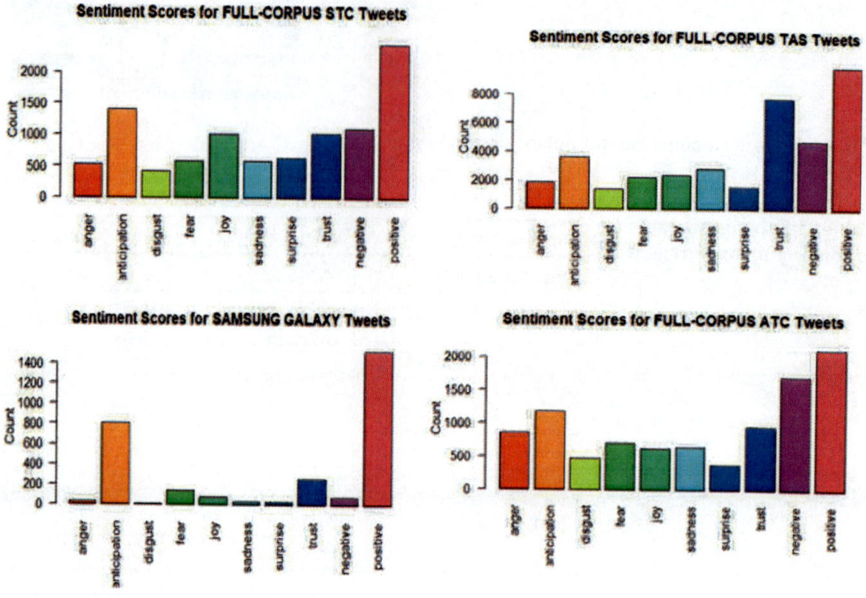

Fig. 6 Sentiment score graph for benchmark tweets

13 Performance Analysis

The eight basic emotions are anger, anticipation, fear, joy, disgust, sadness, surprise and trust. The emotions such as anger, anticipation, fear, disgust and sadness are grouped as negative emotions and joy, surprise and trust are grouped together as positive emotions. Figure 7 shows graph for sentiment score (SC) of positive and negative emotions for benchmark datasets.

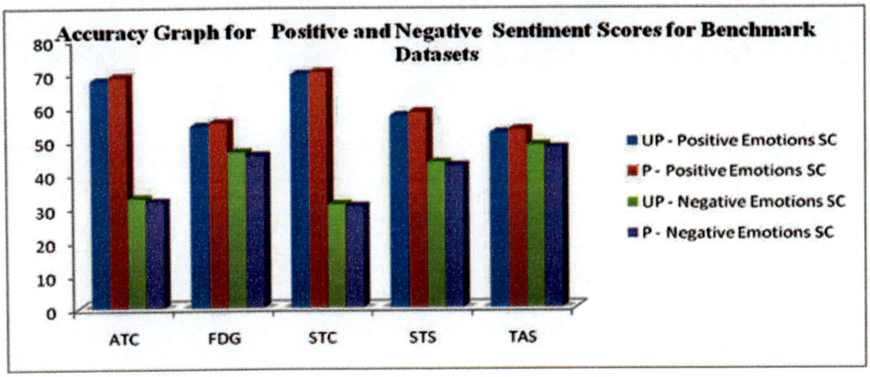

Fig. 7 Benchmark tweets accuracy graph of positive and negative emotions

14 Aspect Term Extraction

The input for aspect term extraction is preprocessed tweets and POS tagging method used for aspect term extraction is given in Table.2 and also sentiment score for individual aspect terms is given in Table.3. The aspect term generated for multiple mobile brands such as Samsung Galaxy, Oppo, IPhone and Oneplus are given below. Performance measures can be evaluated to assess the performance of supervised classifiers. Performance metrics used are recall (R), precision (P) and F1 scores which are calculated based on a confusion matrix.

Table.2. Showing aspect term in tweets

S. No.	Preprocessed tweets—IPhone	Aspect terms
1	Hey I am having some battery issues on my iP…	[Battery, issues]
2	Change atticus Iphone cameraS	[Iphone, cameraS]
3	How do i update my iphone plus ios is not supported	[Iphone, ios]
4	Iphone support	[Support]
5	Iphone pair bluetooth headphones guess marks philistine…	[Iphone, betooth, headphone]

Table 3 Sentiment score for aspect terms

Aspect	No. of positive mentions	No. of negative mentions	No. of neutral mentions
Battery	10,556	3018	1508
Camera	4524	0	15,080
Headphone	0	0	40,718
Hardware	0	3016	0
Processor	0	0	1508
System	1508	0	1508

15 Conclusion

Tweets preprocessing is the most important part of sentiment analysis. It is carried out on the raw tweets collected from Twitter API. The final preprocessed tweets will convert unstructured tweets to structured tweets. Visualization of word cloud depicts the term frequency of most important words in the tweets. Sentiment score analyzes the emotion score of real-time and benchmark tweets and also identifies the positive and negative sentiments. The preprocessed real-time tweets are taken as input and fed into the proposed system. The aspect terms are extracted based on noun phrases using the NLTK library. Then aspects are clustered into aspect categories. The generated aspect categories sentiments are identified as positive, negative and neutral for each mobile brand. Sentiment analyzer aims to identify sentiment about aspect terms present in given textual data and opinion words. Based on the polarity score, sentiments are identified as positive, negative and neutral. The overall polarity of the real-time tweets can also be identified.

The integration of automatic tweets data collection and tweets preprocessing is to get the structured tweets data. Based on Aspect Terms Extraction techniques and applying polarity Classification for given the sentiments of aspects are identified. The overall polarity prediction is done for mobile phone tweets using sentiment models. The system produces sentiment both at the emotional level and polarity levels for mobile phone tweets.

References

1. Charbuty B, Abdulazeez A (2021) Classification based on decision tree algorithm for machine learning. J Appl Sci Technol Trends 2(01):20–28
2. Poria S, Hazarika D, Majumder N, Mihalcea R (2020) Beneath the tip of the iceberg: current challenges and new directions in sentiment analysis research. IEEE Trans Affect Comput: 120–130
3. Mitra (2020) Sentiment analysis using machine learning approaches (lexicon based on movie review dataset). J Ubiquitous Comput Commun Technol 2(3):145–152
4. Sharma P, Agrawal A, Alai L, Garg A (2017) Challenges and techniques in preprocessing for twitter data. Int J Eng Sci Comput 7(4):147–159

5. Ghabayen AS, Ahmed BH (2020) Polarity analysis of customer reviews based on part-of-speech subcategory. J Intell Syst 29(1):1535–1544
6. Ganganwar V, Rajalakshmi R (2020) Implicit aspect extraction for sentiment analysis: a survey of recent approaches, 1877-0509 © 2020. Elsevier B.V. International conference on recent trends in advanced computing 2019, pp 105–115
7. Dosoula NJ, Senarath Y, Ranathunga S (2018) Aspect extraction from customer reviews using convolutional neural networks. In: Proceedings of 18th international conference on advanced ICT for emerging regions (ICTer), Sept 2018, pp 215–220
8. Chatterji S, Feng J, Sun X, Liu Y (2020) Co-training semi-supervised deep learning for sentiment classification of MOOC forum posts. Symmetry 12(1):1–24
9. Mohammed B, Hajar EH (2019) Using synonym and definition WordNet semantic relations for implicit aspect identification in sentiment analysis. In: Proceedings of 2nd international conference on network, information system secure, pp 1–5
10. Dang TV, Nguyen VD, Kiet NV, Ngan NLT (2021) A transformation method for aspect-based sentiment analysis. J Comput Sci Cybern 34(4):323–333
11. Hamad A, Abdullah T (2020) Recent trends and advances in deep learning-based sentiment analysis In: Deep learning-based approaches for sentiment analysis. Springer, pp 33–56
12. Khattak AM, Batool R, Satti FA, Hussain J, Khan WA, Khan AM, Hayat B (2020) Tweets classification and sentiment analysis for personalized tweets recommendation. 3(8):45–55
13. Athar A, Butt WH, Anwar MW, Latif M, Azam F (2021) Exploring the ensemble of classifiers for sentimental analysis. Proceedings of the 9th international conference on machine learning and computing—ICMLC 2021, pp 410–414
14. Bansal B, Srivastava S (2019) Hybrid attribute based sentiment classification of online reviews for consumer intelligence. Appl Intell 49(1):137–149
15. Kumawat D, Jain V (2019) POS tagging approaches: a comparison. Int J Comput Appl 118(6):32–38
16. Yiran Y, Srivastava S (2019) Aspect-based sentiment analysis on mobile phone reviews with LDA. In: Proceedings of 4th international conference on machine learning technology, pp 101–105

Microbial Metabolites and Recent Advancement

Prakash Garia, Kundan Kumar Chaubey, Harish Rawat, Aashna Sinha, Shweta Sharma, Urvashi Goyal, and Amit Mittal

Abstract Microbes have attracted a lot of attention due to their potential in the development of bioprocess technologies for the limitless manufacture of foods and supplements, which could help to meet the ever-increasing demands of the world's population. Additionally, since microorganism-based technologies do not constitute a sizable source of pollution, they are preferred options for resolving large environmental problems caused by conventional chemical treatments. Enzyme inhibitors are used to treat infectious and deadly diseases all over the world. These inhibitors are produced by antibiotics, anticancer medications, immune suppressants, alkaloids, and microorganisms. Additionally, they greatly influence human life expectancy and the reduction of mortality. To meet the demands of a population that is constantly growing, microbial enzymes (Enz) provide immense potential for a variety of industries, including medicine, cosmetics (CSM), food, feed, beverages, detergents, leather processing, and paper and pulp; use of Enz has been developing steadily. Microbial-derived growth regulators and pesticides have demonstrated considerable potential for the development of sustainable agriculture, and as a result, a greener world. It

P. Garia (✉)
School of Management, Graphic Era Hill University, Bhimtal 263139, Uttarakhand, India

K. K. Chaubey · A. Sinha
School of Applied and Life Sciences, Uttaranchal University, Dehradun 248007, Uttarakhand, India
e-mail: kundan2006chaubey@gmail.com

H. Rawat
Department of Botany, Dhanauri P. G. College Dhanauri, Haridwar 247667, Uttarakhand, India

K. K. Chaubey
School of Applied and Life Sciences, Sanskriti University, Mathura 281401, Uttar Pradesh, India

S. Sharma
College of Biotechnology, DUVASU, Mathura, Uttara Pradesh, India

U. Goyal
Sanjay Gandhi Postgraduate Institute of Medical Sciences, Lucknow 226014, Uttar Pradesh, India

A. Mittal
Department of Allied Science, Graphic Era Hill University, Bhimtal 263139, Uttarakhand, India

S. Kumar et al. (eds.), *Fourth Congress on Intelligent Systems*, Lecture Notes in Networks and Systems 868, https://doi.org/10.1007/978-981-99-9037-5_14

should be noted and understood microbes are a source of a range of essential micronutrients and natural compounds. In brief, microbes produce two types of metabolites, primary and secondary whose uses are discussed later. This review discusses and explains various metabolites produced by microbes and their use in daily life.

Keywords Microorganisms · Metabolites · CSM · Fermentation · Wastewater treatment · Antibiotics · Antimicrobial advancement

1 Introduction

Microorganisms have a profound impact on almost every element of human, plant, and animal existence and are a potential source of a wide variety of natural substances. Natural compounds derived from bacteria have proven useful in food production, agriculture, and medicine. Examples of primary metabolites that are utilised in the biotransformation of raw materials to create industrial commodities as well as dietary supplements include amino acids, Enz, vitamins, organic acids, and alcohol when compared to secondary metabolites, which are organic compounds mostly derived from tissues or plants. They are mostly used in the biopharmaceutical industry because they can decrease infectious diseases in people and animals, hence increasing life expectancy. Additionally, the development of sustainable agriculture is unavoidably greatly influenced by microbes and their by products [1]. The vast majority of microorganisms have adapted to many dominating physicochemical circumstances, as evidenced by the omnipresent microbial variety on earth [2]. There are microbes in a variety of natural settings, such as soil, harsh conditions, seas, glaciers, and ponds. The richness of the chemicals found in communities of bacteria, actinomycetes, archaea, lichens, and fungus is reflected in their variety. A total of 30 phyla make up the microbial population, the bulk of which are non-cultivable and found in various habitats [3]. The microorganisms can also adapt to and endure circumstances that are different from those in other habitats, and they can be mass special bioactive substances that are not present in other creatures. Business Communication Company (BCC) predicts that the market for microorganisms and microbial products will expand from \$186.3 billion in 2018 to \$302.4 billion in 2023, at a compound annual growth rate (CAGR) of 10.2% from 2018 to 2023 [4]. Due to technological and financial advantages, new technologies for the manufacture of microbial products are replacing synthetic manufacturing methods. These include medicinal goods, organic acids, agriculturally significant metabolites, Enz, flavouring compounds, and nutritional supplements including vitamins and amino acids [1]. Antimicrobials that are used in production of livestock are also manufactured from some microbes; and this helps in maintaining productivity and health. Drug-resistant infections in cattle and people are a concern to human health as a result of this culture. It was estimated that 63,151 (1560) tonnes of antibiotics would be used in the production of food animals worldwide in 2010, and that number is forecast to rise by 67% to 105,596 (3605) tonnes by 2030. An increasing number of animals are being raised, which accounts for the majority (66–67%) of antimicrobial consumption. By 2030, more animals will likely be raised in intensive farming systems; hence the remaining third (34%) of the increase can be attributable to modifications in agricultural practices. Researchers hypothesise that by 2030, solely in Asia, there will be an upsurge in consumption of antimicrobials. With the help of their present industrial position, we are clearly exhibiting in this communication, the functions of microbial metabolites in CSM, wastewater treatment, fermentation, medicine, and nutrition. Although it is believed widely that microorganisms are the causative agents for various diseases, they are highly useful in medicinal, research and industrial purposes. This review explains and encourages the use of various metabolites primary and secondary in day-to-day life. Research and development has made the most use of the microbial metabolites and now they are used in textile, leather, beverages, paper, and cosmetic industries. Putting metabolites from microbes to positive use also reduces environmental stress yielding sustainable results.

2 Microbial Metabolites

A crucial part of microbial metabolism is played by metabolites, which are chemical molecules with a thin molecular structure (1000 Da) (chemical conversion) [5]. Recent studies have emphasised that the chemicals that microbes produce, rather than the microbes themselves, cause potential and specific health effects [6]. Numerous microbial produced chemicals have been found to have important effects on human health and disease, with some, like butyric acid, offering protection against cancer and inflammation and others, like indoxyl sulphate, promoting inflammation and having negative effects on the kidneys, heart, and brain [7]. The total collection of metabolites that a specific bacterium produces is referred to as its "metabolome" [8]. The term "metabolome" was first put forth in 1997 [9] and can be divided into three different matrices: the interior of the cell, the extracellular medium, and the culture head space (Fig. 1). The metabolome of an organism can reveal information about its biological function since it is almost certainly related to its phenotype [10]. Metabolites can be consumed, produced, degraded, or ejected because of their infinitely long half-lives [11]. On the other hand, metabolites are not just produced during metabolic processes; they are also changed into other compounds. The ratio of metabolite synthesis to conversion rates determines a microbial cell's concentration of each metabolite [5]. Bacterial growth involves the constant synthesis of primary metabolites. Important cellular functions including growth, development, and reproduction depend on them; therefore, they are necessary for survival. Auxotrophy will occur if microorganisms are unable to synthesise these essential metabolites [12], and if the metabolite is not exogenously provided, the organism will perish. All phyla and kingdoms share the same primary metabolites [13], because of this, most microbial species create primary metabolites that are identical to one another [14]. They frequently only exist in trace levels inside the cell because of their high turnover rates. Because of this, intermediary metabolites of the primary [1] metabolism are abundantly produced by microorganisms and then excreted into the extracellular medium. This leads to a general decrease in the ratio between the concentrations of 1° and secondary [2] metabolites inside the cell. But this is not always the case [5]. The precise control over the synthesis of a given metabolite is provided by nutrient availability, growth conditions, feedback regulation, enzyme stimulation or inactivation, and other factors. Certain RNA, elements, and transmitted gene products produced during post-exponential development, as well as some low molecular mass substances, commonly influence their control. Recent research has revealed that chromosomal DNA clustered genes, as opposed to plasmidic DNA, mostly code for the manufacture of particular metabolites. A fresh theoretical perspective for enzymology academics, regulation, and differentiation is presented by the fact that the linked pathways are yet not completely understood [15]. Using these three principles, all microbial metabolisms can be divided into three classes (Table 1).

In actual life, these expressions are commonly combined. Here are a few typical instances:

- **Chemolithoautotrophs:** acquire carbon from the fixing of carbon dioxide and energy from the oxidation of inorganic substances. Sulphur-oxidising microorganisms Examples include nitrifying bacteria, iron-oxidising bacteria, and knell gas bacteria (Fig. 2).
- **Photolithoautotrophs**: Using reducing equivalents from inorganic molecules, fix carbon from carbon dioxide and obtain energy from light. Examples include Cyanobacteria [which uses water (H_2O) as a hydrogen donor], Chromatiaceae Chlorobiaceae [which uses hydrogen sulphide (H_2S) as a donor], and Chloroflexus [which uses hydrogen (H_2) as a donor].

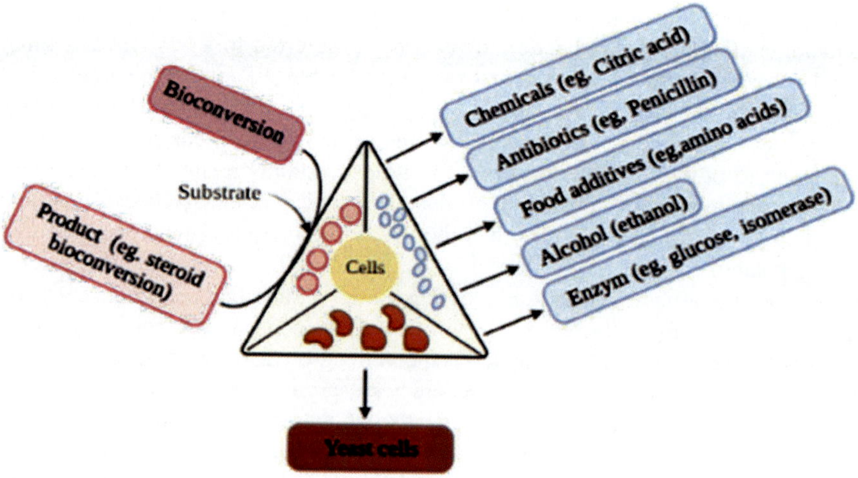

Fig. 1 Microbial metabolites

Table 1 Production of energy from different sources by microorganisms

From carbon			By electron reduction		From light and chemicals	
Auxotrophic	Heterotrophic	Mixotrophic	Lithotrophic	Organotrophic	Phototrophic	Chemotrophic
Carbon dioxide produced from carbon [CO_2]	Organic molecules are a source of carbon	Organic molecules and fixing carbon dioxide are two ways to get carbon	Inorganic compounds are used	Organic compounds are used	Energy can be produced from light	External chemical molecules are used for energy production

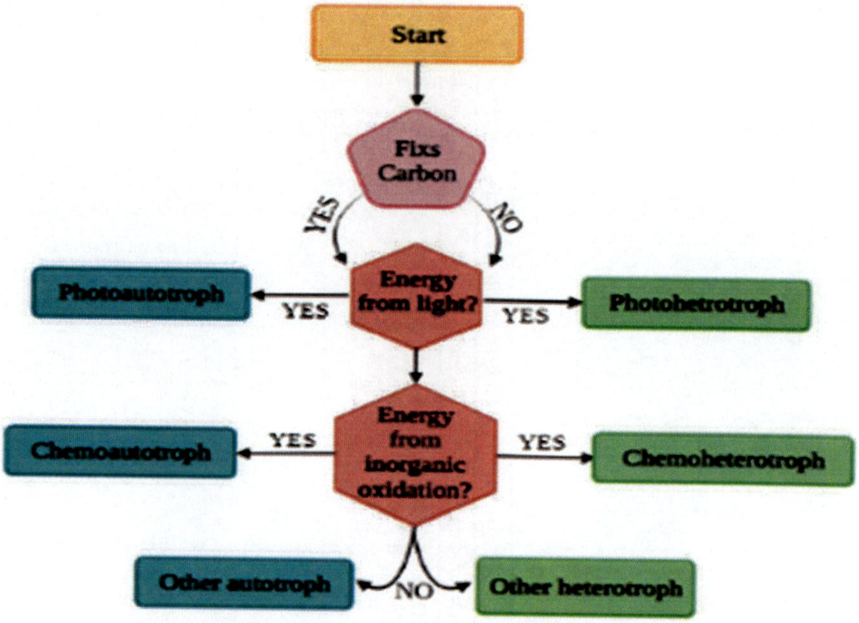

Fig. 2 Flowchart for identifying the metabolic traits of microbes

- **Chemolithoheterotrophs**: produce energy by oxidising inorganic materials but are unable to repair carbon dioxide (CO_2). Sulphate-reducing bacteria, Knall gas, certain Nitrobacter, Thiobacilus, Beggiatoa, and Wolinella (using H_2 as a reducing equivalent donor) are a few examples.
- **Chemoorganoheterotrophs:** the conversion of organic molecules into energy, carbon, and hydrogen for biosynthetic processes. Examples include most bacteria, including Escherichia coli, many Bacillus species, and Actinomycetota.
- **Photoorganoheterotrophs:** obtain energy for organic compound biosynthesis reactions from light, carbon, and reducing equivalents. While many organisms can fix carbon dioxide and are mixotrophic, some are wholly heterotrophic As an alternative to photolithoautotrophy using hydrogen, examples include Chloroflexus, Rhodopseudomonas, Rhodospirillum, Rhodomicrobium, Heliobacterium, Rhodomiccyclus, and Rhodobacter.

2.1 Primary Microbial Metabolites

The microbial synthesis of primary metabolites has a considerable impact on the level of quality of life we currently experience. Microorganisms are thought to develop properly when essential metabolites like amino acids, nucleotides, and by products of fermentation like ethanol and organic acids are present. The simplicity with which microbial synthesis has emerged as the preferred and most efficient approach for producing amino acids because it is possible to create affordable, environmentally acceptable, and enantiomerically pure amino acids [16]. When microorganisms reach the end of or are just about to enter the stationary phase of growth, secondary metabolites, which are organic compounds unrelated to growth, development, and reproduction, begin to form [17]. Primary metabolites are currently produced in the business using strains created through significant mutagenesis followed by screening, or selection/screening for overproducers. Such initiatives frequently begin with organisms that have some ability to synthesise the desired product, but high productivity cannot be achieved until many mutations lead to deregulation in a specific biochemical pathway. Mutants that are auxotrophic can be highly helpful. Sequential mutations make sure that nutrients are efficiently directed to the right products with little to no diversion to alternative pathways. These mutations affect the generation of pathway precursors and intermediates as well as the release of feedback controls. Surprisingly, this strain improvement technique has been successful in producing organisms that generate large amounts of primary metabolites for industrial use. Recombinant DNA technology is being used in more current methods to create strains that can produce too many main metabolite pathway precursors and intermediates [18]. Microbial synthesis is emerging as the most effective and preferred approach for synthesising enantiomerically pure amino acids at low cost and with ecological acceptability [19].

2.2 *Secondary Microbial Metabolites*

Secondary metabolites are organic molecules with multiple possibilities that are found in bacteria, fungus, or plants. These metabolites, which contain several antibiotics, anticancer agents, and therapeutic chemicals, are essential for biotechnological and biomedical advancements. In traditional biotechnology, terrestrial plants were seen as a source of secondary metabolites [20]. Additionally, reports of secondary metabolites that act as pesticides, herbicides, and growth promoters for plants have been made. A few metabolites, including mithramycin, adriamycin, bleomycin, and daunomycin, have been employed as anticancer agents [1]. Also used as anaesthetics, anti-inflammatory drugs, blood thinners, hemolytics, anabolics, hypocholesterolemics, and vasodilators are secondary metabolites [21]. Endophytic microorganisms are important candidates of secondary metabolites. Mefteh et al. showed that, in contrast to healthy plants, plants under biotic pressure offered new and distinctive endophytes with a variety of bioactivities. According to Shrama et al., the endophytic fungus Colletotrichum gloeosporioides produced more cryptic and bioactive metabolites with anti-oxidant and antibacterial potential when dietary ingredients like grape skin and turmeric extract were used. In addition, it was discovered that the endophytic fungus Chaetomium globosum, which was isolated from Egyptian medicinal herbs, had anti-rheumatic action [22]. There have been varieties of approaches investigated for the efficient and excessive synthesis of primary metabolites, where modifications in genetics and physiology have been important. These techniques include the taking metabolites out of the culture continuously, over-expression of related coEnz, deletion of genes related to metabolite deterioration, and over-excited of genes engaged in metabolite synthesis [23]. Recent researches have shown that secondary metabolites have potential role in the area of cosmetics, food, and chemical synthesis. One of the major known such metabolites is kojic acid which is produced largely at industrial level by *Aspergillus flavus*. Kojic acid is also generally called as a tyrosinase inhibitor, that plays a major role in skin whitening [24].

3 Microorganism Growth

3.1 *Lag Phase*

The number of cells does not grow when microorganisms reach a new habitat till the metabolic system has adapted to the environment. It is the early period in bacterial life cycle, when the cells are accustomed to the new surroundings before entering exponential phase. Antibiotics and unfriendly temperatures are only two examples of externally harmful factors for the cells. Anabolism is underway in cells, and the amount of RNA rises. Germination, which includes activation, budding, and growth, is required for spores®. The variation in the activity of individual cells is often connected to the microbial population during the lag period. A number of mathematical models, including individual-based modelling, particular instantaneous simulation, and one-step kinetic analysis, allow us to more extensively study the behaviour of individual cells and help us better understand cell response and growth mechanisms in novel circumstances. Because of their high activity, the metabolites produced during the lag period can easily start the synthesis of numerous Enz (Fig. 3).

3.2 *Logarithmic (Log) Phase*

In this stage, microorganisms display traits including the fastest growth possible, balanced cell development, active enzyme systems, and active metabolisms. With the onset of log phase, microbial growth starts, and there is a supply and demand for nutrition and energy. To meet this chain, extracellular Enz are produced and given in vivo to break down substrates into more readily digestible and active molecules like monosaccharides, active molecules, and polypeptides (polypeps) (Fig. 3). These essential metabolites impart a pivot role in the development, growth, and reproduction of the generating organism [25].

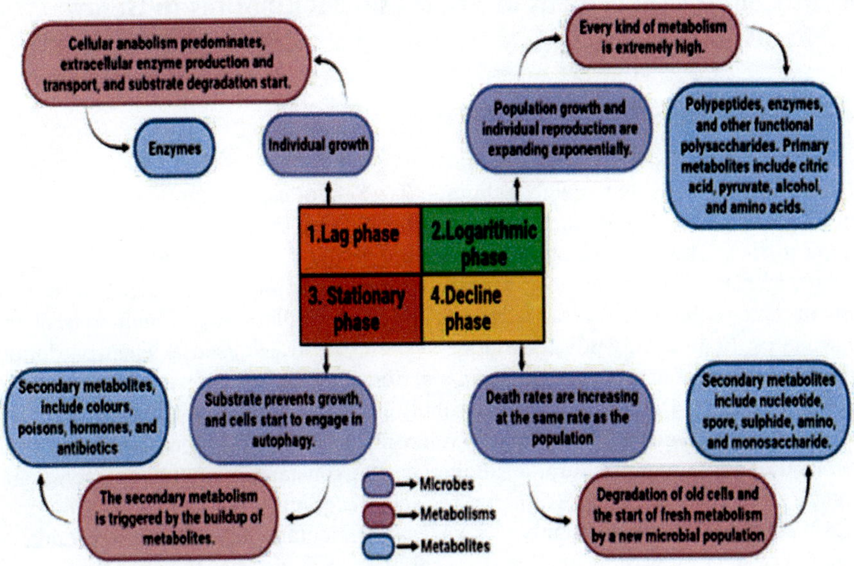

Fig. 3 Effect on microbial metabolites at different growing phases

3.3 *Stationary Phase*

When cells reach the stationary phase also known as the plateau stage, they start to build fat, transfection granules, and glycogen (Fig. 3). The number of freshly replicating cells matches the quantity of dying cells at this stage, and the cell growth rate is zero. Bacilli often start producing spores in the stationary phase. The production is greatly influenced by the stationary phase's regular growth pattern [26].

3.4 *Decline Phase*

The entire population of microorganisms is in a condition of negative growth because the mortality rate of each individual exceeds the division rate. Consequently, the population starts to dwindle (Fig. 3). Cell morphology starts to alter; for instance, expansion takes place or atypical degenerative forms are generated. Some cells start to lyse themselves. The generation that can adjust to the new environment progressively gains dominance and restarts metabolism as the decline period comes to an end [26].

4 Recent Advancement of Microbial Metabolites in Below Field

4.1 Wastewater Treatment (WWT)

The traditional activated sludge technique (CAS) produces a significant quantity of surplus sludge as a byproduct when microorganisms break down contaminants to support growth and reproduction, because it contains a lot of volatile solids (VS), a lot of water, and dangerous substances including heavy metals, pathogens, and persistent organic pollutants, excess sludge has become a challenging problem to treat and dispose of. In an usual activated sludge system, substrates used as electron donors are fully oxidised and broken down in a series of biochemical processes known as catabolism (Fig. 4). The substrate distribution and energy conversion are essential for cell formation during the process of microbial metabolism. More importantly, the removal and treatment of surplus sludge needs a substantial amount of energy and chemical agents, which raises the wastewater treatment process's carbon footprint and resource usage significantly. As an environmentally friendly and sustainable method of treating wastewater, the disposal of extra sludge has thus become a challenge [19, 27]. Core microbial communities in effective activated sludge WWTPs were discovered through molecular investigations; their absence may indicate a problem with the wastewater treatment system. Notwithstanding the presence of the microbial genera *Zoogloea, Dechloromonas, Prosthecobacter, Caldilinea, and Tricoccus*in all WWTPs, it was observed that geographic location affected the species make-up of the biomass [28]. The impact of geography on species composition is influenced by the fact that wastewater treatment temperatures vary based on location [29].

4.2 Fermentation

The ancient industrial technology of fermentation has been used extensively ever since human history began. Early fermentation systems relied on conventional strain breeding, which included extensive rounds of random mutagenesis, screening, and/or selection, to generate productive strains [18]. Since 4000 B.C., vinegar has been made microbiologically. The ideal bacteria to use for vinegar fermentation are *Gluconacetobacter* and *Acetobacter* species [30]. A remedy of ethanol is attacked to produce an acetic acid solution, which contains 12–17% acetic acid after 90–98% of the ethanol has been consumed. Acetic acid concentrations have been measured at 53 g/L in genetically modified *E. coli* [31], and 83 g/L in a mutant of the bacteria *Clostridium thermoaceticum* [32] and 97 g/L by an engineered strain of *Acetobacter aceti* ssp. Xylinium [33]. The overproduction of primary metabolic products has more recently been addressed using molecular genetics techniques. Modern molecular biology technologies have made it possible to enhance strains using methods that are more logical.

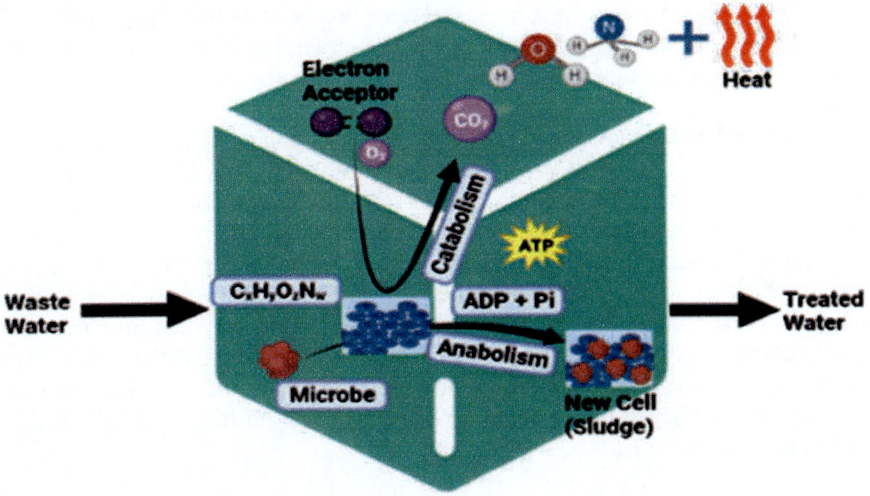

Fig. 4 Diagram of microorganism metabolism in wastewater treatment system

In order to find novel and significant target genes and quantify metabolic activities necessary for further strain development, techniques of transcriptome, proteome, and metabolome analysis, as well as metabolic flux analysis, have recently been developed [18]. The laboratory used the incredibly effective culture flask ventilation and air fibre filtration sterilising technique after the discovery of penicillin and the development of large-scale production. The growth of the antibiotic business has not only made it possible to use microbial technology in the pharmaceutical sector, but it has also encouraged the growth of the industrial aerobic microbial fermentation sector. Biosynthetic metabolisms are now the focus of microbial engineering rather than decomposing metabolisms.

4.3 *Cosmetics*

Traditional sources of new compounds continue to be biologically derived substances from plants and numerous other organisms. Microbes are one of the least expensive, most sustainable, most innovative producers of any chemicals among the many sources of life that exist. Despite the diversity of microorganisms present in nature, relatively few of them are used in the CSM business on a commercial scale. In light of this, the enormous yet untapped biodiversity represents a potential possibility for future biotechnological and aesthetic applications [2]. FA, Enz, peps, vitamins, lipopolysaccharides, and pigments with advantageous qualities for aesthetic procedures are abundant in microbes [34] (Fig. 5). Additionally, special compounds with significant uses in the beauty business, such as ceramides, mycosporine-like amino acids, carotenoids, and omega-3, 6 and 9 FA, are produced by bacteria [35] (Fig. 4). The CSM industry has been driven to investigate microbiological sources due to rising customer demand for biological components and cosmetic goods [33]. Cosmetics companies are doing R&D to develop new novel active compounds for cosmetics goods and explore the biodiversity of those chemical compounds. These compounds are used in various cosmetic products and are used either for beautification or in health industry to preserve diversity, tap market potential, and increase competitive advantage. Many biological molecules have contributed significantly to the development of a wide range of chemicals, including esters, fragrance compounds, and active agents, which are widely employed in the CSM industry. The primary advantage of using microbial components is their biocompatibility, but they can provide additional benefits, such as a streamlined production process, enhanced and consistent product quality, and a less environmental impact. Numerous bacteria and microorganisms secrete a large number of physiologically active substances with significant commercial worth [33].

- The cyclic oligosaccharides known as cyclodextrins, which contain a [1,4] linked glucopyranose moiety bonded together in a ring, play a notable role in the formulation of CSM [36]. Most commonly, cyclodextrin is used to lessen the volatility of esters in fragrances and gel room fresheners. They are also often utilised in detergents to provide a consistent and long-lasting release of scent [37].
- Due to its several functions as a detergent, foaming agent, emulsifying agent, and skin hydrator, biosurfactants are utilised in the creation of many cosmetic

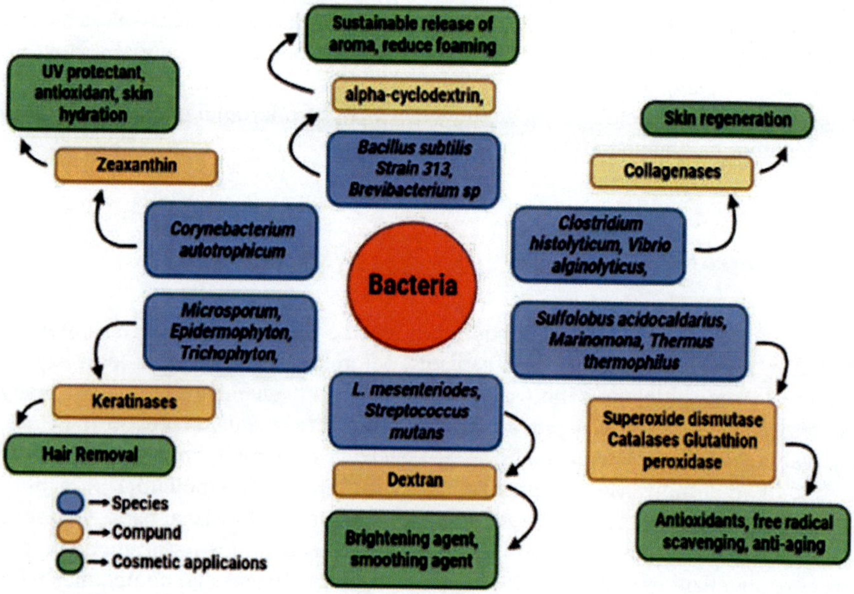

Fig. 5 Bacterial application in CSM

products. Biosurfactants are also biodegradable and relatively non-toxic. In nature, bacteria create the majority of the biosurfactants, followed by fungus and other microorganisms. The majority of biosurfactants are neutral lipids, glycolipids, FA, and lipopeps. Additionally, the US EPA has authorised the use of biosurfactants in food goods, CSM, and medicines as safe [38].

- Dextran is utilised as a skin-brightening and smoothing agent because it improves skin radiance, smoothness, and appearance of wrinkles. Dextrans also has anti-inflammatory properties since it enhances blood flow and increases the production of nitric oxide [NO] in human epidermal keratinocytes cells [33].
- Together, the Enz superoxide dismutase (SOD), catalase, glutathione peroxidase, and lactoperoxidase serve as an exfoliant. When administered to the skin's surface, these Enz act as free radical scavengers and shield the skin from ultraviolet radiation [39]. Another comparable enzyme is lactate dehydrogenase (LDH), which may catalyse the reduction of pyruvate and NADH to produce lactate and NAD + as the end product. When exposed to UV, the aforementioned response is lessened, but when LDH is present, the subunits in the cells are still there and enable regular cell activity [39].
- A glycosaminoglycan (GAG) called hyaluronic acid (HA) is made up of the amino acids -4-glucuronic acid (GlcUA) and -3-Nacetylglucosamine (GlcNAc) [25]. In cosmetic surgery, HA is frequently utilised as a dermal filler. Additionally, because it encourages moisture retention, reduces wrinkles, and boosts skin firmness and elasticity, sodium hyaluronate is a frequently used active ingredient in skin lotions and serums [33].
- Lactic acid is often used in cosmetic skin creams to keep skin moisturised and give it a smooth, supple appearance. It has a high concentration (up to 12%) and is used in skin peeling cream as an exfoliating agent, for skin whitening, and to lessen the occurrence of acne [40].
- Ceramides are employed as skin moisturisers in cosmetic products. Because ceramides are abundant in the stratum corneum of the human epidermis. Only ceramides are present in eukaryotic cells and are animal-derived (e.g., cows). However, the search for other sources of ceramides has been prompted by worries about infectious illnesses. Additionally, plant-derived ceramides have distinct structural characteristics. This restricts its application in CSM because it is derived from animal ceramides. Ceramides have been found in a variety of fungus species [41].

The improvement in microbial strains that are used in the production of cosmetic products can be done by engineering microbe's system biology, and this in turn can raise the productivity of cosmetics [42].

4.4 Antibiotics and Antimicrobial

An unknown fungus's production of penicillin served as a signpost, the start of the era of antibiotics in 1929. The 1940s saw the heyday of penicillin *notatum* manufacture commercially [43]. Later, a variety of antibiotics derived exclusively from fungi and actinomycetes were found in an effort to find more potent pharmacological effects and to fight off novel diseases. A number of antibiotics were discovered as a result of ongoing study in this field, including cephalosporin, macrolids, tetracycline, chloramphenicol, ansamacrolids, aminoglycosides, peptide inhibitors, antifungal antibiotics, anthracyclins, and glycopeps. Some microbial species have been found to be capable of manufacturing a variety of antibiotics. According to published data, between 2000 and 2010, the amount of antibiotics used by humans globally rose by 36%. 10 Brazil, Russia, India, China, and South Africa (BRICS) countries were responsible for three-quarters of the increase although making up just 40% of the global population as a whole. Among these nations, India accounted for 23% of the retail sales volume despite lax enforcement of laws governing the over-the-counter selling of antibiotics. The National List of Essential Medicines [NLEM] has been recommended by the WHO, many national commissions, and reports from India as a vital instrument to promote health equity [40]. Antimicrobials are frequently utilised in food animals to promote growth and prevent illness. An estimated 80% of the country's yearly antimicrobial use in the USA is attributed to the use of antibiotics in food animals [41], a sizeable portion of which contains antimicrobials crucial to human medicine in the treatment of typical infections as well as essential to carry out treatments like major surgeries, organ transplants, and chemotherapy [44]. Countries like the BRICS have shifted toward vertically integrated, highly efficient intensive livestock production systems to meet this need. Rising earnings in transitional countries are actually fuelling an increase in antimicrobial consumption. In the meanwhile, BRICS nations have discovered multiresistant ARBs in food animals [45, 46], especially in the underdeveloped world, where the utilisation of antibiotics for the purpose of promoting growth is still mainly uncontrolled [47]. Since they are among the most commonly prescribed medications in healthcare and can save lives, antibiotics should only be used sparingly when treating infections. However, the greatest significant contributor to antibiotic resistance globally is improper and inappropriate administration of medicines without a valid prescription. Stopping the rise of drug resistance is possible by immunisation, hand washing, clean food preparation, and the use of antibiotics on a proper prescription only when necessary [48]. Recent researches have revealed that, when probiotics are active in the intestine, substances called postbiotics are produced from the composition and metabolism of these bacteria. Postbiotics are a class of drugs with a wide range of therapeutic uses, particularly antibacterial properties, that are produced by numerous methods. Because they do not encourage the growth of antibiotic resistance and do not contain ingredients that could make it worse. These compounds have been shown to have antimicrobial effects and thus are highly beneficial in field of medical [49].

4.5 Marine Carbon Cycle

The activities of marine phytoplankton, bacteria, grazers, and viruses result in the production of a variety of short-lived saltwater metabolites that quickly cycle through one-fourth of the carbon on earth that is obtained through photosynthesis [50]. Half of all biological carbon fixation on earth occurs in marine habitats. However, photosynthetically fixed carbon must be transported to and deposited in deep ocean waters and sediments for long-term carbon sequestration. This work is aided by the biological carbon pump (BCP), which moves particulate organic carbon (POC) from the ocean's surface to its interior and influences the climate throughout geological time periods [51, 52]. Another biological process for sequestering carbon is the microbial carbon pump (MCP) [53]. Refractory dissolved organic carbon (RDOC) is created when microorganisms transform labile dissolved organic carbon (LDOC) into a more stable form, which may survive in the ocean for a long time—possibly millennia—without further biological deterioration [54, 55]. More than 35 years ago, the BCP idea was first developed [56], and since then, study on this topic has been quite busy [6, 52]. Despite the fact that MCP is a relatively recent idea, its study is gaining popularity and pace [57]. Both MCP and BCP have made major research efforts, but neither fully understand their separate quantitative contributions to climate modulation or the environmental and biological factors that may have an impact on these contributions and dynamics [58].

The basis for BCP and MCP is provided by dissolved organic matter [DOM], mostly of which is produced by 1° producers in the ocean surface. Marine primary production, which is subject to top-down and bottom-up regulations[57, 59], is thus a key factor influencing BCP and MCP. The nature, biomass, productivity, and portioning of the organic matter produced by the 1 producers are affected by zooplankton grazing and virus lysis [53, 60]. Through the integration of chemical composition assessments with microbiological investigations, novel complete insights on the relationships and mechanisms governing the microbial metabolism in the ocean would be possible [61] (Fig. 6). The composition, abundance, and productivity of the marine photosynthetic microbial communities are significantly influenced by the availability and chemical speciation of nutrients, and different phytoplankton may have different potentials for exporting carbon [62, 63].

4.6 Microbial Metabolites in Development of Gut Plethora

Metabolites derived from microbes bring about the formation of microbiome communities and are responsible for structuring interactions of hosts with diverse microbes. The advances in computational and biotechnological strategies have now enabled researchers to explore the role of microbes in growth and development of the immune, nervous and endocrine systems. The healthy gut microbiome thus has a preeminent effect on both early and for persistent health in the whole lifespan of

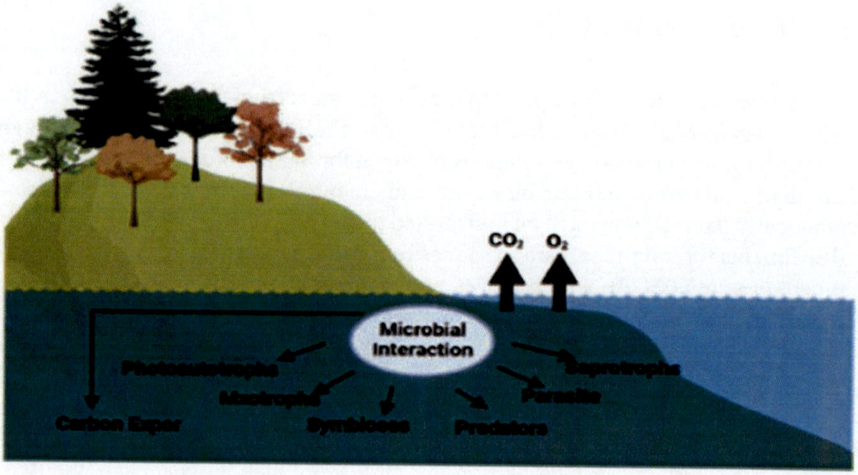

Fig. 6 Interaction of microbial

humans [64, 65]. Early nutrition transitioning from breastfeeding to solid foods, along with an increase in the variety of the microbiome and metabolome, causes alterations in the metabolism of the gut bacteria. High quantities of lactate and acetate as well as aromatic lactic acids (such as phenyllactic acid, 4-hydroxyphenyllactic acid, and indolelactic acid) are produced during breastfeeding because human milk oligosaccharide (HMO) degrading Bifidobacterium species dominate the newborn gut [65]. Refs. [66–70] are also important related to different applications. As the child's diet progresses, the complexity of the diet rises and more fibre and indigestible proteins end up in the colon. Short-chain fatty acids (SCFA, such as acetate, propionate, and butyrate) and gases (such as hydrogen and methane) are consequently produced as a result of alterations in intestinal fermentation. Additionally, proteins are broken down into amino acids, which are then fermented by the gut's resident microbes into branched SCFAs (such as isobutyrate, isovalerate, and 2-methylbutyrate), amines (such as histamine, dopamine, tyramine, -aminobutyric acid (GABA), and tryptamine) [71].

5 Conclusion

All life forms depend on microorganisms as their primary supply of nutrients, and they are effectively used in healthcare, agriculture, and nutrition. In addition to being utilised as feed additives, these small creatures are used in the preparation 2*17 of many different cuisines. Worldwide, microorganisms create antibiotics, anticancer medications, immune suppressants, alkaloids, and enzyme inhibitors that are used to treat infectious and deadly diseases. Microbes through metabolism that begin with cross-talk between microbial cell wall and external environment facilitate global biogeochemical cycles. The essentials required for growth for all heterotrophs and many autotrophs exist in the dilute and heterogeneous mixture of compounds leading to the formation of dissolved organic matter [DOM]. In a nutshell, the relation between microbe and molecule is one of the vital reactions for carbon cycle globally. Additionally, they have a considerable impact on human life expectancy and death rates. Many industries, including the pharmaceutical, food, feed, beverage, detergent, leather processing, paper and pulp, and so on, have considerable potential for using microbial Enz; their use has been steadily expanding to satisfy the demands of the world's population, which is expanding at an alarming rate.

References

1. Singh R, Kumar M, Mittal A, Mehta PK (2017) Microbial metabolites in nutrition, healthcare and agriculture. 3 Biotech 7(1):15
2. Oren A (2002) Diversity of halophilic microorganisms: environments, phylogeny, physiology, and applications. J Ind Microbiol Biotechnol 28(1):56–63
3. Kijjoa A, Sawangwong P (2004) Drugs and cosmetics from the sea. Mar Drugs 2(2):73–82
4. Rathinam NK, Bibra M, Rajan M, Salem D, Sani RK (2019) Short term atmospheric pressure cold plasma treatment: a novel strategy for enhancing the substrate utilization in a thermophile, Geobacillus sp. strain WSUCF1. Bioresour Technol 278:477–480
5. Pinu FR, Villas-Boas SG, Aggio R (2017) Analysis of Intracellular Metabolites from Microorganisms: Quenching and Extraction Protocols. Metabolites 7(4):53
6. Wishart DS, Oler E, Peters H, Guo A, Girod S, Han S, Saha S, Lui VW, LeVatte M, Gautam V, Kaddurah-Daouk R (2023) MiMeDB: the Human Microbial Metabolome Database. Nucleic Acids Res 51(D1):D611–D620
7. Scharlau D, Borowicki A, Habermann N, Hofmann T, Klenow S, Miene C, Munjal U, Stein K, Glei M (2009) Mechanisms of primary cancer prevention by butyrate and other products formed during gut flora-mediated fermentation of dietary fibre. Mutation Research/Reviews in Mutation Research 682(1):39–53
8. Vanholder R, Schepers E, Pletinck A, Nagler EV, Glorieux G (2014) The uremic toxicity of indoxyl sulfate and p-cresyl sulfate: a systematic review. J Am Soc Nephrol 25(9):1897
9. Adesso S et al. (2018) AST-120 reduces neuroinflammation induced by indoxyl sulfate in glial cells. J Clin Med 7(10):365
10. Aldridge BB, Rhee KY (2014) Microbial metabolomics: innovation, application, insight. Current Opinion Microbiology. 19(1):90–96
11. Beale DJ, Karpe AV, Ahmed W (2016) Beyond metabolomics: a review of multi-omics-based approaches. Microb Metabolomics: Appl clin Environ Ind Microbiol 289–312
12. Macintyre L et al. (2014) Metabolomic tools for secondary metabolite discovery from marine microbial symbionts. Mar Drugs 12(6):3416–3448
13. Villas-Bôas SG (2007) Part Ii discovery of new metabolic pathways in saccharomyces. Anal An Introduction, 191–202
14. Pande S, Kost C (2007) Bacterial Unculturability and the Formation of Intercellular Metabolic Networks. Trends Microbiology 25(5):349–361
15. Karlovsky P (2008) Secondary metabolites in soil ecology. In: Secondary metabolites in soil ecology. Springer, Berlin, Heidelberg, pp. 1–19
16. Vaidyanathan S (2005) Profiling microbial metabolomes: what do we stand to gain? Metabolomics 1(1):17–28

17. Jeong Y et al. (2020) Current status and future strategies to increase secondary metabolite production from cyanobacteria. Microorganisms 8(12):1849
18. Sun X et al. (2015) Synthesis of chemicals by metabolic engineering of microbes. Chem Soc Rev 44(11):3760–3785
19. Demain AL (1999) Pharmaceutically active secondary metabolites of microorganisms. Appl Microbiol Biotechnol 52:455–463
20. Choi KR, Lee SY (2023) Systems metabolic engineering of microorganisms for food and cosmetics production. Nature Reviews Bioengineering 1–26 (2023)
21. Demain AL, Fang A (2000) The natural functions of secondary metabolites. History mod biotechnol I:1–39
22. Tilman D, Balzer C, Hill J, Befort BL (2011) Global food demand and the sustainable intensification of agriculture. Proc Natl Acad Sci 108(50):20260–20264
23. US Food and Drug Administration (2021) FDA releases annual summary report on antimicrobials sold or distributed in 2020 for use in food-producing animals
24. Laxminarayan R et al. (2013) Antibiotic resistance—the need for global solutions. Lancet Infect Dis 13(12):1057–1098
25. Silva NCC, Guimarães FF, Manzi MP, Budri PE, Gómez-Sanz E, Benito D, Langoni H, Rall VLM, Torres C (2013) Molecular characterization and clonal diversity of methicillin-susceptible Staphylococcus aureus in milk of cows with mastitis in Brazil. J Dairy Sci 96(11):6856–6862
26. Zhu YG, Johnson TA, Su JQ, Qiao M, Guo GX, Stedtfeld RD, Hashsham SA, Tiedje JM (2013) Diverse and abundant antibiotic resistance genes in Chinese swine farms. Proc Natl Acad Sci 110(9):3435–3440
27. Maron DF, Smith TJS, Nachman KE (2013) Restrictions on antimicrobial use in food animal production: An international regulatory and economic survey. Global Health 9(1):1–11
28. Sanchez S, Demain AL (2009) Microbial primary metabolites: biosynthesis and perspectives. Encycl Ind Biotechnol: Bioprocess, Bioseparation, and Cell Technol, 1–16
29. Singh R, Kumar M, Mittal A, Mehta PK (2017) Microbial metabolites in nutrition, healthcare and agriculture. 3 Biotech 7:1–14
30. Reddy S, Sinha A, Osborne WJ (2021) Microbial secondary metabolites: recent developments and technological challenges. Volatiles and Metabolites of Microbes, 1–22
31. Bentley R (1997) Microbial secondary metabolites play important roles in medicine; prospects for discovery of new drugs. Perspect Biol Med 40(3):364–394
32. Abdel-Azeem AM, Zaki SM, Khalil WF, Makhlouf NA, Farghaly LM (2016) Anti-rheumatoid activity of secondary metabolites produced by endophytic Chaetomium globosum. Front Microbiol 7:1477
33. Tamano K (2014) Enhancing microbial metabolite and enzyme production: current strategies and challenges. Front Microbiol 5:718
34. Sharma S, Singh S, Sarma SJ (2023) Challenges and advancements in bioprocess intensification of fungal secondary metabolite: kojic acid. World J Microbiol Biotechnol 39(6):140
35. Sze JH, Brownlie JC, Love CA (2016) Biotechnological production of hyaluronic acid: a mini review. 3 Biotech 6:1–9
36. Feng R, Chen L, Chen K (2018) Fermentation trip: amazing microbes, amazing metabolisms. Annals of Microbiol 68:717–729
37. Guo JS, Fang F, Yan P, Chen YP (2020) Sludge reduction based on microbial metabolism for sustainable wastewater treatment. Biores Technol 297:122506
38. Cydzik-Kwiatkowska A, Zielińska M (2016) Bacterial communities in full-scale wastewater treatment systems. World J Microbiol Biotechnol 32:1–8
39. Zhang T, Shao MF, Ye L (2012) 454 Pyrosequencing reveals bacterial diversity of activated sludge from 14 sewage treatment plants. ISME J 6(6):1137–1147
40. Wang X, Hu M, Xia Y, Wen X, Ding K (2012) Pyrosequencing analysis of bacterial diversity in 14 wastewater treatment systems in China. Appl Environ Microbiol 78(19):7042–7047
41. Deppenmeier U, Hoffmeister M, Prust C (2002) Biochemistry and biotechnological applications of Gluconobacter strains. Appl Microbiol Biotechnol 60:233–242

42. Ansorge-Schumacher MB, Thum O (2013) Immobilised lipases in the cosmetics industry. Chem Soc Rev 42(15):6475–6490
43. Parekh SR, Cheryan M (1994) High concentrations of acetate with a mutant strain of *C. thermoaceticum*. Biotechnol lett 16:139–142
44. Beppu T (1993) Genetic organization of Acetobacter for acetic acid fermentation. Antonie Van Leeuwenhoek 64:121–135
45. Hellawell JM (Ed.) (2012) Biological indicators of freshwater pollution and environmental management. Springer Science & Business Media
46. Brunt EG, Burgess JG (2018) The promise of marine molecules as cosmetic active ingredients. Int J Cosmet Sci 40(1):1–15
47. Gupta PL, Rajput M, Oza T, Trivedi U, Sanghvi G (2019) Eminence of microbial products in cosmetic industry. Nat Prod Bioprospecting 9:267–278
48. Del Valle EM (2004) Cyclodextrins and their uses: a review. Process Biochem 39(9):1033–1046
49. Buschmann HJ, Schollmeyer E (2002) Applications of cyclodextrins in cosmetic products: A review. J Cosmet Sci 53(3):185–192
50. Nitschke M, Costa SGVAO (2007) Biosurfactants in food industry. Trends in Food Sci Technol 18(5):252–259
51. Falkowski PG, Raven JA (2013) Aquatic photosynthesis. Princeton University Press
52. Le Moigne FA (2019) Pathways of organic carbon downward transport by the oceanic biological carbon pump. Front Mar Sci 6:634
53. Jiao N, Herndl, GJ, Hansell DA, Benner R, Kattner G, Wilhelm SW, Kirchman DL, Weinbauer MG, Luo T, Chen F, Azam F (2010) Microbial production of recalcitrant dissolved organic matter: long-term carbon storage in the global ocean. Nat Rev Microbiol 8(8):593–599
54. Ogawa H, Amagai Y, Koike I, Kaiser K, Benner R (2001) Production of refractory dissolved organic matter by bacteria. Sci 292(5518):917–920
55. Jiao N, Robinson C, Azam F, Thomas H, Baltar F, Dang H, Hardman-Mountford NJ, Johnson M, Kirchman DL, Koch BP, Legendre L (2014) Mechanisms of microbial carbon sequestration in the ocean–future research directions. Biogeosciences 11(19):5285–5306
56. Volk T, Hoffert MI (1985) Ocean carbon pumps: analysis of relative strengths and efficiencies in ocean-driven atmospheric CO_2 changes. Carbon Cycle Atmos CO_2: Nat Var Archean to present 32:99–110
57. Zhang C et al. (2018) Evolving paradigms in biological carbon cycling in the ocean. Natl Sci Rev 5(4):481–499
58. Dang H (2020) Grand challenges in microbe-driven marine carbon cycling research. Front Microbiol 11:1039
59. Siegel DA, Buesseler KO, Behrenfeld MJ, Benitez-Nelson CR, Boss E, Brzezinski MA, Burd A, Carlson CA, D'Asaro EA, Doney SC, Perry MJ (2016) Prediction of the export and fate of global ocean net primary production: The EXPORTS science plan. Front Mar Sci 3:22
60. Sime-Ngando T (2014) Environmental bacteriophages: viruses of microbes in aquatic ecosystems. Front Microbiol 5:355
61. Kujawinski EB (2011) The impact of microbial metabolism on marine dissolved organic matter. Ann Rev Mar Sci 3:567–599
62. Herndl GJ, Reinthaler T (2013) Microbial control of the dark end of the biological pump. Nat Geosci 6(9):718–724
63. Richardson TL (2019) Mechanisms and pathways of small-phytoplankton export from the surface ocean. Ann Rev Mar Sci 11:57–74
64. Roager HM, Stanton C, Hall LJ (2023) Microbial metabolites as modulators of the infant gut microbiome and host-microbial interactions in early life. Gut Microbes 15(1), 2192151(2023)
65. Ratsika A, Codagnone MC, O'Mahony S, Stanton C, Cryan JF (2021) Priming for life: early life nutrition and the microbiota-gut-brain axis. Nutrients 13(2):423
66. Singh A, Kumar N, Joshi BP, Singh BK (2018) Load frequency control with time delay in restructured environment. J Intell Fuzzy Syst 35:4945–4951
67. Singh A, Kumar N, Joshi BP, Vaisla KS (2018) AGC using adaptive optimal control approach in restructured power system. J Intell Fuzzy Syst 35:4953–4962

68. Singh BK, Kumar N, Singh A, Joshi BP (2018) BBBC based frequency controller for hybrid power system. J Intell Fuzzy Syst 35:5063–5070
69. Joshi BP, Pandey M, Kumar S (2016) Use of intuitionistic fuzzy time series in forecasting enrollments to an academic institution. Adv Intell Syst Comput 436
70. Joshi BP, Kharayat PS (2016) Moderator intuitionistic fuzzy sets and application in medical diagnosis. Adv Intell Syst Comput 380
71. Milani C et al. (2017) The first microbial colonizers of the human gut: composition, activities, and health implications of the infant gut microbiota. Microbiol Mol Biol Rev 81(4):10–128
72. Lods, Dres, Scholz, Brooks (2000) The future of enzymes in cosmetics. Int J Cosmet Sci 22(2):85–94
73. Hyde KD, Bahkali AH, Moslem MA (2010) Fungi—an unusual source for cosmetics. Fungal diversity 43:1–9
74. Gao JM, Zhang AL, Chen H, Liu JK (2004) Molecular species of ceramides from the ascomycete truffle Tuber indicum. Chem Phys Lipid 131(2):205–213
75. Moran MA, Kujawinski EB, Schroer WF, Amin SA, Bates NR, Bertrand EM, Braakman R, Brown CT, Covert MW, Doney SC, Dyhrman ST (2022) Microbial metabolites in the marine carbon cycle. Nat Microbiol 7(4):508–523
76. Ozma MA, Moaddab SR, Hosseini H, Khodadadi E, Ghotaslou R, Asgharzadeh M, Abbasi A, Kamounah FS, Aghebati Maleki L, Ganbarov K, Samadi Kafil H (2023) A critical review of novel antibiotic resistance prevention approaches with a focus on postbiotics. Crit Rev Food Sci Nutr 1–19

Design of a 3D-Printed Accessible and Affordable Robotic Arm and a User-Friendly Graphical User Interface

Daniel Bell and Emanuele Lindo Secco

Abstract This paper presents the design, manufacturing and software integration of a 3D-printed robotics arm. The robot design is made of a rotational base combined with a five degrees of freedom arm with a gripper. An intuitive and user-friendly Graphical User Interface is also implemented and integrated with the robotic device. Preliminary tests are performed showing that the system has the potential to be used in different contexts and applications, as well as manufacturing environments where pick and place tasks could be performed in a low-cost fashion.

Keywords 3D-printed prototyping · 3D-printed robotics · Human–robot interface · Intuitive interface · User-friendly human–robot interaction

1 Introduction

Since the 1960s, there has been concern surrounding automation and robots' effect on the manufacturing industry. One of the biggest worries is the impact on workers' jobs. Automation is the *substitution of work activities undertaken by human labor with work done by machines, with the aim of increasing quality and quantity of output at a reduced unit cost* [1]. As technology continues to improve, automation will increase. In the upcoming five years, officials in Zhejiang, an eastern province of China known for its manufacturing industry, are planning to allocate approximately 500 billion Yuan (i.e., $ 82 billion) to facilitate the transition of 5,000 companies annually from human labor to automated systems [2].

Based on findings from the International Federation of Robotics, it has been noted that several nations, including Brazil, Republic of Korea, Germany, China,

D. Bell · E. L. Secco (✉)
Robotics Lab, School of Mathematics, Computer Science and Engineering, Liverpool Hope University, Liverpool, UK
e-mail: seccoe@hope.ac.uk

D. Bell
e-mail: 20210890@hope.ac.uk

S. Kumar et al. (eds.), *Fourth Congress on Intelligent Systems*, Lecture Notes in Networks and Systems 868, https://doi.org/10.1007/978-981-99-9037-5_15

and the USA, have witnessed a growth in paid employment. In contrast, Japan has encountered a decrease in employment [3]. The rise in automation not only leads to job displacement but also supports new employment opportunities. The global robot industry alone generates around 170,000–190,000 jobs. Additionally, there is a need for support staff and operators, which contributes to a similar number of job opportunities. Apart from direct employment, robotics also plays a role in job creation and preservation in some other scenarios. For example, there are tasks that can only be effectively performed by robots, ensuring precise and consistent production while maintaining cost-efficiency.

Also, robotics can be employed in situations where the current working conditions are unsatisfactory or even illegal in developed countries, providing an alternative means of operation [4]. Opponents of the automation boom argue that the rapid technological progress occurring is eradicating jobs that were traditionally held by the middle class. While elevator operators and highway toll collectors have already been replaced by robots and automated systems, the impact of automation extends beyond low-skilled positions. It is encroaching upon higher-skilled job roles as well, raising concerns about the potential for substantial job losses among human workers in the long run [5]. According to the Organization for Economic Co-operation and Development in 2019, it was projected that around 14% of existing jobs could be at risk of disappearing due to automation. Additionally, automation was predicted to have a significant impact on approximately 35% of jobs [6].

Automation, particularly using robotics, often involves the replacement of human labor with automated systems, resulting in the displacement of workers from their place of work. Therefore, the effective effect of automation versus employability is under discussion and may have an effective impact on jobs [7].

The literature shows that there is debate on whether robots' effect on the manufacturing industry is positive or negative. Workers are concerned about being replaced by robots; however, there is evidence that shows the increase in automation can support jobs.

In this context, we want to provide the design of a robotic system which could support human activity and manufacturing, according to the following motivations:

- To make robotics technology more accessible and affordable for individuals, hobbyists, and small businesses.
- To enable the development of new applications and uses for robot arms, such as in education, research, and industry.
- To encourage innovation and experimentation in the field of robotics, by providing a low-cost platform for designers and engineers to build upon.
- To reduce the cost and complexity of existing robot arm designs, making them more practical and efficient.
- To help advance the field of robotics by providing a platform for researchers and developers to test and refine their ideas and algorithms.

We also aimed at taking inspiration from the biological system and, in particular, from the human arm, an articulated structure performing a wide range of movements through the shoulder, elbow, and wrist joints [8].

2 Materials and Methods

This section presents the *Design of the Robotic Arm* (Sect. 1), the *3D Printing Manufacturing Process* (Sect. 2), and the implementation of the *Software Interface* (Sect. 3).

2.1 *Robotics Arm Design*

The proposed design of the robotic arm consists of an articulated system with a set of rotational joints between the links of the robot. Precisely, the project aims at designing a *six actuated degrees of freedom robotic arm* with a *base*, an anthropomorphic *shoulder*, an *elbow*, a *wrist*—performing twist and rotation—and an opening and closing *gripper*. Some parts of this design are taken from *Giovanni Lerda's Kauda*, a printable and open-source robotic arm [9]: in particular, the base joint is taken from this source as it allows full 360-degree rotation, whereas the remaining design of the arm has been developed with the *Autodesk Fusion 360* software. This is a 3D CAD and CAM software widely used in a variety of manufacturing industries.

To design each component, a sketch that defines the basic outline and shape of the part can be initially prepared. Then, extrude tool can be applied to turn the 2D sketch into a 3D part. After this, features are added such as holes and fillets to finalize the part and refine it. Once the part is finished, it can exported as an STL file (i.e., *STereolithography*) for manufacturing. This process is shown for one component in Fig. 1 (left panel) where a final rendering of the section is also shown in the same Fig. 1 (right panel). Due to the nature of the 3D printing process, once the part is exported, the STL file has to be manipulated with a slicing software. This further step is shown in Fig. 2 (left panel). The slicing software, which allows the preparation of the instruction for the 3D printer (i.e., the G-Code), is *Cura Slicer*. This slicer allows the user to set up a variety of parameters such as the height of the layer of materials, the infill density, as well as to optimize the speed of the printing.

2.2 *Arm Manufacturing*

In order to manufacture the arm, the parts were 3D printed. The *Prusa MK 3D Printer* works by building up layers of material to produce a three-dimensional object, as shown in Fig. 3 (left panel). The 3D printer uses *PolyLactic Acid* (PLA) filament to build the object. PLA is an extensively used material with ecological benefit since it is a thermoplastic obtained from renewable sources [10].

Once the 3D printing of the object is completed, the part may require a further sanding, polishing, or painting phase, in order to improve the quality of its surface.

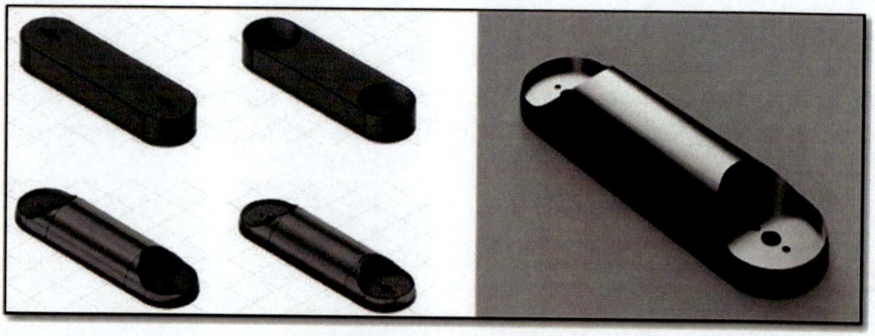

Fig. 1 Design and rendering of the parts with the Autodesk Fusion 360 software (left and right panels, respectively)

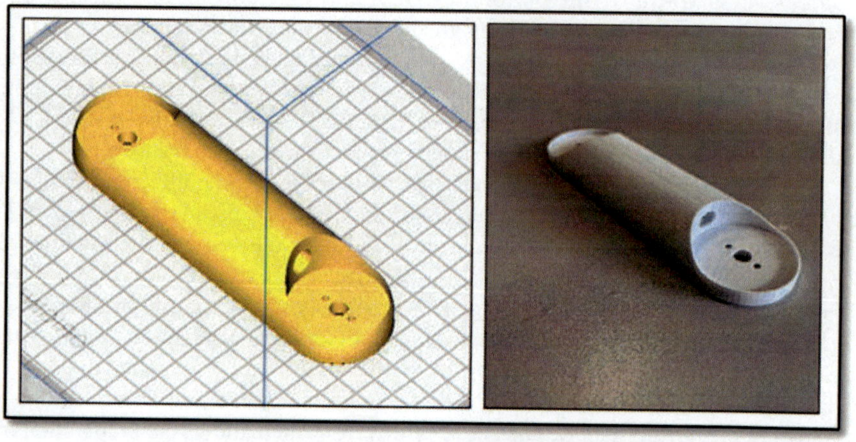

Fig. 2 Preparation with *Cura Slicer Software* and manufacturing of the part (left and right panels, respectively)

The result of this process is, for example, the arm section shown in Fig. 2 (right panel) and Fig. 3 (right panel).

Once all the parts are printed, the arm can then be assembled: the first step requires to assemble the motors with the parts. Then ball bearings are added to the base joint as shown in Fig. 4. After this step, the limbs are assembled as well.

Finally, in order to control the robot and the joint actuators, a *low-cost Arduino board* is selected and connected to a breadboard interface between the robot controller and the actuators. The electronics are wired up and connected to the Arduino, as shown in Fig. 5. Then the breadboard is connected to six different motors, namely one *gripper actuator* and other *five motors*, performing the movement of the *base* and of the robotic arm, namely the *shoulder*, the *elbow*, the *wrist's twist,* and the *wrist's rotation.*

Fig. 3 Manufacturing of the robotics arm parts with the *Prusa MK 3D Printer* and PLA material

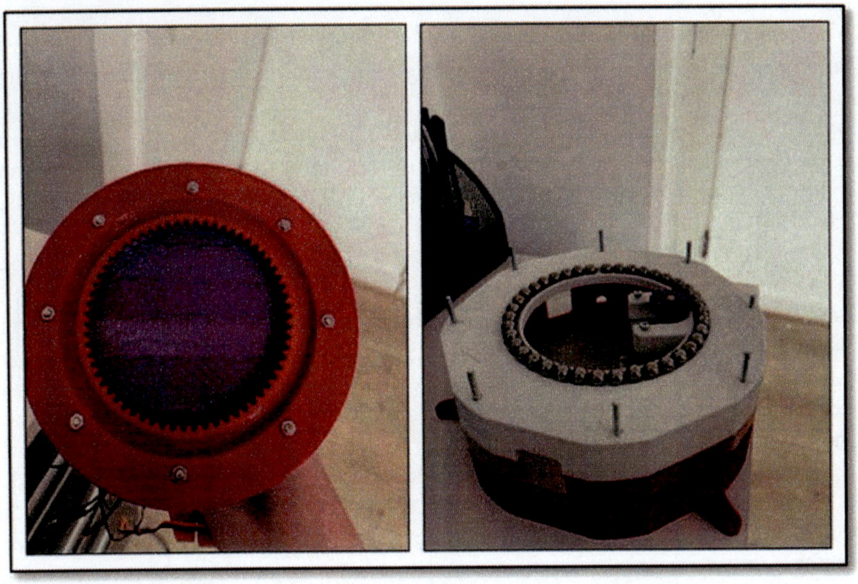

Fig. 4 Manufacturing of the *base joint* and integration with the ball bearings

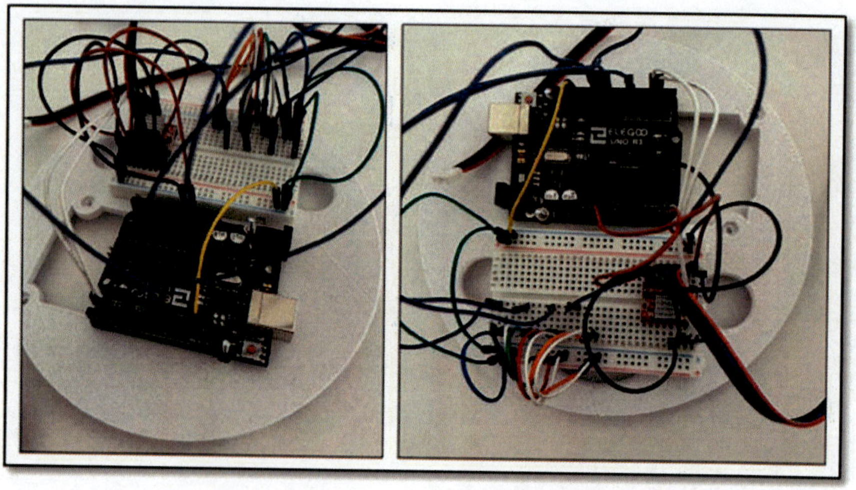

Fig. 5 Integration of the Arduino board and of the breadboard within the base

2.3 *Software Implementation*

To control and move the robotic arm, a software interface has been designed in *Java Programming Language* with the *Processing* programming environment. Processing is a software environment which allows the design of *Integrated Development Environment* (IDE) for animations, and interactive applications. This IDE has been adopted because it is supported by several libraries and functions that simplify the process of designing a *Graphical User Interface* (GUI) with elements such as buttons, sliders, and drop-down menus.

Serial communication is a straightforward method of transferring data. The *Processing.serial.* library* allows the Processing IDE to send and receive information to and from microcontrollers, sensors, and other devices. The library supports several data formats, such as the communication of numerical values and strings.

According to these features, a GUI has been designed as it is shown in Fig. 6: this interface is characterized by six controlling sliders for the six different motors of the robotic arm. For each motor, there is a slider which allows the user to set the angular displacement of each motor. The angle is also returned to the user and displayed. For the stepper motor, there are also two buttons which control these motors in both directions. The color of these buttons goes to grey when the button is pressed in order to inform the end-user.

The Arduino board is connected to 6 actuators, namely a *set of servomotors performing the arm movement* and a *stepper motor* controlling the rotation of the robotic base. The stepper motor requires the integration of a controller between the Arduino board and the robot itself. When a button is pressed on the GUI, an integer value is passed into the board by means of the communication protocol. The Arduino

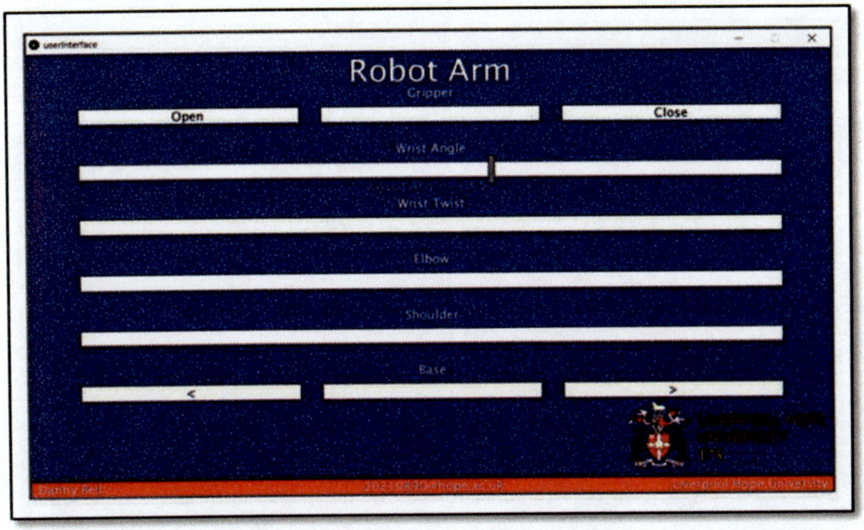

Fig. 6 GUI for the control of the six actuators of the robot by means of sliding commands and interactive buttons; angular position of the actuator is also displayed

is programmed to read the integer and move the correct motor accordingly. An *A4988 motor driver* is used to interface the Arduino board with the microcontroller and the stepper motors.

3 Results

A low cost six degrees of freedom robotic arm has been developed. The arm has been designed to be easy to assemble, and affordable, with a focus on using commonly available materials and 3D-printed components. The final prototype is made of five servo motors and one stepper motor plus an Arduino board as reported in Fig. 7. The servo motors are attached to a series of 3D-printed limbs, allowing an easy assembly and disassembly of the arm.

The base joint uses a stepper motor, which allows a full 360° rotation. The stepper motor embeds a gear on its shaft which engages with another internal gear connected with the base. Metal ball bearings help the arm to smoothly rotate. The shoulder and elbow joints use *MG996R Servo Motors*, which have a 180° rotation range and a torque of up to 11 kg/cm at 6 V thanks to metal gearing. The wrist joints and the gripper use *SG90 Servo Motors*, which also have a 180° rotation range with 1.8 kg/cm of torque. The control system of the robot arm is made of an Arduino board connected to a computer via USB, and a custom user interface developed in Processing. This

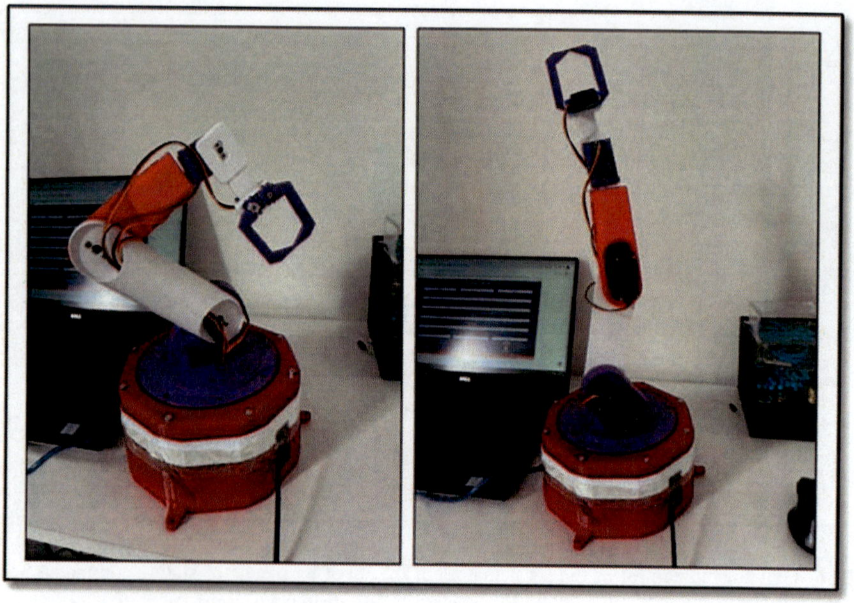

Fig. 7 Robotic arm

interface allows the end-user to control the arm by means of sliders and buttons via serial communication.

Before testing the system, the arm has been calibrated to ensure that it correctly operates. The calibration requires setting the correct angles for each servo motor and checking that the arm moves smoothly. A 47 μF capacitor has been introduced in parallel with the power supply at this stage to stabilize the power and reduce electrical noise in the circuit, which in turn, could cause motor's shaking.

The robot arm's performance has been evaluated through two types of tests or experiments. Firstly, the accuracy of the arm's movement has been tested, by asking the robotic arm to perform movements towards specific points (or marks) within its workspace (i.e., movement accuracy test).

3.1 *Testing Positioning's Accuracy*

The movement accuracy test has been performed by defining the test points, namely the specific points in space that the robot arm has to reach. The arm has been positioned at a designated starting point and calibrated in order to be ready to move.

Next, the control software is used to move the arm towards each designated test point. The arm is moved slowly and precisely to ensure accuracy. After the robot arm has reached each test point, the actual arm position—in relation to the designated location—is recorded. A ruler is used to measure the distance between the end effector and the defined point.

Once the results are collected, they are analyzed. The differences between the actual and designated positions are calculated and assessed. The testing process is shown in Fig. 8, and the results are shown in Table 1.

This test has been designed in order to validate the performance of the arm, as well as to detect any possible issue in terms of movement capability. According to the results, the device shows a proper accuracy vs positioning toward the targets.

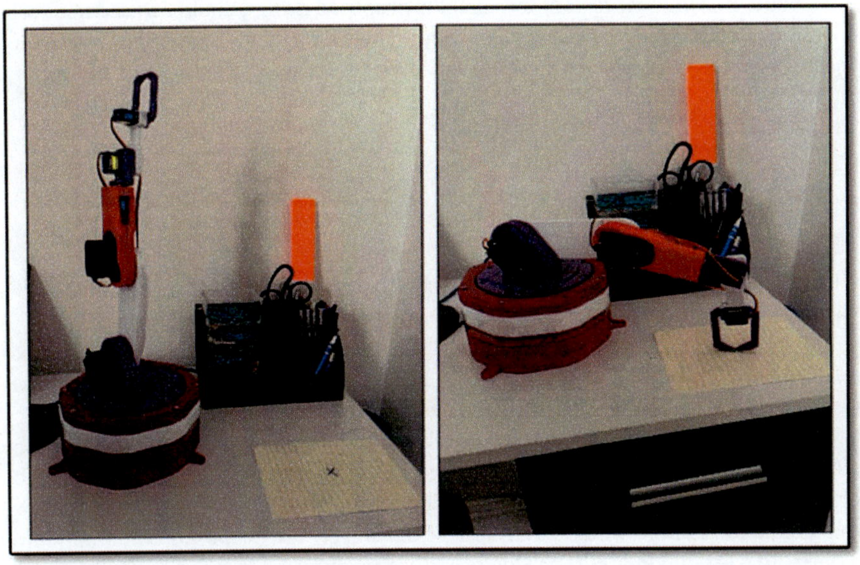

Fig. 8 Testing the positioning's accuracy versus targeting marks

Table 1 Movement accuracy test results

Trial	Distance from base (mm)	Distance from ref marker (mm)	Percentage error (%)
1	250	5	3.3
2	300	6	2.0
3	200	3	1.5
4	220	3	1.4

3.2 *Testing Manipulation Capability*

We have then performed a second type of test where the device has to move a different set of objects of different size and shapes. Here, we have used a set of daily life objects such as a pen and small objects like a screw and a nut. The robotic arm has been positioned on a flat surface, and then a set of moments are performed while grasping these objects. The movements of the device are controlled by means of the GUI, the grasping capability and manipulation capability are observed, and then results are analyzed to detect possible improvements vs the proposed design.

The outcome of these tests shows that the proposed design could be used in a set of laboratory experiments where objects need to be moved and manipulated, provided that the weight of these objects is not too high vs the overall weight of the robotic arm [11].

4 Discussion and Conclusion

In this work, we have presented a 3D-printed design of a robotic arm which shows potential for the low-cost manufacturing of robotic devices [12, 13].

The arm has been integrated with a GUI in order to be used by non-expert end-users on daily basis. The integrated system (i.e., hardware and software design) has been also tested to evaluate performance and capability.

There is room for improving the overall set-up in term of 'sensorization' on the system, given that a better set of transducers could be also integrated within the robot and combined with a suite of other software features, making the device more adaptable and reactive vs the environment. However, the proposed combination of 3D-printed parts and low-cost hardware and software seems to be a significant benefit of this project [14].

A further set of tests should be considered in order to systematically validate the prototype. Integration of another human–robot interface could also be considered to make the system more 'human-like' [15–17].

Acknowledgements This work was presented in dissertation form in fulfilment of the requirements for the BEng in Robotics for the student Daniel Bell from the Robotics Lab, School of Mathematics, Computer Science and Engineering, Liverpool Hope University.

References

1. Muro M, Maxim R, Whiton J (2019) Automation and artificial intelligence: How machines are affecting people and places. Think Asia. https://think-asia.org/handle/11540/9686. Accessed 15 Dec 2022

2. Purnell N (2013) A Chinese province is trying to solve its labor problems with robots. Quartz. https://qz.com/147887/a-chinese-province-is-trying-to-solve-its-labor-problems-with-robots. Accessed 20 Dec 2022
3. Owais Qureshi M, Sajjad Syed R (2014) The impact of robotics on employment and motivation of employees in the service sector, with Special reference to health care. Safety and Health at Work. https://www.sciencedirect.com/science/article/pii/S2093791114000511#bib5. Accessed 20 Dec 2022
4. Gorle P, Clive A (2020) Positive impact of industrial robots on employment. https://robohub.org/wp-content/uploads/2013/04/Metra_Martech_Study_on_robots_2013.pdf. Accessed 20 Dec 2022
5. "Are robots killing jobs or creating them?," Thomasnet® - Product Sourcing and Supplier Discovery Platform—Find North American Manufacturers, Suppliers and Industrial Companies. https://www.thomasnet.com/insights/imt/2013/02/05/are-robots-killing-jobs-or-creating-them/. Accessed 20 Dec 2022
6. Teresa Ballestar M, Díaz-Chao A, Sainz J, Torrent-Sellens J (2020) Impact of robotics on manufacturing: a longitudinal machine learning perspective. Technological Forecasting and Social Change. https://www.sciencedirect.com/science/article/pii/S0040162520311744?via=ihub#cebibl1. Accessed 15 Dec 2022
7. Automation. Encyclopædia Britannica. https://www.britannica.com/technology/automation/Consumer-products. Accessed 09 Jan 2023
8. Guguloth S (2023) Design and development of robotic mechanisms for upper extremity ... https://www.researchgate.net/publication/330522238_DESIGN_AND_DEVELOPMENT_OF_ROBOTIC_MECHANISMS_FOR_UPPER_EXTREMITY_REHABILITATION. Accessed 17 Jan 2023
9. Lerda G (2023) 010-kauda. DIY Tech. https://www.diy-tech.it/010-kauda. Accessed 7 Feb 2023
10. What is PLA? (Everything you need to know). TWI. https://www.twi-global.com/technical-knowledge/faqs/what-is-pla. Accessed 6 May 2023
11. Procter S, Secco EL (2022) Design of a biomimetic BLDC driven robotic arm for Teleoperation and Biomedical Applications. J Human Earth Future https://hefjournal.org/index.php/HEF/article/view/108. Accessed 8 May 2023
12. Manolescu VD, Secco EL (2023) Development of a 3D printed biologically inspired monoped self-balancing robot. Intl J Robot Control Syst ASCEE 3(1):84–97. https://doi.org/10.31763/ijrcs.v3i1.841
13. Manolescu VD, Secco EL (2022) Design of a 3-DOF robotic arm and implementation of D-H forward kinematics. In: 3rd Congress on Intelligent Systems (CIS 2022), vol 1(42), pp 569–583. https://doi.org/10.1007/978-981-19-9225-4_42
14. Tharmalingam K, Secco EL (2022) A surveillance mobile robot based on low-cost embedded computers. In: 3rd International Conference on Artificial Intelligence: Advances and Applications, 25, (ICAIAA 2022). https://doi.org/10.1007/978-981-19-7041-2
15. Howard AM, Secco EL (2021) A low-cost human-robot interface for the motion planning of robotic hands, intelligent systems conference (IntelliSys). In: Advances in Intelligent Systems and Computing, Lecture Notes in Networks and Systems, vol 3(30). Springer, Berlin, p 296
16. Ormazabal M, Secco EL (2021) A low cost EMG Graphical User Interface controller for robotic hand. In: Future Technologies Conference (FTC 2021), Lecture notes in networks and systems, vol 2. Springer, pp 459–475. https://doi.org/10.1007/978-3-030-89880-9
17. Secco EL, McHugh DD, Buckley N (2022) A CNN-based computer vision interface for prosthetics' application. In: EAI MobiHealth 2021–10th EAI International Conference on Wireless Mobile Communication and Healthcare, pp 41–59. https://doi.org/10.1007/978-3-031-06368-8_3

Profit Maximization of a Wind-Integrated System by V2G Method

Gummadi Srinivasa Rao, M. Prem Kumar, K. Dhananjay Rao, and Subhojit Dawn

Abstract An essential component of a power system network is profiting maximization for both power producers and consumers. When demand is low, electricity on the electrical grid is easily available. The electrical system can be supported during periods of high demand by storing any excess energy in storage units. Thus, the grid's stability and safe functioning are guaranteed. The requirement for an energy storage system in a power system that runs on renewable energy is increased due to the unpredictable nature of renewable energy sources. Recently, a method of employing renewable resources to preserve power system stability has surfaced: vehicle-to-grid (V2G) technology. The V2G storage devices can be used to make more money during periods of high demand. This effort's primary goal is to determine where in a power network the V2G system will be most profitable and cost-effective. A deregulated electrical environment's whole economy is taken into account in this study together with the effects of V2G integration. To replicate the suggested work, MATPOWER software is used using the IEEE 14-bus system.

Keywords Competitive market · Wind power · Imbalance cost · System revenue

1 Introduction

Willet Kempton and Steven E. Letendre pioneered the V2G technology in the year 1997. This notion addresses the plausibility of utilizing plug-in electric vehicles (PEVs) as decentralized power generators. During the charging process of an electric vehicle (EV), power is drawn from the electrical grid. The V2G technology, however, reverses this procedure by allowing electricity to be transmitted to the grid from EVs through the discharging operation. By giving the power network more power, this situation has the ability to raise the system voltage to a safe level and

G. Srinivasa Rao (✉) · M. Prem Kumar · K. Dhananjay Rao · S. Dawn
Department of Electrical and Electronics Engineering, Velagapudi Ramakrishna Siddhartha Engineering College, Vijayawada, India
e-mail: vasu1in@vrsiddhartha.ac.in

S. Kumar et al. (eds.), *Fourth Congress on Intelligent Systems*, Lecture Notes in Networks and Systems 868, https://doi.org/10.1007/978-981-99-9037-5_16

maintain the grid frequency. Virtual-to-grid (V2G) technology is the process of transferring energy from an electric vehicle's batteries to the grid while the vehicle is in park mode. A smart grid, which governs energy usage via the use of information technology, would not be possible without this strategy. There are two main benefits that come with electric cars. First and foremost, the addition and clever design of a building's energy management system can be made possible by the use of bidirectional charging and V2G, all without the need for a major overhaul and significant capital expenditures. Second, operating electricity costs are decreased when bidirectional charging is implemented.

Several lessons have been learned in the areas of energy storage applications and renewable energy sources. Because renewable energy is unpredictable, there are significant limitations on its use in power networks. Currently, wind energy is the most important and rapidly growing renewable energy source [1]. About 3% of the planet's total energy production capacity, or 125 GW, can currently be produced by the grid energy storage system. An energy storage technology that is most common and well-liked is the pumped storage hydroelectric plant [2]. Load shedding and variable renewable energy (VRE) have been enhanced by the installation of a battery-based energy storage transport (BEST) system in the power network. A viable solution to the issue of transmission line congestion is also being considered: BEST [3]. More and more wind, solar, and variable electricity is being produced, which means that additional services are required to manage the increased volatility and unpredictable nature of this kind of power [4]. Investigating how grid-scale renewable energy may support economic growth while meeting the growing demand for electricity is one of the ambitious goals that developing countries have set for themselves [5]. The authors of [6] focused on how to minimize generating costs in order to maximize profit and social welfare in an integrated renewable system. The concept of bidding from both sides was examined in this study.

Energy storage solutions are becoming more and more necessary for power grid systems, as evidenced by the recent introduction of special battery capacities for power grid deployment [7]. This work suggests a transmission expansion planning model that spans multiple years and combines four distinct grid pricing techniques in an effort to reduce system costs overall. The model takes into consideration fluctuations in the output of solar and wind energy [8]. Because conventional frequency control approaches and other storage technologies are more affordable, using lithium-ion batteries in these applications is significantly hindered [9]. Few locations with abundant wind energy can generate electricity economically enough to compete with conventional sources. This investigation covers a number of topics, including wind turbine lifespan, operating costs, and maintenance costs [10]. Apostolaki-Iosifidou et al. [11] states that a vehicle-to-grid (V2G) system's effectiveness can range from 60 to 70%, though this figure is dependent on a number of variables. To deliver electricity to thermal power plants while minimizing societal costs, paper [12] suggests integrating energy storage devices with renewable sources in a competitive power system.

Regarding the advancement of renewable energy in a competitive electric system, a number of challenges, regulations, and incentives are covered in Ref. [13]. The objective of the study is to develop an optimization model that accounts for both power generator routing and load scheduling from the perspective of electric vehicle consumers [14]. The analysis suggests that how local resources are used to meet local power needs and how an alternative power system is used in times of crisis will have an impact on how the events play out. Plug-in electric cars (PEVs) can function as mobile power producers in addition to distributed generators (DGs) like solar or wind farms [15]. The following are the outcomes of the research:

Taking into consideration the price difference, the primary objective of this investigation is to optimize the financial benefit of a wind-related competitive power system. The price imbalance that results from wind energy's unpredictability has a direct effect on the financial gain of the system. The aim of this study is to reduce the price gap and boost the economic benefit by utilizing the most advantageous V2G technique operation. The vehicle-to-grid (V2G) technology reaches price equilibrium by executing power operations that flow in both directions between electric cars and the power grid, contingent upon the grid's demand for power.

2 Mathematical Design

This section provides a thorough examination of the vehicle-to-grid (V2G) system, the thyristor-controlled series compensator (TCSC) static model, and mathematical modeling for wind farms.

Wind Farm

Based on the wind speed in a particular area, the wind farm operates. Initially, the wind speed data is collected in Vijayawada, Andhra Pradesh, India. Reference [6] is consulted in order to calculate the capacity of wind power generated and the associated costs.

Power from V2G System

Power transmission in a V2G system occurs bidirectionally. Consequently, regulated converters are incorporated in the majority of V2G systems to facilitate this process. Electricity is transported from the power grid to the battery of an electric car during the charging phase. The enhancement of the system's economic viability is exemplified in (1).

$$P_{\mathrm{Dm}} = p_{\mathrm{dm}} + C_{\mathrm{EV}} \tag{1}$$

Under this context, pdm indicates the power requirement, while P_{Dm} indicates the total power requirement at the mth bus. Additionally, C_{EV} refers to an electric vehicle's (EV) battery capacity that needs to be charged. When an EV battery is being discharged, power is transferred from the battery to the electrical grid. Preventing the battery from fully discharging is crucial because of efficiency constraints. Equation (2) illustrates how the battery's discharge into the grid is influenced by the average efficiency of the EV battery and converter.

$$P_{\text{Dm}} = p_{\text{dm}} - (C_{\text{EV}} * E_{\text{B}}) \tag{2}$$

The battery's overall output efficiency is denoted by EB. Different battery types are used in the field of electric vehicles (EVs) [16]. In the previously mentioned article, utility companies and consumers assessed the performance and profit optimization of lead-acid and lithium-ion batteries. Furthermore, supercapacitors were used as a comparison due to their 97.94% efficiency [17]. The overall efficiency was calculated by averaging the efficiencies of the different batteries and the system's efficiency, which was 70% [11].

3 Objective Function

A future potential generation scenario must be supplied by the renewable power plant to the market operators. This has led to the beginning of power scheduling based on the provided data. The unpredictable nature of renewable sources, however, prevents renewable energy facilities from meeting the necessary power generation targets. Price imbalance is a concept that arises from this circumstance and harms the system's profitability. In the event that renewable energy providers supply the grid with more power than what is specified, they are compensated with a positive imbalance price. Conversely, the grid's incapacity to obtain sufficient electricity from renewable sources is the cause of the negative imbalance pricing. The three variables that comprise the system profit at time h, represented as SEP(h), are the revenue SRC(h), generation cost SGC(h), and imbalance price SIP(h). In this study, the expansion issue (3) deals with profit maximization.

$$\text{Maximize, } \text{SEP}(h) = \text{SRC}(h) + \text{SIP}(h) - \text{SGC}(h) \tag{3}$$

The revenue from thermal power plants SRC_Th (h) and revenue from wind farms SRC_W (h) make up the two components of the total revenue expenditure, as shown in Eqs. (4) and (5).

$$\text{SRC}(h) = \text{SRC}_{\text{Th}}(h) + \text{SRC}_{\text{W}}(h) \tag{4}$$

$$\text{SRC}_{\text{Th}}(h) = \sum_{k=1}^{N} P_{\text{Th}}(k, h) \cdot \text{LMP}(k, h) \tag{5}$$

P_{Th} (k, h) is the thermal power generated at time h. The LMP of the kth thermal generator at time h is expressed using LMP(k,h). The system contains N generators. The price of system imbalance is computed using the current wind power (CW) and the predicted wind power (PW). The imbalance cost of an electrical network powered by wind is shown in (6).

$$\mathrm{SIP}(h) = \sum_{k=1}^{N}\left(\mathrm{ER}(h) + \mathrm{SR}(h)\left(\frac{\mathrm{PW}(k,h)}{\mathrm{CW}(k,h)}\right)^2\right)\cdot(\mathrm{CW}(k,h) - \mathrm{PW}(k,h))\mathrm{IC}(t)$$

$$= \sum_{i=1}^{NG}\left(\mathrm{SCR}(t) + \mathrm{DCR}(t)\cdot\left(\frac{P_p(i,t)}{P_a(i,t)}\right)^2\right)\cdot\left(P_a(i,t) - P_p(i,t)\right) \quad (6)$$

$$\mathrm{SR}(h) = (1+\propto)\cdot\mathrm{LMP}(k,h),\quad \mathrm{ER}(h) = 0 \quad \text{if } \mathrm{PW}(k,h) > \mathrm{CW}(k,h) \quad (7)$$

$$\mathrm{ER}(h) = (1-\propto)\cdot\mathrm{LMP}(k,h),\ \mathrm{SR}(h) = 0 \quad \text{if } \mathrm{PW}(k,h) < \mathrm{CW}(k,h) \quad (8)$$

$$\mathrm{ER}(h) = \mathrm{SR}(h) = 0 \quad \text{if } \mathrm{PW}(k,h) = \mathrm{CW}(k,h) \quad (9)$$

where ER(p) and SR(p) are the wind plant's extra and shortage rates at time p.

4 Results and Discussions

An IEEE 14-bus test system comprising five generators with 20 lines and 14 buses has been chosen in order to confirm the viability, utility, and efficacy of the suggested technique. Firstly, there is the slack bus. Data from [18] was used to create the system under examination. In assessing the competitive power network related to renewable energy, the economic analysis must take imbalance pricing into account. Due to the unpredictable nature of renewable resources, the renewable power plant is unable to meet the predetermined power needs, posing a significant risk to the electrical grid. Two approaches to addressing the problem of price imbalance are the application of energy storage devices and efficient forecasting technologies. Furthermore, there are a number of limitations with the first method that prevent it from providing a practical solution. Thus, scheduled electricity can be maintained and the risk of financial loss can be reduced by using energy storage systems. The data collected on the survey day is referred to as current wind speed (CS), while the data collected the day before is referred to as predicted wind speed (PS). Wind generation is impacted by wind speed variation, which leads to a discrepancy between anticipated power generation and, eventually, profitability. The difference, or gap, between actual and anticipated data is what leads to the development of imbalance pricing. For the city of Vijayawada, real-time data on actual and predicted wind speeds is obtained (as indicated in Table 1). Because wind speed fluctuates so much, we use hourly intervals. Wind speeds are measured at a height of ten meters. The likely wind speed at a height of 120 m—the height of a wind turbine hub—was first evaluated in order to compute wind generation at various hours.

Table 1 Real-time PS and CS of Vijayawada (in Km/h)

Hour	Projected wind speed (PS)	Current wind speed (CS)	Hour	Projected wind speed (PS)	Current wind speed (CS)	Hour	Projected wind speed (PS)	Current wind speed (CS)
00	5	7	08	8	8	16	11	12
01	6	7	09	9	8	17	10	10
02	6	8	10	10	9	18	8	8
03	7	7	11	11	10	19	8	8
04	8	6	12	12	12	20	7	8
05	8	6	13	12	13	21	7	8
06	8	7	14	12	14	22	6	8
07	8	8	15	12	13	23	6	8

The producing companies are compensated in accordance with the award if the current wind speed is higher than the predicted wind speed. If the wind speed at any given time is less than the predicted wind speed, businesses that produce electricity will be fined for failing to deliver the contracted amount of power. Repetitive situations have been omitted since it is evident from the collected wind speed that some repeating condition has occurred. As decided in Step 2, the wind farm is placed on bus number 14, and the imbalance cost is computed by factoring in both the wind power that is expected and the wind power that is currently available. Table 2 shows the impact of imbalance pricing on profit, with real profit representing profit after taking the imbalance price into account and projected profit representing system profit prior to doing so. Step 3 above's imbalance cost can be reduced and similar situations can be avoided by using V2G technology. Power grids use vehicle-to-grid (V2G) technology to use the energy saved in optimal electric vehicles (EVs) to get around problems that arise during peak demand. In times of low demand, EVs use the grid to obtain electricity. Electric cars (EVs) have the capacity to draw energy from the grid and use it to meet demand when it increases. An example of how uneven pricing affects actual profitability is provided in Table 2. A twenty-four-hour time limit has been placed on the proposed methodology. When similar wind speed scenarios and duplicates are removed, there are 17 unique cases left. Bus 14, the most profitable bus in the deregulated system, is connected to the wind farm, as was already mentioned.

This illustrates how the imbalance affects profit fluctuations. More is lost than made in penalties. Due to the fact that a higher penalty entails greater losses, this highlights the necessity of eliminating the imbalance. The integration of the V2G system helps prevent this from happening. The battery-powered nature of V2G technology should be noted. The 3 MW EV fleet under discussion is its capacity. It all depends on the battery what is the ultimate efficiency. It is our ability to prevail that determines the EV's output power. Efficiency raises the output power, which raises profit as well. In the event that there is a pricing imbalance, the EV fleet must supply

Table 2 Imbalance prices effect on system profit

Sl. No	Time (24 h)	PS (Km/h)	CS (Km/h)	Imbalance price ($/hr)	Estimated profit ($/h)	Actual profit ($/h)
1	00	5	7	0.1182	1930.490	1930.6078
2	01	6	7	0.1182	1931.350	1931.4687
3	02	6	8	0.2688	1931.350	1931.6188
4	03	7	7	0	1932.190	1932.18
5	04	8	6	−5.2751	1933.569	1928.293
6	06	8	7	−2.9847	1933.569	1930.5827
7	07	8	8	0	1933.569	1933.5685
8	09	9	8	−3.0506	1935.309	1932.2579
9	10	10	9	−4.5144	1937.317	1932.8019
10	11	11	10	−6.0353	1939.451	1933.4151
11	12	12	12	0	1942.469	1942.469
12	13	12	13	0.414	1942.469	1942.883
13	14	12	14	0.7729	1942.469	1943.241
14	16	11	12	0.3431	1939.451	1939.794
15	17	10	10	0	1937.317	1937.317
16	20	7	8	0.1549	1932.190	1932.345
17	22	6	8	0.269	1931.350	1931.619

the additional power to correct it. In such cases, the profit is computed using the power that remains produced by the EV fleet after the power produced by the EV fleet has been subtracted. By using V2G technology, the profitability of the current power system could be raised. Lithium-ion batteries are utilized in modern V2G technologies. Here, we investigated how system profit was affected by integrated lead-acid, lithium-ion and super-capacitor batteries. Table 3 shows the effects on the electrical grid of lithium-ion, lead-acid, and super-capacitor profit. Given its higher profitability compared to the regulated system, the system being considered in this scenario is deregulated.

In this case, the gain in the charging scenario is the same under all conditions since the battery is fully charged. Depending on the situation, different gains can be made in discharge. This is due to the fact that different batteries have different output efficiencies. Supercapacitor was the most efficient of the three battery options that were examined. However, the installation cost of the supercapacitor is high. In this scenario, the electric vehicle is discharged close to the generator bus. The battery pack in electric cars is related to this V2G concept. It is possible to discharge it on other buses as well as charge it from one. The electrical system may therefore be spared a disastrous breakdown thanks to this technology.

Table 4 illustrates that the integration of the V2G system results in significant system benefits, despite any imbalance present. Power system gains are enhanced by

Table 3 System performance with V2G and without SIP (in $/h)

	SRC	SGC	SEP
Lead-acid			
Base scenario	9431.116	7279.111	2152.005
Lead-acid charging	9563.468	7393.381	2170.087
Lead-acid discharging	9556.777	7296.361	2260.416
Lithium-ion			
Base scenario	9431.116	7279.12	2151.996
Li-ion charging	9563.468	7393.39	2170.078
Li-ion discharging	9556.393	7296.37	2260.023
Super-capacitor			
Base scenario	9431.116	7279.12	2151.996
Super-capacitor charging	9563.468	7393.39	2170.078
Super-capacitor discharging	9556.52	7292.33	2264.19

the presence of V2G systems. For ease of understanding, Fig. 1 presents a graph of comparative improvements.

The entire task was finished using the Mi Power software. The regulated and unregulated system environments were taken into consideration when designing the first two stages of the project. The remainder of the work was only done in the deregulated environment after it was determined that the unregulated system was more profitable than the regulated one.

Table 4 System performance with V2G and without SIP, when imbalance present (in $/h)

	SRC	SGC	SEP
Lead-acid			
Base scenario	9465.178	7527.865	1937.313
Lead-acid charging	9591.531	7646.115	1945.416
Lead-acid discharging	9460.51	7436.405	2024.105
Lithium-ion battery			
Base scenario	9465.178	7527.874	1937.304
Li-ion charging	9591.531	7646.124	1945.407
Li-ion discharging	9460.329	7433.574	2026.755
Super-capacitor			
Base scenario	9465.178	7527.874	1937.304
Super-capacitor charging	9591.531	7646.124	1945.407
Super-capacitor discharging	9463.541	7432.364	2031.177

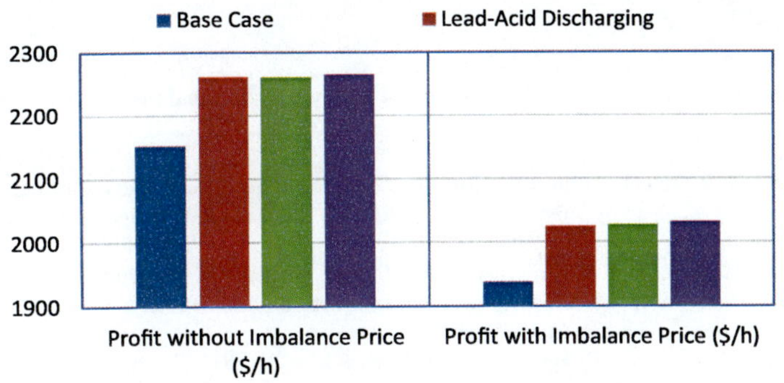

Fig. 1 Profit assessment with and without SIP for the V2G approach

5 Conclusions

An analysis of the impact of wind turbine integration at various grid points on voltage, LMP, generation costs, and profitability in regulated systems is conducted using the IEEE 14-bus system. The addition of renewable energy sources has made the system more profitable as a consequence. Enhancing the profitability of the energy system can be achieved through deregulating the environment. Competitive power networks, as opposed to regulated ones, offer higher returns and employ demand-side bidding. When renewable energy sources are used, the benefits of the power system are maximized. Its instability, which could lower system profit overall, is these sources' greatest drawback. Here, we are describing an imbalance price. To optimize a power system's advantages, pricing must be balanced. Producers receive rewards or penalties based on the difference between the data for projected and actual wind speeds. Due to their intermittent nature, renewable energy sources result in uneven pricing and systemic inconsistencies. To lessen this issue, consider using electric cars (EVs). Based on recent research, vehicle-to-grid (V2G) technology can reduce imbalance prices, increasing the profitability of the electrical system. When EVs are discharged during peak hours, grid blackouts can be prevented. Connecting an EV with a high LMP to the load bus benefits the power system the most. To maximize profit in every situation, the TCSC placement has also been completed. Through the use of V2G and TCSC technologies, this study shows how wind farms can lessen the reliance of power networks on fossil fuels while preserving grid stability. In order to maximize profits while maintaining a steady voltage profile, deregulated market practices collaborate with these technological solutions. For the entire task, the Mi Power program was employed. Considering both regulated and uncontrolled system contexts, the project's first two phases were designed. When the unregulated system proved to be more profitable than the regulated one, only then was the remaining work completed in the deregulated environment.

References

1. Karanja JM, Hinga PK, Ngoo LM, Muriithi CM (20220) Optimal battery location for minimizing the total cost of generation in a power system. In: 2020 IEEE PES/IAS Power Africa
2. Lawder MT, Kumar B, Suthar PW, Northrop C, De Sumitava C, Hoff M, Leitermann O, Crow ML, Santhanagopalan S, Subramanian VR (2014) Battery energy storage system (BESS) and battery management system (BMS) for grid-scale applications. Proc IEEE 102(6):1014–1030
3. Pulazza G, Zhang N, Kang C, Nucci CA (2021) Transmission planning with battery-based energy storage transportation for power systems with high penetration of renewable energy. IEEE Trans Power Syst 36(6):4928–4940
4. Bhatnagar D, Currier AB, Hernandez J, Ma O, Kirby B (2013) Market and policy barriers to energy storage deployment: a study for the energy storage systems program. Sandia National Laboratories, SAND2013–7606. https://www.sandia.gov/ess-ssl/publications/SAND2013-7606.pdf
5. Lee N, Flores-Espino F, Cardoso De Oliveira RP, Roberts BJ, Brown T, Katz JR (2019) Exploring renewable energy opportunities in select southeast Asian countries: a geospatial analysis of the levelized cost of energy of utility-scale wind and solar photovoltaics. National Renewable Energy Lab (NREL), No. NREL/TP-7A40–71814
6. Dawn S, Tiwari PK (2016) Improvement of economic profit by optimal allocation of TCSC and UPFC with wind power generators in double auction competitive power market. Intl J Electr Power Energy Syst 80:190–201
7. Versteeg T, Baumann MJ, Weil M, Moniz AB (2017) Exploring emerging battery technology for grid-connected energy storage with Constructive Technology Assessment. Technol Forecast Soc Chang 115:99–110
8. Bravo D, Sauma E, Contreras J, de la Torre S, Aguado JA, Pozo D (2016) Impact of network payment schemes on transmission expansion planning with variable renewable generation. Energy Econ 56:410–421
9. Swierczynski M, Stroe DI, Lærke R, Stan AI, Kjær PC, Teodorescu R, Kær SK (2014) Field experience from Li-ion BESS delivering primary frequency regulation in the Danish Energy market. ECS Trans 61(37):1–14
10. Morthorst PE (2004) Wind energy the facts. Risoe National Lab., Roskilde (Denmark). The European Wind Energy Association, Brussels (Belgium), vol 2, Costs and prices, pp 95–110
11. Apostolaki-Iosifidou E, Codani P, Kempton W (2017) Measurement of power loss during electric vehicle charging and discharging. Energy 127:730–742
12. Chakraborty MR, Dawn S, Saha PK, Basu JB, Ustun TS (2022) A comparative review on energy storage systems and their application in deregulated systems. Batteries 8(9):124
13. Randhir Singh YR, Sood N, Padhy P (2009) Development of renewable energy sources for Indian power sector moving towards competitive electricity market. In: 2009 IEEE Power and Energy Society General Meeting
14. He J, Hu Z (2014) Hourly coordinated operation of Electric vehiclescharging/discharging and wind-thermal power generation. In: 2014 IEEE Conference and Expo Transportation Electrification Asia-Pacific (ITEC Asia-Pacific)
15. Hussain SS, Aftab MA, Ali I, Ustun TS (2020) IEC 61850 based energy management system using plug-in electric vehicles and distributed generators during emergencies. Intl J Electr Power Energy Syst 119, 105873
16. May GJ, Davidson A, Monahov B (2018) Lead batteries for utility energy storage: a review. J Energy Stor 15:145–157
17. Zhong Y, Zhang J, Li G, Liu A (2006) Research on energy efficiency of supercapacitor energy storage system. In: Power System Technology, PowerCon 2006. International Conference on North China Electric Power University
18. Das SS, Das A, Dawn S, Gope S, Ustun TS (2022) A joint scheduling strategy for wind and solar photovoltaic systems to grasp imbalance cost in competitive market. Sustainability 14(9):5005

Detection of Partially Occluded Area in Images Using Image Segmentation Technique

Jyothsna Cherapanamjeri and B. Narendra Kumar Rao

Abstract Computer vision is subfield of artificial intelligence. Detection of occluded area in images is one of the challenging tasks in the computer vision. In this paper, we concentrate on occluded area detection in images using image segmentation techniques. Occlusion is nothing but one object in image is hidden by another object. Detection of occluded area in images using image segmentation techniques. Image segmentation is the process of dividing the images into different regions based on the characteristics of the pixels in the original image and reduces the complexity of analysis. When determining the number of objects in a scene, instance segmentation is an excellent choice. But in semantic segmentation algorithms if all that is required is to group items that belong to the same class. Image segmentation has many applications in a number of fields, including gaming, robotics, autonomous vehicles, agriculture, object detection, and pedestrian detection. In this research work, we concentrate on detection of occluded area in images using Mask R-CNN. The experiment will show better segmentation results corresponding to the manually labeled occluded area.

Keywords Artificial intelligence · Machine learning · Deep learning · Computer vision · Image segmentation · Occluded images

1 Introduction

Image segmentation is a technique that is frequently used in computer vision, image analysis, and digital image processing. Each segment is easier to interpret and analyze since the complexity of the image is reduced. Technically identifying objects, people, or other important details in an image by assigning labeling to specific pixels. Object detection is a common application of image segmentation. It is utilized in a wide

J. Cherapanamjeri (✉)
Jawaharlal Technological University, Anantapur, India
e-mail: jyothsnamtech@gmail.com

B. Narendra Kumar Rao
Sri Vidyanikethan Engineering College, Tirupathi, India

S. Kumar et al. (eds.), *Fourth Congress on Intelligent Systems*, Lecture Notes in Networks and Systems 868, https://doi.org/10.1007/978-981-99-9037-5_17

range of real-world scenarios, including facial recognition for video surveillance, medical image analysis [1], and computer vision for self-driving cars.

1.1 Image Segmentation Techniques

There are five different types of techniques in image segmentation (1) edge-based segmentation, (2) threshold-based segmentation, (3) region-based segmentation, (4) cluster-based segmentation, and (5) artificial neural network-based segmentation [2]. Finding edges in an image is an essential first step in interpreting visual features. This process is known as edge detection. It is thought that edges keep significant information and relevant characteristics. An image's edges are identified by edge-based segmentation algorithms based on differences in brightness, saturation, contrast, color, texture, and other elements. Supplemental processing procedures must then be performed to concatenate all of the edges into edge chains that better match the borders in the image, as shown in Fig. 1, in order to improve the results [3].

In threshold-based segmentation [2], foreground and background can be distinguished using the threshold segmentation procedure. Since threshold segmentation recovers the foreground mostly based on gray value information, as seen in Fig. 2, it is very useful for segmenting images with a strong contrast between the foreground objects and background.

There are three types of thresholds (1) simple thresholding, (2) Otsu's binarization, and (3) adaptive thresholding. In region-based segmentation technique for determining regions directly, the splitting and merging-based segmentation methods to segment an image combine the two primary approaches of region splitting and region merging. While merging involves bringing together adjacent parts that are accurately similar to one another, splitting includes repeatedly dividing an image into regions

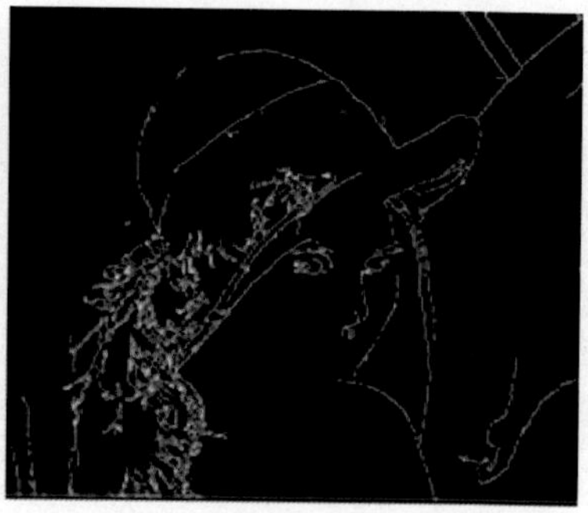

Before Edge Segmentation — After Edge Segmentation

Fig. 1 Edge-based segmentation

Fig. 2 Threshold-based segmentation

with identical features as shown in Fig. 3. Region growing, region splitting, and region merging are three techniques in region-based segmentation.

Clustering algorithm is used to group closer the properties of each object form a group. Figure 4 shows similar group of objects based on properties. Here cows are similar group of objects.

There are two important clustered-based algorithms, namely (1) k-means and (2) fuzzy c-means. In artificial neural network-based segmentation [2], AI is used to process and recognize portions of pictures including objects, faces, text, handwritten

Fig. 3 Region-based segmentation

Fig. 4 Cluster-based segmentation

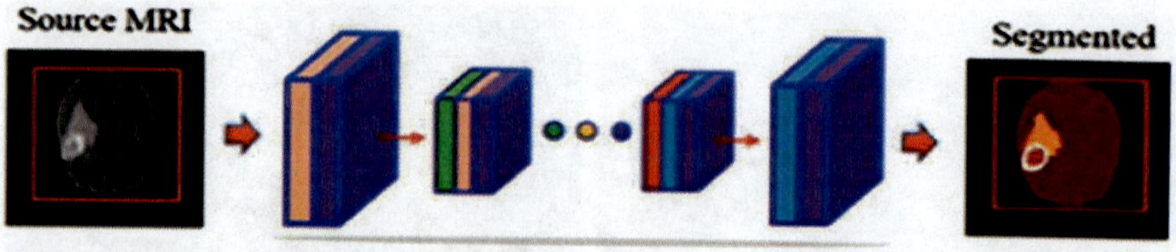

Fig. 5 Artificial neural network-based segmentation

text, etc., automatically. In view of their design to identify and handle high-definition visual data, convolutional neural networks are specifically used in this kind of task. Artificial neural network-based segmentation is shown in Fig. 5.

1.2 *Types of Image Segmentation*

Segmentation can be divided into two basic categories: instance segmentation and semantic segmentation.

1.2.1 Semantic Segmentation

A class label is given to each pixel in the image as part of the semantic segmentation process. In semantic segmentation, the labels are shared by all the objects in the same class. Each pixel in an image has a label attached to it or category using semantic segmentation. It is used to recognize groups of pixels which fall into different categories [4]. For instance, a self-driving vehicle needs to be able to identify other vehicles, pedestrians, traffic signs, roadways, and other portions of the route. There are three steps in segmentation: (1) classifying a certain visual object in the image, (2) locating the object and then drawing a box around it, (3) generating a segmentation mask to group the pixels in a localized image. It is also known as the pixel-level categorization of images. Semantic segmentation cannot distinguish between different instances in the same category; i.e., all sheep and persons are marked as same class as shown in Fig. 6 [5].

1.2.2 Instance Segmentation

The process of classifying and recognizing multiple kinds of objects in an image is known as instance segmentation. It generates a segment map for each category and instance of that class, instance segmentation produces output in a richer format. In instance segmentation, each detected object receives its unique label. Instance segmentation is the process of locating, classifying, and segmenting each individual

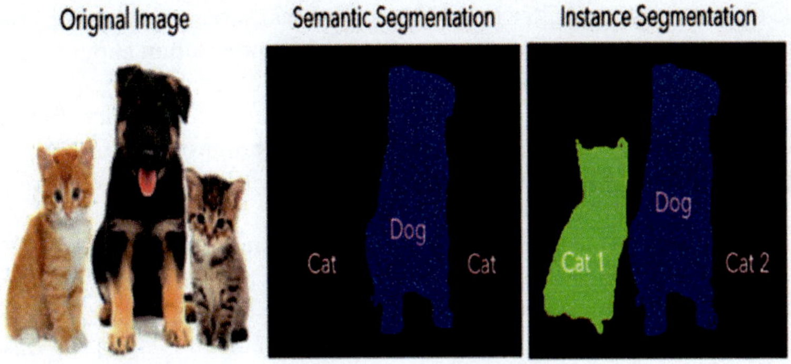

Fig. 6 Differentiation between semantic segmentation and instance segmentation

object in the image [6]. By combining semantic segmentation and object identification, instance segmentation adds the capability of distinguishing distinct instances of any given segment class to the basic segmentation mask. Compared to both semantic segmentation networks and object detection networks, instance segmentation generates outputs that are richer. Instance segmentation can distinguish between different instances of the same categories; i.e., different chairs are distinguished by different colors as shown in Fig. 6 [7].

2 Related Work

Literature survey reveals that the image segmentation method has developed rapidly for recent years such as increasing accuracy and speed. The following describes recent research on image segmentation.

1. Shen et al. proposes backbone network, ResNet 350-FPN-ED, was proposed to improve the mask R-CNN model by introducing an ECA module in the backbone network [8].
2. Gite et al., title is "Enhanced lung image segmentation using deep learning" 2021, in this study four benchmark neural network architectures: U-Net, FCN, SegNet, and U-Net++ [9].
3. Zuo et al., the title is "A method of crop seeding plant segmentation on edge information fusion model," 2022. When extracting features, the backbone network, which is composed of the UNET network, is instructed to perceive the information about the plant [3].
4. Kaihan Lin, the title "Face Detection and Segmentation Based on Improved Mask R-CNN" proposes enhanced Mask. In this proposed method, ResNet-101 is used to extract features, RPN is used to generate RoIs, and RoIAlign faithfully preserves the exact spatial locations to generate binary mask through fully

convolution network (FCN). R-CNN name G-Mask integrates face detection and segmentation into one framework to obtain more information about faces [2].
5. Yi-Ting Chen, the title "Multi-instance Object Segmentation with Occlusion Handling" proposes MCG and SDS CNN architectures extract features for each object hypothesis. Top-scoring segmentation proposals are utilized to infer occluding regions, and categorized segmentation hypotheses are employed to create class-specific probability maps.
6. Sehar and Naseem, the title "How deep learning is empowering semantic segmentation" proposes CNN-LSTM models for classifying the objects belonging to the forest regions; the tree is your required masked area [4].
7. Laith Abualigah, the title "A Novel Instance Segmentation Algorithm Based on Improved Deep Learning Algorithm for Multi-object Images" proposes ResNet with SENet; the multi-object image instance segmentation technique proposed in this research has three phases. In the first stage, a unique backbone approach enhances the image recognition algorithm by extracting low and high characteristic levels from the provided images. Each ResNet block connects to the squeeze-and-excitation network (SENet), which is the essential building block. COCO dataset is being used here for model training.

3 Methodology

The effective method for locating and isolating items or regions of interest in an image or video is image segmentation. The capacity to precisely extract pertinent information from an image is one of the main advantages of image segmentation. Based on the content of the image, predictions are made using this information. For object detection and tracking, image segmentation is essential. Medical imaging, autonomous vehicles, robotics, agriculture, and detecting systems are among the major uses of picture segmentation. Existing architectures to implement image segmentation are (1) encoder and decoder architectures, (2) U-Net, (3) U-Net++, (4) Fast FCN, (5) Gated SCNN, (6) Deep Lab, (7) Mask R-CNN, (8) V-Net (9) SegNet. In this study, we will explore how to use Mask R-CNN [2]. It automatically segments and creates pixel-wise masks for each object in an image using Mask R-CNN. Mask R-CNNs will be used to process both video and image streams. In this research, we process only images [10–12].

3.1 Understanding Mask R-CNN Basic Architecture

The main components in these architectures are backbone network, region proposal network, region of interest, and finally object detection branch as shown in Fig. 7. The following is the description of components in basic Mask R-CNN architectures [13, 14].

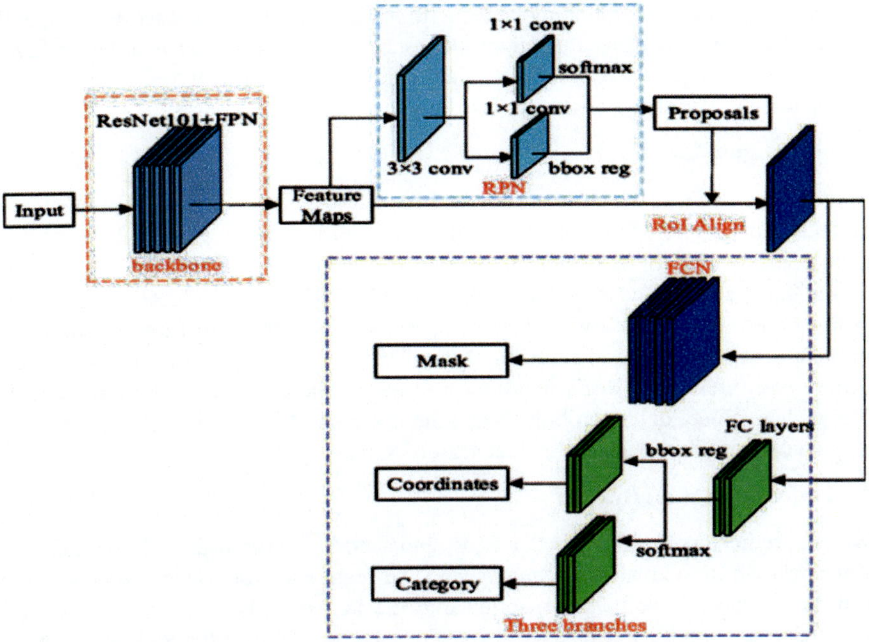

Fig. 7 Basic mask R-CNN architecture

1. **Backbone**

The key feature extraction component of Mask R-CNN is a backbone. Residual networks (ResNets) with or without FPN are frequently used for this component [15]. For simplicity, we take ResNet with FPN as a backbone. A feature map is created when data from a raw image is fed into a ResNet backbone after it has passed through many residual bottleneck blocks. The final convolutional layer of the backbone's feature map comprises abstract picture data, such as instances of various objects, their classes, and their spatial characteristics. The RPN is then fed using this feature map [16].

2. **RPN**

RPN stands for region proposal network. A convolutional layer processes the feature map and creates a tensor with c channels for each spatial vector connected to an anchor center. Given a single anchor center, a collection of anchor boxes with various scales and aspect ratios is constructed. These anchor boxes are distinct areas that totally surround the image and are uniformly distributed throughout. The c-channel tensor is then processed by two sibling one-by-one convolutional layers. One is a binary classifier. It predicts whether each anchor box will contain an item. Each *c*-channel vector is converted to a *k*-channel vector, which represents *k* anchor boxes with various scales and aspect ratios that are centered on the same anchor. The other is an object bounding box regressor. Each *c*-channel vector is converted into a 4k-channel vector. We identify the bounding boxes with the highest abjectness score out of any overlapped bounding boxes that might suggest the same object and discard the others. The non-max suppression process is what it is. Finding each RoI's precise location on the feature map is the next step. It is known as RoIAlign [17–19].

3. **RoIAlign**

In order to prepare feature vectors for further operations, in accordance with RPN's classification of RoI, region of interest alignment, or RoIAlign, pulls feature vectors from a feature map. By scaling [20], we match RoI with their corresponding regions on the feature map. These regions can be found in a variety of places, shapes, and aspect ratios. We sample over relevant aligned sections of the feature map in order to obtain feature tensors of uniform shape. Then, in order to extract features even further, we insert each RoI's feature map into a collection of residual bottleneck blocks. The object detection branch and the mask creation branch will process the results, which reflect each RoI's finer feature map.

4. **Object Detection Branch**

We can predict an object's object category and a more precise instance bounding box if we have an individual RoI feature map. The feature vectors of the last n classes and $4n$ instances have been converted into the bounding box coordinates of this branch, which is a fully connected layer. We successively feed the RoI feature map to a convolutional layer and a transposed convolutional layer on the mask generating branch. The network on this branch is fully convolutional. For one class, only one binary segmentation mask is constructed. Next, based on the object detection branch's class prediction, the output mask is produced. This is one way that per-pixel's mask prediction might avoid conflict between different classes [21].

5. **Mask Generation Branch**

It is final branch of mask R-CNN. It consists of fully convolutional neural network. It generates binary mask for each instance in an image. Instance mask generation is achieved by combining bounding box object detection and binary mask generation for each class.

4 Experiments and Results

The Mask R-CNN model yields accurate results when tested on own dataset. To properly identify instances of digital images during segmentation tasks, the Mask R-CNN architecture has been frequently used. A third branch is added to the faster R-CNN architecture of a Mask R-CNN model, a region-based convolutional neural network, which generates object masks concurrently with the existing branch for bounding box recognition.

4.1 *The Steps Involved in Mask R-CNN for Partially Occluded Area in Images*

Step 1: Data Collection and Cleaning

In this step, we collect raw data from Google images. Here our experiment is occluded area detection in face images. Here the occluded area is face area covered by masks, sunglasses, scarves, etc. In this research, we will take masked face images as shown in Fig. 8. Here we are taking own dataset that dataset named celebrities.

Step 2: Image Annotation

The technique of labeling images to describe the desired qualities of your data is known as image annotation. Depending on the quality of your data, the outcome is then utilized to train a model and complete computer vision tasks with the desired level of accuracy. In this research, we are using VGG image annotator to target our occluded area in images. In dataset folder create two folders train and validation folders, put training images into train folder and validation images into the validation folder along with their annotations.

Step 3: Model Training (Bounding Box Annotation and Single Class Classification)

After collecting the data, we need to train our model using Mask R-CNN algorithm as discussed in Sect. 3.

Fig. 8 Celebrities' dataset

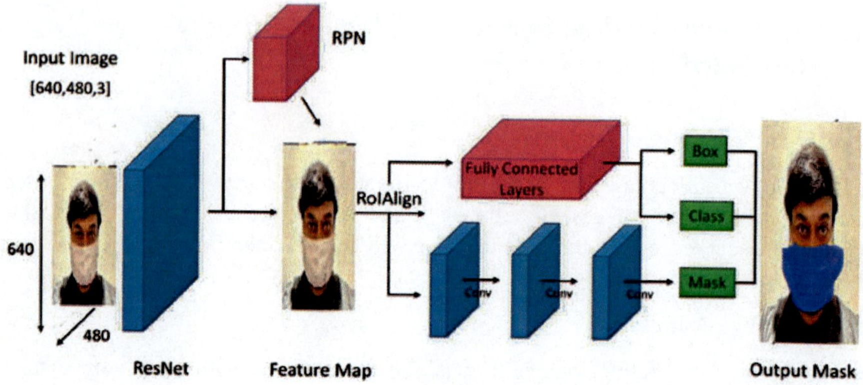

Fig. 9 Testing the model based on new input image

Step 4: Model Testing and Evaluation

After training the model, we need to test our model. For testing, we have to take test images other than training as shown in Fig. 9. After testing, you get accurate results based on the training model.

Step 5: Predicted Results

After testing the model, we get the accuracy of our model and print as our desired output.

5 Conclusion

Image segmentation is important step to reduce the chance of data loss in images or videos. Image segmentation is very efficient technique to retrieve information from visual scenes. In this paper, we explained different types of image segmentation and different techniques used in image segmentation. The main objective of this paper is to classify occluded area in images based on Mask R-CNN image segmentation technique and own dataset named as celebrities' dataset. This architecture achieves satisfactory results.

References

1. Fei C, Chun T, Qi Z, Qin S, Yaqian W, Hongjun T (2021) Application of image segmentation technology in medical image processing. Sci Technol Wind 11(36):70–72. https://doi.org/10.19392/j.cnki.1671-8007341.202136024

2. Yu Y, Wang C, Fu Q, Kou R, Huang F, Yang B, Yang T, Gao M (2023) Techniques and challenges of image segmentation: a review. Electronics. 12(5):1199
3. Zuo X, Lin H, Wang D, Cui Z (2022) A method of crop seedling plant segmentation on edge information fusion model. IEEE Access 10:95281–95293
4. Sehar U, Naseem ML (2022) How deep learning is empowering semantic segmentation: Traditional and deep learning techniques for semantic segmentation: a comparison. Multimed Tools Appl 81(21):30519–30544
5. Zhang X, Yao QA, Zhao J, Jin ZJ, Feng YC (2022) Image semantic segmentation based on fully convolutional neural network. Comput Eng Appl 44:45–57
6. Zhang H, Sun H, Ao W, Dimirovski G (2021) A survey on instance segmentation: recent advances and challenges. Int J Innov Comput Inf Control 17:1041–1053
7. Dai J, He K, Li Y, Ren S, Sun J (2016) Instance-sensitive fully convolutional networks. In: ECCV
8. Shen L, Su J, Huang R, Quan W, Song Y, Fang Y, Su B (2022) Fusing attention mechanism with Mask R-CNN for instance segmentation of grape cluster in the field. Front Plant Sci 13:934450
9. Gite S, Mishra A, Kotecha K (2022) Enhanced lung image segmentation using deep learning. Neural Computing and Applications, pp 1–15
10. Sehar U, Naseem ML (2022) How deep learning is empowering semantic segmentation
11. Minaee S (2021) Image segmentation using deep learning: a survey
12. Minaee S, Boykov Y, Porikli F, Plaza A, Kehtarnavaz N, Terzopoulos D (2022) Image Segmentation Using Deep Learning: A Survey. IEEE Trans Pattern Anal Mach Intell 44:3523–3542
13. He K, Gkioxari G, Dollr P, Girshick R (2017) Mask R-CNN. ArXiv
14. Jiale L, Kai X, Hongping J (2019) Research on image segmentation method based on Python language. Electron World 24:64–65
15. Lin T, Dollr P, Girshick R, He K, Hariharan B, Belongie S (2016) Feature pyramid networks for object detection. In: CoRR
16. Yuting X (2020) Research on segmentation, classification and recognition methods of cytopathological images. School Comput Sci. https://doi.org/10.26918/d.cnki.ghngc.8812020.000859
17. Dai J, Li Y, He K, Sun (2016) R-FCN: object detection via region-based fully convolutional networks. In: NIPS
18. Ren S, He K, Girshick R, Sun J. Faster R-CNN: towards real-time object detection with region proposal networks. In: NIPS
19. Li Y, Qi H, Dai J, Ji X, Wei Y (2017) Fully convolutional instance-aware semantic segmentation. In: CVPR
20. Guiyan T, Chunliang G (2021) Pantograph image segmentation based on multi-scale self-attention mechanism. Inf Technol Inf 32(12):164–167
21. Tianshu L, Jiandong F, Yudong Z (2022) Comparative study on target detection methods based on image segmentation. Comput Age 25(1):14–18. https://doi.org/10.16644/j.cnki.cn33-1094/tp.2022.01.004

Application of IP Network Modeling Platforms for Cyber-Attack Research

Ivan Nedyalkov and Georgi Georgiev

Abstract This work proposes the use of IP network modeling platforms to study cyber-attacks. For the purpose of this work, the GNS3 platform is proposed to be used due to the range of advantages and functionalities it offers. In this work, a model of an IP network is created in which voice streams are exchanged (VoIP connections are established) between users. Kali Linux is used as the attacker because of its capabilities and tools it offers. The device under attack is the Asterisk Free PBX. It is subjected to various TCP flood attacks (TCP SYN, TCP FIN, and TCP RST). The goal is to check how Asterisk reacts when using the different TCP flood attacks and in which attack, the consequences will be the most severe. Ports 80 and 5060 are under attack. A characterization of the traffic and studying of the impact of the applied cyber-attacks on the exchanged voice streams during the attacks are done. Well-known traffic monitoring and delay measurement tools were used during the study.

Keywords Asterisk · Cyber-attacks · DoS attack · GNS3 · IP modeling · Kali Linux · Traffic characterization · VoIP

1 Introduction

With the development of the technologies, almost any device now can be connected to the Internet to be controlled or monitored remotely. Thus, all such devices can be victims of cyber-attacks [1–7]. Therefore, it is advisable to subject a device to a cyber-attack beforehand in order to be able to test different hypotheses, as well as to observe how the device will react when it is subjected to the respective attack: will it

I. Nedyalkov (✉) · G. Georgiev
South-West University "Neofit Rilski", 66 Ivan Mihailov Str., Blagoevgrad 2700, Bulgaria
e-mail: i.nedqlkov@swu.bg

G. Georgiev
e-mail: g.georgiev@swu.bg

S. Kumar et al. (eds.), *Fourth Congress on Intelligent Systems*, Lecture Notes in Networks and Systems 868, https://doi.org/10.1007/978-981-99-9037-5_18

be able to continue functioning normally, will it be able to be accessed despite being subjected to the attack(s), and other hypotheses.

To be able to realize this, it is necessary to make an isolated experimental network. The option of connecting the device to an already existing IP network and subjecting it to various attacks there is unfeasible and strictly forbidden because the cyber-attack(s) are likely to affect the operability of the network. When it is not possible to build a physical experimental IP network due to lack of relevant network equipment or due to inability to purchase it, the use of IP network modeling platforms is the solution to this problem [8–16].

The aim of this work is to demonstrate the capabilities of the IP network modeling platform GNS3 in the study of cyber-attacks. For the purpose of this work, a model of an IP network is created in which there is an IP telephone exchange, users connected to it, and an attacking machine that implements a DoS attack against the telephone exchange. Characterization of the network traffic under attack has been done. In this way, it will be verified under different DoS attacks (attacking with different TCP packets) what will be the reaction of the PBX.

2 Related Work

In [17], the authors propose the use of the GNS3 platform to create an isolated virtual laboratory for monitoring and studying DoS attacks. In this isolated platform, attacks will be isolated from the "outside world" and can be more easily controlled. Thus, students will be able to run various experiments and observe the network response during the attacks.

In [18], the authors study the impact of cyber-attacks on the communication part of the smart grid. Using the GNS3 platform, the authors model the communication part of the smart grid. Thanks to the GNS3 they can observe what happens in the communication network during a cyber-attack and determine what the attack is.

In [19], the authors present the use of GNS3 to create a distance learning cyber security platform based on the Cisco platform Certified CyberOps course. Through this platform, students will be able to observe the processes in a network during a cyber-attack, experiment, manage network security, and many other capabilities. The platform made using GNS3 is closed and isolated from the "outside world" making it ideal for the purpose of the work.

Further work on the subject can be found in [20–24].

As can be seen from the presented review of related work, it is evident that IP network modeling platforms are ideal for the study of cyber-attacks applied to IP networks. The platforms offer a "closed" and control-manageable environment in which any experiments can be translated. Current work continues and further develops existing studies. What is new in this work is the use of network monitoring tools to assess the impact of the cyber-attacks on the performance of the attacked device.

3 Platform, Research Methodology, and Used Tools

3.1 Used Platform

The used platform for modeling of IP networks is GNS3 [25]. This platform has a number of advantages and capabilities that make GNS3 the preferred choice:

- it works with disk images of real operating systems, of real network devices such as routers, switches, Layer 3 switches, and much more. The models thus created will most closely resemble real, physical IP networks built with such network devices;
- integration with network monitoring tools;
- ability to connect the modeled network to a real IP network;
- completely free, unlike other platforms where some of the functionality must be paid for.

3.2 Study Methodology

The research methodology is as follows. Telephone conversations are continuously built up between users. During these conversations, the Asterisk Free PBX is continuously subjected to DoS attacks with different types of TCP packets, attacking port 80, the port through which Asterisk is remotely accessed through a browser and configured and port 5060 as well. The packet size, in all attacks, is 1024 bytes. The goal is to verify for which TCP packet types, the DoS attack is most effective and for which it has no effect while the call is in progress—no drops, drops, and for which TCP packets access to the Asterisk Free PBX management, monitoring and configuration menu is possible and for which TCP flooding attacks access is not possible.

3.3 Used Tools

The used tools for monitoring the network and characterizing its traffic are:

- Network Protocol Analyzer—Wireshark [26]. It is used to monitor the traffic in all nodes of the experimental IP network.
- Two-way delay measurement tools—Colasoft ping tool [27] and Solarwinds Traceroute NG [28] were used. The Colasoft ping tool was used to track the change in round-trip delay (RTD) over the entire study period, and Solarwind Traceroute NG was used to track the instantaneous RTD values.
- Network analyzer—Capsa Enterprise Trial [29]. This tool is used to observe the traffic in the modeled network and its characterization.

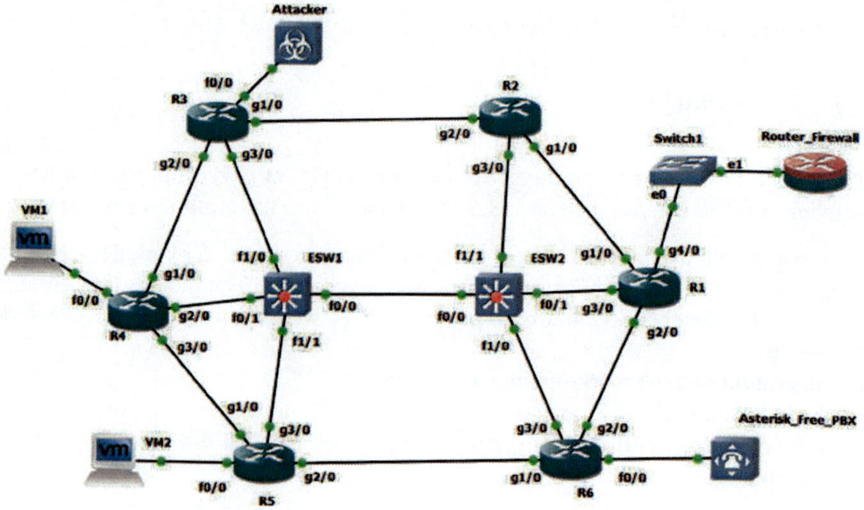

Fig. 1 Topology of the modeled experimental IP network

Additionally, mathematical distributions are used for further analysis [30, 31].

4 Topology of the Experimental Network

Figure 1 presents the topology of the modeled network.

The IP network model is composed of six routers (R1–R6) and two Layer 3 switches (ESW1 and ESW2), but operating only as switches and one unmanaged switch Switch1. Router_Firewall is a module implemented by the pFSense platform and through it the access to the Internet is provided. VM1 and VM2 are Asterisk Free PBX users. These are virtual machines. At-tacker is the virtual machine through which the DoS attack to Asterisk is executed. This is Kali Linux. In the modeled network, the EIGRP dynamic routing protocol is used along with the MPLS technology.

5 Results

Figure 2 shows what the Asterisk network load looks like before a DoS attack has been applied to the PBX. As it can be seen the load is even in both directions. The result is for a phone call between the two subscribers VM1 and VM2. This result will be used as a baseline against which to compare the load when the PBX is under attack.

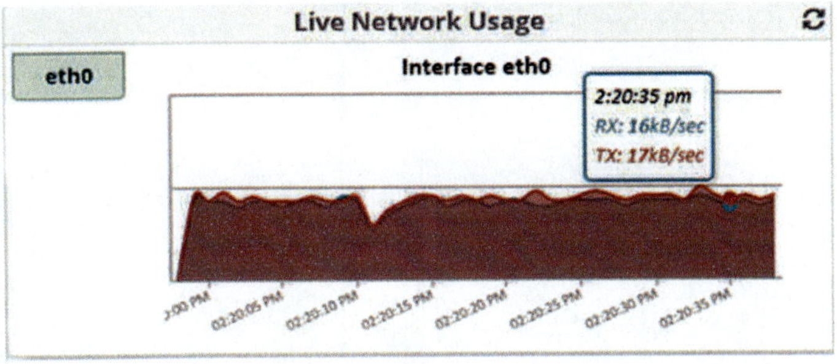

Fig. 2 Normal network usage

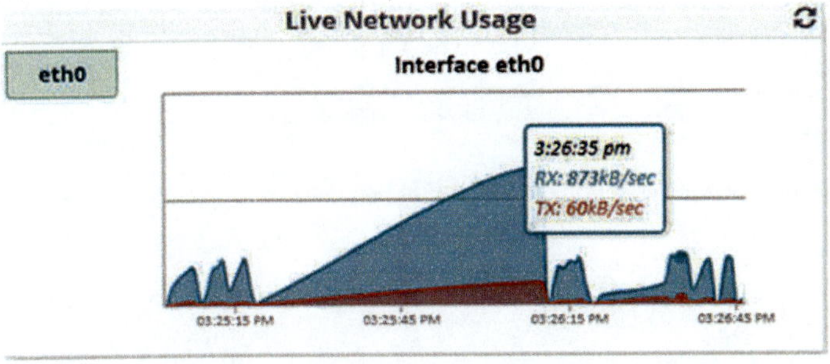

Fig. 3 Network usage during TCP SYN flood attack

Figure 3 presents the network load when Asterisk is subjected to a DoS attack via TCP SYN flood. As it can be seen from the figure, in the receive direction Asterisk is much more loaded than in the transmit direction due to the DoS attack. Asterisk is accessed through a browser, difficult but it is accessed.

Figure 4 shows the summarized results for a voice stream that is exchanged during the TCP SYN attack. The results are from Wireshark. In the direction from VM1 (192.168.200.2) to the Asterisk (192.168.40.2), the voice stream parameters are within the norm, but in the other direction the impact of the TCP SYN attack affects the voice stream parameters: there are 21% lost packets, with 1% lost packets allowed, the maximum jitter values are outside the norm, but the mean jitter value, that is most important, is below 30 ms [32, 33]. This is due to the improved voice stream parameters in the direction from VM1 to the Asterisk. Conversations are choppy and intermittent, but not falling apart.

Figure 5 presents the Colasoft Capsa results, which show the TCP SYN packets generated by Cali Linux. As it can be seen, the TCP SYN packets are many more than the TCP SYNACK packets because Asterisk fails to respond to all TCP SYN packets because it has been flooded. This figure shows that the DoS attack is successfully executed—flooding the Asterisk.

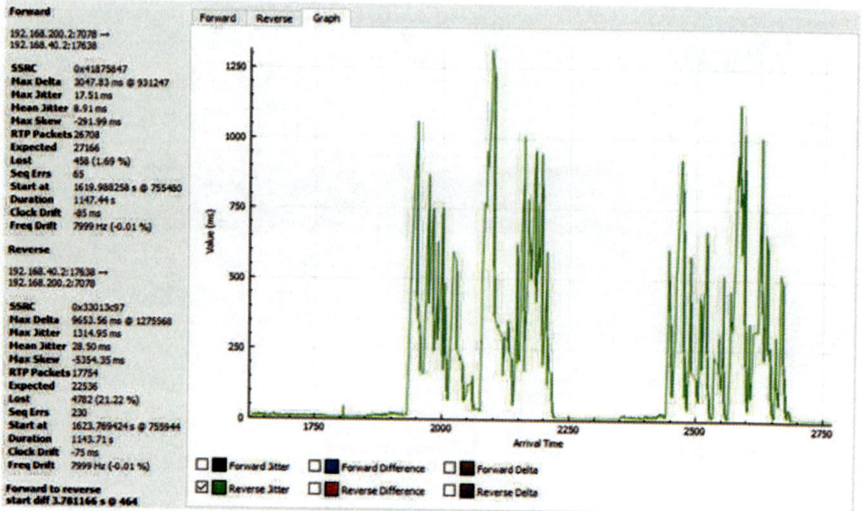

Fig. 4 Voice flow parameters during the TCP FIN flood attack

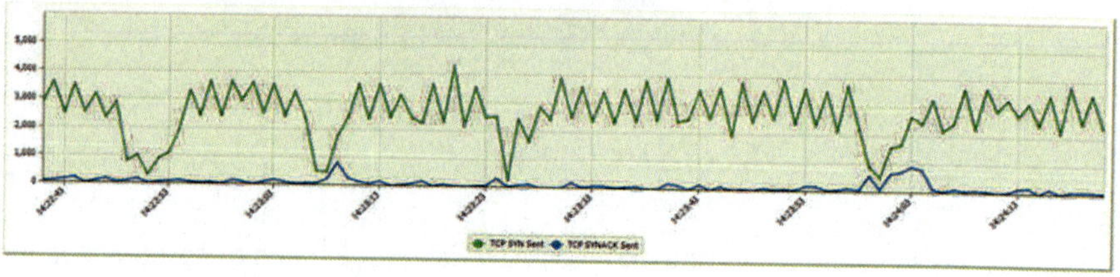

Fig. 5 TCP SYN send TCP SYN ACK send

Figure 6 presents the network load result when Asterisk is subjected to a DoS attack via TCP FIN flood. As it can be seen from the figure in the receive direction Asterisk is again much more loaded than in the transmit direction due to the DoS attack. Here the IP PBX is busier in receive direction than during the TCP SYN flood attack. Asterisk can still be accessed through a browser, but more difficult.

Figure 7 shows the summarized results for a voice stream exchanged during the TCP FIN flood attack. Again, in the direction from VM1 (192.168.200.2) to the Asterisk (192.168.40.2) the voice stream parameters are within the norm, but in the reverse direction the impact of the TCP FIN attack affects the voice stream parameters: there are 20% lost packets, with an allowable 1%, the maximum jitter values are outside the norm, but the average value is again below 30 ms, which is due to the better voice flow parameters in the direction from VM1 to Asterisk. Calls continue to chop and drop, but not falling apart.

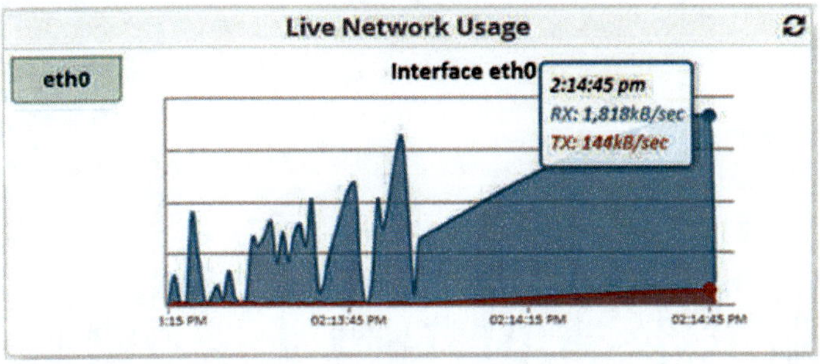

Fig. 6 Network usage during TCP FIN flood attack

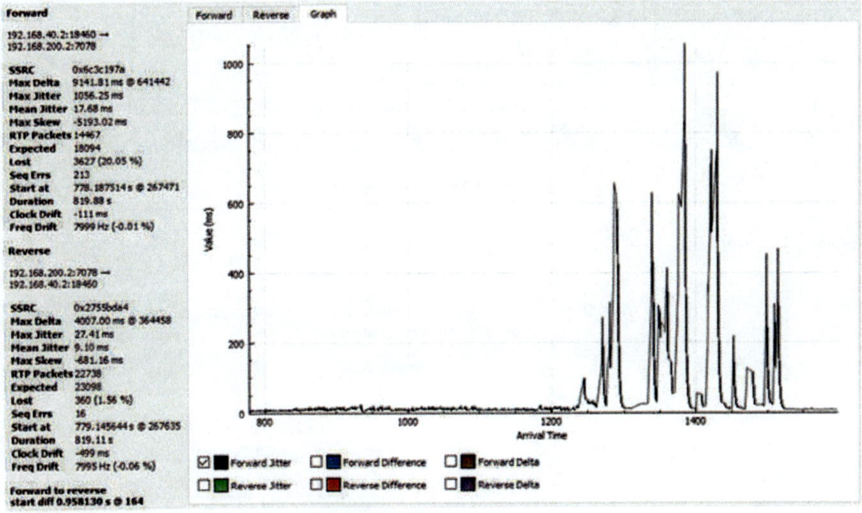

Fig. 7 Voice flow parameters during the TCP FIN flood attack

Figure 8 presents the Colasoft Capsa results, which show the TCP FIN packets generated by Cali Linux. This figure shows that the DoS attack was successfully executed—flooding Asterisk.

Figure 9 presents the network load when the Asterisk is subjected to a DoS attack via TCP RST flood. As it can be seen from the figure in the receive direction Asterisk is again much more loaded than in the transmit direction due to the DoS attack. Asterisk continues to be accessed through a browser.

Figure 10 presents the summarized results for a voice stream exchanged during the TCP RST flood attack. In the direction from VM2 (192.168.20.2) to the Asterisk (192.168.40.2), the voice stream parameters are in the normal range. In the reverse direction, the impact of the TCP RST attack affects the voice stream parameters: there are 12% lost packets, the maximum jitter values are still outside the norm, the

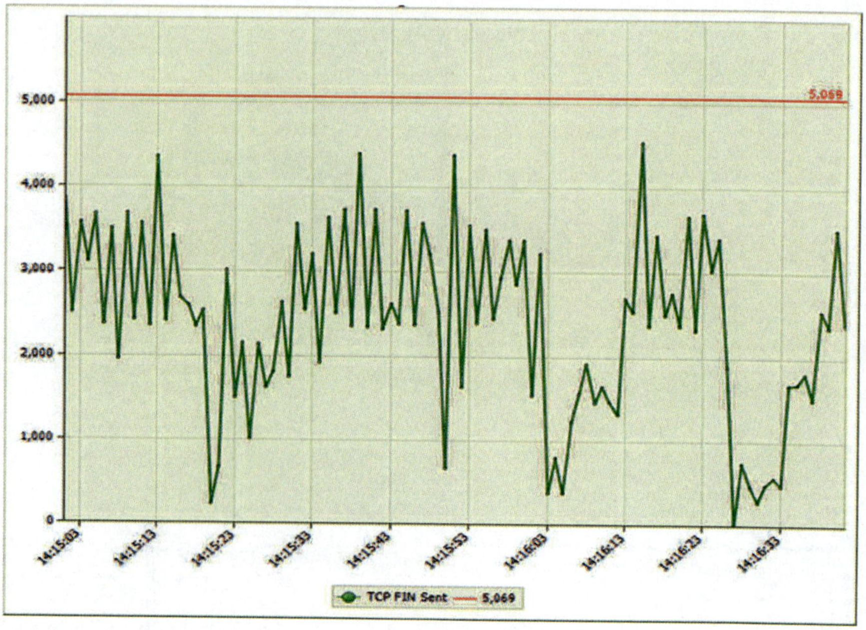

Fig. 8 TCP FIN send

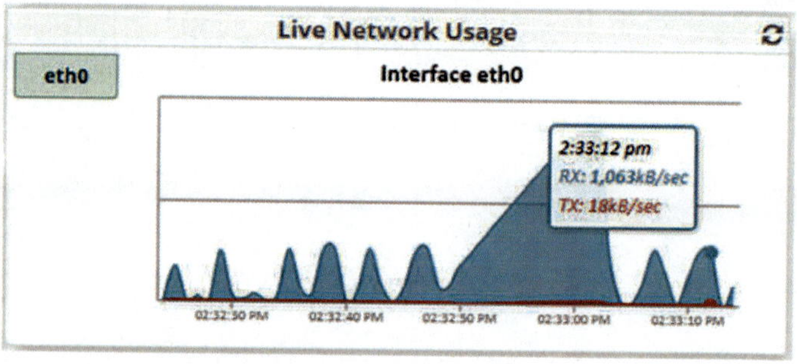

Fig. 9 Network usage during TCP RST flood attack

average value is still below 30 ms, which is due to the better voice flow parameters in the direction from VM2 to Asterisk. Calls continue to drop in and out but do not break up.

Figure 11 presents the Colasoft Capsa results showing the TCP RST packets generated by Cali Linux, indicating that the DoS attack was successfully executed-flooding Asterisk.

Figure 12 presents the network load when the Asterisk is subjected to a DoS attack via TCP SYN flood, but port 5060 is attacked, not as in the previous results where port 80 is attacked. As it can be seen from the figure in the receive direction Asterisk

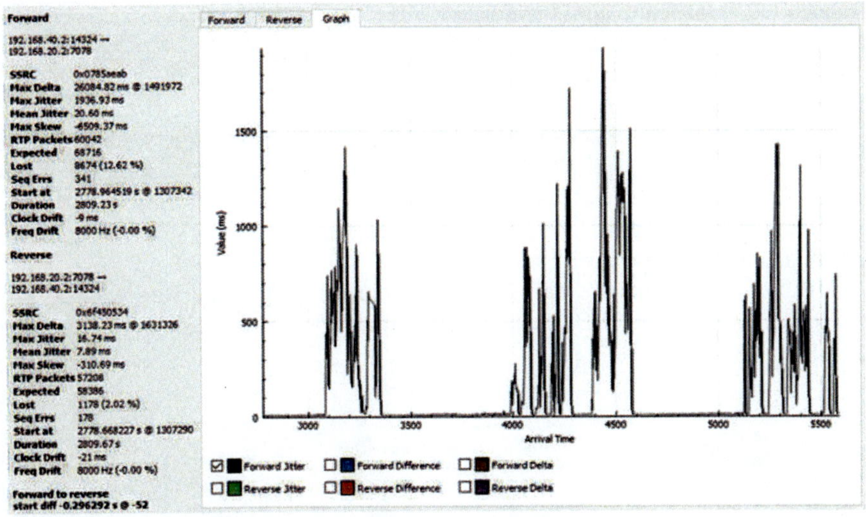

Fig. 10 Voice flow parameters during the TCP RST flood attack

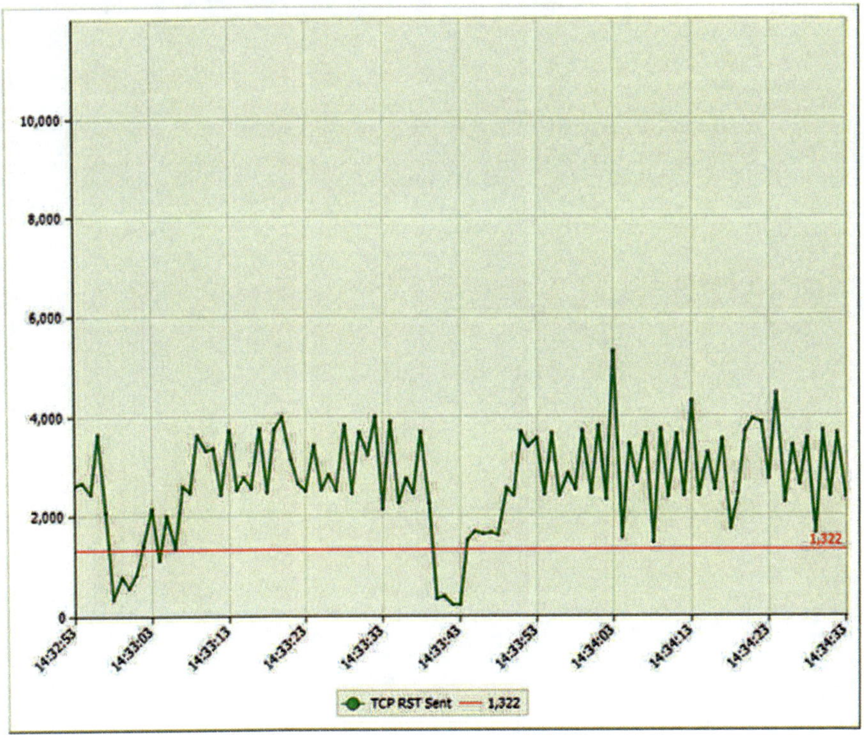

Fig. 11 TCP RST send

is again much loaded than in the transmit direction due to the DoS attack. Asterisk continues to be accessed through a browser.

Figure 13 shows the summarized results for a voice stream exchanged during the TCP RST flood attack. In the direction from VM2 (192.168.20.2) to the Asterisk (192.168.40.2), the voice stream parameters are in the normal range. In the opposite direction, the impact of the TCP SYN attack affects the voice flow parameters, but not as much as the previous DoS attacks on port 80: there are 7% lost packets, the maximum jitter values are still outside the norm, the average is still below 30 ms. Conversations continue to be choppy and dropped but do not break up.

Figure 14 represents the traffic generated during each of the DoS attacks. The results are from Colasoft Capsa, and the sample interval is set to be 1 s to get a

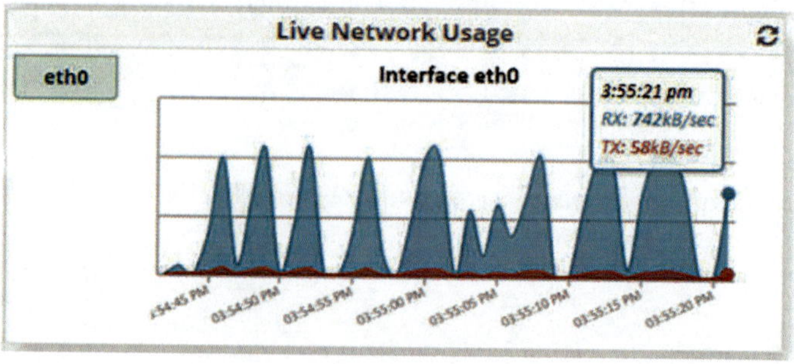

Fig. 12 Network usage during TCP SYN flood attack on port 5060

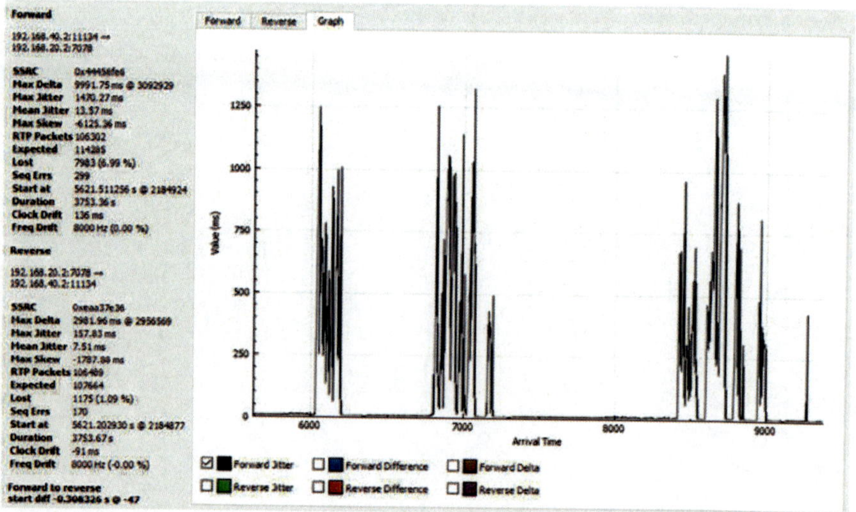

Fig. 13 Voice flow parameters during the TCP SYN flood attack on port 5060

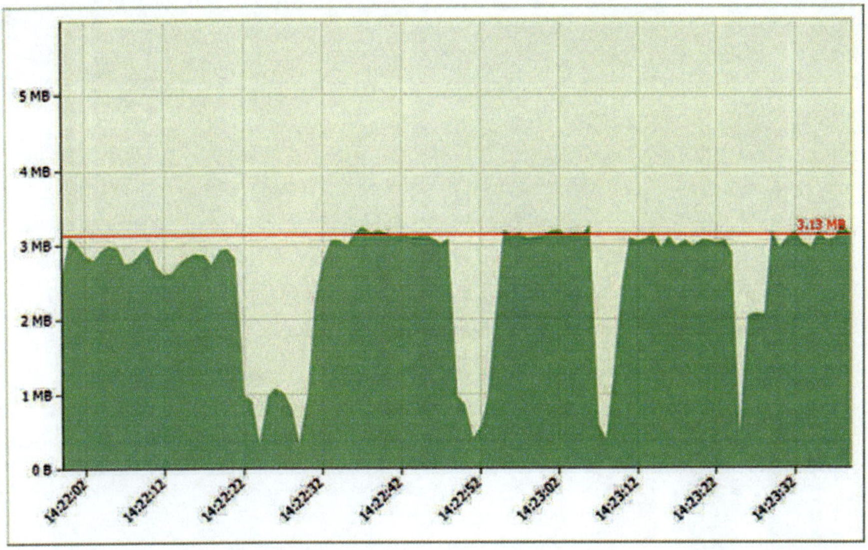

Fig. 14 Generated traffic during a DoS attack

more accurate measurement. As it can be seen from the result the generated traffic is significant, which is normal to be able to flood the attacked device.

Figure 15a represents the traffic generated by Kali Linux during all DoS attacks for the entire study period, and Fig. 15b presents the traffic handled by the Asterisk, again for the entire study period. As it can be seen from the results, the generated traffic by the Kali Linux is significant, which is normal because the network is flooded by this traffic, and it must be significant for the DoS attack to accomplish its goal. For the Asterisk results, there are peaks where the traffic is significant—these are the times of the DoS attacks. When the PBX is operating normally, the traffic it handles is very small because the traffic is only from the phone calls, which do not generate much traffic, unlike the DoS attacks, where a lot of traffic is generated to be able to flood the attacked target.

Figure 16 shows which ports generated the most traffic at the Asterisk. As it can be seen from the result these ports are: TCP port 80, which is used for DoS attacks; UDP port 7078, which handles WebSocket Server requests and is active because of the DoS attacks. Port 5060 is used both by SIP to establish phone connections and exchange signaling information in VoIP calls and for DoS attacks. Because of that the traffic generated by this port is significant. If there were no DoS attacks targeting this port, the traffic would be very small because under normal conditions the signaling information that is exchanged with SIP is very small. The other UDP ports are used to carry the speech information (RTP packets) that are exchanged between VM1, VM2, and the Asterisk.

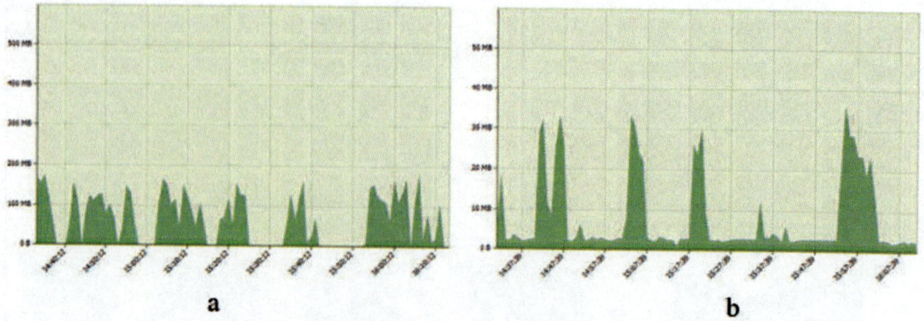

Fig. 15 Total generated traffic from Kali Linux (**a**) and Asterisk (**b**)

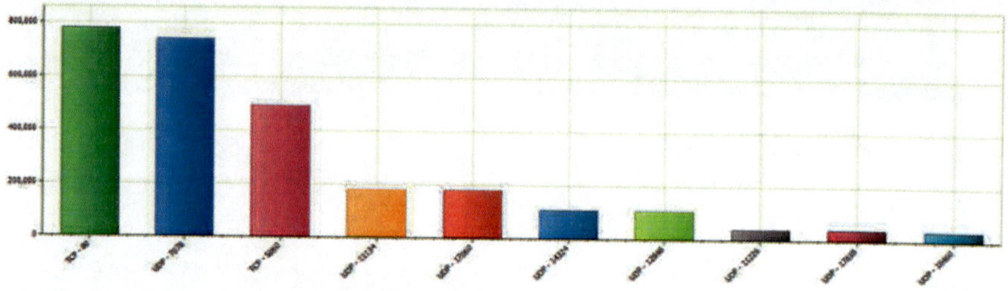

Fig. 16 Top ports by total traffic at Asterisk

Figure 17 shows which ports generated the most traffic at the Cali Linux. As it can be seen from the result, these are TCP port 80 and port 5060, which are used for DoS attacks.

Figure 18 presents the variation of the RTD between VM1 and the Asterisk over the entire study period. The graph was obtained from the Colasoft Ping Tool. As it can be seen from the graph, except for a few spikes, the round-trip delay is around 200 ms. Such a delay value is high, but within the norm, because in VoIP calls the delay in one direction should not exceed 150 ms, i.e., 300 ms in both directions [32, 33]. Because of the constant DoS attacks to the Asterisk, the delay values are inflated.

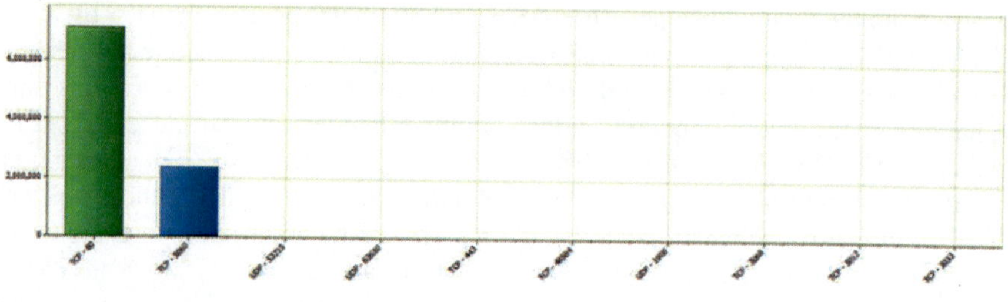

Fig. 17 Top ports by total traffic at Kali Linux

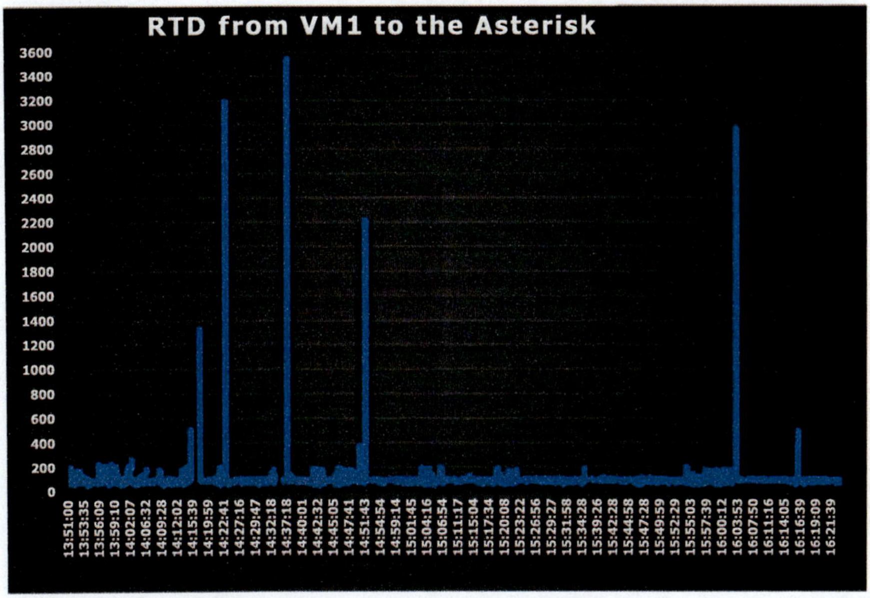

Fig. 18 RTD between VM1 and the Asterisk for the whole study period

Figure 19 represents the instantaneous values of the RTD between VM1 and the Asterisk. The result is from Solarwinds Traceroute NG. 192.168.200.1 is the IP address of port f0/0 of R4 which is the gateway address of VM1. 192.168.40.2 is the IP address of the Asterisk. The other IP addresses are the IP addresses of the ports through which the packets pass. As it can be seen from the result, the values are within normal limits, even well below the 150 ms per direction limit. This shows that the DoS attacks do not affect the network latency.

Figure 20 presents a mathematical distribution with Pareto approximation for the exchanged packets by size between Kali Linux and Asterisk. Using this distribution, one can get an idea of the size distribution between the exchanged packets. As it can be seen from the graph, there are only packets of two sizes: those with a size of 1024 bytes are the packets that are used for the DoS attacks (TCP SYN, TCP FIN, TCP RST, and TCP 5060), the remaining packets with a size of up to 75 bytes are

Hop	IP	PL (%)	Now (ms)	Avg (ms)	min Latency max
1	192.168.200.1	0.0	3.8	3.8	==================
2	10.0.6.2	0.0	34.2	34.2	==================
3	10.0.7.2	0.0	66.8	66.8	==================
4	192.168.40.2	0.0	87.9	87.9	==================

Fig. 19 Instantaneous values of the RTD between VM1 and the Asterisk

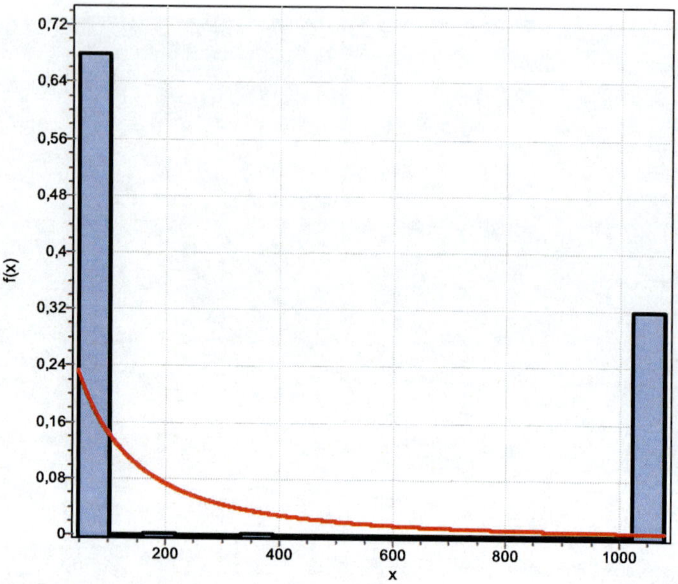

Fig. 20 Mathematical distribution for packet length

service packets that are exchanged between network devices, EIGRP packets and ACK packets from Asterisk to some of the TCP packets used for DoS attacks.

6 Conclusions

The GNS3 platform was chosen and a working IP network model was created. The network has two users and one IP PBX. Voice streams are exchanged between the users.

DoS attacks have been successfully carried out by an attacking user—the Kali Linux. The successful execution of the DoS attacks is evident from the network monitoring results. Different DoS attacks were used on port 80 and port 5060 using different TCP flood packets: TCP SYN, TCP FIN, and TCP RST. For all three TCP flood attacks, no difference in the response of the Asterisk has been observed. During all three types of DoS attacks, the performance of Asterisk has been hampered. The system can be accessed, but with difficulty: every second or third access attempt through a browser is successful. Asterisk is difficult to access, the configuration menus can be operated, but it is difficult. Disconnection and reconnection are observed as access is not completely blocked, but is quite difficult.

In terms of making phone calls during the attacks to port 80, calls are choppy, get dropped in the call, but do not break up. This is evident from the analysis of the voice streams during the attacks, which shows that the parameters of the streams are severely degraded, but not to the point of being disrupted by the Asterisk.

The carried-out study demonstrates the capabilities of the GNS3 to create a model of an experimental IP network in which various cyber-attacks can be studied and the response of the attacked devices can be observed. Also, the processes taking place in an IP network where cyber-attacks take place can be observed. Thanks to the capabilities of GNS3, different user devices can be subjected to different cyber-attacks and their reaction to the attacks can be monitored. Thus, it will be possible to take precautions to protect the examined devices and reduce the probability of the device being subjected to cyber-attacks.

Future work is considering studying the impact of other cyber-attacks on the PBX performance as well as voice call quality. It is considered to make a comparison between the methods of protection from DoS attacks and to propose the most appropriate method where the best ratio between ease of implementation (time of commissioning), use of computational resources, and attacks prevented is available.

References

1. Bharathi V, Kumar CV (2022) Enhanced security for an IoT devices in cyber-physical system against cyber attacks. In: 2022 International Conference for Advancement in Technology (ICONAT), Goa, India, pp 1–5. https://doi.org/10.1109/ICONAT53423.2022.9725884
2. Antzoulis I, Chowdhury MM, Latiff S (2022) IoT security for smart home: issues and solutions. In: 2022 IEEE International Conference on Electro Information Technology (eIT), Mankato, MN, USA, 2022, pp 1–7. https://doi.org/10.1109/eIT53891.2022.9813914
3. Hristova V, Cherneva G, Borisova D (2021) Radio communication system with a high degree of protection of information against non-allowed access. In: Electro-Optical and Infrared Systems: Technology and Applications XVIII and Electro-Optical Remote Sensing XV, vol 11866, pp 267–271. SPIE
4. W Dimitrov B Jekov P Hristov 2021 Analysis of the cybersecurity weaknesses of DLT ecosystem R Silhavy Eds Software engineering and algorithms, CSOC 2021 Lecture Notes in Networks and Systems 230 Springer Berlin
5. Dimitrov W, Dimitrov G, Spassov K, Petkova (2021) Vulnerabilities space and the superiority of hackers. In: 2021 International Conference Automatics and Informatics (ICAI), Varna, Bulgaria, pp 433–436. https://doi.org/10.1109/ICAI52893.2021.9639579
6. Hajamohideen F, Karthikeyan S (2020) Cyber threats detection in the smart city using bigdata analytics. In: 3rd Smart Cities Symposium (SCS 2020), Online Conference, pp 233–238. https://doi.org/10.1049/icp.2021.0872
7. Choudhary A, Chaudhary A, Devi S (2022) Cyber security with emerging technologies and challenges. In: 2022 4th International Conference on Advances in Computing, Communication Control and Networking (ICAC3N), Greater Noida, India, pp 1875–1879
8. Durmuş Ö, Varol A (2021) Analysis and modeling of cyber security precautions. In: 2021 9th International Symposium on Digital Forensics and Security (ISDFS), Elazig, Turkey, pp 1–8. https://doi.org/10.1109/ISDFS52919.2021.9486345
9. Cherneva G, Dimkina E (2013) Simulation and examination of a signal masking chaotic communication system, based on the duffing oscillator. In: Communications–Scientific Letters of the University of Zilina, vol 15(2A), pp 6–10. https://doi.org/10.26552/com.C.2013.2A.6-10

10. Rahim K, Khaliq H (2021) Modeling and simulation challenges for cyber physical systems from operational security perspective. In: 2021 International Conference on Cyber Warfare and Security (ICCWS), Islamabad, Pakistan, pp 63–69
11. TD Tashev MB Marinov RP Tasheva AK Alexandrov 2021 Generalized nets model of the LPF-algorithm of the crossbar switch node for determining LPF-execution time complexity AIP Conf Proc 2333 1 090039
12. Sun Z, Zhang S (2021) Modeling of security risk for industrial cyber-physics system under cyber-attacks. In: 2021 4th IEEE International Conference on Industrial Cyber-Physical Systems (ICPS), Victoria, BC, pp 361–368
13. TD Tashev AK Alexandrov DD Arnaudov RP Tasheva 2022 Large-Scale computer simulation of the performance of the generalized nets model of the LPF-algorithm I Lirkov S Margenov Eds Large-scale scientific computing, LSSC 2021 Lecture Notes in Computer Science 13127 Springer Cham
14. Tashev TD, Marinov MB, Arnaudov DD, Monov VV (2022) Computer simulations for determining of the upper bound of throughput of LPF-algorithm for crossbar switch. AIP Conf Proc 2505(1)
15. F Sapundzhi 2019 Computer simulation and investigations of the roof mount photovoltaic system Intl J Online Biomed Eng (iJOE) 15 12 88 96 https://doi.org/10.3991/ijoe.v15i12.10869
16. FI Sapundzhi MS Popstoilov 2020 Maximum-flow problem in networking Bulg Chem Commun 52 192 196
17. Al Kaabi S, Al Kindi N, Al Fazari S, Trabelsi Z (2016) Virtualization based ethical educational platform for hands-on lab activities on DoS attacks. In: 2016 IEEE Global Engineering Education Conference (EDUCON), Abu Dhabi, United Arab Emirates, pp 273–280
18. Alrashide A, Abdelrahman MS, Kharchouf I, Mohammed OA (2022) GNS3 communication network emulation for substation goose based protection schemes. In: 2022 IEEE International Conference on Environment and Electrical Engineering and 2022 IEEE Industrial and Commercial Power Systems Europe (EEEIC/I&CPS Europe), Prague, Czech Republic, pp 1–6
19. Uramová J, Segeč P, Papán J, Brídová I (2020) Management of cybersecurity incidents in virtual lab. In: 2020 18th International Conference on Emerging eLearning Technologies and Applications (ICETA), Košice, Slovenia, pp 724–729
20. Hajdarevic K, Kozic A, Avdagic I, Masetic Z, Dogru N (2017) Training network managers in ethical hacking techniques to manage resource starvation attacks using GNS3 simulator. In: 2017 XXVI International Conference on Information, Communication and Automation Technologies (ICAT), Sarajevo, Bosnia and Herzegovina, pp 1–6
21. Abisoye OA, Shadrach Akanji O, Abisoye BO, Awotunde J (2020) Slow hypertext transfer protocol mitigation model in software defined networks. In: 2020 International Conference on Data Analytics for Business and Industry: Way Towards a Sustainable Economy (ICDABI), Sakheer, Bahrain, pp 1–5
22. RE Pérez Guzmán M Rivera PW Wheeler G Mirzaeva EE Espinosa JA Rohten 2022 Microgrid power sharing framework for software defined networking and cybersecurity analysis IEEE Access 10 111389 111405
23. Taib AM, Abdullah AA-S, Ariffin MAM, Ruslan R (2022) Threats and vulnerabilities handling via dual-stack sandboxing based on security mechanisms model. In: 2022 IEEE 12th International Conference on Control System, Computing and Engineering (ICCSCE), Penang, Malaysia, pp 113–118
24. Dhivvya JP, Muralidharan D, Raj N, Kumar BK (2019) Network simulation and vulnerability assessment tool for an enterprise network. In: 2019 10th International Conference on Computing, Communication and Networking Technologies (ICCCNT), Kanpur, India, pp 1–6
25. Getting Started with GNS3. https://docs.gns3.com/docs/. Accessed 8 Aug 2023
26. Wireshark. https://www.wireshark.org/docs/wsug_html_chunked/. Accessed 8 Aug 2023
27. Colasoft ping tool. https://www.colasoft.com/ping_tool/. Accessed 8 Aug 2023
28. Traceroute NG. https://www.solarwinds.com/free-tools/traceroute-ng. Accessed 8 Aug 2023
29. Capsa Enterprise. https://www.colasoft.com/capsa/. Accessed 8 Aug 2023

30. MB Marinov N Nikolov S Dimitrov T Todorov Y Stoyanova GT Nikolov 2022 Linear interval approximation for smart sensors and IoT devices Sensors 22 949 https://doi.org/10.3390/s22030949
31. MB Marinov N Nikolov S Dimitrov B Ganev GT Nikolov Y Stoyanova T Todorov L Kochev 2023 Linear interval approximation of sensor characteristics with inflection points Sensors 23 2933 https://doi.org/10.3390/s23062933
32. Cisco-understanding delay in packet voice networks, white paper. https://www.cisco.com/c/en/us/support/docs/voice/voice-quality/5125-delay-details.html
33. Tim S, Christina H (2004) End-to-end QoS network design: quality of service in LANs, WANs, and VPNs. In: Part of the Networking Technology Series. Cisco Press, Indianapolis, Indiana

Enhancing Information Integrity: Machine Learning Methods for Fake News Detection

Shruti Sahu, Poonam Bansal, and Ritika Kumari

Abstract With growing and advancing technology, people have access to the Internet easily which leads to the availability of news online and can reach from one part of the world to another. Internet is the ideal environment for the growth and dissemination of malevolent and fake news. Fake news detection has become a serious problem in the recent era. Due to this, it has become important to detect whether a floating news is fake or real as it may have a serious negative impact on individuals and society. In this study, we use WELFake and Real and Fake News Classification Dataset from the Kaggle Repository for fake news classification using five machine learning algorithms: Naïve Bayes (NB) classifier, decision tree (DT), random forest (RF), support vector machine (SVM), and K-nearest neighbor (KNN). Five metrics are used: accuracy, precision, recall, $F1$-score, and AUC for comparing the model's performances. We notice that the best-performing models are DT and RF yielding an accuracy of 93.92% and 91.18% respectively.

Keywords Fake news classification · Decision tree · Random forest · Classification · Support vector machine

S. Sahu (✉) · P. Bansal · R. Kumari
Department of Artificial Intelligence and Data Sciences, IGDTUW, New Delhi, India
e-mail: shrutisahu2501@gmail.com

P. Bansal
e-mail: poonambansal@igdtuw.ac.in

R. Kumari
e-mail: ritikakumari@igdtuw.ac.in

R. Kumari
USICT, Guru Gobind Singh Indraprastha University, New Delhi, India

S. Kumar et al. (eds.), *Fourth Congress on Intelligent Systems*, Lecture Notes in Networks and Systems 868, https://doi.org/10.1007/978-981-99-9037-5_19

1 Introduction

With the emergence and improvement in recent technologies, people have become more aware about their surroundings through vast range of social media platforms. A vast variety of social media platforms as well as news websites have emerged. Nowadays, people share every other news without realizing its authenticity which may affect the people, brand, etc., by creating a wrong opinion. For example, On January 9, 2018, the South African president, Jacob Zuma, was falsely reported to have resigned by the website AllAfricaNews. For a brief period, this led to a rise in the value of the Rand before other news organizations exposed the story as bogus [1]. Events with high media coverage lead to widespread fake news such as the demonetization process in India, 2016 US presidential election, 2019 Indian General Election, and 2014 General Election in India [2]. During COVID-19, a lot of false information about home remedies and unverified medicines was spread which was misleading and unsafe. Fake news can also disrupt the human minds and alter their course of actions and growth which is not desirable for a country's growth. Thus, it has become important to detect the fake news, which eventually will help people to have positive perspectives and avoid unnecessary engagement in false rumors.

The main objective of this study is to identify the best method for categorizing fake news in the real-world scenario. In this study, we perform fake news classification on two datasets taken from the Kaggle Repository using machine learning techniques NB, DT, RF, SVM, and KNN. We observe DT and RF to be the best performers. The paper is organized as follows: Literature survey is discussed in Sect. 2. The classifiers studied are described in Sect. 3. Dataset description and performance metrics are mentioned in Sects. 4 and 5. The proposed methodology is discussed in Sect. 6. Experiment and results are shown in Sect. 7, and Sect. 8 shows the conclusion.

2 Literature Survey

In the literature, several studies conducted for the fake news detection are discussed below:

Okuhle Ngada and Bertram Haskins in 2020 investigated the study using machine learning algorithms: AdaB, DT, KNN, RF, SVM, and XGBoost on two datasets and achieved highest accuracy of 99.4% and 99.2% with SVM and AdaB, respectively [1].

Rohit Kumar Kaliyar et al. in 2019 conducted the study using machine learning algorithms: NB, KNN, DT, gradient boost, RF, and SVM on FNC-based dataset and achieved highest accuracy of 86% with Gradient [2].

Julio C. S. Reis et al. in 2019 conducted the study using machine learning algorithms: KNN, NB, SVM, RF, and XGBoost on BuzzFeed dataset and achieved highest accuracy of 85% with random forest [3].

Boosting Lilapati Waikhom and Rajat Subhra Goswami in 2019 analyzed the study using machine learning algorithms: bagging, AdaBoost (AdaB), RF, extra trees, and XGBoost on LIAR dataset and achieved highest accuracy of 70% with bagging and AdaB [4].

Arvinder Pal Singh Bali in 2019 analyzed the study using machine learning algorithms: NB, DT, RF, XGBoost, AdaB, and gradient boosting on Jruvika fake news dataset and achieved highest accuracy of 88% with gradient boosting [5].

Rahul R. Mandical et al. in 2020 investigated the study using machine learning algorithms: NB, deep neural network, and passive aggressive classifier on seven datasets [6].

Supanya Aphiwongsophon and Prabhas Chongstitvatana in 2018 analyzed the study using machine learning algorithms: NB, SVM, and neural network (NN) with SVM and NN on a dataset made by collecting information from various sources and achieved highest accuracy of 99.9% with SVM and NN [7].

Thomas Felber in 2021 performed machine SVM on normally annotated fake news dataset and achieved highest accuracy of 95.70% [8].

Hazif Yasir Ghafoor et al. in 2022 investigated the study using machine learning algorithms: passive aggression, NB, and DT on three datasets and achieved highest accuracy of 93.7% with passive aggression [9].

Smitha N. and Bharath R. in 2020 analyzed the study using machine learning algorithms: SVM, RF, linear regression, XGBoost, DT, and RF on a dataset from Kaggle using word embedding, TF-IDF vectorizer, and count vectorizer. SVM gave highest accuracy of 94% with TF-IDF vectorizer, NN gave highest accuracy of 94% with count vectorizer and NN gave highest accuracy of 90% with word embedding [10].

Vanya Tiwari in 2020 conducted the study using machine learning algorithms: KNN, linear regression, RF, and DT on two datasets from Kaggle and achieved the highest accuracy of 71% with linear regression [11].

Sairamvinay Vinjayaraghavan et al. in 2020 investigated the study using machine learning algorithms: artificial neural network (ANN), long short-term memory (LSTM), SVM, linear regression, and RF using word2vect, count vectorizer, and TF-IDF with ANN giving highest accuracy of 93.06% with word2vect, LSTM giving highest accuracy of 94.88% with count vectorizer and linear regression giving highest accuracy of 94.79% with TF-IDF [12].

Arun Nagaraj et al. in 2021 investigated the study using machine learning algorithms: SVM and NB and achieved highest accuracy of 74% with SVM [13].

Nihel Fatima Baariri and Abdelhamid Djeffal in 2021 analyzed the study using machine learning algorithm: SVM on a dataset [14].

Iftikhar Ahmad et al. in 2020 conducted the study using machine learning algorithms: linear regression, KNN, RF, SVM, and classification and regression tree (CART) on four datasets and achieved highest accuracy of 99% with RF and linear SVM [15].

Akshay Jain and Amey Kasbe in 2018 investigated the study using machine learning algorithm: NB and achieved an AUC score of 0.807 for title and 0.939 for text [16].

Jasmine Shaikh and Rupali Patil in 2020 investigated the study using machine learning algorithms: SVM, NB, and passive aggressive classifier and achieved the highest accuracy of 95.05% with SVM [17].

Karishnu Poddar et al. in 2019 analyzed the study using machine algorithms: SVM, NB, logistic regression, DT, and ANN. SVM gave highest accuracy of 92.8% with TF-DIF vectorizer and logistic regression gave highest accuracy of 91.6% with count vectorizer [18].

Awf Abdulrahman and Muhammet Baykara in 2020 conducted a study using machine learning algorithms: RF, KNN, linear SVM, logistic regression, NB, recurrent neural network with long short-term memory, convolutional neural network with long short-term memory, AdaB, XGBoost, and ANN. AdaB gave highest accuracy of 93.24% and 96.13% with count vectorizer and character level vectorizer, respectively, and linear SVM gave highest accuracy of 91.28% with N gram vectorizer [19].

Welin Han and Varshil Mehta in 2019 investigated the study using machine learning algorithms: NB, convolutional neural network, and recurrent neural network [20].

Bashar Al Asaad and Madalina Erascu in 2018 analyzed the study using machine learning algorithms: NB and linear SVM. Linear SVM gave the highest accuracy of 94% with TF-IDF model [21].

Noman Islam et al. in 2021 conducted the study using machine learning algorithms: DT, RF, SVM, and logistic regression and achieved highest accuracy of 93.15% with SVM [22].

Mykhailo Granik and Volodymyr Mesyura in 2017 investigated the study using machine learning algorithm: NB and achieved highest accuracy of 74% [23].

Pawan Kumar Verma et al. in 2021 analyzed the study using machine learning algorithms: NB, SVM, KNN, BT, bagging, and AdaB on WELFake dataset and achieved highest accuracy of 96.73% with SVM [24].

3 Classification Models

We use five machine learning algorithms in this study which have been described below:

Naïve Bayes. It is a supervised machine learning algorithm which uses Bayes theorem and is used to compute the conditional probability. It explains the likelihood of occurrence of an event happening given an event that has already happened. On the basis of likelihood, the prediction takes place [16–21, 23, 24].

Support Vector Machine. It is a supervised machine learning algorithm for classification and regression problems. The training data is first labeled and after that the algorithm defines the best hyperplane for classifying the unseen data [18, 19, 21, 22, 24].

Decision Tree. It is a machine learning algorithm that uses a tree-like structure to make predictions. Each branch represents a possible decision. In a good DT, the splits decrease the randomness of data and the clarification of information received increases [22, 24].

K-Nearest Neighbors. It is a nonparametric, supervised machine learning algorithm that makes a prediction based on the similarity of neighbors. Since the principle of similarity depends on the value of K, cases are classified by the majority of the votes of their neighbors. This occurs because similar cases are close to each other. It is utilized when the dataset is compact, tagged, and noise-free [15, 16, 19, 24].

Random Forest. It is a supervised machine learning algorithm that works on the concept of ensemble learning which is a process of combining multiple classifiers to solve a complex problem and to improve the performance of the model. It consists of multiple decision trees working together and making an individual prediction. The class with the highest votes becomes the final model prediction. This technique reduces errors as it gathers a number of individual predictions before making a final prediction [16, 19, 24].

4 Dataset

We use two datasets: WELFake and Real and Fake News Classification Dataset from the Kaggle Repository [25–27]. WELFake dataset consists of 72,134 news articles with 35,028 real and 37,106 fake news. The authors have merged data from Kaggle, Mclntire, Reuters, and BuzzFeed Political [1]. The class label 0 represents fake news and 1 represents real news. Real and Fake News Classification Dataset consists of 3729 instances with 1850 real and 1871 fake news. In the class label, fake news is represented by 0 and real news is represented by 1. The Real and Fake News Classification Dataset consists of eight null values so they have been dropped and not considered in our research [2]. Table 1 shows the description of the datasets as shown below:

Table 1 Description about the datasets

Name of dataset	Number of instances	Fake	Real
WELFake	72,134	37,106	35,028
Real and fake news	3729	1871	1850

5 Performance Metrics

Five metrics are used for comparing the performances of the models as discussed below [2, 7, 15, 17–19, 27]:

Accuracy. It refers to the number of true predictions made by the machine learning classifier as shown in Eq. (1):

$$\text{Accuracy} = \frac{\text{TP} + \text{NP}}{\text{TP} + \text{FP} + \text{TN} + \text{FN}} \tag{1}$$

Precision. It is defined as the number of true positives present out of the total positives in the dataset as shown in Eq. (2):

$$\text{Precision} = \frac{\text{TP}}{\text{TP} + \text{FP}} \tag{2}$$

Recall. It is defined as the number of data samples that the machine learning algorithm correctly identifies as belonging to the given class as shown in Eq. (3):

$$\text{Recall} = \frac{\text{TP}}{\text{TP} + \text{FN}} \tag{3}$$

AUC. It is defined as the area under the Receiver Operating Characteristic (ROC) curve and is calculated as shown in Eq. (4):

$$\text{AUC} = \frac{\text{TP rate} + \text{TN rate}}{2} \tag{4}$$

***F*1-score**. It combines precision and recall as shown in Eq. (5):

$$F1\text{ - score} = \frac{2 * (\text{precision} * \text{recall})}{\text{precision} + \text{recall}} \tag{5}$$

TP = true positives, TN = true negatives, FN = false negatives, FP = false positives.

6 Proposed Methodology

The proposed methodology used in this study is discussed as shown in Fig. 1.

1. Feed the Dataset.
2. Preprocess the Dataset.
    a. Label Encoding.
    b. Addressing null values.
3. Train the dataset.
4. Apply the Machine Learning Algorithms.
5. Evaluation of various Performance Metrics.
6. Best Model.

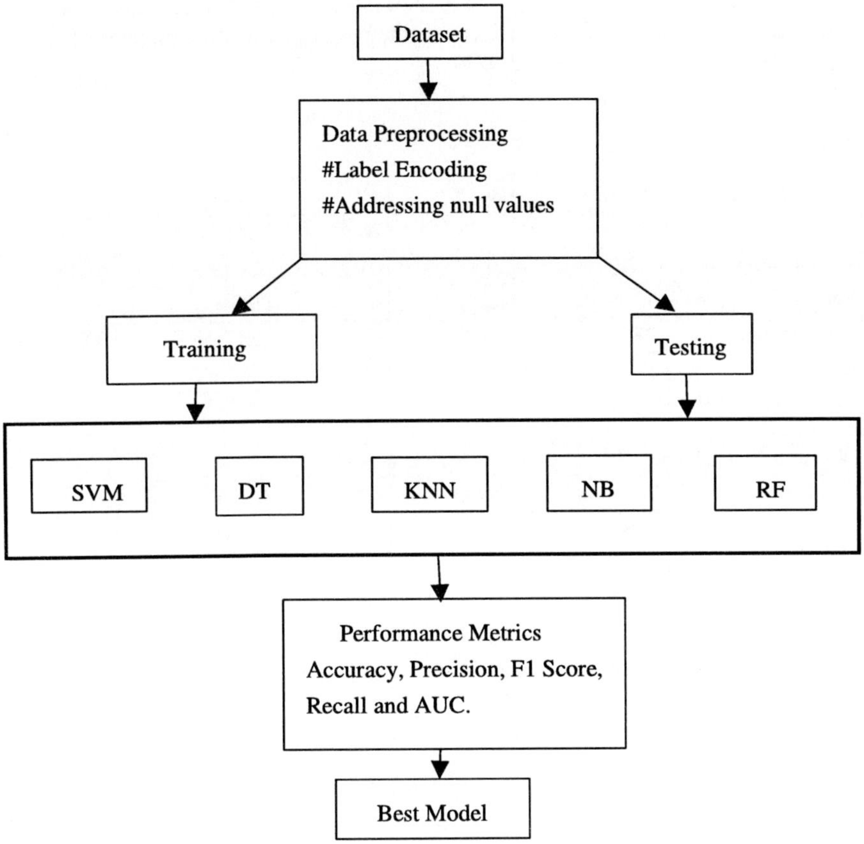

Fig. 1 Proposed methodology

7 Experiments and Results

We experiment with five machine learning classifiers on two real-world datasets for fake news classification taken from Kaggle Repository: NB, DT, RF, SVM, and KNN. Models are compared using different performance metrics as shown in Table 2.

On the WELFake dataset, the best-performing classifier is RF giving an accuracy of 91.18%. SVM is the least-performing classifier giving an accuracy of 67.02%.

On the Real and Fake News Classification Dataset, DT, RF, and KNN are best-performing models giving an accuracy of 93.92%, 93.71%, and 93.38%, respectively, whereas SVM performed the least giving an accuracy of 78.53% as shown in Table 3. Figure 2a and b shows the performances of different machine learning in terms of accuracy.

Comparison of our work with previous studies. As shown in Table 4, we compare our work with the previous work in the area of Fake News Classification. In comparison with previous studies, in our work decision tree surpassed other machine learning algorithms with an accuracy of 93.92%.

Table 2 Results of WELFake dataset

Classifier	Precision	Recall	*F*1	Accuracy	AUC
SVM	69.82	67.55	66.21	67.02	67.55
Random forest	91.17	91.18	91.17	91.18	91.18
Naïve Bayes	61.68	61.57	61.36	61.41	61.57
Decision tree	90.32	90.19	90.23	90.25	90.19
KNN	85.68	85.52	85.56	85.56	85.52

Table 3 Results of real and fake news classification dataset

Classifier	Precision	Recall	*F*1	Accuracy	AUC
SVM	78.52	78.56	78.52	78.53	78.56
Random forest	93.75	93.71	93.73	93.71	93.71
Naïve Bayes	78.78	77.38	76.72	76.92	77.38
Decision tree	93.92	93.90	93.91	93.92	93.90
KNN	93.37	93.38	93.37	93.38	93.38

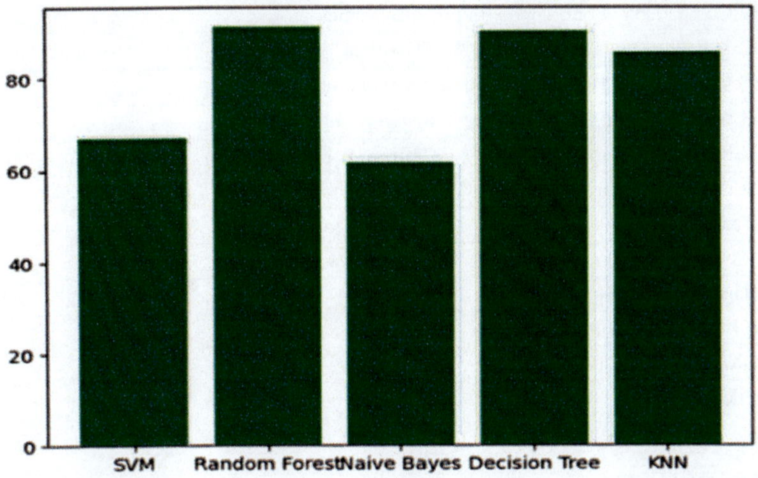

(a) WELFake Dataset

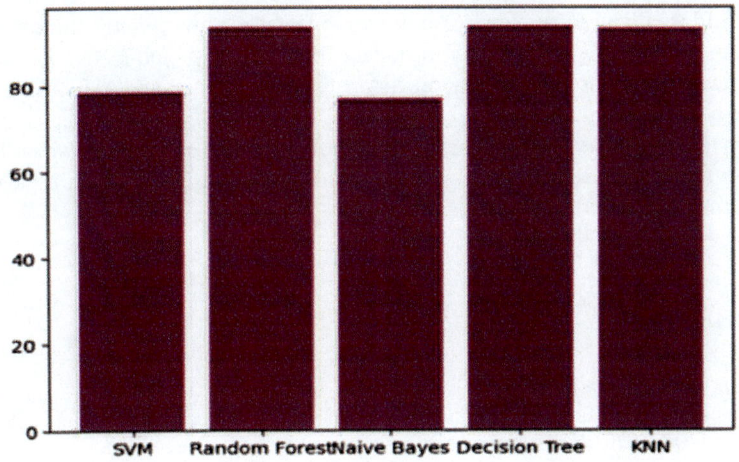

(b) Real and Fake News Classification Dataset

Fig. 2 Performance of different machine learning techniques in terms of accuracy

Table 4 Comparison of our work with previous studies

Source	Classifiers	Accuracy
Julio C. S. Reis et al. [3]	Random forest	85
Lilapati Waikhom et al. [4]	Bagging and AdaBoost	70
Rohit Kumar Kaliyar et al. [2]	Gradient boost	86
Arvinder Pal Singh Bali [5]	Gradient boosting	88
Hazif Yasir Ghafoor [9]	Passive aggression	93.7
Vanya Tiwari [11]	Linear regression	71
Arun Nagaraj et al. [13]	SVM	74
Noman Islam et al. [22]	SVM	93.15
Mykhailo Granik et al. [23]	Naïve Bayes	74
Our work	Decision tree	93.92

8 Conclusion

In this study, we performed the fake news classification on two datasets: WELFake dataset and Real and Fake News Classification Dataset using five machine learning algorithms: Naïve Bayes, random forest, decision tree, SVM, and KNN. Five performance metrics were used: accuracy, precision, recall, $F1$-score, and AUC. We noticed that the best-performing models were random forest (WELFake dataset) and decision tree (Real and Fake News Classification Dataset). For future work, we will work on more related datasets and will conduct studies with different machine learning algorithms.

References

1. Ngada O, Haskins B (2020) Fake news detection using content-based features and machine learning. In: 2020 IEEE Asia-Pacific conference on computer science and data engineering (CSDE), Dec 2020. IEEE, pp 1–6
2. Kaliyar RK, Goswami A, Narang P (2019) Multiclass fake news detection using ensemble machine learning. In: 2019 IEEE 9th international conference on advanced computing (IACC), December 2019. IEEE, pp 103–107
3. Reis JC, Correia A, Murai F, Veloso A, Benevenuto F (2019) Supervised learning for fake news detection. IEEE Intell Syst 34(2):76–81
4. Waikhom L, Goswami RS (2019) Fake news detection using machine learning. In: Proceedings of international conference on advancements in computing & management (ICACM), Oct 2019
5. Bali APS, Fernandes M, Choubey S, Goel M (2019) Comparative performance of machine learning algorithms for fake news detection. In: Advances in computing and data sciences: third international conference, ICACDS 2019, Ghaziabad, India, 12–13 Apr 2019. Revised selected papers, Part II 3. Springer, Singapore, pp 420–430
6. Mandical RR, Mamatha N, Shivakumar N, Monica R, Krishna AN (2020) Identification of fake news using machine learning. In: 2020 IEEE international conference on electronics, computing and communication technologies (CONECCT), July 2020. IEEE, pp 1–6

7. Aphiwongsophon S, Chongstitvatana P (2018) Detecting fake news with machine learning method. In: 2018 15th international conference on electrical engineering/electronics, computer, telecommunications and information technology (ECTI-CON), July 2018. IEEE, pp 528–531
8. Felber T (2021) Constraint 2021: machine learning models for COVID-19 fake news detection shared task. arXiv preprint arXiv:2101.03717
9. Ghafoor HY, Jaffar A, Jahangir R, Iqbal MW, Abbas MZ (2022) Fake news identification on social media using machine learning techniques. In: Proceedings of international conference on information technology and applications: ICITA 2021. Springer, Singapore, pp 87–98
10. Smitha N, Bharath R (2020) Performance comparison of machine learning classifiers for fake news detection. In: 2020 second international conference on inventive research in computing applications (ICIRCA), Jul 2020. IEEE, pp 696–700
11. Tiwari V, Lennon RG, Dowling T (2020) Not everything you read is true! Fake news detection using machine learning algorithms. In: 2020 31st Irish signals and systems conference (ISSC), June 2020. IEEE, pp 1–4
12. Vijayaraghavan S, Wang Y, Guo Z, Voong J, Xu W, Nasseri A, Cai J, Li L, Vuong K, Wadhwa E (2020) Fake news detection with different models. arXiv preprint arXiv:2003.04978
13. Nagaraja A, Soumya KN, Sinha A, Rajendra Kumar JV, Nayak P (2021) Fake news detection using machine learning methods. In: International conference on data science, e-learning and information systems, Apr 2021, pp 185–192
14. Baarir NF, Djeffal A (2021) Fake news detection using machine learning. In: 2020 2nd international workshop on human-centric smart environments for health and well-being (IHSH), Feb 2021. IEEE, pp 125–130
15. Ahmad I, Yousaf M, Yousaf S, Ahmad MO (2020) Fake news detection using machine learning ensemble methods. Complexity 2020:1–11
16. Jain A, Kasbe A (2018) Fake news detection. In: 2018 IEEE international students' conference on electrical, electronics and computer science (SCEECS), Feb 2018. IEEE, pp 1–5
17. Shaikh J, Patil R (2020) Fake news detection using machine learning. In: 2020 IEEE international symposium on sustainable energy, signal processing and cyber security (iSSSC), Dec 2020. IEEE, pp 1–5
18. Poddar K, Umadevi KS (2019) Comparison of various machine learning models for accurate detection of fake news. In: 2019 innovations in power and advanced computing technologies (i-PACT), Mar 2019, vol 1. IEEE, pp 1–5
19. Abdulrahman A, Baykara M (2020) Fake news detection using machine learning and deep learning algorithms. In: 2020 international conference on advanced science and engineering (ICOASE), Dec 2020. IEEE, pp 18–23
20. Han W, Mehta V (2019) Fake news detection in social networks using machine learning and deep learning: performance evaluation. In: 2019 IEEE international conference on industrial Internet (ICII), Nov 2019. IEEE, pp 375–380
21. Al Asaad B, Erascu M (2018) A tool for fake news detection. In: 2018 20th international symposium on symbolic and numeric algorithms for scientific computing (SYNASC), Sept 2018. IEEE, pp 379–386
22. Islam N, Shaikh A, Qaiser A, Asiri Y, Almakdi S, Sulaiman A, Moazzam V, Babar SA (2021) Ternion: an autonomous model for fake news detection. Appl Sci 11(19):9292
23. Granik M, Mesyura V (2017) Fake news detection using naive Bayes classifier. In: 2017 IEEE first Ukraine conference on electrical and computer engineering (UKRCON), May 2017. IEEE, pp 900–903
24. Verma PK, Agrawal P, Amorim I, Prodan R (2021) WELFake: word embedding over linguistic features for fake news detection. IEEE Trans Comput Soc Syst 8(4):881–893
25. Kaggle Repository for WELFake dataset. https://www.kaggle.com/datasets/saurabhshahane/fake-news-classification. Accessed 20 June 2023
26. Kaggle Repository for Real and Fake news classification dataset. https://www.kaggle.com/datasets/imbikramsaha/fake-real-news. Accessed 20 June 2023
27. Kumari R, Singh J, Gosain A (2023) SmS: SMOTE-stacked hybrid model for diagnosis of polycystic ovary syndrome using feature selection method. Expert Syst Appl 225:120102

Optimum Selection of Virtual Machine in Cloud Using Improved ACO

R. Jeena, G. Soniya Priyatharsini, R. Dharani, and N. Senthamilarasi

Abstract This research presents a novel strategy for selecting the optimal virtual machine (VM) in a cloud environment. The approach employs an improved version of the ant colony optimization (ACO) algorithm. Cloud computing, which offers scalable resources and services, has become an essential component of modern IT design. These resources must be used effectively in order to optimize performance and cost-effectiveness. The proposed improved ACO algorithm features enhancements that allow for more precise solution space exploration, which improves VM selection results. Extensive experiments and comparative research have demonstrated the efficiency of the improved ACO algorithm. This study advances cloud resource management methods by assisting in the optimal distribution of virtual machines to meet stated performance and cost targets.

Keywords Cloud data centre · Virtual machine · Load balancing

1 Introduction

Cloud computing is a distributed system that leverages the Internet to provide resources such as servers, software, databases, and storage. The digital era advances on a daily basis. Cloud computing has lately begun to be used by all businesses,

---

R. Jeena (✉)
Department of CSE, Saveetha School of Engineering, Saveetha Institute of Medical and Technical Sciences, Chennai, India
e-mail: rjeenait@gmail.com

G. Soniya Priyatharsini
Department of CSE, Dr.MGR Educational and Research Institute, Chennai, India

R. Dharani
Department of IT, Panimalar Engineering College, Chennai, India

N. Senthamilarasi
Department of CSE, Sathyabama Institute of Science and Technology, Chennai, India
e-mail: senthamilarasi.n.cse@sathyabama.ac.in

S. Kumar et al. (eds.), *Fourth Congress on Intelligent Systems*, Lecture Notes in Networks and Systems 868, https://doi.org/10.1007/978-981-99-9037-5_20

which is a new phenomenon. Many aspects of daily human life have been changed and revolutionized as a result of this ingenious and imaginative digital solution. The term "cloud computing" refers to resource management by a third-party service provider. The end user does not want to be concerned about software or hard drives.

As a result of the emergence of cloud computing, the industry is beginning to pay more attention to virtualization technologies. Virtualization technologies improve the security of cloud computing. The term virtual machine refers to the digital version of a physical computer.

This virtual machine of software can run programs and operating systems, store data, and connect to the network. This, however, demands maintenance in order to monitor and update the system. Multiple VMSs can be managed using virtual machine software on a single machine and most often a server [1].

Each VM type is intended to run a specific task. GPU-based virtual machines, for example, can be utilized for demanding graphics rendering and video editing.

Balancing load in the cloud environment is a little bit crucial and requires some attention. The load described here is the CPU load, memory capacity, and network load. To reduce job resource time and utilization, load can be distributed to the systems across the data centre.

1.1 Need of LB in the Environment

Load balancing refers to the process of dynamically distributing extra burden across all nodes. This undoubtedly leads to increased customer satisfaction and, as a result, greater resource utilization. A poor load-balancing system may allocate resources in an inefficient manner, resulting in resource waste. As a result, proper load-balancing techniques are required to reduce resource consumption [2].

2 Related Works

VM technologies are appealing because they cut energy costs and boost resource utilization in data centres. Certain researches are carried out by the researchers in the efficient utilization of resources and the selection of VM. Consolidation of virtual machines is an NP-hard issue [3].

Tawfeek et al. [4] proposed the ACO algorithm for resource sharing in the cloud. The objective function in this study is makespan reduction. By doing experiments, the constraint satisfaction technique is used to assign VMs to incoming batch jobs. Physical machines (PMs) are represented by a vertex or node, and edges represent VM migration from one physical machine to another. The graph theorem is used. This rule is ineffective because it does not address the issue of transition loops, which causes uniformity to be slow [5].

Mondala et al. presented a soft computing load-balancing technique. Stochastic hill climbing is used to assign jobs to virtual machines (VMs). The performance of stochastic hill climbing algorithms is examined using CloudAnalyst. The round-robin and first-come, first-served methods are compared here [6].

A set of Pareto solutions is obtained through a modified ACO method proposed by Chaharsooghi et al. The authors utilized a tree-search technique based on ACO. By simplifying the probability computation and updating the pheromone rule, the ant's learning ability is improved [7].

Malik et al. suggested a method for prioritizing the multiple tasks assigned to the virtual computer. This method is based on parameters such as the number of required resources or processors, the number of users, the type of task, the user, and the cost [8]. This method is known as the priority-based round-robin load-balancing technique. This strategy enhances the system's capability by optimizing characteristics such as overhead, scalability, resource utilization, and reaction time. The CloudSim is the testing tool here [9].

Qiang et al. proposed the multi-objective ACO (MO-ACO) algorithm, which featured objective functions such as makespan minimization, cost reduction, and load balancing [10]. This protocol fails to communicate the dependency between tasks, which is utilized to measure customer satisfaction.

Gang Li et al. [11] present the SWIM ACOTS LB based on LB. The suggested ant colony task scheduling load balancing (ACTS LB) method outperforms the standard ACO, PSO, and min-min algorithms. The experiment is carried out utilizing the NS-3 simulator [12]. This simulation outperforms existing load-balancing strategies. This is not truly simulated in real time in a large SWIM environment.

Md Hasanul Ferdaus et al. addressed the scalability issue with a hierarchical, decentralized dynamic VM consolidation system. This methodology localizes migration-related network congestion and network cost reduction. This dynamic VM consolidation approach employs ACO metaheuristics in conjunction with the suggested VM migration. The suggested technique reduces overall power usage by 47%, resource waste by 64%, and migration overhead by 83% [13].

Xiao-Fang Liu et al. used evolutionary computing to save energy and reduce the number of physical servers. The ACS technique is offered to achieve the virtual machine placement (VMP) goal. This ACS is used with the order exchange and migration (OEM) search strategy to form OEMACS [14].

The authors proposed a report that contained the effects of the COVID-19 pandemic, as well as how quickly cloud computing data centres are growing in size. This is due to a mix of country-mandated healthcare application adoption and remote working, learning, and entertainment [15, 16].

The aforementioned issues generate a poor convergence speed, which results in a longer time for resource allocation, which indirectly leads to resource waste and decreases overall system performance.

3 Background Formulation

3.1 Virtual Machine Placement (VMP)

Virtualization is a critical component in cloud setup. When the cloud data centre accepts a customer's invitation, a VM is built to host the application. This VM creation is based on resources like CPU, RAM, and storage. The VM is mapped to a free server based on the VM approach [14]. The suggested architecture is depicted in Fig. 1.

Let N be the number of virtual machines and M be the number of servers. Let V = 1, 2, 3, 4, …, N represent the group of virtual machines and P = 1, 2, 3, 4, …,

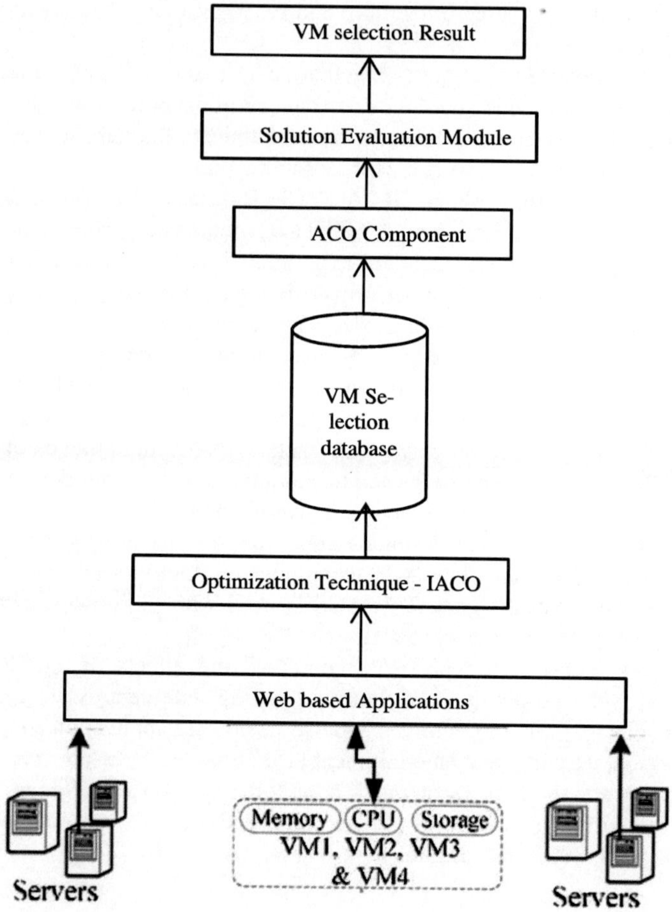

Fig. 1 Proposed architecture

Fig. 2 Construction scheme with $N = 4$, $M_t = 3$

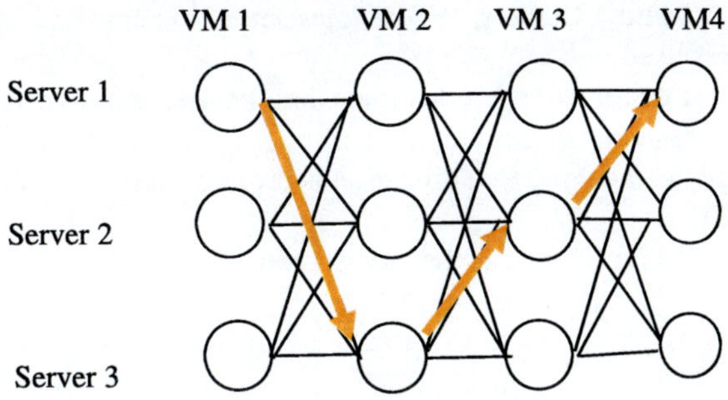

M represent the group of servers. A single server's capability is sufficient to allocate the resources. The VMP problem of minimizing the number of dynamic servers is shown.

$$\text{Minimize } f(S) = \sum_{i=1}^{M} y_i \quad (1)$$

where *S* represents solution. Figure 2 shows the construction scheme with $N = 4$, $M_t = 3$.

The achievement of a minimum number of active servers improves resource utilization and energy consumption. Each VM is in one-to-one manner, i.e. VM → One server. *X* is the zero–one adjacency matrix, where x_{ij} is VM_j to server *i*. $x_{ij} = 1$, then means that VM_j is placed on server *i*. Otherwise $x_{ij} = 0$.

$$x_{ij} = \begin{cases} 1, \text{ if VM}_j \text{ is assigned to server } i \\ 0, \text{ otherwise} \end{cases} \quad \forall_i \in P \text{ and } \forall_j \Big\} \quad (2)$$

$$y_i = \begin{cases} 1, \text{ if } \sum_{j=1}^{N} x_{ij} \geq 1 \\ 0, \quad \text{otherwise} \end{cases} \Bigg\} \; \forall_i \in P \quad (3)$$

$$\sum_{i=1}^{M} x_{ij} = 1 \;\; \forall_j \in V \quad (4)$$

$$\sum_{j=1}^{N} \text{vc}_j \cdot x_{ij} \leq \text{PC}_i \cdot y_i \;\forall_i \in P \quad (5)$$

$$\sum_{j=1}^{N} \text{vm}_j \cdot x_{ij} \leq \text{PM}_i \cdot y_i \;\forall_i \in P \quad (6)$$

In Eq. (3), the $y_i = 0$ denotes server i is not used. Whereas $y_i = 1$ denotes server i is used.

Equation (4) demonstrates that the VM is assigned to the server.

Equations (5) and (6) ensure that the server satisfies the VM's resource. The percentage of CPU utilization is proportional to the amount of power utilized by the CPU.

The following is the power model:

$$P(u) = k \cdot P_{\max} + (1 - k) \cdot P_{\max} \cdot u \tag{7}$$

where $P_{\max}$ denoted maximum power utilized by completely loaded server,

K—fraction of power utilization in idle stage,

$u(u \in [0, 1])$—CPU utilization of the server.

Equation (7) is used to calculate the consumed power.

3.2 Ant Colony System

The ants can recognize the direct path between the shell and the source of food. Pheromone is the component that allows this to happen. This actually inspires the use of an ant colony approach for the selection of VMs. This was initially used by travelling salesman problem. Later, the pheromone properties make it useful to other applications as well. Ants normally construct the optimal solution by visiting each nearby node one by one until all are visited. Pheromones play an important role in determining the next node. This construction phase also applies to VMP. [14]

3.3 Ant Colony System—Pheromone

The pheromone's accumulated history aids in the search for the shortest path. The VM serves as the U vector, whereas the server serves as the V vector. Figure 3 depicts a bipartite graph.

The linked VMs can be assigned to the same server. Figure 4 depicts the VM server mapping formation.

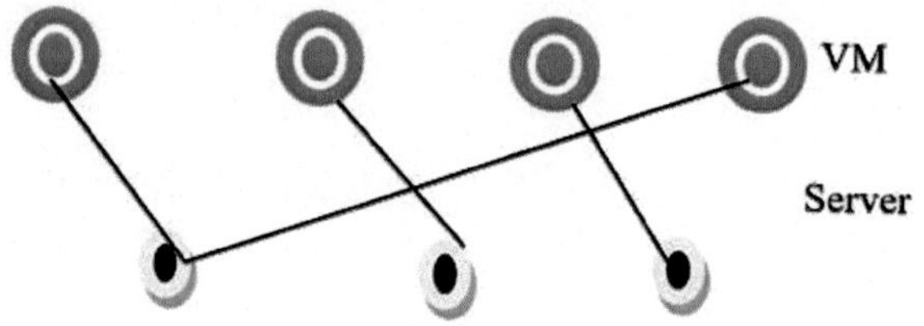

Fig. 3 Virtual machine placement–bipartite graph of VMP

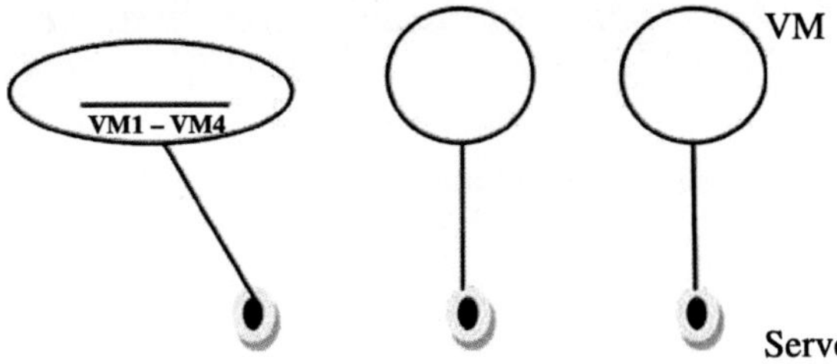

Fig. 4 Mapping VMs with server

4 Implementation and Results

In the cloud environment where the apps are running, there is a pool of servers. Virtualization is an option for the servers. The dimensional vector packing problem arises as a result of server virtualization, specifically the utilization of resources [17]. Typically, academics focus on two-dimensional problems such as CPU and memory. Assume there are two VMSs on a single server. The server's CPU utilization can be determined by utilizing,

CPU consumption = Cumulative of CPU consumption of the 2 VMs.

Consider the example,

Table 1 shows how the total number of VMs affects overall CPU and memory utilization. The total CPU utilization here is 55%, which equals half a century. A threshold can be established in each server to avoid 100% CPU utilization and RAM adaptation. The 100% utilization may degrade overall system performance.

While the ant is seeking for food, it leaves behind a unique chemical compound. This pheromone, for example, can be employed in graph construction. A graph can be built with (*N*, *E*), where *N* denotes the number of nodes, i.e. VMs and tasks.

The *E* in this case represents the edges, or the connections between tasks and VMs. At first, the ants look for food in a haphazard fashion. When the ant discovers the food, it begins to leave a chemical compound known as pheromone along the road. With the help of this material, the follower ant follows the successor ant [18].

CloudSim simulator is used for this research. The CloudSim simulator is essential for load balancing.

Table 1 Overall CPU and memory utilization

VM1		VM2		Total	
CPU	Memory	CPU	Memory	CPU	Memory
25%	35%	30%	45%	55%	80%

4.1 Pseudocode of the Projected Improved ACO Algorithm

Input: List = {Tasks, VMs}
Output: the finest solution for jobs distribution on VMs
Steps:
1. Initialize:
Set Present_iter T = 1.
Set Present_Opt_Sol = null.
Make an initial value τij(T) = c
2. Reside the m ants on the initial VMs randomly.
3. For s: = 1 to m do
Reside the starting VM of the sth ant in tabuk.
Ensure ants_tour while all ants don't finish their tours
To execute the next job, every ants selects the VM based on Eq. (8).
Tabuk -> chosen VM insertion.
Do End
4. For s = 1 to M do
sth ant pheromone is used to compute the length of Ls of the trip based on Eq. (10).
Increase the present_Opt_Sol with the optimized solution.
5. Use the formula (11) to put the local pheromone row each edge (I, j).
6. Calculate global pheromone based on formula (12).
7. Increase Present_iter_T by 1.
8. If (present_iter_T < Tmax)
Tabu -> {empty}
Jump to step two.
Else
Print present_Opt_Sol
End If
Return

4.2 Pseudocode of the Scheduling-Based ACO Algorithm

Input: List of incoming Cloudlets and VMs
Output: print "Completing"
Step by step:
Place CLET_List = NULL and dummy_List_of_CLET = NULL
Get any CLETs in CLET List in order of their arriving time
Do ACO_P iterate up to CLET_List not empty or more incoming CLET
Set r = VMs list size
If (size of CLET_List > r)
Handover the first arrived r CLETs from CLET_List and place on dummy_List_of_CLET
Else

Handover CLet from CLet List and place on dummy_List_of_CLET
IF END
Implement the process of ACO with the involvement of dummy_List_of_CLOUDLET
and r
Stop Do
Display "Completing"
End

The probabilistic transition rule can be applied by iterating the ant $s(1, 2, \ldots, m)$ which builds the number of tasks(r). m represents number of the ants.

The probability that s-ant chooses virtual machine (j) for the following task i is as follows:

$$\mathrm{Prob}_{ij}^{s}(T) = \left\{ \frac{[\tau_{ij}(T)]^{\alpha}.[\vartheta_{ij}]^{\beta}}{\sum_{q \in \mathrm{allowed}\ s}[\tau_{is}(T)]^{\alpha}.[\vartheta_{is}]^{\beta}} \right\} \tag{8}$$

$\tau_{ij}(T)$—pheromone concentration at time T between task i and VM_j,
allowed$_s$—$\{0, 1, \ldots, r-1\}$—tabu$_s$,
$\vartheta_{ij} - \frac{1}{d_{ij}}$ denotes the prominence for the moment(t).
d_{ij} is the tasks expected execution time and transfer time

$$d_{ij} = \frac{\mathrm{TL\_Task}_i}{\mathrm{Processor\_num}_j * \mathrm{Processor\_\ mips}_j - \mathrm{VM}_j} + \frac{\mathrm{InputFile\ Size}}{\mathrm{VM\_bandwidth}_j} \tag{9}$$

$$L^{s}(T) = \arg\max_{j \in J}\left\{\mathrm{sum}_{i \in IJ}\left(d_{ij}\right)\right\} \tag{10}$$

$$\tau_{ij}(T) = (1-\rho)\tau_{ij}(T) + \Delta\tau_{ij}(T) \tag{11}$$

$$\tau_{ij}(T) = \tau_{ij}(T) + \frac{Q}{L+} \text{ if } (i, j) \in TT \tag{12}$$

Let us assume the tasks are independent and pre-emptive in nature. The experiment is executed with 12 data centres with 60 virtual machines and 100–1500 tasks under the CloudSim platform. Let 1000 MI to 20,000 MI can be the length of the task. Table 2 shows the parameters α, β, ρ with number of ants.

- Adaptive parameter is denoted as Q
- M denotes the number of ants
- α, β—weight of pheromone trail control parameters
- ρ—trail decay $0 < \rho < 1$
- $T_{\max}$—maximum number of iterations.

Table 2 α, β, ρ with number of ants

α	β	ρ	M
0	0	0	1
0.1	0.5	0.1	5
0.2	1.5	0.2	8
0.3	2	0.3	10
0.4	2.5	0.4	15
0.5	3	0.5	20

Time is computed using the CloudSim simulator's makespan variable. Figure 5 depicts the makespan time for FCFS, RR, and improved ACO as a graph. Table 3 displays the adaptive parameter and the maximum number of iterations.

When compared to FCFS and RR, the enhanced ACO executes in the shortest time span.

The overall performance of cloud computing is determined by throughput and response time. The response time is denoted by the makespan time in Fig. 6. Figure 7 depicts the throughput of the cloud environment by displaying the number of VMs assigned to the server.

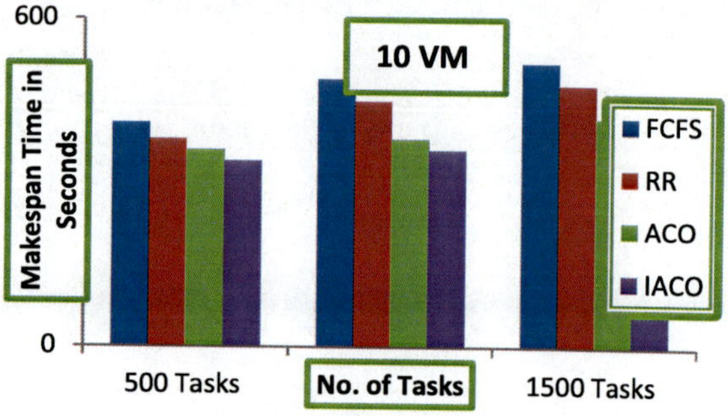

Fig. 5 Comparison of makespan time for FCFS, RR, ACO, and improved ACO (10 VM)

Table 3 Adaptive parameter and maximum of iterations

Q	T_{max}
1	50
100	75
500	100
1000	150
1500	200

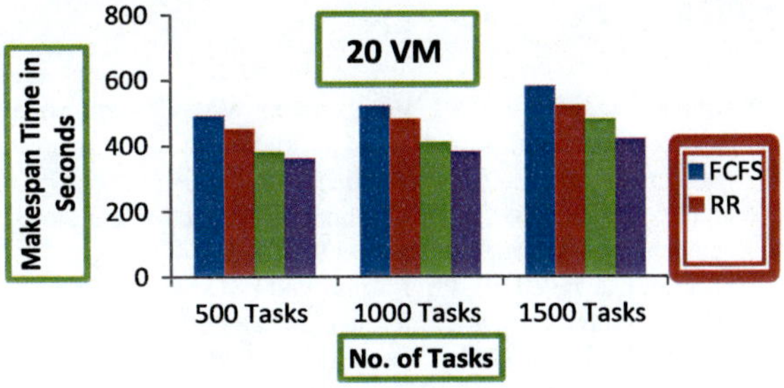

Fig. 6 Comparison of makespan time of FCFS, RR, and improved ACO (20 VM)

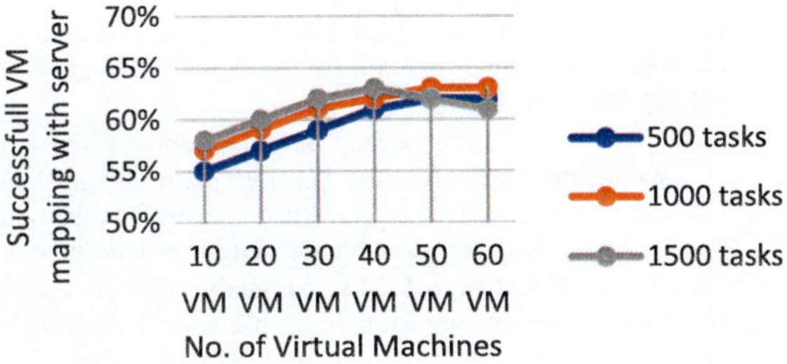

Fig. 7 Throughput based on the number of virtual machines

5 Conclusion and Future Work

Finally, the study detailed in this paper improves the issue of cloud resource management by providing a specialized improved ACO algorithm for the optimal VM selection. By successfully utilizing the power of nature-inspired optimization, this technique aids in the realization of efficient and cheap cloud deployments. The current work provides vital new insights into the possibility of enhanced ACO, but it also indicates new research areas. Future study could focus on the algorithm's scalability to larger cloud settings, flexibility to dynamic workloads, and the incorporation of real-time performance data. Finally, optimizing VM selection will remain critical for defining the future of cloud computing, and the improved ACO method offers a significant step forward in this direction. Existing systems such as FCFS and RR are compared to enhanced ACO and achieve superior results. The revised ACO algorithm is utilized to improve the outcome. In future, there will be various soft computing approaches that can be used to achieve the best results, such as the particle swarm optimization algorithm and the stochastic hill climbing algorithm.

References

1. Shabeera TP et al (2017) Optimizing VM allocation and data placement for data-intensive applications in cloud using ACO metaheuristic algorithm. Eng Sci Technol 20(2):616–628. https://doi.org/10.1016/j.jestch.2016.11.006
2. Fu X (2015) VM selection and placement for dynamic consolidation in Cloud computing environment. Front Comput Sci 9:322–330
3. Luo Z, Liu H (2019) IPSO: improved PSO based TS (task scheduling) at the cloud data center. In: 15th international conference on semantics, knowledge and grids (SKG), pp 139–144
4. Tawfeek M et al (2015) Cloud task scheduling based on ant colony optimization. Int Arab j. inf technol 12(2):129–137
5. Garg SK, Versteeg S, Buyya R (2013) A framework for ranking of cloud computing services. Future Gener Comput Syst
6. Dubey AK, Kumar A, Agrawal R (2021) An efficient ACO-PSO-based framework for data classification and preprocessing in big data. Evol Intell 14(2):909–922. https://doi.org/10.1007/s12065-020-00477-7
7. Li G, Wu Z (2019) Ant colony optimization task scheduling algorithm for SWIM based on load balancing. Future Internet 11(4). https://doi.org/10.3390/fi11040090
8. Rashedi E, Zarezadeh A (2014) Noise filtering in ultrasound images using GSA (Gravitational Search Algorithm). In: Iranian conference on intelligent systems, ICIS, 2014
9. Chaudhary D, Kumar B (2014) An analysis of the load scheduling algorithms in the cloud computing environment. In: IEEE conference, ICIIS. IEEE, pp 1–6
10. Mishra K, Majhi SK (2021) A binary Bird Swarm Optimization based load balancing algorithm for cloud computing environment. Open Comput Sci 11(1):146–160. https://doi.org/10.1515/comp-2020-0215
11. Gupta A, Srivastava S (2020) Comparative analysis of ant colony and particle swarm optimization algorithms for distance optimization. Procedia Comput Sci 173(2019):245–253. https://doi.org/10.1016/j.procs.2020.06.029
12. Zhou Z, Hu Z, Li K (2016) Virtual machine placement algorithm for both energy-awareness and SLA violation reduction in cloud data centers. Sci Program 2016(i). https://doi.org/10.1155/2016/5612039
13. Azad A et al (2019) Comparative evaluation of intelligent algorithms to improve adaptive neuro-fuzzy inference system performance in precipitation modelling. J Hydrol 571:214–224. https://doi.org/10.1016/j.jhydrol.2019.01.062
14. Deng W et al (2019) A novel intelligent diagnosis method using optimal LS-SVM with improved PSO algorithm. Soft Comput 23(7):2445–2462. https://doi.org/10.1007/s00500-017-2940-9
15. Alashhab ZR, Anbar M, Singh MM, Leau YB, Al-Sai ZA, Alhayja'a SA (2021) Impact of coronavirus pandemic crisis on technologies and cloud computing applications. J Electron Sci Technol 19:Art. no. 100059
16. Alhomdy S, Thabit F, Abdulrazzak FH, Haldorai A, Jagtap S (2021) The role of cloud computing technology: a savior to fight the lockdown in COVID 19 crisis, the benefits, characteristics and applications. Int J Intell Netw 2:166–174
17. Hu H, Keqin L (2016) VM placement algorithm for both energy-awareness and SLA violation reduction in cloud data centers. Resource Manag. Virtual. Clouds, Article ID 5612039
18. Jeena R, Logesh R (2022) Optimum selection of virtual machine using improved particle swarm optimization in cloud environment. Int J Comput Netw Appl (IJCNA) 9(1):125–134

Data Imputation Using Artificial Neural Network for a Reservoir System

Chintala Rahulsai Shrinivas, Rajesh Bhatia, and Shruti Wadhwa

Abstract This study provides a comprehensive comparison of the different algorithms implemented on a reservoir system, and the results are statistically analyzed from the results of other machine learning algorithms. Different weights and activation methods have been used to obtain the results. The algorithms implemented on the data of reservoir system are generative adversarial networks, synthetic model, non-dominated sorting genetic algorithm 2. Later on, we have done comparisons and visualization on the data obtained We have attempted to implement generative adversarial networks on a reservoir system that is in the time series representation and the data values are from June 1, 1989, to May 1, 2016. Data was collected from the reservoir authorities, and they did not have the records for some of the months. The target is to regenerate that empty values and find out what could be the next data value in the upcoming months.

Keywords Generative adversarial network (GAN) · Data imputation · Synthetic model · Non-dominated sorting algorithm 2 (NSGA2)

1 Introduction

A reservoir is a human-made or natural body of water used to store and manage water resources. Reservoirs are of great importance for various reasons:

- Water supply: Reservoirs are used to store water for human consumption, agricultural irrigation, and industrial purposes. They ensure a constant water supply even during droughts or periods of low rainfall.
- Flood control: Reservoirs can regulate water levels by releasing water during periods of high rainfall or snowmelt, preventing floods downstream.
- Hydropower generation: Reservoirs can be used to generate electricity by utilizing the gravitational potential energy of water stored in them.

C. R. Shrinivas (✉) · R. Bhatia · S. Wadhwa
Punjab Engineering College (Deemed to be University), Chandigarh 160012, India
e-mail: iamrahulsai@gmail.com

S. Kumar et al. (eds.), *Fourth Congress on Intelligent Systems*, Lecture Notes in Networks and Systems 868, https://doi.org/10.1007/978-981-99-9037-5_21

A major part of accuracy of any machine learning model depends upon its dataset and its use. A machine learning model loses its reliability when the null records are not handled properly. There are numerous techniques that may be used to deal with missing data, including:

- Deleting the missing values: One approach is the removal of records which are null. This process is advisable only if the dataset to be trained is large so that even if the null values are deleted, then there are enough number of data records for training and testing.
- Imputation: This method is used to replace the null values with suitable values like mean of the attribute or by evaluating the median of the attribute and sometimes regression imputation is used to replace the null values.
- Specific algorithms were designed to replace the missing values with the results obtained from them. Decision tree helps to replace the missing values without any preprocessing of the dataset.

2 Related Work

Missing data is a very common problem prevalent across fields. Therefore, for this study, results might be derived from various applications. For instance, Nisha et al. proposed several methodologies in their work for vehicle detection system. If the vehicle detection is weak, significant processes could be used especially at the night due to complex illumination conditions [1].

Dubey and others [2] have used a generative adversarial network (GAN) for true radar detection instead of multiple interferences, noises, and reflections. This model shows better results than some of the state-of- the-art radar detection techniques.

Often the training period of GANS is more than that of traditional machine learning algorithms but Licheng Chen et al. have proposed how this problem can be minimized. The authors proposed a new algorithm, SE-GAN that includes multiple GANs that learn in parallel [3]. In their work, the authors also discussed how the training phase of multiple generative adversarial networks may be improved.

Generative adversarial network was also used by Bhagyalakshmi A et al. to establish a virtual dressing room system for the customers [4]. This model also deals with the numerous stances a person creates for trying out the clothes.

Optical character recognition (OCR) has also been attempted using GANs. Often there is a problem regarding the dataset in the OCR because it requires a dataset for training and a dataset for validation which is hard to get in this problem. Sinha et al. came up with a solution to this by taking the whole dataset as training data and for validation, they have referenced a study that is generating input image from a softmax layer which is present in the OCR model [5].

A remote controlling system has been implemented using GANs in some part of the model due to which the bit error rate (BER) is reduced in the dynamic scenarios [6]. Many activation functions were used to achieve the optimized results, and at last it reached the maximum of 2.5 m of communication distance. The program for the receiver has been implemented in Python 3,8.

A breast cancer image classification was done by Djouima et al. [7]. The authors have implemented deep convolution generative adversarial network (DCGAN) to give consistence in the majority and minority classes. Further, data augmentation was also applied in order to generate efficient number of images to train the model.

Shaojie Li et al. have proposed the use of GANs and variational autoencoder in MRI segmentation [8].

3 Methodology

3.1 Generative Adversarial Network

Generative adversarial networks (GAN) use two neural networks, a discriminator and a generator, to produce new data that is similar to the training data. The discriminator network learns to separate the created data from the actual data while the generator network learns to create new data. Figure 1 depicts the entire working of the generative adversarial networks, and its process is run through two modules.

In the process, two networks are trained together and the generator tries to create data that will deceive the discriminator and the discriminator tries to properly identify whether the data is real or fake.

GANs can be difficult to train and require large amounts of data and computing power. However, they have the potential to revolutionize many fields.

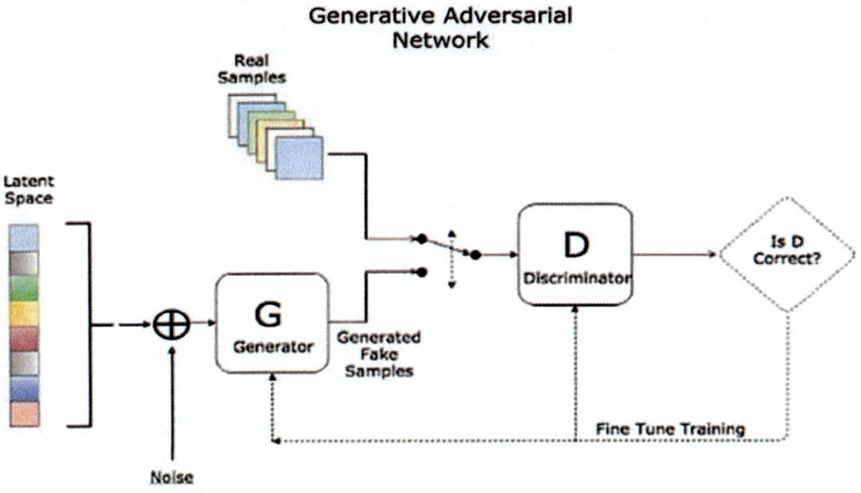

Fig. 1 Schematic diagram of generative adversarial network (GAN) [9]

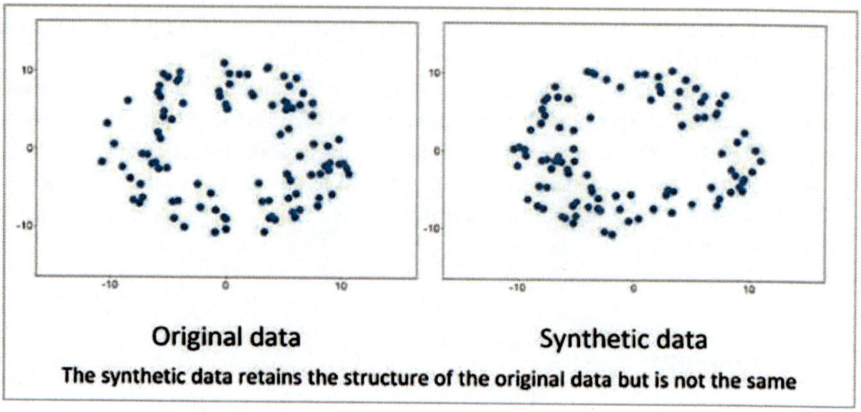

Fig. 2 Scatter plot of original data and synthetic data [10]

3.2 *Synthetic Model*

A machine learning model called synthetic model can be used to create synthetic data, which is data that resembles real data. Data augmentation, which is the process of expanding a training dataset by producing new data samples that are variants of the original data, frequently makes use of synthetic models.

Machine learning models like generative adversarial networks (GANs) and variational autoencoders (VAEs) are examples of generative models that can be the foundation for synthetic models. These models can create new data samples that mimic the original data after learning the underlying patterns and distributions of the training data. Figure 2 shows the similarity between the original data and synthetic data and how the synthetic data takes the visual structure similar to that of the original data. The data shown in the figure is of original outflow values and the synthetic outflow values.

Larger training datasets can help machine learning models perform better by increasing the size of the training dataset through the usage of synthetic models. This is because models with larger datasets often perform better.

Machine learning models can also be tested using fictitious data. By doing so, you can assist make sure the models are accurate and unbiased.

3.3 *NSGA2*

NSGA2 is one of the most popular multi-objective machine learning algorithms in engineering and the fields of computer engineering. It is an evolutionary machine learning algorithm, which mostly obtains desired results with multiple objectives and numerous datasets. It uses a population-based study to derive its output.

It produces a population of numerous solutions in order before prioritizing them using a non-dominated sorting process. The non-dominated sorting algorithm process prioritizes the generated solutions according to the level its dominance, which is a standard of how appropriately they completed each of the tasks successfully.

The solutions, which were not dominated, by different solutions are known as Pareto optimal solutions. NSGA2 algorithm implements a selection strategy to select the best-required solution from the population, and after that, it implements a crossover and mutation procedure to produce a new population of suitable solutions.

The method iterates until a desired requirement is satisfied, such as a minimum fitness value or a maximum number of generations.

4 Material

This research work is done with the objective of utilizing the available data of reservoir system for training and testing the genetic algorithms. One of the primary objectives of this project is data generation for flood control, irrigation supply, and industrial supply.

There are mainly six attributes involved while implementing the algorithm. Attributes used are inflow, initial level of water, evaporation rate, outflow of water, gross storage at start, and gross storage at the end. The parameters can differ in other algorithms because we needed to figure out data for industrial release values and domestic release.

5 Data Source and Study Area

Monthly dataset has been taken for the years 1990–2015. Data analysis is done on the data provided, and output has been generated. After getting the raw data, we obtained the preprocessing to generate reliable forecasts. Few values of some features are missing, and those values need to be regenerated. The main task is to regenerate the missing values with appropriate data which can be obtained from the three models explained earlier in the methodology part of this paper. Majority of this study revolves around the models like generative adversarial networks, synthetic model, and NSGA2 implementation on the dataset.

Tables 1 and 2 describe the different datatypes such as int64 and float64 are present in the dataset. There are 324 rows consisting of different values. Ravishankar Reservoir, also known as Ravi Shankar Sagar Dam, is a reservoir located in the Indian state of Chhattisgarh. The reservoir is formed by the Ravi Shankar Sagar Dam, which is built across the Mahanadi River in the Dhamtari District of Chhattisgarh.

The primary purpose of the reservoir is to provide water for irrigation and hydroelectric power generation. The dam has a height of 54 m and a length of 934 m, and its construction was completed in year 1977. The reservoir has a storage capacity of

Table 1 Dataset information

Column name	Count	Dtype
Year/month	324 non-null	object
Initial level	324 non-null	float64
Gross storage at start (MCM)	324 non-null	float64
Inflow (MCM)	324 non-null	float64
Evaporation (MCM)	324 non-null	float64
Outflow (MCM)	324 non-null	float64
Gross storage at end (MCM)	324 non-null	float64

Table 2 Detailed information about data

Methods	Initial level	Gross storage at start (MCM)	Inflow (MCM)	Evaporation (MCM)	Outflow (MCM)
Count	324	324	324	324	317
Mean	162	344	570	8	150
Std	93	4	216	7	234
Min	1	334	140	0	0
25%	81	341	363	2	14
50%	162	344	588	6	45
75%	243	346	734	11	180
Max	324	388	940	36	1538

1.37 billion cubic meters and a surface area of 47.85 km^2. Figure 3 shows a scatter plot of inflow values and outflow values before implementing the machine learning models. Figure 4 shows the outflow values of each month from Jun 1989 to May 2016. The outflow peak values occur during the monsoon season.

A Python 3.8 script was used to visualize the dataset. The minimum value of a few outflow values is zero, which indicates that these are missing values in the dataset. These missing values need to be regenerated by the machine learning models.

6 Results and Discussion

The monthwise data was implemented on generative adversarial network algorithm, synthetic model, and NSGA2. Newly regenerated data is obtained. Figure 5 shows a bar plot that depicts the different outflow values from before and after the implementation of GAN. It shows the comparison of original outflow and regenerated outflow values; the root mean squared error (RMSE) from generative adversarial networks is about 271.54. RMSE from synthetic model is about 332.41. RMSE from NSGA2 is about 812.57.

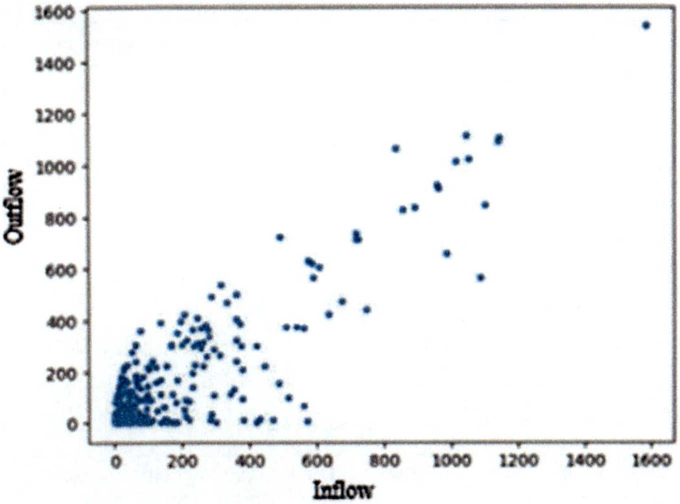

Fig. 3 Inflow–outflow scatter plot

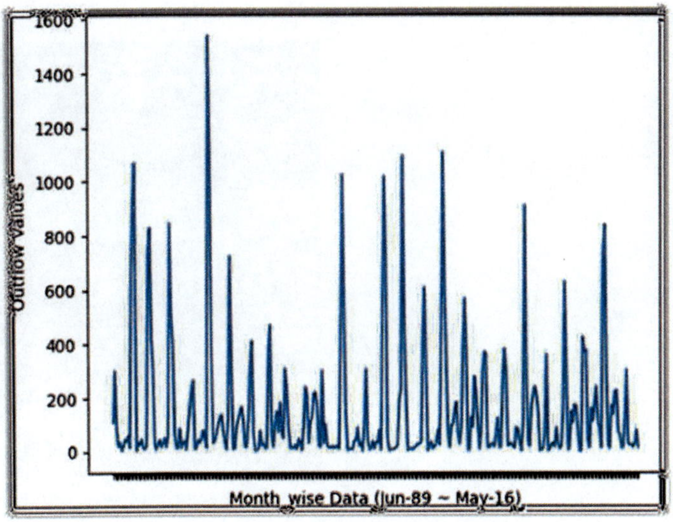

Fig. 4 Outflow between June 1989 and May 2016

Figure 6 shows the results of synthetic model after preprocessing as there were many values missing in the dataset. In the end result, we obtain the 324 regenerated outflow values which are having similar kind of distribution as that of the original values.

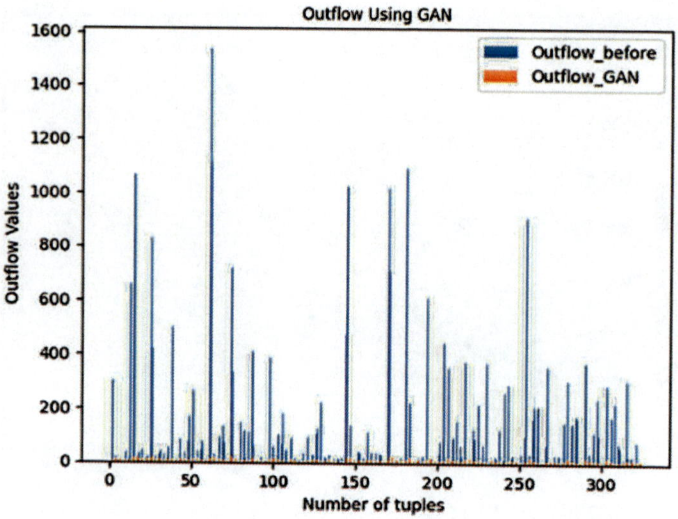

Fig. 5 GAN outflow

Fig. 6 Synthetic model output

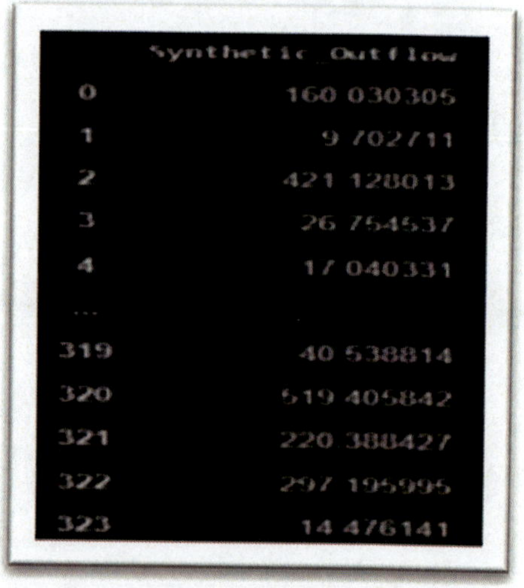

	Synthetic_Outflow
0	160.030305
1	9.702711
2	421.128013
3	26.754537
4	17.040331
...	
319	40.538814
320	519.405842
321	220.388427
322	297.195995
323	14.476141

Figure 7 shows how NSGA2 algorithm requires sensitivity analysis to determine the population size of the model. From this, the best-suited value of population size can be obtained and it is used further in the NSGA2 implementation. Figure 8 shows the output results of the regenerated outflow and storage values through NSGA2 algorithm.

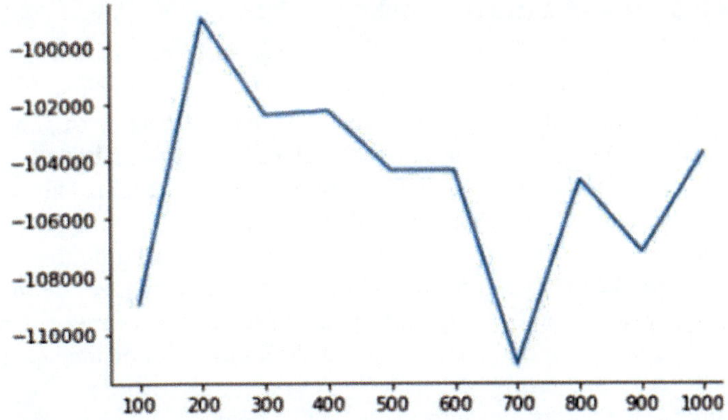

Fig. 7 Population objective function value NSGA2 sensitivity analysis

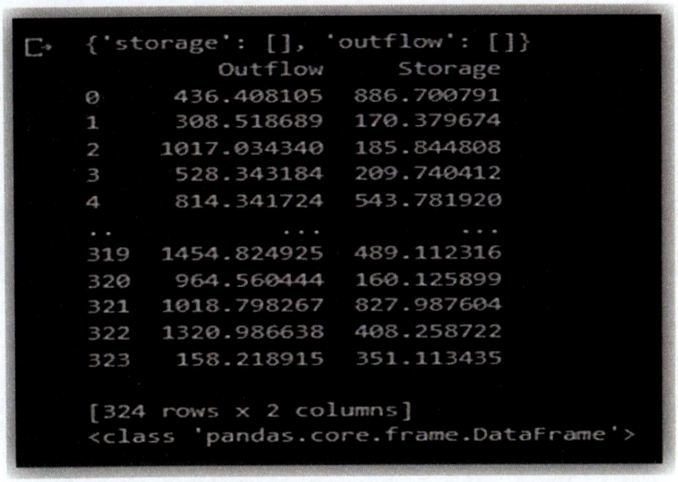

```
{'storage': [], 'outflow': []}
          Outflow     Storage
0      436.408105  886.700791
1      308.518689  170.379674
2     1017.034340  185.844808
3      528.343184  209.740412
4      814.341724  543.781920
..            ...         ...
319   1454.824925  489.112316
320    964.560444  160.125899
321   1018.798267  827.987604
322   1320.986638  408.258722
323    158.218915  351.113435

[324 rows x 2 columns]
<class 'pandas.core.frame.DataFrame'>
```

Fig. 8 NSGA2 outflow and storage values

On analyzing the MSE error of each of these three models, it was concluded that generative adversarial networks machine learning algorithm is comparatively better suited to regenerate data for a time series dataset.

7 Conclusion and Future Work

Following significant technological advancements, the generative adversarial network is a breakthrough innovation in regenerating lost data or to generate new data instead of simply replacing the data with null or mean values. It just regenerates new data which is passed on from the discriminator of the model. The continuous monthwise data was provided as input in this model and the results have not been that much accurate and different results were obtained by changing the parameters of the model. The key parameters include the values of batch size, the number of epochs, and activation function. This method has been used in a variety of areas. We conclude that data generation for time series dataset might not be the best option to obtain new accurate and precise data.

Following are the points for future scope in this field:

- If the discriminator is too strong than the generator in generative adversarial network, it retaliates all the generated data from the generator. Therefore, other machine learning models can be implemented to get better results.
- Maximize the gross storage at the end with constraints so that the reservoir system is protected from flood.

References

1. Nisha UN, Ranjani G (2022) Deep learning based night time vehicle detection for autonomous cars using generative adversarial network. In: 2022 international conference on augmented intelligence and sustainable systems (ICAISS), Trichy, India, 2022, pp 336–341. https://doi.org/10.1109/ICAISS55157.2022.10010725
2. Dubey A, Fuchs J, Lübke M, Weigel R, Lurz F (2020) Generative adversial network based extended target detection for automotive MIMO radar. In: 2020 IEEE international radar conference (RADAR), Washington, DC, USA, 2020, pp 220–225. https://doi.org/10.1109/RADAR42522.2020.9114564
3. Shen L, Yang Y (2019) SE-GAN: a swap ensemble GAN framework. In: 2019 international joint conference on neural networks (IJCNN), Budapest, Hungary, 2019, pp 1–8. https://doi.org/10.1109/IJCNN.2019.8851684
4. Arjovsky M, Chintala S, Bottou L (2017) Wasserstein gan. arXiv preprint arXiv:1701.07875
5. Sinha A, Jenckel M, Bukhari SS, Dengel A (2019) Unsupervised OCR model evaluation using GAN. In: 2019 international conference on document analysis and recognition (ICDAR), Sydney, NSW, Australia, 2019, pp 1256–1261. https://doi.org/10.1109/ICDAR.2019.00-42
6. Ahmed MF, Pervin N, Bhowmik S, Rashid MA, Kuwana A, Kobayashi H (2023) Design and implementation of remote controlling system using GAN in optical camera communication. IEEE Photonics J 15(2):1–10, Art no. 7301510. https://doi.org/10.1109/JPHOT.2023.3250455
7. Djouima H, Zitouni A, Megherbi AC, Sbaa S (2022) Classification of breast cancer histopathological images using DensNet201. In: 2022 7th international conference on image and signal processing and their applications (ISPA), Mostaganem, Algeria, 2022, pp 1–6. https://doi.org/10.1109/ISPA54004.2022.9786028
8. Li S, Zhang Y, Yang X (2021) Semi-supervised cardiac MRI segmentation based on generative adversarial network and variational auto-encoder. In: 2021 IEEE international conference on

bioinformatics and biomedicine (BIBM), Houston, TX, USA, 2021, pp 1402–1405. https://doi.org/10.1109/BIBM52615.2021.9669685
9. Synthetic data generation: 3 key techniques and tips for success. Datagen. https://datagen.tech/guides/synthetic-data/synthetic-data-generation/
10. Synthetic data vs real data: which is a better option? Labellerr, 15 Nov 2022. https://www.labellerr.com/blog/synthetic-data-vs-real-data-which-is-a-better-option-for-your-projects-2/

Depth Multi-modal Integration of Image and Clinical Data Using Fusion of Decision Method for Enhanced Kidney Disease Prediction in Medical Cloud

Tatiparti B. Prasad Reddy and Vydeki

Abstract Due to the shortage of nephrologists and the scarcity of diagnostic labs in rural areas, chronic kidney disease (CKD) has recently emerged as a major health problem. In order to detect illness from clinical and radiological pictures, automated diagnostic models are necessary. Previous research on CKD diagnosis has mostly focused on the independent use of clinical data and CT scans to create AI algorithms. Incorporating clinical data with computed tomography (CT) images, this research seeks to create a Deep Multi-modal Fusion approach with a late fusion mechanism for diagnosis. Clinical data were used in the tests with CT scans of the kidney to ensure accuracy. Clinical data and image-extracted features are kept separate in the proposed model by a process termed late fusion. The model scored 99% accuracy, 98.6 recall, 97.8 precision, and 98.2 F-score, showing that combining clinical data with the CTD increases diagnosis accuracy to that of a human expert. An additional method of verification was comparing the proposed system's results with those of a human expert. In addition, the findings validate the feasibility of the proposed approach as a diagnostic aid for chronic kidney disease (CKD) for medical professionals.

Keywords Chronic kidney disease · CT · Precision · Accuracy · Precision · F-score

1 Introduction

The term "Internet of Things" is used to refer to a system in which a group of medical equipment are linked together so that they may exchange and store patient data. IoMT refers to "the Internet of Medical Things," which "becomes a specialized, extended

T. B. Prasad Reddy · Vydeki (✉)
SENSE, Vellore Institute of Technology, Chennai, India
e-mail: vydeki.d@vit.ac.in

T. B. Prasad Reddy
e-mail: tatiparti.bprasadreddy2017@vitstudent.ac.in

S. Kumar et al. (eds.), *Fourth Congress on Intelligent Systems*, Lecture Notes in Networks and Systems 868, https://doi.org/10.1007/978-981-99-9037-5_22

branch of IoT that incorporates all connected devices used to provide timely assistance" [2]. Wireless or wired connections are also possible for inter-device communication. One of the primary advantages of IoMT-related remote health monitoring is the freedom to go about one's normal activities while still obtaining monitoring of health [1]. The size of the modules attached to the body and the need for regular battery replacement or charging made the conventional remote monitoring techniques look cumbersome to the patients [3]. By developing small, ultra-low-power sensor devices and lightweight gearbox protocols, the IoMT revolution is able to address the aforementioned issues and provide viable solutions. Portable patient monitoring units (PPMUs) in ambulances and residences are used for remote health monitoring [4]. A decision support system is used in hospitals for real-time monitoring. Combining data or information from several sources with different formats and structures is known as "data fusion" [5]. The concept of data fusion is simplified when the human perceptual system is used as an example. The human perceptual system is shown here. The five senses (sight, sound, smell, taste, and touch) and prior experiences serve as building blocks for a person's mental representation of the world. The use of data fusion methods to the human body exemplifies the value of this strategy. In addition to its usage in geospatial systems, defense systems, and intelligence services (for a review, see [7]), data fusion and its related approaches offer a platform for building improved decision support and smart patient monitoring systems in healthcare contexts [6]. Combining data (COD) and combining interpretation (COI) may both lead to successful data fusion [8]. The next sections dissect both strategies in detail.

To begin, we aggregate characteristics from all of our data sources into a single decision model (or classifier) for training, as illustrated in Fig. 1.

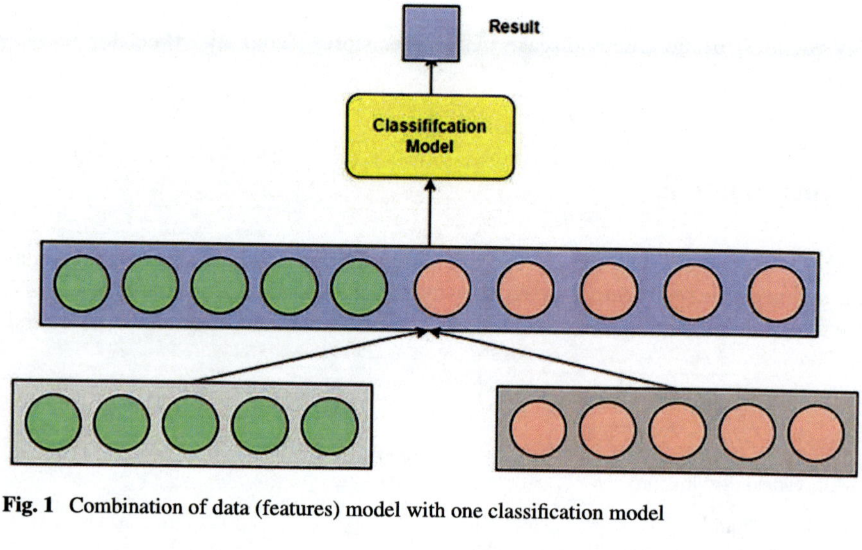

Fig. 1 Combination of data (features) model with one classification model

Second, a combiner (sometimes called a meta-decision model or meta-classifier) synthesizes the results from various decision models using numerous inputs in a process known as a combination of interpretation (COI). Stacking is a common method used for building ensembles of classifiers, which has conceptual ties to COI [9]. The basic premise of the COD model is shown in Fig. 2. Although both COD and COI have shortcomings, the latter is proven to be sub-optimal since it cannot preserve links between data from diverse sources. In order to address these concerns, Lee et al. [10] developed a General Fusion Framework (GFF) that positions COI and COD on opposite extremes of a spectrum.

While both COD and COI have inherent drawbacks, COI has been shown to be sub-optimal due to its inability to maintain connections between data from different sources. The authors [10] created a GFF to deal with these issues; in this paradigm, COI and COD are considered as two ends of a continuum. Numerous encouraging research [7–10] have looked at the use of deep learning models to automate diagnosis in medical imaging. Previous work has proven the practicality of effective automated image analysis based on imaging data alone, in contrast to normal clinical practice which depends on the interpretation of medical imaging in combination with relevant clinical data to indicate a good diagnosis. In radiology, having access to relevant background information at the moment of image interpretation is vital since obtaining

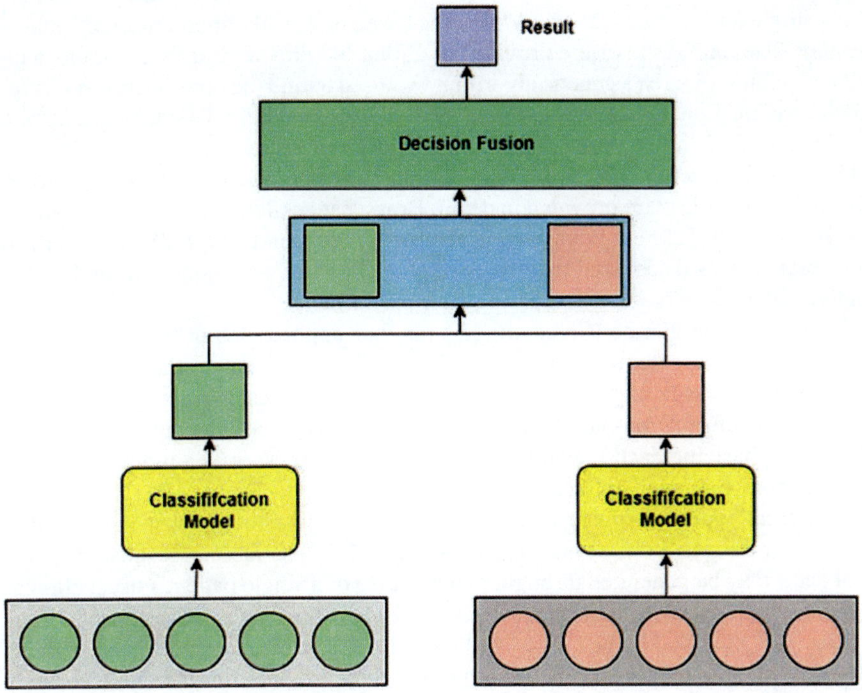

Fig. 2 Combination of interpretation holding two classification algorithms

a proper medical diagnosis on imaging typically relies on pre-test probability, past diagnosis, clinical and laboratory data, and previous imaging. Eighty-five percent of radiologists feel that clinical context is crucial for evaluating imaging data [14]. The use of multi-modal data fusion for automated clinical outcome prediction and diagnosis has increased during the last three years.

2 Related Work

References [5, 8–12] provide an overview of the research on COD and COI techniques. In [8], the authors demonstrated the COD method by predicting the roles of yeast proteins using a support vector machine (SVM). They propose to use a set of kernels to combine data on protein structures, amino acid sequences, and the genes that code for them. Researchers in [9] used machine learning techniques for cancer detection and prognosis, using inspiration from the COD approach to develop multiple categorization models. According to the COI technique published by Rajpurkar et al. in [13], a breast cancer diagnosis is reached using a combination of radiologist interpretation and objective evidence gathered from mammograms and patient history. Methods from detection theory were used in the development of the classifiers. In order to attain high accuracy, we built many binary classifiers for a single data source, each of which was based on the likelihood ratio of a single feature. The final results were provided as a joint likelihood ratio for all of the decision variables. Finally, a genetic algorithm was used to find the optimal thresholds for each classification. Ensemble classifiers, like COI (particularly the stacking scheme), are investigated in [14] and need the development of a large number of individual classifiers. Instructions for building ensemble classifiers and combining the results of several classifiers are provided in detail. Breast cancer detection model developed by Huang et al. [15] employing an ensemble of decision trees, SVMs, and neural networks. It was discovered that the ensemble classifier was more precise than the individual classifiers when comparing their performance.

In [16], the author constructs models for four applications in biological image processing and assesses their COI and COD. Possible uses include atlas-based picture segmentation, tissue segmentation based on average images, multi-spectral classification, and deformation-based group morphometry. The author compared COI with COD, emphasizing each method's strengths and weaknesses as a model generator for the aforementioned applications.

Different types of image data (such as pathology photographs, radiographic images, and camera shots) and non-image data (such as lab test results and clinical data) may be generated throughout the course of a single patient's usual clinical sessions. It is possible that medical diagnoses and prognoses might benefit from the insights provided by heterogeneous data [17–19]. One major drawback of these decision-making approaches is the room that they provide for qualitative analysis and personal preference [20, 21]. With the rapid development of AI in recent years, a plethora of deep learning-based solutions for multi-modal learning in medical applications have emerged. Multi-modal fusion using deep learning is an exciting new field of study because it may give a high-level abstraction of complicated processes hidden inside high-dimensional data [22, 23]. Research in deep learning has shown promise for single-modality diagnosis and forecasting [24–26]. Since different clinical modalities may include diverse information (complementary information of a person) and have different data formats, successful integration of the multi-modal data is not a simple process when creating solutions.

A diagnosis and prognosis are made based on the patient's image and non-image data. Photographs might be as straightforward or intricate as X-rays or pathology slides. Imaging data are said to be pixel-aligned if and only if the pixels in successive photos are in the same physical position and pixel-not-aligned if the pixels in subsequent photographs are in different physical places. Results from demographic laboratory tests, such as blood samples, are examples of data that are not a picture. Input to multi-modal learning systems is complex since it might include both images and text. Two-dimensional pathology pictures show the micro-level morphology, whereas three-dimensional computed tomography (CT) or magnetic resonance imaging (MRI) scans may show the macro-level and geographical information about a cancer. Together with clinical data and laboratory test findings, the molecular, biological, and chemical characteristics shown by structured DNA and mRNA sequences play a crucial role in clinical decision-making. Images are often huge and dense (millions of pixels), as opposed to sparse and low-dimensional, which makes them unique among data sources. To effectively capture the common and complementary information for improved diagnosis and prognosis, many preprocessing and feature extraction procedures (e.g., for varied dimensions, pictures, free text, and tabular data) are necessary.

3 Data

First, we provide a brief overview of the data collection that was analyzed in this research. The assessment matrix and methods employed will be discussed thereafter. DVR diagnosis in Hyderabad provided the data for this analysis. In order to accurately diagnose kidney illness, it is necessary to collect data from a wide variety of sources. The sections that follow will detail the various types of information that were collected for this study.

3.1 Clinical Data

The dataset contains the fields, their titles, brief definitions, and data types that make up the chronic kidney disease dataset. There are a total of 25 characteristics and 400 samples in the dataset. Class is the area of focus for disease forecasting, and it has been implemented. Eleven are numerical and fourteen are category, making up the remaining 24 attributes.

Clinical Data Preprocessing

Preprocessing steps include looking for and eliminating outliers, validating and normalizing noisy data, and checking for missing values. The patient evaluation reveals several missing or inaccurate estimates. The planned research endeavor includes the following preprocessing measures to account for this.

(a) *Handling of missing values*

The simplest approach to dealing with missing values is to ignore the records, although this is impractical for smaller datasets. The dataset is checked for missing attribute values as part of the data production process. By using the statistical technique of mean imputation, we may assess the value of missing data in numerical systems. When filling up missing values for minor characteristics, the mode technique is employed.

(b) *Categorical encoding of data*

Since most deep learning methods only take numbers as input, the category values must be converted into numbers. Yes and no are only two examples of the types of things whose attributes are represented by the binary integers 0 and 1.

(c) *Transformation of data*

Data transformation refers to the act of changing numbers on a small scale such that one variable does not dominate the others. Otherwise, regardless of the unit of measurement, the learning methods will consider larger values to be more advanced and smaller ones to be less advanced. For further processing, the data modifications adjust the values in the dataset. This study employs a data normalization technique to improve the precision of deep learning methods. Transformed information has a standard deviation of 1, a mean of 0, and an intermediate value of − 1.

The standardization can be stated as,

$$s = \frac{(v - \overline{v})}{\sigma}. \tag{1}$$

In the above equation s denotes the standardized score, ν is the observed value, the mean is denoted as V^1, and σ signifies the standard deviation.

Outlier detection

In statistics, outliers are single observations that cannot be explained by the rest of the data. The estimate of experimental variability or signal inaccuracy might generate an outlier. Outliers may skew and mislead a deep learning algorithm's training data. Longer training times, less reliable output, and reduced model accuracy are all consequences of outlier inclusion. This study employs an IQR-based technique to filter out extreme cases before feeding the data into the learning algorithm.

When a dataset is divided into quartiles, the interquartile range (IQR) may be used as a measure of variability. The numbers that separate each section are known as the first, second, and third quartiles, and the formula for determining the interquartile range (IQR) is as follows:

$$\mathrm{IQR} = V_3 - V_1, \tag{2}$$

where V_3 represents the median of the first half of the ordered data set and signifies the median of the second half. The interquartile range is the range between the first quartile V_1 and the third quartile V_3. Data points outside the thresholds are potential outliers.

Imaging Data

Computed tomography (CT) scans were used to acquire the imaging data. The collection includes several different kinds of pictures. The pictures were taken using standard techniques and saved in DICOM format for uniformity and ease of analysis.

Data Preprocessing

Extensive preprocessing processes were conducted on the clinical and imaging data to assure quality and compatibility.

First, the input CT images are class-labeled and then preprocessed to remove noise. To restore integrity to characteristics of an image that have been degraded by noise, a preprocessing technique is presented for use in noise reduction. Here, adaptive filtering is employed to remove noise from specific regions of a picture. A corrupted image may be described as one in which $\hat{I}(x, y)$, the noise difference across the whole image is indicated by σ_y^2, the local mean provided by μ_L. Generally, a pixel window and local variance window are provided by $\hat{\sigma}_y^2$.

3.2 Feature Extraction and Classification

Clinical Data

The GFR is very important in many contexts, including as public health, medical treatment, and scientific inquiry. Clinical labs play a crucial part in determining GFR and diagnosing chronic renal disease. Serum creatinine measurement and GFR estimation are advised as the first steps in the GFR assessment. Here, the filtration rate is determined using age and serum creatinine (SC) level, and the GFR is used to categorize the five phases of CK illness.

GNNs are the glue that holds the interdependence of graph nodes together through inter-node communication. The GNN operates on the graph to characterize nearby data via a series of random phases. This makes GNN a useful resource for wireless networks with complex characteristics that cannot be extracted in a closed form, as is the case with many of these systems. In the suggested study, a GNN-based method is used to achieve weights of the each node. The Q-function is taught in a deep Q-learning procedure by observing examples of cell and entity placement. The Q-function's main benefit is that it allows us to create GNNs that are scalable across multiple sizes, each of which may collect constrained network characteristics with a unique number of cells and entities. Learning the proper Q-function is necessary for producing the best possible choices. Learning the GNN parameters, in this case, the Q-function, requires the progressive accumulation of new cell entity connections across a partially connected network, where the nodes represent cells and the relationships between them represent entities.

Image Data

Here, preprocessed CT scans have been fed into VGGNet-16 and Inception-v4 feature extractors for the purposes of COVID-19 prediction and classification, respectively. The classifier receives a combined set of extracted characteristics from these models.

The convolutional neural network (CNN) model uses a series of convolution layers to identify preexisting picture patterns. CNN's strength is in its ability to train an extremely deep network with just a small number of parameters. It makes the training process easier and takes less time. In addition to the convolutional layer, activation layer, pooling layer, fully connected layer, and SoftMax layer, the CNN consists of several more layers.

VGGNet-16 model: In 2014, the Oxford Visual Geometry Group introduced VGG-16, a popular CNN technique with 16 layers that has shown standard results across a variety of image processing applications (Xu et al., 2019). When increasing a system's depth, VGG-16 swaps out large convolution filters for small ones. The improvement in classification accuracy is mostly attributable to CNNs with very small filters. The VGG-16 CNN approach utilized in this study was pre-trained on the ImageNet dataset, and its front layers are a low-level universal feature suitable to typical image processing applications. New version of Inceptions are utilized throughout several training stages to break up repetitive blocks into smaller networks to displaying a whole model in memory. Therefore, Inception modules are simply tweaked, representing the potential of altering the number of filters from exclusive layers, without affecting the pre-eminence of the trained network. The training time may be reduced by fine-tuning the layer size to strike a balance between the various sub-networks. The newest Inception models were developed using TensorFlow, and unlike previous models, they do not include any redundant segmentation. This may be due to the fact that activating tensors, which play a crucial role in calculating gradient and approximating bounded values, are a function of contemporary memory optimization for Back Propagation (BP). Furthermore, Inception-v4 is planned to remove duplicated work across all grid sizes for Inception blocks (Shankar et al., 2020).

Residual Inception blocks: The filter-expansion layer, which is used to increase the filter bank's dimensionality, applied Inception blocks in this model before calculating the input depth. These are Important for making up for the reduction in dimension necessitated by the Inception block. Inception-v4 is mild because to the inclusion of several levels, and it is only one of many Inception variants. The supplementary shift between residual and non-residual forms of change for typical layers, we make use of something called Batch Normalisation (BN). Since the BN model in TensorFlow uses more memory, it is important to restrict its use to certain situations and reduce the total number of layers.

Scaling of the residuals: Here, if the number of filters is more than 1000, the network is terminated during the first stage of training, which is the destination layer before the pooling layer activates to construct zeros from different iterations, since residual techniques reveal its instability. As a result, restricting training methods would not be enough to get rid of it. The learning method is also confirmed to be effective when the constrained measurements added before the activation layer are used. Accumulated layer activations are typically scaled using factors between 0.1 and 0.3.

The suggested ensemble deep learning model that makes use of the Internet of Things is illustrated below in stepwise. In the beginning, medical IoT devices get the necessary scan of a patient at a nearby medical facility. After a scan is completed, it is sent to the IoT framework's storage layer over a communication medium. The resulting scan is then processed using the ensemble deep learning model. Once the results have been received, they are saved in the database. Medical specialists, physicians, and patients are just some of the IoT users who may have access to data stored in the cloud. Figure 3 depicts the suggested ensemble deep learning model. It makes it quite apparent that we will be training the pre-trained models independently at first. The final ensembled framework for automated screening of CKD suspicious instances is obtained by a majority vote. The remainder of this article will focus on the sequential ensemble model.

(1) Initially, the abdomen CT image data is procured.
(2) Divide the dataset into two subparts, i.e., 60% and 40%, respectively.
(3) They are represented as

$$[C_{\mathrm{tr}} C_{\mathrm{ts}}] = \mathrm{Tf}(\mathrm{Ds}). \tag{1}$$

Here, C_{tr} stands for CT scan training data, C_{ts} for testing data, Tf for tenfold cross-validation, and DS for the aggregated CT scan dataset with four classes.

Tf for 10-fold cross-validation, and DS for the aggregated CT scan dataset with four classes.

(4) The deep learning models, i.e., ResNet152V2, DenseNet201, and IRNV2 are applied on the testing dataset (C_{ts}) as

$$\begin{aligned} \mathrm{RS} &= \mathrm{TL}(R, S), \\ \mathrm{DS} &= \mathrm{TL}(D, S), \\ \mathrm{IS} &= \mathrm{TL}(I, S). \end{aligned} \tag{2}$$

S represents the SoftMax function, R represents the ResNet152V2 network, D represents the DenseNet201 network, and I represents the IRNV2 network. TL stands for the deep transfer learning model.

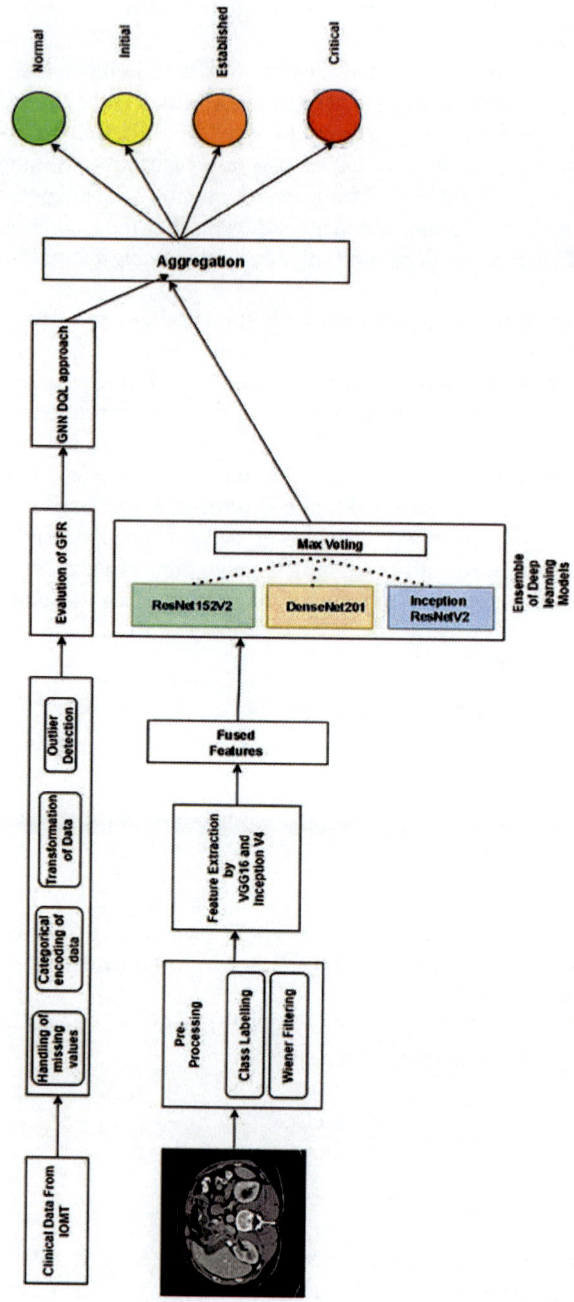

Fig. 3 Multi-modal deep fusion binary classification model for CKD

(5) The trained individual deep transfer learning models can be defined as

$$\begin{aligned} \text{RS} &= \text{MB}(\text{RS}, C_{\text{tr}}), \\ \text{DS} &= \text{MB}(\text{DS}, C_{\text{tr}}), \\ \text{IS} &= \text{MB}(\text{IS}, C_{\text{tr}}). \end{aligned} \quad (3)$$

(6) Here, MB defines the model-building process. Finally, ensembling is achieved by using majority voting as

$$E_c = E_m(\text{RS}, \text{DS}, \text{IS}), \quad (4)$$

where E_c is the trained ensemble model and E_m defines the majority voting ensemble model.

3.3 Fusion Methodology

In order to improve the precision and reliability of kidney disease detection, the fusion technique used in this study combines clinical and imaging data in a complementary fashion. The fusion methods and algorithms explored in this study are outlined below.

Combining the predictions or choices generated by separate models trained on clinical and imaging data is the subject of decision-level fusion. Using methods of ensemble learning including majority voting and weighted average, this study fuses individual decisions. The final determination is based on the combined predictions of many classification models that were trained independently on clinical data and imaging data. Combining these verdicts strengthens the renal disease detection method by compensating for the shortcomings of individual models as shown in Fig. 3.

4 Results

Models have been trained on an NVIDIA GeForce RTX-2060 SUPPER with 8 GB of memory, and the minimum CPU required for augmentation and classification operations is an Intel i7-9700F with 16 GB of RAM. The study is implemented in Python 3.8.5 and makes use of the TensorFlow 2.5 library for DL models. The confusion matrices for clinical and imaging data are shown in Figs. 4 and 5. Table 1 displays the computed performance of two different classifiers—one for picture data and another for clinical data—both of which do well when compared to their predecessor methods.

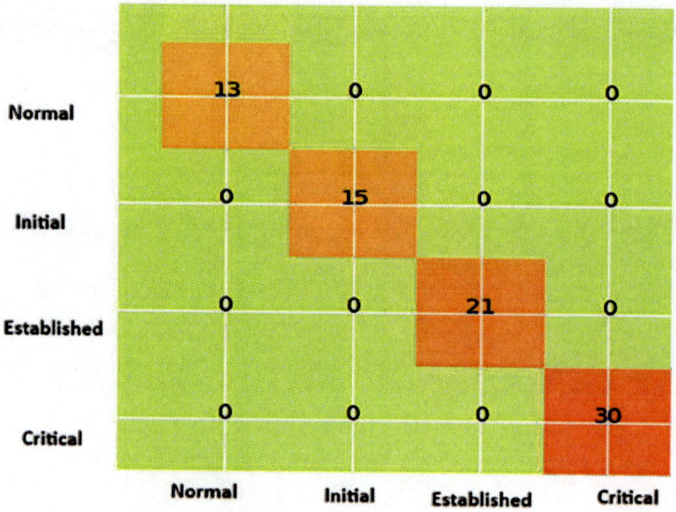

Fig. 4 Confusion matrix for the clinical data model

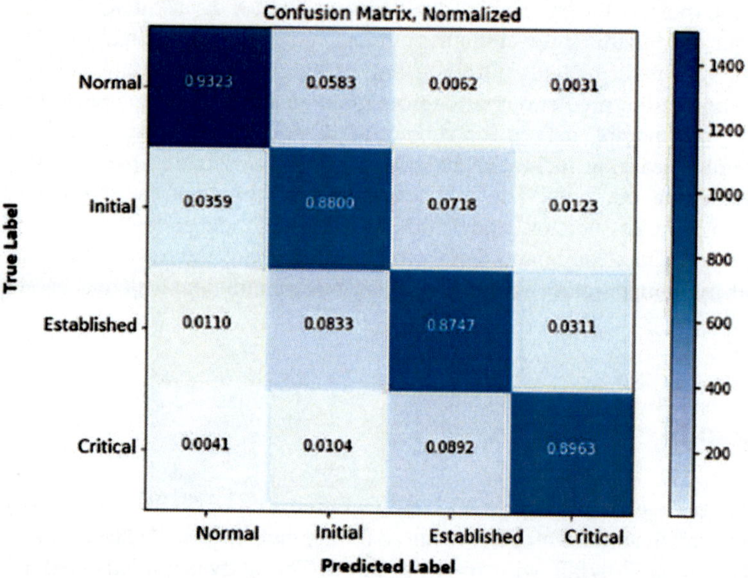

Fig. 5 Confusion matrix for the image data model

Table 1 Metrics of performance

Models	Performance outcomes (%)			
	Accuracy	F1-score	Recall	Precision
GNN-DQL (clinical data)	99.93	99.86	99.86	99.86
Ensemble, model (image data)	98.05	98.24	98.05	98.43

Table 2 Benchmarking with a human expert

Evaluation factors	Human expert	Proposed model
Accuracy	0.960	0.960
Recall	1.000	0.917
Precision	0.923	1.000
F1-score	0.960	0.957

5 Conclusions

Table 2 shows that when clinical data are combined with the CTD, the system outperforms a human expert in making diagnoses. The suggested system will not be able to diagnose CKD on its own, but it will help physicians in countries where it is difficult to get medical professionals. Despite the fact that the suggested approach has yielded very substantial outcomes. There are flaws in the research as well. The validation was carried out on a sample size of 250 patients since only one expert was available and a significant number of patient data were not available.

References

1. Tai Y, Gao B, Li Q, Yu Z, Zhu C, Chang V (2021) Clinic data access based on X-ray imaging and deep neural networks for reliable and intelligent COVID-19 diagnostic iomt. IEEE IoT J 8(21):15965–15976
2. Elbasi, Zreikat (2021, May) Disease prognosis and early detection utilising machine learning on IoMT data. In: 2021 IEEE AI-Internet-of-Things (AIIoT) world congress. IEEE, pp 0155–0159
3. Akter H, Islam H, Fahim WA, Sarkar PR, Ahmed (2021) Comprehensive evaluation of deep learning models for chronic kidney disease risk prediction and early detection. IEEE Access 165184–165206
4. Yu CS et al (2020) A machine learning strategy for the design of an online health care assessment for preventive medicine. J Med Internet Res 22(6):e18585
5. Mitchell HB (2014) Data fusion: concepts and ideas. Springer, Heidelberg. https://doi.org/10.1007/978-3-642-27222-6
6. Lahat D, Adali T, Jutten C (2015) Multimodal data fusion: a review of current approaches, future directions, and obstacles. Proc IEEE 103(9):1449–1477. https://doi.org/10.1109/JPROC.2015.2460697. Integration of health records: a real-world example 301
7. Castebedo F (2007) Data fusion: a literature survey. 2013 Sci World J. https://doi.org/10.1155/2013/704504

8. Rohlfing T, Pfefferbaum A, Sullivan EV, Maurer CR Data fusion vs. interpretation fusion in biological image analysis. In: Christensen GE, Sonka M (eds) IPMI 2005. LNCS, vol 3565. Springer, Heidelberg, pp 150–161. https://doi.org/10.1007/1150573013
9. Ponti MP Jr (2011) Creating ensembles and fusing decisions via the combination of classifiers. In: Tutorials for the 24th annual symposium on instruction in computer graphics and related art forms
10. Preliminary findings of combining mass spectrometry and histology to predict prostate cancer recurrence. Lee G, Madabhushi A A knowledge representation framework for integration, categorization of multi-scale imaging and non-imaging data. In: From Nano to Macro: IEEE's international symposium on biomedical imaging. John et al (2019) CheXpert is a big database of chest X-rays that includes uncertainty labels and a comparison to the opinions of experts. Cs Eess ArXiv 107031
11. Rajpurkar P et al (2017) To diagnose pneumonia on chest X-rays at the level of a radiologist, developed CheXNet, a deep learning system. Cs Stat. arXiv:1710.5225 ArXiv
12. Bien N et al (2018) The creation and retrospective validation of MRNet for deep learning-assisted diagnosis of knee MRI. 15. PLOS Med e1002699
13. Rajpurkar P et al (2020) AppendiX Net: deep learning for appendicitis diagnosis from a small dataset of CT exams with video pretraining. Sci Rep. 10.3958
14. Yang X et al (2019) The identification of pulmonary embolisms in CTPA pictures using a two-stage convolutional neural network. IEEE Access 7
15. Huang et al (2020) Developed PENet, a scalable deep-learning network to automatically diagnose pulmonary embolisms using volumetric CT imaging. NPJ Digit Med 3(1):1–9
16. Tajbakhsh N, Gotway MB, Liang j (2015) Automatic diagnosis of pulmonary embolisms in medical images using convolutional neural networks and a unique vessel-aligned multi-planar picture representation. In: Navab N et al (eds) Medical image computing and computer-assisted intervention—MICCAI 2015. Springer, Berlin, pp 62–69
17. Leslie A, Jones AJ, Goddard PR Clinical information's impact on radiologists' CT report writing. Brit J Radiol 4:1052–1055
18. Zhang SM, Wang YJ, Zhang ST (2021) Accuracy of artificial intelligence-assisted detection of esophageal cancer and neoplasms on endoscopic images: a systematic review and meta-analysis. J Dig Dis 22(6):318–328. https://doi.org/10.1111/1751-2980.12992
19. Date C, Jesudasen SJ, Weng CY (2019) Applications of deep learning and artificial intelligence in Retina. Int Ophthalmol Clin 59(1):39–57. https://doi.org/10.1097/IIO.0000000000000246
20. Cui H, Xuan P, Jin Q, Ding M, Li B, Zou B (2018) Co-graph attention reasoning based imaging and clinical features integration for lymph. 4. You may get a copy from Springer International Publishing. https://doi.org/10.1007/978-3-030-87240-3
21. Neumann M, King D, Beltagy I, Ammar W (2020) ScispaCy: fast and robust models for biomedical natural language processing. 0:319–327
22. Chauhan G et al Joint modelling of chest radiographs and radiology reports for pulmonary Edoema assessment. In: Lecture notes in computer science (including subseries for the whole paper, titled "BERT: pre-training of deep bidirectional transformers for language understanding"
23. Devlin J, Chang MW, Lee K, Toutanova K (2019) In: The 2019 Conference of the North American chapter of the association for computational linguistics, human language technologies (HLT 2019), vol 1(Mlm), pp 4171–4186
24. Cohan A, Lo K, Beltagy I (2020) SCIBERT: a pretrained language model for scientific text. In: 2019 Conference on empirical methods in natural language processing and 9th international joint conference on natural language processing (EMNLP-IJCNLP 2019), pp 3615–3620. https://doi.org/10.18653/v1/d19-1371
25. Parisot S et al (2018) Disease prediction using graph convolutional networks: application to autism spectrum disorder and Alzheimer's disease. Med Image Anal 48(1):117–130. https://doi.org/10.1016/j.media.2018.06.001
26. Yan R et al (2021) Richer fusion network for breast cancer classification based on multimodal data. BMC Med Inform Decis Making 21(1):1–14. https://doi.org/10.1186/s12911-020-01340-6

An Efficient Prediction of Obstructive Sleep Apnea Using Hybrid Convolutional Neural Network

N. Juber Rahman and P. Nithya

Abstract Obstructive sleep apnea (OSA) represents a severe sleep disorder, which exhibited during the interruption of the breathing while sleeping. It causes as a result of inadequate supply of oxygen to both the brain and the physical body. In this research work, Convolutional Neural Network (CNN) has been used for feature extraction and selection and the Artificial Neural Network (ANN) has been used to predict sleep apnea. A flattened layer, fully connected networks, and feature extraction layers make up the suggested CNN model. LSTM model is used for classification, composed of Softmax classification layer to identify disease classification and to compute its association with cardiovascular diseases. Experimental analysis of the proposed framework toward the prediction of obstructive sleep apnea classification has been done using the Physionet apnea dataset. This dataset aims to evaluate the effectiveness of the proposed representative framework against the conventional approaches to analyze in various deep learning architectures. The proposed framework achieves accuracy of 99% on optimal feature classification for disease prediction against machine-based existing classification approaches.

Keywords Sleep apnea · Convolutional Neural Networks · Artificial Neural Networks · Electrocardiogram · LSTM

1 Introduction

Sleep apnea is a medical condition identified by repeated interruptions in breathing while sleeping, which delays the body from receiving sufficient oxygen. It will lead to restraint in airflow because the higher respiratory procedure's pathway is pinched or narrowed off for the period of sleep. On short span of time, an apnea air is limited

N. Juber Rahman (✉)
Computer Science, PSG College of Arts and Science, Coimbatore, India
e-mail: jubijbr@gmail.com

P. Nithya
Networking and Mobile Application, PSG College of Arts and Science, Coimbatore, India
e-mail: nithi.selva@gmail.com

S. Kumar et al. (eds.), *Fourth Congress on Intelligent Systems*, Lecture Notes in Networks and Systems 868, https://doi.org/10.1007/978-981-99-9037-5_23

from relocating beyond the obstruction reducing sufficient blood circulation to the mind. This triggers the brain to awaken from sleep to signal the physique that it needs to breathe. It is frequently accompanied by loud gasping, choking, or snoring sounds as the individual takes a deeper breath than before obstruction. As soon as a normal breath is resumed, the mind returns to sleep, and the system starts unevolved again. Challenges of the existing approaches on processing the segment of ECG signals toward sleep apnea detection and its association to cardiovascular disease on computation of R-peak in the QRS are highly complex. ECG signals which suffer from the power line noise, muscular noise, electrode contact noise have difficulties in detecting the R-peak detection.

Accurately identifying the peaks of individual waves such as P and T, as well as deviations, in ECG signals represented by QRS complexes poses a significant challenge. ECG signal processing containing several segments is highly complex in terms of spectral localization, shift sensitivity, and phase information on the non-uniform signals to detect the R-peak. Further, minute details of the abnormalities of signal contraction can be missed on existing analysis. The detection of this particular feature is challenging due to the presence of both high- and low-frequency disturbances in the ECG signal as well as sudden changes in the signal's configuration. The prior sleep apnea in the ECG segment has an effect on the current sleep apnea. The ECG segment's signal processing needs extra modification for normalization. Therefore, the focus of this research is to provide R-peak detection in QRS complex of ECG signal on various aspects of the challenges defined previously. In this work, a primary objective is analyzing the ECG signal with time window for sleep apnea detection efficiently. The proposed model will be detailed in upcoming sections.

2 Literature Review

RR intervals and EDR signals may provide important clues about the development of OSA. RR intervals and EDR signals are thus extracted from the original ECGs using a preprocessing approach. The time interval between successive peaks of QRS complexes (Q wave, R wave, and S wave) is used to calculate the RR interval, which indicates the length of a heartbeat cycle. An external module sourced from the BIOSIG toolbox is used to find these QRS complex peaks, taking use of the physiologically consistent points that are present within the generated RR intervals. Thus, proves this work achieves better evaluation than the other schemes. However, this work utilizes another tool named BIOSIG toolbox, which utilizes longer heartbeat cycle than usual. This pitfall makes the system to analyze and report for shorter, but in longer intervals the system fails due to longer waiting period [1].

The noise in biological signals needs to be filtered out while preserving the desired information. In particular, the QRS complex of the ECG signal has a significant impact on noise filtering, often resulting in the generation of undesirable large impulses [2]. This filtering process is crucial for detecting the apnea state of RR intervals. Although the variation changes in RR intervals during apnea are not readily apparent, an automated segmentation of RR intervals based on these variation changes can effectively eliminate most misdirected cases. This approach is user-friendly and easily comprehensible [3]. In order to achieve more accurate OSA detection using ECG signals, a hidden Markov model (HMM) is implemented. HMM is a statistical Markov model consisting of two sequences: the unobservable Markov states and the observations. It excels in capturing the temporal dependencies of segmented signal sequences and accounts for the differences among individual subjects. For successful state prediction in hidden Markov models, the transition probabilities as well as the features must be subject-specific, and these aspects should adhere to specific distributions to avail analytical solutions [4].

The segment's state determines all of the features that are derived from ECGs, which is consistent with the methodology used by the majority of segment-based classifiers. Various subjects may display various transition patterns, and the subject-specific nature of the transition probabilities between states allows for the identification of subject-to-subject differences. Support vector machine (SVM) assigns fitness values to each ECG feature attribute, which are then utilized to select an optimal subset of features. This approach focuses on classifying features based on their respective weights. The underlying concept of SVM involves mapping the data to a higher-dimensional space, and the algorithm operates by finding the hyperplane that maximizes the distance between the training data and the hyperplane [5].

SVM can be described as a system composed of linear functions within a feature space of high dimensionality. These functions are trained through an optimization-driven learning algorithm that incorporates a learning bias derived from statistical learning theory. SVM has gained significant popularity as a highly effective classifier in various domains of classification, especially when dealing with large volumes of data [6].

Even though the above works develop a model to identify the sleep apnea, but fail in correlating hyperplane and sleep apnea, lack of distribution in obtaining analytical solutions, utilizing optimal features and subsets, validating and obtaining precision QRS values, and so on. Thus, the proposed work focused in solving the issues and providing a better working model to identify sleep apnea.

3 Methodology

Respiratory information, or signals obtained from ECG-derived respiration (EDR), can be retrieved from the ECG using a variety of signal processing algorithms by examining the morphological changes brought on by the impacts of respiration on the recorded ECG signals. Numerous ECG-derived factors related to respiration (EDR signals) and HRV signals have been used in existing literature in numerous methods for OSA detection based on ECG. These parameters are used to extract time-domain, frequency-domain, and various nonlinear features, and these features are then used as inputs within classifiers in a black box for the decision-making process.

3.1 Signal Preprocessing

In the preliminary stage of signal processing, HRV signal and EDR signals of each subject recording are segmented on minute basis. Those segmented signals are employed to noise reduction on occurrence of interference such as in the signal using various filtering techniques. Noise-filtered segment is employed to Pan–Tompkins's algorithm [7] to determine the R-peaks. The series of RR intervals is generated by computing the time intervals between consecutive R-peak points. To account for poor signal quality and errors in the R-peak detection technique, the RR interval series is adjusted and corrected.

3.2 Signal Segmentation

To address the incomplete understanding of the physiological processes connecting ECG changes to OSA events, the initial single-lead ECG recordings are partitioned into 1-min segments.

3.3 Noise Filtering Using Two-Stage Median Filter

A median filter was used for removing physiologically uninterruptable points. The median filter M is an operator that relies on data and extracts a value from a signal in matrix form, which is obtained from the matrix data extracted from the Physionet database. Prior to obtaining a clean signal, the matrix data undergo filtration. The presence of a significant number of outliers in the sequence has minimal impact on the median selection process, particularly if the sequence exhibits symmetry around a certain point [8].

3.4 Feature Extraction

The respective features are obtained from a refined version of the EDR and RR series, which incorporates a R-peak detector and utilizes the Hilbert transform. Additionally, various methods for estimating EDR are introduced and assessed for their alignment with the reference respiration signal. These approaches are evaluated before being employed as surrogate respiration signals in OSA detection. The most important features of ECG for apnea detection are HRV parameters.

3.5 Minute-By-Minute Classification (Segment by Segment Classification)

Feature classification of the extracted features from the EDR and HRV signals containing R-peak detection on RR intervals for each minute segments is carried out. In this classification, Convolutional Neural Network is employed to select the optimal characteristics of the combined EDR and HRV signals. Selected feature is classified using multilayer perceptron-based Artificial Neural Network. Artificial Neural Network acts as backpropagation technique to the minute-by-minute-based classification. ANN can be considered as unsupervised learning model to one-minute segment and remaining segment can be classified as supervised learning model considering the backpropagation operation of the model. Hyperparameter tuning of the model is represented in Table 1.

3.5.1 Convolutional Layer

The convolutional layer consists of many filters designed to filter out the low-level feature of HRV signals and EDR signals represented as input vector. Convolutional layer processes the feature selection with respect to activation function.

Convolutional operations are performed as mathematical operation in each layer by 20 filter (kernels) with sizes of 10, 20, 15 using stride of 1 to produce the feature map. The feature map of the convolutional operations is represented in Fig. 1. The feature map on the convolutional operations is provided by the convolutional layer. To generate a linear feature map, the convergence is performed by employing epoch and increasing the feature production on normalization of the activation function, which is commonly represented as ReLU. The cosine distance measure is used to calculate the distance between these characteristics [9].

The evaluation of the cosine distance of feature maps is performed as Eq. 1,

$$\mathrm{Cf} = y(\mathrm{mtft} + c). \tag{1}$$

Table 1 Non-parameterized component for the Convolutional Neural Network

Hyperparameter	Values
Learning rate	0.04
Feature extracted	45
Epoch	50
Activation function	Rectified linear units (ReLUs)
Dropout layer	0.2
Loss function	Cross-entropy

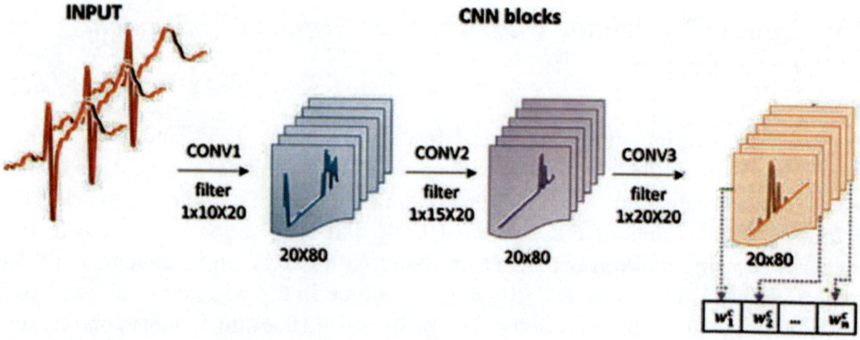

Fig. 1 Feature selection using the convolutional layer of CNN

3.5.2 Pooling Layer

The pooling layer extend diminishes the features of the features extracted from the convolutional layer. Hence, it is considered as max pooling of the features representing as high-level features. This procedure, also known as feature down-sampling, involves lowering the dimension of the feature by maintaining just the selected weighted features. The amplitude of the selected weighted characteristics is the highest. The Max Pooling layer classifies the features based on their amplitude, tying them together. For obstructive sleep apnea, max pooling is used to evaluation the largest number of features within each subgroup. It also helps to increase model generalization [10].

3.5.3 Fully Connected Layer

The fully connected layer is made up of batch normalization, flattening of the EDR and HRV features, a Softmax layer for sleep apnea classification, an activation function layer that uses the ReLU function, and a loss layer that uses cross-entropy limitations to process the feature map containing the relations of the EDR and HRV features. The fully connected layer, as a layer, uses an activation function for feature normalization or flattening to reduce nonlinearity and overfitting problems in the features.

Within the fully connected layer, the Softmax layer is used to produce illness classes by translating the feature vector into a disease class vector reflecting the obstructive sleep apnea classes {Mild, Moderate, and Severe}. In addition, a loss layer is added into the fully connected layer to reduce overfitting and underfitting difficulties in EDR and HRV feature classes.

The class coefficients for feature clarification are determined by the class objective function which is expressed as Eq. 2.

$$Y = \beta_0 + \beta_1 X. \tag{2}$$

The class coefficients are denoted as follows:

$$\beta_1 = \frac{n\sum xy - \sum x \sum y}{n\sum x^2 - \left(\sum x\right)^2}. \tag{3}$$

The deviations of the class objective function with respect to the class coefficients are obtained by evaluating the Error Sum of Squares (SSE). The Softmax layer is defined by the loss function of the hyperparameters, which follows the Delta rule. This layer's aim is to compute linear weights for various spatial features. Additionally, the feature weights can be calculated iteratively as in Eq. 4.

$$\Delta W_i = C(t - \text{net})x_i, \tag{4}$$

where '***c***' represents the learning rate, '***x***' denotes the input associated with that weight.

The Delta rule will be updated in order to reduce SSE and eliminate classifier loss. Algorithm 1 describes how the suggested classification model works.

Algorithm 1: Sleep apnea classification Input: ECG and EDR signals

Output: Obstructive Sleep Apnea classification

Process

Signal Preprocess

Signal Transformation()

Segmentation of signal ()

Minute Segment = Minute by Minute based signals

Signal Filtered SF= Bandpass filter (ECG and EDR Signals)

QRS complex = Differentiation (SF)

R Peak Detection using Hilbert Transform

Feature Extraction

ST_{HRV}=Extract Statistical features of RR interval of HRV

TFF_{HRV}=Extract Time frequency features of RR interval of HRV

NLF_{HRV}= Non Linear feature of RR interval of HRV

HRV feature = { ST_{HRV}, TFF_{HRV}, NLF_{HRV}}

ST_{EDR}=Extract Statistical features of RR interval of EDR

TFF_{EDR}=Extract Time frequency features of RR interval of HRV

NLF_{EDR}= Non Linear feature of RR interval of HRV

EDR feature = { ST_{EDR}, TFF_{EDR}, NLF_{EDR} }

Feature Selection &Feature Classification of CNN to Minute segments ()

Convolutional Layer(Kernel, Stride)

Feature map = low level feature (HRV +EDR)

Max pooling layer (stride, Kernel, pool size)

feature map = High level feature (HRV+ EDR)

Fully connected layer ()

Activation Layer (ReLu)

Flatten layer of high level feature of HRV+EDR

Softmax layer (Multilayer Perceptron)

Event of the segments= { Normal, Mild, Moderate , Severe}

3.5.4 Activation Layer

ReLU activation function, which imparts nonlinearity to the max-pooled feature vector and improves training to reduce mistakes, is a feature of the architecture. After each activation function, a batch normalization step reduces overfitting and improves system generalization by resampling the activation function's output using convolutional processes to account for illness stage labels. It detects the disease of the obstructive sleep apnea [11].

3.5.5 Softmax Layer

Softmax module is employed to classify the flattened feature map using multilayer perceptron to class the disease of the sleep apnea. The Softmax classifier identifies the features classes of the feature output in an N-channel image of probabilities and identifies segments related to the class with the maximum probability of every feature on the backpropagation function constructed using association rules in the MLP model [12].

4 Results and Discussion

4.1 Dataset Description

The data used for the analysis are sourced from the Physionet Apnea-ECG database [13]. Each dataset comprises single continuous ECG signals that span approximately 8 h, obtained from full PSG recordings. These ECG signals were recorded at a sampling rate of 100 Hz, with 16-bit resolution, and using a modified lead V2 electrode configuration. These recordings were gathered from a diverse group of subjects.

4.2 Performance Metrics

Performance of the obstructive sleep apnea classification technique using deep learning architectures and machine learning model is evaluated with the confusion matrix of the validation data. The confusion matrix provides insights into elements such as true positives, false positives, true negative, and false negative to the class results.

4.3 Performance Analysis of Sleep Apnea Classification

In this section, we assess the performance of obstructive sleep apnea classification by combining convolutional neural network and artificial neural network in a hybrid method. Initially, preprocessing of the signal is carried out using the median filter to eliminate the noise, and noise-eliminated signal is segmented into subjects

Table 2 Performance analysis of minute-by-minute classification of OSA

Technique	Accuracy (%)	Sensitivity (%)	Specificity (%)
Multilayer perceptron	91.25	86.56	95.45
Deep Convolutional Neural Network	96.25	91.16	98.14
Hybridization Convolutional Neural Network and Artificial Neural Network	98.87	93.78	98.79

and minute segments for recordings. Segmented minute-based signal is employed for QRS detection and R-peak detection using CNN algorithm. R-peak detection contains RR interval of EDR signal and HRV signals which are subjected for feature extraction. Feature of the EDR signal and HRV signal is extracted using nonlinear and linear feature extraction technique mentioned as discrete wavelet transform on time-domain feature and frequency-domain features and statistical features. Extracted feature is employed to the convolutional neural network for optimal feature selection and selected feature is classified into apnea and normal class using Apnea and Hypopnea Index. The table labeled as Table 2 represents a comparison of minute-by-minute classification outcome for sleep apnea detection between the proposed method and the traditional method, specifically for every HRV features.

The evaluations were repeated on the identical datasets, varying only the window length. The final result of minute-by-minute classification employing the weight calculation algorithm on the apnea ECG database exhibits a notable high level of performance, particularly with the proposed Hybridization Convolutional Neural Network and Artificial Neural Network model. It should be noticed that the proposed work results with 98.87%, respectively, which is considered to be a satisfactory performance. Figure 2 provides the performance evaluation of the obstructive sleep apnea classification using proposed classifier and conventional classifier to minute-by-minute classification.

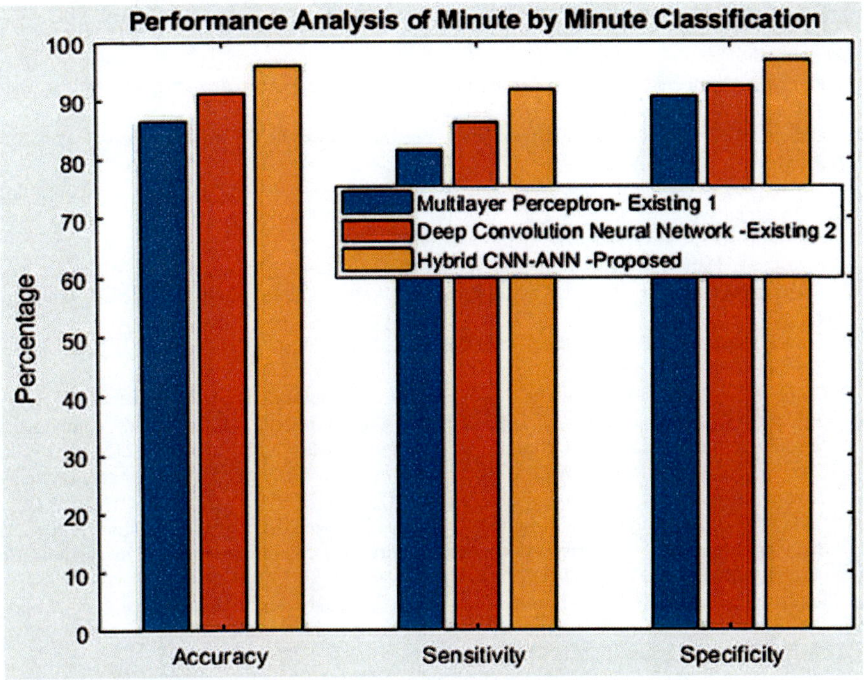

Fig. 2 Performance evaluation of the minute-by-minute classification model

5 Conclusion

Experimental analysis of the proposed framework toward the obstructive sleep apnea classification to the EDR signal and HRV signal has been carried out using ECG and ECG-derived respiratory dataset to evaluating the performance of the proposed representative framework using available optimal features indices against the conventional approaches has been analyzed in detail on various deep learning architectures. Proposed framework achieves accuracy of almost 99% on optimal feature for disease classification against machine-based existing classification approaches. In future, effective feature extraction, data preprocessing method, and data classification can also be enhanced using noise filtering and signal smoothening, an important preprocessing step in the signal classification to achieve good performances. Effective computation of data has become as dominant in the disease classification which becomes the focus of the current research in this field.

References

1. Parhi KK, Ayinala M (2014) Low-complexity Welch power spectral density computation. IEEE Trans Circ Syst I Regul Pap 61(1):172–182
2. Cokelaer T, Hasch J (2017) 'Spectrum': spectral analysis in Python. J Open Sour Softw 2(18):348
3. Lin R, Lee R, Tseng C, Zhou H, Chao C, Jiang J (2006) A new approach for identifying sleep apnea syndrome using wavelet transform and neural networks. Biomed Eng Appl Basis Commun 18:138–143. Sensors 2020, 20:4157 15 of 15
4. Hassan A, Haque A (2017) An expert system for automated identification of obstructive sleep apnea from single-lead ECG using random under sampling boosting. Neurocomputing 235:122–130
5. Penzel T, Moody GB, Mark RG, Goldberger AL, Peter JH (2000) The apnea-ECG database. Comput Cardiol 27:255–258
6. Goldberger AL, Amaral LAN, Glass L, Hausdor JM, Ivanov PC, Mark RG, Mietus JE, Moody GB, Peng C-K, Stanley HE (2003) PhysioBank, PhysioToolkit, and PhysioNet: components of a new research resource for complex physiologic signals. Circulation 101:e215–e220
7. Chen W, Wang Z, Xie H, Yu W (2007) Characterization of surface EMG signal based on fuzzy entropy. IEEE Trans Neural Syst Rehabil Eng 15(2):266–272
8. Porta A, Baselli G, Liberati D, Montano N, Cogliati C, Gnecchi-Ruscone T, Malliani A, Cerutti S (1998) Measuring regularity by means of a corrected conditional entropy in sympathetic outflow. Biol Cybern 78(1):71–78
9. Roche F, Duverney D, Court-Fortune I, Pichot V, Costes F, Lacour JR, Antoniadis A, Gaspoz JM, Barthelemy JC (2002) Cardiac interbeat interval increment for the identification of obstructive sleep apnea. Pacing ClinElectrophysiol 25(8):1192–1199
10. Sadr N, De Chazal P (2014) Automated detection of obstructive sleep apnoea by singleleadecg through elm classification. In: Computing in cardiology conference (CinC). IEEE, pp 909–12
11. Sharma M, Acharya UR (2018) Analysis of knee-joint vibroarthographic signals using bandwidth-duration localized three-channel filter bank. ComputElectrEng 72:191–202. https://doi.org/10.1016/j.compeleceng.2018.08.019
12. Sharma M, Achuth P, Deb D, Puthankattil SD, Acharya UR (2018) An automated diagnosis of depression using three-channel bandwidth-duration localized wavelet filter bank with EEG signals. CognSyst Res 52:508–520
13. Selim B, Won C, Yaggi HK (2010) Cardiovascular consequences of sleep apnea. Clin Chest Med 31:203–220

Bearing Fault Diagnosis Using Machine Learning and Deep Learning Techniques

N. Sai Dhanush and P. S. Ambika

Abstract Machine Fault diagnosis plays a vital role in ensuring the reliability and efficiency of industrial systems. Among various components, roller element bearing is prone to failures due to critical function in supporting rotating machinery. This paper proposes a machine fault diagnosis approach specifically tailored for roller element bearings. The methodology combines vibration analysis, signal processing techniques, and machine learning algorithms to accurately classify bearing faults. Firstly, vibration signals are acquired from the machine using accelerometers, and relevant features are extracted using time-domain, frequency-domain, and statistical methods. Subsequently, autoencoders are also used to extract more features with the aid of existing features. Finally, few state-of-the-art machine learning algorithms such as support vector machines, random forests are trained to classify the fault types. Experimental results on a simulated dataset and real-world scenarios illustrate effectiveness and accuracy of proposed approach in diagnosing rolling bearing faults. This developed methodology offers a practical solution for conditional monitoring and predictive maintenance, enabling timely detection and mitigation of bearing faults, thereby enhancing system reliability, minimizing downtime, and reducing maintenance costs.

Keywords Machine fault diagnosis · Autoencoder · Machine learning · Deep learning

N. Sai Dhanush (✉) · P. S. Ambika
Center for Computational Engineering and Networking (CEN), Amrita School of Artificial Intelligence, Amrita Vishwa Vidyapeetham, Coimbatore, India
e-mail: saidhanush.n@gmail.com

P. S. Ambika
e-mail: ps_ambika@cb.amrita.edu

S. Kumar et al. (eds.), *Fourth Congress on Intelligent Systems*, Lecture Notes in Networks and Systems 868, https://doi.org/10.1007/978-981-99-9037-5_24

1 Introduction

The process of analyzing the reason behind failure of machine (mechanical) components is known as machine fault diagnosis. Due to intricate and harsh circumstances like heavy loads, high temperatures, and high speeds, the intricate elements of mechanical equipment would inevitably give rise to diverse faults of varying magnitudes.

Machine fault diagnosis is a critical aspect of modern industrial systems, aiming to ensure the reliable operation and longevity of machinery. With the increasing complexity and integration of machines, the detection and identification of faults have become essential to prevent costly breakdowns, minimize downtime, and optimize maintenance activities [4]. Machine faults can arise from various sources, including mechanical wear, electrical malfunctions, lubrication issues, and environmental factors. These issues may appear in various ways, including unusual vibrations, changes in temperature, atypical noises, or a decline in performance. Consequently, a successful fault detection approach should possess the capability to scrutinize and comprehend a variety of signals, enabling precise identification of the underlying causes of these problems.

Conventional fault diagnosis methods typically involve manual inspections and scheduled maintenance, which can be both time-consuming and labor-intensive, and might not effectively detect early-stage issues. Nevertheless, progress in sensor technology, data acquisition systems, and computational methods has opened doors to more streamlined and automated fault diagnosis approaches. Vibration analysis stands out as a notable technique in the realm of machine fault diagnosis, as it frequently provides crucial insights into the state and performance of machine components [5]. By analyzing the frequency, amplitude, and other characteristics of these signals, it is possible to detect the presence of faults and determine their severity. Other techniques, such as acoustic analysis, thermal imaging, oil analysis, and electrical measurements, can also provide valuable insights into machine health.

In recent years, the field of machine fault diagnosis has witnessed significant advancements with the integration of machine learning and artificial intelligence algorithms [6]. These approaches harness the capabilities of data-driven models to autonomously grasp patterns and associations within extensive sensor datasets. Through training these models with historical data that contains recognized fault patterns, they become capable of promptly categorizing and identifying faults in real-time. The advantages of proficient machine fault diagnosis extend widely. It facilitates proactive maintenance strategies like condition-based and predictive maintenance, which optimize maintenance schedules, reduce expenses, and curtail unexpected downtime. Furthermore, early fault detection aids in mitigating potential safety risks and enhances overall machinery reliability and productivity.

In this article, we offer an extensive examination of methods for diagnosing machine faults and delve into machine learning algorithms [1]. We will delve into their practical uses, advantages, and constraints across various industrial contexts. Through a grasp of these techniques and their application, engineers and maintenance experts can put into action effective and precise fault diagnosis strategies, resulting in heightened system efficiency, lowered maintenance expenditures, and improved operational safety.

2 Related Work

This section provides an in-depth exploration of traditional approaches employed in diagnosing machine faults with a focus on the IMS dataset. It covers various signal processing techniques, including time-domain analysis, frequency-domain analysis, and the extraction of statistical features. Furthermore, it discusses classical machine learning algorithms like decision trees, support vector machines, and k-nearest neighbors within the context of IMS dataset analysis.

In light of the advancements in machine learning and data-driven methods, this section also investigates the application of advanced algorithms for machine fault diagnosis using the IMS dataset. It delves into the utilization of artificial neural networks, deep learning models, and ensemble methods to achieve precise fault detection and classification. Additionally, the section explores feature extraction and selection techniques specifically tailored for effective use with the IMS dataset.

Adapted weighted Signal Preprocessing technique for Machine Health Monitoring, which uses Signal Processing methods as preprocessing techniques for conditional health monitoring with an accuracy of 98% [2]. A fault diagnosis technique for induction motor bearing has been developed, utilizing cepstrum-based fault preprocessing and ensemble learning algorithms, achieving an impressive accuracy rate of 99.58% [3].

Fang et al. [7] second step in the fault diagnosis is feature extraction, and there are many feature extraction techniques including statistical, time-domain, frequency-domain, wavelet features and pattern recognition models. There have been limited research studies on the topic of bearing fault diagnosis. One such study explores the application of deep learning extraction techniques combined with handcrafted feature extraction in the time and frequency domains. This approach achieves a commendable accuracy rate of 95.8%. An another study which implements Singular value decomposition technique as a feature extractor which achieved an accuracy of 98.33%.

Final step in the fault diagnosis is fault classification, and once input preprocessing and feature extraction is done, faulty and healthy data is segregated and then machine learning algorithms were used classification. This can be posed as a multi-class classification problem, with the four classes being healthy (which means under normal working condition without any flaws), faults in outer race region of bearing, faults in inner race region, and faults in rolling elements. A study used combined approach using SVM and CNN for fault classification which achieved an accuracy of 98.9% and another study which uses ensemble models like random forests and XGBoost for classification with an accuracy of 99.3%.

3 Proposed Work

3.1 Feature Extraction

Each file in three datasets contains one-second signal snapshots recorded. We are converting this huge data in a single file into a data with features with that time stamp.

So each time stamp represents one-second signal snapshots, and useful features are extracted from those data for fault diagnosis. Time-domain and statistical features like peak value, minimum value, mean value, root mean square value, standard deviation, kurtosis, skewness, crest factor and form factor [8]are used for feature extraction. This useful information is extracted and represented as a single row in the pandas data frame. In such manner, we have represented data as a single row for all available files provided in the dataset with the extracted features.

Figure 1 is an example which illustrates the time-domain plot of bearings for one-second signal snapshot.

Features extracted from the signal snapshots are peak value, minimum value, mean value, root mean square value, standard deviation, kurtosis, skewness, crest factor, and form factor. The importance of these features are listed in Table 1.

Buchaiah and Shakya [11] once the feature extraction is done, we are supposed to detect fault degradation which is possible with the help of features extracted and all possible plots are investigated for each test setup to know where the fault occurs and types of faults.

Here are few images which illustrate the behavior of bearings with respect to features we extracted (for second test setup).

Figure 2 clearly depict the behavior of bearings at different timestamps with respect to different features extracted.

3.2 Methodology

Albasi et al. [9] fault degradation is detected manually with the help of above plots, ands as we inspect those plots, it is very obvious that bearing 1 in the second test setup is behaving abnormally toward the end of the plot, from which we can infer that bearing 1 is undergoing fault detection from timestamps towards the end of plot [10]. So in this second test setup, we can create data for healthy data and data with fault in outer race. Similarly for all three tests the same proedure is followed and data is segregated into healthy and non-healthy, where non-healthy data refers to data in which faults in outer region of bearing race, inner race and faults in rolling elements in the bearing.

Thus, the data is labeled manually with the help of timestamps available, and now we have labeled data which is posed as a supervised learning problem. With four different target classes, and posed as a four class classification problem. Data is split into 70% training and 30% for testing [12]. Now we implemented machine

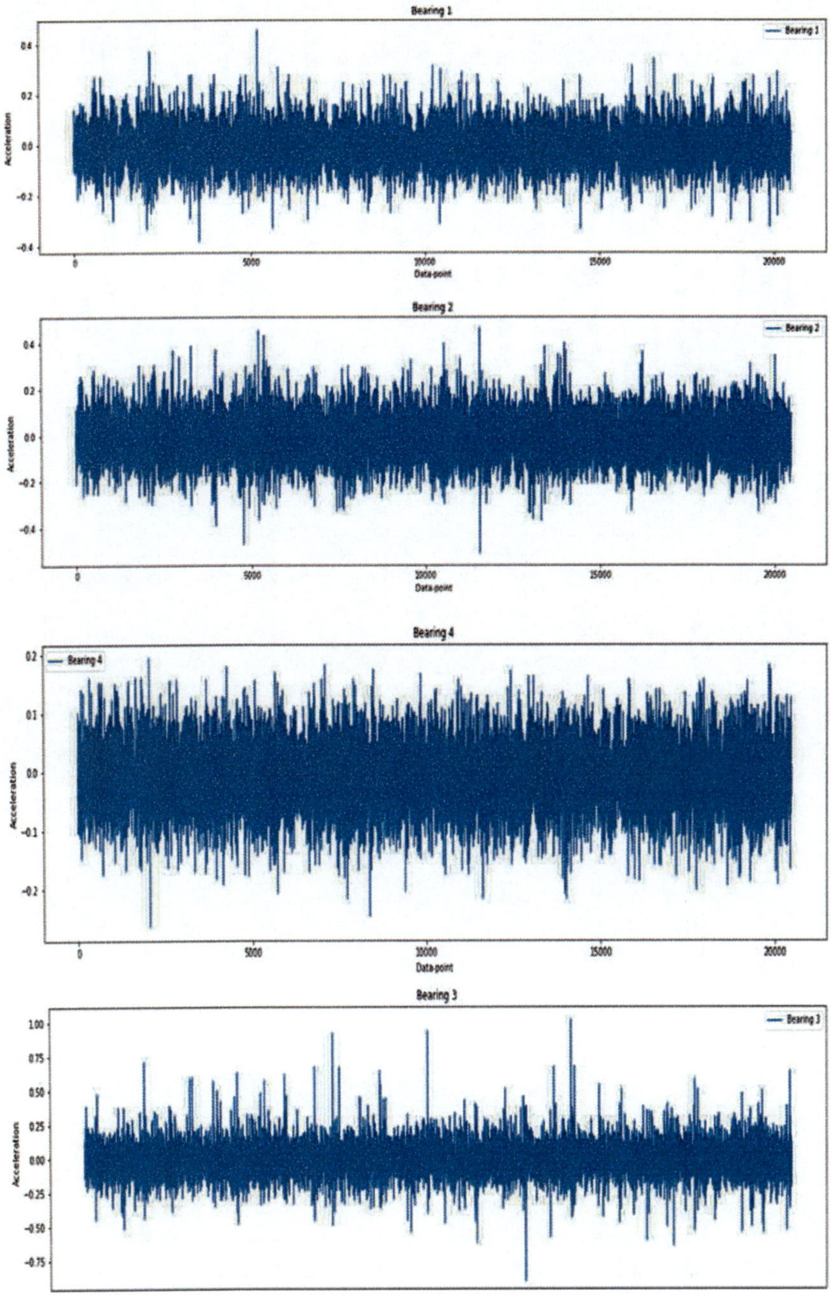

Fig. 1 Time-domain plot for all four bearings for one-second snapshot for test setup

Table 1 Description of features extracted from the signal snapshots

Feature	Mathematical formula	Description
Maximum value	$\max(x) = \max\{x_1, x_2, x_3, \ldots, x_n\}$	The maximum value in a time series is obtained by compressing each sampling observation
Minimum value	$\min(x) = \min\{x_1, x_2, x_3, \ldots, x_n\}$	The minimum value in a time series is determined by compressing each sampling observation
Mean value	$\text{mean}(x) = ((x_1 + x_2 + x_3 + \cdots + x_n))/n$	The average value in a time series is calculated by compressing each sampling observation and taking the simple average
Root mean square (RMS) value	$\text{RMS}(x) = \surd(((x_1^2 + x_2^2 + x_3^2 + \cdots + x_n^2))/n)$	The measure quantifies the dispersion of predicted values around the regression line. In this context, '*n*' corresponds to the sample size, indicating the number of timestamps acquired after preprocessing
Standard deviation	$\sigma = \text{sqrt}\big((1/N) * \Sigma(x - \mu)^2\big)$	Obtained by taking the square root of the variance, it signifies the distribution of a reduced time-series sample in relation to its mean
Kurtosis	$\text{Kurt}(X) = (1/n) * \Sigma\big((x - \mu)^4\big)/\sigma^4$	It quantifies the deviation of the tail of a distribution from the normal distribution and is defined as the fourth moment of the distribution
Skewness	$\text{Skew}(X) = (1/n) * \Sigma\big((x - \mu)^3\big)/\sigma^3$	It quantifies the divergence of a distribution from the normal distribution, either toward the left or right. This measure is defined as the third-order moment of the distribution

(continued)

Table 1 (continued)

Feature	Mathematical formula	Description
Crest factor	Crest Factor = Peak Value/RMS Value	Crest factor is a measure used in signal processing to quantify the peak-to-average ratio of a waveform. It provides information about the amplitude variations or "peaks" in a signal relative to its average value
Form factor	Form Factor = RMS Value/Average Absolute Value	It quantifies the ratio between the root mean square (RMS) value and the average absolute value of a signal, providing information about the symmetry and smoothness of the waveform
Clearance factor	$\text{Clearance Factor} = \text{Maximum Value} / \text{Squared Mean of Squared root of absolute amplitudes} of Maximum Value$	The peak value is determined by dividing the squared mean value of the square roots of the absolute amplitudes by the peak value itself. This characteristic is maximized in the case of healthy bearings and gradually diminishes for faulty bearings

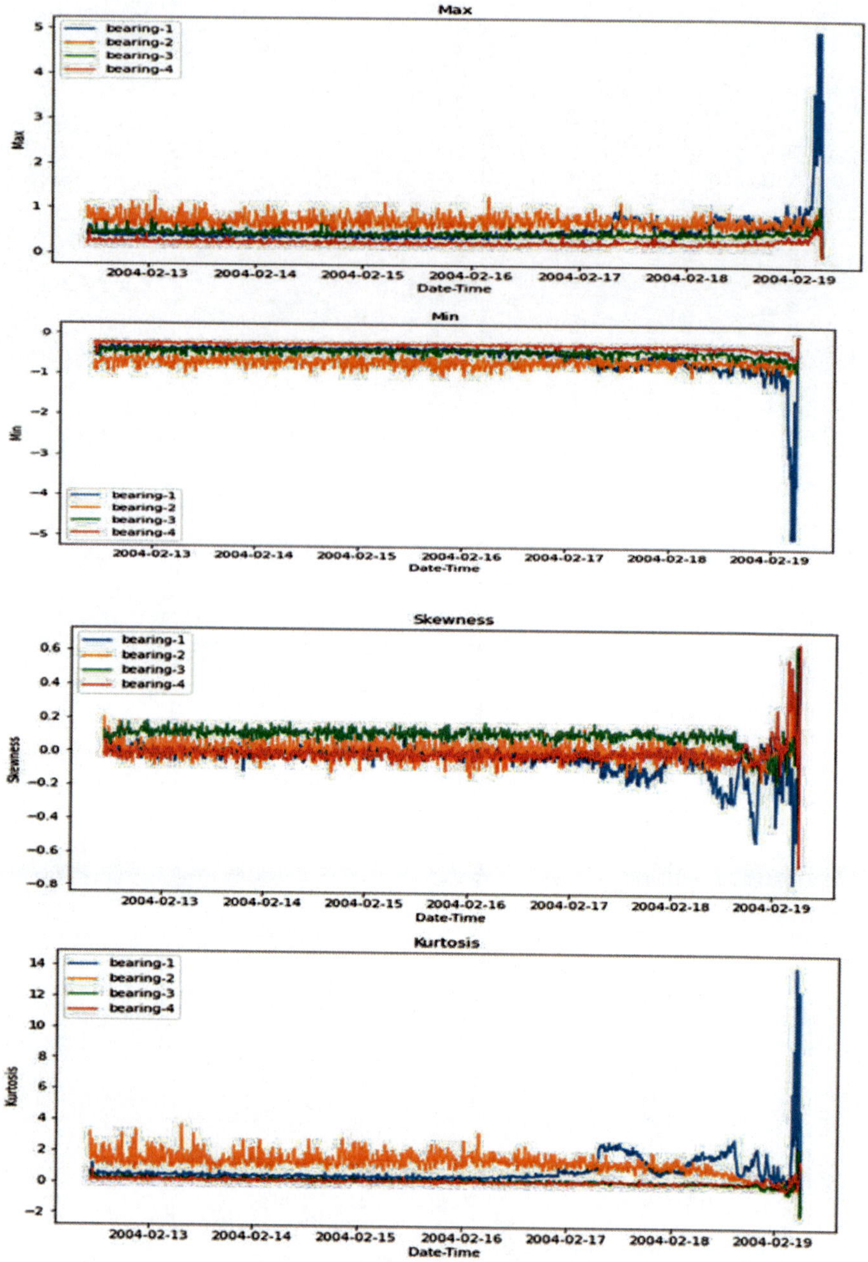

Fig. 2 Behavior of bearings with respect to extracted features for second test setup

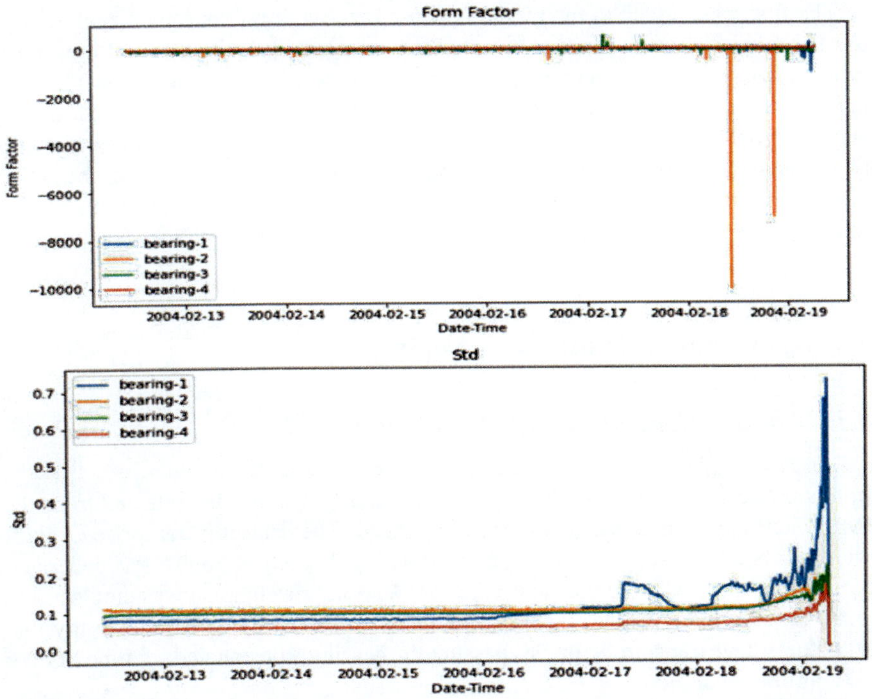

Fig. 2 (continued)

learning algorithms to tackle this problem, several algorithms such as K-Nearest Neighbors, Support Vector Machines, Random Forest, Decision Trees, CatBoost, LightGBM, and other techniques have been employed. In the process of training, neither overfitting nor underfitting is observed.

Wu et al. [13] besides this, we have incorporated the same methodology by adding new other features which were obtained from autoencoder architecture. Earlier we have only ten features, now with the aid of autoencoder, 256 new features were extracted and concatenated with the ten time-domain features we had, and once again, it is posed as a four-class classification problem [14]. And the same machine learning algorithms were addressed this problem and hyperparameter tuning is also done to achieve better results.

3.3 Deep Learning-Based Feature Extraction

The code implementation for extracting deep learning-based features makes use of deep learning model using Sequential API from Keras, which allows linear stack of layers.

The first layer added to the model is dense layer with 64 neurons, which makes use of "relu" activation function to introduce nonlinearity, which is followed by a dropout layer which aids in regularization; in this case, 50% is dropped out. The next layer is another dense layer with 128 neurons and "relu" activation function, and another dropout layer is added after second dense layer. The model continues with two more dense layers of 256 and 512 units. The model is compiled with the aid of Adam Optimizer and loss function as "Categorical Cross Entropy" which suits for multi-class classification problem. The model will also compute the accuracy metric.

4 Experimental Results and Analysis

4.1 Dataset Collection and Description

In this particular research study, the NASA bearing dataset, also referred to as the IMS (Intelligent Machines' Dataset), was utilized. The dataset is comprised a shaft with four bearings, subject to a consistent rotational speed of 2000 RPM generated by an AC motor linked to the shaft via rubs. A spring mechanism was employed to apply a radial load of 6000 pounds to the shaft and bearing. It is noteworthy that all failures have transpired after surpassing the bearing's designated lifetime of over 100 million revolutions.

The dataset is structured as follows: It is provided in a zip format with a size of approximately 1.6 GB. Within the dataset, there are three individual datasets, each describing a test-to-failure experiment. Each dataset comprises multiple files, where each file represents a one-second snapshot of vibrational signals recorded at different time intervals. The experimental setup is illustrated in Fig. 3.

The sampling rate for the signals was set at 20 kHz. For further details regarding the dataset description and other relevant information, please refer to the table presented Table 2.

4.2 Result and Discussion

The proposed method is simulated in Python, running on a 64-bit Windows 11 platform with Intel Core i7-11800H @ 2.30 GHz and 32 GB RAM.

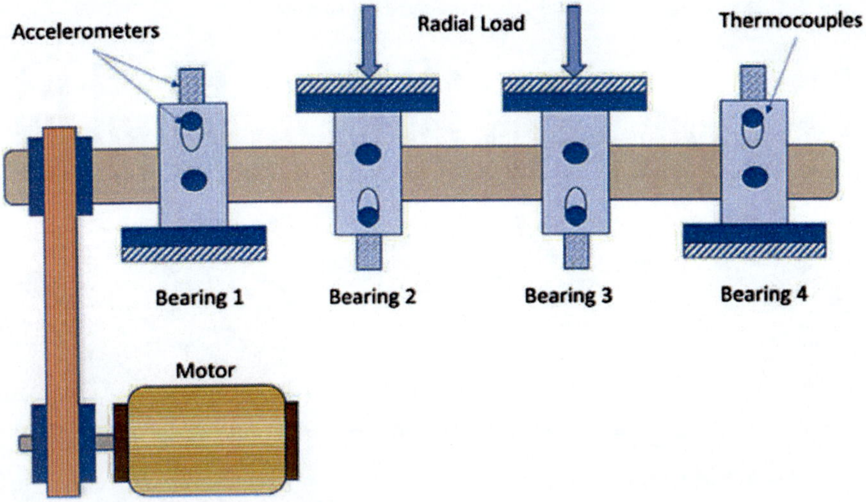

Fig. 3 IMS dataset experiments were conducted using a bearing test rig, and careful consideration was given to the placement of sensors

Table 2 Intelligent machine systems (IMS) dataset description

Description	Dataset 1	Dataset 2	Dataset 3
No. of files	2156	984	4448
No. of channels	8	4	4
File recording interval (in minutes)	Every 10 min interval	Every 10 min interval	Every 10 min interval
Failure description (occurred during end-to-end failure test)	Inner race defect in Bearing 3, roller element defect in Bearing 4	Outer race failure occurred in Bearing 1	Outer race failure occurred in Bearing 2

5 Conclusion

In this work, performance analysis of our work is compared with only statistical features and the other being statistical and deep learning-based features. When implementation is done only with statistical features, highest accuracy of 96.05% is achieved with Gradient Boosting Algorithm, whereas when the implementation is done with both statistical and deep learning-based features, an highest accuracy of 96.05% is achieved with LightGBM algorithm. All the results obtained are maximum possible results as all algorithms are fine tuned for better results with the help of hyperparameter tuning. These results are illustrated in Figs. 4 and 5.

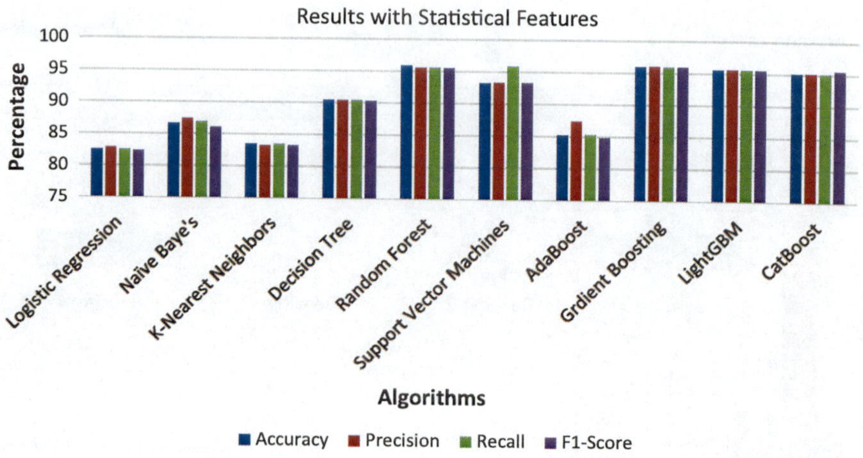

Fig. 4 Results with statistical features

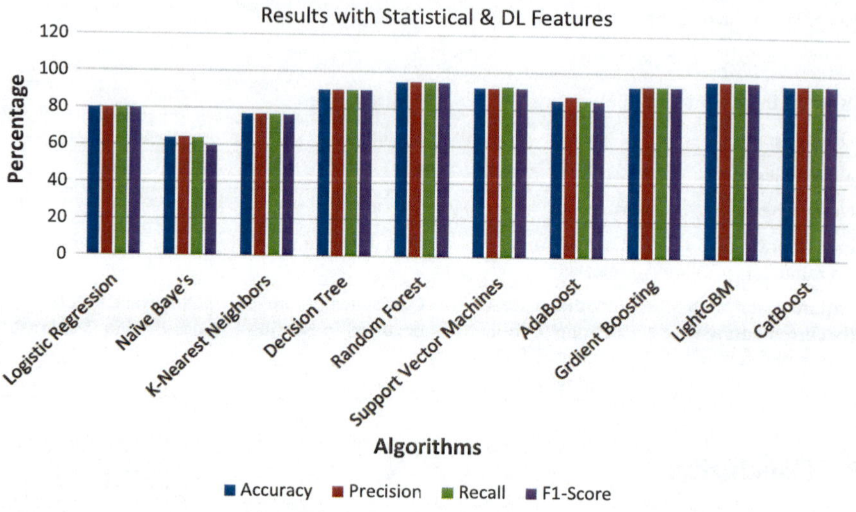

Fig. 5 Results with statistical and deep learning features

Acknowledgements Authors thank Prof. K. P. Soman, Head, Center for Computational Engineering and Networking (CEN) at Amrita Vishwa Vidyapeetham, Coimbatore, Tamil Nadu.

References

1. Gupta M, Wadhvani R, Rasool A (2022) A real-time adaptive model for bearing fault classification and remaining useful life estimation using deep neural network. Knowl-Based Syst 259:110070. ISSN 0950-7051. https://doi.org/10.1016/j.knosys.2022.110070. https://www.sciencedirect.com/science/article/pii/S0950705122011637
2. Hou B et al (2021) Adaptive weighted signal preprocessing technique for machine health monitoring. IEEE Trans Instrum Meas 70:1–11. https://doi.org/10.1109/TIM.2020.3033471
3. Bhakta K et al (2019) Fault diagnosis of induction motor bearing using Cepstrum based preprocessing and ensemble learning algorithm. In: 2019 International conference on electrical, computer and communication engineering (ECCE), pp 1–6. https://doi.org/10.1109/ECACE.2019.8679223
4. Sikder N et al (2019) Fault diagnosis of motor bearing using ensemble learning algorithm with FFT based preprocessing. IEEE Xplore
5. Zhang Z et al (2019) General normalized sparse filtering: a novel unsupervised learning method for rotating machinery fault diagnosis. Mech Syst Sig Process 124:596–612. ISSN 0888-3270. https://doi.org/10.1016/j.ymssp.2019.02.006. https://www.sciencedirect.com/science/article/pii/S0888327019300895
6. Zhang K et al (2020) A compact convolutional neural network augmented with multiscale feature extraction of acquired monitoring data for mechanical intelligent fault diagnosis. J Manuf Syst 55:273–284. ISSN 0278-6125. https://doi.org/10.1016/j.jmsy.2020.04.016. https://www.sciencedirect.com/science/article/pii/S0278612520300601
7. Fang H et al (2021) LEFE-Net: a lightweight efficient feature extraction network with strong robustness for bearing fault diagnosis. IEEE Trans Instrum Meas 70:1–11. https://doi.org/10.1109/TIM.2021.3067187
8. Zhu H et al (2021) Bearing fault feature extraction and fault diagnosis method based on feature fusion. Sensors 21(7). ISSN 1424-8220. https://doi.org/10.3390/s21072524. https://www.mdpi.com/1424-8220/21/7/2524
9. Albasi M et al (2020) Bearing fault diagnosis using deep learning techniques coupled with handcrafted feature extraction. J Vibr Control
10. Han T et al (2021) Rolling bearing fault diagnosis with combined convolutional neural networks and support vector machine. Measurement 70. Elseiver
11. Buchaiah S, Shakya P (2022) Bearing fault diagnosis and prognosis using data fusion based feature extraction and feature selection. Measurement 188:110506. ISSN 0263-2241. https://doi.org/10.1016/j.measurement.2021.110506. https://www.sciencedirect.com/science/article/pii/S0263224121013889
12. Toma RN, Kim JM (2020) Bearing fault classification of induction motors using discrete wavelet transform and ensemble machine learning algorithms. In: MDPI, July 2020
13. Wu X et al (2021) A hybrid classification autoencoder for semi-supervised fault diagnosis in rotating machinery. Mech Syst Sig Process 149
14. Chai Z, Zhao C (2020) Enhanced random forest with concurrent analysis of static and dynamic nodes for industrial fault classification. IEEE Trans Ind Inform 16(1):54–66. https://doi.org/10.1109/TII.2019.2915559

A Transfer Learning Approach to Mango Image Classification

Abou Bakary Ballo, Moustapha Diaby, Diarra Mamadou, and Adama Coulibaly

Abstract Côte d'Ivoire is one of West Africa's leading mango exporters, significantly contributing to its economy. With an annual production of around 180,000 tons, fresh mango is the third most exported fruit from Côte d'Ivoire. The country is the third biggest supplier to the European market, with 32,400 tons a year. However, to maximize profits, farmers must accurately assess mangoes' ripening stage. This paper proposes a CNN-based system using transfer learning to accurately detect mango ripeness. Our study focuses on developing a system using renowned CNN algorithms such as VGG19, ResNet101, and DenseNet121. The model was first trained on a dataset of 996 mango images. Next, it was refined on a smaller dataset specific to the task. Finally, it was evaluated on a test dataset. The DenseNet121 model performed best, with an accuracy of 97.50%. These machine learning techniques were chosen because they can improve the accurate identification of mango maturity. The contributions of our research are manifold: not only have we proposed an accurate solution for detecting mango ripeness, but we have also evaluated the performance of different models of CNN algorithms. This system will strengthen Côte d'Ivoire's position as a leader in the African mango industry by reducing product losses associated with mango harvesting and increasing export volumes.

Keywords Mango · CNN · Ripeness

A. B. Ballo (✉) · D. Mamadou
LaMI, Université Felix Houphouët-Boigny, Abidjan, Côte d'Ivoire
e-mail: bakary.abou@ufhb.edu.ci

A. B. Ballo
LMI Université Péléforo Gon Coulibaly, Korhogo, Côte d'Ivoire

A. Coulibaly
LaMA Université Felix Houphouët-Boigny, Abidjan, Côte d'Ivoire

M. Diaby
Lastic, Ecole Supérieure Africaine des Technologies de l'Information et de la Communication, Abidjan, Côte d'Ivoire

S. Kumar et al. (eds.), *Fourth Congress on Intelligent Systems*, Lecture Notes in Networks and Systems 868, https://doi.org/10.1007/978-981-99-9037-5_25

1 Introduction

Côte d'Ivoire, a West African country, is one of the continent's leading mango exporters. Indeed, the country far surpasses other West African nations as the leading exporter of this delicious, juicy fruit. Mango exports significantly contribute to Côte d'Ivoire's gross domestic product (GDP), accounting for up to 4% of its economy. Mango is the third most exported fruit in Côte d'Ivoire [1–3], after bananas and pineapples, demonstrating its economic importance. However, to make the most of their investment, farmers need to assess the ripening stage of their produce accurately. The harvest timing is essential in determining the quality and shelf life of the mango. Harvesting an overripe mango risks rotting the fruit before it can be marketed. It is, therefore, crucial to find solutions to accurately detect the degree of ripeness of mangoes to reduce product losses while increasing export volumes.

We propose a CNN-based system using transfer learning and proven image processing techniques. We aim to accurately detect the ripeness of mangoes, offering an essential solution for preventing product losses.

The study focuses on developing a system based on convolutional neural networks using transfer Learning, capable of accurately identifying the ripeness of mangoes. This system will help to reduce product losses associated with mango harvesting and boost exports of this precious fruit, thereby strengthening Côte d'Ivoire's economy as a leader in the African mango industry.

The contributions of this research are manifold:

Firstly, we explored machine learning techniques to detect mango ripeness using renowned CNN algorithms, such as VGG19, ResNet101, and DenseNet121. This approach takes advantage of the advanced capabilities of these techniques to improve identification accuracy.

In addition, we evaluated the performance of the different algorithm models used, comparing them with other existing approaches. The results demonstrate our approach's relevance and effectiveness in detecting mango ripeness with high accuracy.

2 Related Work

Machine learning and image processing techniques have made a breakthrough in agriculture in recent years, as they have been widely applied. Some studies have focused on automated detection of fruit ripeness. We can mention just a few of them. Jasman Pardede et al. proposed a technique using the VGG16 model without replacing the top layer by adding a multilayer perceptron (MLP) block. The MLP block contains a flattening layer, a dense layer, and a regularization layer to classify mango maturity. The results of this hybrid model gave an accuracy of 90% [4]. Their study is an important contribution to the literature on mango maturity classification. The study results show that the hybrid model has high accuracy and can be used to classify mango maturity with high precision. However, the technique must be evaluated on a larger dataset and under various weather conditions to confirm the study results.

Similarly, Limsripraphan et al. presented a Bayesian approach to mango ripeness classification. Thresholding and image labeling are used to identify defects in the mango skin. Color features are then extracted from the RGB images using statistical calculations. Finally, a Naive Bayes classifier is used to classify the color features and identify the type of defect. The experimental results of their proposed method gave an accuracy of 83% [5]. Nevertheless, the study has some limitations that need to be considered in future work, such as using a different machine learning model, such as a convolutional neural network (CNN), to evaluate the model over more categories of mango maturity. Kusrini et al. [6] compared five deep learning models for mango pest and disease recognition: VGG16, ResNet50, InceptionResNet-V2, Inception-V3, and DenseNet. The VGG16 model achieved the highest validation and testing accuracy (89% and 90%, respectively) of all the models tested despite being the smallest model with only 16 layers. Testing 130 images took 2 s longer than the rest of the experiment. Their approach is a promising first step toward developing deep learning models for recognizing mango pests and diseases. However, the study may have limitations, such as assessing the accuracy of the models in detecting different types of pests and diseases and the time needed to train and test the models. Similarly, Selly Anatya et al. studied a classification application using convolutional neural networks to classify five classes of fruit: starfruit, mango, melon, banana and tomato. Each class is again divided into 52 sub-classes, including fruit type and ripeness level, with 5030 training data images. The results of this work gave an accuracy of 88.93% [7]. However, the model may be too slow to be used in real-time applications.

Ayllon et al. presented a system for determining the ripeness of banana (Cavendish), mango (Carabao), calamansi/pineapple, and pineapple fruits using convolutional neural networks [8]. The best accuracy for RGB is banana, with a maximum accuracy of 97%, while the best result for grayscale is calamansi, with a maximum accuracy of 89%, but the weaknesses of their study are the total number of fruit studied and the study period. Gururaj N. et al. presented a system based on mango maturity stage, shape, textural features, color, and defects to identify the mango variety and classify it according to its quality. The mango maturity stage is extracted using a convolutional neural network (CNN). The extracted features are fed into the Random Forest classifier to identify the mango variety and classify mango quality into three categories: poor, average, and good. The results of this work gave a score of 93.23% for variety recognition and 95.11% for quality classification [9]. Using a different machine learning algorithm for the variety recognition task can be developed into a more robust system capable of handling variations in lighting, background, and other factors to make the system more user-friendly.

Fig. 1 **a** Ripe mango; **b** unripe mango

3 Methodology

3.1 Dataset

Mango images used in the study were obtained from a plantation in Nafana, a village in northern Côte d'Ivoire's Ferkessédougou department, Savanes region.

Figure 1 shows images of mango fruits collected on a plantation in Nafana using an Infinix Hot Lite camera in controlled and uncontrolled environments. The images were divided into 80% for training and 20% for testing. We trained our model on a machine with an Intel(R) Core™ i7-8650U processor, 16 GB RAM, and an NVIDIA GeForce® graphics card. We used Python software with the Keras 2.4 libraries, Tensorflow 2.4 backend, and CUDA/CuDNN dependencies for GPU acceleration to deploy our models.

3.2 Classification Algorithm

The choice of an appropriate classification algorithm is important for the performance of a prediction model. In our study, we will use VGG19, ResNet101, and DenseNet121.

- **VGG19**

VGG19 is a deep convolutional neural network (CNN) with 19 layers, including 16 convolutional layers and three fully connected layers. It uses small 3×3 filters in each convolutional layer to reduce the number of parameters in the network. The network is trained to classify images into 1000 object categories, and it features two fully connected layers with 4096 units each [11]. The VGG19 model won classification and localization at the ImageNet large-scale visual recognition competition (ILSVRCRC) with 138 million parameters in 2014 [12].

- **ResNet101**

ResNet101 is a convolutional neural network (CNN) developed by Microsoft researchers in 2015. ResNet101 is a convolutional neural network (CNN) with 101 layers. A pre-trained version of the network, trained on over a million images from the ImageNet dataset, can be loaded [14]. The pre-defined training system classifies images into 1000 object categories, such as keyboard, mouse, pencil, and several animals. As a result, the lattice was trained on multiple rich claim feature representations for various images. The lattice is designed with an image input size of 224 $\times$ 224. ResNet101 is a CNN architecture that was developed to improve the accuracy of image classification tasks. It achieved state-of-the-art performance on the ImageNet dataset when it was first introduced and has since been widely applied and studied in computer vision [13].

- **DenseNet121**

DenseNet is a special type of CNN originally proposed by Huang et al. [15]. The DenseNet121 convolutional neural network uses connections between layers via dense blocks. Each layer is connected to all the others. Each layer, the set of feature maps from all previous layers is deployed as input. This applies to all other layers. Each layer has immediate access to the original input image and loss function gradients.

3.3 Methods

The architecture of our method is based on several key steps, each of which brings significant benefits.

- **Step 1: Data preprocessing**

The images in the database were organized and classified into two folders, namely ripe and unripe mangoes. Each image was resized to a size of 150 $\times$ 150 pixels. We also applied extraction techniques to target specific areas of the fruit in our images. After resizing, the training and test images were converted into arrays, and the values assigned to the pixels in these arrays were adjusted to the interval [0, 1].

- **Step 2: Feature extraction**

The second step is to extract the features from each image. To do this, we exploit advanced techniques such as convolution and pooling. We use the DenseNet121, VGG19, and RESNET101 algorithms for independent variable extraction. These operations are applied according to an architecture specific to each algorithm. Next,

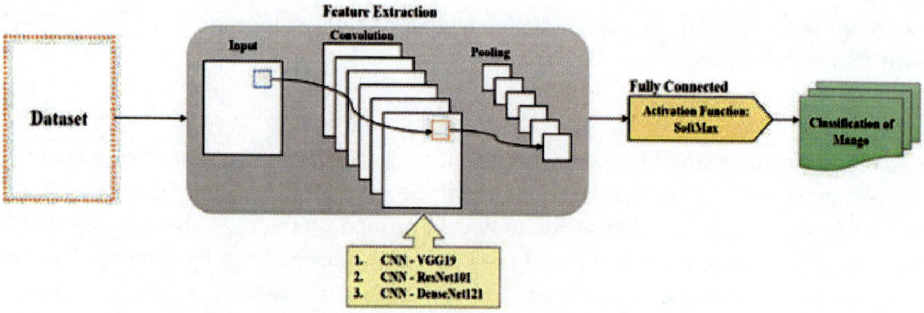

Fig. 2 General architecture of the study

we divide our data into training, validation, and testing. We used the Softmax classifier, a powerful machine learning tool for multi-class classification, to classify the vectors obtained after feature extraction.

- **Step 3: Learning with validation**

Finally, in the third step, we train our model using the validation of each algorithm. This approach enables us to test different algorithms and select the best performance for classifying mango fruits according to their condition. We can perform accurate and efficient classification using the features extracted from each image. This feature-based approach enables us to distinguish between different mango varieties more reliably and robustly.

Figure 2 shows the general architecture of our study.

4 Performance Measurement Phase

At this stage, the proposed approach is evaluated. Six measures are used to evaluate the proposed approach, namely accuracy, precision, F1-score, recall, Matthew correlation coefficient (MCC), and mean square error (MSE). We also use the confusion matrix.

- Accuracy measures how often a system correctly classifies data into the right class.
- Precision shows how good a classifier is at not calling negative samples positive.
- The mean squared error (MSE) measures the average squared difference between a quantity's estimated and actual values. It is a risk function that measures the expected value of the squared error loss. The MSE is always non-negative, and lower values indicate better performance.
- Recall is the proportion of positive cases correctly identified by a classifier.
- The F1-score can be interpreted as a weighted average of precision and recall, where the F1-score reaches its best value at one and worst value at 0.

- The Matthews Correlation Coefficient (MCC) is a machine learning metric that measures the quality of classifications. It ranges from − 1 to 1, with higher values indicating better performance. A MCC of 1 represents a perfect prediction, a MCC of 0 represents a random prediction, and a MCC of − 1 represents an inverse prediction. The MCC is also known as the phi coefficient.

The mathematical formula for these valuation measures is defined as follows:

$$\text{Accuracy} = \frac{\text{TP} + \text{TN}}{\text{TP} + \text{FP} + \text{TN} + \text{FN}}, \tag{1}$$

$$\text{Precision} = \frac{\text{TP}}{\text{TP} + \text{FP}}, \tag{2}$$

$$\text{F1 - Score} = 2*\frac{\text{Precision*Recall}}{\text{Precision} + \text{recall}}, \tag{3}$$

$$\text{Recall} = \frac{\text{TP}}{\text{TP} + \text{FN}}, \tag{4}$$

$$\text{MCC} = \frac{\text{TP} * \text{TN} - \text{FP} * \text{FN}}{\sqrt{(\text{TP} + \text{FP})(\text{TP} + \text{FN})(\text{TN} + \text{FP})(\text{TN} + \text{FN})}}, \tag{5}$$

$$\text{MSE} = \frac{1}{n}\sum_{i=1}^{n}(Y_i - \hat{Y}_i)^2. \tag{6}$$

With TP= True positive, TN= True negative, FP= False positive, FN= False negative.

5 Results and Discussion

Table 1 provides performance results for three convolutional neural network (CNN) models: VGG19, ResNet101, and DenseNet121. Each model is evaluated in accuracy, precision, F1-score, recall, Matthews Correlation (MCC), and mean square error (MSE).

Table 1 General results of the various methods

CNN	Accuracy	Precision	F1-score	Recall	MCC	MSE
VGG19	96.50	96.59	96.48	96.5	92.88	0.035
ResNet101	90.00	90.85	89.79	90.0	80.06	0.10
DenseNet121	**97.50**	**97.60**	**97.48**	**97.5**	**94.95**	**0.025**

The data in bold in the table represent the model with the highest score from our study.

Results for VGG19 were high overall. Accuracy is 96.50%, and precision is 96.59%. F1-score and recall are also high, indicating a good balance between precision and recall. However, the MCC is very low at 0.035, suggesting a low correlation between model predictions and actual observations. The root mean square error (RMSE) is also low at 0.035, indicating low error dispersion.

ResNet101 results are slightly lower than those of VGG19. Accuracy is 90.00%, and precision is 90.85%. The F1-score is also lower, at 89.79%. The recall is 90.0%, suggesting that the model may miss some true positives. However, the MCC is high (80.06), indicating a relatively strong correlation between predictions and actual observations. The root-mean-square error (RMSE) is 0.10, indicating greater error dispersion compared with VGG19.

The results for DenseNet121 are the highest of the three models evaluated. Accuracy is 97.50% and precision is 97.60%. F1-score and recall are also very high, indicating a good ability to identify true positives. The MCC is high at 94.95, suggesting a strong correlation between model predictions and actual observations. The root-mean-square error (RMSE) is 0.025, indicating even lower error dispersion than other models. DenseNet121 achieved the highest performance among the three models evaluated across all metrics. VGG19 also achieved good performance, while ResNet101 had the lowest performance. In summary, the results of this study suggest that DenseNet121 is the best-performing model for sediment image classification. It has the highest accuracy, precision, F1-score, recall, and MCC, as well as the lowest error dispersion.

Graphs of each model and their confusion matrix are shown below:

- **The case of VGG19**

Figure 3 shows the accuracy and loss curves of the VGG19 model during training.

Figure 4 shows the VGG19 confusion matrix during training.

- **The case of ResNet 101**

Figure 5 presents the accuracy and loss curves of the ResNet101 model during training.

Figure 6 shows the ResNet101 confusion matrix during training.

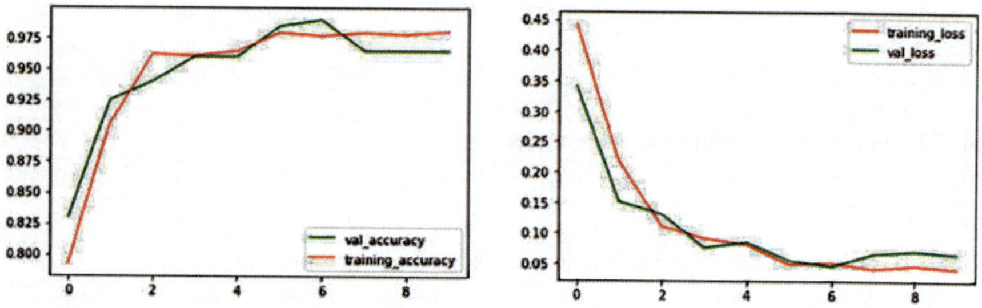

Fig. 3 VGG19 learning and testing graph

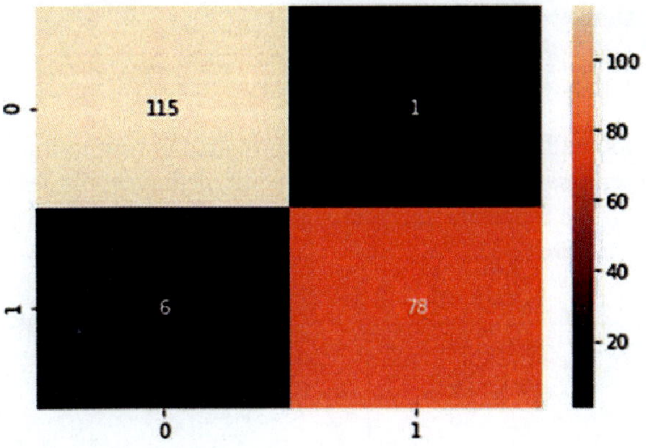

Fig. 4 VGG19 confusion matrix

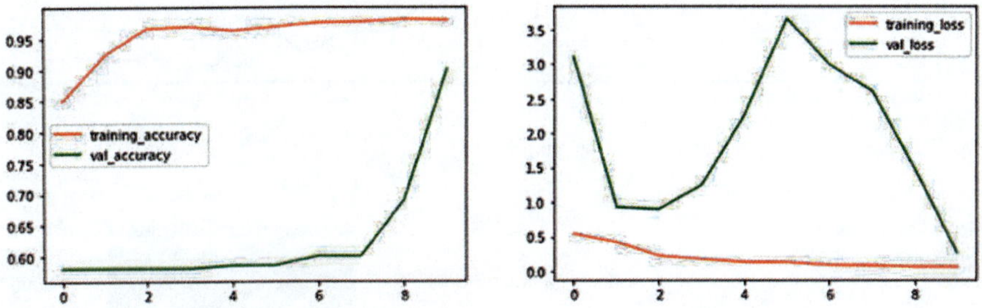

Fig. 5 ResNet101 learning and testing graph

Fig. 6 ResNet101 confusion matrix

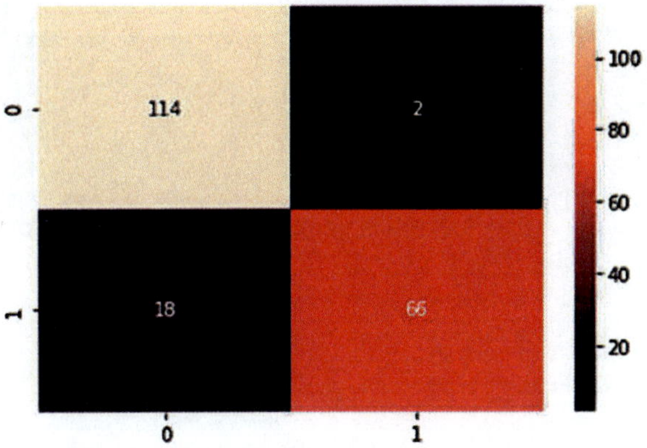

- **The case of DenseNet121**

Figure 7 presents the accuracy and loss curves of the DenseNet121 model during training.

Figure 8 shows the DenseNet121 confusion matrix during training.

We will now compare our study with the state of the art. Table 2 compares our model with that of the literature and shows that the results of our study are superior to those of the literature.

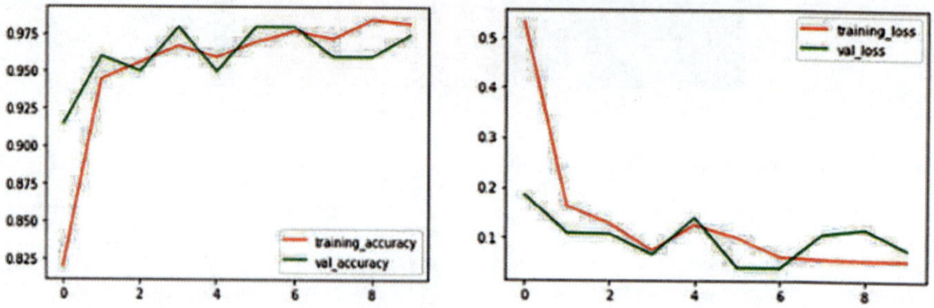

Fig. 7 DenseNet121 learning and testing graph

Fig. 8 DenseNet121 confusion matrix

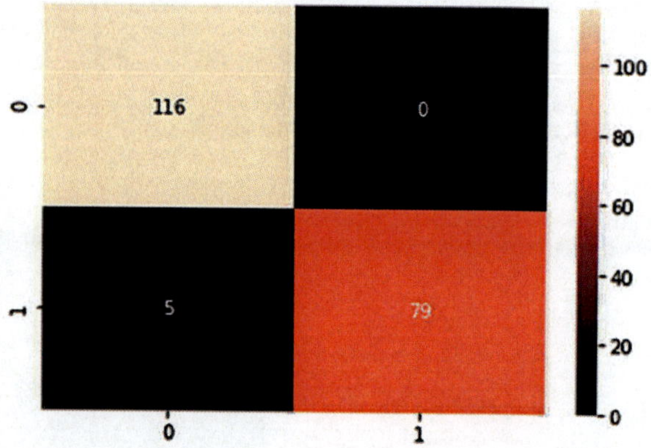

Table 2 Comparison with existing results

Methods	Accuracy
Pardede et al. [1]	90.00
Limsripraphan et al. [2]	83.00
VGG19	**96.50**
ReseNet101	**90.00**
DenseNet121	**97.50**

The data in bold in the table represent the model with the highest score from our study.

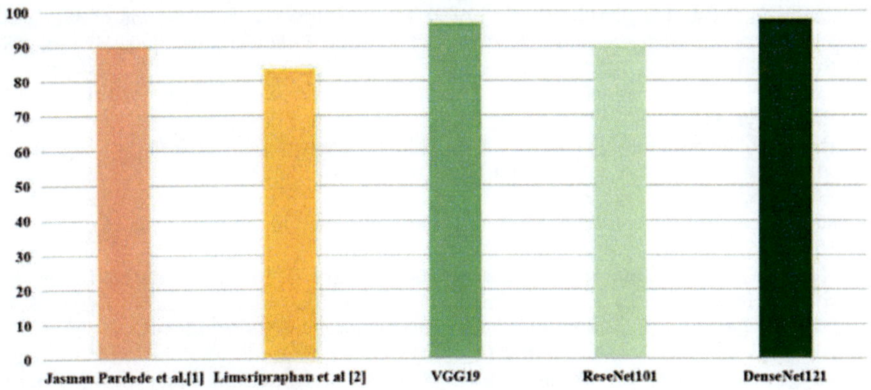

Fig. 9 Histogram of metric values

Figure 9 shows the histogram of the accuracies of our methods compared with those of other authors in the state of the art.

6 Conclusion

Our study demonstrates that using convolutional neural networks (CNNs) such as VGG19, ResNet101, and DenseNet121 in a mango ripeness detection system achieves high and accurate performance. By comparing these models with other existing approaches, we proved the effectiveness of our method. The results of our research are promising for the mango industry in Côte d'Ivoire. By improving farmers' ability to assess the stage of ripeness of mangoes accurately, our system will help to reduce product losses at harvest and increase export volumes, strengthening Côte d'Ivoire's position as the leading mango exporter in West Africa and having a significant impact on its economy. Our research opens up new perspectives for the mango industry using advanced machine learning techniques. It demonstrates the importance of using CNN models to detect fruit maturity and highlights the economic and commercial benefits that can be derived from adopting such technologies in the agricultural sector.

References

1. Coulibaly A, Minhibo MY, Soro S, Dépo ORN, Goran AN, Hala NF, Barnabas M, Djidji H (2019) Effectiveness of weather ants (Oecophylla longinoda), bait application (GF-120) and neem oil (Azachdirachta indica) combination in the control of fruit flies in mango orchards in Northern Côte d'Ivoire

2. Tenon C, Gondo DB, Bertille KEAA (2021) Mango cultivation practices and termite pest attacks: case of mango orchards in Northern Côte d'Ivoire. J Entomol Zool Stud 9(4):150–155
3. Sylvain TBC, Senan S, Lombart K, Lucie Y, Bissiri YHT, Kolo Y, Yao T (2019) Ants assemblage method according to an age gradient of Mango Orchards in Korhogo (Côte d'Ivoire). Adv Entomol 8(1):56–71
4. Pardede J, Sitohang B, Akbar S, Khodra ML (2021) Implementation of transfer learning using VGG16 on fruit ripeness detection. Int J Intell Syst Appl 13(2):52–61
5. Limsripraphan P, Kumpan P, Sathongpan N, Phengtaeng C (2019) Algorithm for mango classification using image processing and naive bayes classifier. Ind Technol Lampang Rajabhat Univ 12(1):112–125
6. Kusrini K, Suputa S, Setyanto A, Agastya IMA, Priantoro H, Pariyasto S (2022) Une étude comparative de la reconnaissance des ravageurs et des maladies des mangues. TELKOMNIKA (Électronique et contrôle informatique des télécommunications) 20(6):1264–1275
7. Anatya S, Mawardi VC, Hendryli J (2020, December) Fruit maturity classification using convolutional neural networks method. IOP Conf Ser Mater Sci Eng 1007(1):012149. IOP Publishing
8. Ayllon MA, Cruz MJ, Mendoza JJ, Tomas MC (2019, October) Detection of overall fruit maturity of local fruits using convolutional neural networks through image processing. In: Proceedings of the 2nd international conference on computing and big data, pp 145–148
9. Gururaj N, Vinod V, Vijayakumar K (2022) Deep grading of mangoes using convolutional neural network and computer vision. Multimedia Tools Appl 1–26
10. Randrianarivony MI (2018) Détection de concepts et annotation automatique d'images médicales par apprentissage profonde. Doctoral dissertation, Université d'Antananarivo
11. Ballo AB, Diarra M, Jean AK, Yao K, Assi AEA, Fernand KK (2022) Automatic identification of Ivorian plants from herbarium specimens using deep learning. Int J Emerg Technol Adv Eng 12(5):56–66
12. Jha BK, Sivasankari GG, Venugopal KR (2021) E-commerce product image classification using transfer learning. In: 2021 5th international conference on computing methodologies and communication (ICCMC), Erode, India, pp 904–912. https://doi.org/10.1109/ICCMC51019.2021.9418371
13. He K, Zhang X, Ren S, Sun J (2016) Deep residual learning for image recognition. In: Proceedings of the IEEE conference on computer vision and pattern recognition. Las Vegas, NV, USA, pp 770–778. https://doi.org/10.1109/CVPR.2016.90
14. Carranza-Rojas J, Goeau H, Bonnet P, Mata-Montero E, Joly A (2017) Going deeper in the automated identification of Herbarium specimens. BMC Evol Biol 17(1):1–14
15. Albelwi SA (2022) Deep architecture based on DenseNet-121 model for weather image recognition. Int J Adv Comput Sci Appl 13(10):559–565. https://pdfs.semanticscholar.org/4577/136cd4f7603a43a1a9e541798892d51957b0.pdf

A Comprehensive Review on Disease Predictions Using Machine Learning Approaches

Suhail Rashid Wani, Shree Harsh Attri, and Sonia Setia

Abstract There are various machine learning models for medical treatment that are now in use, however they all primarily target identifying a single condition. Consequently, this study has created a mechanism to just one graphical interface can predict multiple illness. This framework is capable of predicting numerous illnesses, including coronary artery disease, arrhythmias, Parkinson, Alzheimer, chronic kidney disease, and polycystic kidney disease. If neglected even when handled such illness are dangerous for people. Due to this, early recognition and treatment of may save countless lives in these conditions. This study makes a proposed approach, and use single model and mixed dataset of heart, brain, and kidney. There are several techniques for classification k-nearest neighbor, logistic regression, decision trees, Random Forest, naïve Bayes to do illness prediction. Every algorithm correctness is verified and in contrast to each other to identify the most precise forecasts. Also, numerous datasets (one for each condition) are employed to ensure that predictions are accurate. The primary objective is to develop a proposed model that can forecast various disease using machine learning including coronary artery disease, arrhythmias, Parkinson, Alzheimer, chronic kidney disease, and polycystic kidney.

Keywords Heart disease · Chronic kidney disease · Brain disease · K-nearest neighbor · Support vector machine · Decision tree · Random forest · Logistic regression · Naïve Bayes

S. R. Wani (✉) · S. H. Attri · S. Setia
Department of Computer Science and Engineering, Sharda School of Engineering and Technology, Sharda University, Greater Noida, UP, India
e-mail: wanisuhail305@gmail.com

S. H. Attri
e-mail: shree.harsh@sharda.ac.in

S. Setia
e-mail: sonia.setia@sharda.ac.in

S. Kumar et al. (eds.), *Fourth Congress on Intelligent Systems*, Lecture Notes in Networks and Systems 868, https://doi.org/10.1007/978-981-99-9037-5_26

1 Introduction

The ability to analyze and forecast various ailments using computational learning frameworks has the potential to revolutionize medical testing and improve patient outcomes [1]. In the past few years, the field of machine learning has experienced incredible developments and use cases in various regions, involving medical care. This study investigates the use of the Support Vector Machines (SVM) model to forecast the occurrence of six common diseases: coronary artery disease, arrhythmia's, Parkinson, Alzheimer, chronic kidney disease, and polycystic kidney [2]. Serious global health problems like coronary artery disease, arrhythmia's, Parkinson, Alzheimer, chronic kidney disease. and polycystic kidney, illnesses place a heavy load on people and healthcare systems around the world [3]. The future of patients and the effectiveness of care are greatly impacted by timely recognition and precise identification of these disorders strategies, and lowering medical expenses [4]. Machine learning, with its capacity to assess complex data and spot intricate patterns, presents potential ways to forecast several diseases. Support Vector SVMs are effective supervised learning machines algorithms are frequently employed for categorization jobs [5]. SVMs seek to discover the best hyperplane that divides various data points of one class to another and increasing the gap between classes in the data. It is possible to manage both linear and nonlinear connections between the data entered characteristics and destination factors allowing it appropriate for many different methods for recognizing illnesses [6, 7]. The ultimate objective of this investigation was to create a multi-disease prediction procedure employing SVMs and assess how well it predicted coronary artery disease, arrhythmias', Parkinson, Alzheimer, chronic kidney disease, and polycystic kidney [8]. A complete collection was created using relevant feature design approaches and freely available datasets, which included pertinent clinical, socioeconomic, and laboratory data [9]. On this dataset, the SVM model was taught to discover the complex connections amongst the provided input attributes treatment and the existence of six illnesses [10]. Machine learning models that accurately forecast disease can help with initial treatments, individualized regimens, and focused treatment techniques [11]. It could help healthcare professionals make better judgments, improve treatment of patients, and better allocate resources within the medical system.

2 Literature Review

The use of various statistical approaches to improve illness detection and forecasting has received a lot of attention lately in study. Despite being hampered by a small dataset, effectively diagnosed Parkinson's disease patients using clustering based on K-means and decision tree methods, reaching accuracy rates of 94.87, 100, and 88%. Similar to this, Asmae et al. [13] used voice assessments and K-nearest neighbors to diagnose Parkinson's illness with high accuracy rates of 97.22 and 97.30%.

Moving on to Alzheimer's illness, Balaji et al. [14] achieved a remarkable 98.5% accuracy utilizing neural networks trained in convolution with MRI and PET scans. Rajendiran et al. [15] investigated Alzheimer's utilizing SIFT and SURF characteristics coupled with Support Vector Machines (SURF), indicating the possibility for improved accuracy using deep learning.

Moving on to heart-related illnesses, Lamalem et al. used OCT and FFR data to analyze coronary artery disease, displaying a range of accuracy but being constrained by a small-scale investigation. With a larger dataset and a research extension, Bauer et al. [17] produced accuracy results with acknowledged limits of 71 and 74%. Pour-Ghaz et al. [18] used the method of principal component analysis to address the arrhythmias disease and attained accuracy rates ranging from 81.2 to 92.5%, however particular restrictions were not mentioned. With an accuracy of 87.02%, Bhatt et al. [19] demonstrated the potential of ensemble approaches in the forecasting of arrhythmias.

Harimoorthy et al. [20] predicted chronic renal disease with a high 98.3% reliability in investigations on kidney conditions without specifying any limits. Despite difficulties caused by a short dataset, Reshma et al. [21] managed to achieve a 98.3% reliability. Goel et al. [22] utilized as a convolutional neural network (CNN) for polycystic kidney disease, attaining 95.3% accuracy despite a small dataset. Using segmentation approaches, Kline et al. [23] exhibited a stunning 99.3% accuracy. These research demonstrate the variety of data-driven methods for disease prediction, in conclusion. Successful results are frequently met by difficulties like constrained collection quantities and potential for improved methodologies. Performance is calculated by different measurements, the amount to which an estimated or determined value resembles the real value is known as accuracy. By comparing the measured value to the true value, it calculates the probability error. The spectrum of such values reveals the measure's precision. The F one outcome, an analytical evaluation instrument rates the precision of a model's predictions. It combines a strategy scores for recall and velocity. The total amount of occasions an equation correctly identified the entire dataset is measured by its reliability score. Resilience is a metric that measures how well a program can predict actual favorable outcomes for every category of data. The specificity metric measures a model's ability to predict actual negatives for each available group. While sharpness can be seen as a measurement of effectiveness, recollection can be seen as a measure of volume. An additional precise method will produce findings that are more pertinent. In Table 1 shows the overall literature survey of multiple disease prediction in which A = Accuracy, F = F1-score, S = Sensitivity, SP = Specificity, P = precision, R = Recall.

Table 2 provided displays the diagnostic reliability of every illness with various algorithmic approaches. The most precise version from each disease's training set of various machine learning algorithms is selected and fed to the given data to trace the best model for multiple disease prediction.

Table 1 Summary of literature survey

Author/year	Disease	Datasets	Method	Performance	Limitations
Shobha Rani et al. [12]	Parkinson's (Brain)	Voice assessments database for biomedicine, 4270 (Oxford)	K-means clustering, decision tree	A = 94.87%, S = 100% SP = 88%	This model is not implemented in real life situation
Asmae et al. [13]	Parkinson's (Brain)	240 voice evaluations comprising 44 characteristics are used	KNN	A = 97.22%, F = 97.30%	No matter if the person has the illness or not
Balaji et al. [14]	Alzheimer's disease (Brain)	Kaggle 512 MRI scans and 112 PET images were utilized as the datasets for training	CNN	A = 98.5%	The conversion time is too large to get the results
Rajendiran et al. [15]	Alzheimer's disease (Brain)	SIFT and SURF feature is used. Open Access Series of Imaging Studies (OASIS)	SVM	A = 86.05%, P = 86.25%, R = 86.75%, F = 86.0%	Techniques based on DL can increase precision in prediction while reducing intricacy
Cha et al [16]	Coronary artery disease (heart)	356 people's OCT and FFR data were collected using these techniques	Logistic regression and supervised learning	A = 91.7%, S = 98.3%, SP = 61.5%	A small-scale investigation for individuals with intermediary infections, and the sample lesions eventually evolved into extremely selective
Bauer et al. [17]	Coronary artery disease (heart)	There were 5457 participants with a total of 18 diagnostic and CCTA-derived factors	Random forest	A = 71% SP = 74%	Single-center investigation with no external verification In individuals, there has been bias in recruitment

(continued)

Table 1 (continued)

Author/year	Disease	Datasets	Method	Performance	Limitations
Pour-Ghaz et al. [18]	Arrhythmias disease (heart)	Clinical manifestations of cardiac amyloidosis	Naïve Bayes, KNN, SVM, Random Forest	A = 81.2% SP = 84% P = 91% R = 92% F = 92.5%	Dataset is small
Bhatt et al. [19]	Arrhythmias disease (heart)	Actual information of 70,000 patient records and 12 attributes	Multilayer perceptron, the random forest, decision trees, and XGBoost (XGB)	A = 87.02%	Only a small number of demographic and clinical factors were taken into account in the study The research investigation only used one dataset, so efficiency wasn't assessed
Harimoorthy et al. [20]	Chronic kidney disease (Kidney)	The application of predictive modeling techniques along with information to the field of healthcare	Decision tree, Random Forest, linear SVM, and quadratic SVM	A = 98.3%	Less number of features used
Reshma et al. [21]	Chronic kidney disease (Kidney)	12 of the finest 24 qualities are selected for forecasting. entrepreneurship at UCI	SVM, regression analysis	A = 98.3%	Small number of data has been taken and less attributes used
Goel et al. [22]	Polycystic kidney disease	20 Dice similarity coefficient (DSC) and. Rogosin Institute ADPKD Repository	Convolutional neural network (CNN), segmentation	A = 95.3%	Analysis showed clear mistakes in a few clinical cases Analyses a model's performance at a specific moment in time
Kline et al. [23]	Polycystic kidney	Medical data, MRI data	PKD, segmentation of the kidneys	A = 99.3%	Renal pelvis delineation appears highly variable

Table 2 Summary of efficiency of various algorithms to predict the diseases

Disease	KNN (%)	SVM (%)	RF (%)	NB (%)	LR (%)	DT (%)
Parkinson's [24]	96	82	86	72	85.72	81
Alzheimer's [25]	81	86.56	80	84	83	78
Coronary artery [9]	90.20	91.70	89.80	87.60	89.69	90
Arrhythmias [9]	81	80.80	79.60	81.20	79	78.60
Chronic kidney [26]	97.60	98.30	96.80	95.20	89	92
Polycystic kidney [18]	96.20	98.30	97.70	96.90	92	89

KNN = A K-nearest neighbor, SVM = A Support Vector Machine, RF = A Random Forest, N = A Naive Bayes, LR = A Logistic Regression, DT = A Decision tree

3 Proposed Methodology

In order to achieve accurate results, the recommended approach for the survey research compares the performance of several training algorithms for disease prediction. Heart, brain, and kidney datasets are used in this proposed effort to train a single model. Figure 1 describes how various illnesses can be predicted using a variety of computational learning technique including Naive Bayes (are referred to as NB), K-nearest neighbor (are referred to ask-NN), Random Forest (are referred to as RF), logistic regression (are referred to as LR), and Support Vector Machine (are referred to as SVM) will help patients and healthcare professionals better communicate, resulting in every individual is free to follow their own goals [27]. The precision for every one the optimal algorithm for prediction is found by validating and contrasting each one. Various datasets are combined to produce results that are most accurately expect a model app has been created to simplify things for end customers, allowing them to make predictions the illness they desire simply by going to the respective values for each property (input) of that particular illness [28]. The recommended mechanism's layout is depicted in Fig. 1. The system in question can forecast a variety of illnesses using many different machine learning approaches, which is just one of its advantages.

- Considerably more data can be analyzed quickly time.
- The best effective solution will be found by comparing them all, every one's efficiency is verified and contrasted to other algorithms as a forecast [22].
- **Datasets**: The initial step is gathering information from many different places.

  Heart disease, brain, and chronic renal disease are just a few of the ailments for which this research collects data from the Kaggle algorithm dataset.
- **Preprocessing**: The preprocessing of information is a collection of methods used to enhance the quality of medical data, including addressing missing values, changing the type of feature, and many more method.

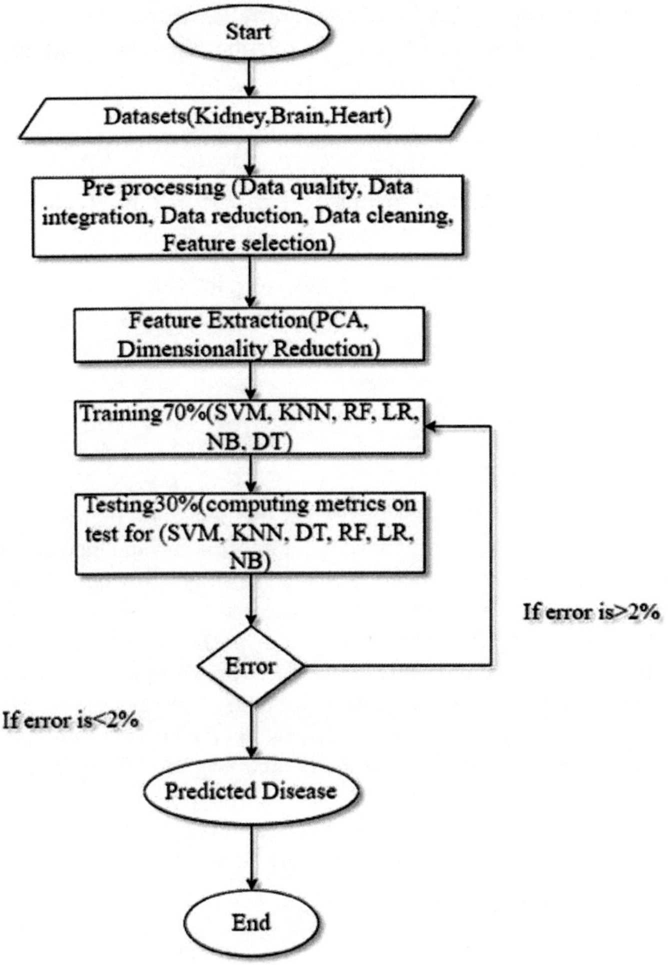

Fig. 1 Flowchart of proposed method

Feature Extraction: In the illness identification assignment, the disease-related characteristics have to be extracted from the rice images' diseased region using a variety of feature extraction approaches, and then key and pertinent features are chosen from the extracted features using the suggested method.

- **Training and Testing**: The training dataset for medical programs taken from patient records, medical pictures, or other pertinent sources. The machine learning model is taught to recognize patterns, correlations, and relationships within the data using the training dataset. The testing dataset is another set of examples that trained the model that has not been seen during the training process. This dataset is employed to evaluate algorithm proficiency in execution and generalization.

- **ML Algorithm**: The algorithm you choose relies on the type of your dataset, the amount of features, the intricacy of relationships, and available computational capacity. It is customary to test various algorithms and assess their efficacy before selecting the best one for a given task of disease prediction. Assessing the approaches using Naive Bayes, neural networks, Random Forest, logistic regression with SVM, and neural networks.
- **Efficiency**: Selecting the best approach. The systems' levels of reliability are evaluated, and the most effective one is chosen for disease prediction.

One model is used in the proposed study to train with heterogeneous datasets from the heart, brain, and kidney. In conclusion, the suggested approach for this survey provides precise illness prediction and offers real-time assistance for health risk effectiveness and decision making.

Machine Learning Models

(1) Statistical Regression (SR)

Machine learning model called logistic regression focuses on binary categorization. This approach might be used to provide dataset information and incorporate the relationship, working similarly to linear regression. An association between two distinct variables and one influencing variable. The easiest way for calculating the threshold value is estimating the likelihood. When there is a linear correlation between specific features and the likelihood of disease occurrence, linear regression may be used to predict diseases. Medical research frequently employs logistic regression to foretell the existence or absence of diseases. It is crucial to take into account variables such model beliefs, variability, choice of features, and proper evaluation metrics when using statistical regression to forecast diseases. Regression model selection is influenced by the type of data being used, the kind of outcome variable being used, and the particular research issue being addressed [18].

(2) K-Nearest Neighbor (KNN)

The KNN approach, often known as lazy optimization, is the most straightforward algorithm for prediction using the Euclidean distance. K determines the neighbor number by taking into account the predictions made by the closest neighbors. It is crucial to properly prepare the data, choose the right characteristics, and think about the consequences of the algorithm's limits and presumptions in a setting of medical diagnosis when using KNN for disease prediction [29].

(3) Naïve Bayes (NB)

The regression and classification strategies used for computational learning are called Naive Bayes. Gaussian Naive Bayes, which depends on the Bayes Theorem, is an augmentation to real-value qualities where recommendations for a Gaussian distribution are made utilizing the information within the dataset. The Naive Bayes classifier performs well in real-world scenarios due to its short database required for approximation and quick computation comparison with additional complex algorithms. The simultaneous presence of numerous illnesses is predicted in this study using Naive

Bayes networks. An assortment of symptoms or medical history are used by the algorithm to predict the likelihood that each disease will manifest. The benefit of employing Naive Bayes for this purpose is that it is simple to construct and can handle huge and complicated datasets [30].

(4) Support Vector Machine (SVM)

The categorization method employed is known as the Support Vector Machine (SVM). Each characteristic of a data point is shown in a space of *n* dimensions here, where *n* is the total quantity of features. The dataset is divided into multiple properties using the hyperplane as a basis. As SVM assists in identifying separable classes from classes that cannot be separated by converting low. It makes it easier to work with even complicated datasets when low-dimensional input is converted to a highly dimensional one. Specifically when faced with complex, unpredictable relationships in the data, SVMs can be useful for disease prediction. They might, however, necessitate careful hyperparameter tweaking and handling of uneven datasets. The precision and clinical value of the SVM-based disease prediction model can be improved by consulting with domain experts and employing feature engineering methods that incorporate medical knowledge [31].

(5) Random Forest (RF)

Random Forest is employed to handle regression and classification issues as it is an ensemble learning technique. This method uses a bootstrapping aggregates methodology and is appropriate for datasets where decision trees are trained. As a result, the approach uses numerous decision trees rather than a single decision tree to derive output. This approach might produce good and efficient outcomes in data mining since it aids in the best or most important characteristic when dividing nodes. Because it can deal with noisy data, multidimensional characteristics, and complex interactions, Random Forest can be a reliable option for disease prediction. To make sure that the model outputs and features are in line with clinical relevance and existing medical knowledge, it is crucial to consult domain experts [32].

4 Evaluation Metrics and Discussion

(A) Performance Assessment: A set of calculations known as assessment of performance are employed to gauge how well a classification algorithm or model performs.

The definitions of a few key words utilized in the outcome assessment equations are given below:

(1) True Positive (are referred to as TP): The individual is physically fit and is assumed to be healthy.
(2) False Positive (are referred to as FP): The individual is normal although the result indicates illness.

(3) True Negative (are referred to as TN): The individual is ill and it was assumed to be ill.
(4) False Negative (are referred to as FN): A person is ill yet is predicted to be in excellent condition.

(B) Confusion matrix: A vital instrument for determining the efficacy of computational learning methods, especially those used to predict heart, brain, and kidney diseases, is a confusion matrix. It offers a success and obvious approach to comprehend how well a model is doing in terms of classification tasks, like predicting diseases. Based on projected and actual classifications, confusion matrices are often divided into four categories shown in.

Several assessment indicators can be obtained from the confusion matrix to evaluate the model's efficiency, especially when considering multiple illness estimations Table 3:

Accuracy: It is determined as (TP + TN)/(TP + TN + FP + FN). It is the percentage of accurate predictions out of all forecasts [33].

Precision: It is determined as TP/(TP + FP), which is the percentage of true positive predictions out of every favorable forecasts [34].

Recall: It (Sensitivity or True Positive Rate) is the ratio of correctly predicted positive outcomes to all true successes. It is computed as TP/(TP + FN) [35].

The percentage of accurate unfavorable judgements compared to all actual negatives, calculated as TN/(TN + FP).

Calculated as 2 * (Precision * Recall)/(Precision + Recall) for the F1-Score, the average of the harmonics of both recall and precision provides the equilibrium among both measurements.

Area under the Receiver's Operating Characteristic (ROC) curve and the curve: The Receiver Operating Characteristic (ROC) curve demonstrates how the two attributes were compromised. The Area Under the Curve is used to determine the machine's overall effectiveness (AUC) [36].

The matrix of uncertainty and accompanying metrics assist evaluate the model's capacity to manage and discriminate among various diseases in the context of heart, brain, and kidney disease forecasts, showing areas where it excels and those that may require work. For the model to be improved and have its efficacy optimized for precise disease forecasting, such data is essential [37].

It's vital to remember that underlying use of evaluation metrics may change based on the clinical situation and the condition being forecasted. For example, in some circumstances, eliminating false negatives (raising recall) could be more crucial than lowering false positives (increasing precision), specifically if ignoring a disease

Table 3 Confusion matrix

	Predicted positive	Predicted negative
Actual positive	True positives (are referred to as TP)	False negatives (are referred to as FN)
Actual negative	False positives (are referred to as FP)	True negatives (are referred to as TN

diagnosis could have serious repercussions. Additionally, explore the ramifications of the disease prognosis in a real-world scenario. Consult with domain experts and to determine the most appropriate evaluation metrics that align with the clinical significance of the predictions. It's typically recommended to employ a combination of these indicators to acquire a thorough knowledge of the model's effectiveness.

5 Conclusion

The precision statistics for several machine learning techniques used to treat various illnesses were reported in the effectiveness Table 2. Here are some considerations to consider if you're seeking for an interpretation or assessment of these findings:

Parkinson's disease KNN had the most accuracy, coming in at 96%. The rates of precision for SVM and logarithmic regression were 82% and 85.72%, respectively. The rates of accuracy for Random Forest and Naive Bayes were 86% and 81%, respectively.

Alzheimer's disease, SVM achieved an accuracy rate of 86.56%. With accuracy levels between 83 and 84 other algorithms performed well as well, such as Naive Bayesian logarithmic regression, and the Random Forest. The accuracy of KNN was 81%.

Cardiovascular disease SVM and logistic regression, with 91.70% and 90%, respectively, had the best accuracy. Accuracy levels for other algorithms were in the 89–90% range. 4. Heart arrhythmias SVM achieved an accuracy rate of 80.80%. The reliability of the remaining algorithms was between 78 and 81%.

Chronic kidney illness SVM had the best reliability, coming in at 98.30%. Reliability rates of 97.60 and 96.80% for KNN and Random Forest likewise shown good performance. The accuracy of Naive Bayes was 89%.

Polycystic kidney disease, or PCK SVM and KNN had the best accuracy, coming in at 98.30%. The precision rates for Random Forest and Naive Bayes were between 97 and 98%. The precision of the logistic regression was 89%.

The outcomes show that each approach performs differently depending on the disease dataset. The sole element to take into account when assessing the effectiveness of a machine learning framework. A more complete picture of a model's performance can be obtained using additional measures such as precision, recall, F1-score, and ROC curve area. Size of the dataset, the complexity of the features, and the availability of processing resources are other variables that may influence the processing method of selection.

References

1. Xie S, Yu Z, Lv Z (2021) Multi-disease prediction based on deep learning: a survey. Comput Model Eng Sci 128(2):489–522. https://doi.org/10.32604/cmes.2021.016728
2. Yahaya L, Oye ND, Garba EJ (2020) A comprehensive review on heart disease prediction using data mining and machine learning techniques. Am J Artif Intell 4(1):20. https://doi.org/10.11648/j.ajai.20200401.12
3. Alanazi HO, Abdullah AH, Qureshi KN (2017) A critical review for developing accurate and dynamic predictive models using machine learning methods in medicine and health care. J Med Syst 41(4):69. https://doi.org/10.1007/s10916-017-0715-6
4. Xiao C, Li Y, Jiang Y (2020) Heart coronary artery segmentation and disease risk warning based on a deep learning algorithm. IEEE Access 8:140108–140121. https://doi.org/10.1109/ACCESS.2020.3010800
5. Gürüler H (2017) A novel diagnosis system for Parkinson's disease using complex-valued artificial neural network with k-means clustering feature weighting method. Neural Comput Appl 28(7):1657–1666. https://doi.org/10.1007/s00521-015-2142-2
6. Kumar KP, Pravalika A, Sheela RP, Vishwam Y (2022) Disease prediction using machine learning algorithms KNN and CNN. Int J Res Appl Sci Eng Technol 10(5):446–450. https://doi.org/10.22214/ijraset.2022.42214
7. Jetti CR, Shaik R, Shaik S (2021) Disease prediction using Naïve Bayes—machine learning algorithm. Int J Sci Healthc Res 6(4):17–22. https://doi.org/10.52403/ijshr.20211004
8. Yu J, Park S, Kwon S-H, Cho K-H, Lee H (2022) AI-based stroke disease prediction system using ECG and PPG bio-signals. IEEE Access 10:43623–43638. https://doi.org/10.1109/ACCESS.2022.3169284
9. Patil DD, Singh RP, Thakare VM, Gulve AK (2018) Analysis of ECG arrhythmia for heart disease detection using SVM and Cuckoo search optimized neural network. Int J Eng Technol 7(2):27. https://doi.org/10.14419/ijet.v7i2.17.11553
10. Beheshti I, Ganaie MA, Paliwal V, Rastogi A, Razzak I, Tanveer M (2022) Predicting brain age using machine learning algorithms: a comprehensive evaluation. IEEE J Biomed Health Inform 26(4):1432–1440. https://doi.org/10.1109/JBHI.2021.3083187
11. Nithya A, Appathurai A, Venkatadri N, Ramji DR, Palagan CA (2020) Kidney disease detection and segmentation using artificial neural network and multi-kernel k-means clustering for ultrasound images. Measurement 149:106952. https://doi.org/10.1016/j.measurement.2019.106952
12. Shobha Rani A, Mutha A, Ranjan A, Gupta A, Mohanty S (2023) Early detection of Parkinson's disease using machine learning. Int Res J Mod Eng Technol Sci. https://doi.org/10.56726/IRJMETS38590
13. Asmae O, Raihani A, Cherradi B, Lamalem Y (2022) Parkinson's disease classification using machine learning algorithms: performance analysis and comparison. In: 2022 2nd International conference on innovative research in applied science, engineering and technology (IRASET), Mar. 2022. IEEE Meknes, Morocco, pp 1–6. https://doi.org/10.1109/IRASET52964.2022.9738264
14. Balaji P, Chaurasia MA, Bilfaqih SM, Muniasamy A, Alsid LEG (2023) Hybridized deep learning approach for detecting Alzheimer's disease. Biomedicines 11(1):149. https://doi.org/10.3390/biomedicines11010149
15. Rajendiran M, Kumar KPS, Nair SAH (2022) Machine learning based detection of Alzheimer's disease in MRI images. J Pharm Negat Results 1615–1625. https://doi.org/10.47750/pnr.2022.13.S08.196
16. Cha J-J et al (2023) Assessment of fractional flow reserve in intermediate coronary stenosis using optical coherence tomography-based machine learning. Front Cardiovasc Med 10:1082214. https://doi.org/10.3389/fcvm.2023.1082214
17. Bauer MJ et al (2023) Prognostic value of machine learning–based time-to-event analysis using coronary CT angiography in patients with suspected coronary artery disease. Radiol Cardiothorac Imaging 5(2):e220107. https://doi.org/10.1148/ryct.220107

18. Pour-Ghaz I et al (2022) A review of cardiac amyloidosis: presentation, diagnosis, and treatment. Curr Probl Cardiol 47(12):101366. https://doi.org/10.1016/j.cpcardiol.2022.101366
19. Bhatt CM, Patel P, Ghetia T, Mazzeo PL (2023) Effective heart disease prediction using machine learning techniques. Algorithms 16(2):88. https://doi.org/10.3390/a16020088
20. Harimoorthy K, Thangavelu M (2021) RETRACTED ARTICLE: multi-disease prediction model using improved SVM-radial bias technique in healthcare monitoring system. J Ambient Intell Humaniz Comput 12(3):3715–3723. https://doi.org/10.1007/s12652-019-01652-0
21. Reshma S (2020) Chronic kidney disease prediction using machine learning. Int J Eng Res 9(7). LBS Institute of Technology for Women, Poojappura, Trivandrum. https://doi.org/10.17577/IJERTV9IS070092
22. Goel A et al (2022) Deployed deep learning kidney segmentation for polycystic kidney disease MRI. Radiol Artif Intell 4(2):e210205. https://doi.org/10.1148/ryai.210205
23. Kline TL et al (2017) Performance of an artificial multi-observer deep neural network for fully automated segmentation of polycystic kidneys. J Dig Imaging 30(4):442–448. https://doi.org/10.1007/s10278-017-9978-1
24. Ouhmida A, Raihani A, Cherradi B, Lamalem Y (2022) Parkinson's disease classification using machine learning algorithms: performance analysis and comparison. In: 2022 2nd international conference on innovative research in applied science, engineering and technology (IRASET), Mar. 2022. IEEE, Meknes, Morocco, pp 1–6. https://doi.org/10.1109/IRASET52964.2022.9738264
25. Usman K, Rajpoot K (2017) Brain tumor classification from multi-modality MRI using wavelets and machine learning. Pattern Anal Appl 20(3):871–881. https://doi.org/10.1007/s10044-017-0597-8
26. Ma F, Sun T, Liu L, Jing H (2020) Detection and diagnosis of chronic kidney disease using deep learning-based heterogeneous modified artificial neural network. Future Gener Comput Syst 111:17–26. https://doi.org/10.1016/j.future.2020.04.036
27. Katarya R, Meena SK (2021) Machine learning techniques for heart disease prediction: a comparative study and analysis. Health Technol 11(1):87–97. https://doi.org/10.1007/s12553-020-00505-7
28. Ferjani MF (2020) Disease prediction using machine learning. https://doi.org/10.13140/RG.2.2.18279.47521
29. Jabbar MA, Deekshatulu BL, Chandra P (2013) Classification of heart disease using K-nearest neighbor and genetic algorithm. Procedia Technol 10:85–94. https://doi.org/10.1016/j.protcy.2013.12.340
30. Langarizadeh M, Moghbeli F (2016) Applying Naive Bayesian networks to disease prediction: a systematic review. Acta Inform Medica 24(5):364. https://doi.org/10.5455/aim.2016.24.364-369
31. Arumugam K, Naved M, Shinde PP, Leiva-Chauca O, Huaman-Osorio A, Gonzales-Yanac T (2023) Multiple disease prediction using machine learning algorithms. Mater Today Proc 80:3682–3685. https://doi.org/10.1016/j.matpr.2021.07.361
32. Paul S, Ranjan P, Kumar S, Kumar A (2022) Disease predictor using random forest classifier. In: 2022 International conference for advancement in technology (ICONAT), Jan. 2022. IEEE, Goa, India, pp 1–4. https://doi.org/10.1109/ICONAT53423.2022.9726023
33. Low JX, Choo KW (2018) IoT-enabled heart monitoring device with signal de-noising and segmentation using discrete wavelet transform. In: 2018 15th International conference on control, automation, robotics and vision (ICARCV), Nov. 2018. IEEE, Singapore, pp 119–124. https://doi.org/10.1109/ICARCV.2018.8581315
34. Tougui I, Jilbab A, El Mhamdi J (2020) Heart disease classification using data mining tools and machine learning techniques. Health Technol 10(5):1137–1144. https://doi.org/10.1007/s12553-020-00438-1
35. Motwani M et al (2016) Machine learning for prediction of all-cause mortality in patients with suspected coronary artery disease: a 5-year multicentre prospective registry analysis. Eur Heart J ehw188. https://doi.org/10.1093/eurheartj/ehw188

36. Yaswanth R, Riyazuddin YMd (2020) Heart disease prediction using machine learning techniques. Int J Innov Technol Explor Eng 9(5):1456–1460. Department of CSE, GITAM University, Hyderabad, India. https://doi.org/10.35940/ijitee.E2862.039520
37. Acharya UR et al (2017) A deep convolutional neural network model to classify heartbeats. Comput Biol Med 89:389–396. https://doi.org/10.1016/j.compbiomed.2017.08.022

Deriving Rectangular Regions Bounding Box from Overlapped Image Segments Using Labeled Intersecting Points

Ganesh Pai and M. Sharmila Kumari

Abstract Numerous neural network models are capable of generating image segments; however, they tend to merge nearby objects or objects situated behind others into a single large region, encompassing multiple overlapped segments. In this paper, we introduce a novel approach for generating individual rectangular overlapped image segments from a larger region, which is produced by applying morphological operations on an image. The proposed algorithm addresses this issue by utilizing labeled intersecting points, enabling the extraction of distinct overlapping segments. To evaluate its effectiveness, we constructed a new dataset specific to this problem domain. The performance of our algorithm is compared with a standard watershed algorithm on this dataset, and our method outperforms with an impressive recall of 96%, F_1-score of 97%, and an average precision of 96.21%.

Keywords Segmentation · Bounding box · Labeled intersecting points

1 Introduction

Image segmentation is essential for identifying various objects in an image, instantiating objects, and distinguishing them from the background. In computer vision, the outcome is critical for a computer to identify and recognize various objects in its environment, followed by tagging, cognitive reasoning of the instances, and reacting to user events. As a result, accurate instance segmentation is essential in computer vision. The overlapped regions sharing common features are frequently a challenge in segmentation. In the literature, several deep learning algorithms have been introduced [1, 2]. Approaches such as fully convolutional network, encoder-decoder-based models, multiscale and pyramid network-based models, R-CNN-based models

G. Pai (✉)
NMAM Institute of Technology-Affiliated to NITTE (Deemed to be University), Nitte, Karnataka, India
e-mail: ganesh.pai@nitte.edu.in

M. Sharmila Kumari
P. A. College of Engineering-Affiliated to VTU, Mangalore, Karnataka, India

S. Kumar et al. (eds.), *Fourth Congress on Intelligent Systems*, Lecture Notes in Networks and Systems 868, https://doi.org/10.1007/978-981-99-9037-5_27

for instance segmentation, attention-based models, generative models, and adversarial training models are in wide use and effective for several applications. Goceri [3] discusses on neural network segmentation approaches trained using weakly supervised, deeply supervised, and transfer learning. It also discusses on the challenges with long training time, over-fitting, and vanishing gradients. Some of the approaches in segmentation of medical images can be found in [4, 8] which covers region-based, cluster-based, and model-based approaches used in segmentation and application in satellite images in [5]. Even with several approaches aiming to address a common issue emerging from a diverse set of domains, there still exists certain issues that will need special attention and may not be handled using in a more generalized approach.

Segmentation of objects of a specific shape is one such issue appearing in most computer vision applications. The issue escalates when they appear overlapped in an image. Instance segmentation segments them in the visible portion of the image and approximates the length of the object extending behind the foreground object. Accurate length of the background object still remains a challenge. The work in this paper tries to address this issue of determining the accurate coordinates of the object for the bounding box using the features available in the visible region. The object of interest may not be a rectangular object but the bounding region is detected as a rectangle. Such images are frequently detected when training a segmentation model with the bounding box of an object. The issue arises when the bounding boxes have to be detected for the overlapped segments of the image. Using pixels relative position, we attempt to address the problem.

This paper is organized as follows. Related works are reviewed in Sect. 2. Section 3 describes in detail the various stages in the proposed methodology and pictorially intermediate stage outcomes. Section 4 initially discusses on the dataset used. Due to the lack of availability of standard dataset to address this problem domain, a new custom dataset is generated for the overlapped and non-overlapped regions and compared the performance with a standard algorithm. This is followed by the experimental results and discussion over the proposed algorithm. Section 5 summarizes and concludes with possible future improvement in the work.

2 Related Works

Determining different objects and distinguishing one from the other is one of the frequent issues in a segmented image. Deep learning segmentation architectures such as U-Net [9] and U-Net++ [10] developed for medical images produces binary targets as segments for a source image. Target objects are often the shape of an object. Ground truth or predicted target objects are often represented as rectangular shaped bounding boxes when the ground truth annotations are rectangular representations. The CNN model learns these bounding boxes based on ground truth bounding boxes of the targets and produces predicted bounding boxes. Consider an example in Fig. 1. Figure 1a shows an original image given as input to a U-Net model, trained to

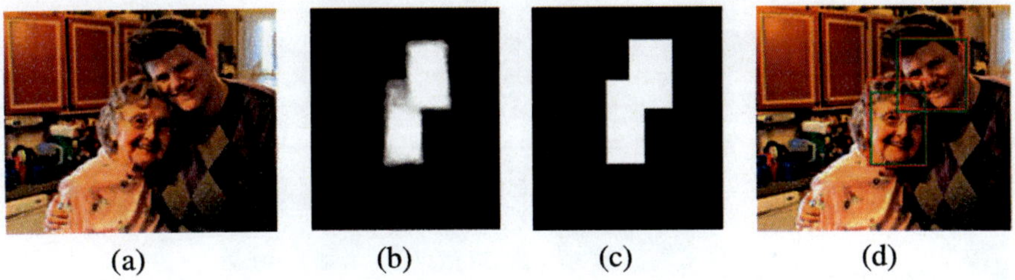

Fig. 1 Issues in the problem domain. Determination of exact boundaries for overlapped image segments in a complex segmentation outcome

produce a binary mask, produces a segmented output image in (b) as predictions. Post-processing produces the image (c). Now observe that the two regions are overlapping. The image in (d) shows the detected bounding boxes for the overlapped region, green being ground truth and red, the detected bounding box. The issue arises when there are multiple overlaps in the predictions. Watershed algorithm [6] is the often used to determine such object boundaries. Recent work on watershed algorithm can be found in [7] that uses image marking method optimization approach. The general approach first uses morphological operations to find background, distance transform to find the foregrounds, and then applies watershed algorithm on the remaining area to compute the image segments. But the approach fails to produce a perfect corner in all cases. It generates an approximation of the region shape that are suboptimal for specific computer vision applications such as in instance segmentation. For applications with need for an accurate rectangular region detection requires algorithm specifically designed for such applications. This paper proposes a novel algorithm to address this issue at a far optimal scale than the standard algorithm.

3 Proposed Methodology

The proposed algorithm is divided into seven steps with each step refining, learning, filtering, and optimizing the candidate points. Figure 2 shows the seven stages of the computation process, beginning from the point where an arbitrary sized segmented binary image is given as input to the system, to the final output image at the seventh stage where the colored bounding box is drawn around each predicted overlapped region. Each step is explained below in detail. The algorithm computes the approximation at the corner areas where the corner edges do not form at ninety degrees. The methodology works around using the pixel intensities and its corresponding coordinates to determine the corners at the outer and overlapped regions. An image may have any number of overlapped or non-overlapped image segments. The algorithm iteratively computes for each region in the input image. For clarity, the image used for demonstration contains only one overlapped region with five overlapped objects on the large rectangular object.

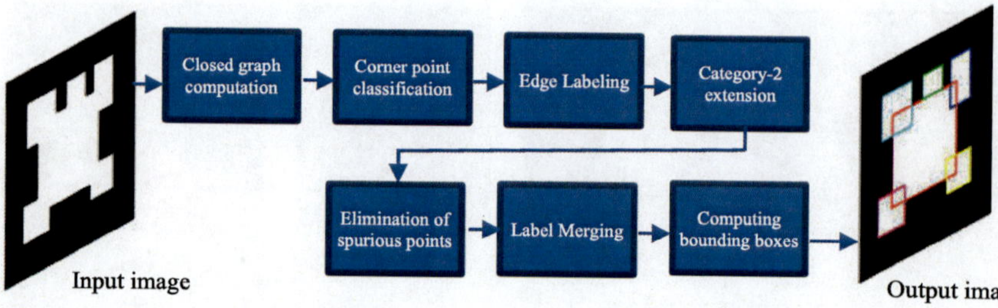

Fig. 2 Seven stages of computation

3.1 Closed Graph Computation

Segmented images are binary images. This is the first step in the process and obtains a graph of all border points of the segmented area. Based on the segmentation structure, there can be a continuous point extraction approach or a corner point approach. As the focus of our problem space is toward rectangular regions, it is relevant to use corner point approach. From every corner of the segments, we obtain a closed graph. An image may have multiple image segments. A distinct graph is extracted for each segment in an image. Figure 3a shows the border points extracted in blue color. For clarity, it shows all the points along the edges and is formulated as

$$S = \bigcup_{i=1}^{n} S_i \text{ where } S_i = \{(x, y) : (x, y) \in E_{\text{seg}}(I)\}, \tag{1}$$

where S_i represents the edges of segment E_{seg} of an image I and S represents the union of border points of all segments in an image. From here on, we focus on only one segment S_i of an image. The process is repeated for all other segments.

3.2 Corner Point Classification

For each point in S_i, point identity is established based on the position of the point in the graph. In this step, each point of the graph is initially classified as top-left, bottom-left, top-right, or bottom-right based on the position of the point relative to its adjacent points in the graph. These points are further classified as category-1 (C_1) or category-2 (C_2) points based on the position of corners around the segment. This classification and its conditions are shown in Table 1. The labels *a*, *b*, *c* are the points corresponding to the intersecting lines and *x*, *y*'s are the coordinates of the respective points. The region of diagonal shades along the line in the figure of Table 1 are the black regions of the image while the opposite side is white. Each computation done are with *b* as the point of interest and points *a* and *c* relatively adjacent to *b*. The

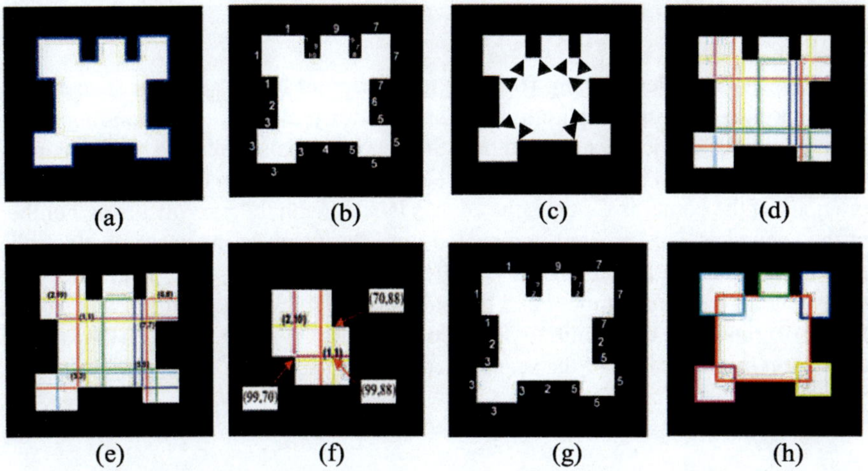

Fig. 3 Computation process. **a** Computed closed graph for the segmented image. **b** Labeled edges. **c** C_2 type points. **d** C_2 lines extended and intersecting. **e** Pair of labeled intersection points (shows limited pairs). **f** C_2 type points and intersecting point. **g** Labeled edges after merging labels. **h** Regions with bounding box (best viewed in color)

direction of traversal of points in the graph is anticlockwise. Points are categorized based on the pixel values along the diagonal coordinates of the midpoint b. Table shows the diagonal coordinates (diamond points in last column) for the top-left corner. The same is applicable to all other corner types. If the difference of pixel value along the diagonal points (top-left minus bottom-right) is negative, the point is C_1 type, and C_2 type otherwise. Diagonal difference for each occurring corner type is computed accordingly and point are classified as C_1 or C_2. This identity of the corner is used in labeling the edges of the segments. This can be formulated as $S_i = C_1 \cup C_2$ and $C_1 \cap C_2 = \emptyset$.

Table 1 Category and corner types

Category/ Corner type	Top-left	Top-right	Bottom-left	Bottom-right	Diagonal points of top-left
Conditions	$x_a < x_c$ $y_a > y_c$ $x_b = x_a$	$x_a > x_c$ $y_a > y_c$ $x_b = x_c$	$x_a < x_c$ $y_a < y_c$ $x_b = x_c$	$x_a > x_c$ $y_a < y_c$ $x_b = x_a$	–
Category-1					
Category-2					

3.3 Edge Labeling

Each corner is labeled starting from the top-left point in the segment sequentially anticlockwise until the graph closes. First label for the edge, corresponding to the first two points, is enumerated with one. Subsequent labels are based on the identity of the corner. If the corner is C_1 type, the edge is labeled with the same label as its previous. If the corner is C_2 type, the edge is labeled with the next occurring number in the sequence. This process is repeated until the graph loops back. To track the points associated with a label $L_i \in L$, a dictionary of labels, D_L, is constructed with the key being the label number and the points belonging to each label as its values. Figure 3b shows the edges initially labeled using this approach. If L_{ab} is the label of the edge *a*–*b*, then labels of the segment can be represented as

$$L = L_{ab} \in N \forall (a, b) \in S_i \tag{2}$$

3.4 Category-2 Extension

When overlapped regions are formed into a segment, they inherently form category-2 intersections. This feature is used to formulate the intersecting coordinates using labeled intersecting points (LIP). LIP's are points in a virtual plane of the image segment. Arrow marks in Fig. 3c shows points formed by C_2 type corner points. LIP is formed by initially extending the edges of each C_2 points inwards and listing each such points, P_L, extending up to the boundary, as shown in Fig. 3d. Candidates of the final points are the intersecting points, P_I along the extended line in P_L, obtained by eliminating the non-intersecting points in P_L and can be formulated as

$$P_I \subset P_L : P_i \text{ is an intersecting point} \in P_L. \tag{3}$$

Each intersecting point internally forms a rectangular virtual plane and is encapsulated with two labels corresponding to the source points of the intersecting edges (hence the name LIP). Figure 3e shows label pairs of some of the intersecting points after extending the edges. Each LIP is made up of three information: source point from where the intersecting line is extended, label number of the extending edge, and the corner type at the source point. For example, consider a LIP of (99, 88): [((70, 88), 1, 'bl'), ((99, 70), 1, 'tr')] occurring with the labeled pair of (1, 1) in Fig. 3e and its close view coordinates highlighted in Fig. 3f. Here, the point of intersection for this label pair is (99, 88); ((70, 88), 1, 'bl') is the parameter of one edge (vertical edge) with the source point at (70, 88), label number of the edge being 1 and formed from a bottom-left (bl) corner. Similarly, ((99, 70), 1, 'tr') indicates, source is at (99, 70), label number is 1 and is formed from a top-right (tr) corner. A dictionary of intersection points with this label pairs as its values are used in all subsequent stages. The above example elaborates one such dictionary entry.

3.5 Elimination of Spurious Points

This step identifies and eliminates spurious points, P_s, out of the candidate intersection points P_I and produces a proposal point list $P_P = P_I - P_s$. Elimination process applies several constraints on the points in P_I there by filtering out the intersections that do not form the corners of the ground truth rectangles. A top-left with bottom-right or top-right with bottom-left corner type combinations in labels of the candidates forms valid intersections. Any other combinations of the corner type are observed to be spurious intersections. Further, points with labels pair framed by adjacent edges and overlapped intersecting lines are also considered invalid. All such points are considered as spurious points and are eliminated from the candidate list. If an intersecting point contains multiple labels with same corner type, invalid labels among them are to be identified and eliminated. Residual points with valid combinations of the labels are refined subsequently. If there are more than two intersecting points, it needs to be validated based on the corner type and eliminate the pairs that forms invalid combination. The eliminated points form the elements of the spurious set P_s.

3.6 Label Merging

A mentioned earlier, label-points dictionary D_L is maintained for each label given to edges, $L \in D_L$ that tracks on the points that are members of a label. The proposal points obtained, P_P, are now merged progressively and conditionally. If the point considered is a midpoint between the two points, it is ignored. First step starts by considering intersection points with same label numbers. These intersections are finalized and added to the corresponding label dictionary. All other intersecting points in P_P that falls along this line of intersections are eliminated. Next step considers intersection points with the distinct label numbers. Recursively label pairs of the intersection points are unified by merging the labels, re-numbering the labels with the smallest number encountered during union and merging its points in the label dictionary. Other points in the line along the finalized intersection point in this step is eliminated from P_P. Further, merging is also constrained by the corners that exists in each label dictionary. A merge is valid only if intersection of corner types in points of two labels results in null set, i.e., $p_l \cap q_l = \emptyset$, where p_l is the labels of point p and q_l, the labels of point q and (p, q) are the source category-2 points of the intersecting point. The end results are the labeled dictionary D'_L with points lying along the boundaries that are sufficient to derive the bounding box coordinated. Figure 3g shows the result of labels merged applying the above conditions.

3.7 Computing Bounding Box Coordinates

Label in D'_L corresponds to a distinct region in the overlapped segment. For each label in D'_L, minimum and maximum of x and y coordinates are computed for the corresponding list of points. They form a list of bounding box coordinates in the form $\langle x, y, \text{width}, \text{height}\rangle$ of each label or region of the segments. The outcome is as shown in Fig. 3h.

These steps are repeated for each segment S_i in the image. The overall algorithm is as given below.

Algorithm getBoundingBox(image):

S = getSegmentCornerPoints() //compute the contours of the image
for each segment s in S:
  label each edge $e \in s$ with a numeric value starting from 1
  construct a list of C_2 points for each $e \in s$
  add e to label dictionary D_L
  for each point p in C_2:
    extend the line inside the region and construct P_L
  store only intersecting points of P_L in P_I
  for each point p in P_I:
    compute the validity of the points and construct a spurious point list P_S
  Compute proposal points as $P_P = P_I - P_s$
 Construct a union of points with matching labels using label matching conditions
  Append the coordinated to respective label in D_L forming the final dictionary D'_L

Compute the min and max x, y coordinates for each label in D'_L that forms the bounding box for each segment.

4 Results and Discussion

This section compares the proposed algorithm with the conventional watershed algorithm against a custom dataset. The algorithm is implemented on an AMD Ryzen 5 3450U with Radeon Vega Mobile Gfx 2.10 GHz processor with 8 GB RAM using python language and opencv2 library. The algorithm took approx. 0.20 s per image when averaged over the dataset.

4.1 Dataset Details

Due to the absence of standardized dataset intended to address the challenge of identifying overlapped rectangular regions, a unique custom dataset is developed and made accessible at https://github.com/paiganesh/rect_segmentation. This dataset is meticulously designed to assess the algorithm's performance using diverse natural overlapping regions found in any segmented outcome of a neural network model. It comprises a collection of 110 basic images, featuring over 800 instances of both overlapped and non-overlapped segments. The dataset encompasses segments of different sizes and relative positions, spanning from zero (discrete independent segments) to seventeen overlapped segments. The samples represent the vast majority of overlapped segments occurring in the standard object segmentation dataset containing rectangular regions. These basic images can be further augmented by 90° rotations along the plane. An annotation file is encapsulated within the dataset containing a filename, followed by number of annotations, followed by bounding boxes for each region precisely annotated in the format $\langle x, y, \text{width}, \text{height}\rangle$.

Performance of the algorithm is evaluated using recall and F_1-scores for five discrete Intersection over Union (IoU's) (0.5–0.9). It can be observed from Table 2 that the algorithm outperforms compared to the standard approach. For an IoU = 0.5, the proposed algorithm gives a recall of 96% against the standard watershed algorithm. For an IoU = 0.9, proposed algorithm maintains its recall quality at 95% against 34% for its counterpart. Table 3 shows the result of applying the two algorithms for few samples with first row containing the samples from the dataset, second row containing the results of watershed algorithm and third row contains results of the proposed algorithm. The bounding boxes are indicated with the colored lines around the segments. For the input image in Table 3a, both algorithms detected the bounding boxes accurately. For the input image in Table 3b, while proposed algorithm detected the bounding box accurately, watershed has considered the two separate segments of different dimensions as one and drawn bounding box around the entire region. In Table 3c, watershed algorithm detected all regions on the boundary but has detected some false boundaries within the regions, while the proposed detected all of them accurately. In Table 3d too is a similar case where the detections are gone beyond the boundary regions by watershed, while the proposed algorithm has detected all boundaries accurately. Comparing the output, we can ascertain that the proposed algorithm has a better capability of determining the boundaries in such situations than the standard algorithm. Table 4 shows some qualitative results of the proposed algorithm. The advantage of this algorithm is that it uses the conventional approach of working on the pixel intensities and does not use any modern CNN approaches thereby reducing the overhead of training the model. Hence it can directly be applied on the output of a segmentation model for prediction. Figure 4 shows the precision–recall curve of the two methods. While average precision with watershed algorithm was 56.44%, with the proposed algorithm we could achieve 96.21%.

Table 2 Recall and F_1-scores for IoU from 0.5 to 0.9

IoU	0.5		0.6		0.7		0.8		0.9	
Algorithm	Recall (%)	F_1 (%)	Recall (%)	F_1 (%)	Recall (%)	F_1 (%)	Recall (%)	F_1 (%)	Recall (%)	F_1 (%)
Watershed	57	62	49	54	45	49	42	46	34	37
Proposed	96	97	96	96	95	96	95	96	95	95

Table 3 Results of few sample images (best viewed in color)

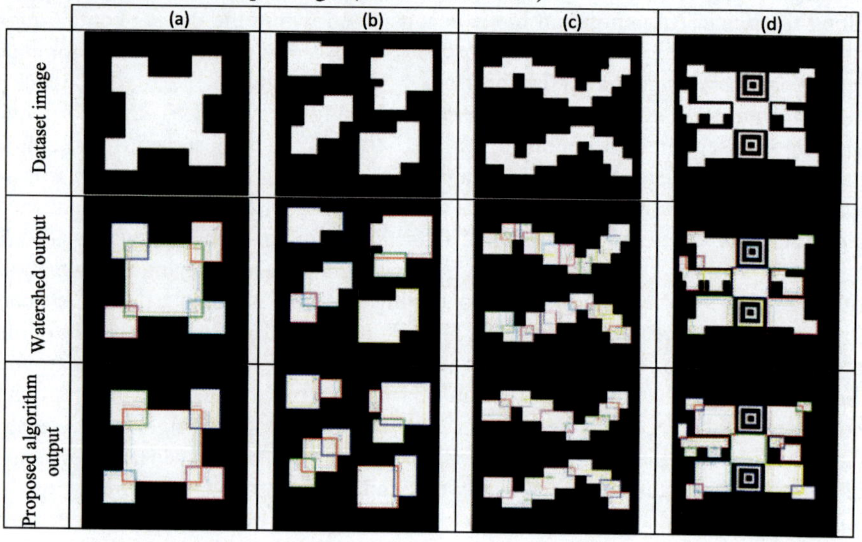

5 Conclusion

Object segmentation is one of the common approaches often used in computer vision that includes instance segmentation. Common issues that emerge out in instance segmentation is when distinct objects are overlapped and instantiating them as separate objects is troublesome when they have close features confidence scores. Standard approaches do distinguish them but have limitations of its own. Watershed algorithm achieves detection of touching and overlapped objects in an image by extracting the contours in an image. The proposed algorithm goes one step further in identifying linear arrangement of points and forming edges of the graph as a means of detecting intersections and refining the intersections to identify those that exists virtually behind the plane of the image segment that forms the plane of the connected rectangular region. The algorithm tries to address the issue for rectangular regions that appear in object segmentation with maximum accuracy as possible. Certain degree of calibration will be needed if extended category-2 points do not coincidence due to difference in pixel position in the initial contour. The algorithm's constraint lies in its

Table 4 Qualitative results (best viewed in color)

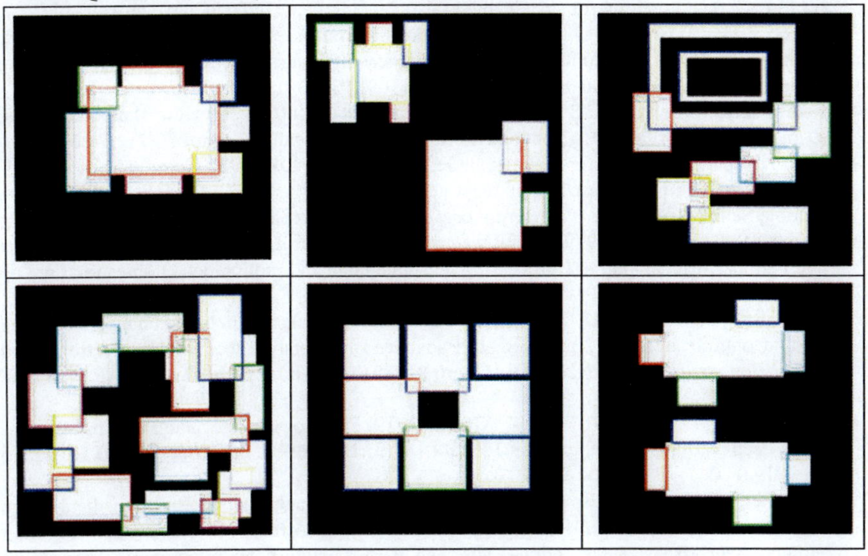

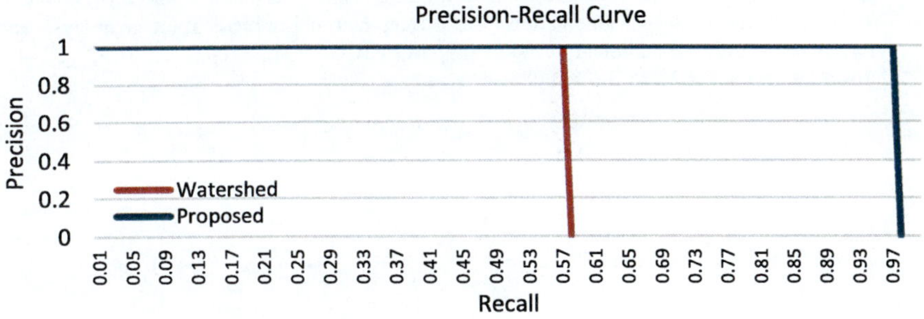

Fig. 4 Precision–recall curve of watershed and proposed algorithm

capability to detect only rectangular regions, making it unable to identify bounding boxes for arbitrary shapes. The study will be further extended to accurately handle elliptical and other related general shapes.

References

1. Minaee S, Boykov Y, Porikli F, Plaza A, Kehtarnavaz N, Terzopoulos D (2022) Image segmentation using deep learning: a survey. IEEE Trans Pattern Anal Mach Intell 44(7):3523–3542. https://doi.org/10.1109/TPAMI.2021.3059968

2. Patel S (2021) Deep learning models for image segmentation. In: 2021 8th International conference on computing for sustainable global development (INDIACom), pp 149–154
3. Goceri E (2019) Challenges and recent solutions for image segmentation in the era of deep learning. In: 2019 Ninth international conference on image processing theory, tools and applications (IPTA), pp 1–6. https://doi.org/10.1109/IPTA.2019.8936087
4. Ramesh KKD, Kumar GK, Swapna K, Datta D, Rajest SS (2021) A review of medical image segmentation algorithms. EAI Endorsed Trans Pervasive Health Technol 7(27):e6–e6
5. Yuan K, Zhuang X, Schaefer G, Feng J, Guan L, Fang H (2021) Deep-learning-based multi-spectral satellite image segmentation for water body detection. IEEE J Sel Top Appl Earth Obs Remote Sens 14:7422–7434. https://doi.org/10.1109/JSTARS.2021.3098678
6. Beucher S, Lantejoul C (1979) Use of watersheds in contour detection. In: Proceedings of the international workshop image processing real-time edge and motion detection/estimation, pp 2.1–2.12
7. Lin H, Song S, Tao S, Liu H (2021) Research on watershed algorithm based on image marking method optimization. In: 2021 IEEE 5th advanced information technology, electronic and automation control conference (IAEAC), pp 811–815. https://doi.org/10.1109/IAEAC50856.2021.9390847
8. Wang S, Yang DM, Rong R, Zhan X, Xiao G (2019) Pathology image analysis using segmentation deep learning algorithms. Am J Pathol 189(9):1686–1698. https://doi.org/10.1016/j.ajpath.2019.05.007; ISSN 0002-9440
9. Ronneberger O, Fischer P, Brox T (2015) U-Net: convolutional networks for biomedical image segmentation. In: Navab N, Hornegger J, Wells W, Frangi A (eds) Medical image computing and computer-assisted intervention—MICCAI 2015. Lecture notes in computer science, vol 9351. Springer, Cham. https://doi.org/10.1007/978-3-319-24574-4_28
10. Zhou Z, Rahman Siddiquee MM, Tajbakhsh N, Liang J (2018) UNet++: a nested U-Net architecture for medical image segmentation BT. In: Deep learning in medical image analysis and multimodal learning for clinical decision support, MICCAI, vol 11045, no 2018, pp 3–11. https://doi.org/10.1007/978-3-030-00889-5

Metaheuristic Optimized Extreme Gradient Boosting Milling Maintenance Prediction

Aleksandra Bozovic, Luka Jovanovic, Eleonora Desnica, Nebojsa Bacanin, Miodrag Zivkovic, Milos Antonijevic, and Joseph P. Mani

Abstract Machining plays a crucial role in modern manufacturing, relying on automated processes to efficiently create complex parts through subtractive like lathe turning and cutting. However, a major concern in this manufacturing process is tool wear, necessitating a robust system for proactive malfunction detection. To keep up with advancements and meet the increasing demands of speed and precision, artificial intelligence (AI) emerges as a promising solution. However, AI algorithms often require fine-tuning of hyperparameters, which poses a challenge. Swarm intelligence algorithms, inspired by collaborative behaviors observed in nature, offer a potential solution. By applying swarm intelligence to hyperparameter optimization, AI algorithms can achieve optimized models that address time and hardware constraints. This work proposes a methodology based on Extreme Gradient Boosting (XGBoost) for forecasting malfunctions. Additionally, a modified optimization metaheuristic is introduced to specifically enhance the performance of this methodology. To evaluate the proposed approach, it has been applied to a real-world dataset and compared

A. Bozovic · E. Desnica
Technical Faculty "Mihajlo Pupin", University of Novi Sad, Đure Đakovića bb, 23000 Zrenjanin, Serbia
e-mail: aleksandra.bozovic@tfzr.rs

E. Desnica
e-mail: desnica@tfzr.uns.ac.rs

L. Jovanovic (✉) · N. Bacanin · M. Zivkovic · M. Antonijevic
Singidunum University, Danijelova 32, 11000 Belgrade, Serbia
e-mail: luka.jovanovic.191@singimail.rs

N. Bacanin
e-mail: nbacanin@singidunum.ac.rs

M. Zivkovic
e-mail: mzivkovic@singimail.rs

M. Antonijevic
e-mail: mantonijevic@singidunum.ac.rs

J. P. Mani
Modern College of Business and Science, Muscat, Sultanate of Oman
e-mail: drjosephmani@mcbs.edu.om

S. Kumar et al. (eds.), *Fourth Congress on Intelligent Systems*, Lecture Notes in Networks and Systems 868, https://doi.org/10.1007/978-981-99-9037-5_28

to several well-known optimizers. The results demonstrate admirable performance, highlighting the potential of swarm intelligence in achieving efficient and effective machining processes.

Keywords Optimization · Predictive maintenance · Forecasting · Machining · Extreme gradient boosting

1 Introduction

Machining is a crucial part of modern manufacturing. Many systems rely on semi or fully-automated processes [28]. Not only does automation reduce the costs associated with manufacturing it decreases margins for error and helps improve standardization across manufacturing. Nevertheless, adequate maintenance is required to maintain continuous operations. The goal of maintenance is to maintain the machine's full operations with minimal possible downtime [7]. Maintenance can include scheduled services and checkups [23]. However, emergency repairs are not uncommon. Additionally, part replacements and realignments are required for proper operations [9].

Subtractive manufacturing processes, such as lathe turning and cutting, remain a dominant method for the production of most complex parts [32]. Milling specifically plays an important role in subtractive manufacturing [34]. One important aspect that needs to be considered with this form of manufacture is tool wear. Despite cutting tools being composed of very durable tungsten-carbide materials depending on the hardness of the material being processed, as well as feed rates, tool quality, and various other settings influence durability [41]. While regular maintenance and careful monitoring help reduce instances of unexpected malfunction, discreet methods often lack the ability to adapt to the dynamic environment and changing demands. Additionally, subtle variations in tool quality are often difficult to account for when scheduling inspections. Unexpected malfunctions can stall the manufacturing process or even damage the workpiece.

With this in mind, the need for a robust system for detecting malfunctions ahead of time is necessary. Furthermore, such a system needs to have the ability to adapt to a dynamic environment and maintain pace with the advancement in modern manufacturing, material sciences as well as increasing demands of speed and precision. One potential solution for such a system is the application of artificial intelligence (AI). However, algorithms are usually designed with good general performance in mind and require additional adjustments of hyperparameters for the best results. Hyperparameter tuning can often become an NP-hard challenge, thus solutions capable of attaining results within realistic time frames and real hardware are needed. Metaheuristic algorithms, specifically swarm intelligence algorithms are a promising contender [1]. By simulating collaborative behaviors in an iterative process, swarm intelligence can be applied to AI algorithms to attain optimized models.

This work applies extreme gradient boosting (XGBoost) [10] for forecasting tool failures in milling processes. Additionally, an altered version of a well-known metaheuristic algorithm has been developed and applied for optimizing XGBoost models to deliver optimal performance. The developed methodology has been applied and evaluated on a real-world dataset and the outcomes are compared to several well-known optimization techniques.

The contributions of this work can be summarized as:

- A proposal for an XGBoost-driven AI approach for maintenance prediction
- An introduction of a modified metaheuristic applied to the optimization of said approach
- The application of the described methodology on a real-world dataset to detect tool malfunction in real-world applications.

The rest of this work obeys the following structure: Sect. 2 covers research relevant to the topic of this paper. Section 3 talks in detail about the methodology proposed by this work. Sections 4 and 5 describe the conducted experimentation and the outcomes of said experimentation respectfully. Finally, Sect. 6 provides a closing word on this work and proposed potential further research.

2 Related Works

The examining needs for industrialization as well as the integration and widespread availability of computing within the industry has led researchers toward exploring novel techniques for process monitoring. Integrated Internet-enabled devices have brought on a revolution in manufacturing and many such devices find their role in monitoring machining tool chains. Temperature and vibration contribute to tool wear as well as increase the risk of damage to the work piece [27]. Researchers have explored the potential of Internet of Things (IoT) devices for monitoring these factors [30]. Collected data could prove useful for more advanced AI-based monitoring systems.

While IoT has merits they often require specialized hardware. A different approach that has great potential is the use of computer vision. A significant advantage of using computer vision is that it does not require specialized hardware and that cameras are relatively cheaper. Researchers have explored computer vision techniques [2] in milling process monitoring and attained promising results. Nevertheless, research on this topic is often difficult due to a lack of well-documented, accessible quality datasets [22].

Several optimization algorithms draw inspiration from a wide range of animal and insect species, aiming to replicate nature's behavioral patterns, which include foraging, hunting, and mating processes. These algorithms utilize mathematical models to capture the essence of these natural behaviors. Notable efficient optimization algorithms in this domain include the Firefly algorithm (FA) [39], Particle Swarm Optimizer (PSO) [36], Whale Optimization Algorithm (WOA) [25], and the well-known Artificial Bee Colony (ABC) [20] algorithm. The genetic algorithm (GA) [26] is also highly effective, drawing inspiration directly from the process of evolution. Additionally, more abstract sources of inspiration have led to the development of powerful methods like the COLSHADE [12] optimization algorithm. The driving factor behind the development of many of these algorithms is the same idea behind the "no free lunch" (NFL) [37] theorem. This theorem proposes that no single method is equally well suited to all challenges.

Due to the excellent performance demonstrated when tackling general optimizations, metaheuristic algorithms have found many applications across several fields. Some interesting examples include computer system security [3, 6, 17, 31], fraud detection [11, 29], tackling complex challenges in emerging industries [8, 14, 24], as well as healthcare [15, 19, 35] applications. Furthermore, metaheuristics have demonstrated admirable performance when tackling optimization related to time series forecasting [16, 18, 33] making them an ideal candidate for tackling solar flair classification. Other examples of metaheuristics and hybrid techniques exist in literature [4, 5, 13].

2.1 XGBoost

With utilizing objective function optimization at the core of the approach, the XGBoost [10] technique employs a progressive training strategy. It considers the results from previous iterations in each step of the optimization process, which subsequently affects the following outcomes. The expression shown in Eq. 1 is utilized to represent the objective function for the tth instance within XGBoost.

$$F_o{}^i = \sum_{k=1}^{n} l\left(y_k, \hat{y}_k^{i-1} + f_i\left(x_k\right)\right) + R(f_i) + C, \tag{1}$$

the loss component of the tth iteration is denoted as l, the constant term is symbolized as C, and the model's regularization parameter is represented by R. The precise definition of the regularization parameter can be found in Eq. 2.

$$R(f_i) = \gamma T_i + \frac{\lambda}{2}\sum_{j=1}^{T} w_j^2 \tag{2}$$

Generally, selecting larger values for the γ and λ parameters results in the generation of less complex tree structures. The mathematical representations for the first derivative (g) and the second derivative (h) of the model are given as follows:

$$g_j = \partial_{\hat{y}_k^{i-1}} l\left(y_j, \hat{y}_k^{i-1}\right) \tag{3}$$

$$h_j = \partial^2_{\hat{y}_k^{i-1}} l\left(y_j, \hat{y}_k^{i-1}\right) \tag{4}$$

The solution can be derived using the following formulas:

$$w_j^* = -\frac{\sum g_t}{\sum h_t + \lambda} \tag{5}$$

$$F_o{}^* = -\frac{1}{2}\sum_{j=1}^{T}\frac{\left(\sum g\right)^2}{\sum h + \lambda} + \gamma T, \tag{6}$$

In this context, $F_o{}^*$ refers to the score of the loss function, while w_j^* represents the solution for the weights.

3 Methods

The following section provides a detailed description of the methods behind this work. The original algorithm is described in detail followed by the observed shortcomings and a solution is proposed. The pseudocode for the introduced algorithm is also provided. Each individual in the population of the metaheuristics represents a potential XGBoost configuration, with hyperparameters encoded as agent characteristics. Thorough successive iterations models are constructed based on solution characteristics then trained and tested. Solutions are improved and an optimized set of parameters for a model is selected.

3.1 Original FA

The FA [40] simulates the mating mechanisms of fireflies. The objective function outcomes determine brightness. Brighter fireflies are more attractive to those with less illumination and all fireflies are attracted to each other. Should no brighter agent exist, movements are randomized. The brightness of a given agent can be according to the inverse square law Eq. 7.

$$I(r) = \frac{I_s}{r^2} \tag{7}$$

in this context, I_s represents the light intensity at the source location. The distance between the source and a particular point is denoted as r, and $I(r)$ represents the light intensity at that specific distance. Moreover, by taking into account the light absorption coefficient of the medium, the calculation of light intensity can be accomplished using Eq. 8.

$$I(r) = I_0 e^{-\gamma r} \tag{8}$$

where the light absorption coefficient associated with the medium is represented by γ. The initial brightness is denoted by I_0, and r indicates the distance. Here the above-described equation is incorporated with the impact of absorption using a Gaussian form, as depicted in Eq. 9.

$$I(r) = I_0 e^{-\gamma r^2} \tag{9}$$

If a slower and more gradual convergence is desired, Eq. 10 can be utilized as a substitute for this approximation.

$$I(r) = \frac{I_0}{1 + \gamma r^2} \tag{10}$$

The degree of attraction between agents is governed by the observed brightness. By considering all these aspects, the attractiveness of a particular agent can be accurately defined using Eq. 11.

$$\beta(r) = \beta_0 e^{-\gamma^2} \tag{11}$$

with β_0 representing an agents highest potential attractiveness should $r = 0$.

Distance between any two fireflies can be calculated using Cartesian distance according to Eq. 12.

$$r_{ij} = \sqrt{(x_i - x_j)^2 + (y_i - y_j)^2} \tag{12}$$

Brighter fireflies are more attractive. For firefly i attracted to firefly j due to higher light intensity, the movement can be computed using Eq. 13.

$$x_i = x_i + \beta_0 e^{-\gamma r_{ij}^2}(x_j - x_i) + \alpha\left(\text{rand} - \frac{1}{2}\right) \tag{13}$$

The second term operates based on attraction mechanisms, while the third term serves to add an element of randomness. The randomization parameter, denoted as α, determines the level of randomization. The variable rand represents a random number selected from a uniform distribution between 0 and 1. Generally, it is assumed that α falls within the range of 0–1, and B_0 is set to 1. However, if necessary, the random distribution can be adjusted to different types of distributions.

The parameter γ governs the variations in attraction, playing a crucial role in determining convergence rates and shaping the behavior of the algorithm. While theoretically, γ can span from 0 to infinity, it is usually limited to approximately 1, as defined by the specific characteristics of the optimization problem denoted by Γ. In most cases, the value of γ ranges between 0.01 and 100.

3.2 Genetically Inspired FA

The original FA demonstrated impressive performance. However, extensive testing with standard CEC functions [21] revealed that certain executions can yield suboptimal results, as the algorithm tends to pay too much attention to less promising areas of the search space. To address this limitation and enhance exploration capabilities, this work introduces the following mechanism.

After each iteration, a new solution is created by combining the current best agent with a randomly selected individual from the population. This combination is achieved using a uniform crossover control parameter, denoted as pc, which has been empirically determined as $pc = 0.1$. The generated solution inherits characteristics from both solutions. Additionally, each parameter in the generated solution undergoes mutation. The mutation process is governed by a mutation parameter, denoted as mp, with an empirically determined value of $mp = 0.1$.

During mutation, a pseudo-random value from the range $\left[\frac{lb}{2}, \frac{ub}{2}\right]$ is either added or subtracted from the corresponding solution parameter. The upper bound is represented by ub, and the lower bound is denoted as lb. The decision of whether to add or subtract the value is determined by an additional control parameter called the mutation direction, denoted as md. A pseudo-random value ψ is generated from a uniform distribution within the range of [0, 1]. If $\psi < md$, subtraction is used; otherwise, addition is employed. In this research, the md parameter is set to $md = 0.5$.

Once the new solution is generated, it replaces the worst-performing solution in the population. However, the performance evaluation of the new solution is deferred until the next iteration. This approach ensures that the computational complexity of the modified algorithm remains on the level of the original FA. The resulting algorithm is referred to as the Genetically Inspired FA (GIFA), as it draws some inspiration from the GA. The pseudocode for the GIFA algorithm is presented in Algorithm 1.

Algorithm 1 Pseudocode for the introduced GIFA

```
Set initial algorithm parameters
Create population P
Evaluate P using objective function
for i = 1 to iteration limit do
  for each solution do
    for each better soltuin do
      if soltuin is beter then
        Determine attraction with respect to distance
        Adjust location towards the better solution
    Evaluate and update solutions in population P
    Generated new solution using genetic crossover mechanism
    Subject new solution to mutation
    Replace the worst-performing agent with a newly generated solution
  Return best-performing solution
```

4 Experimental Setup

To evaluate the performance of the introduced methodology, testing was conducted on a publicly available[1] real-world dataset that consists of relevant machining data and tool status [22]. To detect manufacturing issues, each XGBoost model was tasked with determining a tool failure point based on historical data. Metaheuristics including the introduced algorithm, original version of the FA [38], GA [26], PSO [36], ABC [20], and COLSHADE [12] were tasked with optimizing hyperparameters toward desired performance.

Each metaheuristic was allocated a population size of 10 agents and allowed 10 iterations to improve said population. Additionally, to account for the randomness associated with metaheuristics. The optimization process encompasses the identification of optimized parameter values from constrained ranges for the following parameters: Learning Rate [0.1, 0.9], Minimum Child Weight [1, 10], Subsample [0.01, 1], Colsample by Tree [0.01, 1], Maximum Depth [3, 10], and Gamma [0, 0.8].

Several metrics were considered during experimentation. Due to the imbalance prescient in the dataset Cohen's kappa, described by Eq. 14, was used as the objective function.

$$\kappa = \frac{p_o - p_e}{1 - p_e} \tag{14}$$

in which p_o denotes the observed outcomes and p_e the expected outcomes. Standard classification metrics including precision, recall, F_1-score, and accuracy were tracked for each optimized model.

$$\text{Precision} = \frac{\text{TP}}{\text{TP} + \text{FP}} \tag{15}$$

$$\text{Recall} = \frac{\text{TP}}{\text{TP} + \text{FN}} \tag{16}$$

$$\text{Accuracy} = \frac{\text{TP} + \text{TN}}{\text{TP} + \text{TN} + \text{FP} + \text{FN}} \tag{17}$$

$$F_1 \text{ Score} = 2 \times \frac{\text{Precision} \times \text{Recall}}{\text{Precision} + \text{Recall}} \tag{18}$$

where TP and TN mark true positive and negative respectively, and FP and FN mark false positive and negative accordingly. Finally, the error rate was used as the indicator function determined as 1 − accuracy.

[1] https://www.kaggle.com/datasets/stephanmatzka/predictive-maintenance-dataset-ai4i-2020.

Table 1 Objective function (Cohen's kappa) overall outcomes

Method	Best	Worst	Mean	Median	Std	Var
XG-GIFA	**0.809443**	**0.789226**	**0.799159**	**0.798405**	0.006025	3.63E−05
XG-FA	0.802363	0.771404	0.781398	0.776090	0.010458	1.09E−04
XG-GA	0.800715	0.773448	0.787509	0.788056	0.009145	8.36E−05
XG-PSO	0.789226	0.773448	0.780946	0.781441	**0.005133**	**2.64E−05**
XG-ABC	0.791540	0.757736	0.778439	0.777940	0.009333	8.71E−05
XG-WOA	0.787068	0.769004	0.779910	0.780593	0.006125	3.75E−05
XG-COLSHADE	0.796103	0.766552	0.781235	0.780436	0.007432	5.52E−05

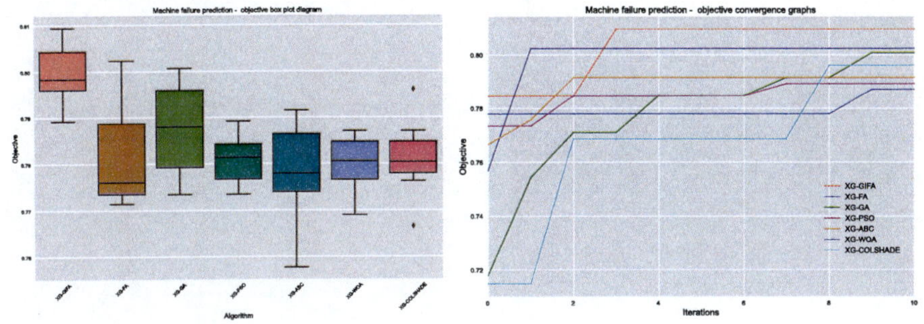

Fig. 1 Objective function distribution and convergence plots

5 Simulation Outcomes

Cohen's kappa objective function outcomes over the 30 independent runs executed during experimentation for the best, and worst-performing models, as well as the mean and median execution, are shown in Table 1. Additionally, standard divination and variance are provided to demonstrate algorithm stability.

As can be observed from the outcomes demonstrated in Table 1, the introduced metaheuristic demonstrates improvements over the base algorithm. Furthermore, the introduced metaheuristic outperformed all competing optimizers. It is also worth noting that the PSO algorithm demonstrated impressive stability despite less favorable outcomes. This is further solidified by the distribution plots shown in Fig. 1. Additionally, improvements to the convergence rate made by the introduced improvement can be observed in the convergence plot.

Aside from the objective for the indicator function, error rate was utilized. Indicator function outcomes over the 30 independent runs executed during experimentation for the best, and worst-performing models, as well as the mean and median execution, are shown in Table 2. Additionally, standard divination and variance are provided to demonstrate algorithm stability.

Table 2 Indicator function (error) overall outcomes

Method	Best	Worst	Mean	Median	Std	Var
XG-GIFA	**0.011333**	**0.012667**	**0.011933**	**0.011833**	0.000416	1.73E−07
XG-FA	0.012000	0.013667	0.013133	0.013333	0.000600	3.60E−07
XG-GA	0.011667	0.013333	0.012733	0.013000	0.000680	4.62E−07
XG-PSO	0.012667	0.013333	0.013067	0.013167	**0.000359**	**1.29E−07**
XG-ABC	0.012333	0.014333	0.013067	0.013000	0.000554	3.07E−07
XG-WOA	0.012333	0.013667	0.013300	0.013333	0.000482	2.32E−07
XG-COLSHADE	0.012000	0.013667	0.013067	0.013000	0.000573	3.29E−07

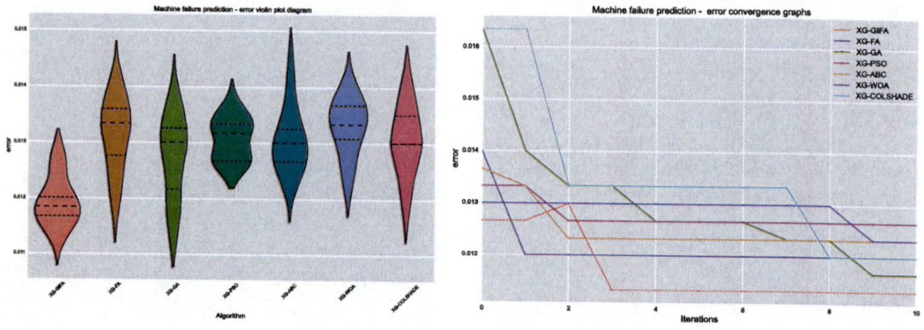

Fig. 2 Indicator function distribution and convergence plots

As can be observed from the indicator function outcomes demonstrated in Table 2, the introduced metaheuristic demonstrates improvements over the base algorithm. Furthermore, the introduced metaheuristic outperformed all competing optimizers. It is also worth noting that the PSO algorithm demonstrated impressive stability despite less favorable outcomes. This is further solidified by the distribution plots shown in Fig. 2. Additionally, improvements to the convergence rate made by the introduced improvement can be observed in the convergence plot.

Detailed metrics for the best-performing models optimized by each metaheuristic are demonstrated in Table 3.

From the outcomes shown in Table 3 it may be deduced that the models optimized by the introduced algorithm attained the best-weighted average for preclusion, recall, and F_1-score. Furthermore, the introduced algorithm attained the best F_1-score for all metrics.

Finally, hyperparameter selections made by each algorithm for their respective best-performing models are visible in Table 4.

Table 3 Detailed metrics for the best-performing XGBoost models optimized by each metaheuristic algorithm

Method	Metric	Non-failure	Failure	Macro avg	Weighted avg
XG-GIFA	Precision	0.990747	0.914634	0.952691	**0.988159**
	Recall	0.997586	0.735294	0.866439	**0.988667**
	F_1-score	**0.994154**	**0.815217**	**0.904686**	**0.988070**
XG-FA	Precision	**0.991078**	0.883721	0.937399	0.987427
	Recall	0.996549	**0.745098**	**0.870824**	0.988000
	F_1-score	0.993806	0.808511	0.901158	0.987506
XG-GA	Precision	0.990072	0.924051	**0.957061**	0.987827
	Recall	0.997930	0.715686	0.856808	0.988333
	F_1-score	0.993985	0.806630	0.900308	0.987615
XG-PSO	Precision	0.990398	0.880952	0.935675	0.986677
	recall	0.996549	0.725490	0.861020	0.987333
	F_1-score	0.993464	0.795699	0.894581	0.986740
XG-ABC	Precision	0.990065	0.901235	0.945650	0.987045
	Recall	0.997239	0.715686	0.856463	0.987667
	F_1-score	0.993639	0.797814	0.895727	0.986981
XG-WOA	Precision	0.989394	**0.922078**	0.955736	0.987106
	Recall	0.997930	0.696078	0.847004	0.987667
	F_1-score	0.993644	0.793296	0.893470	0.986832
XG-COLSHADE	Precision	0.990068	0.912500	0.951284	0.987431
	Recall	0.997585	0.715686	0.856635	0.988000
	F_1-score	0.993812	0.802198	0.898005	0.987297
	Support	2898	102	3000	3000

Table 4 Hyperparameter selections made for the best models optimized by each algorithm

Method	Learning rate	Min child weight	Subsample	Colsample by tree	Max depth	Gamma
XG-GIFA	0.409960	2.619509	1.000000	1.000000	5	0.506223
XG-FA	0.900000	5.444813	0.795093	1.000000	9	0.429615
XG-GA	0.255070	1.076415	0.814741	0.736531	8	0.723615
XG-PSO	0.424214	3.569898	0.877443	1.000000	6	0.323775
XG-ABC	0.427487	4.114513	0.723255	1.000000	5	0.412552
XG-WOA	0.356862	5.637026	0.864202	0.999709	9	0.800000
XG-COLSHADE	0.588753	1.000000	0.935980	0.602135	7	0.800000

6 Conclusion

Due to the crucial role machining plays in modern manufacturing maintaining consistent and optimized processes is essential. However, unexpected variations in materials and tools make it difficult for simple scheduled maintenance insufficient. A need for an advanced system for detecting malfunctions is evident. This work proposes a methodology based on XGBoost for milling maintenance forecasting. Additionally, a modified optimization metaheuristic is introduced to specifically enhance the performance of this methodology. To evaluate the proposed approach, it has been applied to a real-world dataset and compared to several well-known optimizers. The attained outcomes indicate that the modified algorithm shows promise for application to this challenge, outperforming all other competing optimization algorithms evaluated in this work.

Upcoming work will mainly focus on further refining the automated maintenance forecasting approach. Furthermore, additional pressing issues from the real-world will be challenged using the introduced metaheuristic.

References

1. Abdulrahman SM (2017) Using swarm intelligence for solving NP-Hard problems. Acad J Nawroz Univ 6(3):46–50
2. Ahmad MI, Saif Y, Yusof Y, Daud ME, Latif K, Kadir AZA (2022) A case study: monitoring and inspection based on IoT for milling process. Int J Adv Manuf Technol 1–11
3. Al Hosni N, Jovanovic L, Antonijevic M, Bukumira M, Zivkovic M, Strumberger I, Mani JP, Bacanin N (2022) The XGBoost model for network intrusion detection boosted by enhanced sine cosine algorithm. In: Third international conference on image processing and capsule networks: ICIPCN 2022. Springer, Berlin, pp 213–228
4. Bacanin N, Jovanovic L, Zivkovic M, Kandasamy V, Antonijevic M, Deveci M, Strumberger I (2023) Multivariate energy forecasting via metaheuristic tuned long-short term memory and gated recurrent unit neural networks. Inf Sci 642:119122. https://doi.org/10.1016/j.ins.2023.119122
5. Bacanin N, Zivkovic M, Antonijevic M, Venkatachalam K, Lee J, Nam Y, Marjanovic M, Strumberger I, Abouhawwash M (2023) Addressing feature selection and extreme learning machine tuning by diversity-oriented social network search: an application for phishing websites detection. Complex Intell Syst. https://doi.org/10.1007/s40747-023-01118-z
6. Bacanin N, Zivkovic M, Stoean C, Antonijevic M, Janicijevic S, Sarac M, Strumberger I (2022) Application of natural language processing and machine learning boosted with swarm intelligence for spam email filtering. Mathematics 10(22). https://doi.org/10.3390/math10224173; https://www.mdpi.com/2227-7390/10/22/4173
7. Bahga A, Madisetti VK (2011) Analyzing massive machine maintenance data in a computing cloud. IEEE Trans Parallel Distrib Syst 23(10):1831–1843
8. Balaji BS, Paja W, Antonijevic M, Stoean C, Bacanin N, Zivkovic M (2023) IoT integrated edge platform for secure industrial application with deep learning. Hum Centric Comput Inf Sci 13
9. Chavoshi SZ, Goel S, Morantz P (2017) Current trends and future of sequential micromachining processes on a single machine tool. Mater Des 127:37–53
10. Chen T, Guestrin C (2016) XGBoost: a scalable tree boosting system. In: Proceedings of the 22nd ACM SIGKDD international conference on knowledge discovery and data mining, pp 785–794

11. Djuric M, Jovanovic L, Zivkovic M, Bacanin N, Antonijevic M, Sarac M (2023) The AdaBoost approach tuned by SNS metaheuristics for fraud detection. In: Proceedings of the international conference on paradigms of computing, communication and data sciences: PCCDS 2022. Springer, Berlin, pp 115–128
12. Gurrola-Ramos J, Hernàndez-Aguirre A, Dalmau-Cedeño O (2020) COLSHADE for real-world single-objective constrained optimization problems. In: 2020 IEEE congress on evolutionary computation (CEC). IEEE, pp 1–8
13. Jovanovic L, Bacanin N, Antonijevic M, Tuba E, Ivanovic M, Venkatachalam K (2022) Plant classification using firefly algorithm and support vector machine. In: 2022 IEEE Zooming innovation in consumer technologies conference (ZINC). IEEE, pp 255–260
14. Jovanovic L, Bacanin N, Zivkovic M, Antonijevic M, Jovanovic B, Sretenovic MB, Strumberger I (2023) Machine learning tuning by diversity oriented firefly metaheuristics for industry 4.0. Expert Syst e13293
15. Jovanovic L, Djuric M, Zivkovic M, Jovanovic D, Strumberger I, Antonijevic M, Budimirovic N, Bacanin N (2023) Tuning XGBoost by planet optimization algorithm: an application for diabetes classification. In: Proceedings of fourth international conference on communication, computing and electronics systems: ICCCES 2022. Springer, Berlin, pp 787–803
16. Jovanovic L, Jovanovic D, Bacanin N, Jovancai Stakic A, Antonijevic M, Magd H, Thirumalaisamy R, Zivkovic M (2022) Multi-step crude oil price prediction based on LSTM approach tuned by Salp Swarm algorithm with disputation operator. Sustainability 14(21):14616
17. Jovanovic L, Jovanovic D, Antonijevic M, Nikolic B, Bacanin N, Zivkovic M, Strumberger I (2023) Improving phishing website detection using a hybrid two-level framework for feature selection and XGBoost tuning. J Web Eng 22(03):543–574. https://doi.org/10.13052/jwe1540-9589.2237, https://journals.riverpublishers.com/index.php/JWE/article/view/18475
18. Jovanovic L, Milutinovic N, Gajevic M, Krstovic J, Rashid TA, Petrovic A (2022) Sine cosine algorithm for simple recurrent neural network tuning for stock market prediction. In: 2022 30th Telecommunications forum (TELFOR). IEEE, pp 1–4
19. Jovanovic L, Zivkovic M, Antonijevic M, Jovanovic D, Ivanovic M, Jassim HS (2022) An emperor penguin optimizer application for medical diagnostics. In: 2022 IEEE zooming innovation in consumer technologies conference (ZINC). IEEE, pp 191–196
20. Karaboga D, Basturk B (2007) A powerful and efficient algorithm for numerical function optimization: artificial bee colony (ABC) algorithm. J Glob Optim 39:459–471
21. Luo W, Lin X, Li C, Yang S, Shi Y (2022) Benchmark functions for CEC 2022 competition on seeking multiple optima in dynamic environments. arXiv preprint arXiv:2201.00523
22. Matzka S (2020) Explainable artificial intelligence for predictive maintenance applications. In: 2020 Third international conference on artificial intelligence for industries (ai4i). IEEE, pp 69–74
23. Mikic D, Desnica E, Asonja A, Stojanovic B, Epifanic-Pajic V (2016) Reliability analysis of ball bearing on the crankshaft of piston compressors. J Balkan Tribol Assoc
24. Milutinovic N, Cabarkapa S, Zivkovic M, Antonijevic M, Mladenovic D, Bacanin N (2023) Tuning artificial neural network for healthcare 4.0. by sine cosine algorithm. In: 2023 International conference on intelligent data communication technologies and internet of things (IDCIoT), pp 510–513. https://doi.org/10.1109/IDCIoT56793.2023.10053543
25. Mirjalili S, Lewis A (2016) The whale optimization algorithm. Adv Eng Softw 95:51–67
26. Mirjalili S, Mirjalili S (2019) Genetic algorithm. In: Evolutionary algorithms and neural networks: theory and applications, pp 43–55
27. Móricz L, Viharos ZJ, Németh A, Szépligeti A, Büki M (2020) Off-line geometrical and microscopic & on-line vibration based cutting tool wear analysis for micro-milling of ceramics. Measurement 163:108025
28. Novakovic B, Durdev M, Radovanovic L, Speight JG (2018) Optimization of manufacturing processes using modern automated CNC milling machines. Appl Eng Lett J Eng Appl Sci 3(4):124–128. https://doi.org/10.18485/aeletters.2018.3.4.2, https://www.aeletters.com/wp-content/uploads/2019/01/AEL00078.pdf

29. Petrovic A, Antonijevic M, Strumberger I, Jovanovic L, Savanovic N, Janicijevic S (2023) The XGBoost approach tuned by TLB metaheuristics for fraud detection. In: Proceedings of the 1st international conference on innovation in information technology and business (ICIITB 2022), vol 104. Springer Nature, Berlin, p 219
30. Said NHAM, Yusof Y (2022) Applied internet of things (IoT) in temperature and vibration monitoring system for milling machine. Res Prog Mech Manuf Eng 3(1):476–485
31. Salb M, Jovanovic L, Zivkovic M, Tuba E, Elsadai A, Bacanin N (2022) Training logistic regression model by enhanced moth flame optimizer for spam email classification. In: Computer networks and inventive communication technologies: proceedings of fifth, ICCNCT 2022. Springer, Berlin, pp 753–768
32. Sathish K, Kumar SS, Magal RT, Selvaraj V, Narasimharaj V, Karthikeyan R, Sabarinathan G, Tiwari M, Kassa AE (2022) A comparative study on subtractive manufacturing and additive manufacturing. Adv Mater Sci Eng 2022
33. Stankovic M, Jovanovic L, Bacanin N, Zivkovic M, Antonijevic M, Bisevac P (2023) Tuned long short-term memory model for ethereum price forecasting through an arithmetic optimization algorithm. In: Abraham A, Bajaj A, Gandhi N, Madureira AM, Kahraman C (eds) Innovations in bio-inspired computing and applications. Springer Nature Switzerland, Cham, pp 327–337
34. Trung D (2022) Effect of cutting parameters on the surface roughness and roundness error when turning the interrupted surface of 40x steel using HSS-TiN insert. Appl Eng Lett J Eng Appl Sci 7(1):1–9
35. Umapathi K, Vanitha V, Anbarasu L, Zivkovic M, Bacanin N, Antonijevic M (2021) Predictive data regression technique based carbon nanotube biosensor for efficient patient health monitoring system. J Ambient Intell Hum Comput. https://doi.org/10.1007/s12652-021-03063-6
36. Wang D, Tan D, Liu L (2018) Particle swarm optimization algorithm: an overview. Soft Comput 22:387–408
37. Wolpert DH, Macready WG (1997) No free lunch theorems for optimization. IEEE Trans Evol Comput 1(1):67–82
38. Yang XS (2010) Firefly algorithm, stochastic test functions and design optimisation. Int J Bio-inspired Comput 2(2):78–84
39. Yang XS, He X (2013) Firefly algorithm: recent advances and applications. Int J Swarm Intell 1(1):36–50
40. Yang XS, Slowik A (2020) Firefly algorithm. In: Swarm intelligence algorithms. CRC Press, pp 163–174
41. Zhang S, Gong M, Zeng X, Gao M (2021) Residual stress and tensile anisotropy of hybrid wire arc additive-milling subtractive manufacturing. J Mater Process Technol 293:117077

Synthetic Fingerprint Generation Using Generative Adversarial Networks: A Review

Ritika Dhaneshwar, Arnav Taya, and Mandeep Kaur

Abstract In this era of stringent privacy-related laws, the manual collection of fingerprints is a challenging task. This act as a deterrent for collecting large-scale database which is a prerequisite for implementing deep learning approaches in fingerprint-based applications. So, to overcome these challenges researchers came up with various synthetic fingerprint generation approaches using Generative Adversarial Networks (GAN). GANs are deep learning-based generative models which help in generating realistic-looking synthetic data. They help in the generation of fingerprints that are comparative in terms of quality, features and characteristics, with that of manually collected samples. The objective of this paper is to review the existing approaches based on GAN that are used for the synthetic generation of fingerprints. Critical investigation of underlying technological details of various GAN variants for generating synthetic datasets and their comparative analysis aims to assist researchers in the generation of apt data for designing advanced applications. Finally, an appraisal of various performance metrics is presented for the evaluation of the synthetic fingerprint quality to facilitate research.

Keywords Synthetic fingerprints · Enhancement · Reconstruction · Segmentation · Matching · Generative adversarial network

R. Dhaneshwar (✉) · A. Taya · M. Kaur
University Institute of Engineering and Technology, Panjab University, Chandigarh, India
e-mail: ritikadhaneshwar@gmail.com

A. Taya
e-mail: Tayaarnav22@gmail.com

M. Kaur
e-mail: mandeep@pu.ac.in

S. Kumar et al. (eds.), *Fourth Congress on Intelligent Systems*, Lecture Notes in Networks and Systems 868, https://doi.org/10.1007/978-981-99-9037-5_29

1 Introduction

Synthetic data or algorithmically generated data can be defined as artificially manufactured information, which resembles real data in terms of some vital parameters. It is system-generated data that is generally used for validating mathematical models, software testing and quality assurance, safe data sharing, training of machine learning architectures, etc. Today, this synthetically generated data is widely favored over its real counterpart because of its cost-effectiveness, easy tailoring of the data according to the requirement for testing frameworks, anonymization of regulated data, etc. Many data professionals of leading industries are today using this concept of synthetic data for sharing and utilizing sensitive information, of their clients while maintaining their privacy. This synthetic data can be generated by using varied data generative models. GANs are amongst the most widely used architectures for generating realistic synthetic datasets. They are deep learning-based generative models where two contending neural networks, generator and discriminator compete with each other to generate synthetic data. As proposed in 2014, both the generator and discriminator are trained simultaneously, which helps in synthesizing samples by learning from the statistical distribution of training data [1]. A basic flow diagram for GANs is depicted in Fig. 1. The generator generates random output and proposes it to be real, whereas a discriminator detects and labels the data as real/fake. In the scenario where the data is classified as fake, the weights of the network are updated using the back-propagation approach. The training of GANs is defined as a Zero-sum game between the generator and the discriminator, where both try to gain maximum benefits. For a model to be declared as stable we need to ensure that, updation in its weights has no consecutive effect on its loss function. This condition of stability is achieved at a point called Nash equilibrium.

There are various flavors of GANs like CycleGan [2], Deep convolutional GANs (DCGAN) [3], Conditional GAN (CGAN) [4], Semi-supervised GANS (SGAN) [3], etc., with a wide-scale of applications like cloning [5], synthetic data generation [6], handwriting generation [7], and various other audio and speech applications [8]. Today, the generation of synthetic fingerprints using GANs is gaining attraction in the deep learning domain, which ensures both easy data collection and anonymity. Hence, they are mostly exploited in research and related application areas where privacy and anonymity are crucial to preserve. The most common and widely used

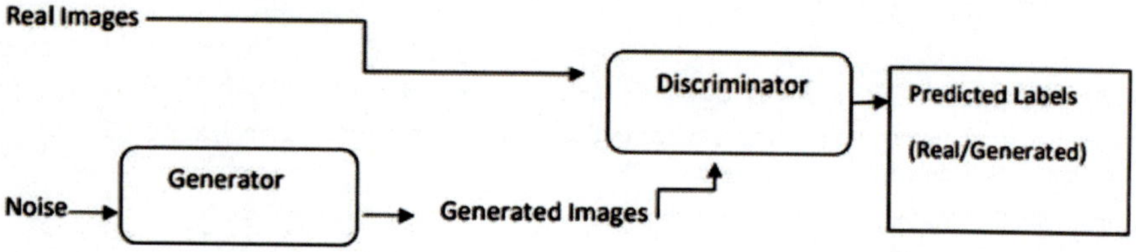

Fig. 1 Basic architecture of GAN

realm is that of fingerprints and latent fingerprints exploited in biometric and forensic applications.

Fingerprints are an important pillar in biometric recognition systems due to two main features, namely, uniqueness and permanence. This has encouraged researchers to promote its usage. Further, due to the easy sample collection of fingerprints as compared to other traits like iris, ear, palm, etc., its usage has been promoted at various levels like Airport authentication systems, criminal conviction, banking operations, etc. Law agencies usually consider fingerprints as important evidence for legitimizing the identity and authenticity of an individual. However, despite having a varied scope of application, many impediments hamper the recognition performance in this area. The foremost issue is to capture and preserve our sample evidence for further investigation. The procedure for the upliftment of a fingerprint depends on many factors like the expertise of the investigating officer, the quality of the surface under consideration, appropriate fingerprint upliftment method selection, etc. [9]. Appropriate reconstruction, enhancement and efficient matching of a large dataset are the other areas of research that need further improvements. So for the majority of the barriers mentioned, upcoming deep learning and machine learning approaches can act as a panacea.

However, recent legal restrictions related to privacy concerns as well as pandemics like COVID have hampered the manual collection of large-scale datasets that are a prerequisite for applying any deep learning approach. Though there are some publically available datasets like NIST27 [10], WVU latent databases [11], FVC 2004 databases [12], IIIT latent fingerprint database [13], IIITD Multi-surface Latent Fingerprint database (IIITD MSLFD) [13], IIIT Simultaneous Latent Fingerprint (SLF) database [13], Multisensor Optical and Latent Fingerprint database [14], Tsinghua Latent Overlapped Fingerprint database [15], ELFT-EFS Public Challenge database [16]. However, they are not large enough to train a deep learning model efficiently. So to overcome this issue, researchers proposed different GAN-based approaches to generate synthetic fingerprints having high quality which are comparable to original images. These approaches would help us in generating large-scale fingerprint databases, which would encourage the community to apply upcoming techniques for improving the recognition performance of biometric systems.

This study is conducted to gain a deep understanding of how we can generate realistic-looking synthetic fingerprints using different GAN architectures. It gives insight into how the quality of output varies with changes in training parameters. A comparison of results may help us in evaluating the suitability of an approach for a particular application model. Therefore in the next section, we would present a detailed review of available approaches, with their comparative analysis.

2 Approaches for Synthetic Fingerprint Generation

A novel approach was proposed by Cao et al., in 2018, where 512 × 512 rolled fingerprint images were generated using the GAN. This approach utilizes the Wasserstein distance for calculating the value function, which leads to better performance than the available generative approaches. The 512-dimensional representations from an input fingerprint were extracted using an encoder component whereas reconstruction of the input fingerprint was done using the decoder [17]. Further fingerprint image quality and diversity were improved by initializing the generator using a discriminator for training the I-WGAN.

In 2018 Minaee et al., proposed a model based on a machine learning framework [18]. As per their proposition, the model was aimed to capture complex texture representations of fingerprint images to ensure the generation of realistic-looking images which helps in overcoming the major drawback of classical statistical models. Further, a suitable regularization term was added to the loss function which would ensure the connectivity of the generated samples. The generator model consists of 5 convolutional layers whereas the discriminator model is having 4 layers. This method has the advantage that it has a less complex gradient update methodology than other similar approaches. An improved WGAN (I-WGAN) architecture was proposed by authors in 2020 for generating synthetic fingerprints [19]. The core contribution of this paper was the use of identity loss which will help in the generation of more distinct images. The first step was to train the convolution auto-encoder using the unsupervised learning method. Further, the training of CAE and I-WGAN was performed to fine-tune the architecture and generate high-quality plain fingerprints. An approach based on the integration of Super-Resolution (SR) methods along with GAN was proposed by the author [20]. The probability distribution of fingerprints was estimated using GANs, whereas to ensure fine-grained textures SR method was used. In the first stage, images were generated randomly by estimating the probability distribution of real images using GAN. ESRGAN architecture with Residual-in-Residual Dense Blocks (RRDB) was the core approach utilized to maintain desired quality images. Further, the embedding of minute details and texture of fingerprint images was done in this phase. Using NIST biometric image software (NBIS) preprocessing tool, image segmentation was performed in the segmentation component to get segmented fingerprints.

The approach proposed by Fahim et al., in 2020 was based on the GANs, where a combination of residual network and spectral normalization was used. Spectral normalization helps to ensure the stability of the network while training the model. Further, the issue of vanishing gradient was handled efficiently using the proposed average residual network [21]. This approach utilizes spectral bounding in the input and the fully connected layers. Also, a conditional loss doping technique was introduced to ensure the continuous flow of the gradient. The author aims to extend his research in future to generate stable 512 × 512 fingerprint patches. Some of the major limitations of this approach are—it is challenging to generate large-size images (i.e., above 128 × 128). Further, some redundancy along with some deformities were also observed in the generated output. The core concept utilized by the author was that generator acts as a black box. It simply means that the weights that are generated for the generator are dependent on the discriminator. So an important reason for the poor fidelity of generated output was that the discriminator is performing improper classification of fake images. To ensure this spectral normalization was used in this approach to control the Lipschitz constant of the discriminator [21]. To differentiate their proposition from other similar approaches, the concept of spectral bounding was utilized in dense and input layers of the discriminator only. Further, the approach of residual connections was used to deal with the issue of vanishing gradient. To overcome the problem of static loss generation that is observed in the traditional model, the author has introduced a loss doping approach. They implemented it by introducing loss augmentations on the generator side if similar loss values are observed for two consecutive epochs.

A Clarkson Fingerprint Generator (CFG) was proposed in 2021, which helps in synthesizing full plain impressions of 512×512 pixels [22]. It was based on progressive growth-based GANs. The concept of multi-resolution training was used for the synthesis of fingerprints in this approach. This training of the generator and discriminator component of the GAN starts with lower spatial resolutions. As the training proceeds, the resolution increases progressively, which has the capability of efficiently capturing the high-frequency components of training data. Seidlitz et al., in 2021, proposed a technique in which privacy-sensitive datasets of real images were converted to privacy-friendly datasets of synthesized fingerprints, which helps overcome the legal restriction imposed on the usage of biometric data [23]. It helps in generating anonymous synthetic fingerprints using a data-driven approach instead of the traditional mathematical modeling technique. In the first step segmentation of the fingerprint was performed to get the core points from the patches using the NIST tool MINDTCT [24]. In the second step augmentation on the fingerprint, patches were performed using StirTrace filters [25]. Further, in another paper, the author has proposed a novel hybrid approach for the generation of realistic synthetic fingerprints. First of all, they acquired the dynamic ridge maps with sweat pores using Anguli [26] fingerprint generator. Using these ridge maps they trained their CycleGan model to synthesize realistic-looking fingerprints. The drawback of this approach was that the manual intervention was still unavoidable and cross-sensor matching was also not supported.

The approach proposed by the author comprises two stages which are summarized in Fig. 2. In the first stage, a detailed procedure was adopted to generate multiple instances of fingerprints or seed images. First of all, master fingerprints were generated by using Anguli and SFinGe. This helps in the creation of ridge maps having a random ridge-flow frequency. Further, with the help of real image distributions, they added pores and scratches. Different seed images for identity were created by cutting the master fingerprint according to the displacement distribution of the same person in a real database. In the second stage, these seed images were converted into realistic fingerprints using CycleGAN. Further, these master prints were flipped horizontally to increase seed images to 5920. After following the above procedure random elastic deformations were applied to get 11,840 distorted images [27]. So

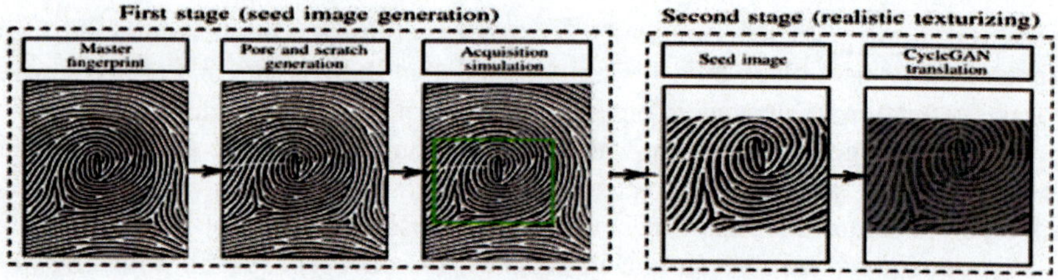

Fig. 2 Flowchart depicting steps for generating synthetic fingerprints using the CycleGan approach [27]

in total, the training seed images obtained after this whole procedure was 23,680 images. Another author proposes a PrintsGAN model which helps in the generation of realistic-looking synthetic fingerprints [28]. The first step in this approach was to generate a binary Master-Print IID using a random noise vector. For the learning of mapping to binary prints, this approach utilizes BigGAN architecture. Further, this IID and warping noise vector was passed to a non-linear TPS warping module and GAN DW to get a warped master print. In the last step, this whole computation was passed to renderer RD to obtain a final fingerprint with textural details.

After having an in-depth discussion regarding the different GAN-based approaches that are available, to the best of our knowledge, Table 1 presents the comparison among these approaches. Parameters like the ratio of input training samples along with their corresponding outputs, and the results generated, would help one to judge the performance of a particular approach.

From the above comparison, it can be concluded that the quality of the input training dataset, number of input images, number of epochs, learning rate, number of layers in an architecture, degree of augmentations, etc., are some of the key factors which govern the performance of a model.

3 Discussion

Synthetic datasets are gaining popularity with the introduction of futuristic machine learning models and architectures. These datasets were created using different generative models. They were defined as the models that ensure the generation of accurate and realistic-looking datasets, by learning from large-scale real datasets. There are different variants of generative models like GANs, Variational Auto Encoders or VAEs, and Autoregressive models. The core difference between autoregressive models and auto-encoding models is at the pre-training stage, the rest of the architecture being the same. A detailed comparison of these models is presented in Table 2. Apart from these deep learning methods, other techniques like Monte Carlo are also available for generating synthetic datasets.

Table 1 Comparison of approaches with results

References	Year	Number of input training images	No. of images generated	Description	Results
[17]	2018	250K fingerprint images from a law enforcement agency were used	12 ms per image	A GAN variant for generating 512 × 512 rolled fingerprint images was proposed	Average NFIQ 2.0 = 62
[18]	2018	Trained two models—1800 images of the FVC2006 fingerprint dataset and 1480 images of PolyU fingerprint databases	16 generated fingerprint images over different epochs (0th, 20th, 40th, 60th, 80th, and 100th)	FVC-2006 fingerprint dataset, and PolyU fingerprint dataset	Frechet inception distance = 70.5
[19]	2020	The model was trained using 280,000 rolled fingerprint images from a law enforcement agency [19] (500 dpi)	100 million fingerprint images	Improved WGAN (I-WGAN) was the backbone of the proposed fingerprint synthesis algorithm	Rank-1 fingerprint search accuracies = 90.35%
[20]	2020	54,000 fingerprint images overall (NIST) in 2009, named special dataset (SD09)	Not specified	GANs and an SR method-based approach were proposed	Classification accuracy of 50.43%
[21]	2020	1000 (real images) captured using Greenbit scanner images	Produce a batch of 36 fake fingerprints in between 4 and 7 s	A deep convolutional generative adversarial network (DCGAN) was used	Multi-scale structural similarity (MS-SSIM) metric = 0.23 for the generated fingerprints
[22]	2021	Trained the model using 72,000 images captured using crossmatch guardian scanner (DB-1)	50K of our synthetic fingerprints	Clarkson fingerprint generator (CFG) was proposed	Mean NFIQ2 score = 54.687

(continued)

Table 1 (continued)

References	Year	Number of input training images	No. of images generated	Description	Results
[23]	2021	NIST special database 27	Progressive GAN generated 266 images and StyleGAN2 191 images	A data-driven approach using GAN was proposed	Bozorth3 scores (s) = 63.53%
[27]	2021	1480 PolyU database (DBII) PG5 relook	7400 images	Obtained dynamic ridge maps with sweat pores by improving the Anguli generator	The equal error rate (EER) (minutiae-based) = 23.84% for real images and 28.12% for synthetic ones
[28]	2022	282K unique fingerprints taken from the MSP longitudinal database used	525K fingerprints (35K distinct fingers, each with 15 impressions)	BigGAN architecture is used to generate synthetic fingerprints	Average realism rating = 2.45 std. dev = 1.47

GANs and VAEs are unsupervised deep learning approaches that generate data by learning from the input data distribution (ignore the density distribution). VAE models learn by comparing their input data with their output. This approach is efficient for learning the hidden data representation. However, due to this approach, the output generated was blurry because of averaging. Hence, this approach is not capable of generating a realistic new sample. On the other hand, GANs are based on an adversarial approach. The discriminator network helps in calculating the similarity/dissimilarity between the generated and the real data. Further, GANs have the capability of being conditioned on different inputs (different classes of data). Therefore, we can conclude that for generating realistic and diverse fingerprint samples GANs are better positioned than VAEs.

4 Performance Evaluation of Synthetic Dataset Generated Through GAN

Holistic evaluation of architecture or validation of its results forms the benchmark for acceptance of any new research model or algorithm. This assessment can be based on a comparison of universally accepted metrics amongst the models or to the ground truth data. In the case of a synthetically generated fingerprint dataset using GANs, we can divide our evaluation into two phases as shown in Fig. 3. In the first stage, we can evaluate the trained model architecture itself. This phase would help us in

Table 2 Comparison of the generative models

Comparison	Generative adversarial networks (GANs)	Variational auto encoders (VAEs)
Components	Simultaneously train two neural networks in an adversarial fashion: a generator and a discriminator	Networks are composed of multiple neural networks: an encoder and a decoder
Working	The generator digests random input from latent distribution and transforms these data points into some other shape without ever directly looking at the original data. The discriminator digests input from the original data and the generator's output, aiming to predict where the input comes from	Two steps working. At first, an encoder network transforms an original complex distribution into a latent distribution. A decoder network then transforms the distribution back to the original space
Input	GANs can be used with discrete latent variables	VAEs can be used with discrete inputs
Advantages	• No need to provide reconstruction error • Works well with unstructured data • To generate realistic new samples of a dataset	• A good approach for solving the transformation problem • Easy to implement and train • Works great for feature extraction • Suitable for compressing data to lower dimensions or generating semantic vectors from it
Disadvantages	• GANs are also more challenging to train than VAEs • GANs require more expertise for training purposes • GANs are prone to the mode collapse phenomenon	• With heterogeneous data, formulation of reconstruction error becomes a challenging task. It sometimes spreads probability mass to unwanted places • As the complexity of the images increases, the image quality decreases
Application	Generate realistic image datasets, image-to-image translation, text-to-image translation, generate new human poses, photo blending and inpainting, 3D object generation, etc.	Used for the reconstruction of input, generating simple synthetic images, text documents, music, etc.

assessing the quality and efficiency of our training process. In the second stage, we can evaluate the quality and the diversity of the data generated.

The performance of generative models, particularly of GANs has improved significantly over time, with the introduction of its different variants and architectures. However, its evaluation remains a challenging task. To the best of our knowledge, there are many parameters for its evaluation, however, none can be labeled as a universally accepted metric. A list of different parameters that are used for evaluating GANs with their analysis is presented [29]. Some of the most widely used metrics are as follows: Inception Score (IS) [30], Fréchet Inception Distance (FID)

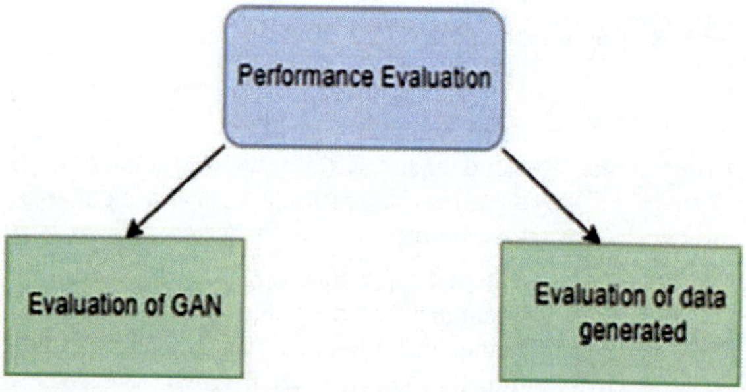

Fig. 3 Two-stage performance evaluation

[31], Mode Score, Maximum Mean Discrepancy (MMD), Birthday Paradox Test, The Wasserstein Critic, etc., [32]. All these methods though give us an insight regarding the model trained, however, they all come with their limitations and disadvantages as mentioned in [32]. In the second phase of the evaluation process, for a synthetically generated dataset, the generated images could be assessed on two broad verticals–the quality of the fingerprint generated and the diversity of the dataset. While evaluating the quality of the images we could focus on the overall quality of an image as well as the nature of features/minutiae generated. For this phase of assessment, the following metrics can be evaluated

1. **Structural Similarity Index (SSIM)**—Structural similarity index is a perception-based model method in which similarity between samples is calculated based on structural information. Structural information refers to the inter-dependent pixels or spatially closed pixels values [33]. It incorporates important human perception-based facts like luminance masking, contrast masking, etc. Luminance masking is the distorted part of an image that is less visible on the edges of an image. Whereas in situations where distortions are less visible in the texture of an image, that term is defined as contrast masking. To calculate it on images, usually, a sliding Gaussian window of size 11×11 or 8×8 is used. It creates a quality map of the image by sliding it pixel-by-pixel. However, in the case of videos, it is suggested to use only subgroups of different window ranges. The range of SSIM lies between -1 and $+1$. It gives a value of 1 in the case where the two images are structurally identical. Hence, the higher the SSIM value, the less the difference between the images. Since SSIM does not satisfy the properties of non-negativity or triangle inequality it is not classified as a distance function. However, it upholds the symmetry properties and the identity of indiscernibles.
2. **Peak Signal-to-Noise Ratio (PSNR)**—PSNR is a measure to compute the quality between the original and the compressed image in decibels. It helps in comparing the images that are reconstructed using different compression algorithms. Further, we can define it as the ratio between the maximum possible value (power) of a signal and the power of distorting noise that affects the quality of its representation. We often use this ratio as a measurement of quality between

the original image and the resultant image. The higher the PSNR, the better the quality of the compressed, or reconstructed image.

3. **NIST Biometric Image Software (NBIS)**—is a standardized software that was developed by NIST for the Federal Bureau of Investigation (FBI) and Department of Homeland Security (DHS) [24]. This software comprises many modules which could help in evaluating the quality of a biometric fingerprint image. First amongst them is the NFIQ score which helps in assigning a quality value between 1 and 5, where 1 represents higher quality [34]. It helps in the numerical calibration of optical and inked 500 PPI fingerprint samples. Further minutiae detector called, MINDTCT could be utilized for automatic detection and marking of fingerprint features. This software helps in evaluating the quality of minutiae points based on local image quality. Matching of fingerprints can be efficiently conducted (both 1 to 1 and 1 to many) by the BOZORTH3 algorithm.
4. **VerifingerSDK**—It is a widely used SDK for fingerprint identification purposes [35]. This system helps in reliable matching (both 1 to 1 and 1 to many) which can help in assessing the diversity of fingerprints generated. Further, it also can detect spoofed and tempered fingerprints. With its added feature of tolerance to translation, rotation and deformation, its reliability and accuracy are increased manifolds as compared to other approaches.
5. **Other Metrics**—Some other standard metrics like True detection rate RT, False detection rate RF, Identification rate (%), Rank-based accuracy, Linear index of fuzziness, Equal error rate, etc., are the other metrics that could help us in analyzing the fingerprint sample efficiently and accurately.

5 Conclusions

In this work, the existing GAN architectures which are available for the generation of synthetic fingerprints were evaluated. This would help the community to generate their own anonymous large-scale fingerprint dataset with realistic-looking features, avoiding time-consuming manual collection. Further, it would encourage us for carrying out research without any hassle of privacy concerns. A brief discussion on performance evaluation metrics would help in quantifying the results accurately. This large-scale dataset would assist researchers to apply upcoming deep learning architectures, which may help in improving the efficiency and reliability of varied biometric recognition systems.

References

1. Gonog L, Zhou Y (2019) A review: generative adversarial networks. In: 2019 14th IEEE conference on industrial electronics and applications (ICIEA), pp 505–510
2. Zhu J-Y, Park T, Isola P, Efros A (2017) Unpaired image-to-image translation using cycle-consistent adversarial networks, pp 2242–2251

3. Zhou K, Diehl E, Tang J (2023) Deep convolutional generative adversarial network with semi-supervised learning enabled physics elucidation for extended gear fault diagnosis under data limitations. Mech Syst Signal Process 185:109772
4. do Lago CAF, Giacomoni MH, Bentivoglio R, Taormina R, Gomes MN Jr, Mendiondo EM (2023) Generalizing rapid flood predictions to unseen urban catchments with conditional generative adversarial networks. J Hydrol 618:129276
5. Golovianko M, Terziyan V, Branytskyi V, Malyk D (2023) Industry 4.0 vs. industry 5.0: co-existence, transition, or a hybrid. Procedia Comput Sci 217:102–113
6. Vega-Márquez B, Rubio-Escudero C, Nepomuceno-Chamorro I (2022) Generation of synthetic data with conditional generative adversarial networks. Logic J IGPL 30(2):252–262
7. Bird JJ, Naser A, Lotfi A (2023) Writer-independent signature verification; evaluation of robotic and generative adversarial attacks. Inf Sci 633:170–181
8. Zhu Q-S, Zhang J, Zhang Z-Q, Dai L-R (2023) A joint speech enhancement and self-supervised representation learning framework for noise-robust speech recognition. IEEE/ACM Trans Audio Speech Lang Process
9. Caterino J, Clark J, Yohannan JC (2019) Analysis of synthetic cannabinoids on paper before and after processing for latent print using DFO and ninhydrin. Forensic Sci Int 305:110000
10. NIST Dataset. https://www.nist.gov/itl/iad/image-group/nist-special-database-2727a. Accessed: 01.09.2019
11. WVU Dataset. https://databases.lib.wvu.edu/. Accessed: 01.09.2019
12. FVC Dataset. http://bias.csr.unibo.it/fvc2004/download.asp. Accessed: 01.09.2019
13. IIIT Delhi Dataset. http://www.iab-rubric.org/resources.html. Accessed: 01.09.2019
14. Sankaran A, Vatsa M, Singh R (2015) Multisensor optical and latent fingerprint database. IEEE Access 3:653–665
15. Tsinghua Dataset. http://ivg.au.tsinghua.edu.cn/dataset/TLOFD.php. Accessed: 01.09.2019
16. ELFT-EFS Public Challenge Database. https://www.nist.gov/itl/iad/image-group/nist-evaluationlatent-fingerprinttechnologies-extended-feature-sets-elft-efs. Accessed: 01.09.2019
17. Cao K, Jain A (2018) Fingerprint synthesis: evaluating fingerprint search at scale, pp 31–38
18. Minaee S, Abdolrashidi A (2018) Finger-GAN: generating realistic fingerprint images using connectivity imposed GAN
19. Mistry V, Engelsma J, Jain A (2019) Fingerprint synthesis: search with 100 million prints
20. Riazi S, Chavoshian S, Koushanfar F (2020) SynFi: automatic synthetic fingerprint generation
21. Fahim MA-NI, Jung HY (2020) A lightweight GAN network for large scale fingerprint generation. IEEE Access 8:92918–92928
22. Bahmani K, Plesh R, Johnson P, Schuckers S, Swyka T (2021) High fidelity fingerprint generation: quality, uniqueness, and privacy
23. Seidlitz S, Jürgens K, Makrushin A, Kraetzer C, Dittmann J (2021) Generation of privacy-friendly datasets of latent fingerprint images using generative adversarial networks, pp 345–352
24. NIST. https://www.nist.gov/services-resources/software/nist-biometric-image-software-nbis. Accessed: 01.09.2019
25. Hildebrandt M, Dittmann J (2015) Stirtracev2.0: enhanced benchmarking and tuning of printed fingerprint detection. IEEE Trans Inf Forensics Secur 10:833–848
26. Anguli. https://dsl.cds.iisc.ac.in/projects/Anguli/. Accessed: 01.09.2019
27. Wyzykowski ABV, Segundo MP, de Paula Lemes R (2021) Level three synthetic fingerprint generation. In: 2020 25th International conference on pattern recognition (ICPR). IEEE, pp 9250–9257
28. Cao K, Jain A (2018) Fingerprint synthesis: evaluating fingerprint search at scale. In: 2018 International conference on biometrics (ICB). IEEE, pp 31–38
29. Borji A (2019) Pros and cons of GAN evaluation measures. Comput Vis Image Underst 179:41–65
30. Barratt S, Sharma R (2018) A note on the inception score
31. Gulrajani I, Ahmed F, Arjovsky M, Dumoulin V, Courville A (2017) Improved training of wasserstein GANs

32. Wang Z, Bovik A, Sheikh H, Simoncelli E (2004) Image quality assessment: from error visibility to structural similarity. IEEE Trans Image Process 13:600–612
33. Niguidula J, Batinggal G (2007) Comparative study on the image compression of photo sharing sites using mean square error, peak signal to noise ratio and data-rate savings
34. NFIQ 2. https://www.nist.gov/services-resources/software/nfiq-2. Accessed: 01.09.2019
35. VeriFinger. https://www.neurotechnology.com/verifinger.html. Accessed: 01.09.2019

A Comparative Analysis of Data Augmentation Techniques for Human Disease Prediction from Nail Images

S. Marulkar, B. Narain, and R. Mente

Abstract Using vast amounts of data can enhance the effectiveness of machine learning algorithms and prevent overfitting issues. It might be difficult and time-consuming to gather a lot of training data for a disease prediction model in the health industry. Without the requirement to gather additional data, data augmentation approaches can broaden the range of data related to training, to overcome this problem. Various methods for enhancing images using deep learning, include color alteration, Neural Style Transfer (NST), image rotation, image cropping, PCA color augmentation, generative adversarial networks (GANs), noise injection, image rotation as well as flipping techniques were used in this study to create enhanced nail image datasets. Modern transfer learning techniques were used to assess the success rate of data expansion approaches, and then the results demonstrated that the supplemented dataset produced by NST as well as GAN methods had higher precision than the primary dataset. The blended approach of artificial intelligence, color, as well as orientation augmentation performed the best across all datasets.

Keywords Deep learning · Data augmentation · Neural style transfer · Medical image processing · Generative adversarial networks

1 Introduction

In health and agriculture image classification and detection, the deep convolutional neural network (DCNN) stays crucial [1]. However, it requires a vast amount of training data and significant computational resources. The positive aspect is that improvements in Graphical Processing Units (GPUs) have taken care of these demands for high-performance computation.

S. Marulkar (✉) · B. Narain
MATS School of IT, MATS University, Raipur, CG, India
e-mail: shwetamarulkar2021@gmail.com

R. Mente
School of Computational Sciences, PAH Solapur University, Solapur, India

S. Kumar et al. (eds.), *Fourth Congress on Intelligent Systems*, Lecture Notes in Networks and Systems 868, https://doi.org/10.1007/978-981-99-9037-5_30

A dataset can be expanded using data augmentation techniques without the need for additional data collection. The original training data is used to create new data using both simple and sophisticated picture alteration techniques. The most popular techniques include image flipping, cropping, rotation, color modification, and noise injection. These can be combined with one or more formatting techniques to produce enhanced images [2, 3].

GANs and NST are advanced image transformation techniques that leverage deep neural networks to generate new augmented images from existing data in the dataset [4, 5]. A generative model called a GAN uses a generator and a discriminator to generate distinct data during training. In an effort to trick the discriminator, which contrasts the created and original data, the generator creates new data. When the discriminator's classification accuracy has been drastically decreased, training is complete [6].

DCGANs are unsupervised learning algorithms that use CNNs to construct both the generator and discriminator [7]. They are commonly used in computer vision applications to produce distinct images and exhibit superior performance compared to traditional GAN networks [8, 9].

The widely used GAN-based learning method known as Wasserstein GANs (WGANs) differs from DCGAN in that it computes loss values during training using a Wasserstein loss function and a function of linear activation in the output layer [10].

Deep neural networks are used in Neural Style Transfer (NST), a complex image alteration method, to manipulate images [11]. The NST creates output images by combining the content and design patterns from a style image and an input content picture [12, 13]. This proposed method is used to efficiently produce augmented data for developing applications.

The paper is structured as follows: Sect. 2 contributes an in-depth analysis of deep learning-based data augmentation methods. While Sect. 3 explains the nail image dataset used for illness identification, Sect. 4 describes how to employ basic image editing methods to build an enhanced dataset, and Sect. 5 describes how to use deep learning-based methods for enhancing data. The conclusion as well as next research are presented in Sect. 7, which is followed by a performance comparison of the derived datasets in Sect. 6.

2 Literature Survey

This survey's main goals are to exemplify the several deep learning approaches used to create intensified images and to look into the significance of the escalation process.

1. A liver lesion classification technique that made use of GAN data augmentation was introduced by the authors in [14]. The system produced a sensitivity of 78.6% and a specificity of 88.4% using the supplemented dataset. However, the classification system achieved a sensitivity of 85.7% and a specificity of 92.4%

by employing GAN data augmentation, which is better than that of the original dataset.

2. In their paper [15], the authors introduced a novel GAN design that enhanced output picture quality during translation between images. The Cycle-Consistency loss function was used by this design, also known as CycleGAN, to stabilize the training of the network.
3. The authors of [16] proposed a different design, the Progressively Growing GANs, which trains the networks using a progressive resolution complexity. With each iteration, the model is trained with higher resolutions. The resolution range, for instance, starts with an input size of 4 by 4 and goes up toward 8 by 8, and so on, until an output determination of 1024 by 1024 is reached.
4. The efficiency of various loss functions for the process of generation as well as discrimination was compared by the authors in [17]. The authors concluded that when the hyperparameters were optimized, the performance of the majority of the loss functions was comparable.
5. The authors in [18] employed GANs to generate authentic data from their simulated data. To prevent class imbalance problems, they created an augmented dataset using CycleGAN, as described in [19], which consisted of a facial expression database that could identify seven distinct emotions. Additionally, [20] evaluated the effectiveness of CycleGANs in recognizing facial expressions for the same task.
6. Conditional GANs is a different GAN architecture that the authors introduced in [21] for picture enhancement. To address concerns about mode collapse, these GANs use a specified scalar in the discriminator as well as generator. The limitations and potential applications of GANs in health imaging applications are covered by the authors in [22]. Similar to this, in order to increase performance, the authors in [23] suggested a liver lesion classification model employing GAN-based data augmentation. In [24], the author suggested a DCNN model for the categorization of liver lesions utilizing DGAN-based data augmentation.
7. The authors of [25] outline the difficulties and disadvantages of enhancing data using GANs through training. Meanwhile, [11] introduces a neural network-based technique called NST, which transfers the manner in which a visual to the input image. Visual Turing Test used to evaluate generated image [26]. For MRI datasets of brain tumors, the authors in [27] developed augmented pictures based on DCGAN and WGAN. The authors use a number of performance indicators to assess the generated photos' quality.
8. The authors of [28] have developed an NST method for making photographs of the real environment. They train using instance normalization rather than batch normalization, which leads to better performance than the traditional NST method. Additionally, [29] has introduced the quickest NST algorithm for generating augmented images. However, the NST algorithm's downside is that it has restricted style transfer.

3 Original Nail Image Dataset

Initial dataset of nail images, which comprises 18,025 color images depicting 09 various classes. The dataset consists of 09 distinct nail images based on color feature. However, the image count per class is uneven, indicating an imbalanced dataset. For instance, the black color nail image class has 2612 images, whereas the purple color nail image class only has 356 images. The classification model's effectiveness may be impacted by the imbalanced dataset. Figure 1 represents samples from original dataset.

Techniques for data augmentation are used to rectify the dataset's imbalance. The strategies for addressing data imbalance issues that are based on basic image alteration are covered in the section that follows. The original dataset's imbalanced structure is depicted in Fig. 2.

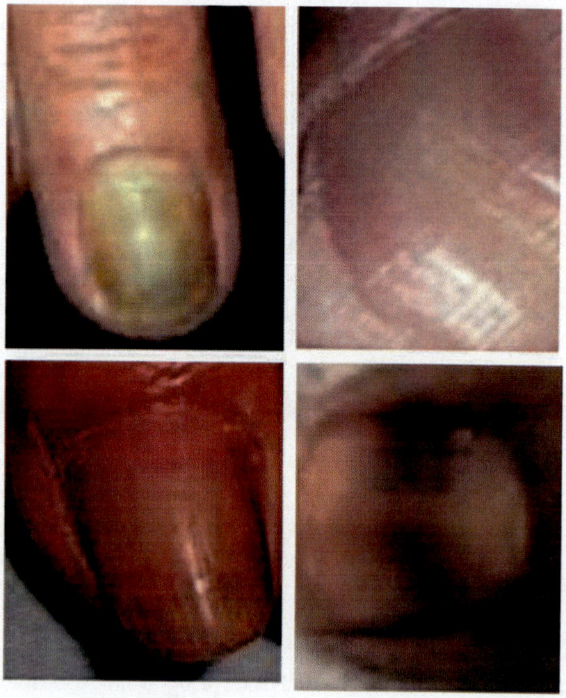

Fig. 1 A sample of the original range of data

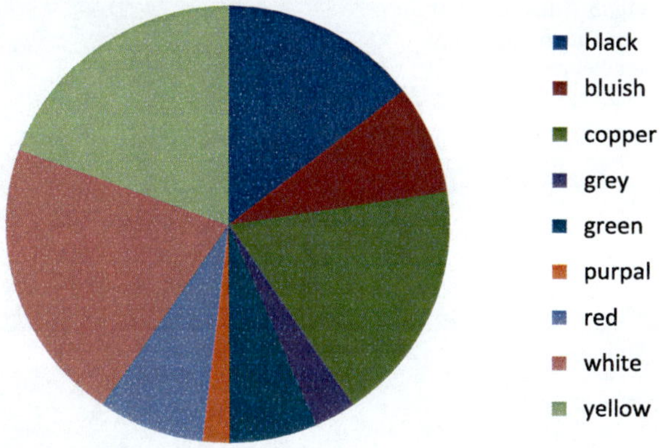

Fig. 2 Original dataset's organizational scheme

4 Augmenting Data Through Fundamental Manipulative Methods

A majority used picture alteration techniques are image noise injection, principal component analysis (PCA), image flipping, image rotation, color transformation, and image cropping. The 36,122 enhanced photos were created using simple image alteration techniques, balancing the size of each category with 2000 visuals from the nail image dataset. Examples of augmented photos created using these fundamental image alteration techniques are shown in Fig. 3. The expanded dataset, which includes 54,147 pictures.

5 Augmenting Data Through Deep Learning Methods

Neural Style Transfer, Wasserstein GAN, and Deep Convolutional Generative Adversarial Networks are the majorly utilized data enhancement method based on deep learning. The conventional WGAN as well as DCGAN were trained using the model of deep learning having batch sizes of 64 with 1000 epochs, respectively.

DCGAN and WGAN were employed to produce 36,122 augmented images, effectively increasing the dimensions of the original set of data while balancing each class size to 2000 images. Figure 4 displays instances of augmented visuals created through DCGAN and WGAN.

5132 images are generated by utilizing the Neural Style Transfer (NST) amplification method, which involves employing a convolutional neural network for constructing the Neural Style Transfer model. Constructed model is then proficient through 3000 training iterations with a mini batch length is 64.

The each class size was standardized to 1000 images by producing an expanded dataset of 42,000 images, attaining equilibrium. The sample at random produced

Fig. 3 Illustration of the basic manipulation of augmented data

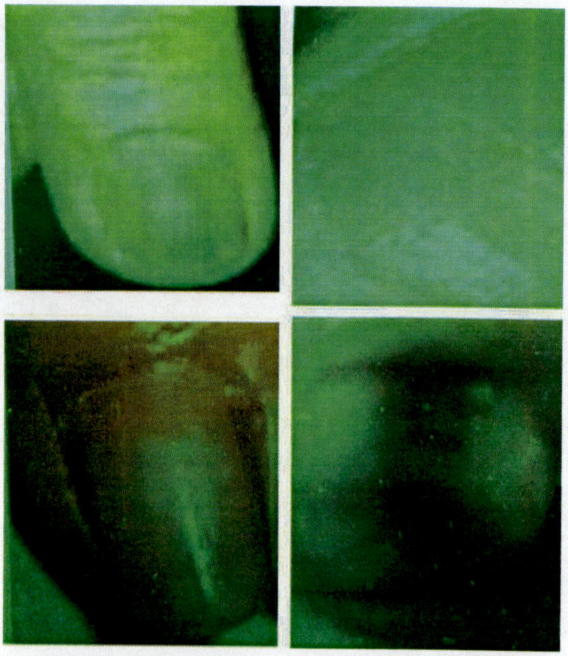

Fig. 4 Sample amplified data by DCGAN and WGAN

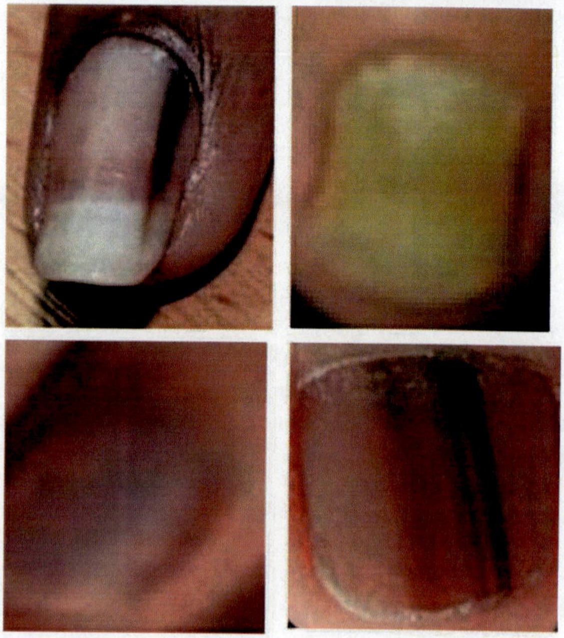

```
import tensorflow as tf
from tensorflow.keras.models import Sequential
from tensorflow.keras.layers import Dense, Reshape, Flatten, Conv2D, Conv2DTranspose
from tensorflow.keras.optimizers import RMSprop
import numpy as np
def build_generator(input_dim, output_shape):
def build_critic(input_shape):
input_dim = 100
output_shape = (64, 64, 3)
generator = build_generator(input_dim, output_shape)
critic = build_critic(output_shape)
critic.compile(optimizer=RMSprop(0.00005), loss=wasserstein_loss)
def train_wgan(generator, critic, epochs, batch_size):

  for epoch in range(epochs):
    for _ in range(batch_size):
      generated_images = generator.predict(np.random.normal(0, 1, (batch_size,) + (input_dim,)))
      critic_loss_real = critic.train_on_batch(real_images, -np.ones((batch_size, 1)))
      critic_loss_fake = critic.train_on_batch(generated_images, np.ones((batch_size, 1)))
        weights = layer.get_weights()
        weights = [np.clip(w, -0.01, 0.01) for w in weights]
        layer.set_weights(weights)
      noise = np.random.normal(0, 1, (batch_size,) + (input_dim,))
      generator_loss = combined.train_on_batch(noise, -np.ones((batch_size, 1)))
      print(f"Epoch {epoch}, Critic Loss Real: {critic_loss_real}, Critic Loss Fake: {critic_loss_fake},
Generator Loss: {generator_loss}")
batch_size = 64
epochs = 3000
train_wgan(generator, critic, epochs, batch_size)
num_augmented_samples = 1000
augmented_data = generator.predict(np.random.normal(0, 1, (num_augmented_samples,) + (input_dim,)))
```

Fig. 5 Sample code

using the enlarged dataset and the Neural Style Transfer method are depicted in Fig. 6.

Figure 5 displays sample code used for implementation of proposed method WGAN (Fig. 6).

6 Results and Discussion

Initially, utilizing the original dataset and a variety of cutting-edge transfer learning algorithms, including ResNet, VGG16, and InceptionV3 the models for diagnosing diseases using nail photos were built.

The dataset was divided into three sets: training, validation, and testing, with relative shares of 90, 5, and 5%. Figure 7 shows the classification accuracy of the models using the original dataset.

Additionally, the pre-trained models were trained with similar hyperparameter values using the dataset that had been enhanced using simple picture modification techniques. The test results showed that the enhanced dataset performed better for classification than the original dataset.

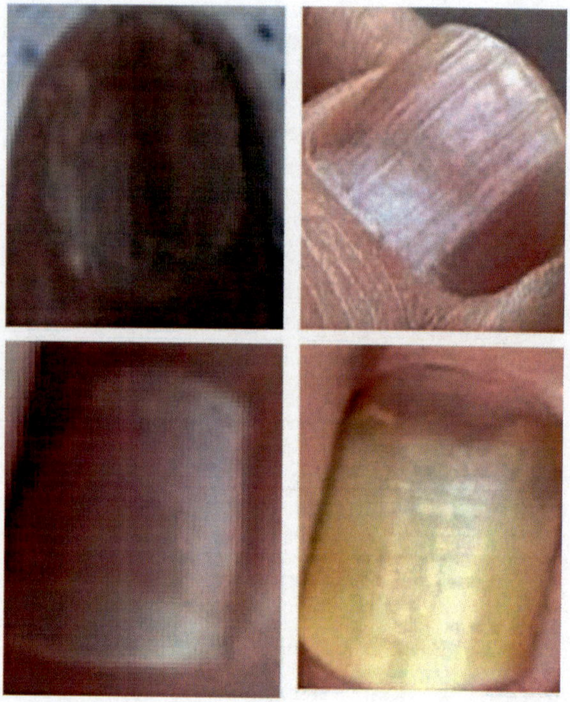

Fig. 6 (NST) sample enhanced data

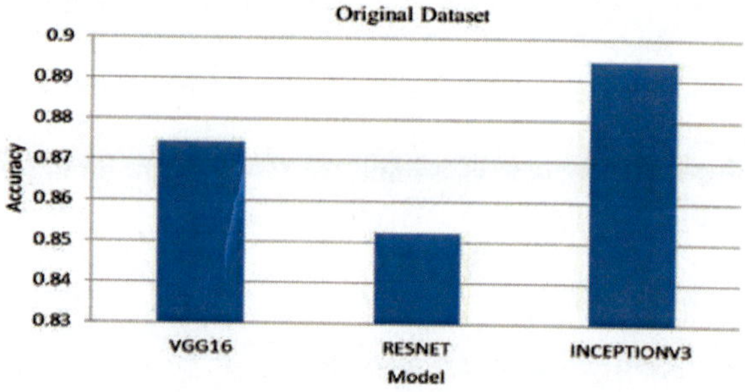

Fig. 7 Accuracy evaluations of various transfer learning techniques

The enactment of various models utilizing the augmented dataset created through basic image manipulation techniques is illustrated in Fig. 8.

Additionally, cataloging prototypes using the expanded DCGAN as well as WGAN datasets were trained using the same hyperparameter settings. These models were then assessed using testing images that had never been seen before. The testing findings showed that cataloging prototypes using the supplemented DCGAN as well as WGAN datasets outperformed models using baseline datasets. Figure 9 shows how well these models performed.

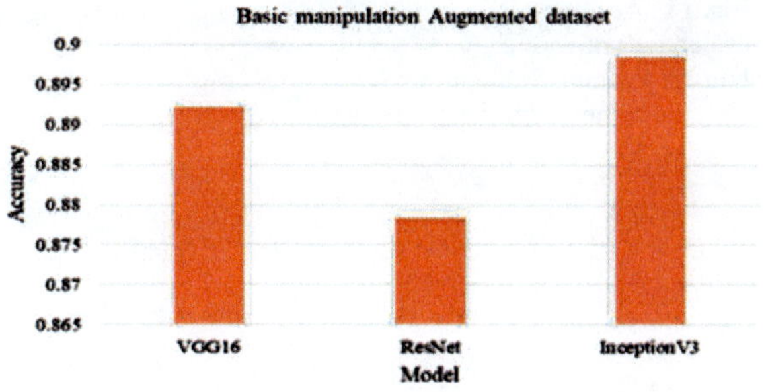

Fig. 8 Accuracy evaluations of various transfer learning techniques with augmented dataset

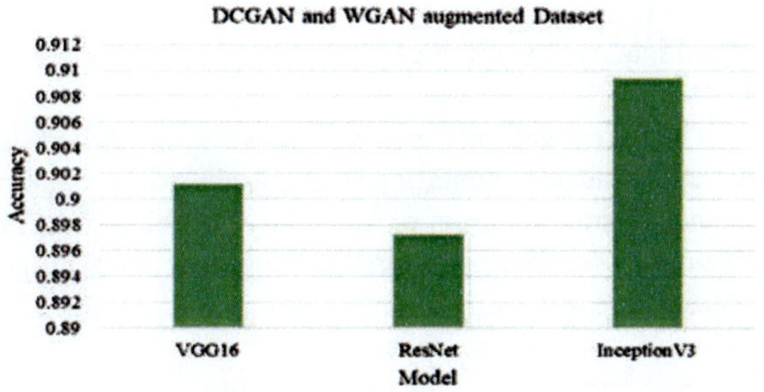

Fig. 9 Accuracy evaluations of various transfer learning techniques with DCGAN in addition to WGAN augmented dataset

The NST enhanced pre-trained models were trained on a dataset in a manner similar to that. The testing outcomes showed that the NST enhanced dataset performed better in terms of precision than both dataset for simple modification and the unique dataset. The NST supplemented dataset, however, did not perform as well as the DCGAN and WGAN enhanced datasets. The dataset's testing results are shown in Fig. 10.

By combining all of the augmentation methods—basic manipulation, DCGAN and WGAN, and NST—a merged dataset was ultimately produced [30]. This dataset was 114,244 bytes large. Compared to other datasets, the pre-trained prototypes employing this database showed considerably improved classification accuracy. The

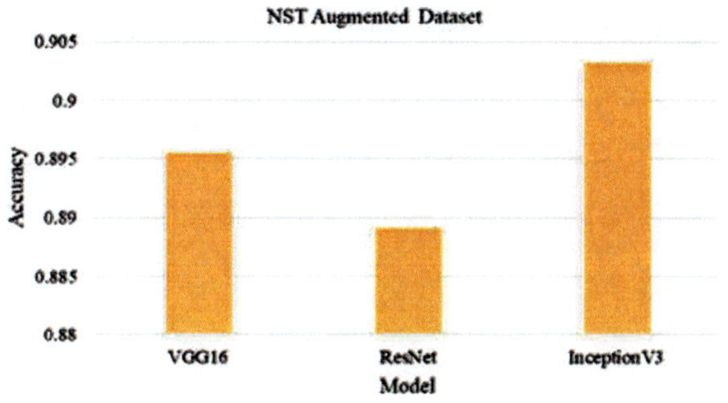

Fig. 10 Accuracy evaluations of various transfer learning techniques with the NST augmented dataset

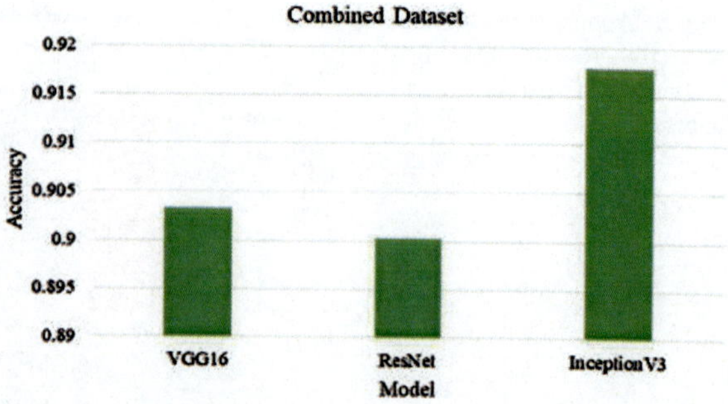

Fig. 11 Accuracy evaluations of various transfer learning techniques with the combined dataset

classification accuracy of the pre-trained prototypes using this dataset is shown in Fig. 11.

The investigational results shows the combined dataset produced by combining different data augmentation strategies performs better than either the original dataset or the separate datasets produced by combining different augmentation approaches having 91% accuracy. Furthermore, deep learning-based datasets outperformed datasets created using simple manipulation methods in terms of performance.

7 Conclusion

In order to solve the classification difficulties associated with disease prediction utilizing nail images, this study suggests four unique image enhanced databases. The DCGAN and WGAN augmented datasets, Elementary image manipulation augmented dataset, NST augmented dataset, and a collective augmented dataset are some of the datasets in this group. Modern transfer learning techniques were trained on these datasets to categorize human diseases. The extensive simulation results provide three key findings that are crucial for creating a powerful model for predicting human disease. First, class size balanced enhanced datasets perform better than the original dataset in terms of classification accuracy. Second, augmentation strategies based on deep learning work better than manipulation techniques. Finally, a combination of datasets based on various augmentation approaches produces results that are superior than those of all other datasets. The nail image dataset may eventually incorporate other picture enhancement methods like feature space augmentation, autoencoders, adversarial training, meta-learning approaches, and random erasure.

References

1. Pandian JA, Geetharamani G, Annette B (2019) Data augmentation on plant leaf disease image dataset using image manipulation and deep learning techniques. In: 2019 IEEE 9th international conference on advanced computing (IACC). IEEE, pp 199–204
2. Stephan RR, Vibhav V, Stefan R, Vladlen K (2016) Playing for data: ground truth from computer games. In: European conference on computer vision (ECCV)
3. Marius C, Mohamed O, Sebastian R, Timo R, Markus E, Rodrigo B, Uwe F, Stefan R, Bernt S (2016) The cityscape dataset for semantic urban scene understanding. In: CVPR
4. Alireza M, Jonathon S, Navdeep J, Ian G, Brendan F (2015) Adversarial autoencoders. arXiv preprint
5. Yanghao L, Naiyan W, Jiaying L, Xiaodi H (2017) Demystifying neural style transfer. arXiv preprint
6. Khizar H (2017) Super-resolution via deep learning. arXiv preprint
7. Pandian JA, Geetharamani G, Annette B (2019) Data augmentation on plant leaf disease image dataset using image manipulation and deep learning techniques. In: 2019 IEEE 9th international conference on advanced computing (IACC), pp 199–204
8. Geetharamani G, Pandian A (2019) Identification of plant leaf diseases using a nine-layer deep convolutional neural network. Comput Electr Eng 76:323–338
9. Radford A, Metz L, Chintala S (2015) Unsupervised representation learning with deep convolutional generative adversarial networks. arXiv preprint arXiv:1511.06434
10. Arjovsky M (2017) Soumith Chintala a Léon Bottou. Wasserstein GAN
11. Gatys LA, Ecker AS, Bethge M (2015) A neural algorithm of artistic style. arXiv preprint arXiv:1508.06576
12. Philip TJ, Amir AA, Stephen B, Toby B, Boguslaw O (2018) Style augmentation: data augmentation via style randomization. arXiv eprints
13. Josh T, Rachel F, Alex R, Jonas S, Wojciech Z, Pieter A (2017) Domain randomization for transferring deep neural networks from simulation to the real world. arXiv preprint
14. Frid-Adar M, Klang E, Amitai M, Goldberger J, Greenspan H (2018) GAN-based data augmentation for improved liver lesion classification
15. Zhu JY, Park T, Isola P, Efros AA (2017) Unpaired image-to-image translation using cycle-consistent adversarial networks. In: Proceedings of the IEEE international conference on computer vision, pp 2223–2232
16. Karras T, Aila T, Laine S, Lehtinen J (2017) Progressive growing of GANs for improved quality, stability, and variation. arXiv preprint arXiv:1710.10196
17. Mario L, Karol K, Marcin M, Olivier B, Sylvain G (2018) Are GANs created equal? A large-scale study. arXiv preprint
18. Ashish S, Tomas P, Oncel T, Josh S, Wenda W, Russ W (2017) Learning from simulated and unsupervised images through adversarial training. In: Conference on computer vision and pattern recognition
19. Goodfellow IJ, Erhan D, Carrier PL, Courville A, Mirza M, Hamner B, Cukierski W, Tang Y, Thaler D, Lee DH et al (2013) Challenges in representation learning: a report on three machine learning contests. In: NIPS. Springer, Berlin, pp 117–124
20. Xinyue Z, Yifan L, Zengchang Q, Jiahong L (2017) Emotion classification with data augmentation using generative adversarial networks, vol abs/1711.00648. CoRR
21. Mehdi M, Simon O (2014) Conditional generative adversarial nets. arXiv preprint
22. Swee KL, Yi L, Ngoc-Trung T, Ngai-Man C, Gemma R, Yuval E (2018) DOPING: generative data augmentation for unsuper-vised anomaly detection with GAN. arXiv preprint
23. Brostow GJ, Julien F, Roberto C (2008) Semantic object classes in video: a high-definition ground truth database. Pattern Recogn Lett 30(2):88–97
24. Maayan F-A, Idit D, Eyal K, Michal A, Jacob G, Hayit G (2018) GAN based synthetic medical image augmentation for increased CNN performance in liver lesion classification. arXiv preprint

25. Tim S, Ian G, Wojciech Z, Vicki C, Alec R, Xi C (2016) Improved techniques for training GANs. arXiv preprint
26. Van der Maaten L, Hinton G (2008) Visualizing data using t-SNE. J Mach Learn Res 9(11)
27. Changhee H, Hideaki H, Leonardo R, Ryosuke A, Wataru S, Shinichi M, Yujiro F, Giancarlo M, Hideki N (2011) GAN-based synthetic brain MR image generation. In: 2018 IEEE 15th international symposium on biomedical imaging (ISBI 2018). IEEE
28. Dmitry U, Andrea V, Victor L (2016) Instance normalization: the missing ingredient for fast stylization. arXiv preprint
29. Johnson J, Alahi A, Fei-Fei L (2016) Perceptual losses for real-time style transfer and super-resolution. In: Computer vision—ECCV 2016: 14th European conference, Amsterdam, The Netherlands, Proceedings, Part II 14. Springer International Publishing, pp 694–711
30. Arun Pandian J, Geetharamani G (2019) Data for: identification of plant leaf diseases using a 9-layer deep convolutional neural network. Mendeley Data, v1

Prediction of Ionospheric TEC Using RNN During the Indonesia Earthquakes Based on GPS Data and Comparison with the IRI Model

R. Mukesh, Sarat C. Dass, S. Kiruthiga, S. Mythili, M. Vijay, K. Likitha Shree, M. Abinesh, T. Ambika, and Pooja

Abstract Total electron content (TEC) is a significant descriptive measure for the ionosphere of the earth. Due to either the sun's activity like solar flare or the positive hall effect caused during earthquake (EQ), the oxygen atoms of the ionosphere split into oxygen ions and electrons increasing the electron content in the ionosphere which causes a rise in the TEC value, thus causing the delay in the signals coming from the satellite to the earth. TEC is associated with the Sun's parameter and geomagnetic indices. In this research, parameters such as planetary K and A-index (Kp and Ap), Radio flux at 10.7 cm (F10.7), Sunspot number (SSN), and IONOLAB true TEC values were collected for the BAKO IGS network station situated in Indonesia (− 6.45° N, 106.85° E) for predicting TEC variations during EQ days occurred in the years 2004 and 2012. A total of three months of TEC data from the BAKO station during the years 2004 and 2012 were used for the developed Recurrent Neural Network (RNN) model in order to predict the TEC before and after the EQ days. For the year 2004, the model has an average Root Mean Square Error (RMSE) and Correlation Coefficient (CC) of 6.79 and 0.90. Also, for the year 2012, during April it has the average RMSE and CC of 8.90 and 0.94. For the same year in August month, the model has the average RMSE and CC of 8.70 and 0.94. The performance of the

R. Mukesh (✉) · M. Vijay
Department of Aerospace Engineering, ACS College of Engineering, Bangalore, India
e-mail: vsmprm@gmail.com

S. C. Dass
School of Mathematical & Computer Sciences, Heriot-Watt University, Putrajaya, Malaysia

S. Kiruthiga
Department of ECE, Saranathan College of Engineering, Trichy, India

S. Mythili
Department of ECE, PSNA College of Engineering and Technology, Dindigul, India

K. Likitha Shree · M. Abinesh · T. Ambika · Pooja
Department of Aeronautical Engineering, ACS College of Engineering, Bangalore, India

S. Kumar et al. (eds.), *Fourth Congress on Intelligent Systems*, Lecture Notes in Networks and Systems 868, https://doi.org/10.1007/978-981-99-9037-5_31

model is also evaluated using linear regression scatter plot. The Pearson's R value calculated from the scatter plot is 0.92, shows that the model has good correlation with the true TEC.

Keywords TEC · BAKO · IONOLAB · RNN

1 Introduction

The TEC is the total number of electrons available along a route between a space-based transmitter and ground receiver. In other words, the number of electrons existing in the ionosphere [1]. TEC is affected due to solar activity and calamities such as solar flares, ultraviolet radiation, geomagnetic storms, EQs, and tsunamis [2–4]. The EQ occurred due to discharge of energy which produces waves that travel in all directions. Earth has many crust levels, the tectonic plates are constantly moving slowly and when they get trapped at their edges because of friction, they produce shock waves which affect the TEC in the ionosphere. TEC anomalies in the ionosphere due to EQ is shown in Fig. 1. Various research was carried out on TEC anomalies during the EQs. In one research work, the authors analyzed the perturbations caused in the ionosphere during two EQ's occurred on 24.6.19 and 19.8.18 in Indonesia. The results show that before EQs there was a good correlation between TEC anomaly and Ozone data obtained from 3 satellites [5]. Dogan et al. [6] monitored the TEC variations on 17.8.99 EQ that happened at Marmara, Turkey by using the GPS data collected from MAGNET. The authors seen that, 3 days before the EQ, TEC anomalies were observed [6]. In another research work, the authors carried out the TEC variations during the EQs that happened on 7.11.2009 and 8.12.2010 in India. They found that the ExB drift and ground waves caused before the EQs caused the TEC anomalies [7]. Three EQs that happened in Indonesia during 25.1.2014, 12.2.2016, and 17.2.2016 were analyzed by the authors based on the TEC anomalies estimated using the GPS data. The results show that TEC anomalies correlated with 2 EQs namely Kebumen and Sumba and it is not matching with Halmahera EQ [8]. Five GNSS receivers were used for estimating the TEC values during the 28.9.2018 EQ and its effect tsunami that occurred over Indonesia. The authors used ray tracing and beamforming methods to find the TTID location and found that ionospheric TEC anomalies can be used to forecast the occurrence of EQ and tsunamis [9]. TEC variations over the Indonesia were analyzed based on the data obtained from DORIS station during the 26.12.2004 EQ [10]. TEC variations before 20 EQs happened during the years 2011–2015 were analyzed based on the data collected from six GNSS stations located in the low-latitude and mid-latitude regions [11]. Ionospheric TEC anomalies were analyzed over the Himalaya region during 5 EQs-based of the data taken from 15 GNSS stations [12]. TEC variations were examined during the 30.10.2020 EQ happened over Turkey according to the data taken from GNSS receivers and RIM-TEC and CODE [13]. Variations in the GPS TEC values were observed based on the ML techniques before the main shocks of Pakistan (Mirpur and Awaran) EQ events [14]. TEC variations were investigated during the 01.04.2015 and 25.04.2015 EQs happened in Himalaya and Nepal based on the GNSS data [15]. Variations in TEC during the 2015 EQs occurred in Nepal were investigated by the authors based on the data collected from GPS and GIMs [16]. Responses of TEC and Rn related to EQs were examined from 2007 to 2009 EQs by using five AI models and found that Rn changes are more related to the seismic event [17]. PCA-NN forecast model was developed to predict the TEC in mid-latitudes [18]. Nearest neighbor method was used by the authors to predict the global TEC during the years 2015 and 2018 [19]. A RNN is a type of neural network (NN), is utilized to forecast TEC to identify potential EQs. This network is made up of interconnected layers and is designed to mimic the structure and functionality of the human brain. Neural networks, including RNNs, are trained using complex algorithms and a large amount of data. The RNN works by using the output of a specific layer as input for predicting future output. To overcome these issues in RNN, Long Short-Term Memory (LSTM) is utilized, which is a unique kind of RNN having the ability of learning long-term dependencies and retaining information for longer periods. LSTMs have a chain-like structure and consist of four layers instead of a single layer [20, 21]. 3 seismic regions Japan, Turkey, and Morocco were considered by the authors for location-dependent EQ forecasting. *K*-means algorithm and Long Short-Term Memory Networks (LSTM) and GRU-based RNN were used to forecast the EQ in advance [22]. Previous studies have explored TEC prediction using neural networks, SWIF, SPM, ARMA, and other techniques [23, 24]. But no significant research has been conducted to predict EQs using TEC anomalies based on AI models for short-term forecasting. In this research the RNN model is adapted and trained using parameters related to ionospheric seismology in order to predict the TEC before the EQ events to act as an EQ precursory indicator. The validation of the proposed model is done by comparing the predicted TEC with International Reference Ionosphere.

2 Methodology

The methodology used in the research work is given as a flow chart in Fig. 2. In this research, we have considered BAKO station which is located in Indonesian low-latitude at − 6.45° N latitude and 106.85° E longitude for analyzing the TEC variations during the three EQs that occurred on 26th of December 2004, 11th of April 2012, and 18th of August 2012. The prediction of TEC is executed by using an AI-based Recurrent Neural Network (RNN) model that uses solar parameters, geomagnetic indices, and TEC data as an input. The RNN model uses previous one month of data as a training data in order predict the TEC values for the next date. The statistical evaluation parameters like Root Mean Square Error (RMSE), Correlation Coefficient (CC), Mean Absolute Error (MAE), and Mean Absolute Percentage Error (MAPE) have been calculated to analyze the performance of the model [25, 26].

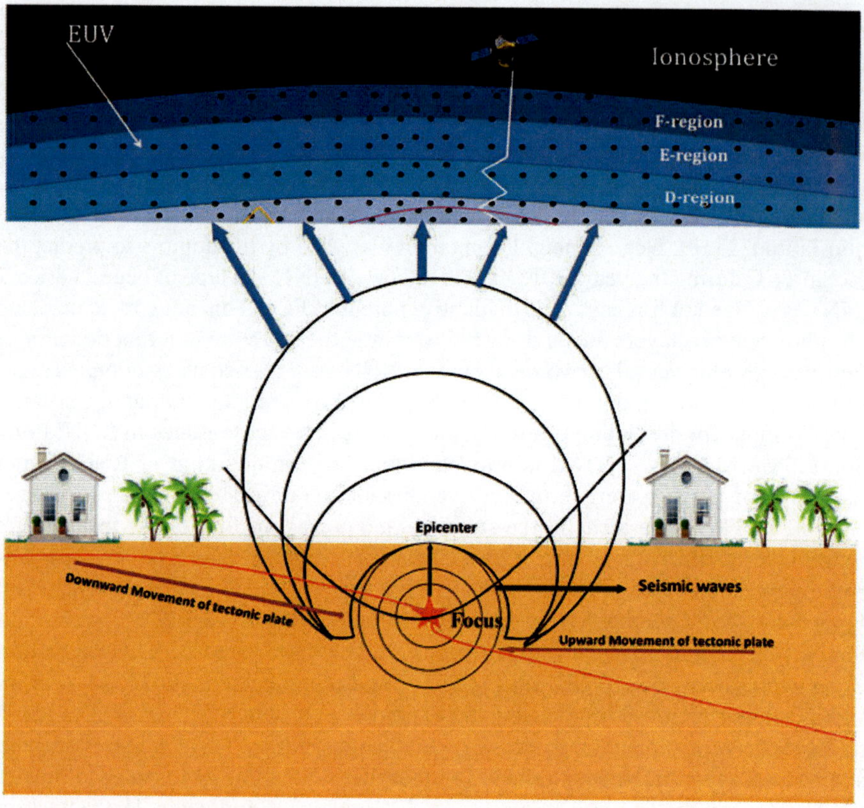

Fig. 1 TEC anomalies in ionosphere due to EQ

2.1 Data Collection

The TEC, solar, and geomagnetic parameters for the BAKO station during the EQ periods were collected from various open source networks. TEC data is collected from the IONOLAB server (http://www.ionolab.org) at a resolution of 576 values of TEC per day and it is down sampled into 24 values per day using the MATLAB program. Solar parameters like SSN, F10.7, and geomagnetic parameters like Ap and Kp indices were collected from OMNIWEB data servers (https://omniweb.gsfc.nasa.gov/).

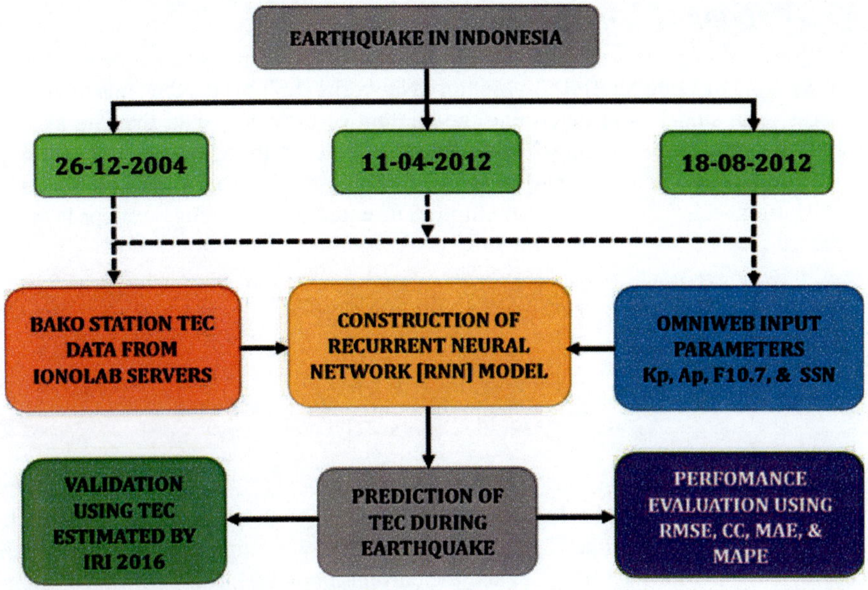

Fig. 2 Flow chart for TEC prediction and comparison

2.2 *Working of Recurrent Neural Network*

The fundamental principle of RNN involves accumulating the output of a certain layer and using it as input to forecast the output of that layer. Nodes in several layers of the NN are combined to one layer of RNNs. The input layer is denoted as "A" and the hidden layer is represented by "h". At each time step "t", the present input is a blend of $A(t)$ and $A(t - 1)$ were utilized to enhance the output. The output at a specified time is then fed back into the network. The process involves taking information from both the current and previous cell, applying the hyperbolic tangent function to it, and passing the output along with the previous information to the next cell, where the same operations are repeated. In the neural network, X is taken as an input layer, passed to the hidden layer, and processing is done there. In the hidden layer, it consists of numerous hidden layers, each with its own weights, biases, and activation functions. The RNN is standardized by these parameters hence each hidden layer will have identical parameters. For RNN's, we have utilized the backpropagation algorithm which has the input as time series data so the inputs from the past and present will be used during the execution.

2.3 Working of LSTM in RNN

The working of Long Short-Term Memory Networks (LSTM) in RNN has three step process. The step-1 is to determine the amount of historical data that the system should retain in memory. Step-2 is to determine which data to let through according to its significance in the present time step and finally step-3: Determine which section of the current cell state is utilized as output. The equation governing the steps is given in Eqs. 1–3.

$$f(t) = \gamma\left(y_f.[n_{t-1}, x_t] + C_f\right) \tag{1}$$

$$\begin{aligned} H_t &= \gamma(y_i.[n_{t-1}, x_t] + d_i) \\ C_t &= \tanh(y_c.[n_{t-1}, x_t] + d_c), \end{aligned} \tag{2}$$

where H_t is the input gate.

The initial stage in the LSTM involves determining which information to exclude from the cell at a given time. This decision is made by the sigmoid function, which considers the earlier state $n(t - 1)$ and the current input $X(t)$, and performs a calculation (C_f). One of the steps in LSTM is to determine the extent to which the current state should be modified. The second layer of the LSTM unit consists of two parts: the sigmoid function and tanh function. The sigmoid function determines whether a particular value should be allowed to pass through (0 or 1). The tanh function governs the level of importance of the values that are allowed to pass, assigning them a value between − 1 and 1. In the third step, which determines the output. The part of the cell state that makes it to the output is decided by the sigmoid layer and the tanh function is used to push the values between − 1 and 1.

$$\begin{aligned} O_t &= \gamma(w_t.[h_{t-1}, x_t] + b_o), \\ H_t &= O_t * \tanh(C_t) \\ O_t &= \text{Output gate.} \end{aligned} \tag{3}$$

3 Result and Discussion

3.1 Prediction of TEC During the August 2004 EQ

The TEC prediction is performed by using RNN over the Indonesia region which is prone to frequent EQs. In this section, the TEC was predicted during the August 2004 EQ and Fig. 3 depicts the comparison of true versus RNN predicted TEC during the 2004 EQ. The RNN model's performance during 2004 EQ is evaluated by parameters such as the RMSE, CC, MAE, and MAPE given in Table 1. The TEC is predicted for two days before, during and two days after the EQ. The EQ occurred on 26-12-2004 in Indonesia. Figure 3 shows a clear assessment of true TEC and forecasted TEC for the BAKO station by using GPS data. In Fig. 3 the predicted TEC values on 24-12-2004 and during that day there was not much changes between the true and predicted TEC values. The RMSE and CC values of the RNN model during that day are 4.24 and 0.96 which shows that the model provides good accuracy. The predicted TEC for 25-12-2004 and 26-12-2004, which is a day before and on the day of EQ shows that during those days our model is performing moderately having RMSE and correlation coefficient values of 9.25, 9.82, and 0.82, 0.84, respectively. The sudden variation in TEC values is due to an EQ that caused sudden disturbances in the ionosphere. The predicted TEC values on 27-12-2004, which is after the day of the EQ. On that day, there is a very small changes between the true and predicted TEC values and the model have RMSE and correlation coefficient of 3.86 and 0.97. From Fig. 3 we can observe that, the later part of the 26-12-2004 EQ event day have drastic TEC variations. Within the observation period, the TEC variations can be seen as alternating peaks and dips at multiple times, even though the exact pattern of variation is not followed by the RNN model, the range of the variation is successfully captured.

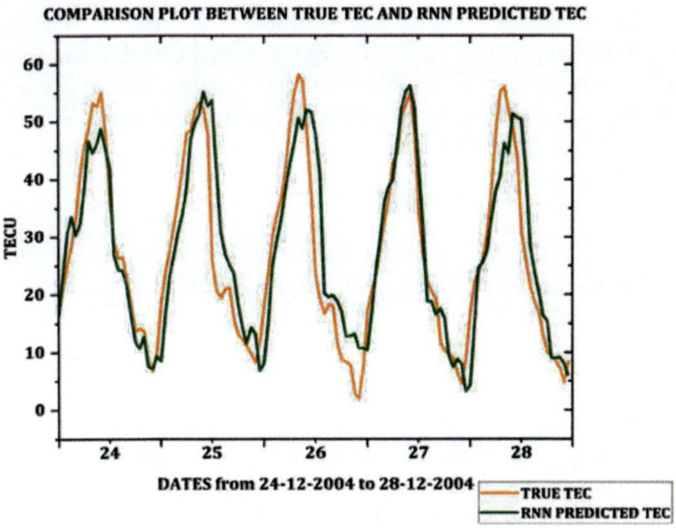

Fig. 3 Prediction of TEC from 24-12-2004 to 28-12-2004

Table 1 Performance analysis of RNN during the 2004 EQ

Date	RMSE	CC	MAE	MAPE
24-12-2004	4.24	0.96	3.39	0.11
25-12-2004	9.25	0.82	6.98	0.33
26-12-2004	9.82	0.84	7.86	0.64
27-12-2004	3.86	0.97	3.25	0.21
Average	6.79	0.90	5.37	0.32

3.2 *Prediction of TEC During the April 2012 EQ*

Figure 4 represents and compare the true with the predicted TEC values for the BAKO station. The EQ occurred on 11-04-2012 in Indonesia. In Fig. 4. TEC prediction for 09-04-2012 is shown where a large difference is observed between the true TEC and predicted TEC, leading to an RMSE of 10.08 and CC of 0.92, indicating that the model is not performing accurately. The TEC predictions during 10-04-2012, 11-04-2012, and 12-04-2012 show that the model performed well on 10-04-2012, with an RMSE of 7.28 and CC of 0.96. On the day of the EQ (11-04-2012) and the next day (12-04-2012), the model's performance was moderate by having two days average RMSE and CC of 9.11 and 0.95, respectively. The performance analysis of the RNN model for the 2012 April EQ is presented in Table 2. The performance of the RNN model during the April 2012 EQ was established by analyzing Fig. 4, the model was able to capture all the maximum TEC within the observed period without any problem. The smaller variation that occurred during the time period of late afternoon and early night on dates such as 09-04-2012 to 11-04-2012 is well correlated by the model.

3.3 *Prediction of TEC During the August 2012 EQ*

The EQ occurred on 20-08-2012 in Indonesia. Here we have analyzed the TEC values two days before, during, and after the EQ day. Figure 5 shows the assessment of true and forecasted TEC for the BAKO station. Table 3 shows the performance analysis of the RNN model for the 2012 August EQ. In Fig. 5 we can see that the predicted TEC on days such as 18-08-2012 and 19-08-2012, our model was predicting well while having the two days average RMSE and CC values of 8.09 and 0.94. In Fig. 5 TEC values were predicted for 20-08-2012 and 21-08-2012 show that our model is predicting moderately with an average RMSE and CC values of 10.40 and 0.95. The prediction on 18-08-2012 two days before the EQ event date is comparatively closer to that of the true TEC and the pattern of prediction during the EQ event date is very close to the true TEC. From Fig. 5, it can be seen that the true TEC slowly lowered

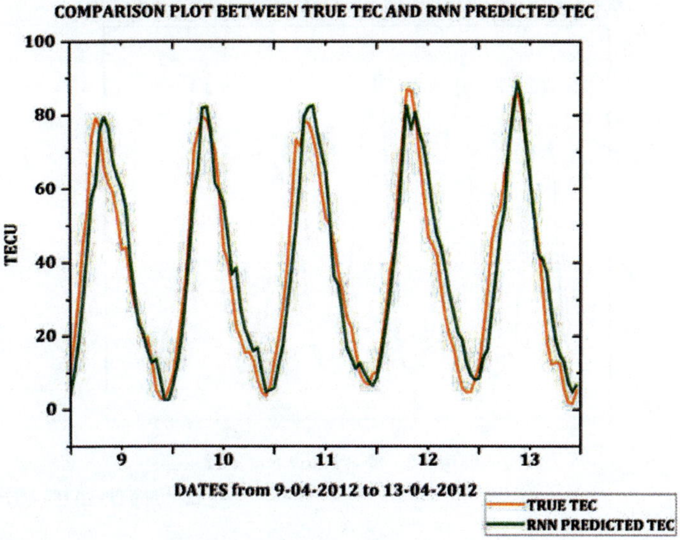

Fig. 4 Prediction of TEC from 09-04-2012 to 13-04-2012

Table 2 Performance analysis of RNN during the 2012 EQ

Date	RMSE	CC	MAE	MAPE
09-04-2012	10.08	0.92	8.32	0.35
10-04-2012	7.28	0.96	6.15	0.30
11-04-2012	9.26	0.94	7.58	0.23
12-04-2012	8.96	0.95	7.83	0.43
Average	8.90	0.94	7.47	0.33

from the high to low TEC values as observed on the event date. It is notable that the model was able to predict those variations with data from previous days and during the EQ day high difference in the maximum TEC shows area of improvements in the model.

3.4 Comparison of RNN with IRI 2012

The RNN and IRI 2012 results were compared to evaluate the TEC prediction accuracy of RNN. For EQ dates, the forecasted TEC using RNN at BAKO station was compared with IRI 2012. Error metrics like RMSE, CC, MAE, and MAPE were used for assessing the performance of the RNN model with the IRI 2012 model during the three EQ dates. Figure 6 shows a comparison plot of the RNN and IRI model for

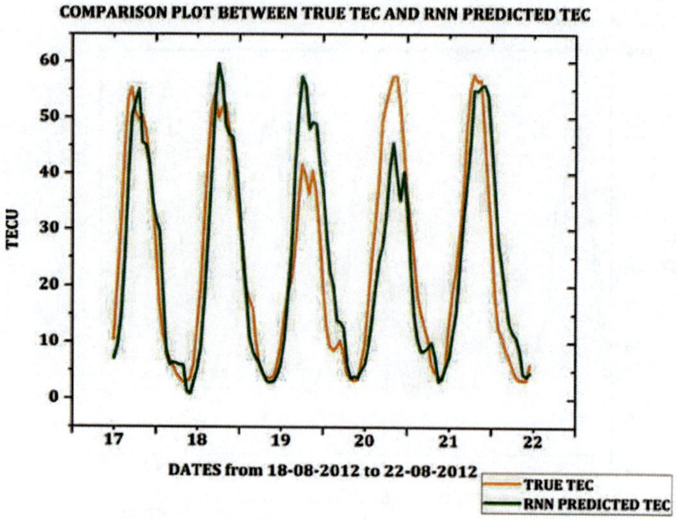

Fig. 5 Prediction of TEC from 18-08-2012 to 22-08-2012

Table 3 Performance analysis of RNN during 2012 EQ

Date	RMSE	CC	MAE	MAPE
18-08-2012	7.51	0.92	5.35	0.34
19-08-2012	8.67	0.95	4.77	0.21
20-08-2012	10.14	0.95	8.12	0.46
21-08-2012	10.67	0.95	8.53	0.32
Average	9.25	0.94	6.69	0.33

the December 2004, April 2012, and August 2012 EQ. According to the comparison results, the RNN TEC prediction plot more closely resembles the actual TEC than the IRI 2012 results. Based on the comparison results given in Table 4, the average RMSE value of RNN is 9.74 TECU, the average CC is 0.91, MAE is 7.85, and the MAPE is 0.44. But the RMSE value of the IRI model is 11.30, the CC is 0.87, the MAE is 8.43, and the MAPE is 0.38. This result indicates that RNN performed better than IRI 2012 during the EQs that happened in Indonesia. A linear regression scatter plot has been plotted between true and RNN predicted TEC values which is shown in Fig. 7. From this figure we can see that the proposed model can predict TEC during EQ with high correlation to the true TEC and very low residuals. The error histogram plot is shown in Fig. 8. The histogram is plotted in MATLAB Neural Network Training module based upon Levenberg–Marquardt back propagation with random data-division. The performance estimation of error is based on Mean Square Error (MAE). The total error range of the model is divided into twenty vertical bars

(20 bins). The histogram plotted using RNN predicted TEC w.r.t true TEC showed errors ranging from − 20.7 to 25.6.

The performance metrics of the RNN model evaluated from the linear regression scatter plot is given in Table 5. The major elements of evaluation are Pearson's R, R-Squared–Coefficient of Determination (COD) and Adjusted R-square. The Pearson's R is having a value of 0.92031 which is closer to 1 showing that the model has good correlation with the system. The remaining two parameters R-square COD and the adjusted R-square have a value of 0.84696 and 0.84654, respectively. These values show the distribution of residuals in the predicted intervals which needs to be improved.

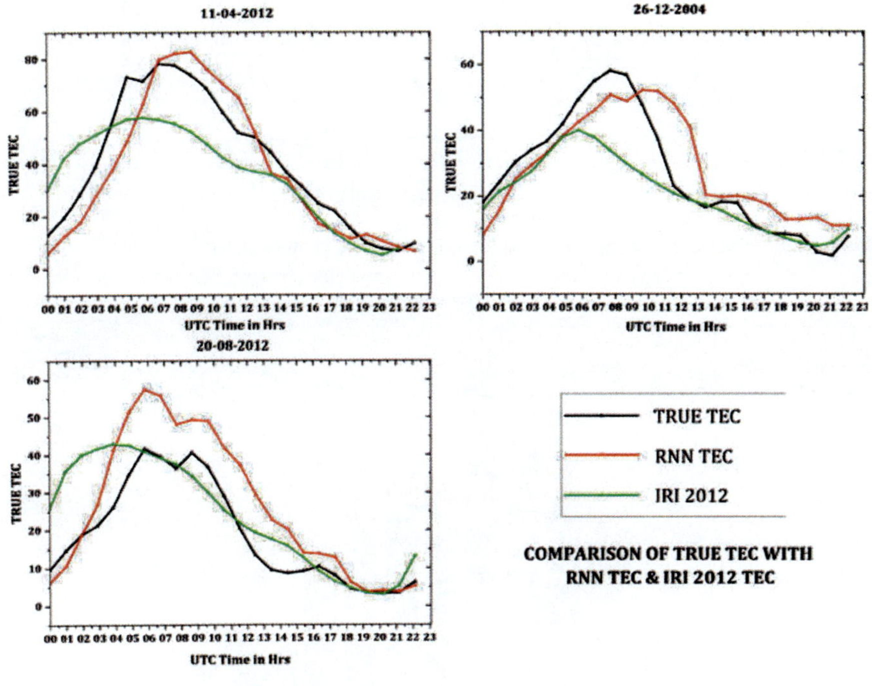

Fig. 6 Comparison of RNN & IRI plot for 2004 EQ

Table 4 Comparison of RNN prediction error metrics with IRI model

Date	RMSE		CC		MAE		MAPE	
	RNN	IRI 2012	RNN	IRI 2012	RNN	IRI 2012	RNN	IRI 2012
26-12-2004	9.82	10.31	0.84	0.93	7.86	6.75	0.64	0.30
11-04-2012	9.26	13.91	0.94	0.86	7.58	11.84	0.23	0.36
20-08-2012	10.14	9.67	0.95	0.81	8.12	6.71	0.46	0.47
Average	9.74	11.30	0.91	0.87	7.85	8.43	0.44	0.38

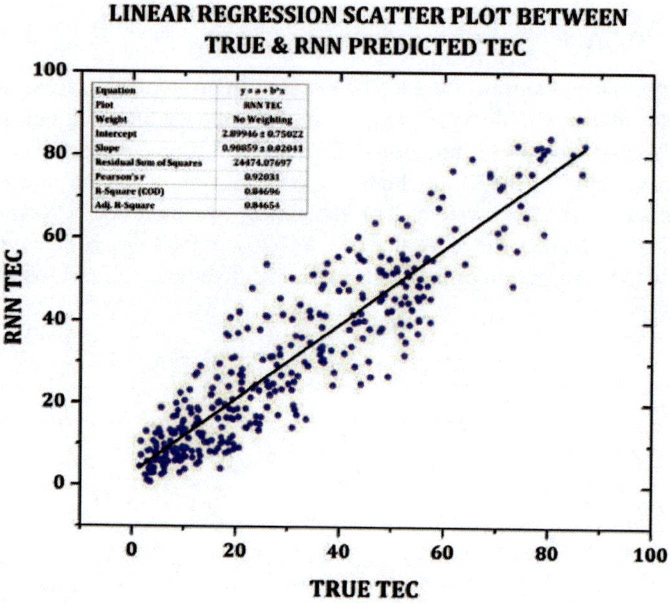

Fig. 7 Linear regression scatter plot between true and RNN predicted TEC

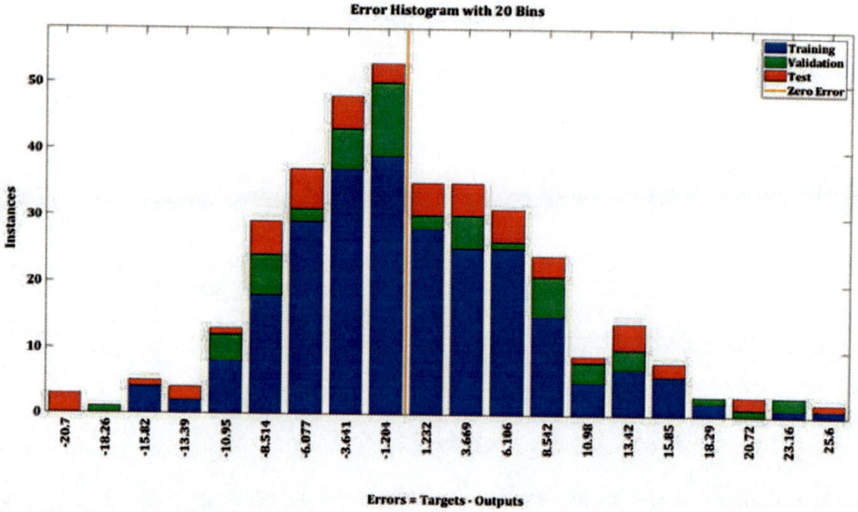

Fig. 8 Error histogram plot

Table 5 Performance of the model evaluated from linear regression scatter plot

Equation	$y = a + b * x$
Intercept	2.89946 ± 0.75022
Slope	0.90859 ± 0.02041
Residual sum of squares	24,474.07697
Pearson's *R*	0.92031
R-square (COD)	0.84696
Adj. *R*-square	0.84654

4 Conclusion

In this paper, we have developed and used the Recurrent Neural Network model based on Artificial Intelligence system to predict the TEC during three EQs that occurred on 26-12-2004, 11-04-2012, and 18-08-2012 in Indonesia. Our model has predicted the TEC during the major EQ that occurred on 26-12-2004 which has a magnitude of 9.1–9.3 Mw. On this day the average RMSE and CC are 6.79 and 0.97. During the EQ that occurred on 11-04-2012 which has a magnitude of 8.6 Mw, our model predicted the TEC with an average RMSE and CC of 8.90 and 0.94. Similarly, during the EQ that occurred on 18-08-2012 which has a magnitude of 6.3 Mw, our model predicted the TEC with an average RMSE and CC of 7.66 and 0.93. From the above three EQs, the RMSE varies from 6.79 to 8.90 and CC varies from 0.93 to 0.97. Finally, we have compared our model results with IRI 2012 and found that the performance parameter results of RNN were better than the IRI model. The linear regression scatter plot comparing the RNN predicted TEC with the true TEC shows that the proposed model has good correlation for ionospheric seismological application. The Pearson's *R* value of 0.92031 and the closely followed values of *R*-square and adjusted R-square clearly signifies the good performance of RNN during EQ. In summary, the proposed RNN model has good correlation with the ionospheric seismological prediction and will be useful to predict the TEC anomalies before the EQ events in order to alert the people living in that area.

Acknowledgements The research work presented in this paper has been carried out under the Project ID "VTU RGS/DIS-ME/2021-22/5862/1," funded by VTU, TEQIP, Belagavi, Karnataka.

Competing Interests The authors declare that there are no financial and non-financial competing interests with respect to the current research work. The authors confirm that there are no known conflicts of interest associated with this publication.

References

1. Shubin VN, Gulyaeva TL (2022) Global mapping of total electron content from GNSS observations for updating IRI-Plas model. Adv Space Res 69(1):168–175. https://doi.org/10.1016/j.asr.2021.09.032

2. Rajana SSK, Shrungeshwara TS, Chiranjeevi G. Vivek, Sampad Kumar Panda, Sridevi Jade., "Evaluation of long-term variability of ionospheric total electron content from IRI-2016 model over the Indian sub-continent with a latitudinal chain of dual-frequency geodetic GPS observations during 2002 to 2019. Adv Space Res 69(5):2111–2125. https://doi.org/10.1016/j.asr.2021.12.005
3. Tsagouri I, Koutroumbas K, Elias P (2018) A new short-term forecasting model for the total electron content storm time disturbances. J Space Weather Space Clim 8:A33. https://doi.org/10.1051/swsc/2018019
4. Davies K, Hartmann GK (1997) Studying the ionosphere with the global positioning system. Radio Sci 32(4):1695–1703. https://doi.org/10.1029/97RS00451
5. Kumar S, Singh PK, Kumar R et al (2021) Ionospheric and atmospheric perturbations due to two major earthquakes ($M > 7.0$). J Earth Syst Sci 130:149. https://doi.org/10.1007/s12040-021-01650-x
6. Dogan U, Ergintav S, Skone S et al (2011) Monitoring of the ionosphere TEC variations during the 17th August 1999 Izmit earthquake using GPS data. Earth Planet Sp 63:1183–1192. https://doi.org/10.5047/eps.2011.07.020
7. Singh OP, Chauhan V, Singh B (2013) GPS-based total electron content (TEC) anomalies and their association with large magnitude earthquakes occurred around Indian region. Indian J Radio Space Phys 42:131–135
8. Sunardi B et al (2018) IOP Conf Ser Earth Environ Sci 132:012014
9. Liu JY, Lin CY, Chen YI et al (2020) The source detection of 28 September 2018 Sulawesi tsunami by using ionospheric GNSS total electron content disturbance. Geosci Lett 7:11. https://doi.org/10.1186/s40562-020-00160-w
10. Li F, Parrot M (2006) Total electron content variations observed by a DORIS station during the 2004 Sumatra-Andaman earthquake. J Geodesy 80:487–495. https://doi.org/10.1007/s00190-006-0053-9
11. Eshkuvatov HE, Ahmedov BJ, Tillayev YA, Arslan Tariq M, Ali Shah M, Liu L (2023) Ionospheric precursors of strong earthquakes observed using six GNSS stations data during continuous five years (2011–2015). Geodesy Geodyn 14(1):65–79. https://doi.org/10.1016/j.geog.2022.04.002
12. Joshi S, Kannaujiya S, Joshi U (2023) Analysis of GNSS data for earthquake precursor studies using IONOLAB-TEC in the Himalayan region. Quaternary 6(2):27. MDPI AG. https://doi.org/10.3390/quat6020027
13. Basciftci F, Bulbul S (2022) Investigation of Ionospheric TEC changes potentially related to Seferihisar-Izmir earthquake (30 October 2020, M_W 6.6). Bull Geophys Oceanogr 63(3)
14. Shah M, Shahzad R, Ehsan M, Ghaffar B, Ullah I, Jamjareegulgarn P, Hassan AM (2023) Seismo ionospheric anomalies around and over the epicenters of Pakistan earthquakes. Atmosphere 14(3):601. MDPI AG. Retrieved from https://doi.org/10.3390/atmos14030601
15. Sharma G, Champati Ray PK, Mohanty S, Kannaujiya S (2017) Ionospheric TEC modelling for earthquakes precursors from GNSS data. Quatern Int 462:65–74. https://doi.org/10.1016/j.quaint.2017.05.007
16. Ulukavak M, Inyurt S (2020) Seismo-ionospheric precursors of strong sequential earthquakes in Nepal region. Acta Astronaut 166:123–130. https://doi.org/10.1016/j.actaastro.2019.09.033
17. Muhammad A, Külahcı F, Birel S (2023) Investigating radon and TEC anomalies relative to earthquakes via AI models. J Atmos Solar-Terr Phys 245:106037. https://doi.org/10.1016/j.jastp.2023.106037
18. Morozova A, Barata T, Barlyaeva T, Gafeira R (2023) Total electron content PCA-NN prediction model for South-European middle latitudes. Atmosphere 14(7):1058. MDPI AG. https://doi.org/10.3390/atmos14071058
19. Monte-Moreno E, Yang H, Hernández-Pajares M (2022) Forecast of the global TEC by nearest neighbour technique. Remote Sens 14(6):1361. MDPI AG. https://doi.org/10.3390/rs14061361
20. Pulvirenti L, Rolando L, Millo F (2023) Energy management system optimization based on an LSTM deep learning model using vehicle speed prediction. Transp Eng 11:100160. https://doi.org/10.1016/j.treng.2023.100160

21. Drewil GI, Al-Bahadili RJ (2022) Air pollution prediction using LSTM deep learning and metaheuristics algorithms. Meas Sens 24:100546. https://doi.org/10.1016/j.measen.2022.100546
22. Berhich A, Belouadha F-Z, Kabbaj MI (2022) A location-dependent earthquake prediction using recurrent neural network algorithms. Soil Dyn Earthquake Eng 161:107389. https://doi.org/10.1016/j.soildyn.2022.107389
23. Song R, Zhang X, Zhou C, Liu J, He J (2018) Predicting TEC in China based on the neural networks optimized by genetic algorithm. Adv Space Res 62(4):745–759. https://doi.org/10.1016/j.asr.2018.03.043
24. Cesaroni C, Spogli L, Aragon-Angel A, Fiocca M, Dear V, De Franceschi G, Romano V (2020) Neural network-based model for global total electron content forecasting. J Space Weather Space Clim 10:11. https://doi.org/10.1051/swsc/2020013
25. Mukesh R, Karthikeyan V, Soma P et al (2020) Forecasting of ionospheric TEC for different latitudes, seasons and solar activity conditions based on OKSM. Astrophys Space Sci 365:13. https://doi.org/10.1007/s10509-020-3730-x
26. Sivavadivel K, Shunmugam M, Raju M et al (2022) Influence of input parameters for prediction of GPS and IRNSS TEC by using OKRSM at Hyderabad stations during solar flare event. Acta Geophys 70:429–443. https://doi.org/10.1007/s11600-021-00712-4

Deep Learning Model for Diagnosing the Severity of Diabetic Retinopathy

Nikitha Reddy Nalla and Ganesh Kumar Chellamani

Abstract Diabetes Mellitus, a condition that affects diabetic individuals, causes diabetic retinopathy. In many nations, it is the leading cause of blindness. A person's likelihood of going blind increases if they have diabetic retinopathy. The failure to properly treat and monitor diabetic retinopathy before it progressed to a severe stage was the root cause of blindness in 90% of instances. In its extreme stages, diabetic retinopathy has no known treatment. However, it can be identified and avoided in its early stages. Consequently, computerized computer diagnosis will help clinicians to find diabetic retinopathy early and more affordably. In health informatics, deep learning is gaining importance. This study uses different deep learning models to create and compare, from which DenseNet169 has been identified as the best deep learning model to categorize the disease into different stages. It may help clinicians to determine the severity of diabetic retinopathy more effectively.

Keywords Deep learning · Diabetic retinopathy · Image processing · Fundus images

1 Introduction

A particularly significant eye consequence of diabetes is diabetic retinopathy. It is caused by excessive blood sugar levels that damage the retinal blood vessels. In its early stages, it could not show symptoms or only minor vision problems; however, later, it can result in blindness. According to NHS Inform, anyone with type 1 or type 2 diabetes has a risk of contracting diabetic retinopathy, and it is estimated that one-third of diabetics have the condition. It is concerning that most people never get their eyes checked for diabetic retinopathy due to a lack of awareness and education.

N. R. Nalla · G. K. Chellamani (✉)
Department of Electronics and Communication Engineering, Amrita School of Engineering, Chennai, Amrita Vishwa Vidyapeetham, Chennai, India
e-mail: c_ganeshkumar@ch.amrita.edu

N. R. Nalla
e-mail: nikitha2241@gmail.com

S. Kumar et al. (eds.), *Fourth Congress on Intelligent Systems*, Lecture Notes in Networks and Systems 868, https://doi.org/10.1007/978-981-99-9037-5_32

Diabetes patients must be aware of their risk factors and see an ophthalmologist to receive treatment for any early signs of retinopathy and prevent additional vision loss.

Diabetic retinopathy (DR) must be detected and treated as soon as possible to reduce adverse effects [1]. The DR diagnosis primarily relies on the competence of a skilled medical professional who can assess retinal images. This time-consuming process, however, delays treatment, particularly in developing nations without the essential resources. Deep learning approaches have demonstrated the ability to automate retinal image analysis, increasing the reach and effectiveness of DR screening programs.

This project aims to create a deep learning model that can detect the severity of diabetic retinopathy from a person's retinal photograph. Previous studies have found that convolutional neural networks (CNNs) are effective methods of classifying DR [1]. In medical practice, the severity of diabetic retinopathy is gauged by analyzing signs and symptoms that include changes in blood vessels, the presence of hard exudates, and the appearance of cotton wool spots. The model will implement multiple classification tasks to correctly identify the severity of DR by recognizing five stages: (0) No DR, (1) Mild DR, (2) Moderate DR, (3) Severe DR, and (4) Proliferative DR. This will enable the necessary medical treatment to be applied accurately.

2 Related Works

The identification and classification of diabetic retinopathy have been built into intelligent systems using a variety of methodologies. This primary topic studies the analysis and evaluation of several publications. Some models are highly accurate but insensitive, whereas others are inaccurate but sensitive. In paper [2], the authors integrate transfer learning [16] intending to extract features for diabetic retinopathy (DR). ResNet50 is a convolutional neural network with a skip connection designed to mitigate the notorious vanishing gradient problem. A Rectified Linear Unit (ReLU) activation function was added to the pre-trained model to simulate the categorization of extracted features. The input images were 512×512 pixels in size. Additionally, the authors used image thresholding to segment the images. Image thresholding was done by creating separate thresholds for the various RGB components of the image, which were then combined using an AND operation. The training and testing datasets had 35,162 images, and the accuracy was 89.4%.

An embedded vision algorithm utilizing a convolution neural network has been developed to differentiate between healthy and diabetic retinopathic images [3]. Here the algorithm was trained using 88,000 high-resolution retinal images from the Kaggle/EyePACS database and had an accuracy of diagnosis of up to 76%. To build the network, Google GoogLeNet, a convolutional neural network with 22 layers, was employed, and its architecture was fine-tuned via a transfer learning approach. During the development process, TensorFlow and NVIDIA Jetson TX2 were used as the underlying platforms to make data processing locally and in real-time, allowing for the classification of retina images without an Internet connection.

In this paper [4], the authors aimed to improve the accuracy of detecting pathologically significant regions within images for diabetic retinopathy detection by combining coarse-grained and fine-grained classifiers. The ensemble's performance was evaluated on public EyePACS and Messidor datasets [15], with successful results in binary, ternary, and quaternary classification, outperforming individual image classifiers and most previously published works. A mix of general and fine-grained deep CNNs, such as residual networks, densely connected networks, NASNet, NTS-Net, and SBS layer, was utilized for predicting automated diagnoses. The authors combined these techniques by training each one independently and integrating the ensemble during inference to take advantage of their combined potential.

This paper [5] focuses on a deep learning approach to identify whether a patient has diabetes. The methodology was to feed a preprocessed image into a convolutional neural network and tested on 30 high-resolution retinal images. The accuracy and sensitivity of this method were found to be 100%. The requisite pre-processing was done using Python's Scipy library. The computational interface between Python's Keras Library and TensorFlow was provided by an Intel(R) Core (TM) i5-4210U processor with a clock speed of 1.70 GHz and 2.40 GHz and 4 GB of RAM. The convolutional neural network was trained using the Adam optimizer.

3 Methodology

3.1 Data Exploration

The project leverages two datasets: (i) the primary dataset used for modeling and evaluation; (ii) the supplementary dataset for pre-training. The primary dataset comprises 3662 labeled retina images taken using a fundus photography technique and labeled by a clinical expert according to their level of DR severity from 0 to 4 [13]. The supplementary dataset carries an additional 35,126 images that have been labeled according to the same severity scale. To ensure accuracy, a test set composed of 1928 images without explicit labels is also included for performance evaluation.

The 3662 colored images included in the labeled dataset contain three RGB channels. As shown in Fig. 1, 49% of the images describe healthy eyes, while the other 51% illustrate varying stages of diabetic retinopathy (DR). Out of this 51%, the least common class is level 3 (Severe), which makes up only 5% of the entire dataset. As the images have been taken with various cameras at various clinics in India, the preprocessing will be required to account for differences in resolution, aspect ratio, and additional variables to ensure an accurate interpretation of the data.

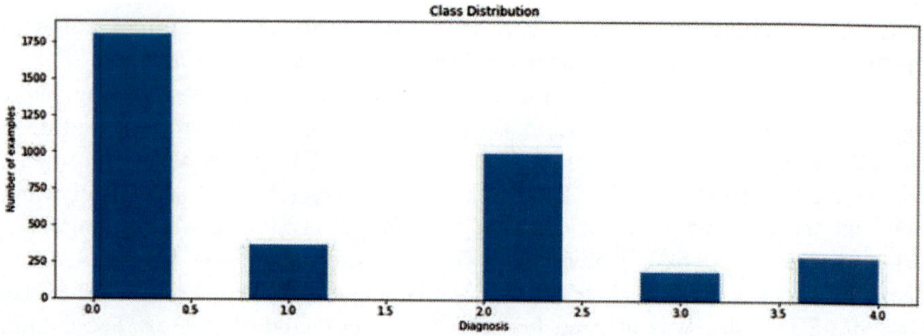

Fig. 1 Class distribution

3.2 Implementation

The implementation of convolutional neural networks like DenseNet169, ResNet50, and VGG16 was executed in Kaggle using the Python language. Various libraries, including OpenCV, have been used for managing the images and preprocessing them. For the mathematical functions, NumPy. TensorFlow and Scikit-Learn are for managing the deep learning models and defining said models. A GPU-enabled device is used to achieve efficient and faster processing.

3.3 Image Pre-processing

The retina photographs are characterized by various image parameters [6, 17]. Standardizing images is essential to ensure the quality of the classification model. To address any discrepancies related to aspect ratios, an image cropping function can be used to adjust the image by converting it to grayscale, identifying intense pixel areas, and removing vertical or horizontal rectangles. Then, the images can be resized to a consistent size of 256 × 256. A Gaussian blur can be applied with a kernel size of 256/6 using the cv2 preprocessing library to correct lighting, color, or contrast discrepancies.

Figure 3 depicts the example batch of eye images after preprocessing. Comparing the images to the ones presented in Fig. 2 clearly shows that the discrepancies between the photographs have been addressed.

3.4 Data Augmentation

After the data analysis, it has been observed that diabetic retinopathy severity picture classes have a highly imbalanced distribution of data, as shown in Fig. 4. In order

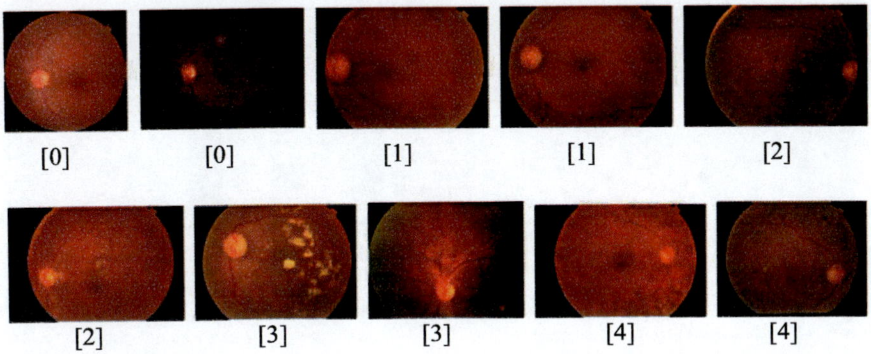

Fig. 2 Data illustration

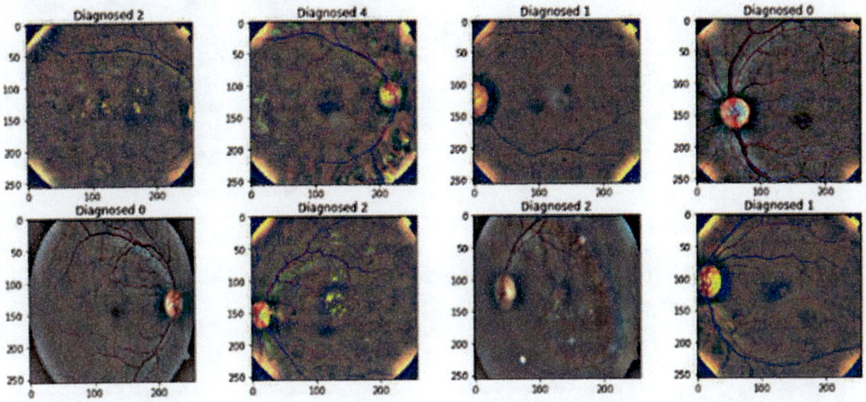

Fig. 3 Images after pre-processing

to balance the severity classifications of diabetic retinopathy, data augmentation was done by bringing the sample sizes of the classes closer together and superimposing the data from the classes with fewer samples onto that of the class with the most samples, as illustrated in Fig. 5. This was implemented using Python's Scikit-Learn and its resampling techniques.

Data augmentation is used during the model training process to improve the generalization capability while avoiding overfitting the model. To achieve this, we use the following augmentations:

- Random rotation by up to 360°
- Random horizontal flip
- Random vertical flip
- Zoom range up to 0.1

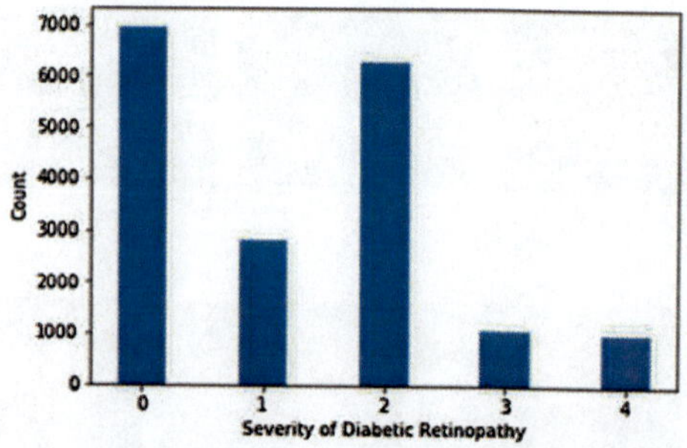

Fig. 4 Imbalanced dataset

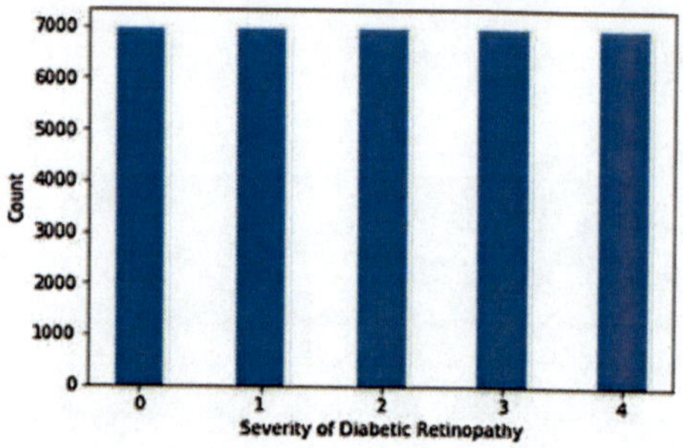

Fig. 5 Balanced dataset after augmentation

Approximately 7000 augmented images were obtained in each class by rotating and mirroring the original images, as shown in Fig. 6.

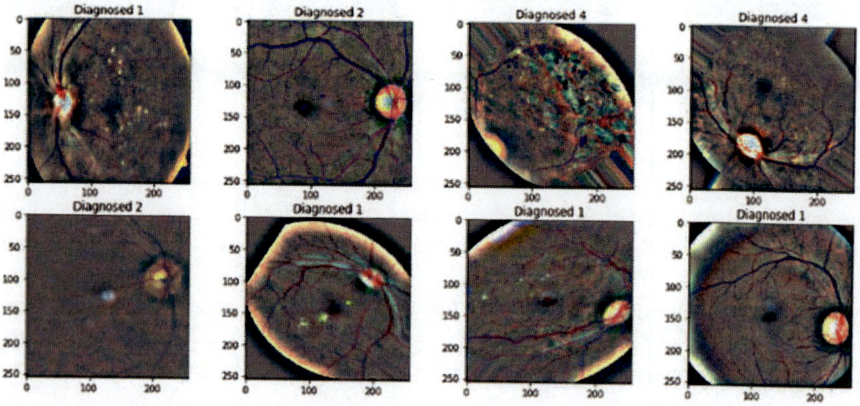

Fig. 6 Images obtained after data augmentation

3.5 Network Architectures

Here, three different networks are taken to compare their kappa score and to decide which network suits the best to the model [7]. Three networks are ResNet50, DenseNet169, and VGG16.

ResNet50—This model incorporates a densely connected layer, a dropout layer with a decay rate of 0.5, a five-unit output layer, and a global average 2D pooling layer at the conclusion of the ResNet50 network [8]. The learning rate is set at an initial value of 0.00005 and is lowered by 0.5 if the model remains static for 15 epochs. The final layer utilizes a sigmoid activation function, while the binary cross-entropy loss was used for optimization. If the model remains at a plateau after five epochs, validation loss is observed to determine if the early stoppage is necessary. An example of a deeper network that applies the same basic ideas is ResNet50 which incorporates over 150 neural network layers [9, 18]. While all algorithms train on the output 'Y' ResNet is distinct in that it is trained on $F(X)$, with the aim of finding $F(X) = 0$ so that $Y = X$.

DenseNet169—The weights of the DenseNet169 model are incorporated into the network, excluding its last layer. The last layer is made up of a dropout layer with a 0.5 probability of decay, global average pooling 2D, and an output layer with five nodes per class. This process works similarly to the 2D average pooling by considering the whole input block size. To counter overfitting, a dropout layer is used. The Adam algorithm is employed to optimize the weights used to train the model. The convolutional layer runs through the fundus images using a series of kernels or filters to evaluate the dot product. These kernels or filters are utilized for deriving a different set of visual properties.

VGG16—VGG16 is a deep convolutional neural network designed to achieve high performance on the ImageNet dataset containing 15 million images divided into 14 million items with 1000 labels [8]. The model is trained with the Adam algorithm, utilizing a global average 2D pooling layer, a dense layer, and a dropout layer with a decaying probability of 0.5. The learning rate is set to 0.00005 and is reduced by half when the model remains static for 10 epochs. The binary cross-entropy loss function is applied to analyze the model's performance, and a sigmoid activation function is used for the output layer. Additionally, the validation loss is continuously monitored in case the model is trapped in a plateau for over five epochs to prevent overfitting.

3.6 Training and Validation

The next stage is pre-training a CNN model on both the APTOS 2019 [13] and 2015 datasets [14] after the image pre-processing and data augmentation. First, we instantiate the model using any of the three networks mentioned in network architectures. The model weights were initialized using pre-trained parameters on ImageNet for the multiple classification problem. The nn.CrossEntropyLoss() function reduces the error between the actual and the estimated values. The Adam optimizer was used with 0.005 learning rate, that decreased by 0.5 for every five epochs. A batch size of 32 and image size of 256 were selected to optimize performance and manage memory consumption. During training, batches of data from the supplementary dataset were sequentially loaded and passed through the network, and the primary dataset was used as a validation sample. If the model did not improve its accuracy over five consecutive epochs, then the training process was halted, and the model weights obtained from the epoch associated with the highest validation accuracy were saved. In total, the network was trained for 15 epochs.

3.7 Testing

The initial model is ResNet50, where weights are initialized with the values pre-trained on ImageNet, and no image preprocessing beyond resizing is applied. Fine-tuning this model on the primary dataset by training the last classification layer achieves a kappa score of 0.819 on the validation.

During the work on the project, iteratively different refinements were tested, and the model performance was tracked within validation to make decisions about including certain modifications. The empirical experiments include:

- Tuning the parameters of image preprocessing functions. That includes cropping and smoothing functions, data augmentations, and image size. The results of the empirical experiments suggest that cropping images and augmentations are crucial for better predictive performance.
- Testing three different network architectures: ResNet50, DenseNet169, and VGG16. The ResNet50 architecture demonstrates the best performance and is selected as a final model architecture. At the same time, model DenseNet169 presents the fastest training and validation, which might be helpful when deploying a model for screening new retina photographs. VGG16 might achieve a better performance, but its convergence speed is slower.
- Testing unfreezing of different layer combinations. It is found that the best performance is achieved when all model weights are optimized during the pre-training stage, and only the last classification layer weights are optimized during the fine-tuning phase. Unfreezing more layer's during fine-tuning does not improve the performance as the sample size is too small.
- Selecting the optimizer and the learning rate. Two optimization algorithms were tried: Adam and stochastic gradient descent (SGD).

The final model is selected based on Cohen's kappa score. The best model achieves a kappa score of 0.83 on the validation.

4 Comparison of Networks

The project considers DR detection as an ordinal classification task. In such a setting, a suitable evaluation metric is Cohen's Kappa, which measures the degree of agreement between the given and the predicted labels [10]. The metric varies from 0 (random agreement) to 1 (perfect agreement).

The Kappa is computed as follows. First, one constructs three matrices: X, which holds the observed scores; M, containing the anticipated results determined by chance agreement, and W, which contains the weights. Next, the Kappa is computed as:

$$k_w = 1 - \frac{\sum_{i=1}^{k}\sum_{j=1}^{k} w_{ij}x_{ij}}{\sum_{i=1}^{k}\sum_{j=1}^{k} w_{ij}m_{ij}}, \quad (1)$$

where x_{ij} are elements in the matrix X and m_{ij} are elements in the matrix M, respectively, while the w_{ij} are elements in W. We will use Kappa with quadratic weights to stronger penalize larger errors.

ResNet50 Model—This study used two datasets to develop a ResNet50 model for diagnosing diabetic retinopathy. Several preprocessing and augmentation techniques were applied to standardize the data and remove unwanted noise.

The graph in Fig. 7 illustrates the accuracy of a ResNet50 model, Fig. 8 depicts the model loss values across the epochs, and Fig. 9 shows the confusion matrix of the ResNet50 model.

This CNN model based on the ResNet50 network reached an accuracy of 99.72% for training and 90.22% for validation, and the Cohen kappa score is 0.8198, as shown in Table 1.

DenseNet169 Model—In this study, a DenseNet169 network was used to develop an early diagnosis model for diabetic retinopathy. Extensive preprocessing and augmentation were performed to standardize the data and remove unwanted noise.

The graph in Fig. 10 provides insight into the model's accuracy, Fig. 11 demonstrates the model's performance in terms of its loss values, and Fig. 12 shows the model's performance with its confusion matrix. This CNN model, which uses the

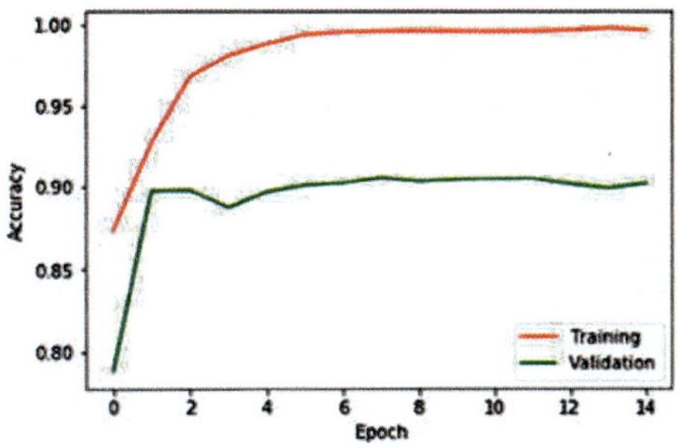

Fig. 7 Training and validation accuracy of ResNet50

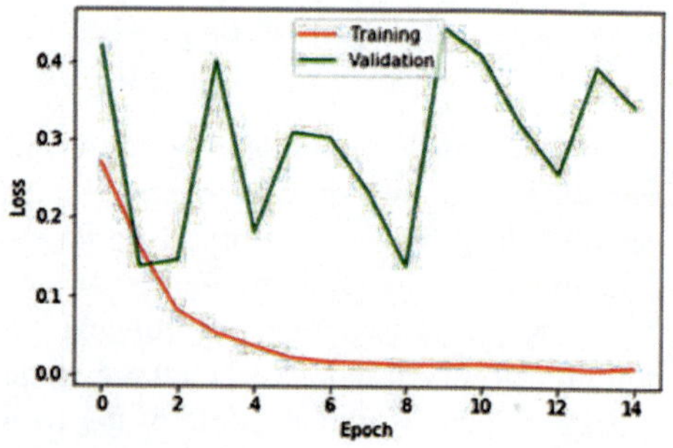

Fig. 8 Training and validation loss of ResNet50

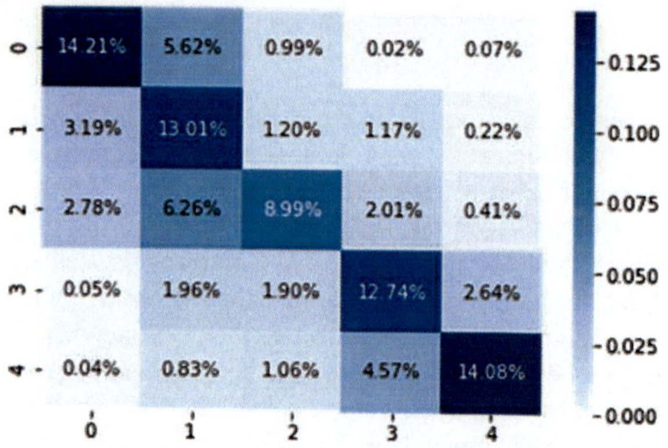

Fig. 9 ResNet50 model confusion matrix

Table 1 Result of ResNet50 model

Accuracy (training)	Accuracy (validation)	k_w
99.7%	90.2%	0.82

DenseNet169 network, achieved an accuracy of 99.52% for training and 90.87% for validation, and the Cohen kappa score is 0.83, as shown in Table 2.

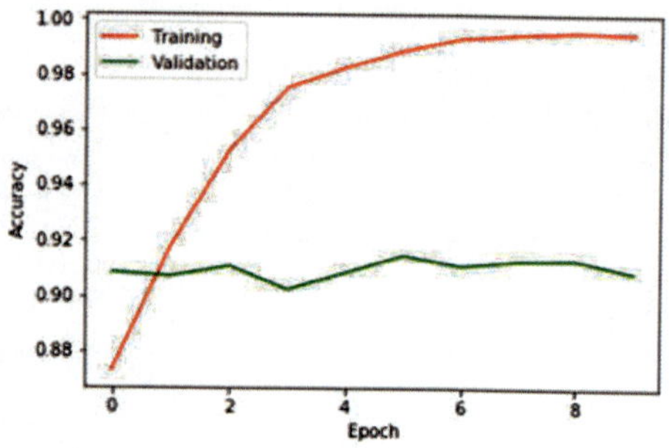

Fig. 10 Training and validation accuracy of DenseNet169

Fig. 11 Training and validation loss of DenseNet169

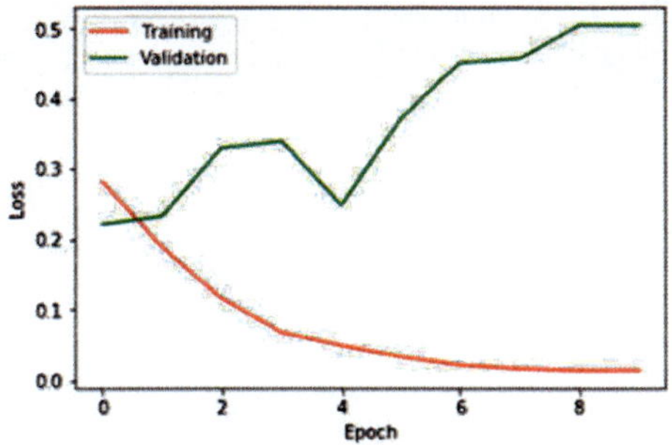

Fig. 12 DenseNet169 model confusion matrix

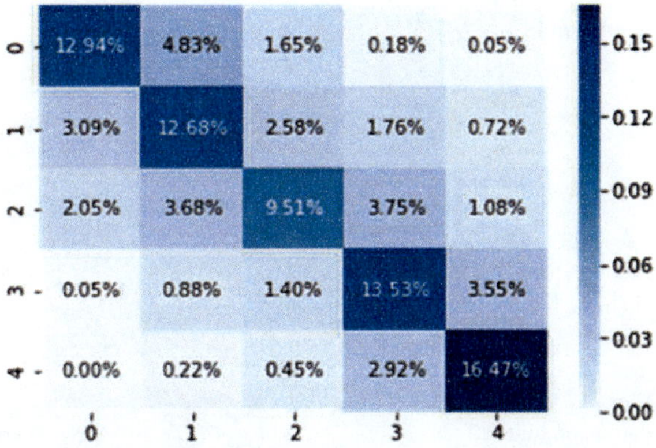

Table 2 Result of DenseNet169 model

Accuracy (training)	Accuracy (validation)	k_w
99.5%	90.8%	0.83

VGG16 Model—The VGG16 network was employed in this model to assist in diagnosing diabetic retinopathy. Extensive pre-processing and augmentation were performed to standardize the data and eliminate the undesired noise.

The VGG16 model performance is evaluated by observing its accuracy and loss of its training and validation set by viewing Figs. 13 and 14, respectively. Figure 15 displays the model's performance with its confusion matrix.

This CNN model utilizing VGG16 yielded an accuracy of 95.15% for training and 90.68% for validation, and the Cohen kappa score is 0.8133 as shown in Table 3.

Comparison of Networks—Here, the accuracy of the different networks when it comes to the classification of different diabetic retinopathy (DR) stages is presented in Table 4. DenseNet169 achieved the best accuracy of 90%, making it the most effective for DR classification with 5 stages.

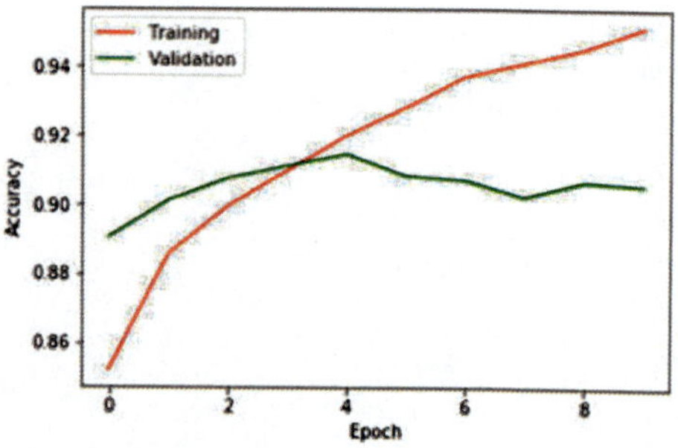

Fig. 13 Training and validation accuracy of VGG16

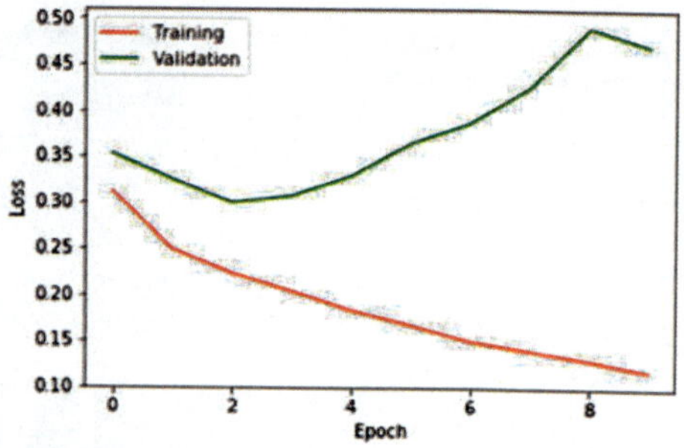

Fig. 14 Training and validation loss of VGG16

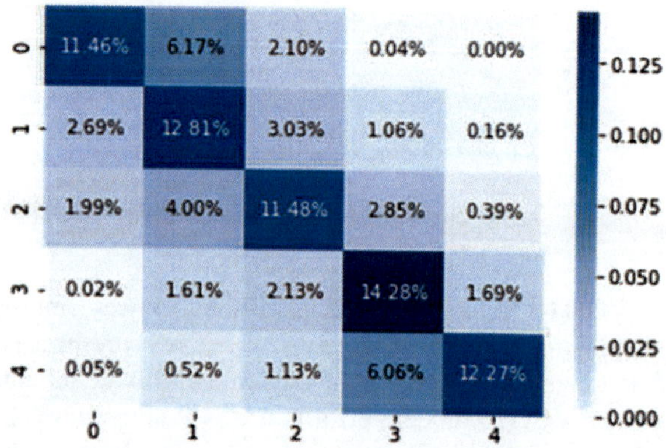

Fig. 15 VGG16 model confusion matrix

Table 3 Result of VGG16 model

Accuracy (training)	Accuracy (validation)	k_w
95%	90%	0.81

Table 4 Comparison of different networks

Networks	Testing accuracy (%)	Stages of DR
SVM [11]	80.6	3
DT [12]	75.1	4
KNN [11]	55.1	3
RESNET50	85	5
DENSENET169	90	5
VGG16	73	5

5 Results and Discussion

To select the optimum architecture for the model, a combination of datasets from Kaggle was used to train the model using DenseNet169, ResNet50, and VGG16. Pre-processing of the dataset was crucial to reduce the noise and enhance the focus on the fundus image. Hence, pre-processing steps included the removal of black borders and corners, downscaling the images to a size of 256 * 256 pixels, and applying Gaussian blur to eliminate Gaussian noise. After pre-processing, it was realized data was highly imbalanced, with mostly no DR cases associated with severity class '0'. Therefore, data augmentation was employed, introducing an additional 7000 images for each class, thus balancing the data. After pre-processing and data augmentation, validation was used to train and validate a set of ResNet50, DenseNet169, and VGG16 models. The validation performance is averaged across the validation to provide a more accurate indication of the model performance. Cross-entropy loss and Cohen's Kappa metric are used to assess a model. The test set that wasn't part of the training data is used for the final performance assessment. Although the labels of the test photos are not explicitly provided, it is still possible to evaluate the model's performance.

6 Conclusion

This study proposes a deep learning approach for diagnosing diabetic retinopathy's severity using the convolutional neural network. With its multiple deep layers, the DenseNet169 technique was more efficient and time-effective than traditional methods for identifying diabetic retinopathy severity. The results of this study demonstrated that DenseNet169 had an excellent performance in detecting retinopathy in retinal images. This strategy was devised after evaluating existing techniques for automatically detecting diabetic retinopathy. Two datasets were combined for this investigation—the Diabetic Retinopathy Detection 2015 Dataset and the APTOS 2019 Blindness Detection Dataset from Kaggle. Extensive pre-processing and augmentation were performed to prepare the data in the required format and eliminate unwanted noise. In addition to the DenseNet169 classifier, the Resnet50, and VGG16 were employed to compare the outcomes. Moreover, the suggested system performed better than machine learning classifiers such as SVM, DT, and KNN, with improved accuracy in classifying images into more classes. The accuracy of the DenseNet169 model was 90%, outperforming the other two models, which produced 85% and 73% accuracy from the ResNet50 model and the VGG16 model, respectively.

References

1. Gulshan V, Peng L, Coram M, Stumpe MC, Wu D, Narayanaswamy A, Venugopalan S, Widner K, Madams T, Cuadros J et al (2016) Development and validation of a deep learning algorithm for detection of diabetic retinopathy in retinal fundus photographs. JAMA 316(22):2402–2410
2. Rajkumar RS, Jagathishkumar T, Ragul D, Selvarani AG (2021) Transfer learning approach for diabetic retinopathy detection using residual network. In: 2021 6th International conference on inventive computation technologies (ICICT), pp 1189–1193. https://doi.org/10.1109/ICICT50816.2021.9358468
3. Vora P, Shrestha S (2020) Detecting diabetic retinopathy using embedded computer vision. Appl Sci 10:7274. https://doi.org/10.3390/app10207274
4. Bajwa MN et al (2019) Combining fine- and coarse-grained classifiers for diabetic retinopathy detection. ArXiv abs/2005.14308
5. Chakrabarty N (2019) A deep learning method for the detection of diabetic retinopathy. https://doi.org/10.1109/UPCON.2018.8596839
6. Wang J, Bai Y, Xia B (2020) Simultaneous diagnosis of severity and features of diabetic retinopathy in fundus photography using deep learning. IEEE J Biomed Health Inf 24(12):3397–3407. https://doi.org/10.1109/JBHI.2020.3012547
7. Classifying diabetic retinopathy using deep learning architecture. Int J Eng Res Technol (IJERT) 5(06). http://www.ijert.org. ISSN: 2278-0181; IJERTV5IS060055 (This work is licensed under a Creative Commons Attribution 4.0 International License.)
8. Bhalekar M, Sureka S, Joshi S, Bedekar M (2020) Generation of image captions using VGG and ResNet CNN models cascaded with RNN approach. https://doi.org/10.1007/978-981-15-1366-4_3
9. Elswah DK, Elnakib AA, El-din Moustafa H (2020) Automated diabetic retinopathy grading using Resnet. In: 2020 37th National radio science conference (NRSC), pp 248–254. https://doi.org/10.1109/NRSC49500.2020.9235098
10. Cohen J (1960) A coefficient of agreement for nominal scales. Educ Psychol Measur 20(1):37–46
11. Gupta S (2015) Diagnosis of diabetic retinopathy using machine learning 3(2). https://doi.org/10.4172/2311-3278
12. Carrera EV, Carrera R (2017) Automated detection of diabetic retinopathy using SVM, pp 6–9
13. Kaggle (2019) APTOS 2019 blindness detection detect diabetic retinopathy to stop blindness before it's too late. [Online]. Available: https://www.kaggle.com/c/aptos2019-blindness-detection/
14. Kaggle (2015) Diabetic retinopathy detection. [Online]. Available: https://www.kaggle.com/c/diabetic-retinopathy-detection/overview
15. Rahul MSP, Mahakalkar NA, Singh T (2019) Novel approach for detection of early diabetic retinopathy. In: 3rd International conference on inventive systems and control (ICISC 2019), JCT College of Engineering and Technology, Coimbatore
16. Aswathi T, SwapnaTR, Padmavathi S (2021) Transfer learning approach for grading of diabetic retinopathy. J Phys Conf Ser
17. Yadeeswaran KS, Mithun Mithra N, Varshaa KS, Karthika R (2021) Classification of diabetic retinopathy through identification of diagnostic keywords. In: 2021 Third international conference on inventive research in computing applications (ICIRCA), Coimbatore, India, pp 716–721. https://doi.org/10.1109/ICIRCA51532.2021.9544621

18. Patra P, Singh T (2022) Diabetic retinopathy detection using an improved ResNet 50-InceptionV3 and hybrid DiabRetNet structures. In: 2022 OITS international conference on information technology (OCIT), Bhubaneswar, India, pp 140–145. https://doi.org/10.1109/OCIT56763.2022.00036

Utilisation of Machine Learning Techniques in Various Stages of Clinical Trial

P. S. Niveditha and Saju P. John

Abstract Clinical trial is a medical approach, which is frequently carried out to analyse the effectiveness of a novel drug or treatment in patients. Clinical trials may also examine different facets of treatment, like enhancing the quality of life for those with long-term ailments. Clinical research as it is now practised is complicated, time-consuming, costly, and sometimes biased, which can occasionally jeopardise its successful application, implementation, and acceptability. Machine learning techniques have become more and more prevalent in the healthcare sector in current era, particularly in fields of study that involve human subjects, like clinical trials, and in which data collection is prohibitively expensive. The role of machine learning in each stage of the clinical trial is different. Machine learning can benefit clinical trials at every stage, from preclinical drug development to pre-trial planning to test conduction to handling and analysing information. This paper conducts an extensive review on the various machine learning approaches that are employed in the various steps of the clinical trial process.

Keywords Clinical trial · Machine learning · Drug development · Health sector

1 Introduction

Healthcare is not an exemption to how machine learning is transforming all fields. These computational algorithms have the capacity to unravel complex data and produce amazing, predicted outcomes if given the proper inputs. Given that conducting and evaluating clinical trials takes up a significant amount of time and money across the entire drug discovery cycle, deep learning could be viewed as beneficial for clinical trials. Since pharmaceuticals are tested on humans to determine their

P. S. Niveditha (✉)
APJ Abdul Kalam Technological University, Thiruvananthapuram, Kerala, India
e-mail: nivedithapsukes@gmail.com

S. P. John
Department of Computer Science and Engineering, Jyothi Engineering College, Cheruthuruthy, Kerala, India

S. Kumar et al. (eds.), *Fourth Congress on Intelligent Systems*, Lecture Notes in Networks and Systems 868, https://doi.org/10.1007/978-981-99-9037-5_33

safety and efficacy, clinical trials are also regarded as the phase with the highest rate of drug molecule screening. Hence there is a higher chance of drug failure. A drug's failure rate is about 57% because of insufficient efficacy [1]. Hence the importance of proper conduction of clinical trial is clear.

The clinical trial process can be divided into four main stages with multiple substages. The main four stages are:

1. Preclinical investigation
2. Protocol optimisation
3. Participant management
4. Data collection and management.

Figure 1 gives the details of processes in each stage of clinical trial where machine learning can be applied. The machine learning techniques used in various stages of the trial process are different.

Some of the most useful machine learning algorithms in the clinical research are discussed below:

Natural language processing (NLP): Natural language processing (NLP) is more particularly the field of 'artificial intelligence' (AI) that is concerned with providing computers the capacity to comprehend written and spoken words in a manner similar to that of humans [2]. NLP blends statistical, machine learning, and deep learning models with computational linguistics rule-based modelling of human language.

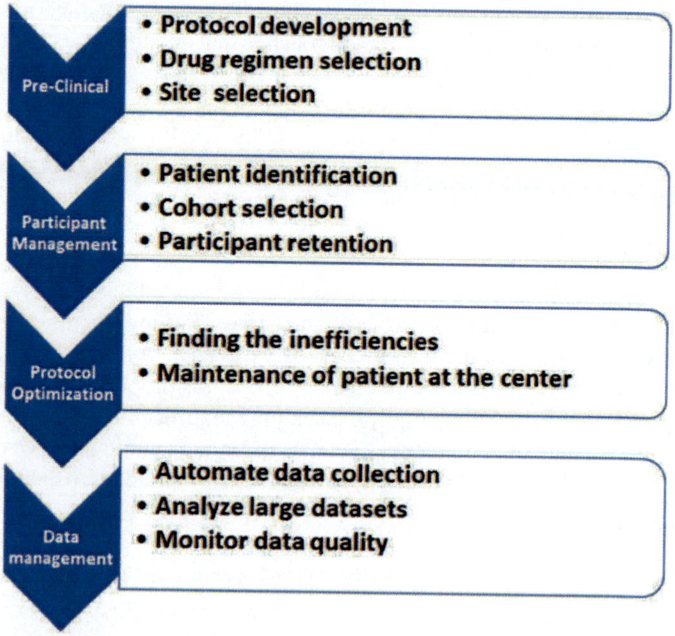

Fig. 1 Stages of clinical trial utilising machine learning

With the use of these technologies, computers are now able to process human language in the form of text or audio data and fully 'understand' what is being said or written, including the speaker's or writer's intentions and sentiments [3]. Large collections of unstructured health data might be analysed using qualitative research methods made possible by NLP, and the resulting relevant insights could be used to enhance patient safety and management. Natural language processing could effectively manage document categorisation, document summarisation, and information extraction.

The ability of NLP to assemble, contrast, and define the best appropriate care recommendations could help to streamline the evaluation of medical policy. NLP replaced the final procedure of double-checking the health record, greatly reducing the workload and time required. Convolutional neural network (CNN) and long short-term memory (LSTM) network could perhaps be modified to produce better outcomes for enhanced sequence modelling extraction [4].

Computer vision: Thanks to its astounding results in the pharmaceutical and healthcare industries, computer vision has had a revolutionary impact. By detecting the items, computer vision aims to teach the machine to mimic human sight. Medical professionals have seen a flowering in this technology as disease detection, particularly radiological imaging, has gotten much faster and more precise. In order to make better medical decisions and to better monitor patient health, doctors use this technology to analyse a broad range of fitness and medical data [5]. By reducing false positive diagnoses, it not only aids in more accurate scan prediction but also in more accurate interpretation for a more accurate diagnosis, which is still a work in progress. Robotic-aided procedures in healthcare have expanded their potential courtesy to computer vision and other deep learning techniques. Another application of computer vision is drug adherence monitoring.

2 Application of Machine Learning Algorithms in Various Stages of Clinical Trial

2.1 Preclinical Study

Preclinical research includes analyses of medication manufacture and purity as well as animal trials. The medicine's safety at dosages that roughly correspond to human exposures is investigated in animal trials, together with its pharmacodynamics and pharmacokinetics components. If the medicine is to be further investigated in human subjects, this information must be presented for IND approval. A clinical trial's first phase is intended to evaluate a drug's safety, maximum tolerated dose (MTD), human pharmacokinetics, pharmacodynamics, and drug-drug interactions. The new experimental drug is being tested in people for the first time during these phase I trials.

Pratik Shah et al. in [6] have described machine learning architectures, for their use in the analysis and learning of publicly accessible medical and clinical trial information sets, practical sensor data, and health records. To find clinically significant patterns in, for example, imaging data, machine learning and computer vision have improved many aspects of human visual perception. Neural networks have been used for a range of tasks, including the creation, categorisation, and forecasting of clinical datasets. Generally, research laboratories, pharmaceutical businesses, and technology firms are investigating the application of artificial intelligence and machine learning in three important areas. Deep learning methods are used on multifaceted data sources, such as integrating genomic and medical data to detect new predictive models, to develop productive algorithms for computational enhancement with current clinical and imaging evidence sets.

Diaconu A. et al. in [7]: With growing complexity over time, AI/ML techniques have become widely employed in drug development, translational research, and the preclinical stage. All of the aforementioned instances highlight a cutting-edge paradigm utilised to enhance disease diagnosis and treatment, urging physicians and researchers to imagine and expect more from AI. The necessity for randomised controlled trials is becoming evident, and as the paradigm shifts, healthcare professionals will be provided with self-improving decision-making tools based on algorithms that use deep learning rather than RCTs. In addition to improving the success and effectiveness of clinical studies and observational studies, ML techniques also enhance fundamental science and translational studies. They are appropriate for planning, study execution, handling of data, and analysis. Although we frequently use AI systems in our daily lives, acquiring, analysing, and using that paradigm to tackle complex clinical problems presents a few difficulties. Its potential to impact the healthcare system, nevertheless, is still developing. This might be partly because of moral and economical restraints, but it could also be because there are not enough substantial studies to confirm their validity.

José Jiménez-Luna et al. [8] suggested that deep learning holds promise for drug development since it allows for sophisticated image interpretation, molecular structure as well as function prediction, and the automated creation of novel chemical entities with specific features. Despite the increasing number of promising future applications, the underlying mathematical models are frequently difficult for the human mind to understand. To meet the need for an innovative machine language of the biological and molecular sciences, there is a desire for 'explainable' deep learning approaches. This paper outlines the key algorithmic ideas of understandable artificial intelligence and dares to predict the prospects, potential uses, and unmet obstacles in future. Users of deep learning techniques have a responsibility to carefully verify and interpret the predictions provided by such a computational model, especially in time- and money-sensitive scenarios like drug discovery. It is reasonable to believe that the continuous development of different approaches that are more understandable and computationally economical will not lose their importance in the light of the potential and constraints of drug discovery.

Andrew W. Senior et al. [9] proposed that protein's three-dimensional shape can be predicted from its arrangement of amino acids using protein structure prediction.

This issue is crucial since a protein's shape significantly dictates how well it performs. However, it might be challenging to discover protein structures through experimentation. Recent years have seen significant advancements due to the use of genetic data. Covariation in homologous sequences can be analysed to determine which amino acid residues are in contact, which helps with the forecasting of protein structures. In this study, the scientists demonstrated that a neural network can be trained to estimate lengths between sets of residues accurately, which provides more structural information than contact predictions. The authors have discovered that a straightforward gradient descent approach can optimise the resulting potential for creating structures without using complicated sampling techniques. Even for repeats with fewer homologous sequences, the resulting technique, known as AlphaFold, achieves high accuracy. High-accuracy structures were produced in the recent evaluation of protein structure forecasting by a blind evaluation of the state of the AlphaFold. A significant improvement in protein structure prediction is AlphaFold. This improved accuracy makes it possible to gain insights into how proteins work and malfunction, particularly when no homologous protein structures have been experimentally identified.

Rory Casey et al. [10] suggested that even if there have been substantial improvements in diabetes mellitus drug discovery over the past few decades, there is still a chance and a need for better treatments. While type 2 diabetes individuals are better able to control their condition, many of the therapies in this field are hormones made from peptides with complex molecular structures that make them difficult to produce and expensive. They have offered novel anti-diabetic peptides that are smaller than 16 amino acids in length using machine learning. In vitro studies have shown these peptides' ability to promote glucose absorption and transporter of glucose type 4 translocation. When compared to therapies already being used, anticipated peptides dramatically lower plasma glucose, decrease glycated haemoglobin, and even ameliorate hepatic steatosis in obese insulin-resistant mice. These unfinished linear peptides are potential prospects for controlling blood sugar levels and need to be further studied. Additionally, this suggests that the authors may have neglected the class of organic brief linear peptide as therapeutic modalities because they often have an outstanding safety profile.

2.2 *Participant Management*

Clinical trial participant management includes both obtaining volunteers and maintaining their interest in the trials. Despite significant resources being spent on participant management, including time, planning, and trial coordinator work, participant dropping out and failure to adhere can occasionally cause trials to go over budget, take longer than anticipated, or fail to deliver usable data. Machine learning (ML)-based strategies could make it easier and more equitable to find, recruit, and keep participants. If specific patient populations are better chosen for trials, a smaller sample size might be required to detect a significant impact. Fewer patients may receive therapies from which individuals are unlikely to benefit because of better patient selection.

Dhaya Sindhu Battina in [11] presented that unsupervised machine learning of clinical groups could discover patterns of patient characteristics that can be used for selecting participants who would benefit most from the course of treatment or intervention, in addition to making it easier to choose patients by quickly analysing large previous study datasets. Unorganised data is essential for phenotyping and choosing target populations, demonstrating the significance of gathering additional data from patients. Diabetes mellitus type II has three subgroups, each with unique phenotypic symptoms and therapeutic requirements. Once a certain cohort has been determined, patients who fit the required phenotypic can be identified using the technique of natural language processing generally necessitating a lot of manual work. It efficiently uses electronic health records to link patients with clinical research by merging written and tabular data from each source into a single latent space. ML-assisted techniques can be used in one of two ways to improve user retention and procedure adherence: as a reminder or as a reward. The first step in identifying study failure to comply risk individuals and acting is to gather and analyse enormous quantities of data using machine learning techniques. The second method, machine learning (ML), aims to reduce participant study burden while enhancing participant experiences.

Bhatt A. [12] have analysed that one of the biggest obstacles to researchers conducting successful clinical studies is patient recruiting. This problem showed up in the identification of patients with early-stage breast cancer. It was suggested that machine learning and artificial intelligence could help increase the total number of people involved and shorten the time it takes to locate and access them. AI was emphasised as having the potential to assist academics in achieving this goal via various sorts of technologies connected to it. It was stated that natural language processing (NLP), a method that recognises written and spoken words to uncover patients who have been identified as having disease, may quickly identify scientists who might desire to take part in such an experiment. The author has described the ideal clinical trial scenario, in which patients should be enrolled utilising genome-patient-specific diagnosis tools and in which patients who would be the best candidates for a medicine in development would have the presence of its target biomarkers. By combining data with medical records stored in the cloud, advanced AI and ML approaches that are now under development may be able to identify individuals or participants in clinical trials who have a particular type of data that might be pertinent to a clinical trial. This method aids in the selection of clinical research trial endpoints that can be accurately monitored to enhance efficacy and results. They increase a researcher's capacity to use AI and ML-based technologies, like optical character recognition tools, to discover and characterise patient subpopulations that are appropriate for particular trials. When these methods are used in clinical research trials, the reading and gathering of evidence pertaining to the studies will be totally automated.

Janette Vazquez et al. [13] suggested that clinical trial (CT) non-participation is a significant obstacle to the assessment of novel drugs and medical devices. In order to identify characteristics of people more likely to express interest in participating in CTs, authors used supervised machine learning methodologies and a deep learning methodology to analyse a set of data from a web-based clinical registry. Using a dataset of 841,377 instances and 20 characteristics, including demographic information, regional restrictions, medical issues, and the visit history, they developed six supervised machine learning algorithms as well as a deep learning technique, convolutional neural network (CNN). Responses indicating particular participant interest in particular clinical trial possibility invitations made up the variable. Based on self-reported medical problems and gender, four subsets of this dataset were individually examined. The machine learning models fell short in comparison with the deep learning model. The findings provide substantial proof that the datasets evaluated with the machine learning models exhibit meaningful correlations between predictor factors and outcome parameters. These methods give hope for finding people who would be more willing to take part in clinical trials when they are presented with the chance.

Jingshu Liu et al. [14]: Although significant progress has been achieved in recent years to optimise patient recruitment for clinical trials, more accurate approaches for predicting patient recruitment are still required to enable trial site selection and determine the proper enrolment deadlines in the trial design stage. The authors have investigated machine learning techniques to estimate the number of patients registered per month at a study's location over the course of a trial's enrolment duration in this research utilising data from numerous previous clinical trials. The authors have demonstrated how these strategies can lower the inaccuracy that is seen with the most recent industry standards and suggested areas for development.

Dr. Vladimir V. et al. [15] put forth the ground-breaking statistical approach for planning and forecasting patient recruitment. They first established the method for modelling recruitment before using it at the prediction stage if we want to have reliable prediction models. The appropriate mathematical frameworks should be founded on statistical concepts because there are various uncertainties in the input data and randomness in the recruitment process at each centre over time. The adoption of a termed Poisson-gamma enrolment model has been suggested by writers as a way to describe the enrolment of patients process in multi-centre clinical trials. This model accurately captures the inherent uncertainties in recruiting statistics that are seen in actual use. This model can be used as a fundamental recruiting model because it exhibits good agreement with actual data, according to statistical analysis of several completed trials. A completely novel data-driven statistical approach for recruiting prediction is created. It enables the estimation of various aspects of the estimated remaining time, particularly credibility intervals, the provision of recommendations on adaptive recruitment adjustment, the evaluation of the bare minimum of centres required to finish in time with a given confidence, study and centre performance, and the resolution of various optimisation problems combining time and costs.

2.3 Protocol Optimisation

The creation of a clinical research protocol is the first step in every clinical trial. The clinical trials protocol is a paperwork that outlines the objectives, design, technique, statistical considerations, and organisational structure of a clinical study. It also ensures the safety of trial participants and the accuracy of the data gathered. A clinical research project's history, reasoning, goals, design, technique, statistical considerations, and organisation are all described in the research protocol.

Marina A. Malikova in [16] has discussed a methodology for identifying planning-stage processes and methods for reducing the risk of financial compliance for clinical trials. Clinical trial management needs thoughtful preparation and effective implementation. It is helpful to establish standardised trial management standards and build a reliable scoring technique for evaluating study protocol complexity in order to achieve timely delivery of significant clinical trials' outcomes. Using a proposed comprehensive scoring model, this review will examine the difficulties clinical teams encounter when creating protocols to make sure proper individuals are enrolled, and appropriate information is gathered to show that a drug is safe and effective. The main things to think about when creating protocols and methods to reduce complexity will be covered. Utilising a complexity scoring model enables clinical research organisations and medical facilities to manage staffing without prejudice, ensure compliance to study protocol as well as processes, allocate resources effectively, and reduce the risk of billing compliance by performing coverage analysis beforehand and in conjunction with study protocol design and complexity.

Kenneth A. Getz et al. [17] proposed that over the past seven years, there has been a noticeable increase in the number of specific procedures and the frequency of those procedures per protocol, with significant variation across therapeutic areas. The typical number of accepted criteria per procedure has significantly increased, while the total number of exclusion requirements has not changed since 1999. In addition, protocols have progressively demanded the use of invasive procedures, X-ray and imaging methods, subjective assessments of research volunteers' hearts, and questionnaires. The frequency of procedures per protocol is much higher in phases 1 and 2. This result is in line with recent Tufts CSDD studies that show a growing dependence by sponsors on early-phase clinical research to decide whether to proceed with larger and more expensive phase 3 studies. The findings of this study offer compelling new understandings of how protocol design changes affect study conduct efficiency and workload. For the research-based pharmaceutical and biopharmaceutical business, enhanced protocol designs may hold the key to obtaining higher levels of efficiency and efficacy in the present drug development environment.

Yizhuo Wang et al. [18] suggested that a good algorithm for predicting a patient's reaction to treatment is a crucial element of precision medicine. To enhance the effectiveness of treatment, it is important to integrate machine learning (ML) techniques into the reaction-adaptive randomisation (RAR) architecture. Such a model guided the selection of treatments and projected the response rate to each medication for each participant. The authors also created an ensemble of these 9 strategies after realising that no one approach may be effective in all studies. The implementation of ML techniques led to more individualised optimal therapy assignments and larger overall acceptance rates among trial participants, according to simulation studies. The ensemble technique outperformed every single ML method in terms of response rate and allocated the greatest proportion of participants to their ideal therapies. The authors have effectively demonstrated the possible enhancements for the practical study if the suggested design had been used in the study.

Yizhuo Wang et al. [19] have proposed a good algorithm for predicting a patient's reaction to treatment which is a crucial element of precision medicine. The goal of the authors is to enhance treatment outcomes by incorporating machine learning (ML) architecture. To simulate the association between patient responses and indicators in a clinical study design, they have included 9 ML algorithms. Such a model guided the selection of treatments and projected the response rate to each therapy for each new patient. They also developed a combination of these 9 strategies after realising that no one approach may be effective in all experiments. By quantifying the advantages for those who participated in the study, such as the total response level and the proportion of patients receiving their optimal therapies, the authors were able to assess their success.

Jian Zhou et al. [20] have proposed the identification of noncoding variations' functional consequences in human genetics. The authors created a deep learning-based algorithmic framework that can forecast noncoding variant impacts de novo from sequence. This framework directly learns a sequence of regulatory code from massive chromatin-profiling information, allowing prediction of chromatin effects of order modifications with single-nucleotide sensitiveness. We also improved the prioritisation of functional variations, such as expression quantitative loci for traits (eQTLs) and disease-associated variants, using this capability. Integrating evolutionary conservation with genomic and chromosomal identifiers at the point of interest has recently resulted in advancements in the prioritisation of functional noncoding variations. In fact, no method has been proven to predict noncoding variant effects on transcription factor (TF) binding, DNA accessibility, or histone marks of sequences with single-nucleotide sensitivity.

2.4 *Clinical Data Management*

Clinical information management is the procedure for gathering and verifying data from clinical trials with the goal of converting it into a digital format for statistical analysis, responding to research questions, and eventually archiving for future research. By appropriately gathering and organising data, eliminating missing data, and improving data quality, this method helps researchers reach the correct conclusions regarding the efficacy, safety, advantages, and potential hazards of the product under study. Creating case report forms, annotating forms, developing databases, entering information, verifying data, handling information discrepancies, managing medical coding, data mining, locking databases, preserving data management processes, and safeguarding data are just a few of the complex steps involved in clinical data management. Clinical data management guarantees the gathering, integrating, as well as accessibility of data at the proper cost and quality. Achieving this objective improves consumer confidence in marketed treatments and protects public health.

Aynaz Nourani et al. [21] defined a clinical information management system as a piece of software that supports the clinical trial's data management process and lowers the possibility of human error. All four key elements were considered when developing the clinical information management framework for the study. The system's capabilities were broken down into five primary categories: administering the research, developing case report forms, managing data, monitoring data quality, and maintaining data confidentiality and safety. The clinical trial manager was assisted by the management section of the system in designing clinical trials, adding researchers and research centres, defining user roles, setting up study groups, randomising participants, and editing trial protocols.

Gazali et al. [22] explained DMP as reviewing the differences, looking into the causes, resolving them further with documentation evidence, and finally declaring them irresolvable are all processes in the disparity management process. Data cleaning and evidence collection for deviations found in the data are the primary duties of discrepancy management. Every CDM software programme has a discrepancy database where all inconsistencies are kept track of and recorded alongside audit trials. Discrepancies are either marked to the investigator for clarification purposes, or they are resolved internally by self-evident corrections (SEC), depending on the nature of discrepancy. Electronic data management systems have replaced paper-based ones to better satisfy the potential in this field. As a result, all recent technological advancements have had a favourable influence on CDM processes and systems, leading to the creation of optimistic outcomes regarding the accuracy and rapidity of the information that has been produced. The CDM experts raise the quality of data by ensuring standards.

Zhengwu Lu et al. [23] discussed that in a cutthroat clinical trial market, most pharmaceutical firms and clinical research organisations find benefits in EDC and e-clinical systems. To encourage the implementation of SDTM and enable electronic submissions to regulatory agencies for providers conducting human drug clinical studies, the FDA has launched a critical path effort. CDISC conceptualised and created SDTM. The industry now has an end-to-end solution to focus on moving data from the stage of record to regulatory delivery thanks to the growing usage of SDTM, the operational information model, data analysis model, case report computations data definition specification define.xml, the testing data model, and growing standards like CDASH and FDA protocols.

E. Mossotto et al. [24] proposed that the prevalence of inflammatory bowel disorder, which includes Crohn's disease (CD), ulcerative colitis (UC), and inflammatory bowel disorder unclassified (IBDU), is rising. A timely and efficient identification of PIBD is required for treatment. This study uses machine learning (ML) to categorise 287 children with PIBD using gastrointestinal and histological data. A

ML model to categorise illness subtypes was developed, trained, tested, and validated using data. While structured clustering found four unique subgroups with varying levels of colonic involvement, unsupervised models showed crossover of CD/UC with broad grouping but no distinct subtype delineation. Three supervised ML models with high accuracy were created using endoscopic data alone, histological data alone, and combined endoscopic/histological data.

Diogo M. Camacho et al. [25] proposed that novel indicators are urgently required to help control the condition due to the shortcomings of the present diagnostic methods for Johne's disease, such as irregular bacterial release or inadequate sensitivity. Here, we investigated the faecal microbiota of cattle with Johne's disease to identify distinctive microbial traits with the potential to serve as novel noninvasive biomarkers. Twelve taxa were chosen as taxonomic signatures to differentiate the disease stage using 16S rRNA genome sequencing along with the machine learning techniques. Additionally, the models demonstrated great accuracy for categorisation, even including animals with preclinical infection, when they were built using relative abundance data of the appropriate species. As a result, the study recommended the use of machine learning and innovative noninvasive microbiological biomarkers for the diagnosis of Johne's.

The comparison of various machine learning techniques in the clinical trial is given as shown in Table 1.

3 Conclusion

Machine learning has eased various tasks by discovering patterns and trends from large volume of data. The medical field which specifically deals with heavy data is also benefited by machine learning. This paper reviews various machine learning techniques which are used in the clinical research activities. Four stages of clinical trial are considered for the research which includes preclinical study, participant management, protocol optimisation, and data management. From the study, it has been coined that machine learning has improved the overall success, generalisability, and the efficiency of clinical research. Based on the requirements in each stage of trial, the appropriate machine learning technique can be opted from the range of approaches available. ML has more potential applications in clinical research than it is currently being used for. This is because change involves time, effort, and collaboration as well as the fact that there are few prospective studies evaluating ML's effectiveness in comparison with conventional methods. To guarantee that machine learning in clinical research is utilised in a fair, moral, and open way that is acceptable to all, open discussion about the potential advantages and disadvantages of machine learning for clinical research and the sharing of best practices must continue, not only in the academic community but also in the public and government.

Table 1 Comparison of various machine learning techniques in clinical trial

Sl. No.	Reference	Objective	Technology employed	Advantages	Stage
1	[6]	To improve medical care for the patients	ML-based computer vision	• ML to predict properties of molecular compounds • Segmentation and pattern recognition • DL technique on clinical data to detect new predictive model	Preclinical study
2	[7]	To improve the cardio renal preclinical model	Deep learning	• Analysis of huge dataset • DL is used in the investigation of haemodynamic parameters • Acts as a very promising tool for the screening of various diseases	
3	[8]	Efficient drug development	Deep learning	• Sophisticated image interpretation • Innovative ML for study of molecular science • Simple and explainable deep learning approaches	
4	[9]	Efficient protein structure prediction	Deep learning	• Faster prediction of 3D shape of protein structure • Neural networks are trained to estimate the lengths between residues accurately • AlphaFold provides an improved protein structure prediction	

(continued)

Table 1 (continued)

Sl. No.	Reference	Objective	Technology employed	Advantages	Stage
5	[10]	Preclinical validation of peptides	Machine learning	• Improved drug discovery for diabetes mellitus • Novel anti-diabetic peptide • Highly improved safety profile	
6	[11]	To identify the patient who would benefit from the trial most	Unsupervised ML	• Simplifies the process of patient selection • Recruiting the patients with needed phenotype • Improved patients' retention, progress tracking	Participant management
7	[12]	To reduce the access time of participants	Natural language processing	• Faster patient recruitment • Maintenance of research integrity • Improved success rate of outcomes • Helps to identify participants with specific characteristics	
8	[13]	To improve clinical trial participation	CNN	• Improved identification of characteristic of people • Relied on self-reporting medical problems • Meaningful coordination between medical factors	
9	[14]	Improved patient recruitment prediction	ML method	• Reduced error with current industry standard • ML methods outperform the existing projections of site • Efficient prediction of enrolment	

(continued)

Table 1 (continued)

Sl. No.	Reference	Objective	Technology employed	Advantages	Stage
10	[15]	Statistical approach for patient recruitment	Prediction model	• Considered the randomness and uncertainties in input data • A well-explained method of modelling of recruitments	
11	[16]	To develop an efficient clinical trial protocol	Seedling model	• A comprehensive scoring system • Clinical institutes can manage staffing accurately • Considers all the regulatory requirements • Upholds scientific integrity	Protocol optimisation
12	[17]	To reduce the procedural frequency	Analysis of data	• Reduced drug development cost • Simplified protocol design • Improved efficiency in study conduction	
13	[18]	To improve response adaptive randomisation design	9 ML algorithm ensemble approach	• Improved treatment outcome • Integration of multiple methods • Personalised optimal treatment assignment • Higher response rate	
14	[19]	A noncoding method to study human genetics	Deep learning	• Direct learning of a sequence of regulatory codes • Integration of evolutionary conversation and genomic identification • Predicts chromatic features	

(continued)

Table 1 (continued)

Sl. No.	Reference	Objective	Technology employed	Advantages	Stage
15	[20]	Reduction of protocol complexity	Complexity ratio model	• Objective method of quantifying clinical trial • Recommendations are precise and accurate • Effective allocation of resources	
16	[21]	To develop an efficient data management system for diabetic clinical trial	Rapid prototyping	• Various components of data management are considered • Usage of data in classified format • Have incorporated multiple inputs to improve the integrity	Data management
17	[22]	Discrepancy management	Artificial intelligence	• The quality of clinical data management is measured based on the standards being followed • Rapid information cleaning process is incorporated	
18	[23]	To develop EDC technology-enabled CDM	EDC	• Efficient path of data from record to regulatory services • Dedicated regulatory compliance • Flexible data format	

(continued)

Table 1 (continued)

Sl. No.	Reference	Objective	Technology employed	Advantages	Stage
19	[24]	Efficient classification of inflammatory disease	Supervised ML	• Improved and personalised treatment of patients • Helps the clinicians to categorise the data • Simple disease classification	
20	[25]	To identify paratuberculosis in cattle	ML-based feature selection	• Precise indicator for diagnostics • Easy and effective analysis of data using machine learning	

References

1. Rghioui A, Naja A, Mauri JL, Oumnad A (2021) An IoT based diabetic patient monitoring system using machine learning and node MCU. J Phys Conf Ser
2. Sundareswaran R, Veezhinathan M, Shanmugapriya M, Dhanush Babu R (2022) Assessment and evaluation of diabetic foot using biothesiometry and artificial neural networks. J Clin Diagn Res 16(11)
3. Mujumdara A, Vaidehi V (2019) Diabetes prediction using machine learning algorithms. In: International conference on recent trends in advanced computing
4. Yin Y, Zeng Y (2016) The internet of things in healthcare: an overview. J Ind Inf Integr
5. Efat MdIA, Rahman S, Rahman T (2020) IoT based smart health monitoring system for diabetes patients using neural network, July 2020
6. Shah P, Kendall F, Khozin S, Goosen R, Hu J, Laramie J, Ringel M, Schork N (2019) Artificial intelligence and machine learning in clinical development: a translational perspective. Scripps Research Translational Institute, 26 July 2019
7. Diaconu A, Cojocaru FD, Gardikiotis I, Agrigoroaie L, Furcea DM, Pasat A, Suciu G, Rezuş C, Dodi G (2022) Expending the power of artificial intelligence in preclinical research: an overview. IOP Conf Ser Mater Sci Eng 1254
8. Jiménez-Luna J, Grisoni F, Schneider G (2020) Drug discovery with explainable artificial intelligence [cs.AI], 2 July 2020
9. Senior AW, Evans R, Jumper J, Kirkpatrick J, Sifre L, Green T, Qin C, Žídek A (2020) Improved protein structure prediction using potentials from deep learning, 15 Jan 2020
10. Casey R, Adelfio A, Connolly M, Wall A, Holyer I, Khaldi N (2021) Discovery through machine learning and preclinical validation of novel anti-diabetic peptides. Biomedicines 9:276
11. Battina DS (2017) The role of machine learning in clinical research: transforming the future of evidence generation. Int J Innov Eng Res Technol (IJIERT) 4(12). ISSN: 2394-3696
12. Bhatt A (2021) Artificial intelligence in managing clinical trial design and conduct: man, and machine still on the learning curve? Perspect Clin Res 12(1):1–3
13. Vazquez J, Abdelrahman S, Byrne LM, Russell M, Harris P, Facelli JC (2020) Using supervised machine learning classifiers to estimate likelihood of participating in clinical trials of a de-identified version of research match, 28 Aug 2020

14. Liu J, Allen PJ, Benz L, Blickstein D, Okidi E, Shi X (2021) A machine learning approach for recruitment prediction in clinical trial design. In: Proceedings of machine learning research LEAVE UNSET, pp 1–7
15. Vladimir V. Recruitment modeling and predicting in clinical trials. Pharm Outsourcing
16. Hope Weissler E, Naumann T, Andersson T, Huang E, Ghassemi M (2021) The role of machine learning in clinical research: transforming the future of evidence generation. Trials
17. Malikova MA (2016) Optimization of protocol design: a path to efficient, lower cost clinical trial execution. Future Sci OA, 12 Jan 2016
18. Getz KA, Wenger J, Campo RA, Seguine ES, Kaitin KI (2008) Assessing the impact of protocol design changes on clinical trial performance. Am J Ther 15:450–457
19. Wang Y, Carter BZ, Li Z, Huang X (2022) Application of machine learning methods in clinical trials for precision medicine. JAMIA Open 5(1)
20. Zhou J, Troyanskaya OG (2015) Predicting effects of noncoding variants with deep learning-based sequence model. Nat Methods 12(10):931–934
21. Nourani A, Ayatollahi H, Solaymani-Dodaran M (2022) A clinical data management system for diabetes clinical trials. Hindawi J Healthc Eng 2022:10, article ID 8421529
22. Gazali, Kaur S, Singh I (2017) Artificial intelligence based clinical data management systems: a review. Inf Med Unlocked 9:219–229
23. Lu Z, Su J (2010) Clinical data management: current status, challenges, and future directions from industry perspectives. Open Access J Clin Trials 2:93–105
24. Mossotto E, Ashton JJ, Coelho T, Beattie RM, MacArthur BD, Ennis S (2017) Classification of paediatric inflammatory bowel disease using machine learning. Sci Rep 7:2427
25. Lee S-M, Park H-T, Park S, Lee JH, Kim D, Yoo HS, Kim D (2023) A machine learning approach reveals a microbiota signature for infection with *Mycobacterium avium* subsp. *paratuberculosis* in cattle 11(1)

Efficient PAPR Reduction Techniques and Performance of DWT-OFDM

M. Thilagaraj, C. Arul Murugan, and R. Kottaimalai

Abstract Wideband digital communication can benefit from the effective wideband OFDM modulation technique, which is used for high-speed data requirement. Some important properties of OFDM are high bandwidth efficiency, robustness to channel fading, capability of handling strong echoes, and immunity to impulse response. High peak-to-average power ratio (PAPR) is the main disadvantage of OFDM, which restricts its use in communication systems. FFT is typically used to implement OFDM. Orthogonal wavelets are being used in recent research to switch from FFT-based OFDM to DWT-based OFDM in order to achieve high performance and high-speed integrated data. By incorporating DWT into OFDM, we can improve spectral efficiency and reduce bit error rates. This work is introduced to compare the performance of DWT-OFDM with traditional FFT-based OFDM examining the performance of PAPR. Additionally, it is suggested that effective PAPR reduction techniques maintain the adverse impacts induced within the allowable limits.

Keywords Wavelet-based OFDM (WOFDM) · Orthogonal · Channel · Modulation · Transforms

1 Introduction

OFDM is a multicarrier modulation scheme that offers selective fading and interference. The use of OFDM is used to achieve a high-rate data stream in UHF and microwave spectrum.

M. Thilagaraj
MVJ College of Engineering, Bengaluru, India

C. Arul Murugan
Karpagam College of Engineering, Coimbatore, India

R. Kottaimalai (✉)
Kalasalingam Academy of Research and Education, Krishnankoil, India
e-mail: rkottaimalai@gmail.com

S. Kumar et al. (eds.), *Fourth Congress on Intelligent Systems*, Lecture Notes in Networks and Systems 868, https://doi.org/10.1007/978-981-99-9037-5_34

1.1 Basic Principles of OFDM

OFDM is a combination of both modulation and multiplexing. The main concept of OFDM is to regenerate the data from carrier which allows simultaneous transmission of subcarriers [1]. OFDM is similar to FDMA, and the total available bandwidth is subdivided into the multiple channels to allocate user for each channel [2]. The signals must be orthogonal and placed in the symbol period between the subcarriers to remove systems error correction. Because of this, there is no overhead, and the channel is time-multiplexed.

1.2 PAPR of OFDM Signals

OFDM has a lot of problems, including a high peak-to-average ratio (PAPR) and transmission to distort the carriers [3]. OFDM has multicarrier which is a sum of complex random variables that produce large output. In some cases it may cancel each other producing zero output. When the signal is stable, and there is no out-of-band interference to the signals [4]. The transmitter side is where this PAPR issue is more problematic [5]. Implementing power amplifier with wild linear range is very expensive and also consumes high power. In this proposed work, the PAPR can be greatly reduced by performing IDWT replacing FFT by wavelet decomposition and IFFT by wavelet reconstruction which gives new dimension in the digital signal processing [6]. Also, an efficient PAPR reduction technique SPI combined with SLM is proposed for better estimation of PAPR.

2 FFT-Based OFDM

OFDM relies on IFFT/FFT for an efficient implementation. OFDM signal could be generated by N point Inverse Fast Fourier Transform (IFFT) to get back the original signal [7]. For an OFDM system with N subcarriers, the channel delay spread baseband signal is given as

$$x(t) = 1/\sqrt{N}\sum_{m=0}^{N-1} X[m_1]\exp(j2\pi\Delta ft), \quad 0 \le t \le T, \tag{1}$$

where $X[m_1]$ is the data symbol carried by the mth subcarrier. Δf is the frequency difference between subcarriers. $x(t)$ is a band limited signal with the Nyquist frequency $fN = N\Delta f$.

3 Wavelet Decomposition Process

Fourier-based conventional OFDM system is employed to improve the spectral efficiency. The channel distortion is decomposed to exploit the orthogonality relationship between modulation and multiplexing schemes [8]. The wavelet transform is achieved by a low-pass and a high-pass filter which results in loss of information. Due to the redundancy, the detail coefficients are obtained from $g(n_1)$ and $h(n_1)$ filters which restrict the data rate [9].

Wavelet decomposition process is very useful for extracting information and approximations. The two sets of coefficients are used to shape the signal of the individual frequencies. A series of filters repeats the decomposition process at different locations in the signal. The reverse operation of this decomposition is difficult to extract the communication system of the signal.

4 Wavelet-Based OFDM

Wavelets have been used for mapping $s(k_1)$ data stream to the symbol stream $x(n_1)$. Once mapping process is performed, these properties change from one variant to the other, which are well localized and faster than any inverse polynomial [10]. Sequentially, frequency selective channel is also analyzed based on the wavelet basis [11].

The filtering processes and up-sampling are applied because half of samples are redundant [12]. The quadrature mirror filter (QMF) is possible to minimize any amplitude distortion in the system [13]. This will occur whenever the system is used to minimize error and optimize the filter [14]. The wavelet decomposition and reconstruction structure are shown in Figs. 1 and 2.

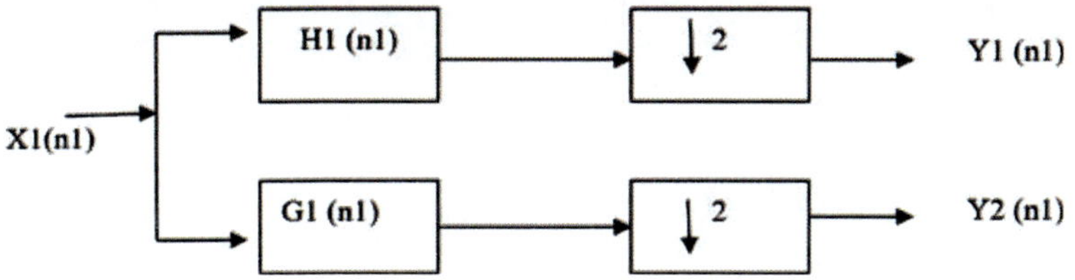

Fig. 1 Wavelet decomposition

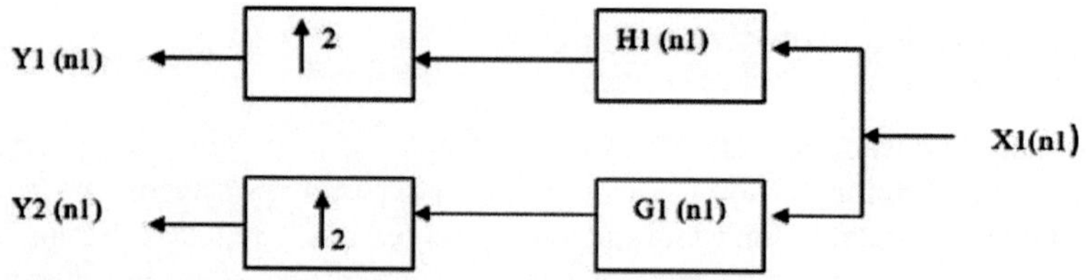

Fig. 2 Wavelet reconstruction

5 PAPR in DWT-Based OFDM and FFT-Based OFDM Systems

An OFDM signal is made up of several individually modulated subcarriers, which put together to provide a high PAPR [15]. Due to use of wavelet transform, it has the ability to perform progressive computations. Increasing the data rate distribution via handheld wireless networks is made possible by the effective approach known as OFDM. FFT was employed in traditional OFDM to convert and map information onto orthogonal subcarriers, but this method had the drawbacks of being highly intricate and rigid by nature. Since it operates in the time–frequency domain, wavelet analysis has significant benefits over Fourier analysis in terms of adaptability and difficulty. The present research compares DWT-based OFDM to traditional FFT-based OFDM with regards to bit error rate (BER) under the combined effects of route loss, multiple path fading, and noisy environments for mobile WiMax.

The power of a sine wave with amplitude provides good performance when compared to FFT-OFDM. A carrier that is unmodulated has a PAPR, so the computational complexity is reduced. The RF power amplifier is decreased by a high PAPR ratio.

A complementary cumulative distribution function (CCDF) curve is given by

$$F(\alpha) = \Pr\{|x_1[n_1]|^2 >\} = e^{-\alpha}. \quad (2)$$

6 Proposed PAPR Reduction Techniques

Since the peak value of the system is quite high, OFDM can be employed to reduce the peak transmitted power [16]. Signal distortion technique leads to the prevention of spectral growth to degrade the system performance. Signal scrambling techniques mostly select the signal with low PAPR to provide information from spectral regrowth. In this investigation, PAPR performances of wavelet-based OFDM are introduced.

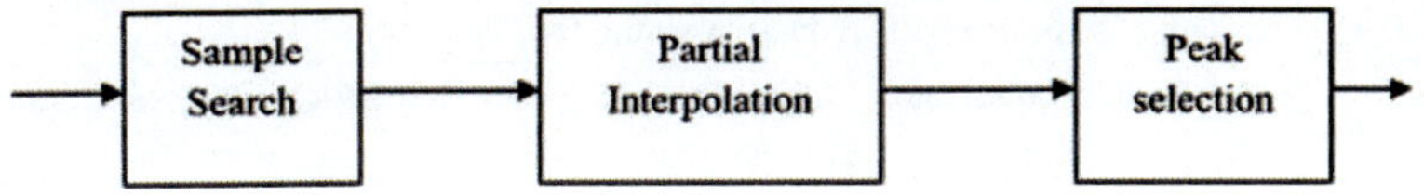

Fig. 3 Block diagram of search and partial interpolation for PAPR estimation

6.1 *Search and Partial Interpolation*

In SPI scheme, the estimation of PAPR is given by $x(t)$ by $x_1(n_1)$ or $x_2(n_2)$. The block diagram of search and partial interpolation for PAPR estimation is depicted in Fig. 3.

6.2 *Selected Mapping Combined with Search and Partial Interpolation*

The input sequences are multiplied by a common phase sequence in SLM, and the IFFT operation is selected by the lowest PAPR. This approach is applicable for all types of modulation as it is evident from the spectral pattern of OFDM systems [17–19].

The diagram of the SLM technology combined with SPI is represented in Fig. 4. By using the SPI technique, the entire bandwidth is occupied by one carrier, and PAPR values are determined. Table 1 compares various PAPR reduction approaches with a wide variety of parameters.

Amplitude clipping is said to be the best method that might be used to reduce PAPR in an OFDM network. In this instance, a threshold amplitude value is applied to restrict the peak envelope of the source signal.

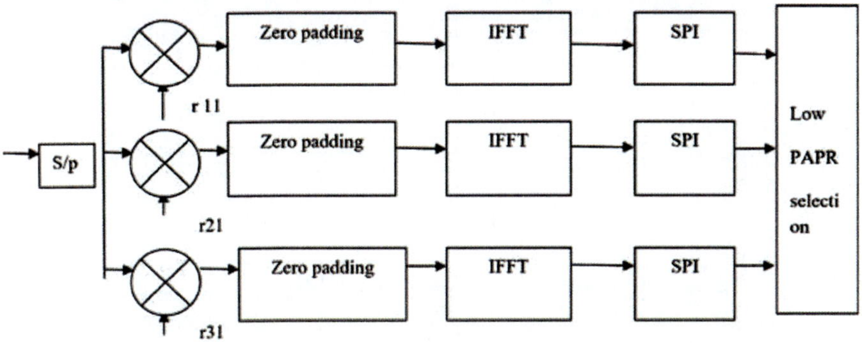

Fig. 4 Block diagram of SLM combined with SPI

Table 1 Comparison of various PAPR reduction methods [17]

Types	Reduced data rate	Distortion less	Boosted BER	Increased transmit power
Tone reservation	√	√	×	√
Clipping and filtering	×	×	×	×
Selected mapping	√	√	√	×
Tone injection	√	√	√	√
Interleaving	×	√	×	√
Coding	√	√	×	×
Partial transmit sequences	√	√	√	×

7 Simulation Results

The CCDF characteristics comparison of PAPR reduction for DWT-OFDM and FFT-OFDM is depicted in Fig. 5.

For an OFDM output symbol block, it is able to reduce the overall computational complexity before and after processing. It is clear that the PAPR is reduced over 5 dB in terms of performance. BER performance for FFT-based OFDM and DWT-based

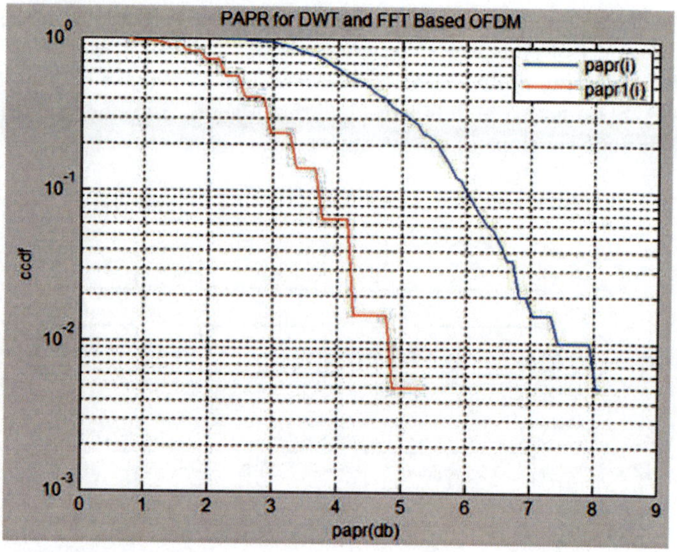

Fig. 5 CCDF characteristics comparison of PAPR reduction

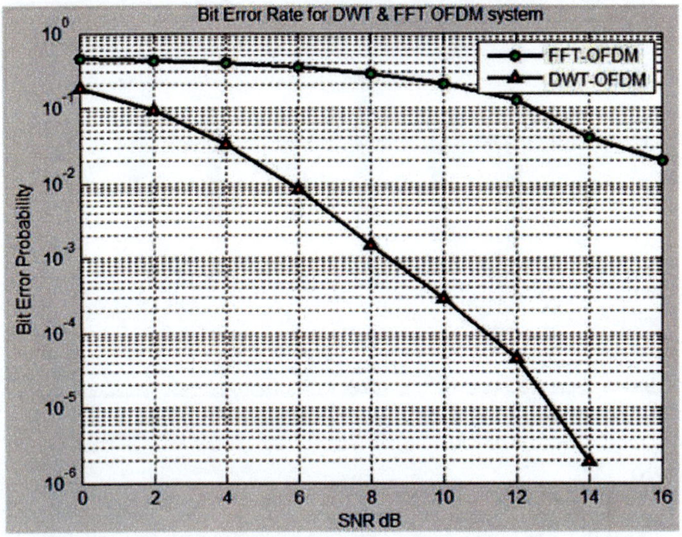

Fig. 6 Performance of FFT in terms of BER

OFDM system is shown in Fig. 6. It can be seen that an excellent PAPR reduction is achieved than existing technique.

Utilizing MATLAB software, the simulation results were obtained to determine the PAPR reduction capabilities. The demonstrated systems primary goal is to compare an existing modulation technique and decrease the PAPR and BER that are shown in Fig. 7a and b. In this approach, the graphical representations to raise bit error rate while simultaneously lowering signal-to-noise ratio are shown.

In Fig. 8, conventional SLM and Base blind SLM technique with number of subcarriers 4, 8, and 16 is represented. The primary goal is to reduce the PAPR by employing modulation techniques. Figure 9 depicts the comparison of BSR and SNR.

In this study, the combined effects of link loss, multipath fade, and distortion in a mobile WiMax setting were used to compare the efficiency of a traditional OFDM against a wavelet-based OFDM system. The experiments revealed that wavelet-based OFDM systems remained highly adaptable and minimal in complexity because they just required simple low-order filters as opposed to sophisticated FFT processing units.

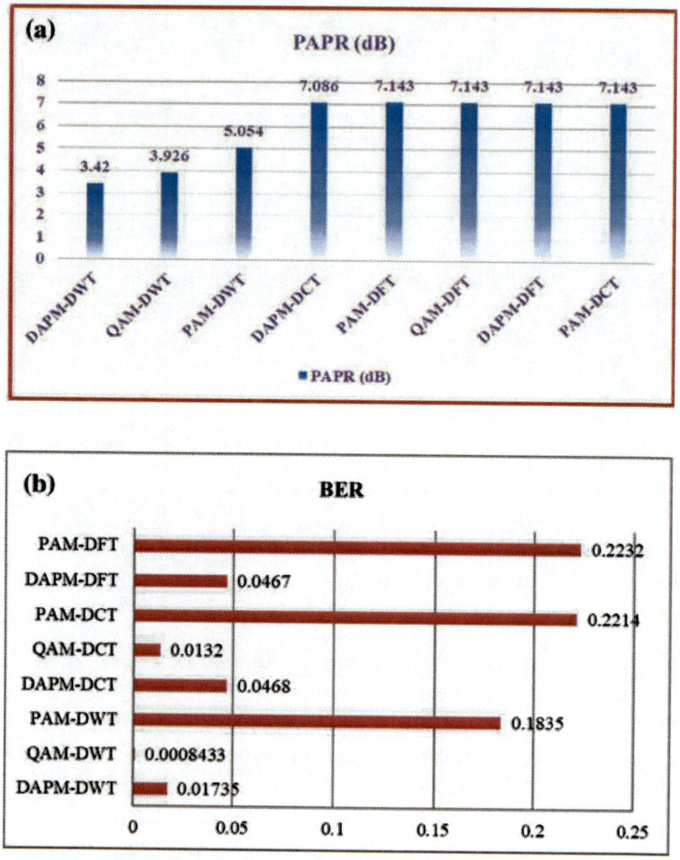

Fig. 7 **a** PAPR results at 10-1 CCDF for various transforms and modulation techniques. **b** Performance of BER for different transforms and modulation technique

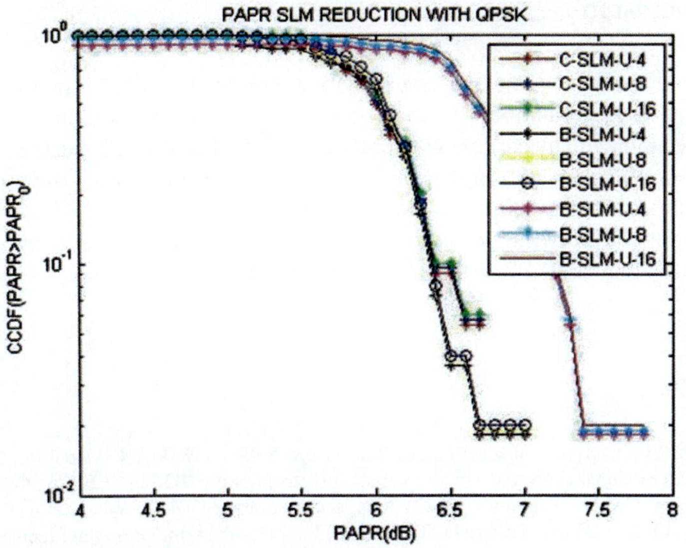

Fig. 8 SLM method for QPSK modulation with PAPR reduction

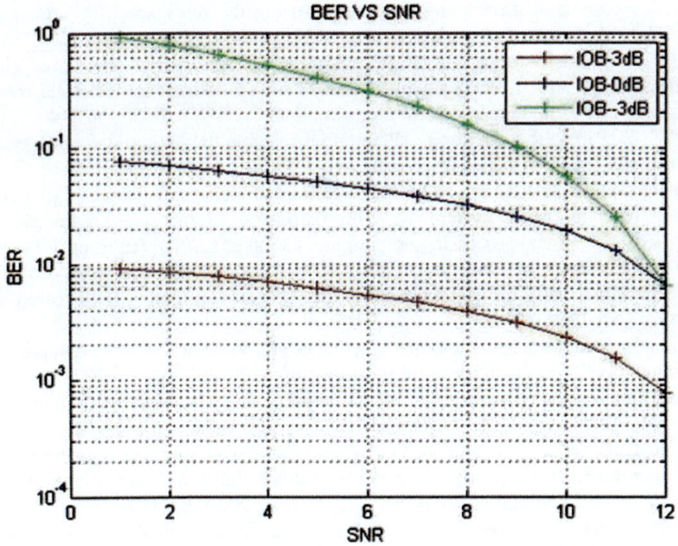

Fig. 9 Comparison of BER and SNR

8 Conclusion

Orthogonal frequency division multiplexing has feature of multicarrier system because of its huge number of subcarriers. OFDM is a very attractive modulation technique among many carriers with each subcarriers to achieve robustness in multi-path channel environment. High peak power is one of the major downsides of OFDM systems. In this investigation, a comparison is shown for the efficient PAPR reduction technique. According to the results of simulations, the wavelet transform can significantly reduce PAPR with suitable PAPR reduction technique.

References

1. Wen Y, Li Y (2018) Cell search algorithms at low SNR for WiMAX system. In: 2018 IEEE 18th international conference on communication technology (ICCT). IEEE, pp 428–432
2. Bebyrahma AMK, Suryani T (2022) Analysis of combined PAPR reduction technique with predistorter for OFDM System in 5G. In: 2022 International seminar on intelligent technology and its applications (ISITIA). IEEE, pp 478–483
3. Puspitasari AA, Nadhiroh UA, Habibah MDN, Palupi GS, Ridwan M, Moegiharto Y (2021) Application of the combination scheme of PAPR reduction and predistortion techniques in cooperative communication with AF protocol using the relay selection strategy. In: 2021 International electronics symposium (IES). IEEE, pp 108–113
4. Al Ahsan R, Wuttisittikulkij L (2021) Artificial neural network (ANN) based classification of high and low PAPR OFDM signals. In: 2021 36th International technical conference on circuits/systems, computers and communications (ITC-CSCC). IEEE, pp 1–4
5. Hu W-W (2019) PAPR reduction in DCO-OFDM visible light communication systems using optimized odd and even sequences combination. IEEE Photonics J 11(1):1–15
6. Sharan N, Ghorai SK, Kumar A (2021) PAPR reduction using blend of precoder and μ-law compander in HACO system. In: 2021 IEEE 2nd international conference on applied electromagnetics, signal processing, & communication (AESPC). IEEE, pp 1–5
7. Liu Z, Hu X, Wang W, Ghannouchi FM (2021) A low-complexity joint PAPR reduction and predistortion based on generalized memory polynomial model. IEEE Microwave Wirel Componen Lett 32(1):88–91
8. Thilagaraj M, Arul Murugan C, Ramani U, Ganesh C, Sabarish P (2023) A survey of efficient light weight cryptography algorithm for internet of medical things. In: 2023 9th International conference on advanced computing and communication systems (ICACCS), vol 1. IEEE, pp 2105–2109
9. Thabet RM, Ali WAE, Mohamed OG (2020) Synchronization error reduction using guard-band allocation for wireless communication systems. In: 2020 International conference on innovative trends in communication and computer engineering (ITCE). IEEE, pp 308–312
10. Bodhe RS, Narkhede SS, Joshi S (2012) Design and implementation of baseband processing for wavelet OFDM. In: National conference e-PGCON2012, Pune
11. Kanti R, Rai M (2013) Comparative analysis of Daubechies wavelets in OWDM with OFDM for DVB-T. Int J Sci Eng Res 4(3)
12. Wu Y, Zou WY (1995) Orthogonal frequency division multiplexing: a multi-carrier modulation scheme. IEEE Trans Consum Electron 41(3):392–399
13. Arul Murugan C, Karthigaikumar P, Priya SS (2020) FPGA implementation of hardware architecture with AES encryptor using sub-pipelined S-box techniques for compact applications. Automatika 61(4):682–693

14. Joo H-S, Kim K-H, No J-S, Shin D-J (2017) New PTS schemes for PAPR reduction of OFDM signals without side information. IEEE Trans Broadcast 63(3):562–570
15. Proakis JG (1995) Digital communications. McGraw-Hill, New York, NY, USA
16. Ramaraj K, Govindaraj V, Murugan PR, Zhang Y, Wang S (2020) Safe engineering application for anomaly identification and outlier detection in human brain MRI. J Green Eng 10:9087–9099
17. Ramaraj K, Govindaraj V, Zhang YD, Murugan PR, Thiyagarajan A (2021) Brain anomaly prediction with the intervention of fuzzy based clustering and optimization techniques for augmenting clinical diagnosis. In: 2021 3rd International conference on advances in computing, communication control and networking (ICAC3N). IEEE, pp 872–877
18. Amiya G, Ramaraj K, Murugan PR, Govindaraj V, Vasudevan M, Thiyagarajan A (2022) A review on automated algorithms used for osteoporosis diagnosis. Inventive Syst Control Proc ICISC 2022:247–262
19. Hu C, Wang L, Zhou Z (2020) A modified SLM scheme for PAPR reduction in OFDM systems. In: 2020 IEEE 10th International conference on electronics information and emergency communication (ICEIEC). IEEE, pp 61–64

Machine Learning-Based Image Forgery Detection Using Light Gradient-Boosting Machine

Meena Ugale and J. Midhunchakkaravarthy

Abstract In recent days, due to the increasing development of digital automation, images have emerged as a significant way to interact as well as transfer messages in our community, and there was a high rise in the number of details transferred in the formation of virtual pictures in the day-to-day life especially with the disclosure of online medias such as Instagram, Twitter, and Facebook. Moreover, uploading pictures on social media and modifying those images with related software apps is considered a common method to do in current days. Even though, if every person does not perform with ominous meanings, but still, there is a noticeable rise in misconduct regarding the malignant image influence as well as updating. This research proposes an image/video forgery identification method by utilizing the light gradient boosting machine (Light-GBM) method to detect the fabrication in the visual data with an increased rate of accuracy. The performance, as well as comparative analysis, is estimated based on the performance metrics such as accuracy at 94.91%, sensitivity at 94.77%, and specificity at 93.26%, respectively, which is superior to the previous techniques.

Keywords Block-wise feature extraction · Feature extraction · Forgery detection · Light gradient boosting machine · Machine learning

1 Introduction

In the current days, trusting the updated pictures on every platform such as web pages, media, social media, as well as publications [1] has become uncertain, particularly regarding popular figures such as entertainers, athletes, and political figures, which is featured to the wide expanse of trespassing by every people including amateurs.

M. Ugale (✉) · J. Midhunchakkaravarthy
Lincoln University College, Petaling Jaya, Selangor, Malaysia
e-mail: meena.u@xavier.ac.in

J. Midhunchakkaravarthy
e-mail: midhun@lincoln.edu.my

S. Kumar et al. (eds.), *Fourth Congress on Intelligent Systems*, Lecture Notes in Networks and Systems 868, https://doi.org/10.1007/978-981-99-9037-5_35

Image manipulation is considered an aspiration for numerous persons [1]; moreover, the pictures are mostly utilized as essential proof to clear situations. Nevertheless, the establishment of several picture-modifying gadgets has let several people conveniently influence the pictures. In addition, the utilization of influenced pictures for malignant reasons could impact negatively the social community, as the legitimacy of the pictures turned skeptical, the detection of the image originality has become an essential problem, and the identification of the fabricated picture is not possible by the people vision directly, and it is necessary to establish convenient techniques for the image fabrication identification. Several types of research have been done on the identification of numerous picture forgeries [2–4], which contain both positive and negative characteristics, especially regarding virtual data safety. However, these researches encourage displaying schemes of image craft as well as enhancing the previous pictures. Furthermore, artificial intelligence (AI), mostly machine learning (ML), influences pictures as well as videos in a certain manner where they are frequently identical from the originals to the naked eyes as well as to the virtual [5]. Several extensive approaches are largely utilized to influence pictures, which includes computer-generated imagery (CGI), as well as content alteration [5].

The main category of image influence attacks people faces in the current days, which was copy-move fabrication, retouching, erase-fill, and image splicing [6]. When the attack includes substituting several phenomena in a virtual picture with an alternate phenomenon, obtained from a varied place, in a similar picture, the attack is considered as a copy-move fabrication. Moreover, the attack in which several current objects from an outside origin are resourcefully enumerated to an original picture along with the malignant purpose of presenting, this compounded picture as an original, along with visibly undetectable evidence of the influence, is called splicing or interlacing attack [6]. Copy-move forgery (CMF) is an often-utilized trespassing technique, where the section of a picture is duplicated from a part of the picture as well as attached in some other part in the exact picture, and the picture could be fabricated to obscure or alter its definition by utilizing the CMF technique. Thus, investigating the dependency of the picture as well as locating the duplicated and progressed regions is essential [2]. Furthermore, DeepFake method based on the deep learning (DL) derived from the terms "deep learning" as well as "fake" is considered a significant competitor between the content-altering video fabricating approaches. The utilization of deep neural networks (DNN) caused a procedure of generating compelling duplicate pictures as well as videos more conveniently and quickly and is a method in which a picture of one human is influenced by the picture of another human by utilizing DL methods [5–7].

The major goal of the research is to identify the images or video forgery, in which the data required for the detection of forgery in pictures/recordings is collected, as well as the preprocessing is functioned to build the image beneficial for detecting the fabrication by utilizing a block-wise feature extraction and an ML technique called the light gradient boosting machine (Light-GBM) method.

Block-wise feature extraction: Block-wise feature extraction is employed to separate the image into blocks and accelerates the extraction process, in which the attributes such as the light coefficients estimate the coherence and sequence of the features as well as pixel flow-based features to standardize the pixel flow bearings, which connect the physical movements to the image movements.

Light gradient boosting machine model: The Light-GBM algorithm is approached for identifying the fabrication in the images or videos, in which the data is performed and the method is also utilized in categorization, reversion, grouping, and assisting effective collateral training.

The image/video forgery identification technique established in the previous research together with its merits as well as demerits is described in Sect. 2. The proposed Light-GBM technique is demonstrated in Sect. 3. The outcomes accomplished are estimated in Sect. 4, and the conclusion is explained in Sect. 5.

2 Literature Review

The review of the related research for image forgery detection is demonstrated in the following section which includes the merits and demerits.

Niyishaka and Bhagvati in [8] developed a convenient and automatically effective image interlacing fabrication identification which regards an exchange among functionalities and price to the end users and includes three stages, and the experimental result states that the developed technique is algorithmically effective and coherence for image trespassing identification. However, the established technique creates lengthy bar graphs that reduce the identification timing, particularly on the high-level face directory.

Al Azrak et al. in [1] developed a convolutional neural network (CNN) technique for copy-move falsification identification on the basis of feature extraction, which is applied along sequence sets of involution and pooling seams. Moreover, the developed technique could as well be utilized for operative fabrication identification due to its hardness to identify the influence of the virtual apex pictures or pictures with signs. But the established technique does not encrypt the arrangement as well as object orientation.

Lee et al. in [2] developed root mean squared energy on the basis of rotation-invariant feature attribute by utilizing using decameter band wavelet quantities, in which a related component is utilized by implementing low attribute reinforcement accelerated by the VGG16 system to achieve the available duplicated and progressed spot sets, and the research outcome states that the established technique is better than the previous approaches. Nevertheless, the perfect identification operation among several techniques is complex as there is a lack of quality instruction as well as test data for differentiation.

Mitra et al. in [5] established an approach based on a novel neural network (NNN) to identify fabricated recordings by implementing a main video-frame eradication approach to minimize the automation in identifying deep fake recordings, as well as a method containing a CNN and a classification algorithm is developed, and the attribute angles from the developed CNN method are employed as the input of the succeeding classification algorithm for categorizing the recordings accomplished an increased rate of accuracy by not instructing the technique with an abundant number of data, as well as minimizes the automation essentially than the previous researches. However, the developed technique normally needed many data when compared to the previous ML methods.

Nath and Naskar [6] developed a blind image interlacing identification approach that utilizes a deep convolutional residual network framework as a background, succeeding by a completely related classification algorithm, and a CNN method is also established to function computerized feature engineering, to save the uninteresting job of choosing the picture attributes, and to evaluate if the picture is original or fabricated, the attribute vector is further transmitted to a classification algorithm, in which the research outcome articulates that the established model outperforms the previous methods. Nevertheless, the developed technique could only identify if a picture is forged or not and not includes the location of the forged area in an image.

2.1 Challenges

The manipulated areas in the fabricated pictures are to be localized to identify the forged images accurately [8].

The memory duration as well as the method range will be further decreased as the memory limit of the DNN is huge to adjust to the sharp mechanics as well as the developed method will be progressed to an extended forum such as national IDs [5].

The evaluation of whether the interfered area of the picture has the ability to be reinstalled from the intelligence of the fabrication placement in an algorithmically realistic manner will be further explored, which paves the way to a large range of investigations in the assigned dominion [6].

The developed ML-based fraud identification technique will be executed and evaluated on an extended size of physical data with various other ML techniques [9].

3 Proposed Light-GBM for Forgery Detection

The major goal of the research is to design and develop the Light-GBM technique for the identification of image/video fabrication. At first, the input image will be put through preprocessing for transforming the raw data into an attainable form and then will be subjected to the feature extraction, which includes the features like block-wise feature extraction (BFE) which separates the image into blocks and accelerates the extraction process, in terms of the statistical features such as variance, median, mean, standard deviation (SD), and root mean square (RMS) and light coefficients-based distant features (LCDF) to estimate the coherence and sequence

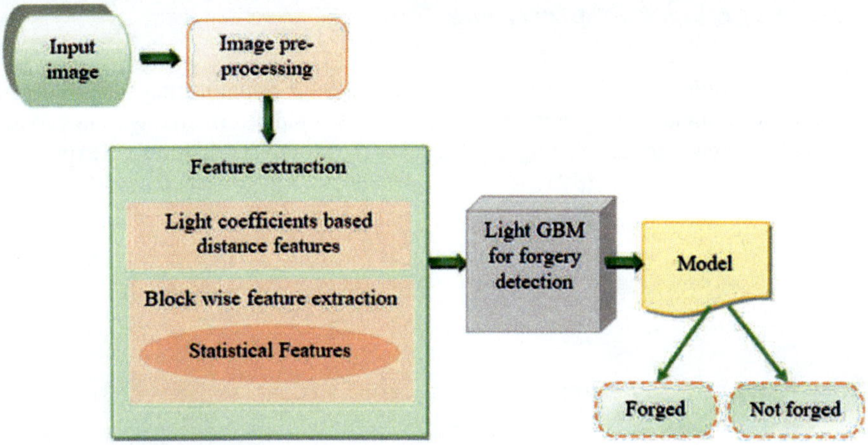

Fig. 1 Diagrammatic representation of the proposed methodology

of the features as well as pixel flow-based features to standardize the pixel flow bearings, which connect the physical movements to the image movements. From these features, the Light-GBM classifier will be employed for identifying the fabrication from the inserted image, which is utilized in categorization, reversion, grouping, and assisting effective collateral training and utilizes the piece-wise sustained trees as well as estimated loss functions along with secondary-level evaluation. Furthermore, differentiation will have functioned between the developed and traditional models, as well as the evaluation will be conducted based on the performance metrics. Moreover, the research will be executed using Python, and efficiency will be estimated by utilizing the accuracy, sensitivity, specificity, and segmentation accuracy, and the articulation of the proposed technique is represented in Fig. 1.

3.1 Input

The input data is collected from the image/video forgery identification dataset (DTS) called DSO-1 and DSI-1 DTS [10] and is numerically demonstrated by the equation as,

$$\text{IN} = \sum I^{d_1} + I^{d_2} \tag{1}$$

where IN is identified as the input and I^{d_1} and I^{d_2} are referred to as the DTSs.

3.2 Image/Video Preprocessing

The preprocessing of the input image is functioned for transforming the unrefined data into an attainable form; therefore, the attributes could be effectively removed for the identification of the image forgery, as well as the preprocessed image is expressed by,

$$P^* = \sum p^{i*} \tag{2}$$

3.3 Feature Extraction

The feature extraction of the inserted input functions is based on the features like LCDF and BFE for achieving a better outcome of the proposed method.

3.3.1 Light Coefficients-Based Distance Features

The distance features based on the light coefficients are measured in a block-wise manner. There is a contrary association between distance and light intensity, in which the distance increases when the intensity of the light decreases. The estimation is conducted on the basis of the distance method such as local gradient pattern (LGP), local optimal-oriented pattern (LOOP), and local directional pattern (LDP) which are explained briefly below.

(a) *Local Gradient Pattern*

The gradient features are fundamental as well as intrinsic properties of human image discrimination, which are frequently in use for differentiating the numerous confined semantic architectures of imagery; they are also physically powerful and have vigorous attributes for quantifying the professed image standard. Moreover, the LGP consists primordial in the main visual cortex, which is directly associated to the professed image standards. Moreover, the local semantic structural alteration can usually reproduce indignity in the image standard. The current research stated that the LBPs can proficiently and efficiently articulate the local semantic structural details of an image as well as can be regarded as the dual estimation of the local semantic structural detail primordial in the main phase of vision, which is expressed as

$$A_\sigma(Z) = \sqrt{[G_{x,\sigma}(z)]^2 + [G_{y,\sigma}, \sigma(Z)]^2} \tag{3}$$

where $G_{x,\sigma}$ and $G_{y,\sigma}$ are defined as the gradient magnitude of the image.

(b) *Local Optimal-Oriented Pattern*

The main drawback of LDP is the subjective series of binarization mass which adds the dependency to direction which also gets led back by the experiential value assignment to the variable of threshold, which provides an ad hoc constraint on the amount of bits allowable to be 1, therefore decreasing the amount of potential words, as stated ahead. The LOOP introduces a nonlinear combination of LBP as well as LDP which overcomes the limitations when protecting the robustness of each by encoding revolution indifference into the major formulation, which is expressed as

$$\text{LOOP}_{(x,y)} = \sum_{n=0}^{7} s(i_n - i_c).2^n \tag{4}$$

where (x, y) is defined as the pixels and i_n, i_c are referred to as the pixel intensity.

(c) *Local Directional Pattern*

The local directional pattern (LDP) is an eight-bit binary code given to every picture element of the inserted data which is measured by differentiating the related edge reaction value of a picture element in the diverse direction, which builds a major solid design in the existence of noise equivalent image later than adding up Gaussian white noise, which is expressed as

$$\text{LDP}_k(x_c, y_c) = \sum_{n=0}^{7} s(m_n - m_k).2^n \tag{5}$$

3.3.2 Block-Wise Feature Extraction

The block-wise feature extraction is functioned by separating the input image into blocks and extracts the attributes from every block which includes the features such as light coefficients-based distant features to estimate the coherence and sequence of the features as well as pixel flow-based features to standardize the pixel flow bearings, which connects the physical movements to the image movements. The inserted image features are removed block-wise from each block row, in which the horizontal blocks are considered at first, and then the block is moved vertically in order to consider the nearby blocks, and the function will be processed until every attribute is extracted, with statistical features such as mean, median, variance, SD, as well as RMS, and the outcome achieved from the technique is trained along the feature parameters which are utilized in identifying the effective parameters in a short time, which is expressed numerically as

$$\text{temp} = \frac{1}{1-3} \log_2 \sum_{a=1}^{18} \sum_{b=1}^{18} \left(\frac{Q(a,b)}{\sum \sum Q(a,b)} \right) \tag{6}$$

where temp is referred to as the temporary parameter and $Q(a, b)$ is signified as the directed weight matrix.

(a) *Mean*

The mathematical mean is the average of the values $[x_1, x_2, \ldots, x_n]$ placed in the time window which is calculated by the given expression

$$\mu = \frac{1}{m}\sum_{i=1}^{m} x_i \tag{7}$$

where μ is defined as the mean and x_i is identified as the average value.

(b) *Median*

Median is the middle number separating the higher part from the lower part, which is used instead of the mean when there is an outlier in the dataset that skews the average as a median is less affected by outliers than the mean, which is expressed as

$$\text{mediam} = \frac{(n/2)\text{th} + (n/2 + 1)\text{th}}{2} \tag{8}$$

where $(n/2)$ is referred to as the rate of observation.

(c) *Standard Deviation*

The standard deviation was calculated by the given equation to measure how the values $[x_1, x_2, \ldots, x_n]$ are spread out, which is expressed as

$$\sigma = \sqrt{\frac{1}{m}\sum_{i=1}^{m} (x_i - \mu)^2} \tag{9}$$

where σ is referred to as SD and μ is defined as the mean value.

(d) *Root Mean Square*

The RMS is the square root of the MS, which is the numerical mean of the values of square groups, which is also known as a quadratic mean as well as a particular case of the general mean, in which the model value is 2 and is also articulated as a changeable process based on a fundamental of the square values which are immediate in a cycle and is expressed mathematically as

$$\text{RMS} = \sqrt{\frac{(x_1^2, x_2^2, \ldots, x_n^2)}{N}} \tag{10}$$

where $(x_1^2, x_2^2, \ldots, x_n^2)$ is defined as the values.

3.3.3 Pixel Flow-Based Features

The pixel flow-based features are extensively utilized for evaluating the pixel movement among double successive frames. The primary requirement for pixel flow is that a point's intensity on an image remains constant over a sufficiently short duration. Let the intensity distribution in a continuous frame is represented as $S_c(a, b, t)$, then the pixel flow condition may be expressed as $\frac{\partial S_c(a,b,t)}{\partial t} = 0$

The pixel flow is evaluated as

$$\varepsilon_{\mathrm{OF}}(a, b, t) = (\nabla S_c(a, b, t)), \nu(a, b, t)) + \frac{\partial S_c(a, b, t)}{\partial t} = 0 \tag{11}$$

where the value of $\nabla S_c(a, b, t)$ is measured by $\left[\frac{\partial S_c}{\partial_n} \frac{\partial S_c}{\partial_y}\right]^{\mathrm{T}}$ and $\nu(a, b, t)$ is measured by $[\nu_1(a, b, t)\nu_2(a, b, t)]^{\mathrm{T}}$.

3.4 *Light-GBM for Forgery Detection*

The Light-GBM algorithm is a sort of gradient boosting decision tree (GBDT) which is commonly utilized in categorization, reversion, grouping, and assisting effective collateral training and utilizes the piece-wise sustained trees as well as estimated functions of loss along with secondary-level estimation. This algorithm is categorized into three major classifications: firstly, feature parallels which are utilized simultaneously in situations with more attributes; secondly, data parallel which is implemented in segments that contain a huge number of data; and thirdly, voting parallels which are implemented in scenes with a large number of attributes as well as votes. While the data which is trained contains more descriptions, there are many vacant arrivals, several personal columns could be integrated with no loss of details, and containing fewer attributes also fasten the function of training, in which the Light-GBM has the ability to surpass many other ML methods in the terms of accuracy and speed and also assist collateral training as well as deductions. Therefore, the proposed Light-GBM classifier gains application in forgery detection. The Light-GBM algorithm contains an integral duplicate for classification adaptable that protects one from constructing them and utilizes data subcase, in which the technique employed is gradient base one-side sampling (GBOSS), as well as utilizes techniques based on bar graphs, which reduces the training capacities that interprets into a less consumption and expressed numerically as

$$K = \frac{f_{aa} f_{bb} - f_{ab}^2}{(1 + f_a^2 + f_b^2)^2} \tag{12}$$

where K is identified as Gaussian curvature and f_{aa}, f_{bb}, f_{ab} are referred to as the first- and second-order partial derivatives.

4 Result and Discussion

The outcome accomplished during image/video forgery detection by utilizing the Light-GBM technique is articulated in the following section.

4.1 Experimental Setup

The proposed Light-GBM technique for image forgery identification is applied in Python as well as the collection included in this method is described as Python 3.7.6 which is compiled in PyCharm 2020 as well as executed in Windows 10 software system.

4.1.1 Datasets

The dataset (DTS) utilized in this research is gathered from the DSO-1 and DSI-1 DTS [9] which contain 200 internal and external images with an image declaration of 2048 × 1536 picture elements.

4.1.2 Evaluation Metrics

(a) *Accuracy*

The image fabrication identification accuracy evaluation is estimated based on the probability of correct prediction to the total detections accomplished from the DTS by using the given equation

$$\text{Accuracy} = \left(n^{cd}\right)/\left(n^{d}\right) \tag{13}$$

here n^{cd} is signified as the correct detection, and n^{d} is identified as the total detection.

(b) *Sensitivity*

The estimation of sensitivity is based on the positive instances identified by the established techniques with the gathered DTS by using the given equation

$$S^{n} = \frac{t^{p}}{t^{p} + t^{n}} \tag{14}$$

here S^{n} is defined as sensitivity, t^{p} is signified as true positive, as well as t^{n} is identified as true negative.

(c) *Specificity*

The specificity of the detection technique is evaluated based on the proportion of true negatives to the total negatives found in the DTS by utilizing the given equation

$$S^p = \frac{t^n}{t^n + t^p} \tag{15}$$

where S^p is referred to as the specificity, t^n is identified as true negative, and t^p is defined as true positive.

4.2 *Performance Analysis*

This part describes the Light-GBM performance for image forgery identification by using their performance metrics on the basis of the DTS [10].

4.2.1 Performance Analysis Based on DTS

The performance evaluation of the developed method is estimated by utilizing the DTS [10], especially in terms of their performance such as accuracy, sensitivity, and specificity, in which the rate of accuracy at the epoch 250 and training percentage (TP) 40 is rated as 88.01%. Similarly, 88.89% at TP 50, 88.98% at TP 60, 89.16% at TP 70, 90.60% at TP 80, and 94.91% at TP 90, whereas the sensitivity value of the proposed technique at epoch 250 and TP 40 is rated as 75.70%, 84.87% at TP 50, 85.03% at TP 60, 85.32% at TP 70, 88.45% at TP 80, and 94.77% at TP 90 as well as the specificity value at the epoch 250 and TP 40 is rated as 65.87%, 70.84% at TP 50, 73.83% at TP 60, 78.12% at TP 70, 85.70% at TP 80, and 93.26% at TP 90. Figure 2 represents the diagrammatic representation of the performance analysis of the developed technique.

4.3 *Comparative Analysis*

The comparative analysis of the proposed technique is measured with various existing techniques for instance support vector machine (SVM) (MD^1) [11], multilayer perception (MLP) classifier (MD^2) [12], *K*-nearest neighbor classifier (KNN) (MD^3) [13], random forest classifier (RF) (MD^4) [14], as well as decision tree classifier (DT) (MD^5) [15].

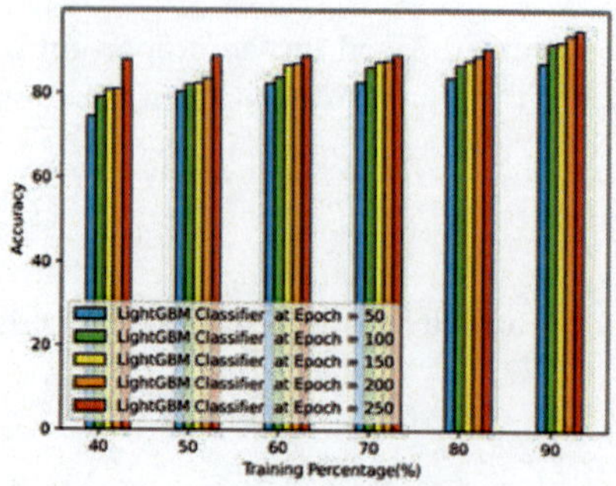

(a) Accuracy

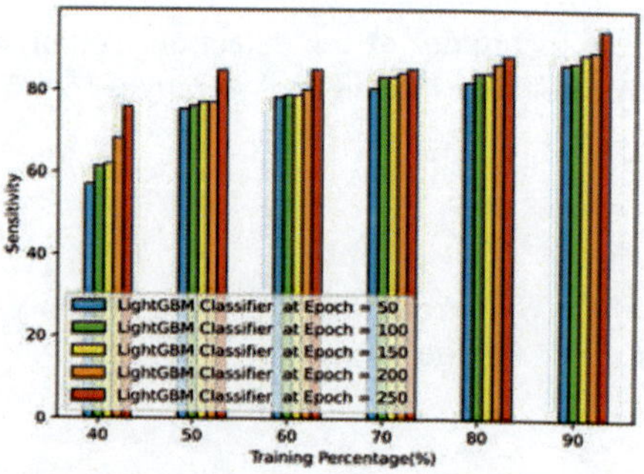

(b) Sensitivity

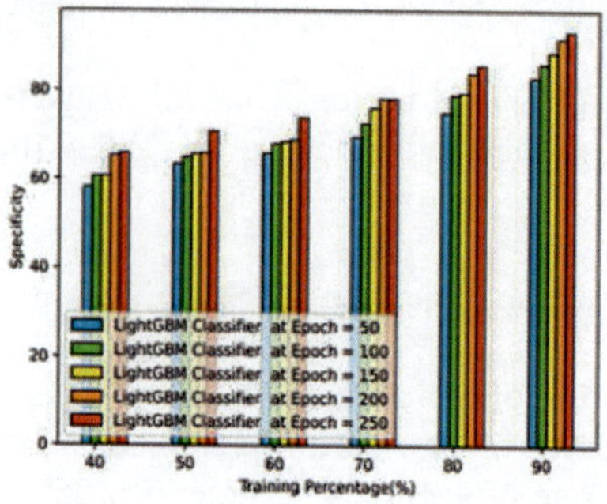

(c) Specificity

Fig. 2 Performance evaluation based on the DTS: **a** accuracy, **b** sensitivity, and **c** specificity

4.3.1 Comparative Analysis Based on DTS

The comparative analysis is measured on the basis of the DTS [10] as well as is achieved based on the accuracy, sensitivity, and specificity, as illustrated in Fig. 3, in which the accuracy value of the proposed technique at TP 90 is esteemed as 94.01% which is increased by 18.09% to the previous MD1, 15.95% to MD2, 11.66% to MD3, 11.07% to MD4, and 3.57% to MD5, whereas the sensitivity value is rated as 95.02% at the TP 90 which is an improvement of 42.78% from the previous MD1, 41.19% from MD2, 34.56% from MD3, 25.95% from MD4, 22.50% from MD5, and the specificity value is rated as 95.01% at TP 90 which is an improvement of 63.64% than the previous MD1, 10.61% than MD2, 5.30% than MD3, and 5.30% than MD4, respectively.

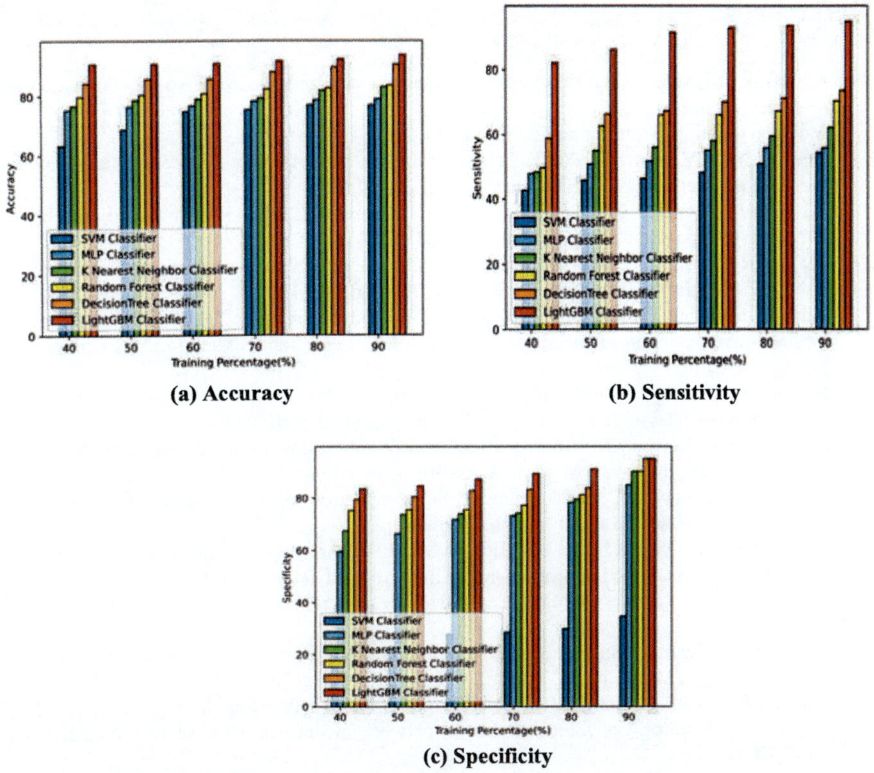

Fig. 3 Comparative analysis based on the DTS: **a** accuracy, **b** sensitivity, and **c** specificity

5 Conclusion

A machine learning-based light gradient boosting machine (Light-GBM) method is proposed in this research to identify image/video forgeries as with the emergence of current virtual automation, the reliability of virtual pictures is developing into extremely endangered in recent days, and therefore it is highly challenging for a people's naked vision to identify the remains of the fabrication. This research also labels the significance of a computerized estimation which can efficiently identify if the inserted image is forged or not. Moreover, the performance and comparative analysis are evaluated based on DTS [10], and the performances metrics like accuracy, sensitivity, as well as specificity and several existing techniques like SVM, MLP, KNN, RF, and DTS were employed for the comparative analysis, in which the proposed model achieved a 94.91% accuracy, 94.77% sensitivity, as well as 93.26% specificity, respectively, which is superior to the previous methods.

References

1. Al_Azrak FM, Sedik A, Dessowky MI, El Banby GM, Khalaf AAM, Elkorany AS, El-Samie FEA (2019) An efficient method for image forgery detection based on trigonometric transforms and deep learning. Multimedia Tools Appl 79:18221–18243
2. Lee SI, Park JY, Eom IK (2022) CNN-based copy-move forgery detection using rotation-invariant wavelet feature. IEEE Access 10:106217–106229
3. Ferreira WF, Ferreira CBR, da Cruz Júnior G, Soares F (2020) A review of digital image forensics. Comput Electr Eng 85, art no 106685
4. Thakur R, Rohilla R (2020) Recent advances in digital image manipulation detection techniques: a brief review. Forensic Sci Int 312, art no 110311
5. Mitra A, Mohanty SP, Corcoran P, Kougianos E (2021) A machine learning-based approach for deep fake detection in social media through key video frame extraction. SN Comput Sci 2:1–18
6. Nath S, Naskar R (2021) Automated image splicing detection using deep CNN-learned features and ANN-based classifier. Signal Image Video Process 15:1601–1608
7. Lourembam A, Kumar KMVM, Singh TR (2021) A robust image copy detection method using machine learning. Malaya J Matematik 23–30
8. Niyishaka P, Bhagvati C (2020) Image splicing detection technique based on illumination-reflectance model and LBP. Multimedia Tools Appl 80:2161–2175
9. Trivedi NK, Simaiya S, Lilhore UK, Sharma SK (2020) An efficient credit card fraud detection model based on machine learning methods. Int J Adv Sci Technol 29(5):3414–3424
10. DSO-1 and DSI-1 Datasets. https://recodbr.wordpress.com/code-n-data/#dso1_dsi1. Accessed on May 2023
11. Dhivya S, Sangeetha J, Sudhakar B (2020) Copy-move forgery detection using SURF feature extraction and SVM supervised learning technique. Soft Comput 24:14429–14440
12. Kolagati S, Priyadharshini T, Mary Anita Rajam V (2021) Exposing deep fakes using a deep multilayer perceptron–convolutional neural network model. Int J Inf Manag Data Insights 2(1):100054
13. Himeur Y, Alsalemi A, Bensaali F, Amira A (2021) Smart non-intrusive appliance identification using a novel local power histogramming descriptor with an improved k-nearest neighbors classifier. Sustain Cities Soc 67:102764
14. Kaur RP, Kumar M, Jindal MK (2019) Newspaper text recognition of Gurumukhi script using random forest classifier. Multimedia Tools Appl 79:7435–7448
15. Zhu E, Ju Y, Chen Z, Liu F, Fang X (2020) DTOF-ANN: an artificial neural network phishing detection model based on decision tree and optimal features. Appl Soft Comput 95:106505

Lecture Notes in Electrical Engineering

Volume 825

The book series *Lecture Notes in Electrical Engineering* (LNEE) publishes the latest developments in Electrical Engineering - quickly, informally and in high quality. While original research reported in proceedings and monographs has traditionally formed the core of LNEE, we also encourage authors to submit books devoted to supporting student education and professional training in the various fields and applications areas of electrical engineering. The series cover classical and emerging topics concerning:

- Communication Engineering, Information Theory and Networks
- Electronics Engineering and Microelectronics
- Signal, Image and Speech Processing
- Wireless and Mobile Communication
- Circuits and Systems
- Energy Systems, Power Electronics and Electrical Machines
- Electro-optical Engineering
- Instrumentation Engineering
- Avionics Engineering
- Control Systems
- Internet-of-Things and Cybersecurity
- Biomedical Devices, MEMS and NEMS

For general information about this book series, comments or suggestions, please contact leontina.dicecco@springer.com.

To submit a proposal or request further information, please contact the Publishing Editor in your country:

China

Jasmine Dou, Editor (jasmine.dou@springer.com)

India, Japan, Rest of Asia

Swati Meherishi, Editorial Director (Swati.Meherishi@springer.com)

Southeast Asia, Australia, New Zealand

Ramesh Nath Premnath, Editor (ramesh.premnath@springernature.com)

USA, Canada:

Michael Luby, Senior Editor (michael.luby@springer.com)

All other Countries:

Leontina Di Cecco, Senior Editor (leontina.dicecco@springer.com)

**** This series is indexed by EI Compendex and Scopus databases. ****

More information about this series at https://link.springer.com/bookseries/7818

Keshav Dahal · Debasis Giri · Sarmistha Neogy · Subrata Dutta · Sanjay Kumar
Editors

Internet of Things and Its Applications

Select Proceedings of ICIA 2020

Springer

Editors
Keshav Dahal
School of Computing, Engineering, Physical Sciences
Artificial Intelligence, Visual Communication and Networks (AVCN) Research Centre
University of the West of Scotland
Paisley, Renfrewshire, UK

Sarmistha Neogy
Department of Computer Science and Engineering
Jadavpur University
Kolkata, West Bengal, India

Sanjay Kumar
Department of Computer Science and Engineering
National Institute of Technology Jamshedpur
Jamshedpur, Jharkhand, India

Debasis Giri
Department of Information Technology
Maulana Abul Kalam Azad University of Technology
Kolkata, West Bengal, India

Subrata Dutta
Department of Computer Science and Engineering
National Institute of Technology Jamshedpur
Jamshedpur, Jharkhand, India

ISSN 1876-1100 ISSN 1876-1119 (electronic)
Lecture Notes in Electrical Engineering
ISBN 978-981-16-7636-9 ISBN 978-981-16-7637-6 (eBook)
https://doi.org/10.1007/978-981-16-7637-6

This Springer imprint is published by the registered company Springer Nature Singapore Pte Ltd.
The registered company address is: 152 Beach Road, #21-01/04 Gateway East, Singapore 189721, Singapore

Preface

This book is the result of the First International Conference of IoT and its Applications (ICIA2020), which was held online at NIT Jamshedpur, India, from 26 to 27 December, 2020. The International Program Chairs of ICIA2020 took on the task of acting as the editorial board to publish the Proceedings of the ICIA2020. The conference received 140 papers, and out of that, only 41 papers were accepted for the presentation and publication following review process by the international program committee. The program committee members were well-known experts from different parts of the world. The authors of those 42 papers presented their papers online to the conference audiences. We have divided the papers into four different fields: IoTs and cloud computing; signal processing and machine learning; wireless network; information security. The editors of this book would like to sincerely thank Prof. Karunesh Kumar Shukla, the director of NIT Jamshedpur for his cooperation to organize this wonderful international event. The Department of CSE at NIT Jamshedpur has played great role in organizing this successful conference. We also appreciate the TEQIP-III for providing the funding.

We would also like to thank all program committee members for reviewing the papers and the session chairs for their contribution to run the conference smoothly. And, of course, we greatly appreciate the authors who submitted their valuable research papers to the conference, which made this book possible.

Paisley, UK — Keshav Dahal
Kolkata, India — Debasis Giri
Kolkata, India — Sarmistha Neogy
Jamshedpur, India — Subrata Dutta
Jamshedpur, India — Sanjay Kumar

Contents

Information Security

About the Editors

Keshav Dahal is a Professor of intelligent systems and the director of the Artificial Intelligence, Visual Communications and Network Research Centre at the University of the West of Scotland, Paisley, UK. He received his master's and Ph.D. degrees from the University of Strathclyde, UK. He also held academic and research positions at the University of Strathclyde and the University of Bradford, UK. His research interests are applied AI, trust, and security modeling in distributed systems, blockchain technologies, and scheduling/optimization problems. He has been the principal investigator or co-investigator on more than 15 externally funded projects and supervised over 20 Ph.D. and postdoctoral researchers. He has published over 150 papers in his research fields with award-winning publications and has been a part of organizing/program committees of over 60 international conferences. He is a Senior Member of IEEE.

Debasis Giri is currently working as an Associate Professor in the Department of Information Technology, the Maulana Abul Kalam Azad University of Technology (Formerly known as the West Bengal University of Technology), West Bengal, India. Before this, he also held academic positions as Professor in the Department of Computer Science and Engineering and Dean in the School of Electronics, Computer Science and Informatics, Haldia Institute of Technology, Haldia, India. He did his master's (M.Tech. and M.Sc.) from IIT Kharagpur, India, and also completed his Ph.D. from IIT Kharagpur, India. His current research interests include cryptography, information security, e-commerce security, and the design & analysis of algorithms.

Sarmistha Neogy is a Professor in the Department of Computer Science and Engineering, Jadavpur University, Kolkata, India. She received her Ph.D. degree in Engineering and Master's degree and Bachelor's in Computer Science and Engineering from Jadavpur University, Kolkata, India. She is a senior member of IEEE and IEEE Computer Society. She has publications in reputed international journals and proceedings of international conferences. She has co-authored four books. Dr. Neogy is a member of the Govt. of India mirror committee for the Joint Technical Committee

(JTC 1). Dr. Neogy's research interests are fault tolerance in distributed systems, reliability and security in wireless and mobile systems, security and privacy, and wireless sensor networks.

Subrata Dutta is an Assistant Professor in the Department of Computer Science and Engineering, National Institute of Technology, Jamshedpur. He has worked as a post-doctoral researcher at the University of the West of Scotland, UK. He received his M.Tech., and Ph.D. degrees from the Jadavpur University, Kolkata, India. His research interests are wireless sensor networks, mobile computing, and machine learning. He has published over 20 articles in international journals and conferences.

Sanjay Kumar is an Associate Professor and Head of the Department of Computer Science and Engineering, National Institute of Technology, Jamshedpur. He received his Ph.D. degree in the safety analysis of computer-based systems from the National Institute of Technology, Jamshedpur, India. He held numerous positions in academic and research activities throughout his career. His research interests are network security, mobile computing, parallel &distributed systems, and artificial intelligence, etc. He is a member of the Association for Computing Machinery (ACM), Computer Society of India (CSI), and Institute for Computer Sciences, Social Informatics and Telecommunications Engineering (ICST).

IoT and Cloud Computing

An Android-Application-Controlled Car for Human Safety Against COVID-19

Jhilam Jana, Sayan Tripathi, Akash Bhattacharya, Ritesh Sur Chowdhury, Angana Karmakar, Deep Ranjan, and Jaydeb Bhaumik

Abstract COVID-19 pandemic has been spread over a large geographical area and has affected a large proportion of the population. Maintaining social distance, quarantine, frequent hand-washing, using alcohol-based hand sanitizer and wearing medical mask are essential to protect us and others from COVID-19. To safe the people and help the government authority, several researchers have also developed many mobile applications, medical masks, drones, ventilation systems, etc. for this purpose. In this paper, an efficient android-application-controlled car for human safety against COVID-19 pandemic has been proposed. The proposed architecture can be useful for patient handling and proper transportation of COVID-19-suspected patients. This approach will be beneficial for the ambulance drivers and other emergency service providers who are playing an important role to protect the community from this infectious disease.

Keywords Coronavirus (COVID-19) · Android application · Arduino Uno · Human safety · Wireless communication

1 Introduction

COVID-19 is a newly identified infectious disease. Day by day, this coronavirus is becoming invincible and difficult to control. The world is now facing a lockdown due to this disease. Doctors, scientists, nurses and other front-line workers are giving their best effort to prevent the spread of this disease. Scientists are trying to find an effective and affordable vaccine for this disease.

This paper aims to design an approach for an android-application-controlled car for human safety against COVID-19 pandemic. The main objectives of the paper are—(i) to protect the ambulance driver carrying a COVID-19-suspected or affected patient from getting affected by coronavirus, (ii) there is no specified operating range

J. Jana (✉) · S. Tripathi · A. Bhattacharya · R. S. Chowdhury · A. Karmakar · J. Bhaumik
Department of ETCE, Jadavpur University, Kolkata, India

D. Ranjan
Indian Institute of Technology (ISM), Dhanbad, India

K. Dahal et al. (eds.), *Internet of Things and Its Applications*, Lecture Notes in Electrical Engineering 825, https://doi.org/10.1007/978-981-16-7637-6_1

in case of controlling the car, (iii) this type of car is controlled manually and as well as partially automatic and (iv) to help the local authority or hospital management systems.

The proposed system consists of a GSM-GPRS-GPS (SIM 808) module, Arduino Uno micro-controller, pyroelectric-infrared (PIR) sensor, web camera, motors, motor driver module, light-emitting diode (LED), buzzer and android application. The PIR sensor is used for motion sensing and detection of living objects. The SIM 808 module is a three-in-one module consisting of GSM, GPRS and GPS modules. This module will provide a good range of operation and send the GPS location of the car. The web cameras are used for detecting any obstacle which comes in the path of the vehicle. The buzzer will alert the driver if any living object comes in front of the car. The motor driver module is used for driving the whole system. This car can be controlled by using an android application which should be installed in driver's smart phone. The proposed system will be beneficial for the protection of the ambulance drivers who have the highest risk of contamination with the COVID-19 patients.

The remaining part of this paper is organized as follows. Section 2 describes the literature survey. Section 3 presents proposed system. Section 4 provides working principle of proposed system. Section 5 presents android app design and PCB layout. Section 6 highlights the features of proposed system. Section 7 presents the conclusion.

2 Literature Survey

COVID-19 pandemic is a threat to the human civilization. However, in this pandemic, the medical front-line warriors including doctors, nurses, medical staff, ambulance drivers are fighting against this disease and are putting their lives at stake to keep the community safe. The ambulance driver carrying a COVID-19-suspected or affected patient has a high chance of getting affected by coronavirus. To reduce the chances of road accident due to human driving error, to help local authority and also to safeguard the human life in this pandemic situation without human to human contact, the robotic car plays an important role nowadays. Several mobile-application-controlled configurations are already introduced in different applications. Winter et al. proposed android-controlled robot. This system transferred the information wirelessly [1]. Jing et al. developed an android remote control car for searching missions. This robot car contains ultrasonic distance sensor, Wi-Fi transmitter, camera, Bluetooth receiver, two Arduino Uno micro-controllers [2]. Tezel and Hangun developed a Bluetooth-controlled robotic car using Arduino. This design is collision free, and the sensor data is obtained using the linear interpolation [3]. Raihan et al. designed and implemented a robotic vehicle with the help of the hand movement [4]. The intelligent vehicle system which navigated autonomously was designed by Fernandes and his co-workers. This autonomous driver assistance system was developed with the help of CaRINA I platform [5]. Goud et al. proposed a pick-and-drop robot which can be very useful in defense for diffusing the bombs, land mines, etc. To avoid human errors, the manual and autonomous driving switching techniques were described in [6]. The smart robot car is developed for not only minimizing the human labors, but also for beneficial rescue operations [7–11]. Wang et al. introduced wheeled robotic car which is based on STM32. This system uses MPU6050 module to adjust the route direction and the turning angle [12].

Chen et al. designed a wireless control robotic system based on Arduino Uno. Arduino Uno micro-controller, Bluetooth module, sound alarm and ultrasonic distance measurement sensor and motor driver were used in this design [13]. The motion-controlled robot car was developed by Xu et al. This car was based on digital signal processing and control system [14]. Tripathi et al. have developed Bluetooth-controlled robot car which can reduce the physical labors of human during material handling [15]. The operating range for the existing robot cars is not sufficient, and also these are not useful for the drivers who are in essential service. So designing of an android application-based system which can assist to run and control a car from a safe location is an important requirement in the present pandemic situation.

3 Proposed Architecture

An android-controlled car for human safety against infectious diseases is presented in this paper. This proposed design can be used for patient handling and proper transportation of COVID-19-suspected or affected patients. In this design, Arduino Uno micro-controller, GSM-GPRS-GPS (SIM 808) module, PIR motion sensor and web camera are employed to run the car. The car can be controlled using smart phone through android application. The driver can control the car from his home or any safe-place. The proposed circuit diagram is presented in Fig. 1. The proposed architecture will play an important role for handling the COVID-19-suspected or affected patients. This design is also cost-effective and easy to handle.

The requisite components of proposed architecture are described as follows.

1. Arduino Uno (IC 1136): Arduino Uno is a micro-controller board which is based on ATmega328P. It comprises 14 digital input/output pins and 6 analog input/output pins. This micro-controller board is cost-effective and provides solutions to many complex embedded problems.

2. PIR Sensor: Pyroelectric-infrared (PIR) sensor basically consists of a pyroelectric sensor, which has the capability to detect different levels of infrared radiations. It can also detect motion, i.e., acts as a motion sensor. When any motion is detected, it will provide a voltage swing as output.

3. Motor Driver Module: The motor driver module (IC L293N) is an integrated monolithic circuit which is embedded in a 15-lead multiwatt package. It can accept standard TTL logic level and can drive inductive load. It has an operating voltage of 20 V.

4. GSM-GPRS-GPS Module (SIM 808): SIM 808 module is a three-in-one function module. It contains global system for mobile communications (GSM) module, general packet radio service (GPRS) module and Global Positioning System (GPS)

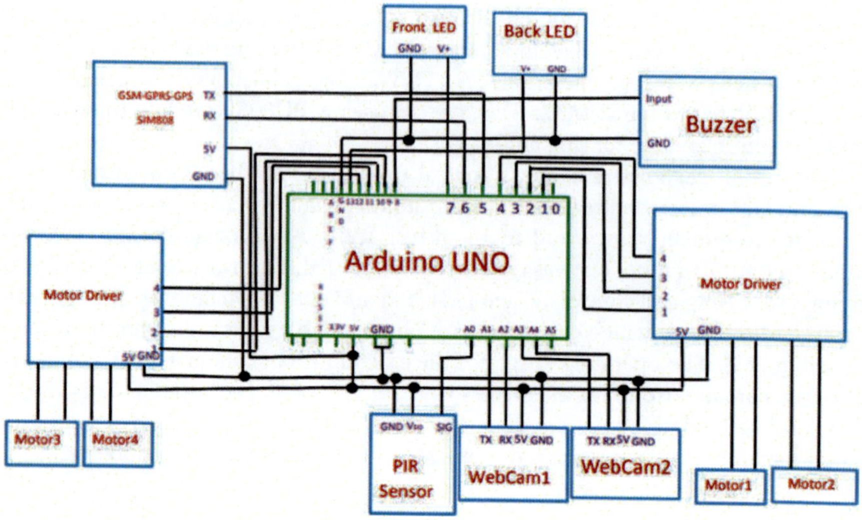

Fig. 1 Circuit Diagram of Proposed Model

module. It supports GPRS/GSM quad-band network and combines GPS technology for satellite base navigation. This module supports ultra-low power consumption in sleep mode and is integrated with a charging circuit for Li-ion battery.

5. Web Camera: The web camera is basically a simple camera which will be used for identifying the obstacles and for comfortable operation of the device. The resolution of the camera can be varied depending upon the requirement.

6. Buzzer: The buzzer module is also used in the proposed design. This module is an audio signaling device. The PIR sensor will confirm the presence of a living object, and then a message will be sent to the Arduino Uno and then buzzer will emit the sound.

4 Working Principle of Proposed Architecture

The proposed architecture is divided into two major units. First is the vehicle unit that is installed in the vehicle itself, and second is the control unit which is an app that can work on any platform or devices like mobile, tablet, laptop or any personal computer. The controller of the proposed design is Arduino Uno micro-controller through which other modules are connected and controlled. The working flow of proposed model is described in Fig. 2.

The vehicle is connected to the app via GSM-GPRS-GPS (SIM 808) module. The Tx and Rx pins of this module are connected to the pin nos. 5 and 6, respectively, of the micro-controller. When an user wants to operate a vehicle from the number of vehicles available, the driver has to select the vehicle from the list available in the app. This list contains all the virtual IDs of vehicles that is provided to every vehicle to

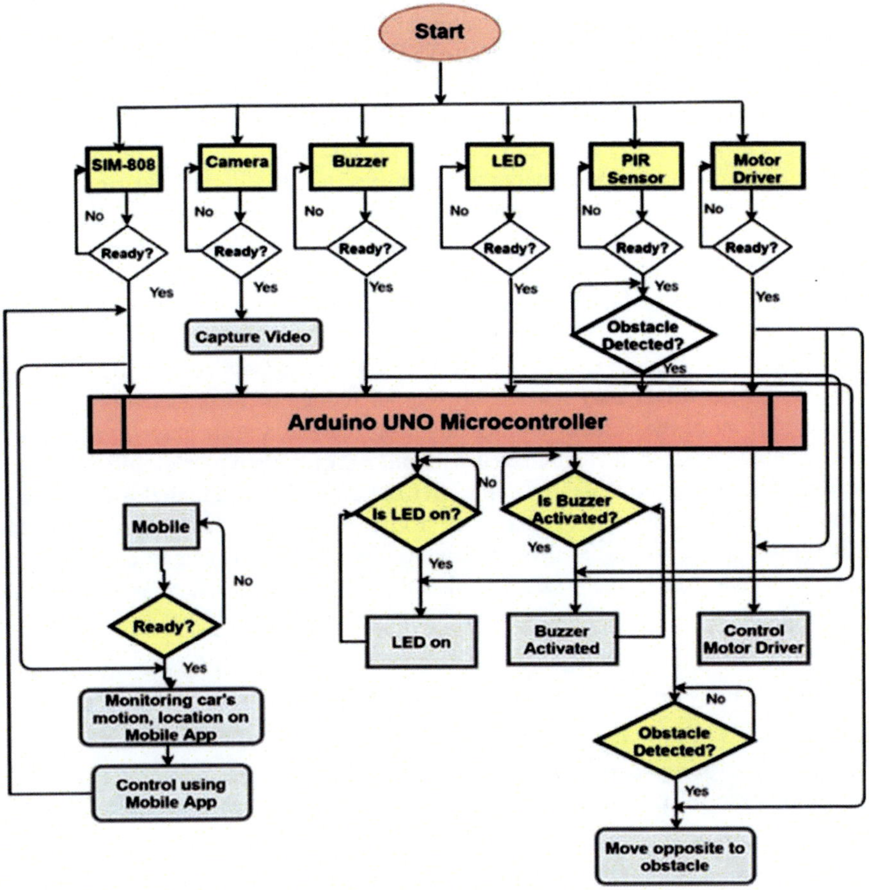

Fig. 2 Flowchart of working principle for proposed model

identify them. In this list, if a vehicle is not connected to any other controlling device, then it appears green otherwise red. When a driver selects a vehicle from the list, then it appears red means not available for further connection to the other drivers, when a secure connection is established between a driver and a vehicle then its current status will start showing on the app and the driver can now operate the vehicle. This module also provides the continuous and uninterrupted connection from the consoling device to the vehicle via cellular tower and also provides the GPS location of the vehicle on the app. By the use of the GPS location, the driver can able to identify where the vehicle is and can also able to get the directions through the map available on the app to reach the destination.

Two web cameras are also installed on the vehicle, one is in the front and other is in the back, which are connected to the micro-controller through A1, A2, A3 and A4 pins. These web cameras provide the live images or videos on the app screen that help the driver to see and handle the vehicle properly. A buzzer is also connected to the micro-controller that can be controlled by the app and driver can use it according to their need. It is connected via pin no. 8 of the Arduino Uno. Two LEDs are also attached in the vehicle which can be controlled via micro-controller, and these LEDs can be used by the driver according to the light condition.

There are two motor drivers which are connected to the Arduino Uno to control the speed of the motors. From each motor driver, two motors are connected. Speed and acceleration of these motors are controlled by the driver with the help of acceleration, direction and brake button that are available on the app screen. The motor drivers are connected with the Arduino Uno with its pin nos. 9, 10, 11, 12 and pin nos. 1, 2, 3, 4, respectively.

A PIR sensor module is also used in this vehicle to provide extra safety. It may happen that the driver may not able to see the sudden or unexpected arrival of a person in front of the vehicle. During this time, the PIR sensor plays an important role to avoid accident. It is connected with pin A0 of the micro-controller. When a person or any other moving thing comes in front of the vehicle, then this sensor will identify it and give signal to the micro-controller. After receiving the signal from the PIR sensor, micro-controller gives instruction to the motor driver to de-accelerate and also activate the buzzer to send an alarm to the person or vehicle in front of it.

5 Android Application Design and PCB Layout

In this section, design of android mobile application and printed circuit board (PCB) layout are described.

5.1 Android Application Design

In proposed model, a car can be controlled by using an android application which has to be installed in the smart phone of the driver. The driver can operate the car from his home or any safe-place. This application can also be run in PC. The commands are received by the GSM-GPRS-GPS (SIM 808) module which are then passed to the Arduino Uno micro-controller. The commands are sent in the form of ASCII characters.The android application's layout of proposed system is presented in Fig. 3.

The android application consists of the following parts:

1. Connection button: This button is used for connecting the android application with the car. If multiple users are there, they can check whether the car is connected to another device or not.

2. Light buttons: The light buttons are used for turning on or off the front lights and the back lights of this car.

3. Acceleration button: This button is used for either accelerating or decelerating the car by varying the magnitude using a scroll bar.

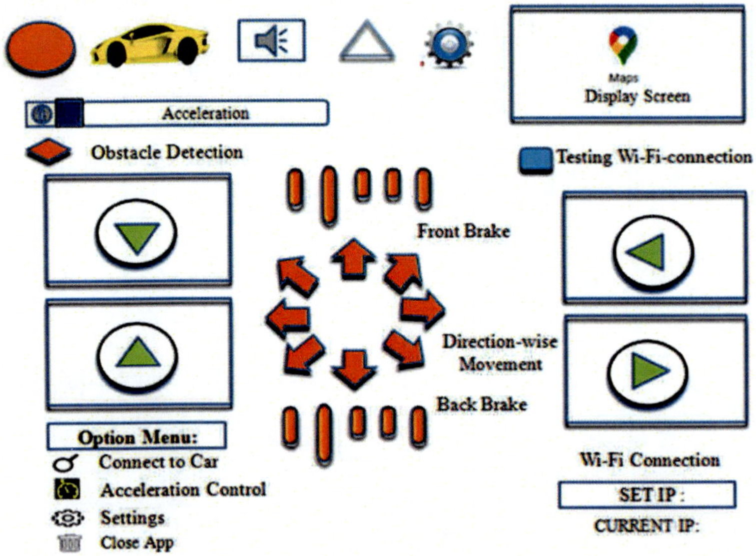

Fig. 3 Layout of Android Application for Proposed Model

4. Obstacle detection button: The obstacle detection is used for detecting any obstacle that comes in the way of the car.

5. Testing Wi-Fi connection button: When the car is connected to the android application, the Internet speed might be low and not up to the mark for controlling the car. So the testing button can be used to test the connection of the device.

6. Control button: The car can be controlled by using the direction buttons provided in the application.

7. Brakes: The front brakes and back brakes are provided separately to facilitate the driver and also to ensure safe driving.

8. Wi-Fi connection section: This section is used for setting up the connection and the IP address of the car which is shown in SET IP and the user's IP is shown in current IP.

5.2 PCB Layout

Printed circuit board (PCB) provides mechanical support to all electronic components. Here, single-sided PCB has been designed where the layout is on one side and the circuit is another side.

The proposed design is represented by using the PCB Wizard 3.50 software. PCB design and PCB solder side artwork layout of proposed system are shown in Figs. 4 and 5 respectively.

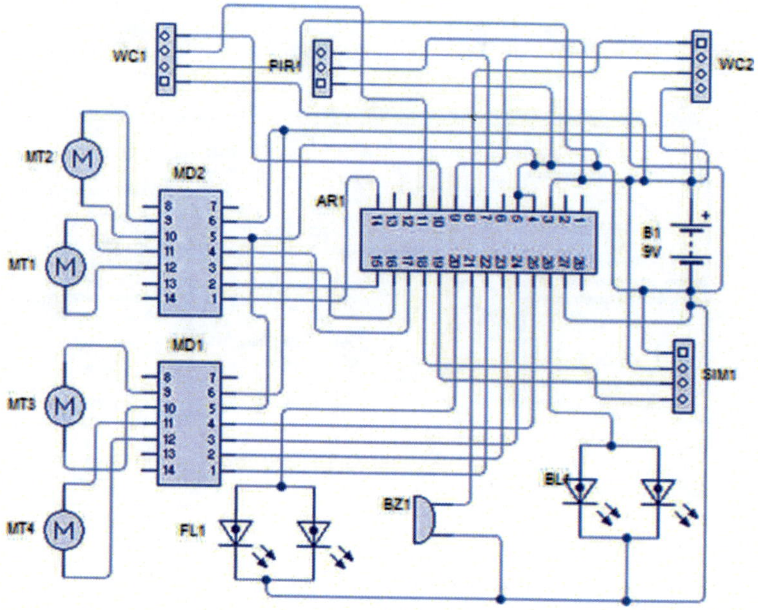

Fig. 4 PCB design of proposed model

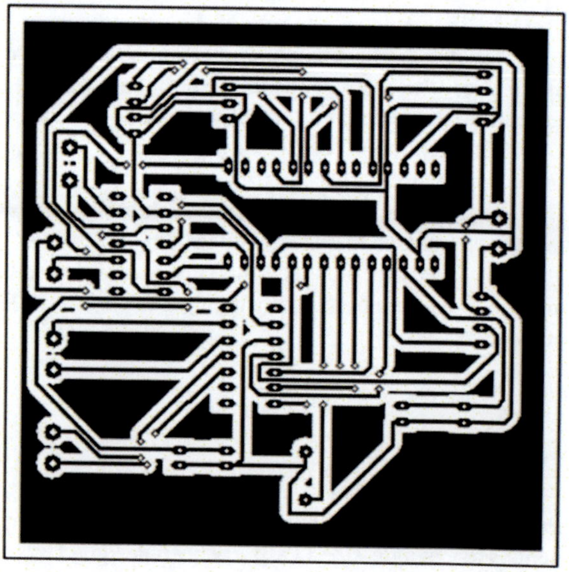

Fig. 5 PCB solder side artwork layout of proposed model

6 Features of Proposed Design

The proposed android-application-controlled car has several notable features which are mentioned as follows.

(i) This car can be operated by a driver from any convenient location without any manual intervention to car.

(ii) Though this car can be operated from any location, it will help the health workers during transportation of COVID-19 patients, without getting in close contact with them. Thus, it will help to eliminate the chance of getting infected to the drivers.

(iii) This type of car is easy to operate, anybody can operate this car after little bit of training.

(iv) The smaller version of this car can be used inside the hospitals to transport the COVID-19 patient from one ward to another.

(v) This type of car can transport the COVID-19 patients up to any distance where the cellular network reaches.

(vi) Proposed system can be accommodated in a small battery-operated car to deliver foods and medicines inside a ward.

7 Conclusion

An android-application-controlled Arduino-based car is proposed to reduce the risk of spreading COVID-19 among the people such as ambulance driver, drivers of other essential services who are in direct contact with COVID-19-suspected or affected people. From this perspective, we have proposed a technique for controlling car from a safe-place, and it can be controlled from any place where network is available. In this car, PIR sensor is used to detect the motion of living object and to reduce the chances of accident. This type of robotic car can also be employed inside the hospital to transfer the COVID-19-affected patient from one ward to another; this is another advantage of the proposed system. We have proposed a model, but it has not been validated yet. The future scope of this architecture is to change the punctured tire using robotic arm and to improve the efficiency by using artificial intelligence and deep learning. This type of robotic car can be employed for home delivery services, industrial services, medical and military fields.

References

1. Kazacos Winter, Jorge. Android controlled mobile robot (2013)
2. Jing Y, Zhang L, Arce I, Farajidavar A (2014) May. AndroRC: an android remote control car unit for search missions. In: IEEE Long Island Systems, Applications and Technology (LISAT) Conference 2014. IEEE, pp. 1–5

3. Tezel C, Hangün B. Design and implementation of bluetooth controlled collision avoidance 4 wheel robot using arduino with linear İnterpolation method for determination. Int J Eng Sci Appl 1(4): 151–156
4. Raihan MR, Hasan R, Arifin F, Nashif S, Haider MR (2019) Design and implementation of a hand movement controlled robotic vehicle with wireless live streaming feature. In: 2019 IEEE International Conference on System, Computation, Automation and Networking (ICSCAN). IEEE, pp 1–6
5. Fernandes LC, Souza JR, Shinzato PY, Pessin G, Mendes CC Osório, FS, Wolf DF (2012) Intelligent robotic car for autonomous navigation: platform and system architecture. In: 2012 Second Brazilian conference on critical embedded systems. IEEE, pp 12–17
6. Goud RK, Kumar BS (2014) Android based robot implementation for pick and retain of objects. Int J Eng Trends Technol (IJETT) 16(3)
7. Birk A, Schwertfeger S, Pathak K (2009) A networking framework for teleoperation in safety, security, and rescue robotics. IEEE Wireless Commun 16(1):6–13
8. Liu Y, Nejat G (2013) Robotic urban search and rescue: a survey from the control perspective. J Intelligent Robot Syst 72(2):147–165
9. Kuhnert KD (2008) Software architecture of the autonomous mobile outdoor robot AMOR. In: Intelligent vehicles symposium, 2008. IEEE, pp 889–894
10. Braun T, Schaefer H, Berns K (2009) Topological large-scale off-road navigation and exploration RAVON at the European Land Robot Trial 2008. In: IEEE/RSJ international conference on intelligent robots and systems, IROS 2009. IEEE, pp 4387–4392
11. Wang Z, Lim EG, Wang W, Leach M, Man KL (2014) Design of an arduino-based smart car. In: 2014 International SoC Design Conference (ISOCC). IEEE, pp 175–176
12. Wang Z, Xie H, Lin Z, Wen T, Zhen Z, Chen H (2020) Design of control algorithm of wheeled robot based on Wi-Fi and inertial navigation. In: IECON 2020 the 46th annual conference of the IEEE industrial electronics society. IEEE, pp 4762–4768
13. Chen L, Zhang J, Wang Y (2018) Wireless car control system based on ARDUINO UNO R3. In: 2018 2nd IEEE Advanced Information Management, Communicates, Electronic and Automation Control Conference (IMCEC). IEEE, pp 1783–1787
14. Xu M, Zhang H, Tang H (2017) Design of motion control system for robot car based on DSP. In: 2017 29th Chinese Control And Decision Conference (CCDC). IEEE, pp 7494–7497
15. Tripathi S, Jana J, Mandal S, Pal D, Das K, Jana AK, Pandit MK (2020) Cost-efficient bluetooth-controlled robot car for material handling. In: Proceedings of the 2nd international conference on communication, devices and computing. Springer, Singapore, pp 343–353

Regular Self-Health Monitoring and Medicine Reminder System with Emergency Alert Messaging Using IoT

Basudeba Behera, Riya Mehta, Prachi P. Fulzele, and Rashmi Sinha

Abstract The main objective of this work is to develop a reliable, low cost, self-health monitoring system using Internet of Things (IoT), so that healthy adults, old people, as well as patients at home can reliably monitor their fundamental health measuring variables. The health-related parameters need to monitor are body temperature, pulse rate, blood glucose level, and ECG on their own. The self-health monitoring system can also be used in rural hospitals or small clinics where the number of doctors per patient is very low. There, the attendants can measure these parameters and reduce the burden on the doctor. The data taken by the device are stored on the cloud through the Internet and therefore will be analyzed using machine learning algorithms to provide real-time online monitoring of the patient by any doctor or a caretaker. If any of the vital parameters exceed the pre-set limit, a danger LED starts to glow, a buzzer beeps, the readings are displayed on the LCD screen, and an emergency message/ email is sent to a close relative/ doctor indicating the reading of the parameter which was above the pre-set limit. Depending on the reading, proper action can be taken, and in extreme cases, an ambulance can be sent instantly to the GPS location of the patient. The device is incorporated with an additional feature of timely reminder for medicines on the LCD screen as well as through the buzzer.

Keywords IoT · Temperature sensor · Pulse rate sensor · ECG sensor · Arduino UNO · Node MCU · Cloud computing

1 Introduction

At present scenario, all kinds of technologies are evolving at an enormous pace, and the application of technology in the field of health, known as e-Health, needs to keep getting better and better. About half of the people in India and over 3/5th of those in rural areas must travel beyond 5 km to reach a healthcare center. People suffering

B. Behera (✉) · R. Mehta · P. P. Fulzele · R. Sinha
Department of Electronics and Communication Engineering, National Institute of Technology, Jamshedpur, Jharkhand, India
e-mail: basudeb.ece@nitjsr.ac.in

K. Dahal et al. (eds.), *Internet of Things and Its Applications*, Lecture Notes in Electrical Engineering 825, https://doi.org/10.1007/978-981-16-7637-6_2

from health problems such as diabetes, high blood pressure, etc., need to visit the doctor regularly, cutting from their precious time.

Health is one of the biggest concerns for everyone. The various fundamental health measuring variables of the human body which indicate health are blood pressure, body temperature, oxygen level, pulse rate, ECG, glucose level, and respiratory rate. This real-time health monitoring system is designed to monitor three of these fundamental parameters of the patient, that is, body temperature, heart rate, and ECG. The device uses simple body sensors to measure the readings of the parameters and displays the readings on an LCD screen. It further acts based on the reading. If the reading exceeds a pre-set value (set according to the normal values of these parameters), then the system lets the patient know through an LED and buzzer sound.

A copy of the reading is sent to the doctor/relative of the patient along with the patient's GPS location, so that an ambulance can be sent to the location if need be. Since the readings are on the cloud, they can be accessed by the doctor/ relative any time. Thus, this device can provide a smart, integrated, and time saving way to tackle the increasing burden of visiting a doctor very frequently even for the measurement of fundamental body parameters. By tracking the fundamental parameters, keeping a note of the readings and detecting abrupt changes, we can even be able to predict health risks in the future. The device will find use in various fields such as the household medical kit, sports fitness, rural area patient management, and many more.

The various features of the system are:

1. Real-time regular health monitoring of various vital physical parameters.
2. Automatic alert messaging system if any vital parameter exceeds the pre-set value. The system will send a message to the doctor/ caretaker along with the current location of the patient with help of Blynk application.
3. Medicine reminder for the patients daily.
4. Emergency switch for the patient to inform the doctor/ caretaker immediately in case they feel uneasy or helpless along with the current location of the user.
5. LED indicator for proper heart rate monitoring and display.

2 Literature Review

The earliest presentation of wearable and implantable wireless body network was done by Wang et al. [1], using a silicon chip taking the mixed signals. This work was done using an 8-bit microcontroller, analog sensors for temperature and PH measurement, data encoding and multiplexing module, and a RF section to limit off the chip components. The silicon chip is tested with all the analog inputs and system fulfills all the requirements. A healthcare system in IoT using RFID was introduced where the system was divided in four layers, namely sensor layer, network layer, Internet layer, and service layer [2]. The combination of microcontroller with the smarts sensors offers advantages like as incorporated precision analog capabilities, small power consumption, and easy for designing GUI's. And finally, the readings of all these sensors are sent to an android mobile that can work with RFID via

Internet. The system also has SMS and email alert facility if any crisis is met. A multidisciplinary outlook for building an effective health monitoring system [3] was reported. They combined IoT, wireless body area network, and cloud computing for the collection, transmission, and analysis of patient data. Bourouis et al. [4] have presented a system that uses the biomedical sensors to obtain, analyses, process, and transmit the data through GPRS/UMTS using a smartphone. The system aims at nursing the patient's mobility, location, and vital bio-parameters such as heart rate and sp02. A Web application is used to stream the data and make it accessible to the reasoning server in real time. The work also presented the overall architecture of UMHMSE and the operation of other parts of UMHMSE. UMHMSE integrates intelligent central node and server as well as wireless body area network (WBAN). The data is monitored using a pulse oximeter based on Python application such as ICN. Another work was reported where they came up with an idea of a Ubiquitous healthcare system (u-Healthcare) that can be accessed from anywhere [5]. The work was implemented using Zigbee. The idea was achieved by connecting the analog sensor for measuring the fundamental health or environment parameters to the mobile systems through wireless personal area network (WPAN) such as Zigbee. Sohail et al. [6] have compared the all existing papers on e-Health monitoring systems and conclude that accurate data collection by sensors is the main problem that needs to be addressed in such systems.

3 Theory and Concepts

A. *Component Description*

Before explaining the working principle, here is brief description of its components.

1. Temperature sensor

LM35 model was used which requires an input of +3.3 to +5 V and produces analog output, such that, 10 mV for raise of every 1 °C. Output can range from −1 V (−55 °C) to 6 V (150 °C). The sensor can easily be connected to any microcontroller that has ADC function or to any development platform like Arduino or Raspberry pi.

2. Pulse rate sensor

Robodo PLSNSR1 model was used which requires an input of +3.3 to +5 V and gives pulse rate in beats per minute as output. Off-the-shelf plug-and-play pulse sensor is used which needs to be tapped onto a fingertip or wrist and plugs right into any development platform. The front side of the pulse sensor has an LED that shines through from the back and just under it is a light sensor. The LED shines light into the fingertip, and sensor counts the light pulses that bounce back to give the reading of the pulse rate.

3. ECG Sensor

   ECG module AD8232 model was used which needs an input of +3.3 V. In terms of output: The ECG module AD8232 sensor is used to measure the electrical activity of the heart which is charted as an ECG or electrocardiogram, while the output is given as an analog reading. It is designed to extract, amplify, and filter small bio-potential signals that are used to diagnose certain heart conditions. It consists of three electrodes of red, green, and blue colors. These electrodes are placed on the patient's right wrist, left wrist, and right ankle.
4. ARDUINO UNO

   Arduino UNO is first in the series of microcontroller boards based on ATmega328P. This microcontroller consists of 14 pins for digital input and output out of which five are used as PWM, six analog pins, one USB connection, an ICSP header, a 16 MHz quartz crystal, and a reset button. It is used to measure and process the input data in a very simple and reliable way. As it takes voltage between 9 and 20 V, it can be powered by a PC or any external source with 9 V. It has a memory unit and a processor to process the data and a reset button of its own.
5. ESP8266

   ESP8266 is a low-cost, self-contained system on chip with TCP/IP protocol stack produced by Espress if systems. It can be connected with minimum seven external components. It requires very low power and does not require a lot of external components. This device can simply be hooked up to the Arduino device as it comes pre-programmed with an AT command set firmware to get the Wi-Fi ability. Due to its wide temperature, operating range is used in many industrial applications. It consists of 32-bit RISC microprocessor core, 32 KB instruction RAM, 32 KB instruction cache RAM, 80 KB user-data RAM, and 16 KB ETS system-data RAM. In this work, it is used to upload the patient's data to the cloud using Wi-Fi module.
6. Piezoelectric Buzzer

   A piezoelectric buzzer is device that is used to make audio signals. It consists of two ends, a positive end for high voltage, and a negative end for low voltage. A piezoelectric element can be driven by devices having piezoelectric audio amplifiers such as oscillating electronic circuits or other signal sources. In this work, the buzzer is used to indicate that the health parameter of the patient has exceeded the normal value in case of emergency and while reminding the medicine time.
7. Global Positioning System (GPS)

   Global positioning system (GPS) is a system based on satellites that use ground stations as well as satellites to measure its position on Earth. Also known as Navigation System with Time and Ranging (NAVSTAR) GPS, the GPS module is connected to the Arduino UNO. The Arduino module using code collects the data and the current location of the patient, which is displayed in terms of latitude and longitude on the LCD at a periodic interval. At least, four satellite's data are needed by a GPS to get accuracy. A GPS module has been used in order to make an inbuilt tracking system for the patient in case of emergency which gives current location of the patient at any instant of time in terms of latitude and longitude.

8. Toggle Switch

The toggle switch we used is an electrical switch which has a handle or lever that can control the flow of current from a power supply to a device or within two devices. The switch can take two positions—ON and OFF. In this work, the toggle switch is used as an emergency switch for the patient. When the patient feels uncomfortable or helpless, he/she can turn the toggle switch ON which in turn sends an email to the care takers.

B. *Function Description*

As stated, the main idea of the system is to monitor various health parameters remotely. For that purpose, the system has been divided into two parts.

The first part which takes the readings of the health parameters will be with the patient/ self-health monitoring individual. The second part of the system will be with the doctor/caretaker. As shown in Fig. 1, the first part consists of the three body parameter sensors, namely temperature sensor, pulse rate sensor, ECG sensor, as well as other components such as emergency switch, LCD, buzzer, GPS, and LED indicator for heartbeat level. These sensors collect the measured parameters from the user and send the data to the connected microcontroller [7].

The microcontroller continuously collects data when the sensors are attached to the user and checks for any anomalies from the maximum pre-set value of each parameter. These values are displayed on the LCD screen, and appropriate LEDs of various colors glow in response. Also, the data are sent at real time to the ThingSpeak and Blynk [8] cloud which generate graphs and appropriate messages based on the recorded data for real-time monitoring.

When any parameter value exceeds the maximum pre-set value, an emergency provision is turned on. This emergency system sends an email to the doctor or care-taker using the Blynk application depending on the reading, so that proper action can be taken, and in extreme cases, an ambulance can be sent instantly to the GPS location of the patient.

The same emergency procedures will take place if the user presses an emergency button installed in the system when he or she suddenly has an attack or simply feels helpless due to any reason. We have also implemented an LED indicator for heartbeat of the user using three LEDs which will glow depending on whether the pulse rate is normal, above average, or critical. A medicine reminder system has also been incorporated in the device which will timely remind the user regarding their medicines on the LCD screen as well as through the buzzer.

The second part of the system consists of the doctors/ care takers who can remotely access all the vital body parameters of the patient using ThingSpeak on a personal computer or a smartphone or a Tablet via the Internet [9, 10], as shown in Fig. 2. Based on the readings, future action for the patient can be decided. And all this was

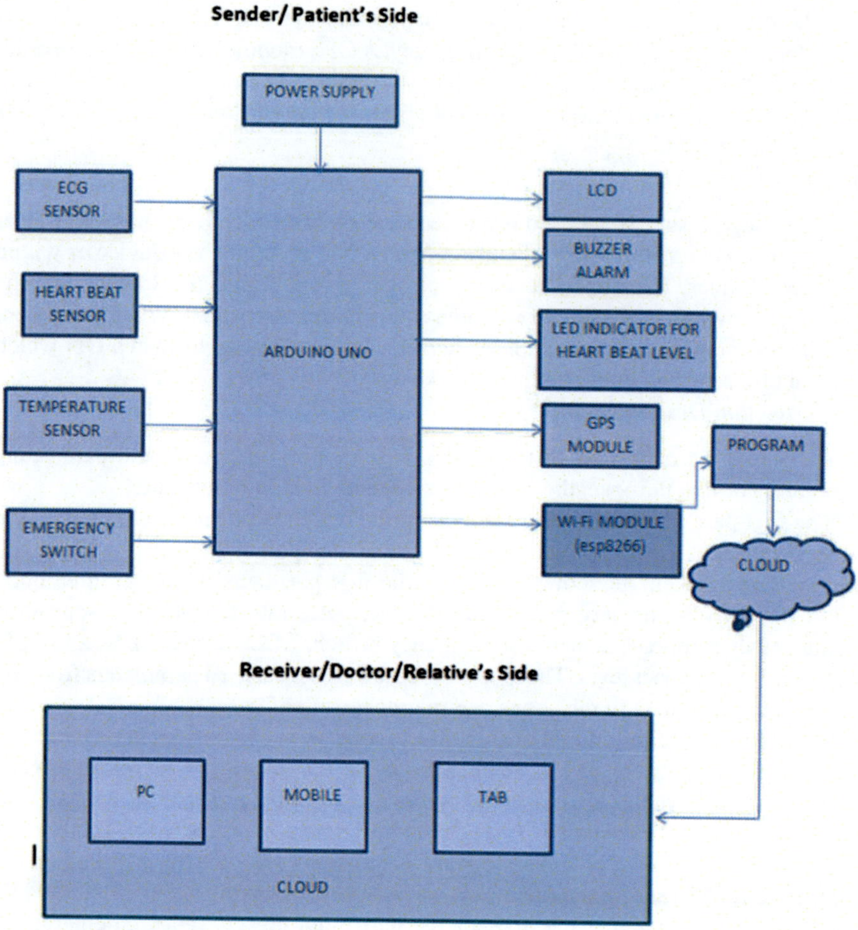

Fig. 1 Block diagram for the system

possible in the comfort of the individual's home without the trouble of going to the hospital in case it is too far or inaccessible. Hence, an efficient and reliable health monitoring system is achieved.

4 Results and Discussion

The system is designed to monitor and analyze the data obtained by the health monitoring sensors that can benefit both patient and doctors/care takers. After getting the data of various health parameters, it is uploaded on ThingSpeak cloud to evaluate

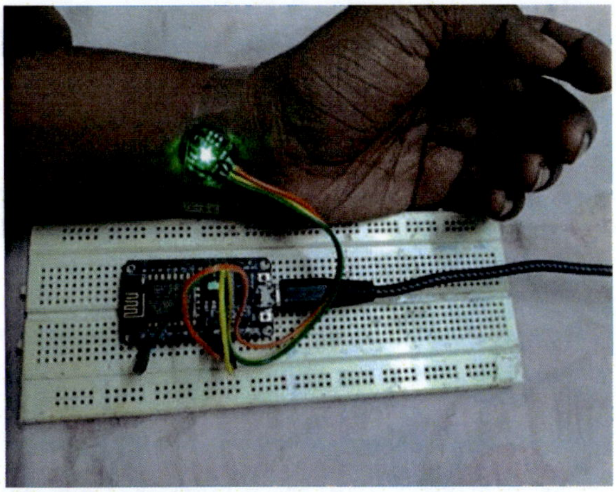

Fig. 2 Pulse rate measurement

and set an emergency alert by sending email to doctor and care takers as well as an ambulance in case of emergency (Fig. 3).

The Web interface is used to catalog the readings and obtain the pictorial representation of the sensor information as well as analyze the information by analysis module to decide the seriousness of the patient.

Figure 4 shown above represents the data analysis of temperature readings of the patient taken by the temperature sensor. The sensor readings are taken from the microcontroller at an interval of 15 s and uploaded to the ThingSpeak cloud via the Wi-Fi module ESP8266. The temperature obtained is in degree Celsius.

The pulse sensor is used to obtain the PPG waves obtained through the microcontroller. When the signal exceeds 50% of the amplitude in fast upward ramp (PT), the timing is taken to measure IBI through microcontroller code manipulation. The

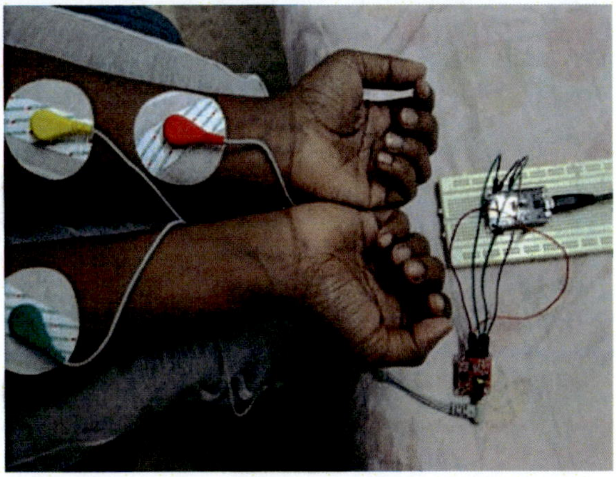

Fig. 3 ECG measurement

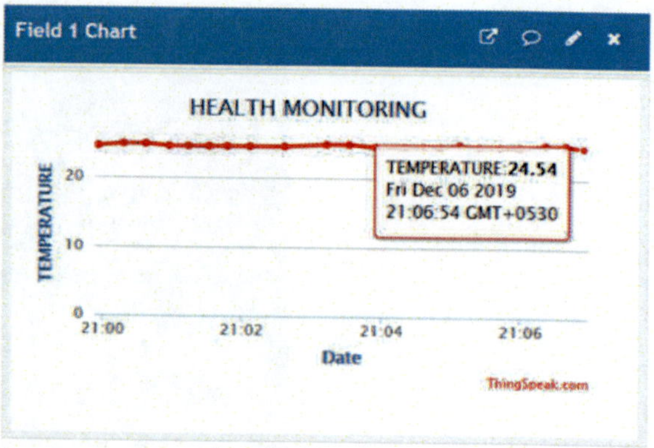

Fig. 4 Graph of temperature measurement generated by ThingSpeak

average of previous 10 IBI beat value is taken in order to calculate the BPM. In order to get the correct value of amplitude, the highest (P) and lowest (T) points of PPG signal are determined. To find the 50% of amplitude, the threshold value is set to 560, and the variation of pulse rating is observed. Before the signal is updated, the time corresponding to 3/5 of IBI value is passed to avoid noise or anomalies in reading. As soon as threshold exceeds 50% signal amplitude and time period is greater than 3/5 of IBI, the ten elements of IBI values are stored. The average of these values is calculated by the formula shown below and BPM is obtained. Figure 5 shown below represents the data analysis of the pulse rate readings of the patient taken by the pulse rate sensor.

$$\text{IBI avg} = (\text{IBI0} + \text{IBI1} + \text{IBI2} + \text{IBI3} + \text{IBI4} + \text{IBI5} + \text{IBI6} + \text{IBI7} + \text{IBI8} + \text{IBI9} + \text{IBI10})/10$$

$$\text{BPM} = 6000/\text{IBIavg}$$

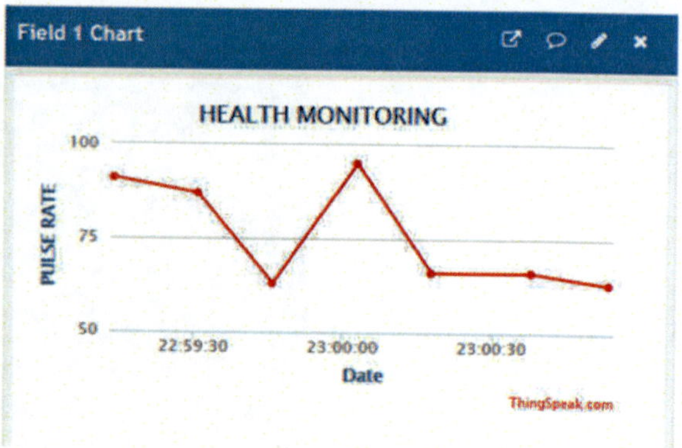

Fig. 5 Graph of pulse rate measurement generated by ThingSpeak

The Arduino sends the ECG data to the ESP8266 Wi-Fi module which then sends it to ThingSpeak cloud in real time [11]. The data is plotted on the ThingSpeak Web site. Figure 5 shown above is the graph obtained by the ECG AD8232 sensor.

The code also displayed the latitude and longitude coordinates of the patient when the temperature reading exceeded the threshold limit set at 41 degree Celsius, Fig. 6. We used the Blynk app to send the location coordinates of the patient to the doctor (Figs. 7 and 8).

To connect our system to Blynk, we installed Blynk libraries and added the Auth token into our code which we received while creating a virtual system on the Blynk app. When the system was connected to the Node MCU GPS Module, it provided the location of the system to Blynk which in turn sent an email containing the latitude and longitude coordinates to the registered email address. A simple medicine reminder system [12] also created using IFTTT (If this then that) Web site and application which sends timely SMSs with the name of the medicine to be taken via the Internet.

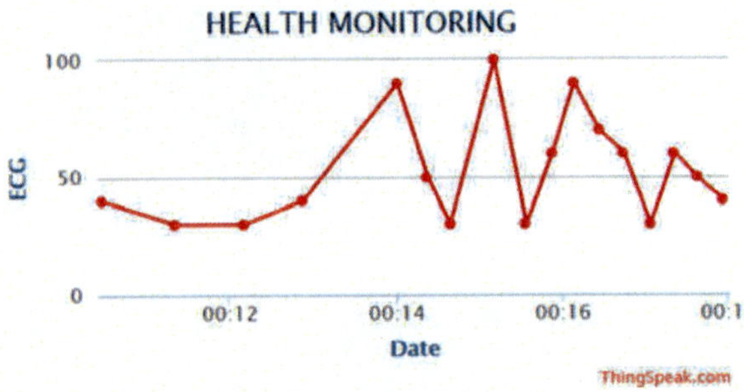

Fig. 6 Graph of ECG measurement generated by ThingSpeak

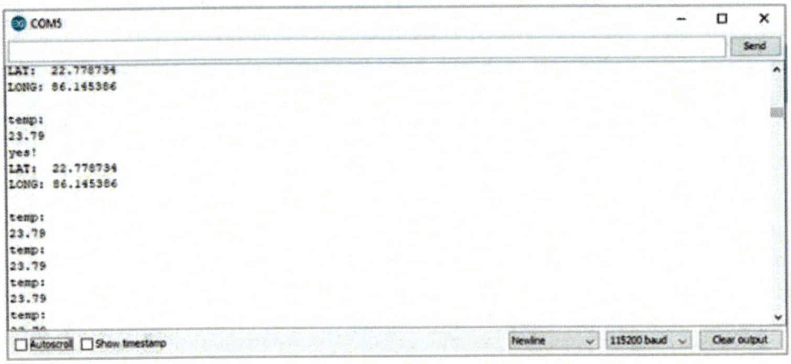

Fig. 7 Location coordinates displayed on serial monitor of Arduino

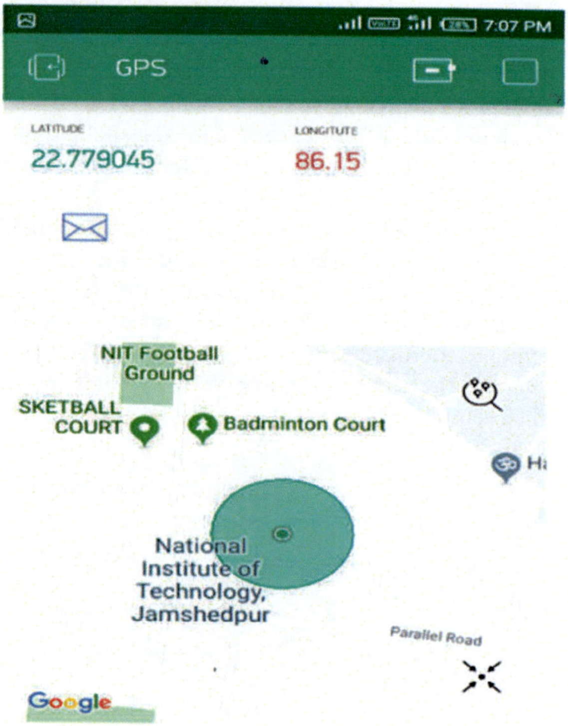

Fig. 8 Location coordinates displayed on the Blynk application

5 Conclusion

Due to technological enhancement, different new features are getting added into such systems every day. This work demonstrated build an integrated e-health monitoring system with variety of features ranging from vital parameter measurement to medicine reminder, and emergency location sharing to real-time data sharing with the doctor. All in all, the system worked well, and we were able to measure the temperature, pulse rate, and ECG reading of the patient and share it over the Internet via the ThingSpeak platform. Further parameters can be added to the present system like blood glucose level, blood pressure, etc., to make a more integrated health monitoring system. Analysis of data collected from several patients can be used for generating a pattern between body parameter readings and certain aliments. This can be done either manually or by incorporating machine learning into the system. Thus, the system will able to give suggestions and predictions based on continuous readings from the patients. As the system deals with sensitive and private data of patients which shall be uploaded to the cloud, there is a lot of scope of work in improving the security of the system.

References

1. Wang L, Yang G-Z, Huang J, Zhang J, Yu L, Nie Z (2010) A wireless biomedical signal interface system-on-chip for body sensor networks. IEEE Trans Biomed Circuits Syst 4(2):112–117
2. Khan SF (2017) Health care monitoring system in internet of things (IoT) by using RFID. In: 2017 6th international conference industrial technology management ICITM 2017, pp 198–204
3. Sawand A, Djahel S, Zhang Z, Na F (2014) Multidisciplinary approaches to achieving efficient and trustworthy ehealth monitoring systems communication. In: China (ICCC), 2014 IEEE/CIC international conference, pp 187–192
4. Bourouis A, Feham M, Bouchachia A (2011) Ubiquitous mobile health monitoring system for elderly(UMHMSE). Int J Comput Sci Info Technol (IJCSIT) 3(3)
5. Lee JW, Jung JY (2007) ZigBee device design and implementation for context-Aware U-Healthcare system. In: The IEEE 2nd international conference on systems and networks communications, Cap Esterel, French Riviera, IEEE Computer Society, pp 22
6. Shaikh S, Waghole D, Kumbhar P, Kotkar V, Awaghade P (2017) Patient monitoring system using IoT. In: 2017 International conference big data, IoT data science BID 2017, vol 2018-January, pp 177–181
7. Thaduangta B, Choomjit P, Mongkolveswith S, Supasitthimethee U, Funilkul S, Triyason T (2017) Smart healthcare: basic health check-up and monitoring system for elderly. In: 20th International computer science engineering conference smart ubiquitos computer knowledge, ICSEC 2016, pp 1–6
8. Taştan M (2018) IoT based wearable smart health monitoring system. *Celal Bayar Üniversitesi Fen Bilim. Derg.*, no. October, pp 343–350
9. Geman O, Chiuchisan I (2016) A health care self-monitoring system for patients with visual impairment using a network of sensors. In: 2015 E-Health bioengineering conference EHB 2015, vol 3. pp 1–4
10. Gupta N, Saeed H, Jha S, Chahande M, Pandey S (2018) Iot based health monitoring systems. In: Proceedings 2017 International conference innovation information, embeding communications systems ICIIECS 2017, January, vol 2018. pp 1–6
11. Kalaivani P, Thamaraiselvi T, Sindhuja P, Vegha G (2017) Real time ECG and saline level monitoring system using arduino UNO processor. 1(2), 160–164
12. Zanjal SV, Talmale GR (2016) medicine reminder and monitoring system for secure health using IOT. Phys Proced 78:471–476

A General Data Retrieval Technique in Remote Healthcare Application

Safikureshi Mondal and Zeenat Rehena

Abstract Mobile cloud platform plays an important role in remote health care monitoring system. Healthcare data is stored in the cloud as semi-structured and unstructured and metadata format which are coming from different heterogeneous sources. Data retrieval technique is a primary and one of the efforts in remote health care monitoring system in which users/doctors do the query through mobile to know the patient information from the cloud. The data retrieval techniques may be computer-intensive task, and it is required much more memory and computational power as its intermediary processing steps. The objective of the paper is to propose a framework for retrieving the patient-centric data from the cloud-assisted semi-structured data/metadata.

Keywords Unstructured data · Keyword search · Offloading · Application partitioning

1 Introduction

The collaboration of mobile and cloud computing services offers a new emerging research area, called, mobile cloud computing(MCC), and there are lot of definitions already done in different research papers [1–3]. The integration of mobile devices and cloud and its research problems can be solved by three approaches [3]:

1. First approach is that user's access applications as services via mobile devices offered by cloud through web browsers which is known as Software-as-a Service(SaaS)cloud.
2. The second approach is that mobile devices worked collaborate as cloud services.

S. Mondal (✉)
Narula Institute of Technology, Kolkata, India

Z. Rehena
Aliah University, Kolkata, India

K. Dahal et al. (eds.), *Internet of Things and Its Applications*, Lecture Notes in Electrical Engineering 825, https://doi.org/10.1007/978-981-16-7637-6_3

3. And third approach is that the user access though mobile devices emerging application and uses cloud storage for big data by offloading and application partitioning approach. This service is called as Platform-as-a-Service.

The work is based on this third approach. Remote health care is one of the emerging research application as mobile cloud like image, multimedia and GPS sensor data. The purpose of using mobile cloud in remote health care is to minimize delay and providing mobile users easy access to medical function efficiently. Already, there are lot of research [4, 5] works done where remote healthcare is implemented by mobile and cloud collaboration. The processing of remote health data is required more energy. But mobile are basically low power, low storage device. Thus, to increase the mobile battery power, mobile computation task can be executed partially or remotely to the cloud.

The remainder of this paper is organized as follows. First, the challenges of data retrieval techniques of remote healthcare are discussed in Sect. 2. Next Sect. 3 describes the review of related work of data retrieval techniques in mobile cloud environment. After this, a data retrieval framework is proposed in Sect. 4 and for understanding the framework, a sequence diagram is also created in this section. Finally concluded in Sect. 5 with future work.

2 Challenges of Data Retrieval in Remote Healthcare

Remote health care, as an application of MCC, it has lot of research challenges which are coming from both mobile as well as cloud side. All the challenges are related or dependencies to each other. Some of the important challenges for data retrieval in remote healthcare are energy efficiency, computation offloading, retrieval algorithm, network connectivity, fault tolerance, etc. which are written as a concise way in the following.

Energy efficiency: In mobile cloud computing, energy efficiency is the very big challenges for low battery mobile devices. Mobile devices have always some challenges like memory, computation, network connectivity, delay, bandwidth and battery power. Cloud computing helps to overcome these mobile constrained to provide better computation and storage. But, the energy power of mobile is remain big challenges. Thus, data retrieval in mobile cloud application remains the big challenge in remote healthcare system.

Computation offloading: Mobile faces the problem of computation power in mobile–cloud communication. Computation offloading is the only best choices for the MCC application design. In remote health care monitoring system, data retrieval techniques also face the problem of low computations if this data retrieval technique is require much more processing power and memory for intermediary steps of processing as it uses mobile devices and the best solution is the computation offloading for data retrieval techniques. Offloading increases the mobile-based application performances, reduces battery power consumptions and provides sufficient resources. Course-gained offloading offload small part of code which is energy hunger by migration and partitions. Course-gained offloading may be offload full application or code. Computation offloading also depends on some parameter. Application migration can be additional overhead to the programmer. Offloading can be measures by threshold value, size of an applications and critical section. Threshold value can be processing time, energy consumption or memory usage. Depending upon size of the code, offloading consumes more or less energy and compared with local processing. Critical section means complex computations which is required more energy for mobile devices. It should be offloaded. Latency, network bandwidth and heterogeneity are the important issues for computation offloading.

Data retrieval Algorithm for Remote healthcare Environment: Due to the heterogeneous patient-centric data is stored in cloud as structured, semi-structured or metadata format. Choices of database for semi-structure or metadata are an another challenging task. Semi-structure or metadata can be stored in the form of XML or JSON in cloud. For this limitation, finding a suitable data retrieval technique in this environment is also a big challenge.

Network connectivity: The main aims of mobile cloud computing are to utilize or increase the capability of resources constraint mobile devices. Mobile always gives long-time services by accessing resources of cloud. Remote health care monitoring system needs an uninterrupted communication between mobile devices and cloud. This continuous and consistent connectivity is called "seamless connectivity". Network connectivity is the important and very challenging issue between mobile and cloud for seamless connection [6]. Due to mobility nature of mobile devices, handover in the network causes service disconnection. There are so many factor of seamless connection such as very low bandwidth, heterogeneity that affects a lot between mobile and cloud. In [7, 8], the author surveys and shows that seamless connection is an open issue of mobile cloud.

Fault-tolerant: Fault-tolerant issue is an important issue in mobile cloud. Fault-tolerant gives the reliability for critical services in mobile cloud platform. It provides mechanism to prevent system failure occurrences [9]. It recovers failure of a system recovery, lower cost and improved various types of metrics. Basically, two types of tolerance can happen in cloud. One is reactive fault tolerance and other proactive technique. There are various techniques such as checkpoint/restart, job migration, SGuard, rescue workflow, replay and retry which is under the category of reactive fault tolerance mechanisms. On the other, category proactive policies recovers by replacing and predicting some component from faults, errors and failures.

Others Challenges: Some other important challenges of data retrieval techniques are trust, security and privacy in data.

3 Related Work

Cloud computing is now a new paradigm for remote healthcare and mobile cloud is an important part of this application. There are lot of works already done which are based on implementation on healthcare using mobile and cloud. Searching techniques in cloud always remain a big challenge for the mobile based application. In this section, some related work is mentioned for the retrieval of data from cloud by the mobile devices in particular healthcare application.

MediAlli is proposed in [10] in where sensors data is collected as medical data in real time using physiological sensors. After it is transferred to the mobile phone in where medical data streams processed on the basis of user specific context rule. In [11], authors proposed a personal healthcare system, named WAITER as emergency aid system which monitor health condition in real time by heartbeat, motion, body sensors etc. and generate alarm in emergency situation. WAITER continuously senses and transmit real data (vital signs) as raw data to the data server in healthcare center via Bluetooth device. Data server generates from mobile device for multiple users. After storing the reports in local database, the caregivers can access the database for users' health status. Also, the server can receive emergency alert from a specific user and then, immediately ask medical aid for the user.

In [12], authors proposed a energy efficient mHealthMon scheme which is work in mobile health platform in a distributed P2P overlay network. The ultimate goal of this research is to maintain the quality and service of mobile health monitoring system. This system modeled by medical sensors in mobile, cloud services and network medium. Authors proposed this scheme to maximize the benefits of parallel offloading scheme in multiple cloud machines.

In [13] research paper, authors proposed a full-text keyword search technique which is built in hybrid approach. The keyword search technique retrieves the matching input keyword locally or remotely. The whole idea is proposed to show the energy efficiency of mobile devices. First, query can be split into two sets. One set is running in mobile device, and another set is offloaded to powerful remote server from mobile. After processing remote server back to the result to the mobile phone. To split this query, author also proposed a SSQ problem as a min-cost flow graph for splitting this query, and remote server solves this problem in offline manner. The authors only concentrated on energy to propose a hybrid keyword search approach but not proposed any keyword search techniques. The input query file collected from MSN and experimented in a given TRACE AP dataset. In [14], authors studied the properties of search queries which is submitted through mobile Yahoo! Search application. The author analyzed the mobile query on various aspects of mobile search with a large-scale study on English mobile queries.

In a [15] research paper, the author discusses the advantages and disadvantages of combining mobile desktop search with cloud-assisted activities such as operation offloading, cloud storage, and cloud-assisted search. The authors analyzed the energy trade-off between local computations and remote services. By this research paper, authors concluded that mobile device should offload the computationally intensive task if mobile devices is connected to network weakly. The authors proposed and evaluated a cloud-assisted search like Dessy desktop search. In this paper, authors shows the offloading of the search for save better energy and performance issues in a multi-user environment.

Authors [16] proposed "Find All" search procedure that is designed specifically for mobile devices with local search engine. This search engine is used to do user search with both memory and energy-intensive. It is worked with retrieve web pages, and also it works if there is no connectivity. This search procedure uses keyword search and implements a full-blown search engine. This system collects all web pages by re-finding procedure which is a combination of some steps like visiting all devices, caching and indexing. Also, author implements this work in a android platform with publicly available Galago search engine.

4 Proposed Framework for Data Retrieval in Remote Healthcare

A data retrieval technique is proposed for remote health care in Fig. 1. In this framework, there are two important tasks required for data retrieval.

1. First, a searching technique is required where users do the query to retrieve data from cloud-assisted semi-structure data through resource-constraint mobile devices and secondly,
2. an offloading technique is required which helps resource-intensive task is offloaded to cloud.

The data retrieval techniques may be required much more processing power or memory for indexing, etc. The particular search procedure or offloading techniques are beyond of this paper. In remote healthcare, user searches the specific information from patient-centric data in the cloud which is stored semi-structured or metadata format. Application module, Profiler, offloading manager, query Engine, queue and connectivity controller are the active module in mobile side. Also cloud side contains *Application*, *Profiler*, *Query Solver*, *Connectivity Controller* and database which contains metadata or semi-structure data as patient health data which is coming from different heterogeneous sources. The database must be contents XML or JSON format, and in this proposed work, we assume JSON is the appropriate data format. Actually it is the general framework for data retrieval technique in the remote healthcare application in MCC environment. Now it is required to discuss every module or functions for both mobile side as well as cloud side.

Application profiler: Actually profiling is the estimation of the execution cost of each module of the application. Often prediction-based and model-based profiling approach is used. For prediction based profile, the estimation is predicted from past historical records. But model based does not directly measure the execution cost but measure the device and network status such as battery level, CPU load, wireless con-

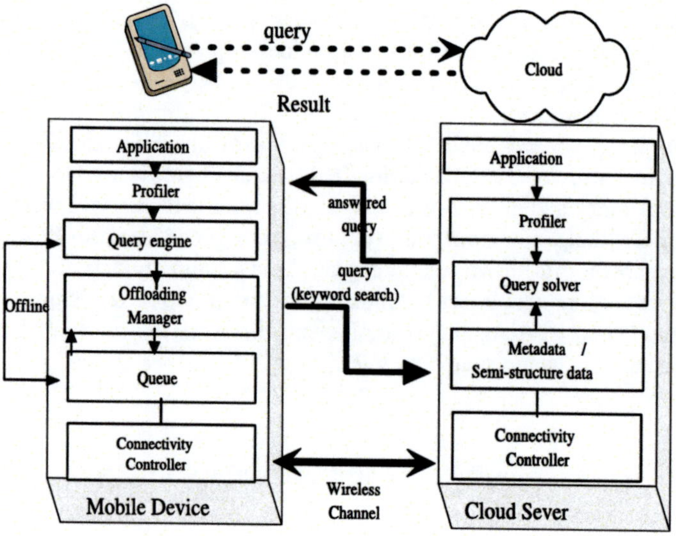

Fig. 1 A proposed data retrieval framework in remote healthcare

nection quality and so on. Odessy [17] uses the prediction-based profiling approach, but MAUI [18], CloneCloud [19] and ThinkAir [20] are used model-based profiling approach.

Offloading manager: It is a decision-maker to make offloading in this framework. Before this decision-making, this module optimizes the partition according to profiling-based execution cost and then offloaded the task. Partitioning decision can be done offline or online. Online means decision will be take place cloud side, and offline is mobile side. Always offline is better for connectivity problem but if this problem persists then obviously extra overhead is not required for mobile device, means online preferable in this case.

Query Engine: This query module is needed for input of the keyword and if required any for pre-processing of this keyword. We assume that query or data retrieval technique is computing-intensive task and processing of this query needed much more memory and computational power for mobile device. The device of appropriate query or data retrieval technique in this scenario is a challenging task.

Queue: Queue module is very important for offline mode of remote healthcare system in MCC environment. If connection is fail due to network problem and user want to search a particular query from cloud, then queue, store the particular keyword in this queue. When connectivity is happened, then this keywords are processed as query sequentially. Actually queue is used mainly for storing keyword.

Connectivity Controller: This module is responsible for connection between mobile and cloud.

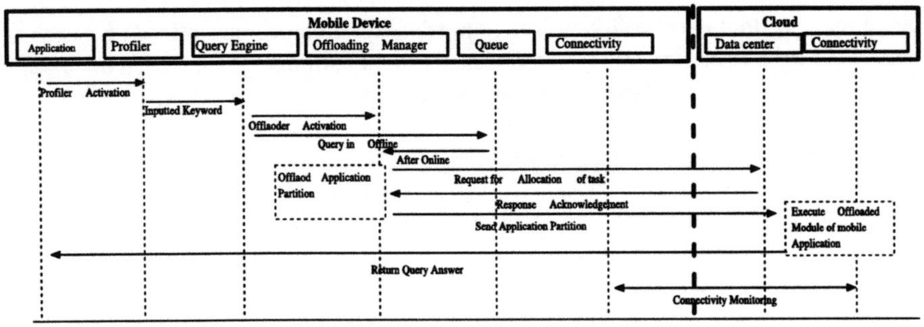

Fig. 2 A sequence diagram for data retrieval framework

Metadata/Semi-structure data: In remote health care system, heterogeneous data is stored from heterogeneous sources in cloud storage. This heterogeneous data is stored in the form of metadata or semi-structure data. Metadata are the descriptive information about data, and data is measured with data's properties. Actually metadata are those elements that concisely identifies the "who, what, when, why, and how" of the data. The main motivation for metadata is storing data in an easily retrievable format. The goal of the metadata is to make it easy for the users to learn, locate and retrieve the desired data. Metadata must also be compatible to add new fields to the metadata, provide user support and maintain software, hardware, and storage and network connectivity. Data should be stored in a format that can be retrievable, and more importantly it should be in a format that will continue to be accessible as technology changes. In this module, data stored in either XML or JSON format. We prefer JSON format.

In Fig. 2, it is given a simple sequence diagram for data retrieval techniques in mobile cloud framework. First profile is activated after application running in mobile cloud, and it is estimate the execution cost of each module of the application. At the meantime query engine module is taking simple query from general user. Depending upon estimation cost of the modules which is predict by the profiler, offloading manager activated and offload application partition to the cloud and offload manager send the application partition and after processing, result of the query is back to the mobile side. And response back to the mobile device. Connectivity always connects mobile device to the cloud. If connectivity fails (offline), then inputted query is store in a queue in the mobile side. In online mode by one by one, inputted query is executed.

5 Conclusion

This paper proposes a framework for data retrieval techniques on semi-structure data /metadata in remote health system where general user or health professional searches with specified simple query from cloud. The query or data retrieval techniques may be computing-intensive task which is require much more computational power and memory for intermediary steps. To do this, some challenges of data retrieval techniques are pointed out. Also a sequence diagram is presented for propose framework. Our future work will be the implementation of data retrieval techniques using appropriate search techniques with existing or new offloading mechanism on remote health care system in MCC environment with care of existing design challenges of MCC for remote health care system.

References

1. Peter M, Timothy G (2009) The nist definition of cloud computing. retrieved from. In: In http://csrc.nist.gov/publications/nistpubs/800-145/SP800-145.pdf, vol 1
2. Wang M, Chen Y, Khan M (2014) Mobile cloud learning for higher education: a case study of moodle in the cloud. Int Rev Res Open Distributed Learning 15(2)
3. Sanaei Z, Abolfazli S, Gani A, Buyya R (2014) Heterogeneity in mobile cloud computing: taxonomy and open challenges. Commun Surv Tutorials IEEE 16(1):369–392
4. Mondal S, Mukherjee N (2016) Mobile-assisted remote healthcare delivery. In: 2016 fourth international conference on Parallel, Distributed and Grid Computing (PDGC), pp 630–635
5. Mondal S, Mukherjee N (2017) A framework for ICT-based primary healthcare delivery for children. In: 2017 9th international conference on Communication Systems and Networks (COMSNETS), Bangalore, pp 525-529. https://doi.org/10.1109/COMSNETS.2017.7945447
6. Joshi SC, Sivalingam KM (2014) Fault tolerance mechanisms for virtual data center architectures. Photonic Netw Commun 28(2):154–164
7. Nguyen H, Shen Z, Tan Y, Gu X, Inc M (2013) Fchain: toward black-box online fault localization for cloud systems. In: ICDCS
8. Bala A, Chana I (2015) Intelligent failure prediction models for scientific workflows. Expert Syst Appl 42(3):980–989
9. Vallée G, Charoenpornwattana K, Engelmann C, Tikotekar A, Leangsuksun C, Naughton T, Scott SL (2008) A framework for proactive fault tolerance. In: Proceedings of the Third International Conference on Availability, Reliability and Security, ARES 2008, March 4–7, 2008, Technical University of Catalonia, Barcelona , Spain, pp 659–664
10. Chowdhury AR, Falchuk B, Misra A (2010) Medially: a provenance-aware remote health monitoring middleware. In: PerCom. IEEE Computer Society, pp 125–134
11. Wu W, Cao J, Zheng Y, Zheng Y (2008) Waiter: a wearable personal healthcare and emergency aid system. In: PerCom. IEEE Computer Society, pp 680–685
12. Ahnn JH, Potkonjak M (2013) mhealthmon: toward energy-efficient and distributed mobile health monitoring using parallel offloading. J Med Syst 37(5):1–11
13. Rao W, Zhao K, Lagerspetz E, Hui P, Tarkoma S. Energy-aware keyword search on mobile phones. In: In Proceedings of the first edition of the MCC workshop on Mobile cloud computing (MCC '12). Association for Computing Machinery, New York, NY, pp 59–64. https://doi.org/10.1145/2342509.2342522
14. Yi J, Maghoul F, Pedersen J (2008) Deciphering mobile search patterns: a study of yahoo! mobile search queries. In: Proceedings of the 17th international conference on world wide web, WWW '08. ACM, New York, NY, pp 257–266
15. Lagerspetz E, Tarkoma S (2011) Mobile search and the cloud: the benefits of offloading. In: PerCom workshops. IEEE, pp 117–122
16. Balasubramanian A, Balasubramanian N, Huston S, Metzler D, Wetherall DJ (2012) Findall: a local search engine for mobile phones. In: Proceedings of the 8th international conference on emerging networking experiments and technologies, CoNEXT '12. ACM, New York, NY, pp 277–288

17. Ra M, Sheth A, Mummert L, Pillai P, Wetherall D, Govindan R (2011) Odessa: enabling interactive perception applications on mobile devices. In: Proceedings of the 9th international conference on mobile systems, applications and services, MobiSys '11. ACM, New York, NY, pp 43–56
18. Cuervo E, Balasubramanian A, Ki Cho D, Wolman A, Saroiu SCR, Bahl P (2010) Maui: making smartphones last longer with code offload. In: Proceedings of ACM MobiSys
19. Chun B, Ihm S, Maniatis P, Naik M, Patti A (2011) Clonecloud: elastic execution between mobile device and cloud. In: Proceedings of the sixth conference on computer systems, EuroSys '11. ACM, New York, NY, pp 301–314
20. Kosta S, Aucinas A, Hui P, Mortier R, Zhang X (2012) Thinkair: dynamic resource allocation and parallel execution in the cloud for mobile code offloading. In: INFOCOM, 2012 Proceedings IEEE, March 2012, pp 945–953

Study on Efficient Service Broker Policy for Data Center Selection Analyst for Education System Using Cloud Computing

Malti Bhardwaj and Kiran Deep Singh

Abstract Cloud computing is an area that is quickly getting differentiation in both shrewd world and industry. Cloud Analyst is huge mechanical gathering to show and separate cloud computing condition and applications before certified relationship of cloud things. Capriciously picked server farm is inclined to give prohibited outcomes in term of reaction time, asset use, cost or different cutoff points. In this paper we propose a need-based Round-Robin administration merchant figuring which passes on the mentioning dependent on the need of server farms and gives ideal execution over the standard Discretionary choice calculation. Education is a need of everybody these days for moderate new development. It anticipates that for the future, an enormous portion could establish a healthy society. Advancement for high-quality education anticipates a main trend. Starting now, cloud computing is the most unexpected and solid asset for exchanging knowledge, as it works in various areas such as clinical management, business, correspondence and more. It also occurs in the scholarly organization, as it offers better benefits on request and numerous other monumental and strong inclinations. Many expert co-ops offer specialized cloud-based software so consumers can work on such apps according to their needs without a completely shocking stretch. In this review article, we will propose a model to educators and understudies with the aid of cloud computing so that teachers can share the course materials over the cloud and understudies can get the information with the help of this model about their assessments, activities and various items. With the aid of this model, guard/gatekeepers may also review the evaluation content, participation records, and so on of their young people. This review paper therefore focuses on how cloud computing can be used in the context of the arrangement to boost the difficulties in the preparation of foundations.

Keywords Cloud computing · Cloud Analyst · Data center · Service broker policy · Round-Robin · Load balancing

---

M. Bhardwaj
Khalsa College of Engineering and Technology, Amritsar, India

K. D. Singh (✉)
IKG Punjab Technical University, Kapurthala, Punjab, India

K. Dahal et al. (eds.), *Internet of Things and Its Applications*, Lecture Notes in Electrical Engineering 825, https://doi.org/10.1007/978-981-16-7637-6_4

1 Introduction

Cloud computing has gotten approval and pace, beginning late through the game plan of particularly depicted administrations in basically every field as shown in Fig. 1. It has emerged with a high pace and almost covered every aspect in the industry. The processing power that was before a need for the customer premises is straightforwardly moving to the cloud specialist co-ops, with customers simply utilizing an essential terminal with a Web program. Notwithstanding, the force and client of the application has not reduced. Without a doubt, administration has improved by moving the weight to the administration provider [1]. The cloud computing model has end up being a guide for end clients, structure bosses, programming originators and even IT purchasers, corporate and government customers. These clients have been drawn by this viewpoint considering the highlights it gives like client self-provisioning, low operational expense and opportunity from capital cost, transparency of a gigantic pool of advantages, multitenant client access, relentlessness, security and evaluated administration. The basic rule that directs the cloud computing point of view is the strategy of processing assets like figuring power, gathering and data transfer capacity as "administrations" on a compensation for each utilization premise. The administrations are given as programming as a service (PaaS), stage as a service (SaaS) and system as a service (IaaS), which might be passed on straightforwardly, private, association or mix structure. The standard element that portrays cloud computing is the use of virtualization (1). Virtual machines are run over the open equipment to address the client needs. A hypervisor is a type of virtualization software that is used in cloud hosting to divide and allocate resources across several pieces of hardware

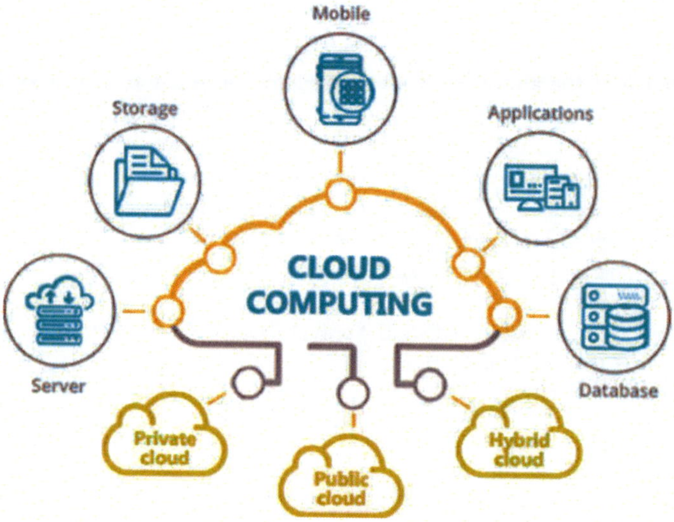

Fig. 1 Cloud computing

[2]. The choice of virtual machines at whatever point the outstanding task at hand is experienced is finished by the load balancer, whose point is to scatter the load, so no virtual machine is overwhelmed by demands reliably while staying inactive at different occasions. Over this level lies another reflection called the administration intermediary, which is the inside individual between the clients of the cloud and the cloud specialist co-ops. It utilizes the current assistance specialist blueprints to emphasis the client mentioning to the most real server farm. Thusly, the confirmation of the best approach picks the reaction time of a specific deals and the productivity of usage of the server farm [3]. Server farms are had and controlled by the specialist organizations at unquestionable zones, and subject to use, a specialist organization may choose to design its server farms with various kinds of equipment. Likewise, the rigging continues changing chance to time as indicated by the client need. Hence, regardless of whether specialist co-ops attempt to keep up consistency in the choice of rigging, they need to expand or rot the measure of machines as shown by the requirements of customers. The current dealer strategies do not base their choice of server farms within equipment game plans, in any case on the region of server farms, reaction time or current execution load. We propose an intermediary strategy that picks a server farm for satisfying its deals dependent on the level of equipment accessible, notwithstanding, when the apparatus open at various server farms is of various course of action [4]. Load balancing works on the principle of closest data center and optimal response time. For maintaining load, the first solution is to find out nearest data center for the user as it will be helpful in finding best internet facilities.

The second efficient way to maintain the load is to check the best response time which would be performed by using different algorithms.

2 VM Load Balancing

1. Cooperative load balancer: Using a simple cooperative search to circulate VMs.
2. Round-Robin Load Balancer: This load balancer aims to maintain relatively exceptional loads on all available VMs.
3. Gagged Load Balancer: This load balancer assures that at any self-assertive time only a pre-depicted number of Web Cloudlets will be allocated to a single VM. In the event that a greater number of packs of deals are available than the amount of accessible VMs on a server farm, an element of the mention should be arranged before the VM is opened [5]. The working of the VM is shown below in Fig. 2.

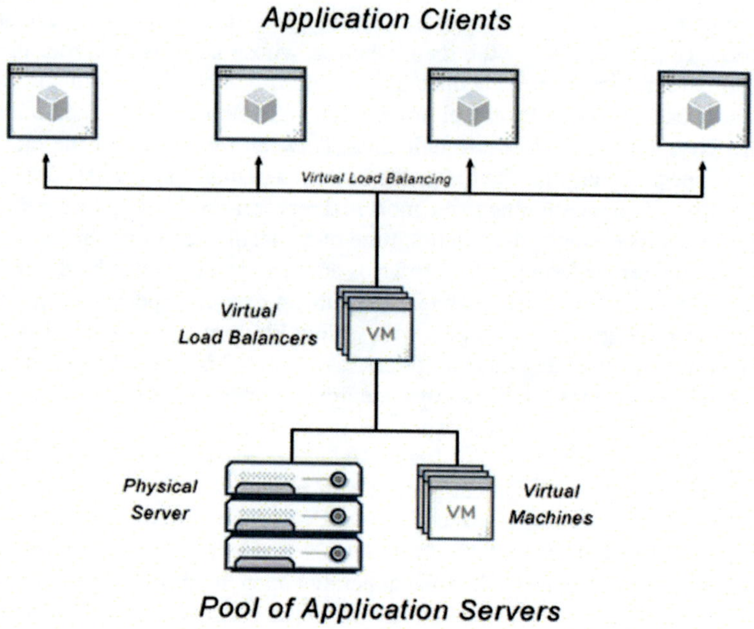

Fig. 2 VM load balancing

3 Cloud Application Service Broker

(1) An assistance representative picks which server farm should offer the assistance to the mentioning beginning from every client base. Also, subsequently, administration intermediary controls the traffic organizing between client bases and server farms. Three assistance representative philosophies are right presently connected with the Cloud Analyst [6].

(2) Administration Region-Based Guiding: In this organizing strategy, administration agent picks the most succinct course from the client base to the server farm, subordinate upon the affiliation dormancy and dependent on that, courses the traffic to the nearest server farm with the possibility of transmission inaction [7].

4 Benefits of Cloud Computing for Education

1. Altered learning: Cloud registration manages transparent portals to make learning decisions more popular in understudies. Using an Internet-related tool, understudies can achieve a wide range of interests and mechanical congregations that fit their learning styles and costs. [8].

2. Lessened Costs: Cloud-based administrations can assist establishments with diminishing expenses and animate the utilization of new headways to meet advancing edifying prerequisites. Understudies can utilize office applications futile without buying, present and keep conscious with the latest on their PCs. It also gives the working environment of pay per use for explicit applications [9].
3. Receptiveness: Availability of the administrations is the most immense and required by the client utilizing the readiness cloud. 24X7 is the accessibility that is required by this framework without thwarted expectation. From any place, one can login and get to the information [10].

5 Literature Survey

Cloud computing is a cutting-edge advancement to improve coordination information structure by utilizing its foundation, stage and programming courses of action through the Web. Cloud computing is skilled to give cloud-based assistance for making money-related and operational administrations in co-arrangements the board. Cloud computing framework as an assistance offers smart assistance to the clients who can spare their expense by utilizing administrations of cloud specialist co-ops or outcast on the rental explanation. There are assorted proposed approaches and structure in a line to understand the coordination for better utilization of coordination information framework over cloud.

Ren et al. proposed a craving-based approach which is called exponential smoothing check to find the weighted least relationship with handle the long association. This check figures the load on the pro from various cutoff points like, CPU use, memory, number of affiliations and size of plate occupation [11].

Peng et al. zeroed in on the effect of cloud computing on information partaking in deftly chain. This paper has systematic assessment of endless information sharing subject to cloud computing by utilizing amusement model to ascertain the commonplace great circumstance of cloud for complex deftly chain [12].

Leukel and Kirn et al. proposed a help organized methodology, got a handle on by cloud computing to give interconnected activities in agilely chain [13].

Sunderaswarn, Squicciarini and Lin et al. (2018) gave a business-based way to deal with oversee cloud administration affirmation. They proposed a novel financier-based education in the cloud, where the cloud merchants is subject for the administration choice and plan [14].

Thomas Schramm et al. dissected cloud computing in easily chain. This paper perceived six solicitations for managers to get a handle on new progression which gives monstrous effect on deftly chain [15].

Jain et al. [16] dissected service broker policy algorithms which enables cost performance and provides gains in service performance.

6 Proposed Work

We can pick server farm in Round-Robin way to deal with fitting solicitations dependably among all the server farms inside a region. It prompts more asset use, in any case server farms may have indisputable arranging speed. Some server farms are sleeker than others, so these server farms should have chosen more often than the milder ones to maximize property execution and usage. Therefore, we have to pick the farm server according to their speed limit. We used preprocessing Fig. 1 to generate the appropriate list at the start of the redirection. Just a single period can be achieved during pre-processing stage. Round-Robin Guarantee Estimate 2 is used after pre-processing for selecting the server farm for each zone.

Calculation 1 Preprocessing

Step 1: do more than one server farm for all locals
2: Find the specifications of each server farm according to its configuration (usually speed)
3: Provides a social log case for which the server farm is used a number of times equal to the needs delegated
4: Store in need list these movements reported by their region
5: Exit to
6: Must return list.
Calculation 2 Need-based Round-Robin Determinatio
Information: Area number, need list
Yield: Objective DC name
Steps:
Dclist ← regionalDataCenterIndex.get(region
if Dclist isn't Invalid, by then.
noDc ← Dclist.size()
if noDC = = 1, by then
DcName ← Dclist.get(0)
else index ← getnextindex(region)
DcName ← Dclist.get(index)
end if
end if
return DcName

This proposed model is for structure planning by the use of distributed computing. Advancement of cloud computing provides flexible tools that can aid in different stages. Basically, we are only concentrating on planning in this investigation report. Notable clients plan cloud fuse employees, understudies, manager workers, endorsement, documents and workplaces for assessment. In this model, we are also making a step for gatekeepers to see their kids' interest records, cost vouchers, exceptional tests, errands, assessment plan and especially the results with the aid of the cloud administration provider's login ID and mystery key. Normally, understudies cannot share their results and class plans with their citizens, but now, with the aid of this model, gatekeepers will certainly review their children's records and needed items using our proposed model. This model also offers office to the employees in order to pass exam content, understudy cooperation and other items through the cloud. In addition, they can reinvigorate their class plans if extra classes are required. In the event that an event of any delay owing to Web or cloud detachment should occur, customers may simply get to the last revived information. They have to keep it together for alteration of the system for extra updating. While this model is being realized, Web fortification network must be accessible to maintain a key good way from any deferrals. Our conceptual model gives all of the accomplices a real situation.

The model proposed has different phases, as shown in Fig. 3. Model creation depicts that the basic stage is certification / records and work environment evaluation where all understudy knowledge is available. This stage shows the option of undergraduate studies in different projects, the concept of undergraduate studies in each program, cost status and understudy test plans. Two mists unambiguously understudy record cloud and EDU cloud which is education cloud are included in the proposed model. Understudy data cloud holds profile of understudies, their classes, records

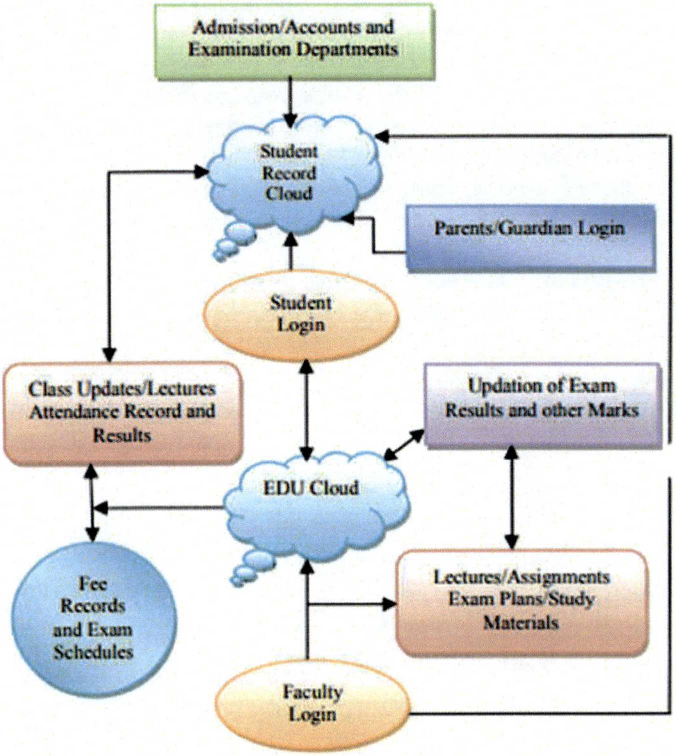

Fig. 3 Education system model using cloud computing

of cooperation and plans for study. Understudies can login via the cloud specialist co-ops using login ID and riddle word techniques, so they can communicate with the education cloud to get their activities, exams, outcomes and backup records as well as getting their conversations from the EDU cloud. On the other hand, we also uphold the job power so that they can login with their given user ID and secret key in the same way to animate their course profile, transfer addresses, study schedules, front line test schedule, assignments, marks and a buzz of additional classes if simple. Educators should also keep records of undergraduate studies so that if any undergraduate studies do not satisfy the minimal participation requirements set by the institution's assessment systems, they may be contacted.

7 Results and Discussion

The proposed policy has been implemented in the Cloud Analyst simulation tool, which is based on the CloudSim (7) extensible simulation toolkit. Simulations run using the same configuration for each of the two existing and the proposed broker policy reveal an improvement in the results. The results include an overall response time summary, response time by region and data center request servicing time. The throttled load balancing policy is being used to manage the load across the VM in the data centers. The summary of the results from the three simulations with three service broker policies, closest data center, optimized response time and the proposed proportion-based algorithm, is shown in Fig. 4:

Figure 4 clearly shows that the values of overall response time and DC processing time for the proposed algorithm is far better than that for the other two existing broker policies.

Figure 5 reveals the DC processing times of all the data centers using the three policies. If we consider the closest data center, the gap of processing time among

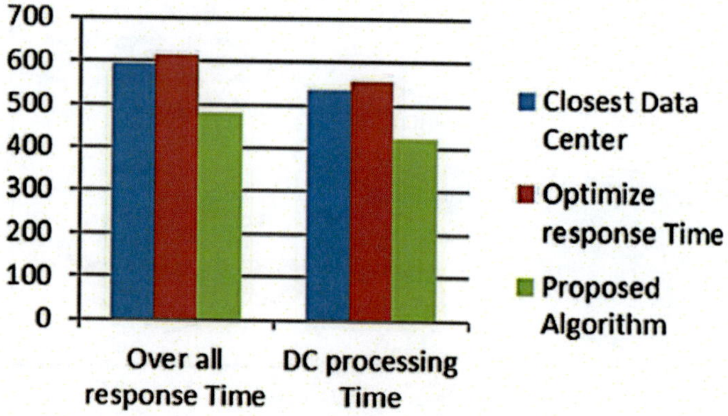

Fig. 4 Bar chart for overall response time and DC processing time for the three broker policies

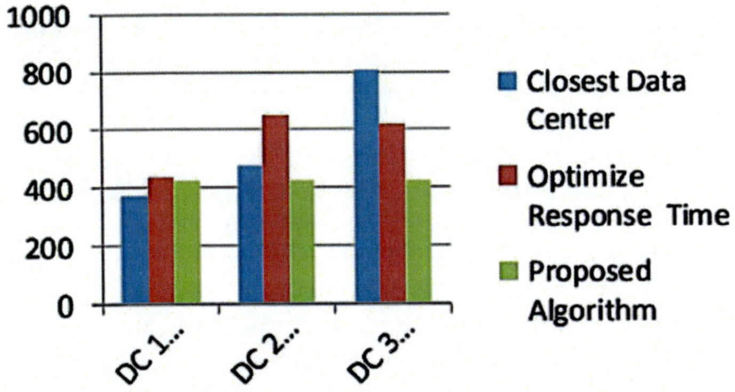

Fig. 5 Bar chart for individual DC processing time

all the data center is much more. When we consider the optimize response time as service broker, the gap of processing time is also available. The data center with high capacity (DC1) finishes their task earlier and remains ideal, and data center with low processing capacity takes more time accordingly. When the proportion-based broker policy is being considered, the values have greatly improved for DC 2 and DC 3. The workload of DC 2 and DC 3 is shifted to DC 1, whose capacity to process the request is more than the others. The DC processing time for DC 1, DC 2 and DC 3 are quite comparable when using the proposed policy, which shows a better utilization of all the data centers. Graph for the same is shown in Fig. 5 below, which is showing the comparable processing time of proposed policies.

8 Contribution of the Present Research

The present research work guaranteed the loss in total cost. It also ensured whenever there is a need, it will be sharing the information. To achieve the work and endure the function the resources should be used effectively. The better response time and the request for data center will enhance the overall service quality.

9 Conclusion

It can be accomplished from potential entertainment outcomes that require Round-Robin confirmation of server farms in intermediate strategy administration works fairly with regard to asset utilization. In addition, our proposed test works successfully to deal with the server farm time and the client's reaction time. The main benefit of cloud computing is the availability, on a pay-per-use basis, of processing and storage capacity in the form of different services. The consumers are now free from the concern of learning about the fundamental hardware that meets their demands. To sustain this function, the available resources must be effectively used. When viewed from the user's point of view, reaction time is the key governing element. A better response time and request for data center duty time accounts for service quality, and relies directly on the option of a suitable data center. Through enforcing the proposed selection policy for data centers, data center resource optimization could be feasible effectively completed. We have got a major increase in the total response time, the service period for geographic requests and the transfer time of the data center under the proposed policy for business brokers.

Future work can be guaranteed in the midst of a decline in the loss in total cost. As we seen that by and by consistently cloud computing offers quicker kinds of help by their foundation. The administration of cloud computing is so relevant in different districts. In addition, it awards us to get to our basic information whenever there is a need, equips to share the specific details with someone else. It also asks us to establish an obligatory condition where you can reliably use the focal points open on cloud computing system. In this assessment paper, we present a staff, understudies and guard structure model. The proposed model gives the associates an easy-to-use condition.

10 Future Scope

We study more properties and attributes for the complex structure of this proposed model for future work.

References

1. Wickremasinghe B, Buyya R (2019) Cloudanalyst: a cloudsim-based tool for modelling and analysis of large-scale cloud computing environments. MEDC Project Report
2. Cloud-analyst can be downloaded from here http://www.cloudbus.org/cloudsim
3. Limbani D, Oza B (2016) A proposed service broker strategy in cloudanalyst for cost-effective data center selection. Int J Eng Res Appl India 2(1):793–797
4. Sarfaraz Ahmed A (2016) Enhanced proximity-based routing policy for service brokering in cloud computing. Int J Eng Res Appl India 2(2):1453–1455
5. Dash M, Mahapatra A, Chakraborty NR (2016) Cost effective selection of data center in cloud environment. Int J Adv Comput Theory Eng (IJACTE) 2319–2526
6. Rawat PS, Saroha GP, Barthwal V (2015) Performance evaluation of social networking application with different load balancing policy across virtual machine in a single data center using cloudanalyst. In: Parallel distributed and grid computing (PDGC), 2012 2nd IEEE international conference on. IEEE, pp 469–473
7. Xu M, Tian W, Buyya R (2017) A survey on load balancing algorithms for virtual machines placement in cloud computing. In: Concurrency computation practice and experience, Wiley Online Library, vol 29(2), March 2017
8. Ghomia EJ, Rahmania AM, Qaderb NN (2017) Load-balancing algorithms in cloud computing: a survey. J Netw Comput Appl 88:50–71

9. Dashti SE, Rahmani AM (2016) Dynamic VMs placement for energy efficiency by PSO in cloud computing. J Exper Theoretical Artif Intell: Adv Appl Swarm Intell 28(1–2)
10. Ghahramani MH, Zhou MC, Hon CT (2019) Toward cloud computing QoS architecture: analysis of cloud systems and cloud services. EEE/CAA J Automatica Sinica 4(1):6–18
11. Ren X (2019) A load balancing method for massive data processing under cloud computing environment. Intell Autom Soft Comput 4:547–553
12. Peng LY (2019) Load balancing strategy of cloud computing based on artificial bee algorithm in computing technology and information management (ICCM), IEEE, Seoul, pp 185–189
13. Leukel J, Kirn S (2018) MQRC: QoS aware multimedia data replication in cloud. Int J Biomed Eng Technol 25(2/3/4):250–266
14. Squicciarini S, Lin (2018) BMAQR: balanced multi attribute QoS aware replication in HDFS. Int J Internet Technol Secured Trans 8(2):195–208
15. Schramm T (2018) An optimized replica allocation algorithm amidst of selfish nodes in MANET. Wireless Personal Commun 94(4):2719–2738
16. Jain R, Sharma T, Sharma N (2017) A review on service broker algorithm in cloud computing. Int J Comput Appl 159(3):19–23

Effective View of Swimming Pool Using Autodesk 3ds Max: 3D Modelling and Rendering

Ganesh Kumar and Debabrata Samanta

Abstract As well as setting up the sources, working with editable poly, information in the interior of the swimming pool, using turbo-smooth and symmetry modifier, this procedure of making a 3D swimming pool model is clarified. The lighting the scene and setting up the rendering, the method in which substances are added to the replica is defined. The methods and techniques of rendering are defined, too. The final rendering is the result of multiple images being drawn. The aim of our research work is to create a swimming pool design with enhancing models with materials affect. The shapes used for that are cylinder, sphere, box, plane and splines. The modifiers are editable poly, editable spline and UVW map. Finally, we used a material editor and target lights for enhancing the model.

Keywords Autodesk · Rendering · 3D modelling · UVW map

1 Introduction

When visualization and visual effects are commonly used, we exist in time. There is no film that does not use extraordinary illustration consequences and no TV that does not show advertising with exceptional illustration consequences. 3ds Max launched the 3D computer graphics upheaval and, relative to the other 3D graphics software packages, is the longest on the market [1, 2]. This long life makes 3ds Max the most popular 3D graphics program still at the top. For creating 3D animations, models, games and pictures, 3ds Max is professional 3D computer graphics software. Autodesk Media and Entertainment is designing and manufacturing it [3–5]. It consists of several features expressly developed to assist in the execution of their designs by artists, architects, engineers and designers in different disciplines [6, 7]. This paper mainly focuses on steps and problems one might face while modelling a swimming pool especially while creating the water texture [8, 9]. We will also have a little bit of glance at the animation part. When we talk about animation the flow of water, waves generated by objects in water [10, 11], its viscosity and tension can be

G. Kumar · D. Samanta (✉)
Department of Computer Science, CHRIST (Deemed to be University), Bangalore, India

K. Dahal et al. (eds.), *Internet of Things and Its Applications*, Lecture Notes in Electrical Engineering 825, https://doi.org/10.1007/978-981-16-7637-6_5

discussed [12–14]. We have also made sure that we use the least amount of geometry and polygons while modelling so that it is easy to render our model finally.

2 Objects Created for Swimming Pool Design

i. **Pool tank:**

Created using 2D circle, modified by editable spline and extruded by converting it into editable poly.

ii. **Pool water:**

Created using a plane and water texture given using a material editor. In the material editor noise is chosen as bump map and realistic water effect has been given by adjusting the reflection and transparency [14].

iii. **Floor:**

Created using box and modified by Pro Boolean modifier and later floor texture was given to it [15].

iv. **Balls and tube:**

Created using sphere and torus and texture have been given using material editor.

v. **Ladders:**

Created using cylinder and extrude by converting it into editable poly and bent using bend modifier.

vi. **Swimming pool chair:**

Created using a box and extruded into the chair by converting it into editable poly.

3 Configuring the Reference Picture

In order to construct a complex 3D model, such as a swimming pool, a technical drawing of at least two views is needed [16–18]. It is also very important to draw these sketches correctly and properly and to be of the same size from all points of view. They are the basis for the development of a precise 3D model. The first step in the creation of a swimming pool car model is to build a basic material with a reference picture on a diffuse diagram. On a plane, this substance is supported and shifted to arrange in a line with the outlook [19–21]. This approach for all views should be replicated. In addition, it is appropriate to build a cube with measurements equivalent to the width of the swimming pool and to increase or decrease all planes with technical drawings to match the size of the cube. This approach helps the swimming pool to be on the actual size [22, 23]

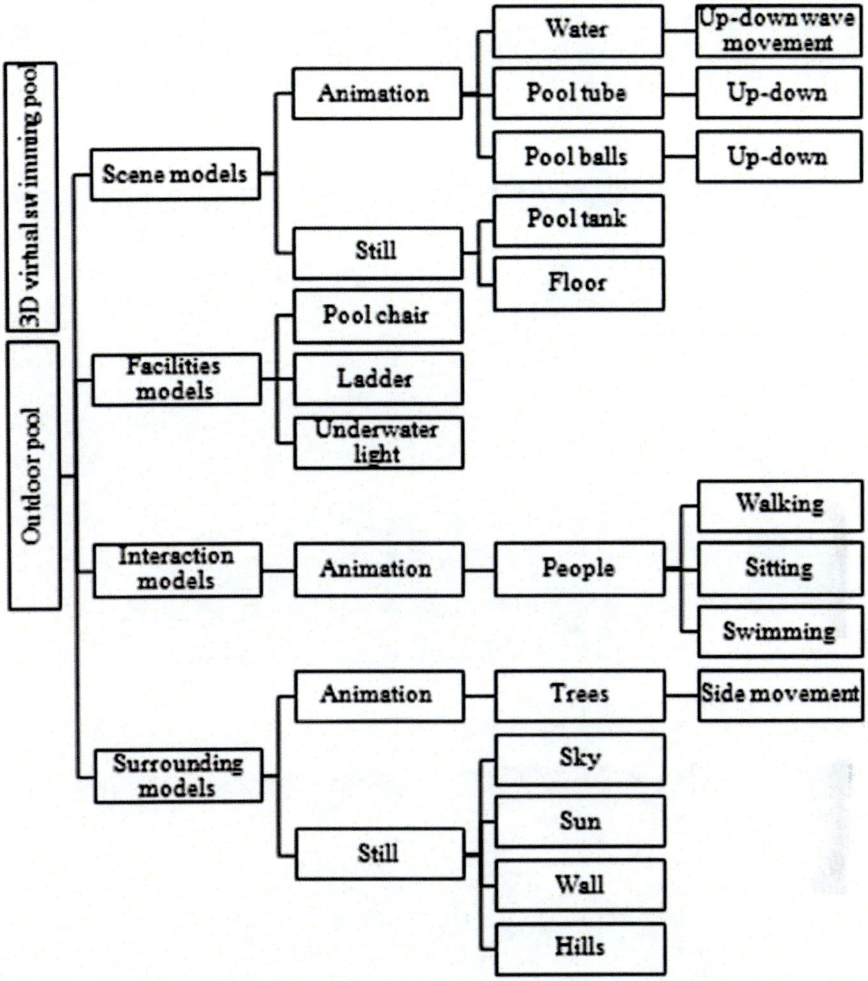

Fig. 1 Virtual swimming pool structure

4 Flow of 3D Model

The final model express in Fig. 1.

5 Rendering Images

The swimming pool interior consists of several items and features modelled as disconnect components [24, 25]. There are several buttons on them with distinct details and indications and they should all be modelled [26–28]. This stage of modelling highlights the specifics of the interior of the swimming pool. Many of the objects are modelled using the same methods as objects in the interior of the swimming pool. The different render images are given Figs. 2, 3 and 4.

Fig. 2 Side view with rendering image

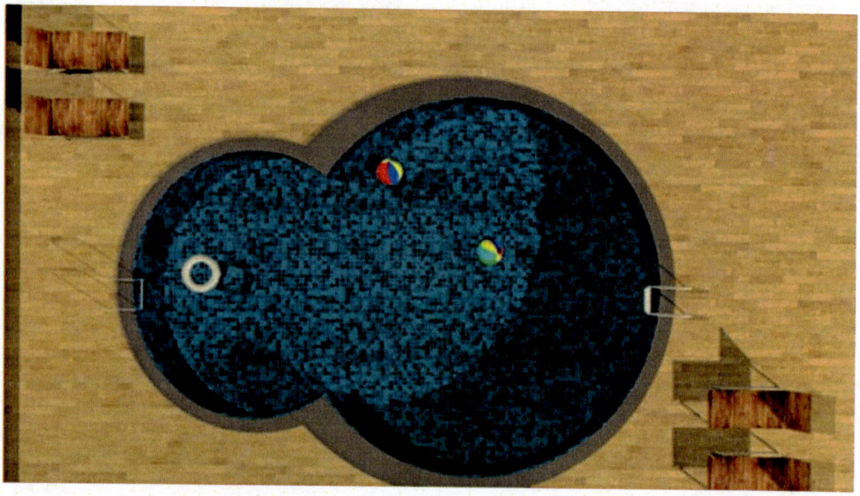

Fig. 3 Top view with rendering image

Fig. 4 Back view with rendering image

6 Conclusion

3ds Max is authoritative computer software that is especially blueprinted to assist designers, architects, engineers and 3D artists in the execution of their projects in different disciplines. It is a complex task to model a swimming pool with 3ds Max, and for this purpose, the mainly significant step is to set up orientation figures from at least two observations. This paper focuses primarily on steps and issues that one could face when modelling a swimming pool, especially when creating the texture of the water. We are also going to take a little bit of a look at the animation portion. The movement of water, waves produced by objects in water, its viscosity and tension can be discussed when we speak about animation.

References

1. Lin TH, Lan CC, Wang CH, Chen CH (2014) Study on realistic texture mapping for 3D models. In: International conference on information science, electronics and electrical engineering (ISEEE) vol 3. pp 1567–1571
2. Ru-chuan W, Deng-yin Z, Chen-yun X (2000) The researche of 3D graphic building technology. Comput Aided Eng Papers 12(4):25–30
3. Yong W (2011) Design and implement of an interactive virtual campus roaming system base on VRML. Jilin University (May 2011)
4. Guha A, Samanta D (2020) Real-time application of document classification based on machine learning. In: Jain L, Peng SL, Alhadidi B, Pal S (eds) Intelligent computing paradigm and cutting-edge technologies. ICICCT 2019. Learning and analytics in intelligent systems, vol 9. Springer, Cham

5. Prieto PA, Wright DK, Qin SF (2003) A novel method for early formal developments using computer aided design and rapid prototyping technology. Proc Instit Mech Eng Part B: J Eng Manuf 217(5):695–769
6. Debevec P, Taylor C, Malik J (1996) Modeling and rendering architecture from photographs: a hybrid geometry-and image-based approach. In: Proceedings of the ACM SIGGRAPH conference on computer graphics, 11–12
7. Ghosh G, Samanta D, Paul M, Kumar Janghel N (2017) Hiding based message communication techniques depends on divide and conquer approach. In: Proceedings of IEEE international conference on computing methodologies and communication, Erode, 18–19 July 2017
8. Wu L, Ru-wei L, Xiao-wei C (2007) Research on realization of dynamic interaction by VRML and 3DS MAX.ComputSimul Papers 24(1):213-216
9. Chakrabarti S, Samanta D (2016) Image steganography using priority-based neural network and pyramid. In: Emerging research in computing, information, communication and applications, pp 163–172
10. Praveen B, Umarani N, Anand T, Samanta D (2019) Cardinal digital image data fortification expending steganography. Int J Recent Technol Eng 7(5S2). ISSN: 2277–3878
11. Althar RR, Samanta D (2021) Building intelligent integrated development environment for IoT in the context of statistical modeling for software source code. In: Kumar R, Sharma R, Pattnaik PK (eds) Multimedia Technologies in the internet of things environment. Studies in Big Data, vol 79. Springer, Singapore
12. Lian Q, Li D-C, Tang Y-P, Zhang Y-R (2006) Computer modeling approach for a novel internal architecture of artificial bone. CAD Comput Aided Design 38(5):507–514
13. Hossain MA, Samanta D, Sanyal G (2012) Statistical approach for extraction of panic expression. In: 2012 fourth international conference on computational intelligence and communication networks
14. Levoy M, Pulli K, Curless B, Rusinkiewicz S, Koller D, Pereira L, Ginzton M, Anderson S, Davis J, Ginsberg J, Shade J, Fulk D (2000) The digital Michelangelo project: 3D scanning of large statues. In Siggraph 2000:131–144
15. Montenegro AA, Carvalho PCP, Velho L, Gattass M (2004) Space carving with a hand-held camera. In: Proceedings of the XVII Brazilian symposium on computer graphics and image processing (SIBGRAPI'04), 396~403
16. Asai T, Kanbara M, Yokoya N (2008) 3D modeling of outdoor environments by integrating omnidirectional range and color images. In: Proceedings of the fifth international conference on 3-D digital imaging and modeling (3DIM'05)
17. Arayici Y, Hamilton A (2005) Modeling 3D scanned data to visualize the built environment. In: Proceedings of the ninth international conference on information visualisation, 509~514
18. Setan H, Ibrahim MS (2004) Close range measurement and 3D modeling. In: Presented at the 1st international symposium on engineering surveys for construction works and structural engineering
19. Kriete A, Breithecker A, Rau W (2001) 3D imaging of lung tissue by confocal microscopy and micro-CT. Proc SPIE—The Int Soc Opt Eng 469~476
20. Tseng Y, Chung S (2014) Profile conversion of a picture into a 3D model reminiscent of low relief for 3D-printing. In: Proceedings of the 33rd Chinese control conference, Nanjing, pp 2953–2957
21. Jones R, Haufe P, Sells E, Iravani P, Olliver V, Palmer C, Bowyer A (2011) Reprap-The replicating rapid prototype. Robotica 177–191
22. Guha A, Samanta D (2020) Hybrid approach to document anomaly detection: an application to facilitate RPA in title insurance. Int J Autom Comput
23. Biswas J, Kureethara JV, Samanta D, Sandhya M (2020) Efficient algorithm for people management in an elevator. TEST Eng Manage 83. Publication issue: March–April 2. ISSN: 0193–4120
24. Samanta D, Galety MG, Shivamurthaiah M, Kariyappala S (2020) A hybridization approach based semantic approach to the software engineering. TEST Eng Manage 83. Publication issue: March–April 2020, ISSN: 0193–4120

25. Gurunath R, Samanta D (2020) Studies on encrypted secret data storage techniques analogous to steganography. Int J Adv Sci Technol 29(2):3705–3711
26. Roy S, Kanti MM, Samanta D, Venkatanagaraju (2020) Awareness with informatics on hypertension and effects on hemoglobin. Int J Adv Sci Technol 29(4):423–433
27. Thomas B, Shwetha P, Dey P, Biswas J, Samanta D (2020) An efficient and holistic approach to reduce output and dependent parameters for multi-output learning. Int J Adv Sci Technol 29(4):25–33
28. Kureethara V, Biswas J, Samanta D, Eapen NG (2019) Balanced constrained partitioning of distinct objects. Int J Innov Technol Explor Eng. ISSN: 2278–3075 (Online)

Applications and Challenges in Internet of Vehicles: A Survey

Surbhi Sharma and Baijnath Kaushik

Abstract The Internet of Vehicles (IoV) is one of the trending concepts in the automotive industry domain. It introduces the idea of smart transportation and smart cities due to the merging of Internet of things (IoT) and vehicular ad hoc networks (VANETs). IoV intends to overcome the existing flaws of VANETs to improve traffic efficiency, safety and ease the driving experience. The main objective of IoV is to strengthen the existing intelligent transportation systems (ITS) by the use of numerous intelligent technologies like cloud computing, intelligent sensing techniques, fog computing, Edge computing, etc. It has achieved a lot of market attention due to reliable Internet connectivity, smart device compatibility, smart decision making, and heterogeneous vehicular network and thus is a better alternative for existing transportation systems. In this paper, an extensive survey is conducted to focus on the comparison of VANETs and IoV networks to give the researchers a clear understanding of the difference between these two networks. Also, a sincere attempt is made to explore the diverse range of applications of IoV in detail before bringing it into actual deployment. In the last, open problems of IoV are also outlined that need to be resolved while deploying the IoV network.

Keywords Applications · Efficiency · Hazards · Intelligence · IoV · Smart cities · Safety · VANETs · Vehicle

1 Introduction

IoV networks have revolutionized the transportation systems by blending the IoT with VANETs, resulting in smart transportation systems. Its main goal is to enhance traffic efficiency, road safety, and ease travellers' driving experience by offering

S. Sharma (✉) · B. Kaushik
School of Computer Science and Engineering, Shri Mata Vaishno Devi University, Katra, India

B. Kaushik
e-mail: baijnath.kaushik@smvdu.ac.in

K. Dahal et al. (eds.), *Internet of Things and Its Applications*, Lecture Notes in Electrical Engineering 825, https://doi.org/10.1007/978-981-16-7637-6_6

infotainment-based features [1]. Each vehicle is equipped with several sensors. On-board unit (OBU) and road-side units (RSUs) are also available for gathering information regarding the environment. It includes five types of communication: (vehicle-to-vehicle (V2V), vehicle-to-roadside unit (V2R), vehicle-to-personal devices (V2P), vehicle-to-sensors (V2S), and vehicle-to-infrastructure of cellular networks (V2I) [2]. Different wireless technologies are used by each communication type, i.e., DSRC.

(Dedicated short range communications), WiMAX, Wi-Fi, etc. A basic structure of IoV is shown in Fig. 1.

By analyzing the data collected from multiple sensors and other vehicles, IoV can make timely decisions to attain intelligent driving assistance. Due to this, traffic congestion and accidents can be prevented; thus, pollution and travel time also reduce. IoV offers numerous benefits, as mentioned below [4]:

(a) Reduces the impact of greenhouse gases to eliminate ecosystem hazards.
(b) Preliminary information regarding vacant parking areas, route optimization, etc., thus, the concept of smart cities is evolved.
(c) Efficient traffic management and toll systems will reduce the waiting time of travelers, so it is time-efficient.
(d) Useful in monitoring the health of patients in case of emergency scenarios.

Fig. 1 Basic structure of IoV network [3]

Section 2 illustrates the comparison of IoV and VANETs based on few parameters. Section 3 elaborates the vast applications supported by IoV and VANETs, i.e., safety-related applications, infotainment applications, traffic efficiency and control, applications of health care. Section 4 discusses the open problems of the IoV network that need to be considered before actual deployment. Section 5 concludes the complete review of this paper.

2 Comparison of IoV and VANETS

Although both VANETs and IoV have a common goal to prevent road mishaps and enhance the traffic efficiency, there are certain distinguishing features between both the networks that need to be clear. In this section, a comparison of IoV and VANETs is done based on numerous parameters and is discussed below:

i. *Purpose*

The primary purpose of designing the VANET network is to prevent road mishaps and enhance traffic efficiency, but it does not offer entertainment-based features while traveling [5]. On the other hand, the primary purpose of IoV is to improvise safety, traffic efficiency, and entertainment features. Entertainment-based features ease the driving experience as passengers are likely to get bored. Hence, such features rejuvenate their mood as they can watch movies, listen to songs, and download any data.

ii. *Smart device Compatibility*

Everything is now reliant on personal smart devices such as tablets, laptops, and smartphones. But smart devices cannot communicate information among other nodes/vehicles in the scenario of VANETs because of the associated divergent network architecture. VANETs, therefore, have compatibility issues with smart devices. Smart devices in IoV are network compliant and can appropriately transmit information among other communicating nodes in case of any emergency scenarios resulting in a friendly atmosphere [6].

iii. *Market-oriented*

Due to various factors such as poor Internet communication, smart devices incompatibility with smart devices and local processing, etc., the desired commercialization of VANETs has not occurred over the past few decades. These factors induced its commercial growth. Hence, its usage tends to decline. The rapid advancement of intelligent and computing technologies in IoV has achieved significant market attention along with good Internet connection and smart device compatibility [7].

iv. *Communication types*

Within VANETs, two kinds of communication are possible: vehicle-to-vehicle (V2V) and vehicle-to-infrastructure (V2I) [8]. In IoV, five kinds of communication occur vehicle-to-smart personal devices (V2P), vehicle-to-vehicle (V2V), vehicle-to-sensors (V2S), vehicle-to-road side units (V2R), and vehicle-to-infrastructure (V2I).

v. *Size of Data*

VANETs have minimal data because it performs decisions locally as well as due to a non-collaborative environment. A vast amount of critical data is generated dynamically in networks like IoV and it involves interaction of different heterogeneous networks [9].

vi. *Decision making*

In VANETs, because of computational and storage constraints, smart decisions are hampered because computations powered by big data mining are missing [10]. In IoV architectures, decisions are focused on artificial intelligence-based data mining and big data computations.

vii. *Internet facility*

Internet access is one of the basic requirements in the modern era. Expanding Internet access in VANETs is a challenge as RSUs are limited in certain places or not appropriately networked [11]. In IoV, vehicles are all time connected to reliable Internet. Faster Internet connections boost the application of the IoV system.

3 Various Applications in IoV and VANETs

Due to the unique characteristics of IoV networks, i.e., dynamic topological structures, massive network scale, non-uniform distribution of nodes, and mobile limitation, etc., such networks support numerous applications that enhance the network's overall performance. IoV applications can be divided into the below-mentioned categories: (a) safety-related applications, (b) infotainment, (c) traffic efficiency, and control, and (d) applications for health care [12].

3.1 Safety-Related Applications

The main goal of IoV and VANETs safety applications is to prevent and minimize road accidents as this entails life risk. Safety applications are further divided into numerous categories and are mentioned below [13]:

(1) *Intersection Collision Avoidance*

Its main objective is to prevent road mishaps by avoiding intersection collisions and thus improvising the road safety. This application has different categories and is described below:

(a) *Traffic signal violation warning*

This IoV application is an example of V2I communications to alert the drivers regarding traffic signals. The driver might miss the red traffic signal and continue to drive, resulting in a collision with other vehicles having a green signal. To avoid such scenarios, this application intends to place RSUs with traffic controllers to alert the drivers in advance about signals so that any signal will not be missed, and ultimately, the network will be secured [14].

(b) *Left turn assistant*

This application aims to avoid intersectional mishaps by guiding the drivers regarding turns at intersections. Sensors placed at RSUs will gather the required information regarding traffic and then forward it to the driver for safe turns, thus preventing road hazards [15].

(c) *Blind merge warning*

This application warns the drivers regarding traffic at junctions where visibility is poor. It aims to prevent crashes at junctions by giving them prior information.

(2) *Public safety-related applications*

Public safety is one of ITS's core priorities. As the delay is not acceptable in such networks, this application intends to ensure drivers' safety by providing emergency vehicles in case of any hazards. Different applications belong to this category are discussed below:

(a) *Approaching emergency vehicle warning*

This application is designed by keeping in mind the concern of emergency vehicles. In this, the emergency vehicle is given as top priority, and all the vehicles on the same route clear the path for an emergency vehicle after receiving the notifications from RSUs.

(b) Emergency vehicle signal preemption

It involves V2I communications, and it aims to reduce traffic delays. Its motive is to help the emergency vehicle reach the destination as fast as possible by changing the traffic signal.

(c) *Post-crash warning*

It intends to alert the approaching traffic about the damaged vehicle, and it involves both V2V and V2I communications. Malfunctioning of vehicles might be due to mechanical breakdown or some other reason. It notifies the approaching traffic regarding the damaged vehicle's location and status to prevent the secondary crash.

(3) Sign extension

It is designed to keep in mind the importance of signboards while traveling, and it should not be ignored in any scenario [16]. Numerous applications belong to this category and are mentioned below:

(a) In-vehicle signage

It is used to warn the approaching vehicles regarding RSUs located along the roadside to alert them about the restricted zones. Few critical signage warnings include hospital zone, school zone, etc., indicating slow driving or no honking. So, such signs should be carefully noticed by drivers.

(b) Low parking structure warning

This application's goal is to assist the drivers regarding parking areas. It relies on RSUs near parking lots to sense the information regarding vacant parking space or timings, etc. Then this gathered information will be forwarded to the driver who is willing to park his vehicle.

(c) Low bridge warning

It is intended to warn drivers on the way regarding bridge with a low height. Then RSUs near the bridge will gather the required information and forward it to the approaching vehicle, then the vehicle will decide whether to go through the bridge.

(4) Information from other vehicles

It relies on V2V and V2I communications. Different applications which belong to this category are discussed below:

(a) Cooperative forward-collision warning

Its goal is to prevent accidents due to collisions at the rear end. V2V communications help to determine the vehicle's velocity, location, etc., to guess the chances of rear-end vehicle collisions and to avoid it.

(b) Vehicle-based road condition warning

It aims to notify the vehicles regarding road status, i.e., the road's condition as road plays a significant role in vehicles' smooth driving. If the road is not good, the driver may choose some alternate route because the damaged road often impacts the vehicle.

(c) Emergency electronic brake lights

It aims to warn the vehicles that preceding vehicles are performing sudden braking. It is quite beneficial in scenarios where visibility is poor due to bad weather conditions. It notifies the drivers in advance about hard braking to avoid the chances of collisions.

3.2 Infotainment Applications

Infotainment applications are intended to ease the driving experience of passengers as it will rejuvenate their moods. For such applications, the fundamental mechanism used for communication is unicast routing. Gaming, file uploading, watching movies, intelligent parking systems, online file sharing, and Internet access are a few examples of such applications [17].

(a) *Peer-to-peer file sharing applications(P2P)*

Such applications of file sharing increase passenger convenience because they can share data with others. It felicitates the sharing of large size media files from one device to another using an Internet connection. In this application, each connecting device is known as a peer, and BitTorrent is one example of the P2P file sharing application [18].

(b) *Intelligent parking navigation system*

Searching for the nearest vacant parking lot during peak hours is a time-consuming and tedious task. An intelligent parking system is designed by authors in [19] to save users' time by suggesting vacant parking areas as fast as possible. Also, the proposed smart system can interact with other vehicles; thus, the system is quite interactive as well.

(c) *Internet service provisioning*

The Internet is the basic necessity in the modern era, and it seems to be impossible to imagine life without the Internet. Thus, the Internet is one of the key requirements while traveling. In such applications, RSUs are always connected to OBUs through the Internet, and due to this, users can access the Internet while traveling. Online video streaming, gaming, watching online movies are a few examples of such applications.

3.3 Traffic Efficiency and Control

These applications tend to regulate traffic flow and to prevent traffic congestion. OBUs and RSUs are connected and disseminates the information regarding any congestion in the network, thus it saves time as well. Few applications which fall under this category are discussed below:

(a) *Intersection management*

The primary purpose of this application is to manage the crossing of vehicles at intersections efficiently. Any ignorance while crossing at intersections may lead to loss of lives. Traffic lights are also one of the current ways to manage ongoing traffic[20].

(b) *Road congestion management*

This application's main objective is to maintain smooth traffic flow and avoid road congestion. It will ultimately enhance the road capacity, as well as traffic jams will be prevented. Such applications use GPS systems to determine the optimal path and notify the other vehicles to change the route if the current route has traffic congestion [21].

(c) *Electronic toll collection*

Fig. 2 Electronic toll collection [23]

Toll payments are time-consuming operations as sometimes there is a long queue waiting near toll gates for payment. Such scenarios are unacceptable in networks like IoV. So, nowadays, the concept of electronic toll is introduced where everything is automatic, and thus a lot of time is saved. In these systems. Figure 2 shows an example of electronic toll collection.

3.4 Applications for Health Care

VANETs and IoV applications are extended to the medical field also to cure the patients. In case of emergency scenarios where a patient cannot visit the hospital, such applications play a significant role. Numerous wireless body sensors are deployed on the human body to detect the patient's health conditions like BP, pulse rate, body temperature, etc. These readings will then be forwarded to medical professionals using different wireless technologies like Wi-Fi, WiMAX, etc. Thus, it is quite useful in curing health issues [22].

4 Open Problems of IoV

Although IoV has numerous applications and has a broader scope, its implementation is still considered to be a critical issue. Few research challenges are mentioned below:

(a) *Network model and service model of human–vehicle*

Convenient human–vehicle network model in IoV is considered an open problem, and it needs to resolve soon. It takes into consideration the optimization of resources, robustness, and stability of the network. The development of a cognitive learning model and the survey of service attributes during the coordination process are vital issues that aim to improve the ability to cope with the complex space–time change in IoV service requirements, but there is no research on this issue [7].

(b) *Location privacy*

Vehicle location privacy is of the highest concern in-vehicle environments as an intruder or hacker may manipulate passenger or driver personal information such as most passengers visited places. Location privacy is preserved by an approach known as Mix Zone and is successful for multilane highways but is not appropriate for one-way roads [24].

(c) *Enhancing communication ability*

Each layer's communication potential in IoV needs improvement due to effective transmission medium, network dimension, and inconsistent user, service, and network optimization objectives. Preservation of bandwidth required for IoV traffic and network traffic minimization are assumed as open problems in IoV [7].

(d) *Location verification*

In ITS applications, location verification systems are quite useful to prevent road mishaps. Because of the lack of trusted authority, location verification is assumed to be a critical challenge in vehicular communication. User privacy, system trust in user correctness, and computational security costs are the three major security positioning problems.

(e) *Lack of standards to enable robust V2V communication*

Open standards are required to focus on improving user experience and services in the IoV ecosystem in an attempt to accomplish efficient communication and information dissemination that assumes transparent and seamless integration with currently closed standards [25].

5 Conclusion

Internet of Vehicles is one of the emerging concepts in the transportation systems due to the incorporation of intelligent technologies. Due to this, the concept of smart cities is evolved, and life-threatening risks will be minimized. IoV assists the ITS in accomplishing its goals, i.e., enhancing traffic efficiency and road safety, parking space and management, ease the driving experience of passengers, less fuel wastage, automated warning systems, etc. It has gained a lot of market attention due to big data technologies like Edge computing, cloud computing, etc. The presence of heterogeneous environments, smart device compatibility, and reliable Internet connectivity, etc.

In this paper, an extensive survey of IoV is provided in which a comparison of IoV and VANETs is done based on a few parameters, although they both the same objective. After this, a plethora of applications of VANETs and IoV are studied in detail and are classified into four major categories—safety-related applications, infotainment applications, traffic efficiency and control, applications of health care. These categories are further divided into numerous applications. In the end, open problems of IoV are outlined, which need to be addressed while deploying the IoV.

References

1. Sharma S, Kaushik B (2020) A comprehensive review of nature-inspired algorithms for internet of vehicles. In: 2020 international conference on emerging smart computing and informatics (ESCI). IEEE
2. Sharma S, Kaushik BJVC (2019) A survey on internet of vehicles: applications, security issues and solutions. 20:100182
3. Storck CR, Duarte-Figueiredo FJS (2019) A 5G V2X ecosystem providing Internet of vehicles. 19(3):550
4. Bagga P et al (2020) Authentication protocols in internet of vehicles: taxonomy. Anal Challenges 8:54314–54344
5. Barba CT et al. (2012) Smart city for VANETs using warning messages, traffic statistics and intelligent traffic lights. in 2012 IEEE intelligent vehicles symposium. IEEE
6. Spelta C et al. (2010) Smartphone-based vehicle-to-driver/environment interaction system for motorcycles. 2(2):39–42
7. Yang F et al. (2014) An overview of internet of vehicles. 11(10):1–15
8. Saini M, Alelaiwi A, Saddik AEJACS (2015) How close are we to realizing a pragmatic VANET solution? a meta-survey. 48(2):1–40
9. Xu W et al. (2017) Internet of vehicles in big data era. 5(1):19–35
10. Bitam S, Mellouk A, Zeadally SJIWC (2015) VANET-cloud: a generic cloud computing model for vehicular Ad Hoc networks. 22(1):96–102
11. Aslam B et al (2013) Extension of internet access to VANET via satellite receive–only terminals. 14(3):172–190
12. Hua LC et al (2017) Social networking-based cooperation mechanisms in vehicular ad-hoc network—a survey. 10:57–73
13. DoT UJDH (2008) Administration, NHTS, others, nd identify intelligent vehicle safety applications enabled by DSRC. 809:859
14. Florin R, Olariu SJVC (2015) A survey of vehicular communications for traffic signal optimization. 2(2):70–79
15. Rabieh K et al (2015) Cross-layer scheme for detecting large-scale colluding Sybil attack in VANETs. In: 2015 IEEE international conference on communications (ICC). IEEE
16. Batish S et al (2015) A comprehensive review on recent issues and applications in VANETs. 2(6):508–512
17. Toor Y et al (2008) Vehicle ad hoc networks: applications and related technical issues. 10(3): 74–88
18. Nandan A et al (2005) Co-operative downloading in vehicular ad-hoc wireless networks. In: Second annual conference on wireless on-demand network systems and services. IEEE
19. Lu R et al (2009) SPARK: a new VANET-based smart parking scheme for large parking lots. In IEEE INFOCOM 2009. IEEE
20. Lu R et al. (2009) SPARK: a new VANET-based smart parking scheme for large parking lots. In: IEEE INFOCOM 2009. IEEE
21. Cunha F et al (2014) Data communication in VANETs: a survey, challenges and applications
22. Vassilaras S, Yovanof GSJWpc (2010) Wireless innovations as enablers for complex and dynamic artificial systems. 53(3):365–393

23. Kumar V et al (2013) Applications of VANETs: Present and Future 5(01):12
24. Ying B, Makrakis D, Mouftah HTJICL (2013) Dynamic mix-zone for location privacy in vehicular networks. 17(8):1524–1527
25. Contreras-Castillo J, Zeadally S, Guerrero-Ibañez JAJIiotJ (2017) Internet of vehicles: architecture, protocols, and security5(5):3701–3709

Industrial IoT: Development of Smart Cooler for Solder Paste Storage and Management

G. Kannan, K. Indra Gandhi, S. Ganesh, S. Priyanka, and A. Anusuya

Abstract Nowadays technology has led to many comforts that would enhance higher effectiveness and increased productivity in short time. One of those technologies is the "industrial Internet of things." The latest improved trend of industrial IoT is industry 4.0 which aims in achieving increased production with less human effort in a short period of time. This paper presents about the system which is mainly designed for the workers of production industry to reduce their work load in storing and maintaining the solder paste that is used in manufacturing processes. The workers in the surface-mount technology (SMT) line of any production factory make use of the solder paste in their production process. To reduce their difficulty in maintaining the solder paste at ambient temperature (25 degree Celsius) before getting it used, a system was designed such that the workers can remotely control the cooler box according to their usage whenever needed using Web page. As a third-eye, the temperature of the cooler box is automatically updated to the end user and thus confirming the working state of the cooler box. In this paper, ThingSpeak cloud platform and HTTP server are used for data visualization.

Keywords Industry 4.0 · Solder paste · Remote sensing and automation · ThingSpeak · HTTP server

1 Introduction

Nowadays industry 4.0 rises over its hand in all the industrial domains. Any production industry makes use of technologies developed across the world. One of the vast developed technologies is the sensing and automation systems. They have many benefits which can replace many efforts of the workers. Any industry will consist of

G. Kannan (✉) · K. Indra Gandhi · S. Ganesh · S. Priyanka · A. Anusuya
Department of Electronics and Communication Engineering, B.S. Abdur Rahman Crescent Institute of Science and Technology, Chennai, India
e-mail: kannan@crescent.education

K. Indra Gandhi
e-mail: indra@crescent.education

K. Dahal et al. (eds.), *Internet of Things and Its Applications*, Lecture Notes in Electrical Engineering 825, https://doi.org/10.1007/978-981-16-7637-6_7

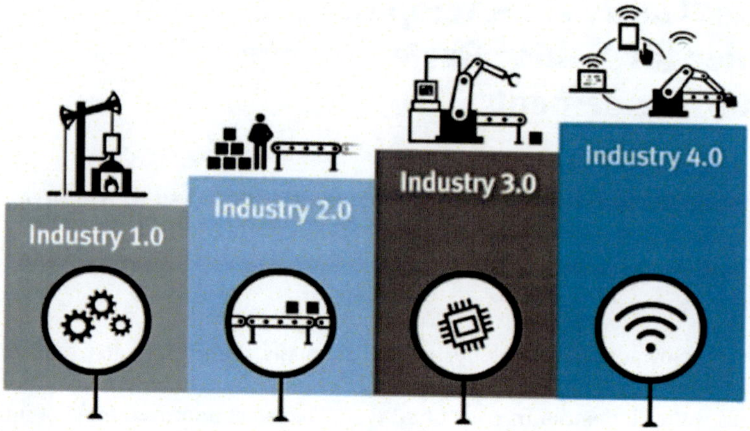

Fig. 1 Industrial developments, [1]

divided workspaces. All the workspaces will be assisted with some latest technologies for sure. For example, "the vision picking system" in warehouses, "safety monitoring systems" in hazardous areas. The various fields in technology provide various usages. The proposed system is also a certain kind of automation and sensing technology. Technologies of various fields have been interconnected in order to provide an expected service to the targeted workspace.

Figure 1 depicts about the various industrial developments throughout a vast period of time. The key tool of production in each developmental stage has been shown. In production industry, while accomplishing SMT work, you can utilize flimsy wire yet in many cases even that is not adequate, you have to utilize solder paste as shown in Fig. 2.

Solder paste comes in tubs of 1/2–1 lb. The solder paste has consistency of smooth nutty spread and is made of wad of patch suspended in motion. As the solder paste is warmed in a broiler, the patch softens and the transition consume with extreme heat leaving a strong weld joint. This issue with solder paste is that the motion can vanish off, leaving the solder paste "old" and "dry." It will not screen-print too—you will experience issues with spans and getting spotless stores. Solder paste ought to be kept cold, however, not freezing. The proposed system in this paper deals with overcoming this issue with less efforts.

2 Related Works

Minkyu Shin et al. have explained about the visualization of the floor plan with the use of different modules. Here, the temperature changes are expressed by the colors. Also they explained how temperature changes in an indoor environment.

Fig. 2 Solder paste and its usage area, [2, 3]

Ibtihaj A. Abdulrazzak et al. have researched about the weather station that gives information on temperature and humidity in a given area and its usage. Dragan Vuksanović et al. have explained about the direction of industry 4.0 and the concepts for future development. Also the evolution of smart factory with the use of digital technology is explained as well. Saima Zafar et al. present the execution and results of an ecological observing framework which utilizes sensors for temperature and dampness of the encompassing zone. Lalbihari Barik has used many different types of sensors to sense the temperature in the environment around them. They also obtained the surrounding temperature and the soil moisture of any plants. K. M. Smruthi et al. have concentrated on structuring and creating android application for controlling independent vehicles. José Ramírez-Faz et al. put a minimal effort IoT arrangement, in light of free equipment and programming, for observing the temperature in refrigerated retail cabinets.

3 Proposed System

In order to overcome the above described difficult situation of making the solder paste available at room temperature before its usage at the SMT line, an individual cooler box which can be operated remotely is designed. The cooler box is controlled by the embedded system, which is already pre-programmed for the required specification.

The customized embedded system controls the cooler box based on the commands received from the end user through the Web page. The Web page is self-developed using the latest Web technology components. The sensed values are fed to the ThingSpeak cloud, and the graphical representation of the data is obtained. Before entering into the real-time implementation, an open-source platform called "Proteus 8 Professional" is used for simulation purposes and its results were obtained. In Proteus, the Arduino is interfaced with the DHT11 temperature sensor and the values sensed by the temperature sensor which is displayed using the virtual terminal. The readings are compared with the real-time values and then verified. Figure 3 shows the block diagram of the proposed system.

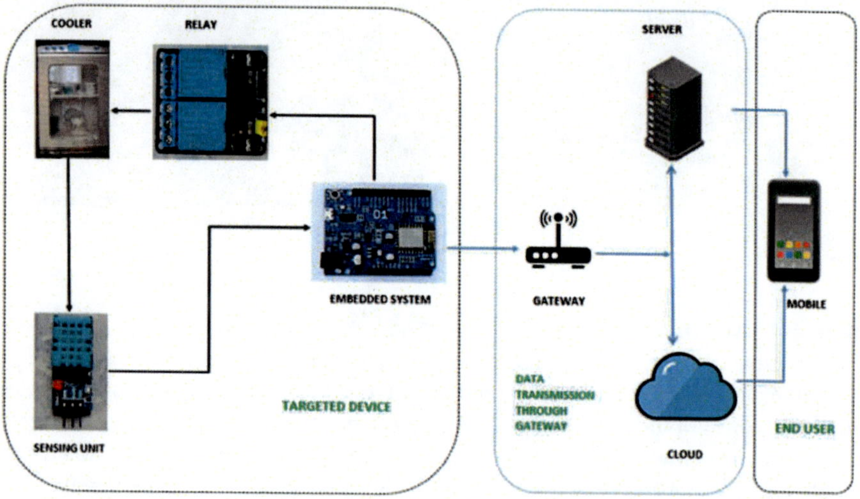

Fig. 3 Architecture of proposed system

Fig. 4 Proteus tool, [4]

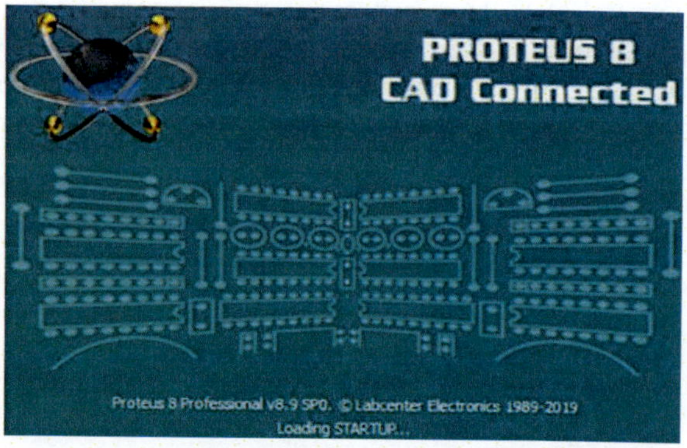

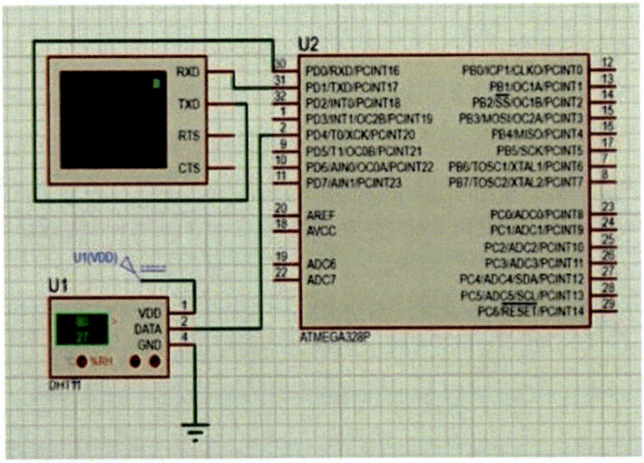

Fig. 5 Circuit built in proteus tool

4 Methodologies and Implementations

4.1 Simulation Using Proteus Tool

The Proteus Design Suite is an exclusive programming instrument suite utilized basically for electronic plan computerization. The product is utilized primarily by electronic structure architects and specialists to make schematics and electronic prints for assembling printed circuit boards. It was created in Yorkshire, England by Labcenter Electronics Limited and is accessible in English, French, Spanish, and Chinese languages. The first form of what is presently the Proteus Design Suite was called PC-B and was composed by the organization executive, John Jameson, for DOS in 1988. Schematic capture support followed in 1990, with a port to the Windows condition presently Figs. 4 and 5.

Blended mode SPICE simulation was first coordinated into Proteus in 1996 and microcontroller reproduction at that point showed up in Proteus in 1998. Shape-based auto-routing was included 2002 and 2006 saw another significant item update with 3D board visualization. All the more as of late, a devoted IDE for reproduction was included in 2011 and MCAD import/trade was remembered for 2015. Backing for a fast plan was included in 2017 (Fig. 6).

After once the virtual circuit is built, the simulation is started and the virtual monitor displays the results as below in Fig. 7. The result shows the temperature,

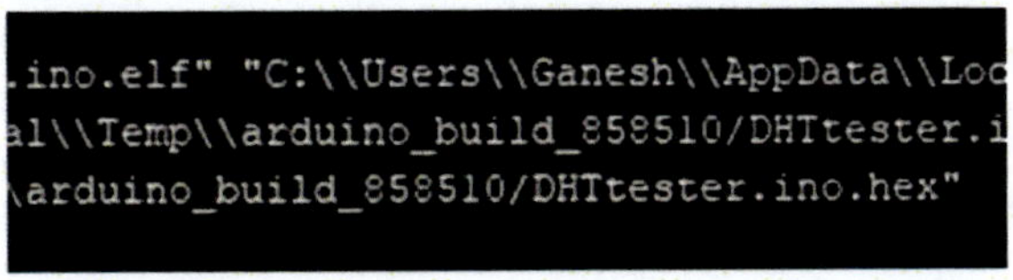

Fig. 6 Generating ".hex" file in IDE

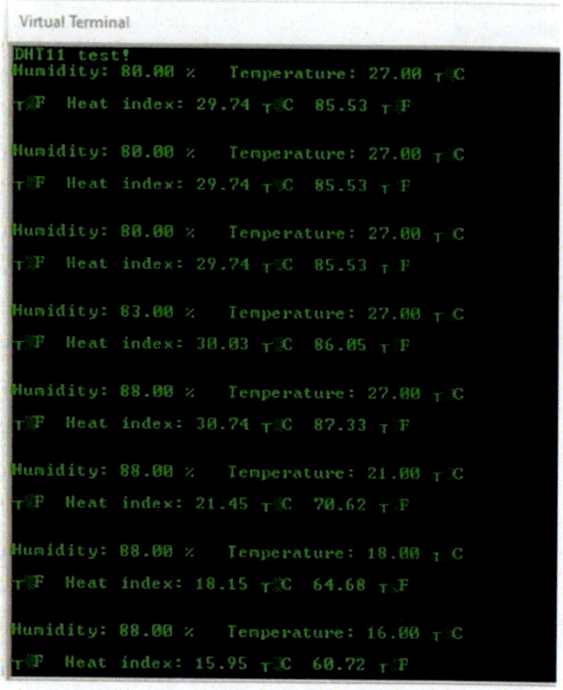

Fig. 7 Simulation output from proteus

humidity, and heat index (both in Celsius and Fahrenheit degree) which are all the properties of a particular temperature zone.

4.2 *Hardware Implementation*

The circuit that is designed holds a major scope of achieving the expected functionality in reducing the efforts of workers in targeted worker's segments. The embedded board plays a vital role as a heart of the product. The HTTP server provides a transmission medium for the data to be transmitted to and from the embedded board. The Web page provides a user-interface for the end user to provide commands as per the needs. The relay module enables the embedded system to control the electrical hardware (i.e., the electric cooler) (Fig. 8).

The work-flow is described into two parts as two different ways of communication.

- The one-way is the user provides commands to be executed through the Web page that provides an easier and interactive UI.
- The commands are transmitted to the embedded system through a transmission path provided by HTTP server.

Fig. 8 Proposed system

- The microcontroller which is embedded inside the embedded board processes the received commands and signals the relay to change its working state.
- The electric cooler which is connected to the relay module responds as per the commands received from the embedded systems.
- As another way of communication, the temperature sensor which is kept inside the electric cooler, senses the temperature inside the cooler, and updates it to the end user. This ensures the successful working of the product.
- The sensed values are fed to the ThingSpeak cloud and the graphical representation of the data is obtained.

4.3 Data Visualization in Thingspeak Cloud

ThingSpeak is an IoT cloud platform and it is open-source. It engages the creation of sensor logging applications, zone following applications, and a causal association of things with status updates. ThingSpeak was at first moved by ioBridge in 2010 as assistance on IoT applications. ThingSpeak has facilitated help from the numerical enlisting programming MATLAB from MathWorks, allowing ThingSpeak customers to examine and imagine moved data using MATLAB without requiring the securing of a MATLAB grant from MathWorks (Fig. 9).

To get into ThingSpeak, a user account is created and the channels for data visualization also created. The respective API key has been generated to write data into channels as per the terms and conditions listed by the ThingSpeak platform.

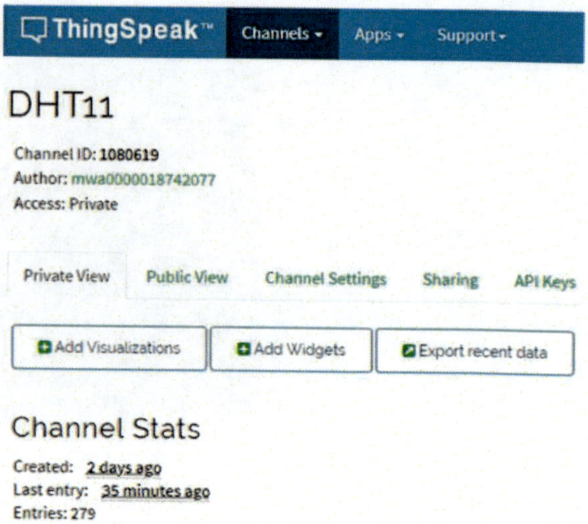

Fig. 9 Creating the channel in ThingSpeak cloud platform

5 Results and Discussion

5.1 Data Monitoring in End User Side

According to the instruction given in the embedded processor, an IP is generated and it is shown below in Fig. 10. The IP thus generated is a key to access the Web page. The Web page is hosted in HTTP server. The processing unit hosts the Web page as per the HTML tags given in its instruction. The developed Web page has been shown in Fig. 11.

As soon as the Web page is accessed, the user now has the capability of controlling the targeted device. As per the instructions given in the code, the processor processes the commands from the Web page and controls the targeted device according to it. The snap of working has been included in Figs. 12 and 13.

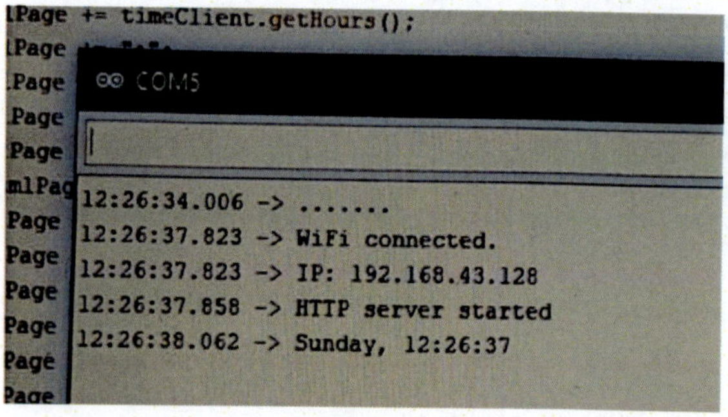

Fig. 10 Generation of IP by the processor

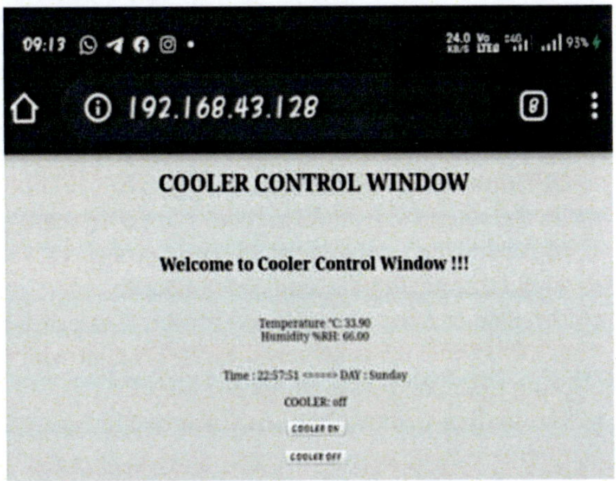

Fig. 11 Web page hosted in HTTP server

Fig. 12 Cooler-ON condition

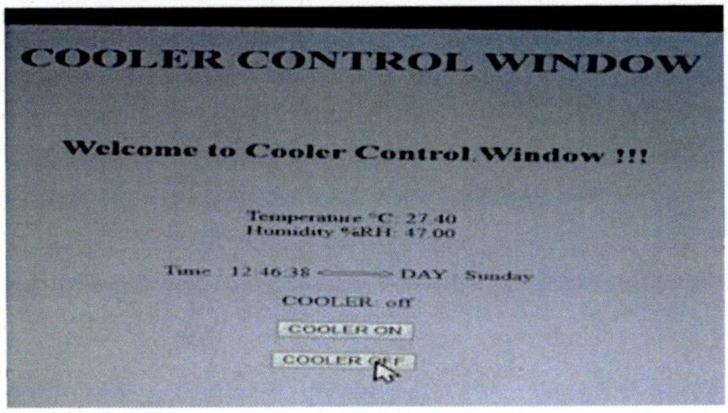

Fig. 13 Cooler-OFF condition

Data from the DHT11 temperature sensor are processed by the processor according to the instructions in the given code and it is transferred to the ThingSpeak cloud through any Internet network. The snaps of data visualization in ThingSpeak cloud platform have been attached below.

Different operations executed

- When the user click on the ON button, the respective commands in the HTML tag which are uploaded to the embedded board is executed and thus the expected ON operation is performed.
- When the user click on the OFF button, the respective commands in the HTML tag which are uploaded to the embedded board is executed and thus the expected OFF operation is performed.
- On the other hand, the term "cooler state" in the Web page is the output displaying terminal. The temperature which is sensed by sensing unit is displayed in this output terminal by the embedded board.
- The date and day are included for the user benefits to keep an eye over the time of ON/OFF. This is achieved by NTP client which is linked along with the code being executed by the embedded board.
- The snap of working has been included in Figs. 12 and 13.

5.2 *Data Visualization in ThingSpeak Cloud*

Data from the DHT11 temperature sensor are processed by the processor according to the instructions in the given code and it is transferred to the ThingSpeak cloud through any Internet network. The snaps of Data visualization in ThingSpeak cloud platform have been attached below. Figure 14 shows the temperature graph. The graph is plotted with temperature against time. At the time 22:27, there will be a decreased point in the curve. This is obtained when the cooler is in ON condition and so the temperature will be low at that time. At time 22:30, the cooler is turned OFF and so the increase in temperature is traced with a rise in curve and the following points were observed.

- In the graph, there would be an empty column as because when the sensing unit does not sense any values or the sensing unit is kept OFF.
- From 23:20 to 23:25, the curve will be reducing down as because the temperature is decreasing. This is because the cooler is kept ON.
- From 23:26 to 23:30, the curve will be increasing up as because the temperature is increasing. This is because the cooler is kept OFF.

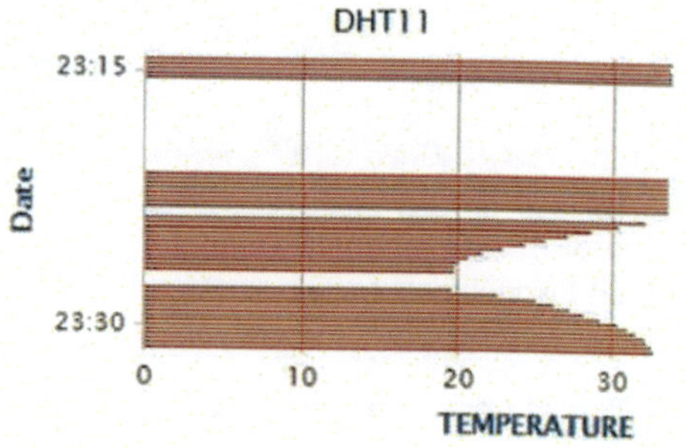

Fig. 14 Temperature values in ThingSpeak sensed by DHT11

Fig. 15 Humidity values in ThingSpeak sensed by DHT11

Figure 15 shows the graph of the humidity values traced over a certain time period. The values over different conditions have been listed below.

- From 23:25 to 23:30, the curve is rising due to increase in humidity when the temperature decreases
- From 23:20 to 23:25, the curve is varying due to variation in humidity but it is comparatively low when the temperature is high.

Figure 16 shows the combinational view of both temperature and humidity at same instances with respect to the particular time zones. From this figure, it is proving that both the temperature and humidity are inversely proportional to each other. Temperature increases with decrease in humidity and vice versa.

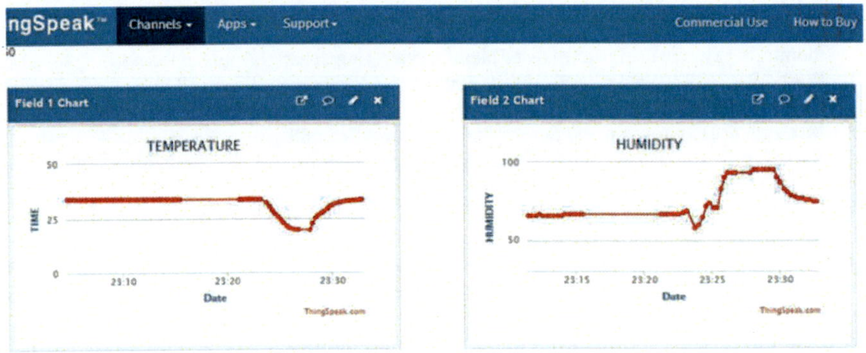

Fig. 16 Data visualization in ThingSpeak cloud

6 Conclusion

In this work, solder paste temperature is continuously monitored using IoT technology. The information obtained from the sensing unit is uploaded in the ThingSpeak cloud and retrieved with Web application for the data visualization. The integration of server and embedded systems is leading as a demanded technology and it will extend its wings toward the growth. The proposed system can be well developed throughout all the domains in the industries to achieve industry 4.0 standards.

References

1. https://images.app.goo.gl/gnVgh4HC9YWCBcmdA, Accessed on 21 June 2020
2. https://rarecomponents.com/store/image/cache/data/0-2011-1-500x500.jpg, Accessed on 23 June 2020
3. https://www.bestpcbs.com/blogimages/2018/12/12/solder-paste-printing.JPG, Accessed on 23 June 2020
4. https://www.labcenter.com/, Accessed on 23 June 2020
5. Shin M, Sang-ik L, Lee H, Lee J-K (2014) Sensing the indoor temperature data on floor plans. Springer Open J. https://doi.org/10.1186/s40327-014-0010-2
6. Abdulrazzak IA, Bierk H, Aday LA (2018) Humidity and temperature monitoring. Int J Eng Technol 7(4):5174–5177
7. Hwang JS (1991) Solder paste technology and applications. Springer Science+Business Media New York 1991
8. Vuksanović D, Ugarak J, Korčok D (2016) INDUSTRY 4.0: the future concepts and new visions of factory of the future development. In: International scientific conference on Ict and E-Business related research, SINTEZA
9. Zafar S, Miraj G, Baloch R, Murtaza D, Arshad K (2018) An IoT based real-time environmental monitoring system using arduino and cloud service. Eng Technol Appl Sci Res 8(4):3238–3242
10. Prasad RP (1997) Solder paste and its application. Chapter-9 , Surface Mount Technology© Chapman & Hall
11. McGrath (2013) Sensing and Sensor Fundamentals. Scanaill
12. Barik L (2019) IoT based temperature and humidity controlling using Arduino and Raspberry Pi. (IJACSA) Int J Adv Comput Sci Appl 10(9)
13. Smruthi KM, Yashwanth KN, Vijayalakshmi MN (2020) Intelligent autonomous vehicle control using smartphone. Springer Nature Comput Sci 1:146
14. Ramírez-Faz J, Fernández-Ahumada LM, Fernández-Ahumada E, López-Luque R (2020) Monitoring of temperature in retail refrigerated cabinets applying IoT over open-source hardware and software. Sensors 20:846.https://doi.org/10.3390/s20030846
15. Mohamed Thameez R, Kannan G (2015) Design and implementation of smart sensor interface for herbal monitoring in IoT environment. Int J Eng Res-Online 3(2):469–475

A Novel Internet of Medical Things Model to Progress COVID-19 Testing

Sakthi Jaya Sundar Rajasekar

Abstract The COVID-19 pandemic has devastated the public health infrastructure of the globe. The crucial strategy has been to carry out aggressive testing, which could be the only way to get back to normalcy. The COVID-19 testing is carried out through the Reverse Transcriptase Polymerase Chain Reaction (RT-PCR), which is considered to be the gold standard. However, these tests are sometimes known to provide inaccurate results which might be due to improper sample storage and transportation techniques. The swab samples transported in a viral transport medium need to be maintained under optimum environmental conditions. The proposed model involves tagging humidity and thermologger devices with the sample container box. This would record the real-time temperature and humidity, which would be stored in the cloud server. This would predict any breakage in the cold chain using AI-powered pattern analysis techniques. This would intimate the authorities of a possible cold chain breakage, thereby assuring the quality of the samples. This would drastically reduce the possibility of false outcomes. This would help the healthcare workers to trace, isolate and treat the right affected individuals, preventing the further spread of the disease. This model could also be used for distribution of COVID-19 vaccines whenever they are available. This could preserve the potency of the vaccines, thereby significantly reducing the wastage of vaccines.

Keywords Artificial Intelligence · COVID-19 · Environmental conditions · Internet of Medical Things (IoMT) · Reverse Transcriptase Polymerase Chain Reaction (RT-PCR) · Transportation

1 Introduction

The COVID-19 pandemic which originated in China has spread across the entire globe affecting over 5.25 crore people leaving over 12 lakh people dead [15]. India has been the worst hit country in Asia and the second-worst hit country in the world

S. J. S. Rajasekar (✉)
Melmaruvathur Adhiparasakthi Institute of Medical Sciences and Research, Melmaruvathur, Chengalpattu District, Tamil Nadu, India

K. Dahal et al. (eds.), *Internet of Things and Its Applications*, Lecture Notes in Electrical Engineering 825, https://doi.org/10.1007/978-981-16-7637-6_8

after the United States. The Health Care Workers (HCWs) form the backbone of the battle against the pandemic. Many drugs are in different phases of trials all around the globe, though none of them has yet been deemed the standard drug of choice. Research teams throughout the world have been working on a safe and effective vaccine against the dreadful virus. Some of these vaccines look promising. However, large-scale production and administration of these vaccines for the entire global population could take significant time. Hence, until then the only crucial weapon we have is to follow strict social distancing, usage of mask and gloves and ensuring highest degree of personal hygiene. Aggressive testing is the only proven way to reduce mortality rates in COVID-19 cases. Early diagnosis and immediate treatment can reduce the mortality rates to significant levels. The COVID-19 testing is carried out using Reverse Transcriptase Polymerase Chain Reaction (RT-PCR), which is considered to be the gold standard. However, these tests are sometimes known to provide inaccurate results. These may be attributed to improper sample storage and transportation techniques. The swab samples collected from the nasopharyngeal and oropharyngeal regions are placed in a viral transport medium. This ensures the viability of the collected sample. These samples need to be transported in perfect environmental conditions, i.e., temperature and humidity conditions. When the samples are exposed to conditions outside the desired ranges, it may degrade the quality of the samples collected. Testing carried out with such samples may not yield accurate results.

2 Related Works

The various uses of Internet of Things technology in the management of the COVID-19 pandemic such as the smart thermometers are discussed. The readings of the thermometers are aggregated through the Internet which helps plot a chart of the regions experiencing high temperatures during a given duration, which help in planning of COVID-19 testing and containment strategies [3]. Various smart solutions to battle the pandemic using the IoT technology ranging from intelligent contact tracing, ensuring adherence to physical distancing, development of smart medical equipment and patient monitoring systems for home quarantined patients are briefed [5]. Mitigating the impact of COVID-19 using the smart city technology like smart delivery technology enabling the contact-less delivery of food, groceries, medicines and smart healthcare technology enabling the healthcare workers to treat the infected patients without any contact, are discussed [8]. Robotic technology and IoT are used to assist COVID-19 affected patients and disabled people. The robotic systems powered with the gesture recognition systems are capable of following the instructions of the patients [11]. Wearables gather range of vital parameters such as temperature and heart rate which are used to develop an edge surveillance model. These help in efficient remote monitoring of the patient round the clock [1]. A novel stochastic model for management of COVID-19 pandemic data using secure communication systems is discussed [13]. A smart IoT model for ensuring cold chain for the transportation of medicines is discussed. Moreover, the cons of the existing cold chain model are studied, and the characteristics which need to improve are highlighted [10]. An intelligent model to detect infection clusters and alert them for mass isolation without compromising the individual's privacy is briefed. This model helps in learning pattern of infection spread and hence devise policies to curb the transmission of the dreadful disease [6]. An IoT-based system comprising an IoT module and a mobile app could effectively monitor the patient's vitals and notify them of their health status through their smartphone. Also, machine learning techniques are applied for analyzing the medical data aggregated and deliver its diagnosis [14]. A humanoid software using the IoT and artificial intelligence is developed. These humanoids undertake medical diagnosis by moving to desired destinations and screening people for COVID-19. This could also make a survey of a society [9]. A wrist wearable which records various bodily parameters and detects if the individual has been infected with the COVID-19 infection much before the symptoms occur is discussed. This model enables early identification of infected individuals, and hence, isolating and treating them at the earliest would not only lead to the protection of the individual from adverse health impacts but also helps break the COVID-19 transmission chain [2]. IoMT powered with blockchain technology to address the safety issues of the IoMT systems is presented. The challenges and future prospects of this model are briefed [4]. iFeliz, a smart model which could analyze the stress levels of individuals, is detailed. This could let the users analyze their stress levels during the given period and helps them relax during appropriate times. This would be particularly useful in times of post-COVID recovery during which the stress levels will be high [12]. The various healthcare technologies using the IoT technology are discussed, and some of the state-of-art models are reviewed. Concerns revolving around the privacy and security issues in the usage of the IoT in health care are briefed [7].

3 Materials and Methods

The paper proposes a novel Internet of Medical Things powered model for preserving the viability of the swab samples, thereby progressing the COVID-19 testing. The architecture of the proposed model has been given in Fig. 1. The proposed model involves a sensor unit which includes humidity and thermologgers attached to the sample containers. Real-time surveillance of these sensor recordings is carried out through the pattern analysis techniques augmented by artificial intelligence. Any breakage in the cold chain would be predicted at the earliest, and hence, mitigatory activities could be planned. This ensures the accuracy of the COVID-19 test results negating the effect of errors owing to improper sample storage and transportation issues. The COVID-19 swab sample containers are well packed in suitable boxes undertaking all the standard precautions. The entire package is tagged with module consisting of humidity and thermologger devices. These devices will be uniquely identified for each of the boxes. These sensors will be connected to the cloud server through the WiFi communication technologies. These are used to provide real-time surveillance. Right before the start of transportation of the samples, the healthcare

Fig. 1 Architecture of the Proposed model

workers will ensure that all the samples are within the desired environmental temperature ranges. If any abnormality is found at this stage, the healthcare workers could collect a fresh sample for the subject again. This could reduce the possibility of false results and unwanted transportation costs. During the transportation of the samples, these environmental conditions could be monitored in real time. Any breakage in the cold chain could be analyzed for this real-time surveillance. The real-time readings are aggregated and processed through pattern analysis techniques powered with artificial intelligence. For instance, if any breakage in the cold chain happens, the temperature of the sample boxes will gradually rise. This gradual rise in temperature is picked up by the AI-powered pattern analysis techniques. Immediate notifications are sent to the driver, the chief healthcare worker at the source and the destination. The driver reviews the situation, and if possible, he fixes the breakage in cold chain. In this case, normalcy would be restored which would be reflected by the real-time humidity and thermologger readings. If the breakage could not be fixed, the higher authorities are notified. They look at other options for preserving the viability of the samples. Firstly, they look to identify suitable cold chain facilities in proximity. The distance of the cold chain facility and the time required to reach it are computed. Through the AI-powered pattern analysis techniques, the remaining time for which the samples could be preserved viable would be analyzed. Depending on these, the authorities make a decision if the samples could be housed in a nearby cold chain facility. Secondly, the authorities dispatch another cold chain transporter vehicle, which would transport the samples from then. The time and distance of approach of both the vehicles will be determined along with the remaining time for the viability of the samples. Thereby, the samples could be well preserved within the desired environmental conditions.

Table 1 Comparison of sample shipment by conventional and the proposed IoMT powered methods

Parameters	Conventional sample shipment	Proposed IoMT powered sample shipment
Real-time surveillance	Not possible	Possible
Accuracy of test results	Chances of false results are significantly higher	Chances of false results are considerably low
Detection of breakage of cold chain, if any	Not possible	Possible
Planning of mitigatory strategies in event of breakage in cold chain	Not possible	Possible
Pros	Low cost	Ability to detect and analyze breakage in cold chain and plan mitigatory activities
Cons	Inability to detect breakage in cold chain, thereby leading to false test results	Slightly higher costs than the conventional method

4 Conclusion

The proposed model looks promising in ensuring the viability of samples during the course of the transportation. This would significantly reduce the possibility of false outcomes, thereby preventing both unwanted panic among the false positive as well as negligence among the false negative cases. Accurate test results would lead to tracing and isolation of the patients, if found to be positive without any further delay. The comparison of sample shipment by conventional method and the proposed Internet of Medical Things (IoMT) powered model is listed in Table 1. The social transmission of the disease could be curbed to a large extent with immediate isolation and treatment of the COVID-19 positive patients. Immediate treatment following diagnosis would reduce the COVID-19 mortality rates. This would be a boon to regions where there are very low number of testing facilities due to which the samples need to be transported for very long distances. Countries have started giving approvals for the usage of COVID vaccines on an emergency usage basis. The same model could be used in the effective distribution of these vaccines to the grassroot level. This would reduce the unwanted wastage of precious life-saving vaccines.

References

1. Ashraf MU, Hannan A, Cheema SM, Ali Z, Jambi KM, Alofi A (2020) Detection and tracking contagion using iot-edge technologies: Confronting covid-19 pandemic. In: 2020 International Conference on Electrical, Communication, and Computer Engineering (ICECCE), pp 1–6
2. Aydemir F (2020) Can iomt help to prevent the spreading of new coronavirus? IEEE Consumer Electronics Mag 1
3. Chamola V, Hassija V, Gupta V, Guizani M (2020) A comprehensive review of the covid-19 pandemic and the role of iot, drones, ai, blockchain, and 5g in managing its impact. IEEE Access 8:90225–90265
4. Dai HN, Imran M, Haider N (2020) Blockchain-enabled internet of medical things to combat covid-19. IEEE Internet Things Mag 3(3):52–57
5. Fadlullah Z, Fouda MM, Pathan ASK, Nasser N, Benslimane A, Lin YD (2020) Smart iot solutions for combating the covid-19 pandemic. IEEE Internet Things Mag 3(3):10–11
6. Garg L, Chukwu E, Nasser N, Chakraborty C, Garg G (2020) Anonymity preserving iot-based covid-19 and other infectious disease contact tracing model. IEEE Access 8:159402–159414
7. Islam SMR, Kwak D, Kabir MH, Hossain M, Kwak K (2015) The internet of things for health care: a comprehensive survey. IEEE Access 3:678–708
8. Jaiswal R, Agarwal A, Negi R (2020) Smart solution for reducing the covid-19 risk using smart city technology. IET Smart Cities 2(2):82–88
9. Karmore S, Bodhe R, Al-Turjman F, Kumar RL, Pillai S (2020) Iot based humanoid software for identification and diagnosis of covid-19 suspects. IEEE Sensors J 1
10. Monteleone S, Sampaio M, Maia RF (2017) A novel deployment of smart cold chain system using 2g-rfid-sys temperature monitoring in medicine cold chain based on internet of things. In: 2017 IEEE international conference on Service Operations and Logistics, and Informatics (SOLI), pp 205–210
11. Niamat Ullah Akhund TM, Jyoty WB, Siddik MAB, Newaz NT, Al Wahid SA, Sarker MM (2020) Iot based low-cost robotic agent design for disabled and covid-19 virus affected people. In: 2020 fourth world conference on smart trends in systems, security and sustainability (WorldS4), pp 23–26
12. Rachakonda L, Mohanty SP, Kougianos E (2020) ifeliz: An approach to control stress in the midst of the global pandemic and beyond for smart cities using the iomt. In: 2020 IEEE International Smart Cities Conference (ISC2), pp 1–7
13. Rana MM, Abdelhadi A, Ahmed MRU, Ali A (2020) Secure iot communication systems for prediction of covid-19 outbreak: An optimal signal processing algorithm. In: 2020 third International Conference on Smart Systems and Inventive Technology (ICSSIT), pp 135–139
14. Vedaei SS, Fotovvat A, Mohebbian MR, Rahman GME, Wahid KA, Babyn P, Marateb HR, Mansourian M, Sami R (2020) Covid-safe: an iot-based system for automated health monitoring and surveillance in post-pandemic life. IEEE Access 8:188538–188551
15. WHO (2020) Who coronavirus disease (covid-19) dashboard

Energy-Efficient Power Allocation for Secure SWIPT in IoT-DAS Using Fractional Optimization

Aaqib Bulla and Shahid Mehraj

Abstract In this paper, we study secrecy and energy-efficiency optimization in a distributed antenna system (DAS)-based IoT network with energy harvesting nodes. Considering simultaneous wireless information and power transfer (SWIPT), we define secure energy efficiency (SEE) as ratio of the achievable secrecy rate to the total consumed power. Our goal is to maximize SEE subject to maximum transmit power constraint of distributed antenna (DA) ports and minimum energy-harvesting requirement of the IoT devices. For a single IoT device and single eavesdropper, maximizing SEE is formulated as a constrained fractional optimization problem, and the optimal solution is derived by solving Karush–Kuhn–Tucker (KKT) conditions. We also consider the case of an energy-harvesting eavesdropper and attempt to restrain it from harvesting the energy from the radio signal. This is achieved by introducing one more constraint into the optimization problem corresponding to the energy harvesting requirement of the eavesdropper. Further, we also consider a general case of multiple IoT devices and multiple eavesdroppers in an *N*-port DAS.

Keywords Wireless power transfer · Distributed antenna system · Physical layer security · Wireless energy harvesting

1 Introduction

The Internet of things (IoT), a massive infrastructure of devices connected over Internet, has been extensively studied as part of research in the field of wireless communication engineering [1]. With the recent advancements in wireless communication technologies, especially 5G/6G [2, 3], backed by significant developments in VLSI and network architectures, IoT is now being studied from a real-world perspective. The realization of such a huge network of IoT devices faces numerous challenges from several different directions, among which the issue of energy consumption is most severe. The demand for higher data rates and exponential growth in the number

A. Bulla (✉) · S. Mehraj
Communication Control and Learning Lab, National Institute of Technology, Srinagar, India

K. Dahal et al. (eds.), *Internet of Things and Its Applications*, Lecture Notes in Electrical Engineering 825, https://doi.org/10.1007/978-981-16-7637-6_9

of devices is responsible for higher energy consumption. Both economic and ecological concerns have thus made IoT a primary focus in the field of energy efficient wireless communication or Green Communications [4, 5].

Distributed antenna system technology, primarily designed for increasing network coverage and data rates, is now being studied in the field of energy-efficient wireless communication [6–8]. In addition to enhancing the system performance, DAS can significantly help in wireless power transmission because of the reduced transmitter–receiver access distances, which in turn can improve the energy efficiency of the system. This technology of simultaneous wireless information and power transfer is believed to be a promising solution for sustainable operation of battery-less IoT devices in energy-constrained wireless communication networks [9]. Optimizing energy efficiency for SWIPT in DAS was recently studied in [10, 11].

In a wireless communication system, since the signals are broadcast over the medium, eavesdroppers can easily intercept the information. Thus security is another major challenge faced in the real world IoT applications. The problem of information security was conventionally dealt with cryptographic algorithms at the higher layers. However, the security provided by cryptographic algorithms is based on the assumption of limited computational power of eavesdroppers. On the other hand, in physical layer security, no such assumptions are necessary. Therefore, security at physical layer has been extensively studied in the literature [12, 13], where several aspects of the radio channel are exploited to ensure communication security. Moreover information security at the physical layer is now being widely studied alongside energy efficient wireless communication [14].

In this paper, we study security and energy efficiency optimization for the wireless information and power transfer in a DAS-based IoT network. The fundamental essence of security at the physical layer is the information-theoretic characterization of secrecy capacity in context of wiretap channels [15]. In case of fading wireless channels, secrecy capacity was characterized in [16]. We use the same secrecy rate metric to define secure energy efficiency (SEE) as a ratio of the secrecy capacity to the total power consumed at DA-ports. Our goal is to maximize SEE subject to per DA-port maximum transmit power constraint and minimum energy harvesting constraint.

The rest of the paper is organized as follows: In Sect. 2 we first discuss the case of a single IoT device in an N-port DAS and we assume that only one eavesdropper is present. The corresponding system model and problem formulation are given in Sects. 2.1 and 2.2, respectively. In Sect. 2.2, we first discuss the case of non-energy-harvesting eavesdropper, and then, we extend the approach to an energy harvesting eavesdropper scenario. In Sect. 3, we consider orthogonal frequency division multiple access (OFDMA)-based general DAS setup with multiple IoT devices and multiple eavesdroppers. Again, the problems for non-energy harvesting and energy harvesting eavesdroppers are formulated separately. The numerical results of all the cases discussed in Sects. 2 and 3 are provided in Sect. 4. Finally, the conclusion of the paper is given in Sect. 5.

2 Single IoT Device and Single Eavesdropper

2.1 System Model

Let's consider a DAS with N DA-ports serving a single legitimate user (Bob) in the presence of an eavesdropper (Eve), where all the DA-ports, the Bob and Eve are equipped with a single antenna. As an example, a 3-port DAS with a common control unit is shown in Fig. 1.

We assume that all the DA-ports communicate with each other via dedicated channels but operate independently, having individual transmit power constraints. Moreover, the instantaneous channel state information (CSI) from DA-ports to Bob and Eve is assumed to be available at the transmitter. The signals received by Bob and Eve in this network are given by;

$$y_b = \sum_{i=1}^{N} \sqrt{P_i S_i^{(b)}} h_i x_i + z_b \tag{1}$$

$$y_e = \sum_{i=1}^{N} \sqrt{P_i S_i^{(e)}} g_i x_i + z_e \tag{2}$$

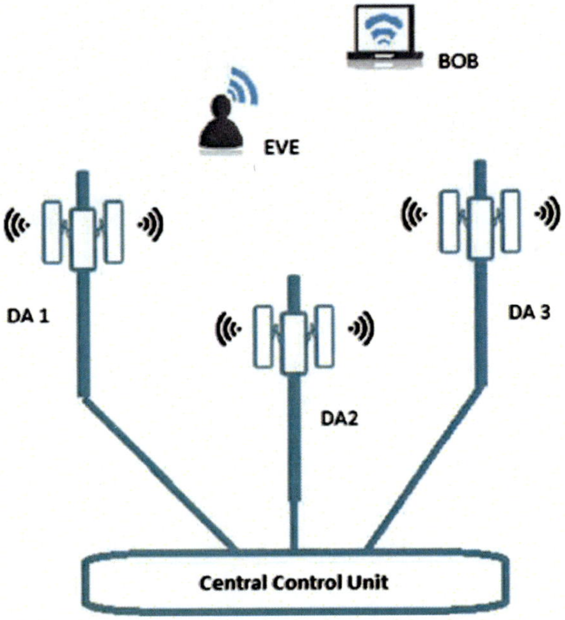

Fig. 1 A DAS model with $N = 3$

for Bob and Eve, respectively: P_i is the transmit power of ith DA-port, $S_i^{(b)} = (d_i^{(b)})^{-\alpha}$ and $S_i^{(e)} = (d_i^{(e)})^{-\alpha}$ denote the propagation path-losses with path-loss exponent α; h_i and g_i denote the independent and identically distributed (iid) circularly symmetric complex Gaussian channel coefficients with zero mean and unit variance; x_i denotes the transmitted symbol from the ith DA-port with an average power $E[|x_i|^2]$ and z_b, z_e denote the additive white Gaussian noise at their receivers with zero mean and variance ${\sigma_b}^2$ and ${\sigma_e}^2$, respectively.

In SWIPT systems, the IoT device has a power splitter which splits received signal power into two parts; one part ($0 \le \Delta \le 1$) for information decoding and the other part ($1 - \Delta$) for energy harvesting (EH). The achievable data rate in this scenario for the Bob is given by [10]:

$$R_b = \log_2\left(1 + \Delta \sum_{i=1}^{N} \gamma_i^b P_i\right) \tag{3}$$

where $\gamma_i^b = \frac{S_i^b |h_i|^2}{\sigma_b^2}$ is the effective channel gain to noise power ratio (CGNR) from the ith DA-port to the Bob. Assuming a linear energy harvesting model having conversion efficiency τ^b ($0 < \tau^b \le 1$), the energy harvested (in Joules) by the device is given by:

$$E = \tau^b (1 - \Delta) \sum_{i=1}^{N} \gamma_i^b P_i \tag{4}$$

Now, in order to ensure the confidentiality of the messages, the transmitter at each DA-port uses Wyner's wiretap coding; therefore, the achievable secrecy rate R_s for Bob is given by:

$$R_s = \left[\log_2\left(1 + \Delta^b \sum_{j=1}^{N} \Gamma_j^b\right) - \log_2\left(1 + \sum_{j=1}^{N} \Gamma_j^e\right)\right] \tag{5}$$

where Δ^b is the PS ratio of Bob, $\Gamma_i^b = \gamma_i^b P_i$, $\Gamma_i^e = \gamma_i^e P_i$ represent the instantaneous SNRs and $\gamma_i^e = \frac{S_i^e |g_i|^2}{\sigma_e^2}$ represents the effective CGNR from the jth DA-port to Eve.

Let's denote $\sum_{j=1}^{N} \Gamma_j^b$ by B and $\sum_{j=1}^{N} \Gamma_j^e$ by E; therefore, Eq. (5) becomes:

$$R_s = \left[\log_2\left(1 + \Delta^b B\right) - \log_2(1 + E)\right] \tag{6}$$

Now, the secure energy efficiency for an N-port DAS, is defined as:

$$\eta_{\text{SEE}} = \frac{R_s}{\sum_{i=1}^{N} P_i + P_c} \tag{7}$$

where P_c denotes the circuit power which represents the total power consumed at the transmitter in different modules during various signal processing operations.

2.2 *Problem Formulation*

2.2.1 Case 1: Non-Energy-Harvesting Eavesdropper

In this case, the objective is to maximize η_{SEE} by varying the transmit power of each DA-port and the power splitting factor Δ^b of the receiver module, subject to the minimum harvested energy requirement $E_{\min}$ for the device and perD -port maximum transmit power constraint $P_{\max,i}$. But, η_{SEE}, being the function of P_i and PS ratio Δ^b, is a non-concave function, and hence, it is difficult to arrive at the optimal solution directly. However, according to Eq. (5), we observe that η_{SEE} increases with the power splitting ratio Δ^b; hence, we can set a bound on Δ^b according to minimum energy harvesting constraint as: $\Delta^b = 1 - \frac{E_{\min}}{\tau^b \sum_{j=1}^{N} \gamma_i^b P_i}$. Since $\frac{E_{\min}}{\tau^b \sum_{j=1}^{N} \gamma_i^b P_i} > 0$ and $0 \le \Delta^b \le 1$, we have:

$$\frac{E_{\min}}{\tau^b \sum_{j=1}^{N} \gamma_i^b P_i} \le 1 \Rightarrow \sum_{j=1}^{N} \gamma_i^b P_i = \sum_{j=1}^{N} \Gamma_j^b \ge \frac{E_{\min}}{\tau^b} \tag{8}$$

Hence, we define the optimization problem as following:

$$P1: \max_{\{P_i\}} \eta_{\text{SEE}}$$

$$\text{s.t}: 0 \le P_i \le P_{\max,i}, \sum_{j=1}^{N} \Gamma_j^b \ge \frac{E_{\min}}{\tau^b}, \quad i = 1, \dots, N \tag{9}$$

where

$$\eta_{\text{SEE}} = \frac{\log_2\left(1 + B - \frac{E_{\min}}{\tau^b}\right) - \log_2(1 + E)}{\sum_{i=1}^{N} P_i + P_c}.$$

Our aim is to optimally allocate the power to the DA-ports such that SEE is maximized under the given constraints. As is customary in physical layer security literature [15, 17], we assume transmitter-legitimate user channel is better than transmitter–eavesdropper channel (that is, $B - \frac{E_{\min}}{\tau^b} > E$). Therefore, R_s given in Eq. (5) is concave. Since, η_{SEE} in (9) is ratio of a concave and an affine function, it is a pseudo-concave function of P_i. Since KKT conditions are necessary and sufficient for optimality of such functions, any maximizer that exists will be the global maximizer [18].

Now, the Lagrangian function for the given problem (*P1*) is written as:

$$L(\{P_i\},\{\lambda_i\},\{\upsilon_i\},\mu) = \frac{R_s}{\sum_{j=1}^{N} P_{j+}P_c} + \sum_{j=1}^{N}\lambda_j P_j + \sum_{j=1}^{N}\upsilon_j(P_{\max,i} - P_j) + \mu\left[\sum_{j=1}^{N}\Gamma_j^b - \frac{E_{\min}}{\tau^b}\right] \quad (10)$$

where $\{\lambda_i\}$, $\{\upsilon_i\}$ are the Lagrange multipliers corresponding to the constraints $0 \leq P_i$ and $P_i \leq P_{\max,i}$, and μ is associated with the constraint $\sum_{j=1}^{N}\Gamma_j^b - \frac{E_{\min}}{\tau^b}$.

According to the KKT conditions, the optimal values $\{P_i^*, \mu_i^*, \lambda_i^*, \nu_i^*\}$ should satisfy the following equations [18]:

$$\frac{\partial L}{\partial P_i} = f_i\left(P_1^*, P_2^*, \ldots P_N^*\right) + \lambda_i^* - \nu_i^* + \mu^*\gamma_i^b = 0$$

$$\lambda_i^* P_i^* = \nu_i^*\left(P_{\max,i} - P_i^*\right) = 0, i = 1, \ldots, N$$

$$\mu^*\left[\sum_{j=1}^{N}\Gamma_j^b - \frac{E_{\min}}{\tau^b}\right] = 0$$

while $P_i^* \geq 0$ and $\left(P_{max,i} - P_i^*\right) \geq 0$ and where;

$$f_i = \frac{\partial L}{\partial P_i} = -\frac{R_s}{\left(\sum_{j=1}^{N} P_{j+}P_c\right)^2} + \frac{1}{\left(\sum_{j=1}^{N} P_{j+}P_c\right)}\left\{\frac{\gamma_i^b}{\left(1 + B - \frac{E_{\min}}{\tau^b}\right)} - \frac{\gamma_i^e}{1+E}\right\}$$

2.2.2 Case 2: Energy-Harvesting Eavesdropper

Now, if the eavesdropper in the system also relies on the energy harvested from the received radio signal, then we can restrain it from charging as well. This goal can be achieved by re-defining the optimization problem with an additional constraint corresponding to maximum energy that the Eve can harvest. However, in this case, we assume that the PS factors of both Bob and Eve are known at the transmitter. This assumption is usually valid because of the standardized receiver modules of the IoT devices. The harvested energy in Joules at the Eve is given by:

$$E^e = \tau^e\left(1 - \Delta^e\right)\sum_{i=1}^{N}\Gamma_i^e \quad (11)$$

where Δ^e is PS ratio of the Eve. Thus, we can re-formulate the problem as:

$$
\begin{aligned}
&P2 : \max_{\{P_i\}} \eta_{SEE} \\
&\text{s.t} : E \geq E_{\min} \text{ and } E^e \leq E^e_{\min} \\
&0 \leq P_i \leq P_{\max,i} \quad i = 1, \ldots, N
\end{aligned} \tag{12}
$$

where $\eta_{\text{SEE}} = \frac{\log_2(1+\Delta^b B) - \log_2(1+\Delta^e E)}{\sum_{j=1}^{N} P_j + P_c}$.

Since in both cases discussed above the objective functions are twice differentiable, we use sequential quadratic programming (SQP) to solve the KKT conditions for the optimal solution [19, 20]. The numerical results of these two cases are discussed in Sect. 4.

3 Multiple IoT Devices and Eavesdroppers

3.1 System Model

In narrow-band IoT standards, due to ease of implementation OFDMA has already been studied for multiple access [21]. In the IoT network under consideration with K_b number of devices, we assume that the entire spectrum is equally segmented into K_b non-overlapping channels, and each device occupies a given channel. In such a scenario, the achievable rate for the kth device is given by:

$$
R_b = \frac{1}{K_b} \log_2 \left(1 + \Delta^b_k \sum_{j=1}^{N} \Gamma^b_{j,k} \right) \tag{13}
$$

where Δ^b_k denotes the PS ratio of kth device and $\Gamma^b_{j,k}$ is the instantaneous SNR from jth DA-port to kth device. We consider two scenarios for eavesdropper:

Case 1: K_b number of IoT devices and a single eavesdropper. In this case, the achievable secrecy rate for the kth device will be:

$$
R^k_s = \frac{\left[\log_2\left(1 + \Delta^b_k B_k\right) - \log_2(1 + E)\right]}{K_b} \tag{14}
$$

where $B_k = \sum_{j=1}^{N} \Gamma^b_{j,k} = \sum_{j=1}^{N} \gamma^b_{j,k} P_{j,k}$.

Case 2: K_b number of IoT devices and K_e number of eavesdroppers. In this case, we approach the problem by considering the worst case secrecy rate for the given user. That is, with K_e number of eavesdroppers present, we define the secrecy rate for the kth user as:

$$R_s^k = \frac{\left[\log_2\left(1 + \Delta_k^b B_k\right) - \max_{l \in K_e} \log_2(1 + E_l)\right]}{K_b} \tag{15}$$

where $E_l = \sum_{j=1}^{N} \Gamma_{j,l}^e$: the instantaneous SNR from jth DA-port to lth eavesdropper.

Now, we define the secure energy efficiency in a multi-user scenario as:

$$\eta_{\text{SEE},K} = \frac{R_{\text{total}}}{P_{\text{total}}} = \frac{\sum_{k=1}^{K_b} R_s^k}{\sum_{k=1}^{K_b} \sum_{i=1}^{N} P_{i,k} + p_c} \tag{16}$$

Since each IoT device can decode the information from a given channel, but can harvest energy from all the available channels, the energy harvested at device k can be expressed as;

$$E_k = \tau_k^b\left(1 - \Delta_k^b\right) \sum_{i=1}^{N} \gamma_{i,k} \sum_{j=1}^{K_b} p_{i,j} \tag{17}$$

where $\sum_{j=1}^{K_b} p_{i,j}$ is the total transmit power of ith DA-port and $\gamma_{i,k} \sum_{j=1}^{K_b} p_{i,j}$ is the received power at kth device from ith DA-port.

3.2 Problem Formulation

Following the same procedure discussed in the single user case, we now define the optimization problem for the multi-user case as:

$$\begin{aligned} &\max_{\{P_{i,k}\}} \eta_{\text{SEE},K} \\ &\text{s.t}: \sum_{i=1}^{N} \gamma_{i,k} \sum_{j=1}^{K_b} P_{i,j} \geq \frac{E_{\min,k}}{\tau_k^b}, \sum_{k=1}^{K} P_{i,k} \leq P_{\max,i}, P_{i,k} \geq 0 \\ &P_{i,k} \geq 0, k = 1, \ldots, K_b, i = 1, \ldots, N \end{aligned} \tag{18}$$

We note that:

For case 1 (Single Eavesdropper):

$$R_s^k = \frac{\left[\log_2(1 + B_k - \beta_k) - \log_2(1 + E)\right]}{K_b}$$

For case 2 (Multi-eavesdropper):

$$R_s^k = \frac{\left[\log_2(1 + B_k - \beta_k) - \log_2\left(1 + \max_{l \in K_e}\{E_l\}\right)\right]}{K_b}$$

where $\beta_k = \delta_k \frac{E_{\min,k}}{\tau_k^b}$ and $\delta_k = \frac{\sum_{i=1}^{N} \gamma_{i,k} P_{i,k}}{\sum_{i=1}^{N} \gamma_{i,k} \sum_{j=1}^{K_b} p_{i,j}} \leq 1$. Again, in a scenario where transmitter-legitimate user channel is better than that of transmitter-eavesdropper (such that, for case 1: $B_k - \beta_k > E$ and for case 2: $B_k - \beta_k > \max_{l \in K_e}\{E_l\}$), $\eta_{\text{SEE},K}$ is a pseudo-concave function of P_i and the KKT conditions are necessary and sufficient for the global optimality [18]. Further, as discussed in Sect. 2, if the eavesdroppers in the system are also energy-harvesting nodes and the PS ratios of the devices are known at the transmitter, it is possible to restrain the eavesdropper from harvesting the energy. In this case, the energy harvested by the eavesdropper is given by $E_l^e = \tau_l^e\left(1 - \Delta_l^e\right) \sum_{i=1}^{N} \gamma_{i,l} \sum_{j=1}^{K_b} p_{i,j}$, and hence, the optimization problem is re-defined as:

$$\begin{aligned} &\max_{\{P_{i,k}\}} \eta_{\text{SEE},K} \\ &E_k \geq E_{\min,k},\ E_l^e \geq E_{\min,l}^e \\ &\sum_{k=1}^{K} P_{i,k} \leq P_{\max,i},\ P_{i,k} \geq 0 \\ &k = 1, \ldots, K_b, l = 1, \ldots, K_e, i = 1, \ldots, N \end{aligned} \tag{19}$$

where for case 1: $R_s^k = \frac{\left[\log_2\left(1+\Delta_k^b B_k\right)-\log_2\left(1+\Delta^e E_l\right)\right]}{K_b}$ and for case 2: $R_s^k = \frac{\left[\log_2\left(1+\Delta_k^b B_k\right)-\max_{l \in K_e}\log_2\left(1+\Delta_l^e E_l\right)\right]}{K_b}$.

4 Results and Discussion

In this section, we present the numerical results of optimization problems defined in Sect. 2 and 3 with the energy conversion efficiency $\left(\tau^b = \tau^e = \tau\right) = 0.75$, $z_b = -90$ dBm and $z_e = -110$ dBm. In all the cases described in Sects. 2 and 3, since objective function is twice differentiable, we use sequential quadratic programming (SQP) to solve the KKT conditions for the optimal solution [19]. In Fig. 2a, we observe that the EE initially improves as the $P_{\max}$ of each DA-port is increased but eventually gets saturated. Also, a comparison is provided with a system without security. As expected, there is an energy efficiency loss when we have security

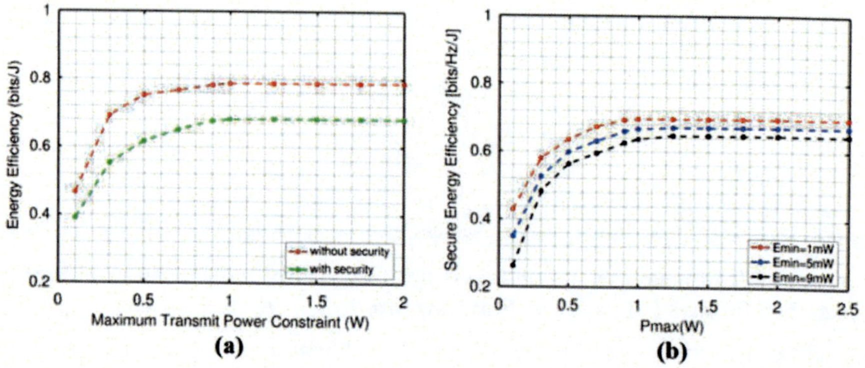

Fig. 2 **a** Energy efficiency w.r.t $P_{\max}$ for $N = 6$ and $E_{\min} = 1\text{mW}$, **b** secure energy efficiency w.r.t $P_{\max}$ for different values of $E_{\min}$

constraint. In Fig. 2b, we plot SEE w.r.t $P_{\max}$ for different values of $E_{\min}$, it is observed that as the energy-harvesting requirement of the device increases, energy efficiency of the system decreases. Figure 3a shows the effect of an energy-harvesting eavesdropper, we observe that in SWIPT environment, it is actually advantageous if the Eve is also energy harvesting node. However, with an energy constraint on the Eve, there is a trade-off between security and energy efficiency.

In Fig. 4a, we plot energy efficiency w.r.t number of DA-ports (N), we observe that as the number of DA-ports increase, energy efficiency improves, and gradually saturates for large N. However, from Fig. 3b, we note that for higher values of maximum transmit power constraint, less number of ports need to be active. In Fig. 4b, SEE versus number of users (K_b) is plotted, we observe that energy efficiency of the system decreases as the number of users in the system increase. However, we can

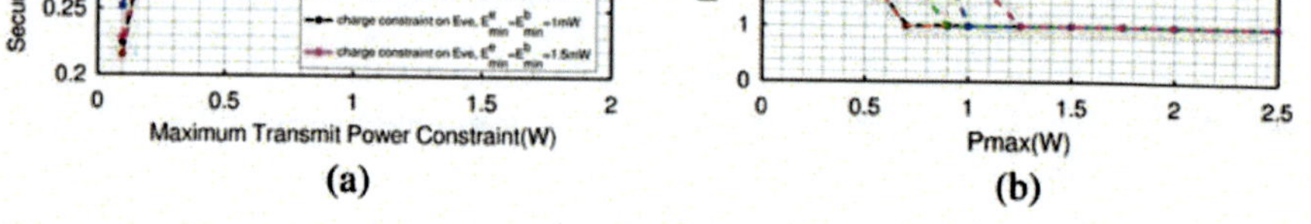

Fig. 3 **a** Secure energy efficiency in case of energy harvesting Eve, **b** No. of active DA-ports w.r.t $P_{\max}$ for increasing $E_{\min}$

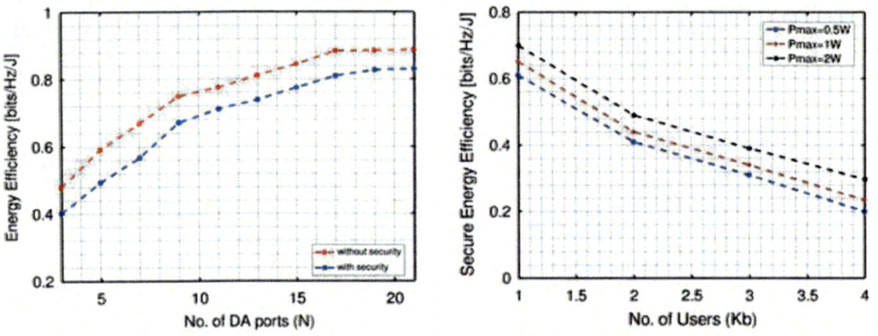

(a) (b)

Fig. 4 **a** Energy efficiency w.r.t N for $P_{\max} = 1$ W and $E_{\min} = 1$ mW, **b** secure energy efficiency w.r.t K_b for different values of $P_{\max}$

still improve the system efficiency if there is a provision for increasing the $P_{\max}$ of the DA-ports.

5 Conclusion

In this paper, we studied secure energy efficiency problem of DAS-based IoT network for simultaneous wireless information and power transfer. We used secrecy rate metric of a wiretap channel model to define secure energy efficiency. The primary objective was to maximize secrecy rates at minimum possible energy consumption. To achieve this goal, we formulated the maximization of SEE as a constrained fractional optimization problem. With the objective function being pseudo-concave, we obtained the optimal solution by solving KKT conditions. We considered several different cases, like, single IoT device and single eavesdropper, Multiple IoT devices with a single eavesdropper and finally multiple IoT devices with multiple eavesdroppers. Moreover, in SWIPT environment, for an energy-harvesting eavesdropper, an attempt was made to restrain the eavesdropper from harvesting the energy by not allowing it to charge. Numerical results reveal that there is a trade-off between security at the physical layer and energy efficiency of the system.

References

1. Atzori L, Iera A, Morabito G (2010) The internet of things: a survey. Comput Netw 54(15):2787–2805
2. Andrews JG, Buzzi S, Choi W, Hanly SV, Lozano A, Soong AC, Zhang JC (2014) What will 5g be? IEEE J Sel Areas Commun 32(6):1065–1082
3. Li S, Da Xu L, Zhao S (2018) 5g internet of things: a survey. J Ind Inf Integr 10:1–9

4. Gandotra P, Jha RK, Jain S (2017) Green communication in next generation cellular networks: a survey. IEEE Access 5:11727–11758
5. Jamil S, Abbas MS, Umair M, Hussain Y et al (2020) A review of techniques and challenges in green communication. In: 2020 international conference on information science and communication technology (ICISCT). IEEE, pp 1–6
6. Zhu H (2011) Performance comparison between distributed antenna and microcellular systems. IEEE J Sel Areas Commun 29(6):1151–1163
7. Chen X, Xu X, Tao X (2012) Energy efficient power allocation in generalized distributed antenna system. IEEE Commun Lett 16(7):1022–1025
8. Kim H, Lee SR, Song C, Lee KJ, Lee I (2014) Optimal power allocation scheme for energy efficiency maximization in distributed antenna systems. IEEE Trans Commun 63(2):431–440
9. Perera TDP, Jayakody DNK, Sharma SK, Chatzinotas S, Li J (2017) Simultaneous wireless information and power transfer (SWIPT): recent advances and future challenges. IEEE Commun Surv Tutorials 20(1):264–302
10. Huang Y, Liu M, Liu Y (2018) Energy-efficient swipt in iot distributed antenna systems. IEEE Internet Things J 5(4):2646–2656
11. Yu X, Chu J, Yu K, Teng T, Li N (2019) Energy-efficiency optimization for iotdistributed antenna systems with swipt over composite fading channels. IEEE Internet Things J 7(1):197–207
12. Wu Y, Khisti A, Xiao C, Caire G, Wong KK, Gao X (2018) A survey of physical layer security techniques for 5g wireless networks and challenges ahead. IEEE J Sel Areas Commun 36(4):679–695
13. Wang N, Wang P, Alipour-Fanid A, Jiao L, Zeng K (2019) Physical-layer security of 5g wireless networks for iot: challenges and opportunities. IEEE Internet Things J 6(5):8169–8181
14. Ng DWK, Schober R (2015) Secure and green swipt in distributed antenna networks with limited backhaul capacity. IEEE Trans Wireless Commun 14(9):5082–5097
15. Wyner AD (1975) The wire-tap channel. Bell Syst Tech J 54(8):1355–1387
16. Gopala PK, Lai L, El Gamal H (2008) On the secrecy capacity of fading channels. IEEE Trans Inf Theory 54(10):4687–4698
17. Leung-Yan-Cheong S, Hellman M (1978) The gaussian wire-tap channel. IEEE Trans Inf Theory 24(4):451–456
18. Zappone A, Jorswieck E (2015) Energy efficiency in wireless networks via fractional programming theory. Found Trends Commun Inf Theory 11(3–4):185–396
19. Nocedal J, Wright SJ (2006) Sequential quadratic programming. Numer Optim 529–562
20. Boyd S, Vandenberghe L (2004) Convex optimization. Cambridge University Press, Cambridge
21. Xu D, Zhu H (2019) Secure transmission for swipt iot systems with full-duplex iot devices. IEEE Internet Things J 6(6):10915–10933

Implementation of IoT-Based Smart Healthcare Monitoring System

Madhumita Sarkar, Shovon Nandi, and Sayamuddin Ahmed Jilani

Abstract The progressive improvement in the medical field shifts the diagnosing system toward automation as manual monitoring sometimes fails to give or produce accurate results. Thus, in our work, we proposed an IoT-based system of health monitoring for smart hospital which will be treated as an advancement toward e-health care. The system has different provisions; the first one has a saline monitoring system using IoT that keeps a check on the saline fluid level of the patient and sends alert for bottle replacement. The second part consists of a vast system that digitally monitors the patient's metabolic activity like pressure, pulse, sugar level, etc. Digitalizing these systems makes it more accurate and efficient and also helps the hospital stuff to look faster and easier. The third portion of this article focused on pee or urine bag monitoring system, where the pee bag is controlled by opening up a valve automatically when it is full for the easy flow of the liquid. This article also delineates how the smooth operation is possible from a Web page or from the mobile application.

Keywords Internet of things (IoT) · Blynk server · Healthcare monitoring system · If this then that (IFTTT) server · ESP8266 · Saline monitoring system · Pee bag monitoring system

1 Introduction

In ongoing period, well-being risk is not an age subordinate factor because of sporadic way of life and occupied timetable. Patients' well-being is procured by different sensors and afterward the information which is put away by the Internet of things (IoT) is shown through the Web site or mobile applications that assists with getting

M. Sarkar (✉)
B. P. Poddar Institute of Management and Technology, Kolkata, India

S. Nandi
Bengal Institute of Technology, Kolkata, India

S. A. Jilani
Maulana Abul Kalam Azad University of Technology, Kolkata, India

K. Dahal et al. (eds.), *Internet of Things and Its Applications*, Lecture Notes in Electrical Engineering 825, https://doi.org/10.1007/978-981-16-7637-6_10

to the remote monitoring. Traditional system of manual monitoring of the status of the saline flow is a difficult job for the hospital staffs for a large number of patients. Also, to accurately displaying the parameters like pressure, pulse, sugar level, etc., is a tiresome job. Sometimes the urine or pee bag monitoring is needed for the critical patients. To counter measure, the above issues a saline monitoring system using IoT are being developed which could help the patient as well as a nurse in the hospital and improve the patient monitoring process. By the same way, we can have benefitted from the properly deploying the sensors of various health measurements parameters. Like this, the digitization process also creates a smooth operation of tracking the data of the patients and helps in properly controlling from the remote end also. In the smart saline bottle, when the fluid level or weight is low, it will alert the observers through the display or/and mobile phone at the control room indicates the room number of the patient for quick assistance. Nowadays, simple electrolytes bottles with no indication parameter are used in most of the hospitals; it may create a problem to patients because if the reverse flow will have started, blood started to flow from body toward the bottle. So to prevent this type of anomaly, our proposed smart saline bottle is very much effective to handle this situation.

2 Literature Survey

The world is changed these days via computerization and Internet of things. Naga Malleswari et al. described in their journal that how smart saline bottle system can be build using IoT [1]. They identify that by Web mode and server mode how it can be possible to monitor saline level, but in their work, they do not use any type of notification system. Just use LED for notifying the nurses or doctors, they just use ThingSpeak. But no one does leave open only ThingSpeak site all day and for the layman they do not understand what is going on unless they do not get proper notification. So, in their system, build-up of email notification is missing which have also identified the time and bed number, and their system can be used just only in hospitals but our prototype also be used in the home.

Debjani Ghosh et al. say in their paper that caretaker's ignorance about monitoring saline bottle can be a cause of critical situation and also mentioned about when saline will be finished and removing of the needle timing will be delayed that time, reverse flow will be started [2]. So, by the notification system, it is good to prevent critical situation but practically no one can have removed the needle at proper time. They do not give any solution for this situation. But in our model it consists of a valve in between the pipe which is on and off by the position of hanging bottle measured by ultrasonic sensor which is shown in Fig. 1a–e. When saline is finished, that time valve will be automatically off and stop the reverse flow of blood.

Anusha Jagannathachari et al. in their paper used a buzzer system for notification in their proposed model which is not actually suitable in hospital environment. Patient may be in panicked by that sound [3]. Also, they did not propose the use of any proper server for sending the notification, whereas in our system we use IFTTT server for

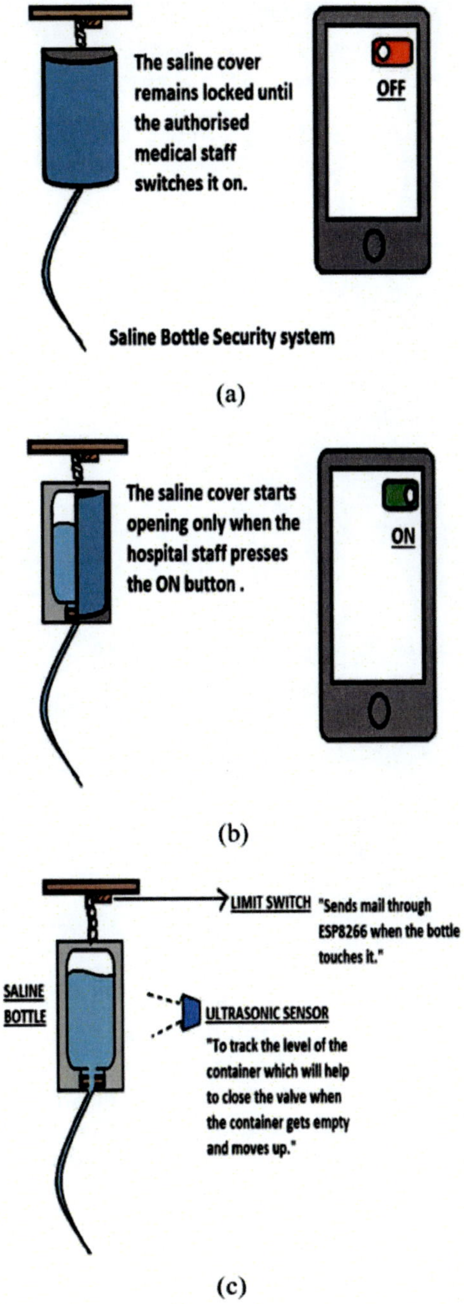

Fig. 1 Smart saline bottle system

Fig. 1 (continued)

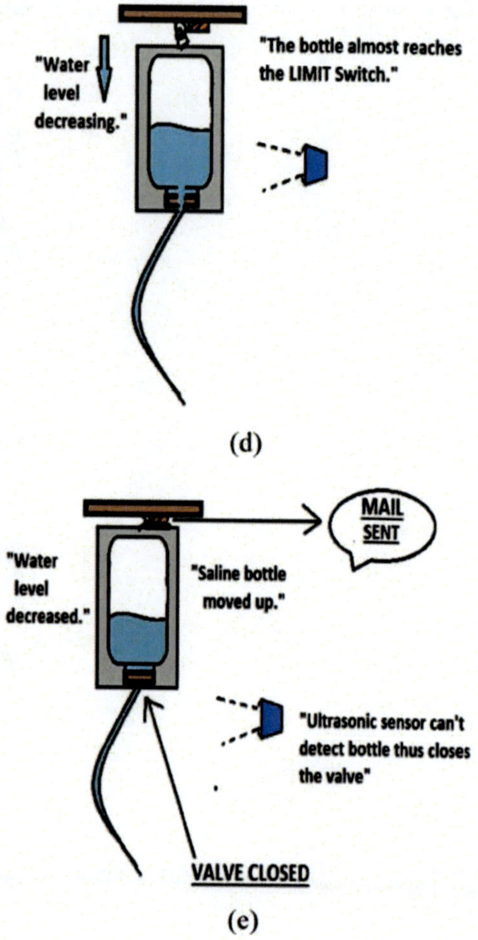

sending mail and anywhere in the world to get notified which help if this system used in home and it is very cost effective also.

Ashika A. Dharmale et al. mentioned about the security system as future development in saline bottle system in his work [4, 5]. We focused on that issue seriously and developed a locking system. In our design, saline bottle is placed hanging in a box arrangement, which has a locking system controlled by servo motor and mobile app. Authorized person can only use that app by a unique mail id and password. After login when that authorized person pressed the unlock button, a mail will be sent with proper date and time with the bed number to administrator's mail id, same for locking time.

Vikramsingh R. Parihar et al. described in their research paper that how smart health monitoring system can be described using IoT. In their system, they have used LCD screen for temperature and blood pressure monitoring (BPM) for patients

[6]. But in modern days, their system can be updated by incorporating mobile app for monitoring. Also, the LCD screen for monitoring patient's health is not suitable for real-time monitoring process. So in their system, they did not used smart phone application for doctor which can monitor the patients in real time from anywhere. But in our system, we used smart phone application for patient monitoring system and doctor can monitor the patient's health in real-time basis from anywhere in the world. So, in their system, they displayed the BPM value only, not able to generate the graph of BPM in their screen. But in our design, doctor or nurse can observe or monitor the patient's temperature, heartbeat in their smart phone from anywhere in hospital or world in real-time basis.

M. Sathya et al. described in their research paper that they did not much worried about patient monitoring in hospital, only focused on fetching the health-related data [7]. They personify that by using Bluetooth sensor or ZigBee for data transmission which is not suitable for large hospital because these types of communication only work for short distances. Also, they did not use any type of notification system for sending alert of the concerned situation of critical patients. But in our system, we use alert system like email and message when temperature and BPM are high enough of patients in real time. Their system can be used in hospital not for outside the hospital.

In our system, low cost and portable patient monitoring system is possible for e-health services in rural areas. IoT-based system has been developed which can be used by paramedics for collecting different sensor data such as ECG signal (electrocardiogram), EEG (electroencephalogram), heart beat signal, temperature, etc., from a patient and send these signals to the specialist doctor also for the sharp observation of the patient. Also, the efficient data transmission using modern communication is also kept in mind [8–10].

Sathish Kumar et al. described in their research paper that how smart urine bag system can be developed using IoT [11, 12]. They defined that, urine coming through the catheter is being deposited in the urine bag. There is a urine meter sensor attached to the urine bag. Until the urine in the urine bag is completely filled the message/notification will not be send to the nurse or the ward boy of that hospital. Then they come and make empty the urine bag. It is harmful because as the critical patient's urine can have accumulated in the urine bag or pee bag, it can spread the infection from the urine bag to the body of the patient and also the medical staffs can be affected. So, this can cause a lot of damage to them.

But we have automated our system or design, because nobody has to come and clean the pee bag for which the infection will not be spreaded. In our system, we have connected a flex sensor to the urine bag and a solenoid valve to the output pipe, which will create a pressure on the flex sensor as urine from the patient's body accumulated in the urine bag causing the solenoid valve to open and urine will be passed out of the urine bag. It is also very safe process and will also save one's life.

As this era is fully IoT-based smart devices, we remove that barrier of using sim & GSM module and connect that device with the Internet which sends notification including time and a health-related message via email. If any old age person feels any health issues, then the family member will be notified immediately.

Lipsa Dash et al. used in their project ARM controller which is very costly regarding Arduino and also they used buzzer system [12]. So our prototype is very cost effective and it also helps hospitals as well as individual patients in home. Also, the security is another key issue [13–16] and this article also portrayed that.

3 System Model

E-health education is characterized as "the capacity to look for, find, comprehend and assess wellbeing data from electronic sources and apply information picked up to tending to or tackling a medical issue" [17]. This article deals with a project work of saline bottle that is placed inside a cover with a lock that grants access to only the authorized officials. The lock works with the help of a servo motor, NODEMCU, and blynk app. The servo motor is used to open and close the lock, and the NODEMCU helps to grant access for the officials using private passcode. On opening the lock, a mail will be send in the hospital control room or if the patient is being treated in the home atmosphere then the notification will be delivered to the authorized person only. Figure 1 pictorially describes the smart saline bottle arrangement.

The saline water flow is also monitored using IoT; here, we use NODEMCU and limit switch that works according to the weight of the bottle. After the loss of certain weight of liquid, the limit switch is triggered and an email will be send in the control room of the hospital with IFTTT server.

In this design, saline bottle is hanging with a spring on top of the bottle cover. Besides the spring a label switch is established. When the weight of the bottle reaches a certain level (this level is measured previously which will be fixed), saline bottle goes up and presses the limit switch. Switch is connected with ESP8266, which will send an alert mail for changing the bottle to provided email address. Also we used this mass reduction technique to stop the reverse flow of blood. Ultrasonic sensor continuously gives sonic wave to the bottle, that sensor is set in a fixed point. When the bottle will be empty, it goes up and this will be sensed by sensor and closed the valve, which is connected in between the channel pipe. Also, we provided a security lock in bottle by servo motor. This lock will be opened by Blynk (app will be operated by authorized person's mail id) when the door will be opened it will transmit a door opening alert via email to the provided email id. Also, wireless body area network (WBAN) for well-being observing is finished by putting sensor nodes (SN) in different pieces of the body with equivalent energy, i.e., 0.5 J. These SNs sense the different natural parameters, for example, ECG, EEG, temperature, and blood pressure [18]. ECG signals check the heart status of the human body and require higher transfer speed contrasted with different sign. The project model implementation of smart saline bottle is shown in Fig. 2. Figure 3 shows how patient's health-related data are sensed and notified to the assigned authorities.

For reducing continuous monitoring by doctor, a monitoring system is developed by Blynk app where doctor can easily monitor patient temperature and pulse by showing the numeric data as well as graph in the app screen. This app will be used

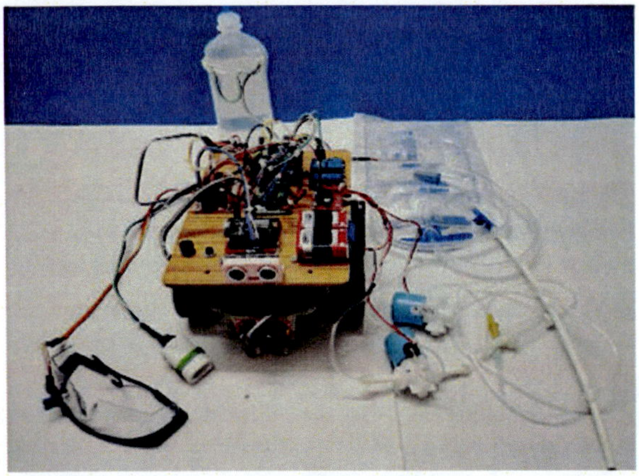

Fig. 2 Implementation of smart saline bottle

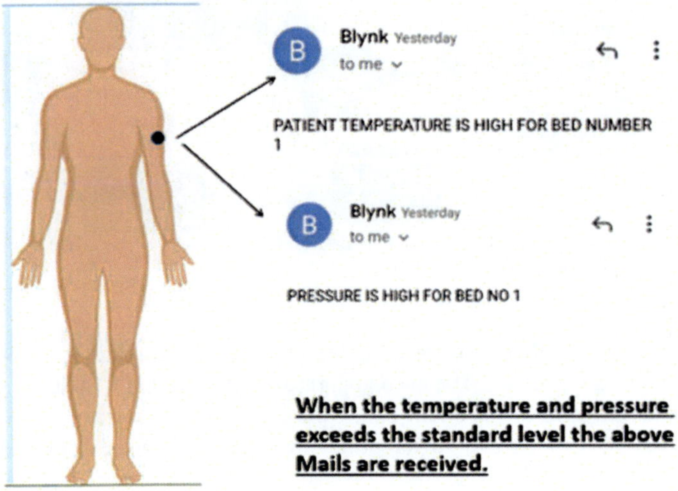

Fig. 3 Notification system of health parameters of the patient

by doctor in anywhere in the hospital. App must be connected with local Wi-Fi server [19, 20]. When the temperature of a person is greater than 100F, it will send an email to doctor to the specified email id. Also, it is possible to send the text message to the approved authority.

Sometime problem happened in pee bag, when it filled-up with urine that time if no one present to change the pee bag it overflows and an unhealthy situation is created. So, by this design when pee bag will be filled that time flex sensor is folded and the solenoid valve will be opened and urine goes out through the pipe. Again,

Table 1 Used apparatus

Node MCU	Arduino uno	Resistor	Blynk (server)
Limit switch	Flex sensor	Voltage regulator	IFTTT (server)
Temperature sensor	Relay	Transistor potentiometer	LCD screen
Servo motor	LED	Push button	Solenoid valve
Ultrasonic sensor	Pulse sensor	Jumper wire	

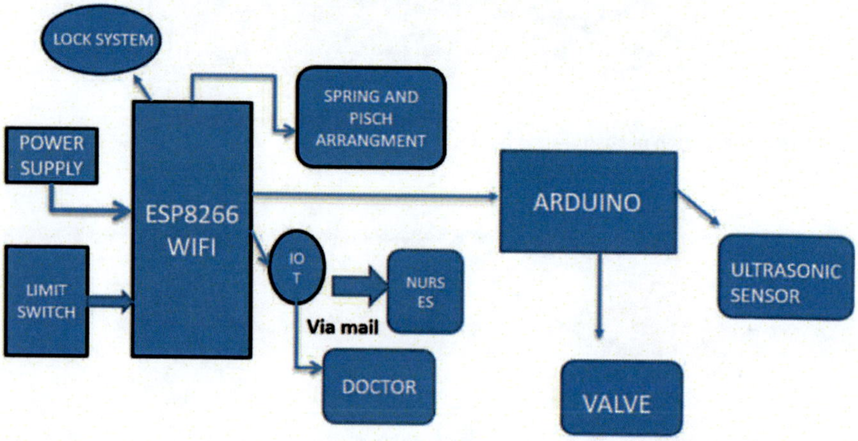

Fig. 4 Block diagram of smart saline bottle

when bag will empty that time solenoid valve will have restricted its flow and wait for filling up of the pee bag. The used apparatuses are shown in Table 1.

The block diagram of the smart saline bottle, smart pee, or urine bag is shown in Figs. 4 and 5, respectively. Different health parameter sensors attached with the Arduino Uno and Blynk server are shown in Fig. 6. To monitor the health-related data, one mobile application is created which is shown in Fig. 7.

4 Conclusion

In this paper, smart patient monitoring system using Internet of things (IoT) has been successfully explained. This work is highly energy efficient as it uses Arduino board having microcontroller which is having low power utilization. The monitoring of the health equipment through Web page or mobile application in this work makes us one step advance toward e-health services. In this work, some of the automated patient monitoring system using IoT like smart saline monitoring system, a temperature and

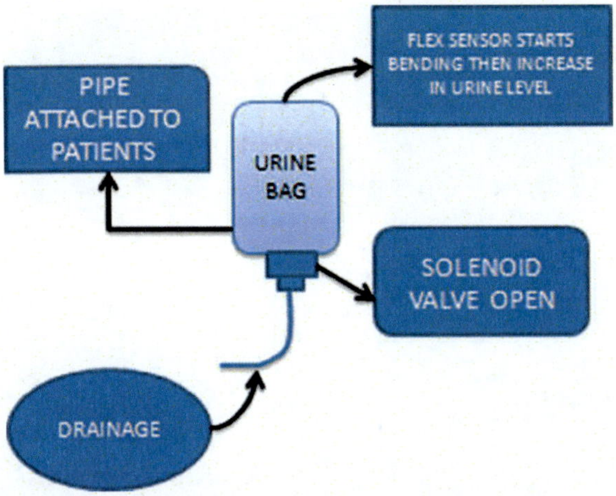

Fig. 5 Block diagram of smart pee or urine bag

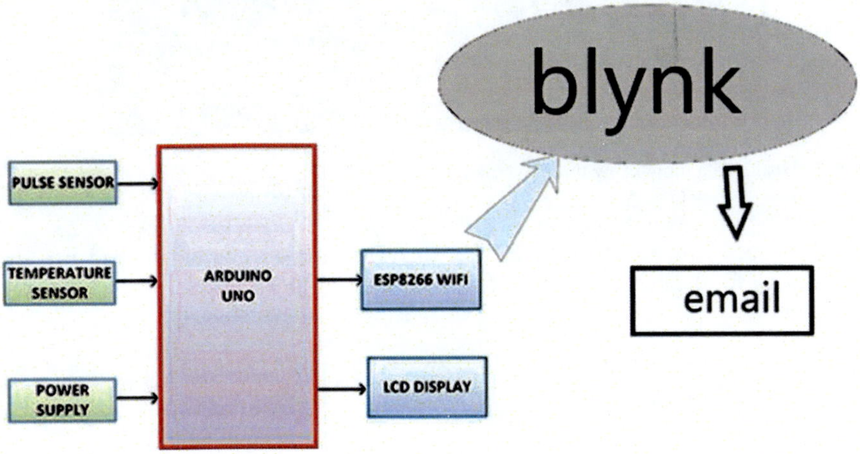

Fig. 6 Block diagram of different health parameter sensors with the Arduino Uno and Blynk server

pulse monitoring system, and a pee or urine bag monitoring system are implemented and explained with maintaining proper guidelines. The temperature and the pulse sensor sends the patient's statistics digitally and the urine bag opens up a valve automatically when the bag is full for the easy passage of the liquid. The main advantage is that the status of a patient's health can be monitored from remote location also. It is a user-friendly system and can be used in a smart hospital as well as in home atmosphere also. The other benefit is that the maintenance of this project is very low.

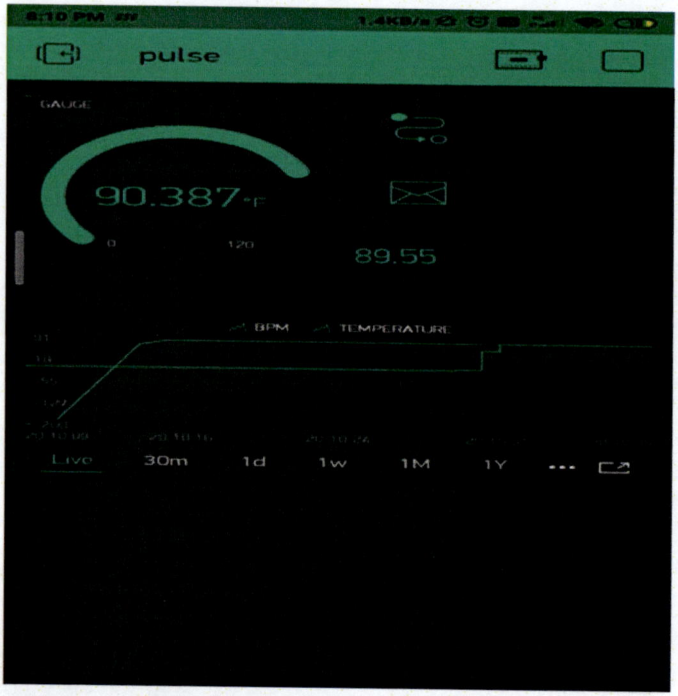

Fig. 7 Health data monitoring through smart phone

This work can be extended further to make smarter equipment and use the technology for the benefit of mankind.

5 Future Scope

The benefit of this work has been utilized in the automation of hospitalization. The deployment of trained medical staffs for patient monitoring can be minimized if we can enjoy the flavor of this work fully. In future, lots of health issues can be controlled electronically and the test data can be employed for diagnosing using machine learning approach.

References

1. Naga Malleswari B, Vijay Varma P, Venkataram ND (2018) Smart saline level monitoring system using IOT. Int J Eng Technol 7(2.7):817–819. doi:http://dx.doi.org/https://doi.org/10.14419/ijet.v7i2.7.10986

2. Ghosh D, Agrawal A, Prakash N, Goyal P (2018) Smart saline level monitoring system using ESP32 and MQTT-S. In: IEEE 20th international conference on e-health networking, applications and services (Healthcom), Ostrava, pp 1–5. https://doi.org/10.1109/HealthCom.2018.8531172
3. Jagannathachari A, Nair AR, Saline level indicator. IOSR J Comput Eng (IOSR-JCE), pp 13–16. e-ISSN: 2278-0661, p-ISSN: 2278-8727
4. Dharmale AA, Mehare RR, Bharti AR, Meshram SR, Deshmukh SV (2019) IOT based saline level monitoring & automatic alert system. IJARCCE 8(4)
5. Jilani SA, Koner C, Nandi S (2020) Security in wireless sensor networks: attacks and evasion. In: 2020 national conference on emerging trends on sustainable technology and engineering applications (NCETSTEA). Durgapur, India, pp 1–5. https://doi.org/10.1109/NCETSTEA48365.2020.9119947
6. Parihar VR, Tonge AY, Ganorkar PD (2017) Heartbeat and temperature monitoring system for remote patients using Arduino. IJAERS 4(5). https://dx.doi.org/https://doi.org/10.22161/ijaers.4.5.10
7. Sathya M, Madhan S, Jayanthi K (2018) Internet of things (IoT) based health monitoring system and challenges. IJET 7(17):175–178
8. Nandi S, Pathak NN, Nandi A (2020) A novel adaptive optimized fast blind channel estimation for cyclic prefix assisted space—time block coded MIMO OFDM systems. Wireless Pers Commun. https://doi.org/10.1007/s11277-020-07629-z
9. Nandi S, Pathak NN, Nandi A (2019) Efficacy of channel estimation and efficient use of spectrum using optimised cyclic prefix (CP) in MIMO-OFDM. IJEAT 9(2), ISSN: 2249-8958 https://doi.org/10.35940/ijeat.B4093.129219, 2019
10. Nandi S, Pathak NN, Nandi A (2020) Avenues to improve channel estimation using optimized CP in STBC coded MIMO-OFDM system-a global optimization approach. In: 5th international conference on microelectronics, computing & communication systems (MCCS-2020), 11–12th July 2020
11. Sathish Kumar R, Rani C, Ganesh Kumar P (2018) IoT based monitoring of container vehicle for secure and reliable delivery of goods. In: Proceedings of the 33rd Annual ACM symposium on applied computing, pp 628–633. doi.org/https://doi.org/10.1145/3167132.3167201
12. Dash L, Arun R, Bhavani R, Rethna Jennifer S (2018) Flexible compartments IoT driven smart pill-box. IJSRCSEIT, 4(5). ISSN: 2456-3307
13. Sarkar M, Sikder S, Ghosh S (2018) Development of architecture for secured data transmission in OCDMA system with designed modified walsh code. WOCN, Kolkata, India
14. Sikder S, Sarkar M, Ghosh S (2015) Theoretical analysis and simulation investigation of designed 1-D and 2-D Modified Walsh Code (MWC) in optical CDMA system. In: Microwave, optical and communication engineering (ICMOCE), IEEE Xplore, pp 7489785. https://doi.org/10.1109/ICMOCE
15. Sikder S, Sarkar M, Ghosh S (2018) Optical network security using unipolar walsh code. AIP Conf Proc 1952:020099. https://doi.org/10.1063/1.5032061
16. Nandi S, Nandi A, Pathak NN (2017) Performance analysis of Alamouti STBC MIMO OFDM for different transceiver system. IEEE conference on ICISS 2017, pp. 883–887. https://doi.org/10.1109/ISS1.2017.8389305
17. Norman CD, Skinner HA (2006) eHEALS: The eHealth Literacy Scale. J Med Internet Res 8(2):e9. https://doi.org/10.2196/jmir.8.2.e9
18. Qu Y, Zheng G, Ma H, Wang X, Ji B, Wu H (2019) A survey of routing protocols in WBAN for healthcare applications. Sensors (Basel) 19(7):1638.https://doi.org/10.3390/s19071638 PMCID: PMC6479667
19. Nandi S, Pathak NN, Nandi A (2020) Analysis of hard decision and soft decision decoding mechanism using viterbi decoder in presence of different adaptive modulations. Int J Future Gen Commun Network 13(3):3002–3012
20. Nandi S, Pathak NN, Nandi A, Implementation of adaptive optimized fast blind channel estimation of MIMO-OFDM System using MFPA. Book chapter, John Wiley & Sons, ISBN: 978-1-119-57138-4

Blockchain-Based Access Control Model for IoT Applications

Ashish Singh, Punam Prabha, and Kakali Chatterjee

Abstract In Internet of things (IoT) domains, safe and easy access to the data and resources is a big challenging issue. The data and resources are also vulnerable to availability, integrity, security, and privacy threats. To overcome such issues, access control models were proposed. But, most of the traditional access control models are not providing complete security in terms of integrity and correctness to the IoT applications as well as the risk of privacy leakage exists. To overcome such problems, a blockchain-based access control model is proposed which can secure the IoT applications (data and resources) as well as provide guarantee the auditability and correctness of access control policies. Smart contracts are deployed in the blockchain that will provide the correct access to correct users in a distributed manner. The security analysis of the proposed model ensures that the blockchain-based access control model will provide safe and easy access to the data and resources to the IoT applications.

Keywords Blockchain technology · Access control · IoT data and resources · Smart contracts · Smart access policies

1 Introduction

IoT is a new paradigm in Internet-based services due to the digitalization and incorporation of new technologies and the development of existing ones. It is composed of physical and virtual objects known as things that have unique identities and are

---

A. Singh (✉)
School of Computer Engineering, KIIT Deemed to be University,
Bhubaneswar, Odisha 751024, India
e-mail: ashish.singhfcs@kiit.ac.in

P. Prabha · K. Chatterjee
Department of Computer Science and Engineering, National Institute of Technology,
Patna, Bihar 800005, India
e-mail: punamp.phd18.cs@nitp.ac.in

K. Chatterjee
e-mail: kakali@nitp.ac.in

K. Dahal et al. (eds.), *Internet of Things and Its Applications*, Lecture Notes in Electrical Engineering 825, https://doi.org/10.1007/978-981-16-7637-6_11

connected with each other using the Internet. The remote accessibility introduced new security and privacy issues such as unauthorized data modification, unauthorized access, integrity, authentication, and many more. With the exponential growth in the IoT-based services and protecting the resources from malicious use, some proper security solution is needed. Access control is such a security solution that restricts unauthorized access. Several access control solutions were present in the past years. But, the modern IoT system devices are vulnerable to several new security attacks. Also, centralized access control may lead to a single point of failure in the case when the access environment is highly malicious. Lack of access transparency, access policy integrity, and static nature of the access control system are some issues. Due to these issues, the traditional access control system is failing to protect the system completely. We have to need a distributed access control system which will ensure the auditability and correctness of access control policies as well as it is immutable and transparent. The access control system over block-chain technology may overcome such access control issues as well as provides the desired security level. A blockchain-based access control system provides (1) auditability and correctness of access control policies (2) immutable and transparent access control policies (3) maintain the integrity of the access control rules (4) decentralized self-evaluating policies.

These studies and features of the blockchain-based access control system gave the guarantee of completely safe and easy access to the data and resources. The discussion also says that smart contracts-based access control rules are secured from unauthorized rule modification. Also, sharing and revoking access permissions to the data happen in a controlled manner. Based on the discussion, we have summarized the contribution of the work:

- This paper summarized the current problems in the traditional access control system and how the blockchain solution can overcome such issues.
- We have proposed a blockchain-based access control system for the IoT applications.
- The concept of a smart contract is used for the generation and protection of access control policies.
- The result and security analysis of the proposed model shows the proposed access control system provides complete access security to the IoT applications.

The rest of the paper is organized as follows: Sect. 2 discussed the background and literature work that are used for the development of the proposed model. Section 3 discussed the proposed access control model. The implementation results of the proposed model are illustrated in Sect. 4. The security analysis of the proposed model is discussed in Sect. 5. Finally, the conclusion is discussed in Sect. 6.

2 Background and Related Works

In this section, we have discussed blockchain taxonomy, blockchain-based access control system, and some security issues that will be considered while we have designed our proposed model.

2.1 *Blockchain Taxonomy*

The blockchain technology is founded by Satoshi Nakamoto in 2009 [1]. Blockchain technology is based on the concept of a decentralization mechanism. It is also known as distributed ledger technology due to the implementation of a distributed public ledger, which is used by each node of the blockchain. With the help of a distributed public ledger, digital records remain unaltered, transparent, and secure. Each block is linked to each other and known as a node. Node has three basic components: data, nonce, and hash function. When blockchain is created first time, the nonce of the first block creates a hash function and form a tree-like structure (Merkle tree). Miner is responsible for creating the block. The created block is attached to the blockchain after successful validation by every node. The hash of the newly created block is attached to the hash of the first block. The first block is named as genesis block. A consensus algorithm is used during the selection of miner [2]. The basic structure of blockchain is shown in Fig. 1.

2.2 *Blockchain-Based Access Control Systems*

Blockchain has desirable features that make the access control system more flexible. Hence, it can verify each access request continuously. In the past years, several access control solutions [3–7] were proposed. Due to the advancement of technology in the IoT-based application, they are not adequate. Hao Wang et al. [8] proposed a blockchain-based data storage model in which healthcare data are stored on the cloud storage in encrypted form. The metadata of the data is stored on a blockchain to

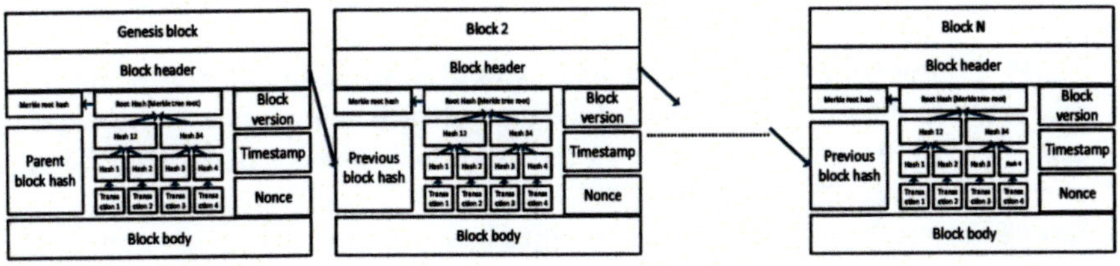

Fig. 1 Blockchain structure

maintain integrity in the data. In 2017, Maesa et al. [9] proposed a blockchain-based access control model by extending the bitcoin concept. The policies are defined based on the attribute-based access control mechanism and eXtensible Access Control Markup Language (XACML). Secure data sharing in IoT-based smart cities are handled by Makhdoom et al. [10]. The proposed access privacy rule is developed with the help of blockchain technology. The access rule uses the concept of smart contracts. The challenges and state of the art of block-chain based access control system is discussed in [11]. In this paper, they have discussed the limitations and problems present in the traditional access control system. Another review paper [12] was found in the literature that discussed the access control issues and challenges in the IoT environment. In the medical sector, an access control model based on the blockchain technology was developed by Thwin and Vasupongayya [13]. Their main aim is to preserve the privacy of the medical healthcare records. In [14], blockchain-based security and privacy issues were discussed. In [15], exploit blockchain technology to proposed an access control system. Access control is considered as a service in [16] for the IoT applications. An access control mechanism is discussed in [17] for the IoT environment. An object model, architecture, and mechanism-based access control model are discussed in [18]. They have also integrated blockchain technology for secure access control in IoT applications.

2.3 Design Consideration Issues

The following issues are considered while we developed the proposed system.

- **Transparency** Transparency provides security mechanisms that are designed for providing in-detectable or hidden from view. The user only sees those works that are available in the foreground [19].
- **Privacy** The ability to secure information and identity in a group is one of the features of privacy [20].
- **Modularity** Modularity minimizes duplication and implements re-usability mechanisms. It is one of the features of object-oriented programming. It divides big problems into different sub-modules, and these modules are linked to each other to operate a single big, complex, and powerful system [21].
- **Delegation** By introducing this concept in a designed system, if any user was not able to do their assigned work, then they can transfer it to other subordinate users [22].
- **Policy Complexity** Complex password is used to make the system secure. For this, a set of rules or policies is defined among each sub-module in presence of authorized third parties. Blockchain technology do this work smartly with the help of smart contract [23].
- **Data Sharing** Shared data may be modified by a any user. Hence, one copy of the original data must be available to each sub-module of the system [24].

- **Immutability** It is different from the module because it can not be modified after creation. When different modules are run on different platforms, they must work on the platform used by the desired systems [25].

3 Proposed Model

The proposed blockchain-based access control model for the IoT applications are discussed in this section. The proposed system has several modules. Each module has some responsibilities and functionalities which are discussed in Sect. 3.1. The working steps of the proposed model are as follows:

Step 1: In the first step, users are sent the access request to the system.
Step 2: These user access requests are accepted by PEP.
Step 3: All the accepted requests are forwarded to the PDP.
Step 4: PDP check the user requests with their credentials and if the access request and credentials are valid then forwarded to the PIP.
Step 5: PIP collects all the user, object, and environment attributes with the help of the attribute manager. All the values of the attribute are stored into blockchain so they are immutable and auditable. All the attribute information are passed to the PDP.
Step 6: After getting all the required attribute information, PDP passes the information to the PAP.
Step 7: PAP consult with the context handler and rule data in which all the access control rules are stored in the form of smart contracts and smart policies. The access request is valid and the user is eligible for getting the resources, then this information is sent back to the PDP.
Step 8: PDP collects all the information received from the PIP and PAP and allow to user to access the IoT application, data, and resources.

Network model of blockchain-based access control system: The proposed network model of blockchain-based access control system is presented in Fig. 2. In this network model, the blockchain component is the main module. The complete security of the proposed model is to depend on the security of the blockchain module. Our proposed model mainly consists of a concept of access control model, IoT users, IoT data, applications and resources, a blockchain component with smart contract and smart access control policies, and resource administrator for defining and controlling the access policies. The details of each component and relation with the blockchain technology are discussed in the next Sect. 3.1.

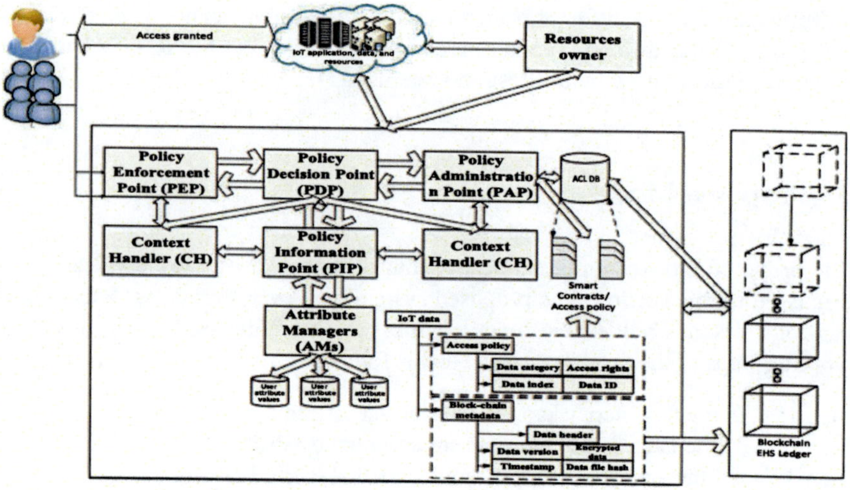

Fig. 2 Blockchain-based access control model

3.1 *Component of Blockchain-Based Access Control Model*

- **Policy Enforcement Point (PEP)**: PEP is one of the components, which is used to protect the resources of the proposed system. With the help of PEP, access request either be accepted or rejected based on access rules defined in the smart contract. Hence, the role of PEP is to collect an access request along with attributes and triggers decisions regarding acceptance and rejection.
- **Policy Decision Point (PDP)**: The input parameters considered by the PDP component are access request, policy, and a set of available attributes. After completing the evaluation process on input parameters, access decision (such as permit, deny, not applicable) is produced as an output.
- **Policy Information Point (PIP)**: PIP is used by attribute manager. In the proposed system, various kinds of attribute managers are available. The available attribute accessed by them depends on their protocol. PIP acts as plugin points of the proposed system. Each attribute manager is connected to PIP for gathering the latest information about attributes.
- **Policy Administration Point (PAP)**: The main role of PAP is to manage access control policies. PAP can also work as a policy repository. Hence, PIP stores policies during the evaluation of access requests. PIP can help those users who want to create and modify policies. It supports complex functions that are used in the phase of policy production and management.
- **Context Handler (CH)**: CH behaves as miner of the blockchain-based access control system. It means CH manages, controls, and monitors entire functionalities, attributes, access requests, and policies used in the proposed system. The decision

prepared by CH depends on the previously described decision used to manage similar work.

- **Smart Access Policies (SAP)**: Access policy is a set of rules which is prepared to grant access request. A smart contract is prepared in XACML language and modified during the occurrence of transactions done by nodes of the blockchain networks.
- **Smart Attribute Manager (SAM)**: Attributes and policies written in a smart contract are managed by the manager smartly. Hence, named as smart attribute manager. The updated result will be automatically updated at the end of the node via DPL.

4 Implementation Results

We have developed a prototype implementation of the proposal to validate the security and proof of concept of the proposed model. The experiment work is divided into three phases: the first phase is to develop smart contracts, the second phase is to develop smart access control rules, and the third is to measure the performance (latency, execution time, and throughput) of the proposed model. We have used International Educational blockchain academic testnet. This tool provides Ethereum-based private testnet blockchain protocol because it is widely used for the development of smart contracts. A JavaScript style language called Solidity as a programming language is used for writing the code of smart contracts. All the access control policies are written in the XACML language. The simulation of the work is tested with the PC configuration of Intel(R) Core(TM) i5-8250U CPU @ 1.60GHz 1.80 GHz; RAM is 8.00GB (7.89 usable), 64-bit Operating System, x64-based processor, 520GB SSD.

To measure the performance of the proposed model, we have measured the latency, execution time, and throughput of the proposed model. Latency is defined in terms of a time delay to send access information from one point to another point. We have measured the latency of the proposed model at a different number of access (transactions). For this experiment, we have considered a maximum of 50 access (transactions). In this experiment, we have achieved average delay of the proposed model 42.8 ms. The achieved experimental result (latency) is shown in Fig. 3.

The second experiment is conducted to achieve the execution time of the proposed model. The execution time refers to the time taken for completion of a task. To achieve the result of this experiment, we have pass 50 numbers of requests at different time intervals. The average execution time is achieved by 103.7 ms. The execution time of the proposed model is shown in Fig. 4.

In the third experiment, we have measured the throughput of the proposed model. The throughput of the proposed model is measured in terms of transactions per second. We have taken multiple time interval (t_1, t_2, t_3, t_4,...,t_{10}) to measure the number of transactions per second. In each time interval, the number of transactions is varied due to the size and complexity of transactions.

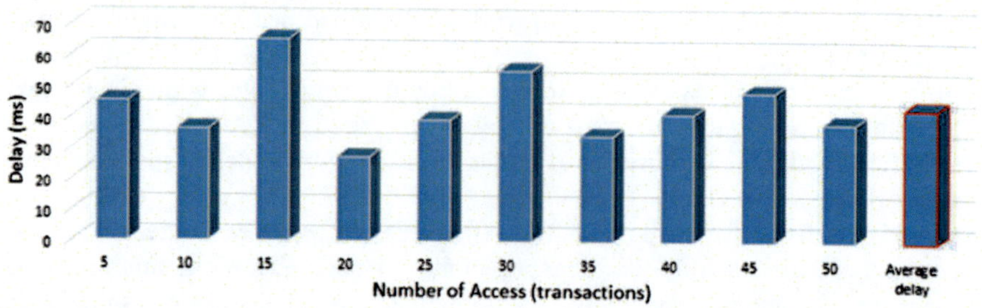

Fig. 3 Latency of the proposed model

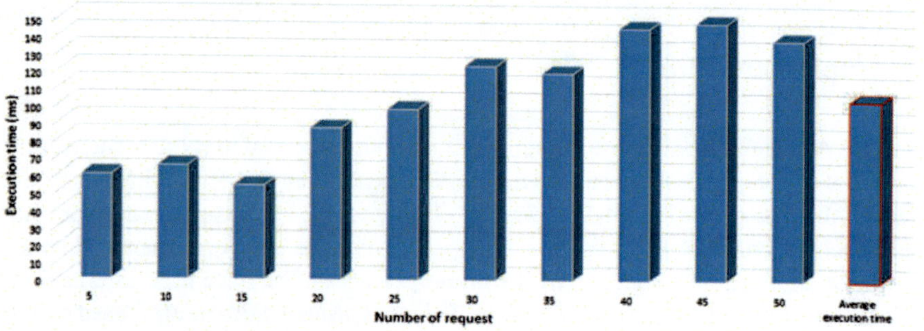

Fig. 4 Execution time of the proposed model

5 Security Analysis

In this section, we have analyzed the security strength of the proposed model. The security effectiveness of the proposed model illustrated that it overcomes all the security problems present in the current system. The following security points are considered to measure security effectiveness.

- **Secure data sharing**: The permission blockchain-based proposed system provides only authorized data access. Distributed public ledger (DPL) shares the same data among every node. Hence, any change in data on any node due to misbehaving and violate permission can be tracked easily.
- **Proof of concept**: Ethereum protocol is used for validation and evaluation by implementing a proof of concept tool in the blockchain technology-based access control model. With the help of a smart contract, the proposed model is deployed globally in a realistic testnet.
- **Security of the access API**: Access rule is defined in a smart contract, which is designed in presence of all nodes. Therefore, it is available on DPL. If an attacker wants to change anything, then installation and initialization of new code are required on all nodes, which is not possible discretely. Therefore, it protects from the hacking of malicious users or servers.

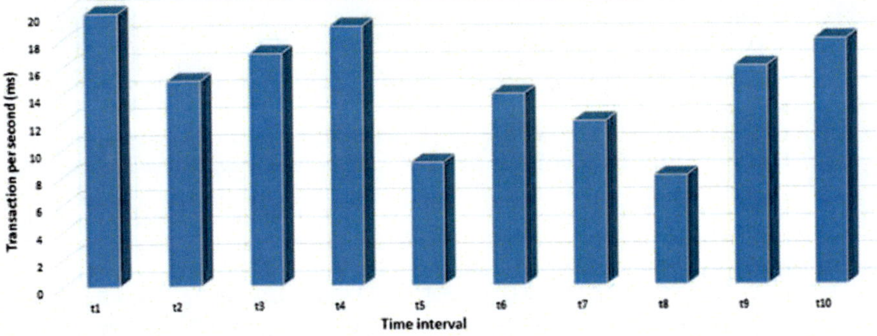

Fig. 5 Throughput of the proposed model

- **Controlled user access**: In case of emergency, if any authorized user has to give access control to other users then, blockchain technology introduces the feature of delegation. Hence, the private record of the user can be accessed by other users for a limited period.
- **Validation of access rule**: In blockchain technology, DPL consists of access rules in encrypted form. Hence, a miner can add a new block in the existing blockchain network, only if the access rule is successfully verified by all nodes. Otherwise, the created block is simply discarded.
- **Integrity and audibility of access rules**: The software (code) and hardware (devices) used for the proposed system are defined in the smart contract before the creation of a single node. Therefore, only predefined code and devices are used for further adding of the new block (Fig. 5).

6 Conclusion and Future Work

In this paper, we have investigated the security issues and the possible solutions in IoT-based applications. Access control is one of the security solution present in the IoT application. But, from the literature work, we have identified several limitations in the access control system. To overcome such issues, we have proposed a blockchain-based access control model which will cover almost all the security problem present in the access control system. The experimental results show the efficiency of the proposed model. Security analysis is also discussed which shows the effectiveness of the proposed model. In the future, we try to implement an access control model where nature and access privileges are dynamic. We have also tried to address more security issues using blockchain technology.

References

1. Nakamoto S (2009) Bitcoin: a peer-to-peer electronic cash system, may http://www.bitcoin.org/bitcoin.pdf
2. Mingxiao D, Xiaofeng M, Zhe Z, Xiangwei W, Qijun C (2017) A review on consensus algorithm of blockchain. In: 2017 IEEE international conference on systems, man, and cybernetics (SMC). IEEE, pp 2567–2572
3. Xu L, Shah N, Chen L, Diallo N, Gao Z, Lu Y, Shi W (2017) Enabling the sharing economy: privacy respecting contract based on public blockchain. In: Proceedings of the ACM workshop on blockchain, cryptocurrencies and contracts, pp 15–21
4. Thwin TT, Vasupongayya S (2018) Blockchain based secret-data sharing model for personal health record system. In: 2018 5th international conference on advanced informatics: concept theory and applications (ICAICTA). IEEE, pp 196–201
5. Ye N, Zhu Y, Wang RC, Malekian R, Lin QM (2014) An efficient authentication and access control scheme for perception layer of internet of things
6. Yao S, Chen J, He K, Du R, Zhu T, Chen X (2018) Pbcert: privacy-preserving blockchain-based certificate status validation toward mass storage management. IEEE Access 7:6117–6128
7. Novo O (2018) Blockchain meets iot: an architecture for scalable access management in iot. IEEE Int Things J 5(2):1184–1195
8. Wang H, Song Y (2018) Secure cloud-based EHR system using attribute-based cryptosystem and blockchain. J Med Syst 42(8):152
9. Maesa DDF, Mori P, Ricci L (2017) Blockchain based access control. In: IFIP international conference on distributed applications and interoperable systems. Springer (2017)
10. Makhdoom I, Zhou I, Abolhasan M, Lipman J, Ni W (2020) Privysharing: a blockchain-based framework for privacy-preserving and secure data sharing in smart cities. Comp Secur 88:101653
11. Rouhani S, Deters R (2019) Blockchain based access control systems: state of the art and challenges. In: IEEE/WIC/ACM international conference on web intelligence, pp 423–428
12. Ouaddah A, Mousannif H, Abou Elkalam A, Ouahman AA (2017) Access control in the internet of things: big challenges and new opportunities. Comput Netw 112:237–262
13. Thwin TT, Vasupongayya S (2019) Blockchain-based access control model to preserve privacy for personal health record systems. Secur Commun Netw
14. Shi S, He D, Li L, Kumar N, Khan MK, Choo KKR (2020) Applications of blockchain in ensuring the security and privacy of electronic health record systems: a survey. Comput Secur 101966
15. Maesa DDF, Mori P, Ricci L (2019) A blockchain based approach for the definition of auditable access control systems. Comput Secur 84:93–119
16. Alonso Á, Fernández F, Marco L, Salvachúa J (2017) Iaacaas: Iot application-scoped access control as a service. Future Internet 9(4):64
17. Cruz-Piris L, Rivera D, Marsa-Maestre I, De La Hoz E, Velasco JR (2018) Access control mechanism for iot environments based on modelling communication procedures as resources. Sensors 18(3):917
18. Ouaddah A, Abou Elkalam A, Ait Ouahman A (2016) Fairaccess: a new blockchain-based access control framework for the internet of things. Sec Commun Netw 9(18):5943–5964
19. Bertino E, Kundu A, Sura Z (2019) Data transparency with blockchain and ai ethics. J Data Inf Q (JDIQ) 11(4):1–8
20. Zhang R, Xue R, Liu L (2019) Security and privacy on blockchain. ACM Comput Surv (CSUR) 52(3):1–34
21. Mattis T, Hirschfeld R (2018) Activity contexts: Improving modularity in blockchain-based smart contracts using context-oriented programming. In: Proceedings of the 10th international workshop on context-oriented programming: advanced modularity for run-time composition, pp 31–38 (2018)

22. Tapas N, Merlino G, Longo F (2018) Blockchain-based iot-cloud authorization and delegation. In: 2018 IEEE international conference on smart computing (SMARTCOMP), pp 411–416. IEEE (2018)
23. Novak M (2019) Crypto-friendliness: understanding blockchain public policy. J Entrepreneurship Pub Pol (2019)
24. Kang J, Yu R, Huang X, Wu M, Maharjan S, Xie S, Zhang Y (2018) Blockchain for secure and efficient data sharing in vehicular edge computing and networks. IEEE Int Things J 6(3):4660–4670
25. Emmadi N, Narumanchi H (20174) Reinforcing immutability of permissioned blockchains with keyless signatures' infrastructure. In: Proceedings of the 18th international conference on distributed computing and networking, pp 1–6 (2017)

An IoT-Based Smart Garbage Segregation System Using Deep Learning

Subham Divakar, Abhishek Bhattacharjee, and Rojalina Priyadarshini

Abstract Deep learning (DL), machine learning (ML), computer vision, and Internet of things (IoT) have played a significant role in innovation of new intelligent and smart systems having digital eyes and brains. These modern systems are capable of taking their decisions on their own. Earlier systems designed were only using one technology but think about the modern IoT systems having the power of computer vision, deep learning, and AI. Our proposed work tends to combine the power of DL and IoT to propose a unique waste segregation technique that comes with the least sensors and fast decision-making capability with easy installation. It easily separates organic, recyclable, and electronic waste, thus making it eco-friendly and contributing toward a greener environment. Our major contribution is the waste dataset which contains waste belonging to three categories organic, recyclable, and electronic waste along with the novel IoT-enabled smart dustbin which uses cloud technology and is power efficient.

Keywords Internet of things · Deep learning · Waste management · Cloud computing

1 Introduction

With the rising population of the world, the waste generated by us also keeps on rising. Waste in general can be classified into organic waste, solid waste, recyclable waste, non-recyclable waste, and e-waste which is the most hazardous and rising day by day. With the rise of smart-phone, laptops, TV, computers, or electronic gadgets users so is the e-waste increasing which needs to be handled carefully. In the USA, there are machines just like ATM machines which collect your old smart phones and

S. Divakar
Persistent Systems, Pune, India

A. Bhattacharjee
Cognizant Solutions, Bengaluru, India

R. Priyadarshini (✉)
C.V. Raman Global University, Bhubaneswar, India

K. Dahal et al. (eds.), *Internet of Things and Its Applications*, Lecture Notes in Electrical Engineering 825, https://doi.org/10.1007/978-981-16-7637-6_12

immediately return some cash amount in exchange for your broken smart phones [1, 2]. This is a new and innovative thinking which reduces e-waste. However, that is not the case for all smart phones and electronic devices. In today's world, numerous campaigns are running to alert people about proper waste collection and segregation followed by waste dumping safely [3]. Even though it is the responsibility of the government to collect waste properly, we face another challenge before this. Waste segregation has to be done properly and carefully before we even hand it over to the people collecting it in our area. This solves two problems. First not only the waste is segregated but also it becomes easier for the authorities to dump them properly. Thus, waste segregation becomes a vital problem that needs to be solved before we dump the waste into the dustbins. However, with the booming market of IoT, AI, ML, and DL we can innovate some smart system that can do this waste segregation itself [4–6]. IoT makes an equipment smart and intelligent, AI gives digital brains and eyes to a system, so why not combine these and solve the problem of waste segregation and technology has advanced so much that nothing seems impossible now. Numerous research works have been done and are also in progress in these fields that will help in maintaining an eco-friendly environment [7]. Our proposed work is an IoT and DL-based smart waste management system that reduces the task for humans to segregate waste by performing automatic waste segregation into three categories—organic, recyclable, and e-waste. This smart system can be used at home or offices or wherever there is a need to segregate waste. The waste segregation part is a deep learning classifier that takes images of waste and classifies them into one of the three categories of recyclable, organic, or e-waste. We used an existing dataset [8] and added a new category of e-waste into it, and thus, the new dataset became ready for use [9]. Using this dataset and using transfer learning approach for training the model, we obtained an accuracy of 88–90% with the pre-trained models. We tested the models against the original dataset and then with our newly created dataset and upon comparison the newly created dataset produced more accurate results than the original dataset. Also, the IoT mechanism proposed in this paper uses least sensors, and this makes our system less complex to install as compared to the existing systems.

The rest of the paper is divided into six sections. The first section is the introduction, the second section contains related works, and third section describes our proposed work in details followed by the fourth section which is implementation. The fifth section deals with results and discussion, and the last section talks about conclusion and future scope.

2 Related Work

We performed some literature survey in which we studied some of the top related works done in the domain of waste management and IoT. Debajyoti Mishra and et al. in their paper [10] demonstrated the use of ultrasonic sensors and gas sensors for the detection of hazardous gases, checking whether the dustbin is full or not. The use of sensors in waste management is studied in this section, which is discussed over here [8]. Abhimanyu Singh et al. in their paper used Raspberry Pi board to collect the status of the dustbins and conveyed it to the corresponding waste managers. The use of hardware like Pi boards and Arduino Boards has been found to be widely used for the purpose of IoT. [11]. In Hong et al. [12], the authors have come up with an IoT-based smart garbage system for food management. The solution focused mainly on the collection of food waste and saving cost. In [13], authors have discussed an IoT approach in which they did similar work as [12] but their focus was dustbins. S. Vinoth Kumar, T. Senthil Kumaran his co-authors in their paper checked the waste level of garbage bins using sensor systems [14]. In Kumar et al. [14] and Bharadwaj et al. [10], the authors prime focus is to come up with systems that can alert when the dustbin is full and used various techniques General Packet Radio Service (GPRS)/Global System for Mobile Communication (GSM) or Message Queuing Telemetry Transport (MQTT) connections for transmitting the data.

Most of the work done tries to find out the condition when the dustbin is full or not and simply alerts when the condition arises. For the purpose of alerts, various techniques like MQTT, Wi-Fi, ZigBee, and other technologies are used to convey this alert. However what we found out missing was that none of the work tried to segregate the waste into several categories of waste. Also, most of the work used IoT alone and did not combine any other technology like cloud computing, deep learning, machine learning to leverage and create a more smart system. Our proposed work solves this issue of automatic waste segregation and performs effectively in real time. It is an end-to-end waste segregation and alert system that uses cloud for communication which is more effective and useful. The key aspect of our work is the automatic waste segregation which is done by the classifier which is present on cloud and which takes images from the dustbin and classifies them into one of the three categories of organic, recyclable, and e-waste.

3 Proposed Work

In this paper, we have proposed a deep learning and IoT-based waste collection and segregation technique which does the following tasks:

- Automatic waste segregation.
- Alert system for full dustbin via android application and notifications to users or authorities.

The workflow diagram depicted in Fig. 1 has 3 major components which explain the entire process from waste collection to alerting system.

- **Cloud**: The cloud is the central part of our proposed system that holds the data and the classifier. The images taken from the camera module are sent to the cloud where the classifier sends the classification result back to the bin. For our implementation purpose, we used AWS cloud. Also, it is the cloud which stores

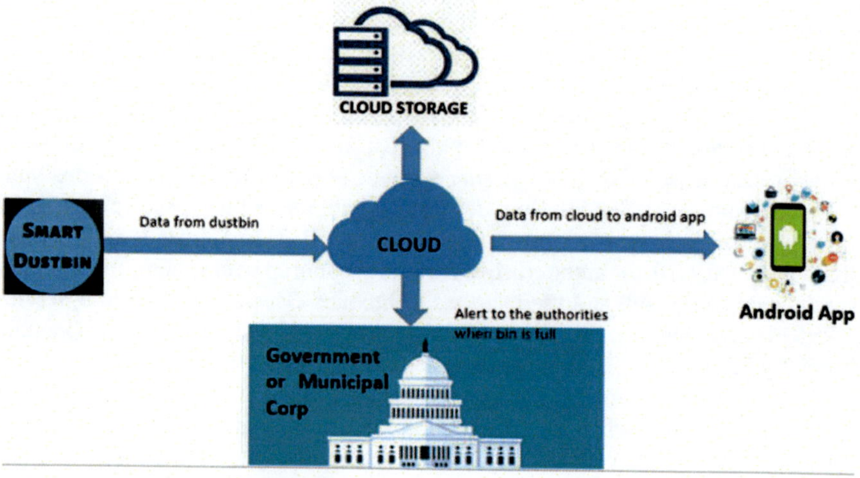

Fig. 1 Data captured by smart dustbin is transported to cloud; after processing, as notification it moves to user through android application or Government Municipal corporation, and it is persistently stored in cloud

the dustbin status whether it is full or not and sends the alert to the government authorities and the android application of the user.

- **Smart bin**: The smart bin contains Raspberry Pi, an ultrasonic sensor to find full dustbin, and a camera module for taking live snapshots of the waste put on the tray of the bin. The camera module picks up snapshots, and the deep learning model classifies the waste and accordingly the tray tilts to the R or O side.
- **Android Application**: The android application developed for us shows the status of the dustbin along with the previous quantity of waste generated by a particular user. It requires Google sign which is mandatory and pulls the data from the cloud storage.

Figure 1 depicts the working of the smart bin from waste collection to segregation. From this, it is clear that the cyclic process starts with the camera module taking the snapshots of the waste which is sent to the cloud via Internet where the classifier is present. Now the classifier predicts the category of the waste, and the result is sent back to the smart bin whose job is now to put the waste in the corresponding area inside the bin. Then the bin checks whether the bin is full or not and sends the alerts accordingly. Thus, this cyclic process continues and the smart bin waits for further tasks, this has been shown in Fig. 2.

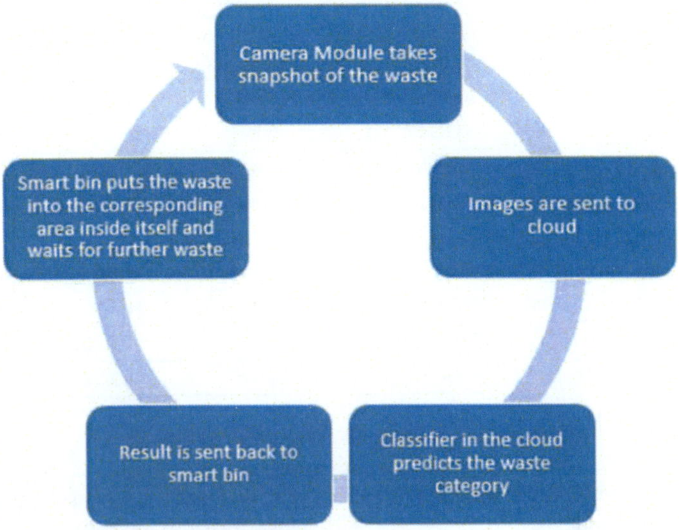

Fig. 2 Flow diagram of proposed smart bin

3.1 Hardware Implementation

Our proposed work tends to reduce complexity thus uses only one Raspberry Pi module, one motor, one camera module, and one ultrasonic sensor. The camera is connected to Raspberry Pi which is connected with the Internet via Wi-Fi module which takes snapshots of the waste and sends it to the classifier which classifies it, and accordingly, the waste is put into proper bin.

3.2 Deep Learning-Based Waste Segregation

Deep learning and IoT have become popular due to the recent advancements in computing power. We in our proposed work have leveraged this fact. For our proposed work, we took the public available dataset from Kaggle [8] which contains two categories of data belonging to O—organic waste and R—recyclable waste only. It contains data already divided into train and test categories. The train part contains 9999 images belonging to O category and 12,565 images belonging to R category. The test part contains 1401 images belonging to O category and 1112 images belonging to R category. First we used transfer learning approach to train four models and generated some results to see which model performed best on this actual dataset and also noted the highest accuracy obtained from this dataset which is presented in Table 1.

Table 1 Accuracy, precision and F1 Score for Inception, VGG 16 and VGG 19 Model On Customized Dataset

Used model types	Accuracy in percentage	precision	F1 score
Inception	0.90	0.90	0.89
VGG19	0.863	0.857	0.8577
MobileNetsV2	0.86	0.876	0.8557
VGG16	0.86	0.858	0.85
On Actual Dataset			
Inception	0.8917	0.8976	0.8903
VGG19	0.8746	0.8802	0.8729
MobileNetsV2	0.8746	0.8802	0.8729
VGG16	0.873	0.8784	0.8713

Now using the actual dataset, we decided to add a new a third category of waste which is very common e-waste. We collected some random e-waste images from Google search and also collected some images from our own houses which are presented in Figs. 3 and 4. In total, we collected 150 images out of which 124 were kept for training and remaining 26 were kept for testing. Thus, we have used train set for training and test set for testing, upon which the accuracy of the classifier is evaluated. We have used the process of transfer learning for training the models because not only it saves time but also it is very efficient in terms of accuracy when the dataset contains images belonging to general category of images. We have also used the process of fine-tuning the layers of the pre-trained models for achieving better accuracy. We have used four pre-trained models:

Fig. 3 Samples from newly created data.

Fig. 4 Samples from newly formed dataset

- Inception.
- VGG 16.
- VGG 19.
- MobileNetsV2.

For this work, we have used the models trained on ImageNets dataset. The results obtained from each model are discussed along with their comparison on our new dataset. Figure 5 shows the various results and outcomes obtained from the training and evaluation of Inception model. Figure 6 shows the various results and outcomes obtained from the training and evaluation of VGG16 model. Figure 7 illustrates the various results and outcomes obtained from the training and evaluation of VGG19 model. Figure 8 shows the various results and outcomes obtained from the training and evaluation of MobileNetsV2 model. These are pre-trained models. For our paper, we have used the models trained on ImageNets dataset. The full code is available on Kaggle and can be accessed from the link [9]. The results obtained from each model are discussed below along with their comparison. Also, we have shown two states of each model and the accuracy after fine-tuning. Figure 5 is showing the plots of training accuracy versus validation accuracy and training loss versus validation loss for Inception model. Likewise Figs. 6, 7, and 8 present the accuracy and loss functions for VGG16 model, VGG19 model, and MobileNetsV2 model, respectively.

4 Results and Discussion

The entire code for classification is written in python code can be found at Kaggle notebook, and new dataset created by for this work is available at url [9].Table

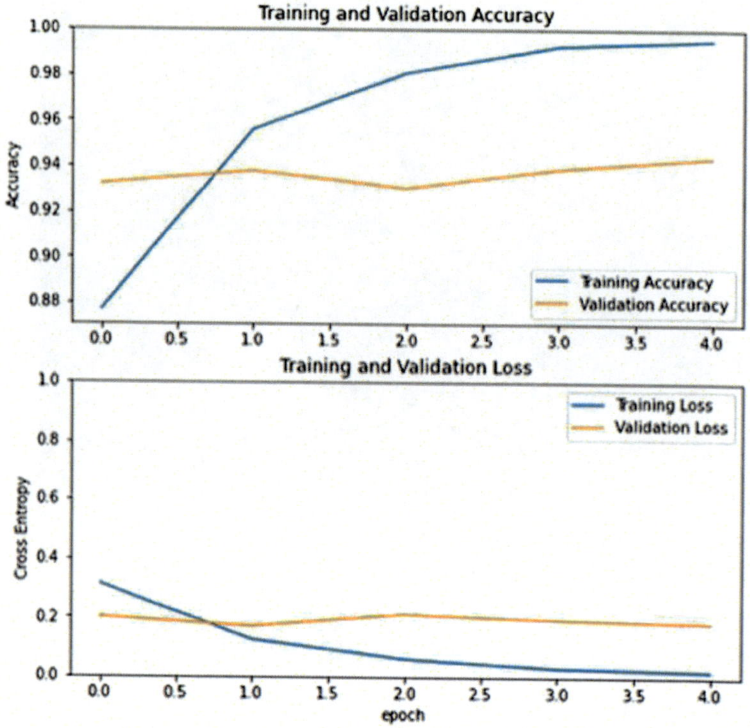

Fig. 5 Plots of accuracy and cross-entropy for Inception model

1 presents the comparison of all four models along with their accuracy, F1score, precision which is sorted in decreasing order of their accuracy. From this table, we can clearly figure out that MobileNetsV2 with accuracy of 89.99% has achieved the best accuracy on the actual dataset. From Table 1, it is evident that MobileNetsV2 is the best model as it has the highest accuracy among all the models. This result has been obtained on a separate test dataset as mentioned in Sect. 3.1. Also, we cannot rule out that this high accuracy is almost similar for the top two models and thus both could perform well in the real scenarios. However for the purpose of implementation, we used MobileNetsV2 since it had the best accuracy. From Table 1, it is evident that Google Inception model is the best model as it has the highest accuracy among all the models. This result has been obtained on a separate test dataset as mentioned in Sect. 3.1. This entire test was performed on the new dataset. Thus, we can clearly see that MobileNetsV2 performed better on the actual dataset and Inception model performed better on the new dataset but the accuracy obtained on the new dataset is slightly greater than the accuracy obtained on the actual dataset.

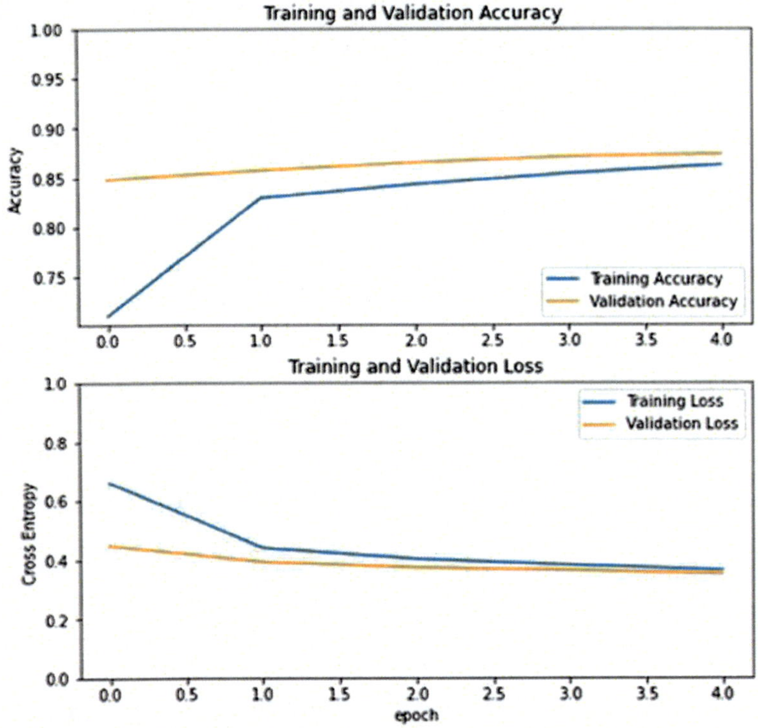

Fig. 6 Plots of accuracy and cross-entropy for VGG16 model

5 Conclusion and Future Scope

In this work, transfer learning technique was applied onto two datasets [5, 8] to come up with classifiers that classify waste into one of the three categories. Initially, four models were trained using transfer learning on the actual dataset that had two categories of waste recyclable and organic waste and the results obtained were mentioned in Table 1 which shows MobileNetsV2 as the best model with an accuracy of 89.993%. We introduced a new category of waste by collecting pictures on our own and named it as e-waste category. Again the same four models were trained on our new dataset, and the results were presented in Table 1. This time Inception model got the highest accuracy of 90%, and this accuracy is slightly greater than the previous highest accuracy of 89.993% by MobileNetsV2. Thus, our proposed system has the capability to perform automatic waste segregation and it also does its tasks accurately as it combines the power of IoT and deep learning which gives it digital eyes and brains.

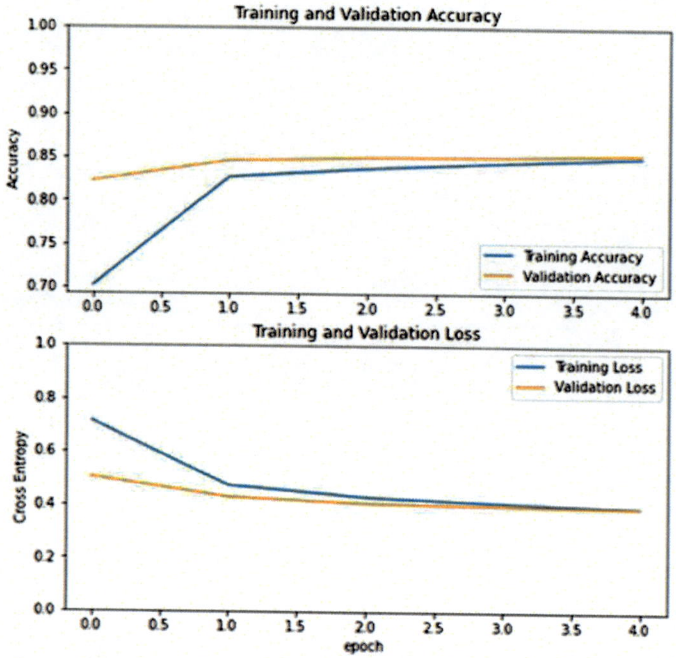

Fig. 7 Plots of accuracy and cross-entropy for VGG19 model

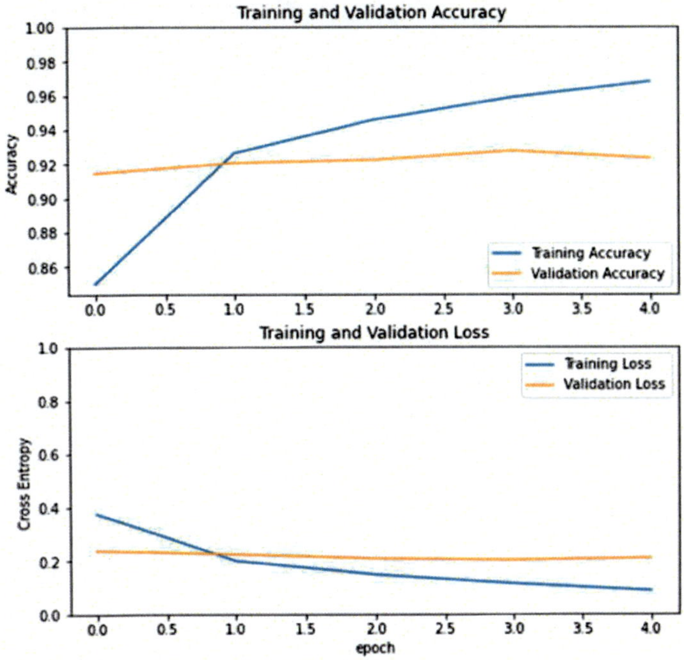

Fig. 8 Plots of accuracy and cross-entropy for MobileNetsV2

References

1. Fan YJ, Yin YH, Da Xu L, Zeng Y, Wu F (2014) Iot-based smart rehabilitation system. IEEE Trans Industr Inf 10(2):1568–1577
2. Roy DS, Behera RK, Reddy KHK, Buyya R (2018) A context-aware fog enabled scheme for real-time cross-vertical Iot applications. IEEE Internet Things J 6(2):2400–2412
3. Marques P, Manfroi D, Deitos E, Cegoni J, Castilhos R, Rochol J, Pignaton E, Kunst R (2019) An Iot-based smart cities infrastructure architecture applied to a waste management scenario. Ad Hoc Netw 87:200–208
4. Pardini K, Rodrigues JJ, Kozlov SA, Kumar N, Furtado V (2019) Iot-based solidwaste management solutions: a survey. J Sens Actuator Netw 8(1):5
5. Pradhan B, Vijayakumar V, Pratihar S, Kumar D, Reddy KHK, Roy DS (2020) A genetic algorithm based energy efficient group paging approach for Iot over 5g. J Syst Archit 101878
6. Priyadarshini R, Barik RK, Panigrahi C, Dubey H, Mishra BK (2020) An investigation into the efficacy of deep learning tools for big data analysis in health care. In: Deep learning and neural networks: concepts, methodologies, tools, and applications. IGI Global, pp 654–666
7. Xu B, Da Xu L, Cai H, Xie C, Hu J, Bu F (2014) Ubiquitous data accessing method in iot-based information system for emergency medical services. IEEE Trans Industr Inf 10(2):1578–1586
8. Misra D, Das G, Chakrabortty T, Das D (2018) An Iot-based waste management system monitored by cloud. J Mater Cycles Waste Manage 20(3):1574–1582
9. Diwakar S, Bhattacharya AR (2020) DataSet and Code Link. https://www.kaggle.com/shubhamdivakar/ai-and-iot-based-system-conference-paper. Accessed 28 Oct 2020
10. Bharadwaj AS, Rego R, Chowdhury A (2016) Iot based solid waste managementsystem: a conceptual approach with an architectural solution as a smart city application. In: 2016 IEEE annual india conference (INDICON). IEEE, pp. 1–6

11. Singh A, Aggarwal P, Arora R (2016) Iot based waste collection system using infraredsensors. In: 2016 5th international conference on reliability, infocom technologies and optimization (Trends and Future Directions) (ICRITO). IEEE, pp 505–509
12. Hong I, Park S, Lee B, Lee J, Jeong D, Park S (2014) Iot-based smart garbagesystem for efficient food waste management. Sci World J
13. Malapur B, Pattanshetti VR (2017) Iot based waste management: an application tosmart city. In: 2017 international conference on energy, communication, data analytics and soft computing (ICECDS). IEEE, pp 2476–2486
14. Kumar SV, Kumaran TS, Kumar AK, Mathapati M (2017) Smart garbage monitoring and clearance system using internet of things. In: 2017 IEEE international conference on smart technologies and management for computing, communication, controls, energy and materials (ICSTM). IEEE, pp 184–189

Container-Based Lab-as-a-Service Application

S. Thiruchadai Pandeeswari, S. Padmavathi, M. Sanjaybabu, S. S. Srilakshmi, and K. Sabari Priya

Abstract In most of the Indian engineering and science educational institutions, variety of courses on problem-solving through programming and computer-aided tools are taught. These courses often have theory and practical components. The practical components are conducted in dedicated laboratories equipped with necessary software and tools. Students need to work in the respective labs to carry out the exercises. Lab setup with diverse requirements restricts the users to depend on the lab facility. In order to allow the users to access the lab environment beyond the working hours and eliminate the dependency on a particular lab facility, number of solutions leveraging virtualization and cloud services have been provided in the past. However, these approaches have few drawbacks. The virtual images are quite heavy. Though portable, it takes quite sometime to configure the virtual images. Similarly, cloud-based solutions may not be affordable for every institution and given the scale of usage, the cloud services may be expensive. In this paper, we propose a simple container-based solution for providing customized lab environments for the users. Containers are lightweight, portable, and easily configurable when compared to virtual image formats. The proposed container-based solution is presented to the users through a simple web portal that allows users to create required lab environment with few clicks. The container-based lab-as-a-service application presented in this paper incorporates necessary settings to save the changes made on the given image by the user and also allows the user to save and retrieve his customized image as and when required. The users do not require knowledge on Docker commands and do not require to use command-line interface for getting the lab setup ready using containers. The web portal's UI allows the user to create lab environments without having to worry about the intricacies of tools required for the lab setup and containers. The proposed solution has been implemented in one of our department labs, and the results are evaluated and presented.

Keywords Virtual labs · Containers · Docker · Lab-as-a-service · Virtualization

S. Thiruchadai Pandeeswari (✉) · S. Padmavathi · M. Sanjaybabu · S. S. Srilakshmi · K. Sabari Priya
Department of Information Technology, Thiagarajar College of Engineering, Madurai, Tamil Nadu, India
e-mail: eshwarimsp@tce.edu

K. Dahal et al. (eds.), *Internet of Things and Its Applications*, Lecture Notes in Electrical Engineering 825, https://doi.org/10.1007/978-981-16-7637-6_13

1 Introduction

Computers and programming have become essential part of learning engineering. Especially the domains such as computer science, information technology, cyber security, data analytics, telecommunications rely severely on the use of computers. Learners have number of practical courses that deal with programming and application development in their curriculum. These practical courses are conducted in designated laboratories that are equipped with necessary computer software and tools. For example, for learning system administration with Linux, a lab with machines that are loaded with Linux operating system is required. Similarly, for learning web development with PHP and JavaScript, machines that are loaded with necessary libraries of PHP is required. The same applies to Python programming and similar other programming courses. In most of the Indian educational institutions, students do hands-on experiments in designated labs. Extensive arrangements in terms of scheduling and equipping the lab with necessary software are carried out, to make the lab available for different sets of students at least for a minimum number of hours every week. This leads to number of challenges such as.

1. Difficulty in setting up labs with diverse requirements as each lab setting requires different set of software tools.
2. Inability to access the lab during non-working hours.
3. CAPex involved grows higher for creating dedicated lab environments for every practical course offered.
4. A laboratory with n number of computers may not be able to support different practical courses if the software requirements of the courses are conflicting with one another.

It is not affordable to create physically dedicated lab environments for every practical course. Number of solutions that leverage virtualization and cloud services has been proposed to overcome the above-mentioned challenges. Working with virtual machines in the given context has been proved viable in the past. In this paper, a novel solution for creating customized programming and application development lab environments especially for educational institutions has been proposed. The solution proposed makes use of containerization, instead of virtualization for creating customized programming and app development environments for individual students. Related works are studied and presented in Sect. 2. Section 3 deals with the design of the solution, and Sect. 4 explains the implementation. The results are shown in Sect. 5.

2 Related Work

The problem of provisioning virtual labs has been solved in number of ways earlier. One of the landmark solutions for provisioning virtual labs is given in [1]. In [1], a web interface is provided for the users to request their images of choice. The image refers to either a virtual image or a bare metal image. The cloud management tool, xCAT is used for loading the requested image by the user on to the servers. Virtual machines are leveraged for creating flexible lab environments at low cost in [2]. VM on-demand tool has been developed to overcome the problems of using multiple virtual machines at a shared computing facility in [2]. The solution proposed in [2] allows a user to walk-in to the lab with the necessary VM image in removable media like USB or DVD. The machines in the laboratory are equipped with customized tool written in POSIX shell command language. This tool guides the user to configure the required environment.

Performance of virtual machines and containers while executing similar workloads have been analyzed in [3]. It is observed in [3] that Docker containers' number of transactions per second is higher than that of virtual machines while executing similar workloads. It is highlighted in [3] that the performance of Docker containers is 30% more than that of virtual machines.

A virtual lab project that provides interactive environment for the user to carry out the tasks that are usually carried out in lab facility is presented in [4]. The project is funded and assisted by MHRD's NMEICT initiative. The projects provide number of lab experiments virtually over Internet. It leverages web technology to provide animation and simulation-based experiments to large userbase.

Usability of the virtual labs has been examined in [5]. Extensive checklists for ensuring technical usability of the virtual lab and pedagogical usability of the lab are presented in [5].

3 Motivations

The solution developed and presented in this paper has the following objectives.

1. Develop an application using containerization tools to deliver various programming and application development lab environments virtually.
2. Remove dependency on a particular lab facility for carrying out programming lab experiments and allow students to carry out experiments from anywhere in the campus.
3. Replace the existing virtual machine-based solutions with lightweight containerization-based **solution.**

4 Design

4.1 Containers versus Virtual Machines

Virtualization has been proved as an effective solution for creating customized execution environments. As explained in Sect. 2, numerous solutions have been proposed in the past leveraging virtual machines for creating execution environments. However with the advent of very advanced containerization technologies, containers have proved flexible and effective in providing customized execution environments. Containers pack all the necessary dependencies into one single unit and allow the applications to run seamlessly. Virtual machines (VM), when used in the context of providing customized lab environments in educational institutions, require virtual machine managers to be installed on all the machines. The VM image can be ported as and when required. The VM images may also be stored and served from a centralized server. However, the size of the VM image is very large when compared to that of images used by Docker engine. Also, the time involved in porting or configuring virtual image on top of a virtual machine manager (VMM) is relatively high. On top of Docker engine, any required image can be pulled and run as containers within few seconds. Also, the time taken to boot is very less for the containers when compared to that of virtual images. Hence, the proposed lab-as-a-service application makes use of Docker images and containerization for creating customized lab environments.

4.2 Application Architecture

The application architecture is explained in top-down fashion. The users can access the application through a web portal which has user-friendly GUI. The users of this application are segregated into classes, viz. admins and students. The admin users are the lab administrators who are responsible for the installation and monitoring of various tools in the lab machines as per lab and user requirements. To begin with, the admin users can create images with necessary software as per the requirements of the lab. The images are pulled from Docker hub, and necessary changes are made on the image. Say for example, for Java programming lab, a base Linux image is pulled from Docker hub and necessary JDk files are added to the image. The image is then made available for students over the local file server. The student users would be able to pull image of their choice or requirement from the server and add their work. They can save the state of their container using the commit options. The user's image would be pruned and saved in the server against the user account. A simple web portal is designed in such a way that the application's user credentials are saved along with their images. From anywhere in the LAN, the users will be able to access the application and pull their own customized images and create their lab environment within few clicks.

In order to make use of the available storage space in the file servers effectively, the images are pruned, compressed into a tar file and stored in the server. Also, each user is given limited amount of storage which is sufficient to cover all the practical course-related images. The unused images are monitored and deleted after a stipulated amount of time. All the activities involved in the workflow can be carried out by choosing necessary options in user interface and clicking few buttons. Users do not need to work on command interface to get the lab setup done. The flow of activities that can be carried out by the admins and students is given in Fig. 1. The figure explains the various activities the admin can carry out on the application for setting up an image for a lab course. Also, the maintenance activities, such as removing the rarely used images, sending out alert messages to the users are also shown in the diagram.

The steps involved in getting the images customized for the lab requirements are shown in the block diagram Fig. 2. The necessary dependencies must be imported into the image by writing appropriate Docker files. By using appropriate Docker commands the images can be instantiated and run as containers. However, to do these steps with Docker, one should have good knowledge on Docker commands and must be comfortable with command-line interface.

In our proposed application, these complex activities like pulling the base image from Docker hub, getting customized images ready by writing appropriate Docker files have already been carried out for a selective set of lab environments and are made available to the admin users through the web console. The admin users can select a lab environment based on the requirements in a given semester and make the corresponding image available to student users.

The student users with their access to the application will be able to choose the necessary lab environment and create them with few clicks. The activities that can be carried out by the admin and student users are shown in Figs. 3 and 4, respectively.

All of the above-mentioned activities can be fired up from the user interface of the application, without the need to use command-line interface and Docker commands. The state of the images as they get manipulated by the users of the application are shown in Fig. 5.

5 Implementation

The web application designed for delivering container-based lab-as-a-service is implemented using Apache Http Server, MySQL database, and PHP for backend scripting. The proof of the concept developed includes provisions for five programming labs for the following courses: Web Development, C programming, C++ Programming. Python Programming, Java Programming.

A server in cloud systems laboratory of the Department of Information Technology in our institution has been used for deploying the lab-as-a-service application. The server has 32 GB RAM and 4 TB hard disk. In order to be able to successfully

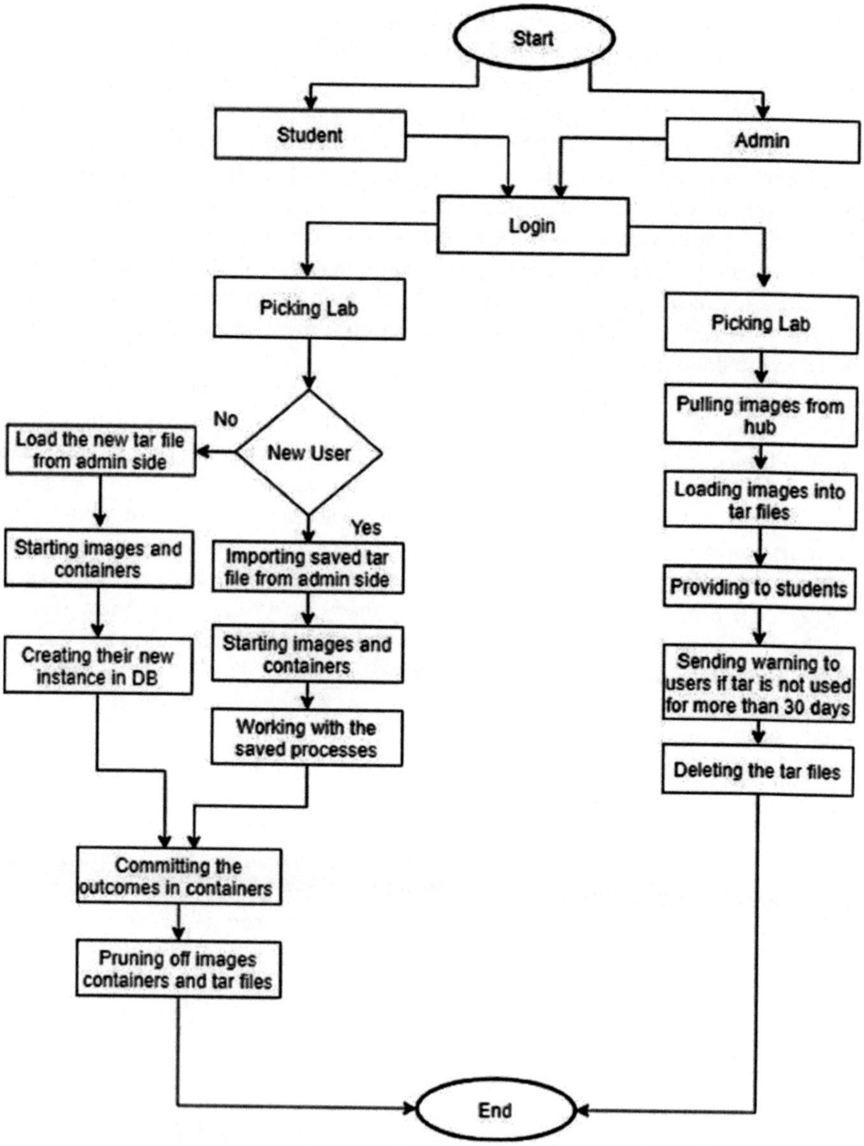

Fig. 1 Workflow of container-based lab-as-a-service application

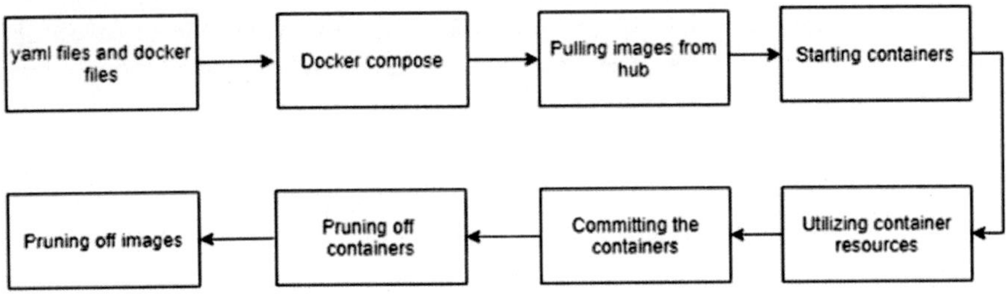

Fig. 2 Steps involved in getting ready-to-use images

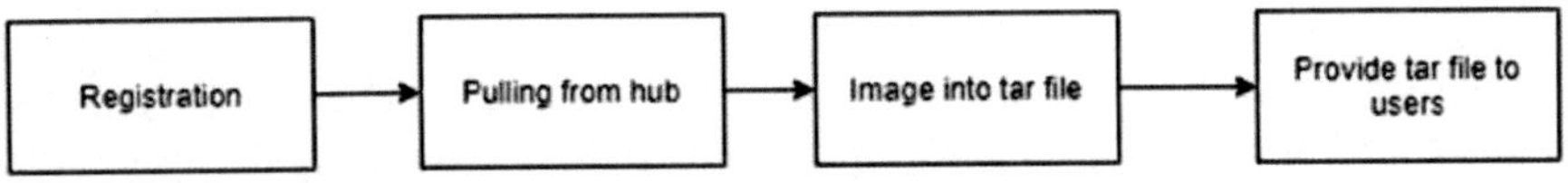

Fig. 3 Activities of admin user

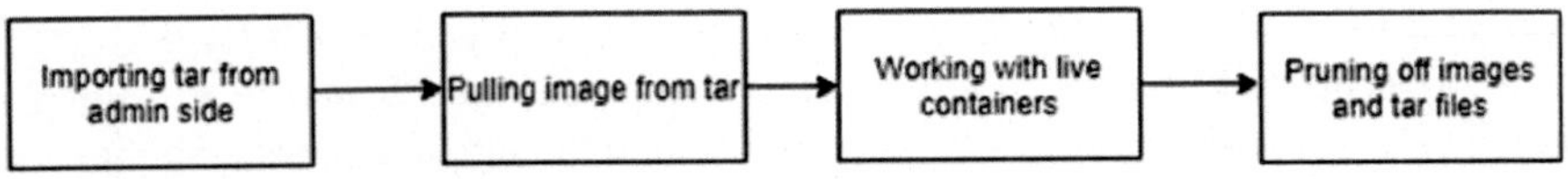

Fig. 4 Activities of student user

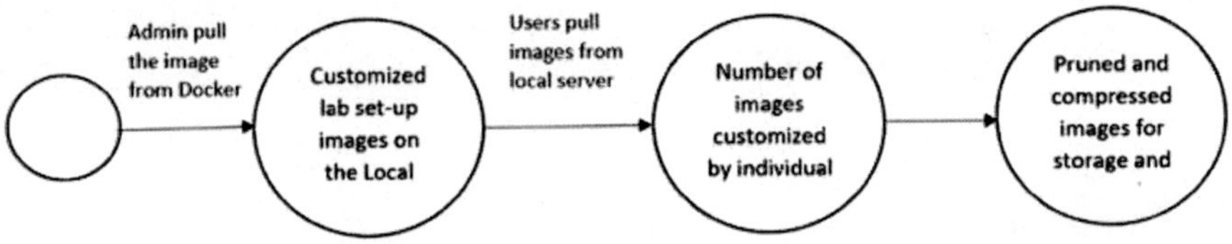

Fig. 5 Transition of images in the application

configure execution environments in few seconds by using the proposed application, all the machines in the cloud system lab were equipped with Docker engine.

Other than the cloud systems laboratory, the information technology department also has seven other labs. Each lab is equipped with 35 computers with 8 GB RAM. These labs are utilized for conducting a minimum 9 practical courses concurrently during a semester. Some labs host more than 2 practical courses for over 280 students spread in eight batches during a semester. Conducting these practical courses requires meticulous planning to avoid overlapping of batches. The machines in the laboratory are also used for other purposes such as projects and other trainings. So the use of labs by students was constrained before the implementation of the lab-as-a-service application.

After the implementation of the application, students were able to access the required lab environment by accessing the application and configuring the images on the available machine as and when needed, without any constraint.

6 Results and Discussion

The UI pages are shown in the following figures below. There are separate logins for admin users and student users. The lab options available for admins for configuration are shown in Fig. 6c. The lab options available for students are shown in Fig. 6d.

The UI page for setting up web development lab environment is shown in Fig. 6c.

The performance of the container-based lab as a service application is measured using the parameters such as size of the image file and CPU utilization. The virtual image size for various programming environments is compared with that of Docker images in the graph given in Fig. 7.

For the sake of comparing the performance of container-based lab setup against the VM-based lab setup on student machines, a laptop with AMD Ryzen5 1600 Six-core processor is taken. The CPU utilization of the machine while executing a simple Hello world printing Java program through the lab-as-a-service application is shown in Fig. 8.

CPU utilization of the same machine, while executing the Java program on a virtualized environment provided by VM, is shown in Fig. 9

Thus, the proposed application is found to reduce the actual size of the image on disk and also the CPU utilization.

7 Conclusion

Container-based lab-as-a-service application is found to reduce the time involved in setting up the required lab environment to a greater extent. It simplifies the application development and deployment environment setup by leveraging containerization for creating the environments. The web console designed as part of the application eliminates the need to use complex commands and allows user to enjoy the benefits of containerization without having to worry about meddling with the command-line interface. It also eliminates the dependency on a particular lab facility for carrying out experiments.

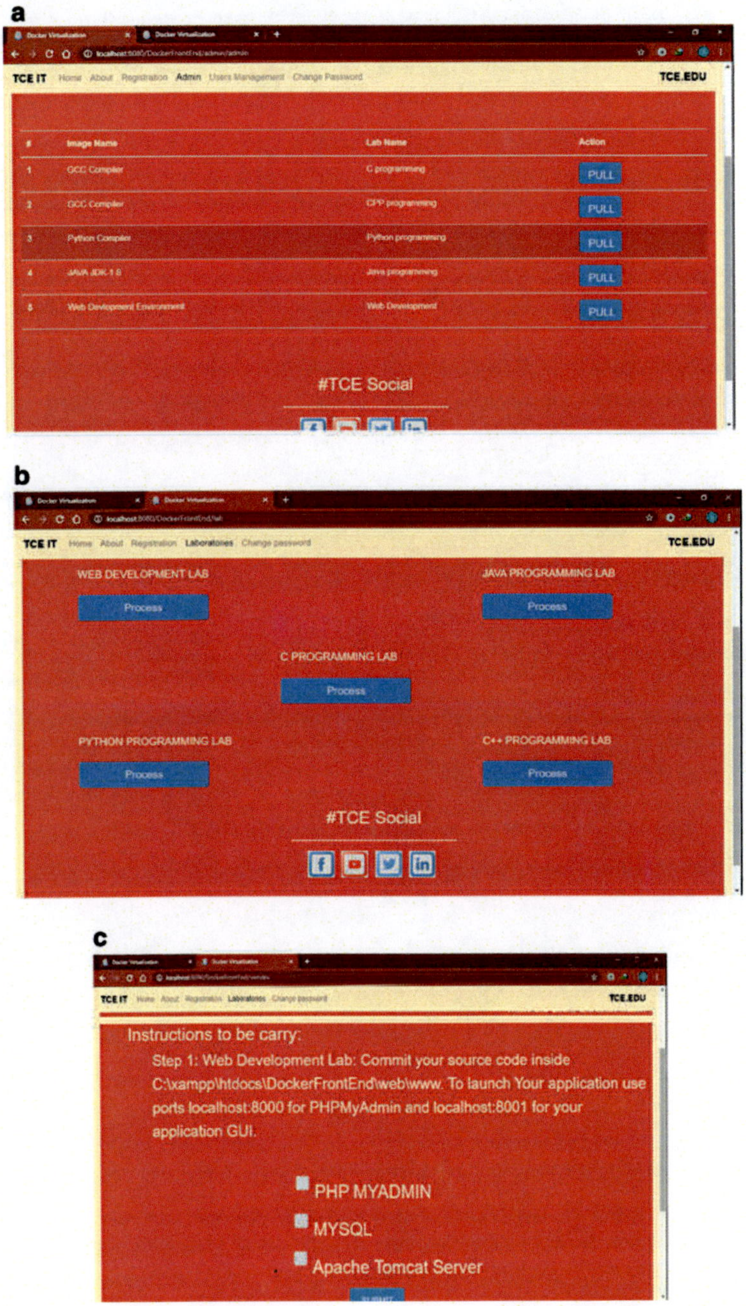

Fig. 6 **a** Lab configuration options available for admin users **b.** Lab setup options available for students. **c** Web development lab setup

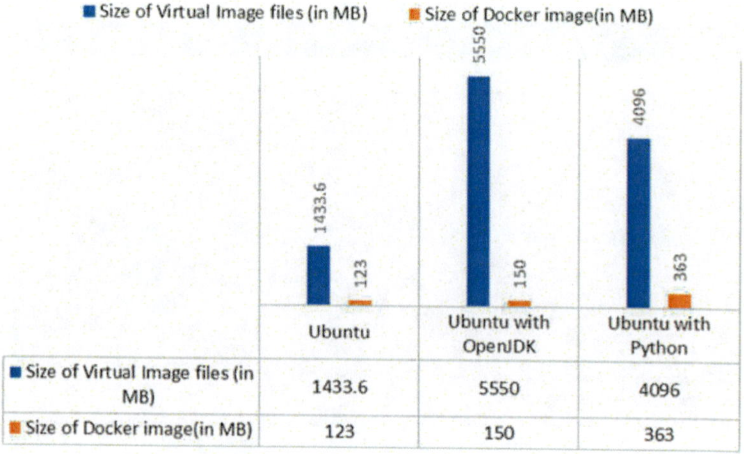

Fig. 7 Comparison of virtual image file size and Docker image file sizes for various lab environments

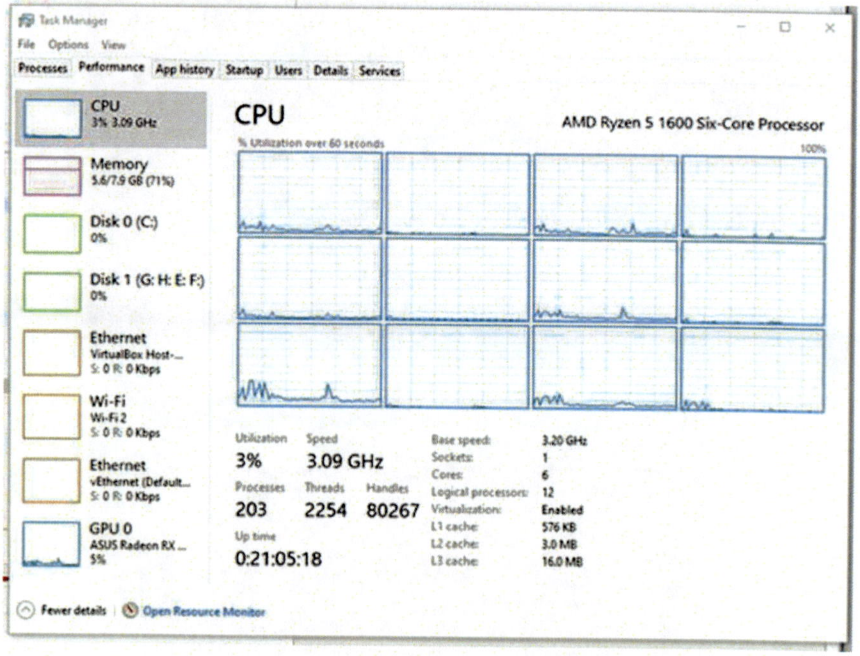

Fig. 8 CPU utilization during execution of Java program through our application

Fig. 9 CPU utilization during execution of Java program on a virtual machine

References

1. Schaffer HE, Averitt SF, Hoit MI, Peeler A, Sills ED, Vouk MA (2009) NCSU's virtual computing lab: a cloud computing solution. 42(7):94–97, Computer
2. Muñoz-Calle J, Fernández-Jiménez FJ, Ariza T, Sierra AJ, Vozmediano JM (2016) Computing labs on virtual environments: a flexible, portable, and multidisciplinary model. IEEE Revista Iberoamericana de Tecnologias del Aprendizaje 11(4):235–241
3. Shirinbab S, Lars L, Emiliano C (2020) Performance evaluation of containers and virtual machines when running Cassandra workload concurrently. Concurrency Comput: Practice Exper
4. Achuthan K et al (2011) The value@ amrita virtual labs project: using web technology to provide virtual laboratory access to students. In: 2011 IEEE global humanitarian technology conference. IEEE
5. Kumar M, Jessica E, Venkatesh C (2018) Usability analysis of virtual labs. In: 2018 IEEE 18th international conference on advanced learning technologies (ICALT). IEEE

Internet of Things: Concept, Implementation and Challenges

Nilupulee A. Gunathilake, Ahmed Al-Dubai, and William J. Buchanan

Abstract Through the technical advancements over five generations, today's digital communication has become much smarter, more intelligent and punctual. This causes a massive amount of continuous data collection in real-time whose analytics are later used to make useful insights, *i.e. prevention of road accidents using vehicular communication applications, fault detection in industrial machineries, etc.*. The means of information reception is usually via sensors. This inter-connectivity of communicating things is basically known as Internet of Things (IoT) which will become a widespread infrastructure of next-generation networking. The devices used in the IoT are physically small and resource-constrained, *i.e. low-end processors, small internal capacities, etc.*. Also, those are operated in small data rates, usually in kbps. Thus, it is unable to adopt conventional security mechanisms which require high-end computational processing. Meanwhile, the low-energy consumption of these networks conducive for green networking requirements offeres the planet a sustainable atmosphere. Due to the wide ranging nature of the subject, existing literature studies often focus on a narrowed-down area. This survey identifies up-to-date information on all IoT-related topics, *i.e. technologies, standardisation, liability, regulations, security, etc.*. This will provide a useful reference for beginners in the field for quick overall comprehension.

Keywords IoT technologies · IoT standards · IoT security · Green networking

This work is supported by the research grants from the School of Computing, Edinburgh Napier University. Any correspondence related to this article can be sent to **(nilupulee.gunathilake@napier.ac.uk)**.

N. A. Gunathilake (✉) · A. Al-Dubai · W. J. Buchanan
School of Computing, Edinburgh Napier University, Edinburgh, UK
e-mail: nilupulee.gunathilake@napier.ac.uk

K. Dahal et al. (eds.), *Internet of Things and Its Applications*, Lecture Notes in Electrical Engineering 825, https://doi.org/10.1007/978-981-16-7637-6_14

1 Introduction

The next-generation technology platforms are mainly based on 5G cellular evolution, big data, industrial 4.0, Internet, machine-to-machine (M2M) communication and Internet of Things (IoT). In contrast, the IoT further introduces several versions known as industrial IoT (IIoT), Internet of Everything (IoE), Internet of Me (IoM) and Web of Things (WoT) [1]. The IoT is a novel approach which is becoming highly successful in terms of smartness, intelligence, autonomy and portability all over the world and beyond.

The core of the IoT purpose is to produce useful insights depending on the nature of data gathered. The vision of the IoT involves many emerging technologies such as artificial intelligence (AI), machine learning and blockchain to be specific. Estimations predict that there may be 200 billion connected devices already in 2020 with an economic impact to be $13 trillion per year by 2025 [2]. The evolution of the IoT is known to start from wireless ad hoc networks that allow direct connectivity between the devices through wireless nodes.

IoT promises to be applied from personal use to applications in space, as in,

- **Low-power applications**: Intelligent transportation systems (ITS) including vehicular communications (VANET, V2X), smart home/ office/ buildings/ cities/ streetlights/ metering/ logistics, disaster rescue missions and intelligent security systems.
- **Sensor-based applications**: Agriculture, health monitoring via wearable devices, climate and weather monitoring and data analytics, factory automation (*i.e. failure predictions, etc.*), ITS (*i.e. automatic pilot, etc.*), machine-to-machine (M2M) communication, natural disaster/status monitoring (i.e. water length in a dam, etc.)
- **Tactical applications**: Mission-critical military use (*i.e. army ad hoc radios, navy ship area ad hoc networks, etc.*), ad hoc robotics, unmanned aerial vehicle (UAV), explosive hopping mines, space shuttle missions, etc.

Among various wired or wireless choices in the IoT, its data flow is recognised to be automatic, dense, unobtrusive and structured. Therefore, dependable pros and cons may exist in budget (CapEx and OpEx), energy drainage and accuracy of the results. For example, contact tracing mobile apps issued by the UK government for COVID-19 pandemic control require each individual's location data in real time. However, turned off phones/GPS or denied permissions by the user would add an unknown tolerance to the produced insight.

1.1 Our Contribution

Innovations and enhancements of IoT applications are greatly being updated. Besides, the opaque transmission of data tends to cause serious privacy violations. For this reason, both professional and non-professional bodies require a thorough awareness of this large area of activity. Available surveys mainly discuss a particular domain, whereas this work summarises up-to-date information of all IoT-related topics.

This paper deals with the IoT communication architecture, propagation techniques, status and challenges in standardisation, law and regulations as well as security. We also consider green networking.

2 Internet of Things

IoT is a complex infrastructure that includes sensing, clusters of data from numerous sources and remote monitoring. The interaction of human-to-human or human-to-computer is not a necessity. The transmission covers four levels: device, edge, fog and cloud as shown in Fig. 1 [3]. The device layer contains sensor nodes. Then, sensed data is processed through edge and fog computing up to the cloud where information is saved. The communication is often wireless as well as full duplex.

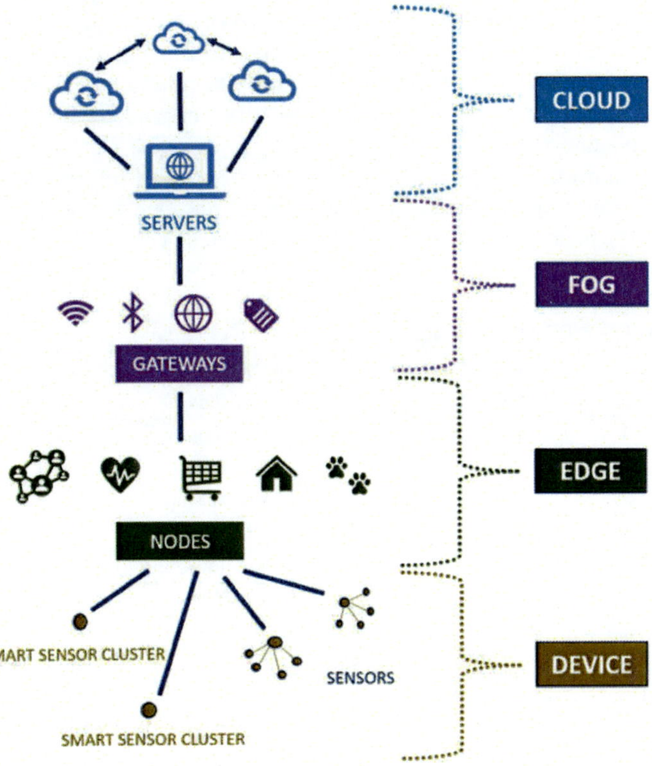

Fig. 1 IoT communication architecture

This extensive area is subjected in different categories to offer the optimised functionalities nationally and internationally. This includes propagation technology development, privacy/safety challenges, standardisation in common platforms technically to avoid translation overheads [4] and law/regulation fixation for liability.

The frequency spectrum used here is the unlicensed ISM (industrial–scientific–medical) band. Depending on future possibilities, the licensed band may be utilised because the existing wired/wireless telecommunication infrastructure operated under the International Telecommunication Union (ITU) regulations also uses IoT benefits.

3 Propagation Technologies

There are numerous transmission protocols used and still being developed that are compatible with efficient IoT communication. Among those, Wireless Fidelity (Wi-Fi), Bluetooth Low Energy (BLE), Narrowband (NB) IoT, Long Range Wide Area Network (LoRaWAN), SigFox, ZigBee and Z-wave are trending.

Wi-Fi is based on the IEEE 802.11 standard that generally consumes 1mW of power. Several versions of it have been introduced subsequently. Their relevant parametric values are in Table 1. The IEEE 802.11p is specifically allocated to vehicular ad hoc networks (VANet) [5] and the standards which have two simple letters after 802.11 are known to be the next-generation Wi-Fi enhancements including IEEE 802.11ba in addition [6].

BLE is a subversion of generic Bluetooth that is explicitly implemented for power-constrained device-to-device (D2D) communication. Due to the low-energy consumption, the battery would gain its lifetime in years. The connection times are in

Table 1 Wi-Fi - IEEE 802.11 standard's versions

Wi-Fi	Frequency	Data rate	Range
IEEE 802.11a	5 GHz	6M-54 Mbps	120 m
IEEE 802.11b	2.4 GHz	1M-11 Mbps	140 m
IEEE 802.11g	2.4 GHz	6M-54 Mbps	140 m
IEEE 802.11n	2.4/5 GHz	288M-600 Mbps	250 m
IEEE 802.11p	5.9 GHz	3M-27 Mbps	1 km
IEEE 802.11ac	5 GHz	346M-3.466 Gbps	70 m
IEEE 802.11ad	60 GHz	Up to 6.7Gbps	1–10 m
IEEE 802.11ah	900 MHz	Up to 347Mbps	1 km
IEEE 802.11aj	45/60 GHz	-	1 km
IEEE 802.11ax	2.4/5 GHz	Up to 10.53Gbps	70–240 m
IEEE 802.11ay	60 GHz	Up to 20Gbps	100 m
IEEE 802.11az	60 GHz	–	–

Table 2 LoRaWAN specifications

Parameter	Value
Frequency band	ISM 433 MHz ISM 868 MHz ISM 915 MHz
Data rate	27 kbps
Range	Urban 2–5 km Clear LoS 15 km
Network topology	Star

a few milliseconds, while the power drainage is in a few microWatts. It has a high data rate, approximately 1 Mbps.

NB-IoT was standardised as the Third Generation Partnership Project (3GPP)'s Release 13 in 2016 [7]. It operates on LTE FDD 180kHz frequency band under three modes of operations which are stand-alone, guard-band and in-band.

LoRaWAN is an enhancement of LoRa protocol that is used to establish direct communication in long distances up to several kilometres. With this, network sessions are handled between nodes and gateways as well as end-to-end encryption at the application level. Moreover, *over the air* registration/activation and multicasting are the main advantages of this low-power WAN. The general specifications are as in Table 2 [8].

Sigfox is a proprietary LPWAN technology operating on ISM 868/902MHz bands. It employs DBPSK and GFSK modulation techniques in a star network topology. Its security mechanisms are tailored via AES-128.

ZigBee is an open standard which is designed to facilitate interoperability between IoT devices due to its affordability, adaptability and deployability. This technology is well suited to sleeping end-devices and energy harvesting applications. It can tolerate up to 65,000 nodes at 250kbps.

Z-Wave is a series of implementations for future proof LPWAN hardware with integrated software tools. It also introduced the SmartStart protocol that empowers pre-configuration of devices to the network by security authorities prior to installation. This evidently reduces time spent on site which would result in minimised CapEx and OpEx as well as maximised RoI. The expected data rates are between 40–100 kbps, and it is capable of allocating up to 232 devices in the network [9].

4 Standardisation

The nature of the IoT standardisation process is complex due to its opaque data flow, continually upgrading hardware, software and network functionalities over billions of connected things. To take forward steps in an interoperable, heterogeneous and secured IoT ecosystem, standardisation is a must. For example, unique standards offered by a number of manufacturers would require extra expense in translation to connect a device with another kind of standard [4]. Work is progressing towards accomplishing this need, as described under the following topics, Sects. 4.1, 4.2 and 4.3.

4.1 Global Standards Development

The authorities relevent to the case are classified to be Standards Development Organisations (SDO), government agencies and industrial contributors. They mainly consider the openness, transparency, mechanisms, power balance/liability issues and due processes to certify adherence to anticipated procedures [10]. The types of committees involved there are as below:

- **Formally recognised fora**
  ITU, International Organization of Standards (ISO)/International Electrotechnical Commission (IEC), European Telecommunications Standards Institute (ESTI), etc.
- **Global fora/consortia**
  Institute of Electrical and Electronics Engineers (IEEE), Internet Engineering Task Force (IETF), Organisation for the Advancement of Structured Information Standards (OASIS), etc.
- **Small/private consortia**

4.2 Standards for Functionality and Compatibility

This includes consensus-driven efforts, private and proprietary standards. The IETF, ISO/IEC and IEEE are the major partners in this. Technically, a vast area of protocols and technologies is defined to handle proper functionalities in the IoT environment, but further optimisation is essential in order to obtain fewer specific compatible standards. Some of the widely used examples are:

- **Communication/Transport**: *Wi-Fi, BLE, LPWAN, etc.*
- **Data protocols**: *Message Queuing Telemetry Transport (MQTT), Advanced Message Queuing Protocol (AMQP), Constrained Application Protocol (CoAP), Websocket, Node, etc.*
- **Device management**: *Technical Report (TR)-069, Open Mobile Alliance Device Management (OMA-DM), etc.*
- **Discovery**: *Physical Web, multicast Domain Name System (mDNS), DNS Service Discovery (DNS-SD), etc.*
- **Identification**: *Electronic Product Code (EPC), uCode, Internet Protocol version 6 (IPv6), Uniform Resource Identifiers (URIs), etc.*

- **Infrastructure**: *6 Low-power Wireless Personal Area Network (6LoWPAN), IPv4/IPv6, Routing Protocol for Low-Power and Lossy (RPL), etc.*
- **Multi-layer frameworks**: *AllJoyn, IoTivity, Weave, Homekit, etc.*
- **Semantic**: *Java Script Object Notation-Linked Data (JSON-LD), Web Thing Model, etc.*

4.3 Standards for Security and Privacy

Security and privacy vulnerabilities mostly depend on a particular operation or application. However, device connectivity to either cloud or fog is an assured task. Thus, the amount of data to be transferred/processed is time-sensitive too. In that case, a quadruple trust, namely a combination of protection, security, privacy and safety, is the aim of the authorities in their efforts via proper security models, *i.e. blockchain, lightweight cryptography, etc.*.

The National Institute of Standards and Technology (NIST) has introduced a cybersecurity framework with five layers, specified as identify, protect, detect, respond and recover. That also includes asset management, access control and detection process to address IoT threats and hazards. Meanwhile, the IEC states in their 62443 release that conformity assessment process and certification thereafter by certification bodies is a promising data safety structure [11].

4.4 Law and Regulations

Transparency, responsibility and liability of IoT data are challenging issues to be addressed due to personal, industrial and governments' overall engagement which could trigger risks individually, locally, regionally and internationally. Hence, power imbalance and possible security breaches must be subjected to introducing or updating law and regulation frameworks specifically for IoT-related cybersecurity issues. Some of the considerations in this are [10, 12];

- Evidence-based support to make cost-beneficial assessment/insurance across affected stakeholders
- Application of product liability for IoT services
- Execution of recommended security standards for integrated IoT features and practices
- Enforcement of minimum-security warranties for data and products
- Transitive liability scheme for supply and service chains
- Compulsory disclosure of IoT security breaches meeting certain thresholds
- Penalty-based regulations if the security of IoT is ignored or for insecure products and practices.

Presently the General Data Protection Regulation (GDPR) brought by the European Union (EU) is implementing strict policies that apply to both personal data and personal-sensitive data in IoT. Besides, ePrivacy regulation of the EU would also smooth up the principles of confidentiality in IoT expansions. The USA mainly follows the Connected Devices Act [13]. Consequently, baseline security standards are required for all connected devices in government that ban the procurement of devices with hardcoded passwords or known weaknesses which are incapable of being updated.

5 Data and Network Security

The most significant difference between the IoT and former Internet technology is that the probabilities of threats and hazards are substantially greater because of [14, 15];

- **More points of exposure**: An exponential increase in connected devices, applications, systems, end-users through billions of billions of communication nodes
- **Creation of new self-attack vector**: Every compromised node becomes a new attack point that may remain unnoticed for a while
- **Risen impacts of attacks**: Due to incompatibility among a number of standards, blind spots may be an advantage to attackers
- **New threats form across the stack**: 'More complexity to sort out' means daily forming new threats where continuous attention of security professionals is high priority.

Blockchain and lightweight cryptography [3, 16] are the major mechanisms in IoT security. The following formula is used to measure cybersecurity risk [17];

$$\text{Cybersecurity Risk} = \frac{\text{Threat Level} \times \text{Probability of Attack} \times \text{Point of Exposure}}{\text{Cybersecurity Measures implemented}}$$

IoT security structures exist under four main layers depending on methods of computation and communication atmosphere, as in Figs. 1 and 2. It is assumed that the highest percentage of security breaches is possible at the cloud level, while it is minimum at the device level.

- **Device**: Hardware level including the elements of physical security, data at rest, chip security, secure booting, device authentication and its identity via edge processing.
- **Communication**: Connectivity networks via fog computing that cover physical layer, *i.e. Wi-Fi, Ethernet, etc.*, network layer, *i.e. IPv6, Modbus, etc.* and application layer *i.e. MQTT, CoAP, etc.*, of the OSI model that is extremely prone to

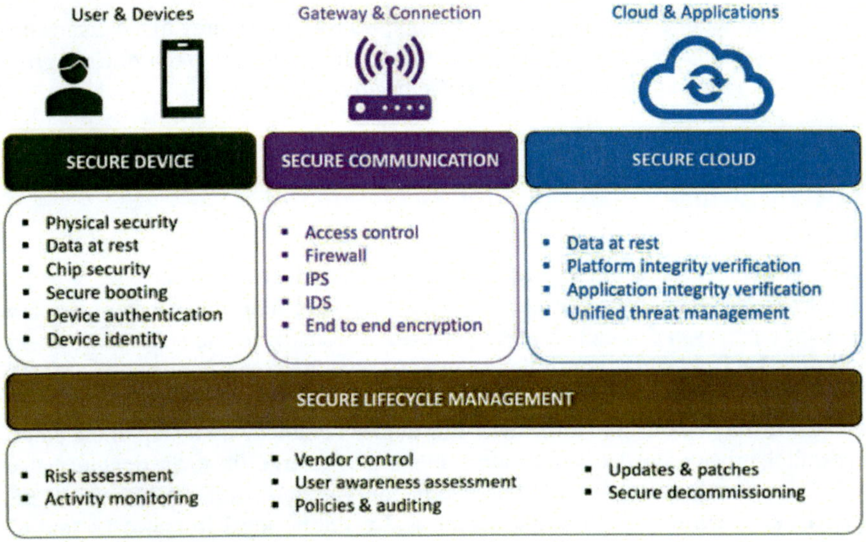

Fig. 2 IoT security architecture

man-in-the-middle attacks. This includes access control, firewall, intrusion prevention system (IPS), intrusion detection system (IDS) [18] and end-to-end encryption to be made sufficiently secure.

- **Cloud**: Software backend where received data is analysed, insights are generated, and useful actions are performed. At this level, components of data at rest, platform/application integrity verification [19] and unified threat management [20] are matters of concern.
- **Lifecycle management**: Handling of continuous processes to keep sufficient security up-to-date. Hence, risk assessment, activity monitoring, vendor control, user awareness assessment, policies and auditing, updates and patches and secure decommissioning should be maintained [17, 19].

In addition, ***processing data locally whenever cloud storage is not necessary is a suitable security mechanism to lower the risks***, *i.e. VPN*. Methods of data wipe out for compromised devices remotely would be an option to avoid spreading vulnerabilities over the entire network.

6 Green Networking

Green networking is a practice of enhancing energy-efficient networking technologies and products and optimally minimising resource use sustainably. Therefore, it is an added advantage of IoT implementations due to its small power consumption and resource limitations. The stack of IoT techniques and procedures that satisfies this scenario is alternatively known as green IoT applications, *i.e. green design, green production, green utilisation and green disposal/recycling*.

7 Conclusions

IoT is a massive infrastructure in the coming generation that is an integration of billions of resource constraint devices. The devices are operated on low data rates, low onboard memory and usually battery-powered. Special technical enhancements are introduced to tackle these constrained computational functions. For example, MQTT is an IoT supportive data protocol where transmission technologies like BLE, NB-IoT and LoRaWAN handle low-power full-duplex communications up to the cloud. In parallel, privacy and data protection complexities emerge due to the opaque nature of IoT data distribution. National and international authorities (*i.e. NIST, ISO/IEEE, the EU, etc.*) are working on introducing and updating legal frameworks for IoT efficiency and liability (*i.e. GDPR, ePrivacy, etc.*) However, adequate IoT security still struggles to provide compatible lightweight primitives to cope with possible and futuristic IoT hazards and threats (*i.e. AI, blockchain and lightweight cryptography*).

References

1. Lueth KL (2014) Why the Internet of Things is called Internet of Things: definition, history, disambiguation. IoT Analytics. https://iot-analytics.com/internet-of-things-definition/
2. Gremban K (2018) Editorial and introduction to the issue: risk and rewards of the Internet of Things. In: IEEE Internet of Things Magazine (IoTM) 1 September (1), 2 (Sep 2018). https://www.comsoc.org/publications/magazines/ieee-internet-things-magazine
3. Gunathilake NA, Buchanan WJ, Asif R (2019) Next generation lightweight cryptography for smart IoT devices: implementation, challenges and applications. In: 2019 IEEE 5th World Forum on Internet of Things (WF-IoT), pp 707–710. https://doi.org/10.1109/WF-IoT.2019.8767250
4. Kranz M (2018) Why industry needs to accelerate IoT standards. IEEE Internet of Things Magazine (IoTM) (1):14–18. https://www.comsoc.org/publications/magazines/ieee-internet-things-magazine
5. Jiang D, Delgrossi L (2008) IEEE 802.11p: towards an international standard for wireless access in vehicular environments, pp 2036–2040. https://doi.org/10.1109/VETECS.2008.458
6. Haiming W, Wei H, Jixin C, Bo S, Peng X (2014) IEEE 802.11aj (45GHz): a new very high throughput millimeter-wave WLAN system. Communications China 11: 51–62. https://doi.org/10.1109/CC.2014.6879003
7. Violette M (2018) Standards matters. IEEE Internet of Things Magazine (IoTM) (2):6–7. https://www.comsoc.org/publications/magazines/ieee-internet-things-magazine
8. LoRaWAN Classes (2020) https://www.thethingsnetwork.org/docs/lorawan/classes.html. Accessed 01 Apr 2020
9. Introduction to Z-Wave SmartStart (Sep 2017). https://www.silabs.com/documents/login/white-papers/introduction_to_z-wave_smartstart_091317.pdf. Accessed 11 Jan 2020

10. Marcus JS (2018) Liability: when things go wrong in an increasingly interconnected and autonomous world (a European view). IEEE Internet of Things Magazine (IoTM) (2):4–5. https://www.comsoc.org/publications/magazines/ieee-internet-things-magazine
11. IEC 62443-4-1 Security for Industrial Automation and Control Systems (Part 4-1: Secure Product Development Lifecycle Requirements) (2018). https://webstore.iec.ch/publication/33615. Accessed 05 May 2020
12. Barrera D, Molloy I, Huang H (2018) Standardizing IoT network security policy enforcement
13. Kenneally E (2018) The TTPs of privacy and security of the IoT. IEEE Internet of Things Magazine (IoTM) (2): 8–11. https://www.comsoc.org/publications/magazines/ieee-internet-things-magazine
14. Gurunath R, Agarwal M, Nandi A, Samanta D (2018) An overview: security issue in IoT network, pp 104–107. https://doi.org/10.1109/I-SMAC.2018.8653728
15. Sundar S, Subramanain S (2018) Security stipulations on IoT networks, pp 289–306. https://doi.org/10.1007/978-3-319-70688-7_12
16. Gunathilake NA, Al-Dubai A, Buchanan WJ (2020) Recent advances and trends in lightweight cryptography for IoT security. In: 16th international conference on Network and Service Management (CNSM 2020). http://dl.ifip.org/db/conf/cnsm/cnsm2020/1570662904.pdf
17. Scully P (2017) Five things to know about IoT security. https://iot-analytics.com/5-things-to-know-about-iot-security/. Accessed 27 Jul 2020
18. Padraig S (2020) Understanding IoT Security—Part 1 of 3: IoT security architecture on the device and communication layers. https://iot-analytics.com/understanding-iot-security-part-1-iot-security-architecture/. Accessed 28 Jul 2020
19. Scully P (2020) Understanding IoT security—Part 2 of 3: IoT cyber security for cloud and lifecycle management. https://iot-analytics.com/understanding-iot-cyber-security-part-2/. Accessed 29 Jul 2020
20. Jadhav P (2018) Cloud unified threat management system. Int J Res Appl Sci Eng Technol 6: 1712–1715. https://doi.org/10.22214/ijraset.2018.4288

Design of Intelligent Transportation System for Smart City

Hrishikesh Ugale, Pushpak Patil, Shubham Chauhan, and Neeraj Rao

Abstract The smart city concept is one of the most emerging applications of the Internet of things (IoT) technology. Paper presents a prototype of a system for intelligent transportation and mobility in a smart city. The system aims to improve the overall vehicular and transportation experience of people in a city by use to technologies to tackle various transportation problems. The proposed intelligent transportation system includes real-time monitoring of the health of vehicles and anomaly diagnosis, efficient traffic management, and dealing with road emergencies. Paper describes the detailed implementation of these features using a smartphone application as an intermediary. Validation of concept is carried out through software simulations and hardware onsite testing, and results are presented here.

Keywords DSRC · Connected vehicles · Android · Smart city

1 Introduction

Recent advancements and rapid evolution in the field of the Internet of things (IoT) are introducing smartness in the working of many traditional systems [1]. IoT-driven devices have the capability to sense the surrounding environment and communicate this information to a suitable destination which helps in achieving better situational awareness and decision making. Vehicular technologies and transportation systems are also undergoing rapid transformation through the involvement of the Internet of things [2, 3]. Researchers in this field are working to develop vehicular systems that would solve various vehicular problems to increase customer satisfaction. Smart city is one of the emerging concepts as an application of the Internet of things. One of the biggest challenges in the development of smart cities is intelligent traffic management [4]. Cities owing to large urban populations have a large number of vehicles. As per the data available [5], an urban city like Pune (a city in India) has around 3.62 million vehicles which is greater than the population of that city. Such a large number of vehicles plying on city roads increases the traffic congestion and transportation time

H. Ugale (✉) · P. Patil · S. Chauhan · N. Rao
Visvesvaraya National Institute of Technology, Nagpur, India

K. Dahal et al. (eds.), *Internet of Things and Its Applications*, Lecture Notes in Electrical Engineering 825, https://doi.org/10.1007/978-981-16-7637-6_15

of people due to poor inefficient management of traffic. This adversely affects the overall transportation experience of people in cities. Clearly, there is a need for a system to tackle this issue using technology in order to develop a truly smart city. The proposed work is related to different aspects of technology to tackle various transportation and vehicle-related problems in the city and to improve the driving experience of people.

The first part of the proposed system is an IoT-based approach for maintenance of the health of the vehicle using the user's smartphone for early diagnostic of faults or anomalies in different parts of vehicles. On-board diagnostic (OBD) is the standardized system [6] that allows external electronics to interface with a car's computer system to diagnose the faults or anomalies in different parts of the vehicle. The proposed system communicates real-time data obtained from OBD sensors to the user's smartphone via Bluetooth interface and uses machine learning to analyze the data coming from the sensors and to identify the anomalies or faults present in different parts using this data.

The second part of the proposed system is an innovative approach for the proper management of traffic in cities to improve the overall driving experience of citizens. Conventional traffic signals can be very inefficient for the management of traffic and can result in large traffic congestion. Traffic intensities on different roads of cities are not uniform throughout. In the proposed system, the behavior of traffic in the city for different times of the day and different weekdays can be mathematically modeled using the analytics data collected from the citizen's smartphone application and can be used to plan traffic management strategies accordingly. Traffic signals in the proposed system are smart traffic infrastructures that communicate with vehicles ensuring a smooth flow of traffic in the city.

Connected vehicles are a newly emerging concept [7] where vehicles have the capability to communicate with each other as well as with traffic infrastructures like signals and poles. For vehicular communication, IEEE has approved the IEEE 802.11p protocol to add wireless access in vehicular environments (WAVE) [8]. Development of the dedicated short-range communication (DSRC) technology for use of vehicle-to-vehicle and vehicle-to-roadside communication is in progress. However, implementation of on-board DSRC units on existing vehicles can be a really costly and time-consuming job since millions of vehicles without DSRC on-board units are plying on roads. The third part of the proposed system attempts to achieve vehicular communication using a smartphone as an intermediary using WiFi peer-to-peer communication technology.

2 Related Work

Researchers have been working in various fields of vehicular technologies for quite a long time. A number of approaches for each of the applications mentioned in Sect. 1 have been developed to date. These approaches have their own pros and cons. This section highlights some of these popular approaches or systems developed by other researchers working in this field.

Various methodologies for anomaly detection in-vehicle health parameters have been proposed and studied till now. In Prytz [9], the author used both Kalman filtering and CNN to recognize and distinguish anomalies in information coming from various sensors installed in the vehicle. A point-by-point conversation of predictive maintenance of the vehicle industry is introduced by Prytz [10] in which different machine learning algorithms are used for the prediction of faults. Consensus self-organized models for fault detection (COSMO) were introduced [11] in which the sensor's information is utilized and testing was carried out on heavy trucks and city buses. In Prytz et al. [12], the author used the random forest for predicting the need for repair in the air compressor of buses and trucks.

Similarly, a large number of approaches for vehicular communication, intelligent transport systems, and traffic management have been proposed. Some solutions rely on sensors mounted on the road for detecting the traffic density and thereby controlling the signal duration [13]. In Mazloumi et al. [14], author proposed a system in which GPS modules were used to track the location of vehicles and to manage the traffic flow. While [15] uses CCTV cameras on roadsides to detect the vehicles through image analysis and computer vision algorithms and thereby manage the traffic flow, an attempt to achieve inter-vehicular communication using an android application was made by Su et al. [16] using the WiFi of the phone.

3 Proposed System

The section describes the prototype of the system for intelligent transportation in a smart city. The prototype consists of an android application that communicates with multiple subsystems [17]. The application communicates with the user's car (vehicle) via Bluetooth interface, with nearby vehicles via WiFi Direct (peer-to-peer WiFi communication), and with the cloud via the Internet through GSM/LTE system. These different modes of communication between various subsystems are required for different applications which are discussed below.

3.1 Vehicle Health and Diagnostics

As discussed in Sect. 1, real-time monitoring of vehicle health and diagnosis of faults or anomalies in various parts of a vehicle is one of the main features of the system. The ELM327 is a programmed microcontroller produced by ELM electronics for translating the on-board diagnostics (OBD) interface found in most modern cars [18]. A portable ELM327 device is connected to the OBD port of the vehicle. An Arduino-based circuit interfaced with Bluetooth module HC05 is connected with this device to receive data from OBD sensors. Data received from OBD sensors is periodically communicated to the android phone via Bluetooth interface at regular intervals. Machine learning models are deployed on the android device that analyzes the incoming data and checks for the anomaly in it. A detailed description of neural network used for this purpose is given below.

OBD parameters, namely barometric pressure, engine coolant temperature, engine rpm, intake manifold pressure, mass airflow (MAF), speed, and throttle position, were considered as the input features for the analysis. The open-source dataset of these parameters was used for training and testing of algorithms. As per the source, this dataset was collected using ELM 327 connected to the OBD port of the car and data logged on mobile phone via Bluetooth. Thus for each point in the dataset, there are 7 parameters (Fig. 1).

Six-layered neural network (NN) was developed for the anomaly detection consisting of an input layer of 7 neurons (1 neuron for each feature) followed by 4 hidden layers and then an output layer. For each neuron, there are a certain number of inputs and the same number of weights. Weights were initialized randomly close to zero. During training, their values were adjusted to new values, and this contributed to deciding the importance of inputs. Three operations are done by a single neuron. First, it calculates the weighted summation of all the input values. Then, it applies an activation function to the weighted summation at last it passes the results to a neuron in the next layer. The rectified linear activation unit, or ReLU, was used because of its computational simplicity and its linear behavior which increases the chances of optimizing the NN. For the output layer, sigmoid activation was used.

The initial information propagates to the hidden neurons at each layer and finally produces the output y, which is a number in the range from 0 to 1. We used the adaptive moment estimation (Adam) optimizer. We used 20 epochs to train the NN with a batch size of 32. The problem of over-fitting was catered by setting the dropout rate to 0.1. Implementation of the neural network model was carried out using TensorFlow.

Once the neural network model was satisfactorily trained and tested, it was then deployed in our proposed android application using tensorflowlite. Input parameters to the model were communicated to phone via Bluetooth interface. Detailed results of implementation are discussed in Sect. 4.

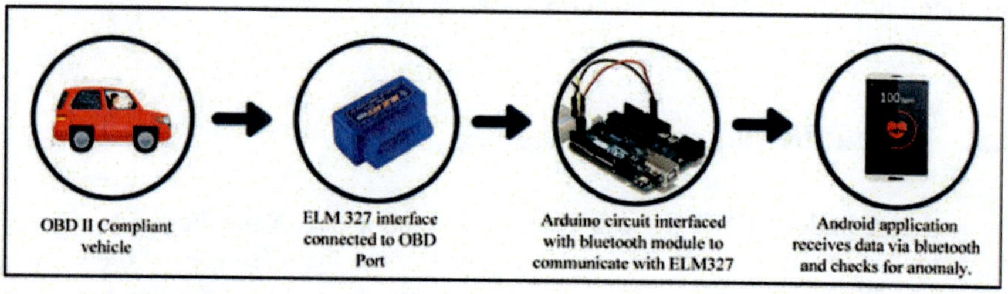

Fig. 1 Anomaly detection in the proposed system

3.2 *Traffic Management*

The proposed system is designed to collect real-time information about traffic conditions in the city. The proposed android application comes with driving mode feature in it which automatically gets triggered when the vehicle starts and then transmits information about the journey of the vehicle to the traffic cloud. Google's Firebase Real-time database service is used as the traffic cloud in the proposed prototype. Analysis of traffic information on different roads of the city collected over a long period of time is used to create a mathematical model of the behavior of traffic intensity. This analysis can provide a good estimate or prediction of traffic intensity at different roads or intersections. For example, roads going toward city school show sharp rise in traffic intensity curve in the morning time when school opens and the evening time when school closes and remains flat for rest of the day. Thus, traffic management strategy at these peak times of day should be different from that for the rest of the day to minimize the traffic congestion (Fig. 2).

The proposed system is designed to deal with huge traffic in big cities. The study of such IoT systems can be achieved through simulations since immediate real-life experimentation of such systems is not practically feasible. These simulations are derived from certain assumptions to simplify the study. A customized simulator for simulating the traffic flow at the intersection was developed using object-oriented JAVA programming language [17], wherein vehicles and traffic signals were modeled as JAVA simulation objects with all the properties that the hardware prototype is to be designed with. These objects are made to act independently on the same timeline object on the common system thread. A customized user-defined scenario is simulated on a timeline [17].

A study of traffic flow at the intersection of two one-way roads was studied using the simulator developed. The model presented in the work is based on the macroscopic traffic model. We assume that all vehicles close to intersection travel

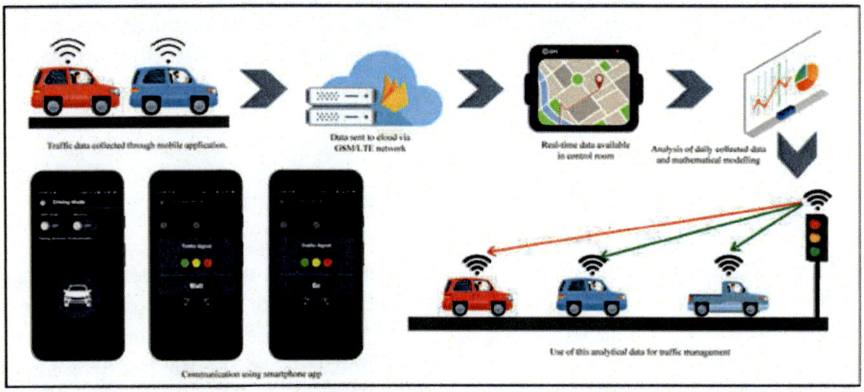

Fig. 2 Traffic management in proposed system

with constant speed v. Also, it is assumed all vehicles are of the same length L and have proposed system deployed in them; i.e., all the vehicles have our driving application installed in the driver's phone. Traffic signals in the scenario are smart traffic infrastructures that can communicate with vehicles via WiFi Direct (peer-to-peer WiFi communication).

Traffic inflow rate φ is the rate at which the vehicle approaches the intersection via a given road per unit time. Since all vehicles are assumed to be moving with same speed, in the case of the road with very high traffic, the maximum traffic inflow rate can be $\varphi_{max} = v/L$, i.e., vehicles moving back to back on highly congested roads. Traffic density ρ is the number of vehicles queued on the road per unit length. For high traffic roads with congestion, its maximum value is given as $\rho_{max} = 1/L$. Traffic outflow rate σ is the rate at which vehicle leaves the intersection or crosses the intersection via the given road. σ is actually controlled by the traffic signal. If the traffic signal is red, no vehicle can cross the intersection, and thus, we have $\sigma_{min} = 0$. On the other hand, when the signal is green for a very long time (let us say infinitely), clearly there would be no traffic congestion at the intersection and thus outflow rate would be equal to the inflow rate, i.e., $\sigma_{max} = \varphi$.

Congestion of traffic on roads is caused due to accumulation of vehicles on road. This rate of accumulation is directly proportional to the difference of inflow and outflow rate, i.e., $(\varphi - \sigma)$. For the intersection of two one-way roads, values of $\varphi 1$ and $\varphi 2$ are obtained through the analysis of data collected through the proposed smartphone application.

Traffic signals in the proposed system have the capability to communicate with the vehicles via WiFi Direct. Instruction regarding whether to cross the intersection or wait is provided by the traffic signal to vehicle via V2I communication. The number of vehicles allowed to cross the intersection from a given road is a function of the ratio of $\varphi 1/\varphi 2$. The aim is to minimize the overall waiting time of vehicles at the intersection and to maintain uniform traffic flow on both the intersecting roads. For example, at a particular instance, the traffic inflow rate of given road A is twice the inflow rate of other road B meeting at an intersection. In order to maintain the uniform traffic flow on both the roads, for every 2 vehicles from road A allowed to cross the intersection, 1 vehicle from road B should be allowed to cross the intersection. Detailed simulation results for such scenarios are presented in Sect. 4.

3.3 Vehicular Communication Using Android Phone

WiFi Direct communication functionality found in most modern android devices enables compatible devices to communicate with each other without connecting to a wireless access point. Due to its easy setup and WiFi frequency band being close to the standard DSRC frequency band (5.9 GHz allocated), an attempt has been made to use WiFi as the substitute for the on-board units for vehicular communication.

WiFi Direct also is known as WiFi peer-to-peer communication functionality is available in all modern-day android devices having an android version above

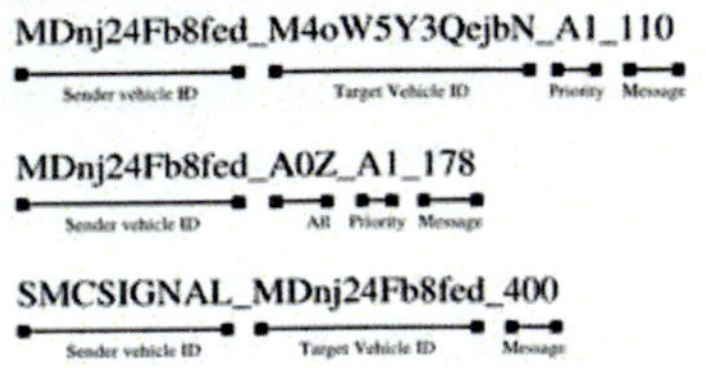

Message Code Lookup

Code	Message
110	Turning Right
178	Slow Down
400	Wait
440	Go

Fig. 3 AP naming convention used in prototype

Android 4.0 (Ice cream Sandwich). The driving mode of the smartphone application of the proposed system is programmed to control the WiFi direct functionality. The device receives messages from other devices in client mode and broadcast messages to other devices in access point (AP) mode. Switching logic between client mode and AP mode is controlled to achieve communication between devices. In order to reduce the delay in establishing a connection with AP, messages with low priority are encoded in a particular format and this encoded string is set as the AP name of the broadcast device. Other devices can thus read this encoded message without actually connecting to the AP and then decode this string to interpret the message. However, for the transfer of messages with high priority, connection between AP and client needs to be established and delay is slightly greater.

Figure 3 shows the AP naming convention used for the communication. A lookup table of message codes is implemented inside the mobile application. A similar procedure was used for the transfer of messages between two vehicles via smartphone. Each vehicle is assigned a unique 12 UTF-8 characters long id. Typical AP name for transfer of message between two vehicles follows the simple convention of sender id followed by receiver id followed by priority of the message and encoded message code. The message is then decoded by using a lookup table implemented inside the application.

Supporting traffic infrastructures like smart traffic signals used in Sect. 3.2 have WiFi AP mode configured in them. Devices connect to these infrastructures in client mode and receive commands from these APs. In the case of multiple APs available at the same time, the priority of AP decides which AP is to connect with. The priority of AP was set according to the urgency of message to be delivered and the device delivering the message and was identified by the clients by decoding the string of AP name. For example, AP names of traffic signals were set higher priority than the AP names of the vehicles. Results of the proposed approach are discussed in Sect. 4.

4 Results

For anomaly detection in vehicle health parameters, neural network algorithms discussed in Sect. 3.1 were implemented and tested using python programming language [17]. Dataset of 12,000 sample points was splitted into 4:1 ratio for training and testing, respectively. A separate test set for testing was used at the end for further validation of results. Performance of model was evaluated in terms of accuracy of recognizing the anomaly point. In order to compare the performance of proposed model, two other well-known anomaly detection algorithms, namely local outlier factor and isolation forest algorithm, were implemented separately using Python and tested against the same test dataset described above. The measured accuracy obtained for the 3 algorithms is given in Table 1.

Clearly, from Table 1, the accuracy of the neural network model was found to be maximum. The proposed android application was also tested with deployed system as shown in Fig. 5 (first application screen).

For the proposed android application, testing of driving mode feature for collection of traffic information was carried out using 5 vehicles with driving mode enabled in their driver's device. When these vehicles were made to drive on roads, traffic information was updated live on the cloud and was observed on the GUI.

Traffic management using control of traffic signals at the intersection was tested using simulations of traffic at intersection of two one-way roads for three different scenarios. In the first scenario, the traffic intensity of both roads was kept low, high and low in the second scenario and both high in the third scenario.

The average waiting time of vehicles and number of vehicles waiting for a green signal at a given time over the simulation duration were recorded and plotted as shown in Fig. 4. It was observed that for all three scenarios, the waiting time of vehicles in the case of conventional traffic signals is quite high, goes up to 4 min, and is dependent on the fixed duration of the red signal. Also, it was observed that conventional traffic signals can lead to long waiting queues of vehicles resulting in traffic congestion. In the case of roads with unequal traffic intensities, roads with higher traffic intensities have very long waiting queues as compared to one with low traffic intensities indicating non-uniformity in traffic flow. On the other hand, for the proposed system, vehicles need to wait for a much lower time. Also, the number of waiting vehicles is found to be much lower. In the case of roads with unequal traffic intensities, the number of waiting vehicles is found to be nearly the same on both sides (with little difference), clearly indicating the much smoother flow of traffic. Thus, the overall experience of people can be clearly improved in the proposed system.

For the vehicular communication system, the proposed application was installed in two mobile devices, and then the driving test was carried out. A prototype of the proposed smart traffic signal was developed using ESP 32 NodeMCU with a WiFi hotspot enabled in it. Test commands like "wait" and "Go" were sent from the traffic signal to the proposed mobile application as shown in Fig. 5. NodeMCU module was programmed to act as a WiFi access point (AP).

Table 1 Anomaly detection algorithm accuracy

Sr No.	Algorithm	Accuracy (%)
1	Local outlier factor (LOF)	62.29
2	Isolation forest algorithm	78.75
3	Neural network	99.79

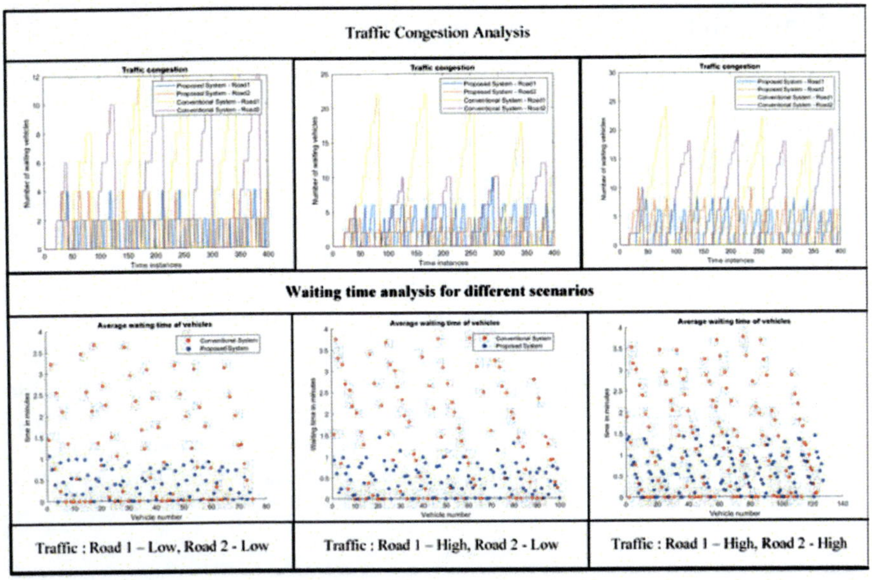

Fig. 4 Comparative simulation results for traffic management system

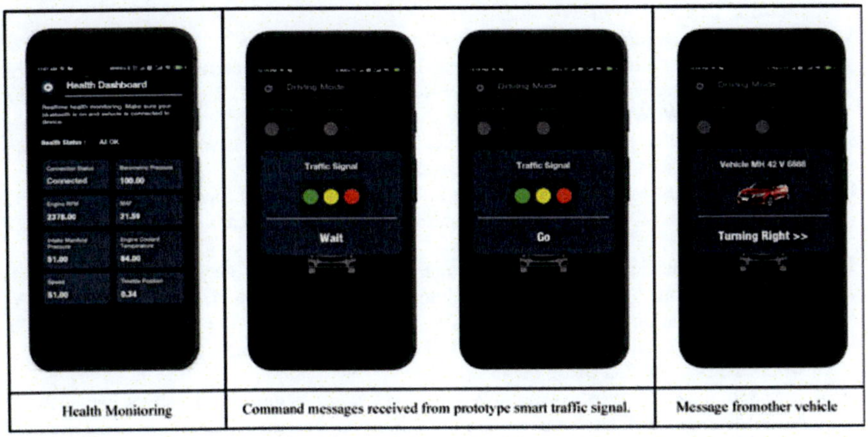

Fig. 5 Testing of proposed android application

When a vehicle with its driver having driving mode enabled in his phone went near the prototypic traffic signal, the application identified the traffic signal by the naming convention of AP and connected with it in the client mode. Commands regarding whether to wait or stop were communicated by the naming convention of the access point (AP).

5 Conclusion

The work proposed in the paper describes the cost-effective design of an intelligent transportation system for a smart city. The prototype of the system was developed, and results of hardware testing and software simulations were presented. Results obtained clearly indicate that such a system can be very beneficial for the maintenance of vehicle health and management of city traffic thereby improving the overall mobility experience and satisfaction of people in city. Since the system is primarily developed using the smartphone as an intermediary, additional hardware required for implementation is quite less, and hence, the system is cost-effective. Some of the drawbacks of the proposed system can be the primary dependency on the user's smartphone, consumption of mobile data, and battery power during the process. But since the use of smartphones is quite common and modern mobile devices come with powerful processors, access to 3G/4G/5G networks, and long battery life. Hence, such a system can be fairly implemented with some advanced modifications in it.

References

1. Nord JH, Koohang A, Paliszkiewicz J (2019) The internet of things: review and theoretical framework. Exp Syst Appl
2. Datta SK, Da Costa RPF, Härri J, Bonnet C (2016) Integrating connected vehicles in Internet of Things ecosystems: challenges and solutions. In: 2016 IEEE 17th international symposium on a world of wireless, Mobile and Multimedia Networks (WoWMoM). Coimbra, pp 1–6. https://doi.org/10.1109/WoWMoM.2016.7523574
3. Wu C (2017) Connected vehicles and Internet of things. In: 2017 2nd international conference on telecommunication and networks (TEL-NET). Noida
4. Saikar A, Parulekar M, Badve A, Thakkar S, Deshmukh A (2017) Smart traffic management for smart cities. In: 2017 international conference on emerging trends and innovation in ICT (ICEI). Pune
5. Number of vehicles in Pune overtakes human population. India Today, 6 April 2018, p. A1. Referred from https://www.indiatoday.in/auto/auto-news/story/vehiclesfigure-in-pune-overtakes-human-population-1206311-2018-04-06
6. "On-board diagnostics" Wikipedia, Wikimedia Foundation, 5 November 2020. en.wikipedia.org/wiki/On-board diagnostics
7. "Connected car" Wikipedia, Wikimedia Foundation, 18 October 2020, en.wikipedia.org/wiki/Connected car
8. Kenney JB (2011) Dedicated short-range communications (DSRC) standards in the United States. Proc IEEE 99(7):1162–1182. https://doi.org/10.1109/JPROC.2011.2132790
9. van Wyk F, Wang Y, Khojandi A, Masoud N (2020) Real-time sensor anomaly detection and identification in automated vehicles. IEEE Trans Intell Transp Syst. https://doi.org/10.1109/TITS.2019.2906038
10. Prytz R (2014) Machine learning methods for vehicle predictive maintenance using of-board and on-board data dissertation. Halmstad University Press
11. Byttner S, Rognvaldsson T, Svensson M (2011) Consensus self-organized models for fault detection (COSMO). Eng Appl Artif Intell 24(5):833–839
12. Prytz R, Nowaczyk S, Rognvaldsson T, Byttner S (2015) Predicting the need for vehicle compressor repairs using maintenance records and logged vehicle data. Eng. Appl. Artic. Intell. 41:139–150

13. Rizwan P, Suresh K, Babu MR (2016) Real-time smart traffic management system for smart cities by using Internet of Things and big data. In: 2016 international conference on emerging technological trends (ICETT). Kollam
14. Mazloumi E, Asce MS, Currie G, Rose G (2010) Using GPS data to gain insight into public transport travel time variability. J Transp Eng 136:623–631
15. Desai G, Ambre V, Jakharia S, Sherkhane S (2018) Smart road surveillance using image processing. In: 2018 international conference on smart city and emerging technology (ICSCET). Mumbai
16. Su K, Wu H, Chang W, Chou Y (2012) Vehicle-to-vehicle communication system through Wi-Fi network using android smartphone. In: 2012 international conference on connected vehicles and expo (ICCVE). Beijing, pp 191–196. https://doi.org/10.1109/ICCVE.2012.42
17. Intelligent Transport System. Github Repository. https://github.com/Hrishi-3331/Intelligent-Transport-System
18. Elm Electronics (2002) "OBD to RS232 Interpreter" ELM327DSJ datasheet

Signal Processing and Machine Learning

Students Performance Prediction Using Educational Data Mining

Anisha Mitra, Aakash Decosta, Nilanjana Roychoudhury, and Anal Acharya

Abstract The goal of educational institutions is to enhance the quality of education and uplift students' academic performance. To attain the highest level of quality, experts need to find out the most influential features affecting academic performances and must attempt to solve weakness of those features. Educational data mining (EDM) tools provide the best measure to achieve so. For educational datasets, quality of prediction results can be upgraded by using feature selection (FS) procedures. Unrelated data and attributes present in dataset can be removed by using proper FS methods which will result in more accurate results in EDM practices. Performance of prediction results can be enhanced by choosing appropriate dataset and attributes. This paper utilizes educational data mining approach through a few chosen feature selection procedures and machine learning algorithms to predict students' final grade. Obtained results attempt to find out influence of different student-related attributes in the prediction.

Keywords Educational data mining · Feature selection · Prediction · Accuracy · Classification · Regression

1 Introduction

Educational upgrades play primary role in the development of a country. Quality of education in a country is the deciding factor in creating influential member of society. Students' data kept in educational repositories need to be explored for getting hidden information which will help in finding main factors improving educational processes. Academic performance of students is observed by using various well-built prediction models which are implemented by EDM techniques. High-dimensional students' data consisting of large number of attributes require going through dimensionality reduction before using it for prediction. This paper consists of two parts

A. Mitra (✉) · A. Decosta · N. Roychoudhury · A. Acharya
Department of Computer Science, St. Xavier's College (Autonomous), Kolkata, India

A. Acharya
e-mail: anal.acharya@sxccal.edu

K. Dahal et al. (eds.), *Internet of Things and Its Applications*, Lecture Notes in Electrical Engineering 825, https://doi.org/10.1007/978-981-16-7637-6_16

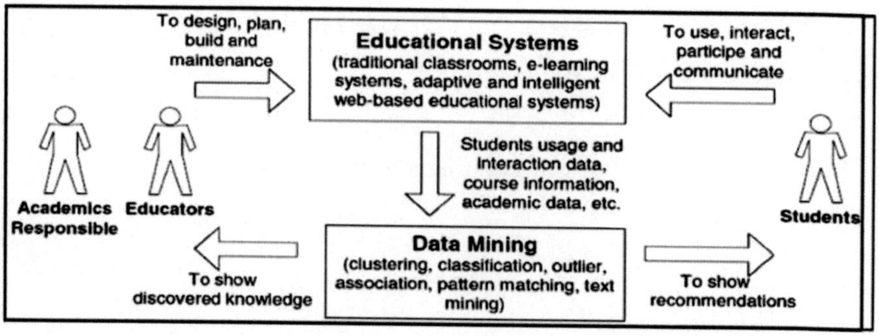

Fig. 1 Performance prediction procedure [1]

including the implementation of different feature selection procedures on the chosen dataset followed by performance prediction. Various well-known feature selection algorithms have been employed on the chosen dataset to reduce dimension to get better results. Later, multiple classification and regression methods are used on the reduced number of features to get the actual prediction output. Most promising result can be detected by comparing accuracy score of individual methods. This paper aims to observe the best combination of feature selection processes and machine learning methods which will give more accurate prediction result. Here, several fundamental algorithms found in the literature are studied to assess their performance in a controlled scenario (Fig. 1).

2 Related Works

The literature review discloses that predicting performance at higher education level has involved substantial attention in the recent past and persists to remain focus of research and discussion. Several works have been done successfully in this domain. Many of those suitable works have been thoroughly examined to gain knowledge in this field.

The study conducted by Asif et al. in [2] shows prediction of graduation performance by only using pre-university marks and marks of first- and second-year courses without requiring any other features with reasonable accuracy. This work includes usage of well-known machine learning algorithms like decision tree, naive Bayer's, neural network, random forest along with feature selection steps and without using feature selection.

In [3], Cortez and Silva have used data collected from two Portuguese schools to predict secondary students' grades of two core classes by using pass school grades, demographic, and social data. This paper consists of implementation of three different data mining goals (binary classification, classification with five levels, and regression) with four well-known data mining methods (decision tree, random forest, naive Bayes method, and support vector machine). The obtained results reveal that it is possible to achieve a high predictive accuracy, provided that the first and/or second school period grades are known. Results obtained also conclude that students' results highly depend on their previous examination performance.

In the research paper [4] by Yassein et al., clustering (using two-phase clustering technique) was done to classify data in to two clusters and then suitable classification algorithm (C5.4 algorithm) was used for prediction. This work establishes the fact that there exists a strong relation between students' attendance in class and their success rate and between practical work and success rate of courses.

Research work [5] published by Hasan et al. used WEKA data mining tool to evaluate decision tree algorithm for discovery of students' performance. A comparison between random forest tree algorithm and decision tree algorithm has been performed here.

Rahman and Islam's research work in [6] has attempted to analyze the effect of students' absence from regular class as well as students' behavioral attributes on their academics. This paper includes application of methods like naive Bayes and artificial neural network. It has also implemented an ensemble filtering method to identify erroneous instances from dataset.

In the domain of students' performance prediction, different kinds of machine learning algorithms have been used. Understanding the impact of various procedures through knowledge gathering on this task, an experimental approach has been taken here. This work tries to find out the effect of well-known feature selection algorithms in a random dataset consisting of students' data and focuses on how different feature selection methods followed by different machine learning procedures can change the prediction result.

3 Proposed Work

The main objective of this research paper is to know in detail the individual effects of various attributes on the students' performance. Educational data mining plays an important role in modern education by helping students as well as teachers to improve overall academic environment. This work tries to find out the effect of well-known feature selection algorithms in a random dataset consisting of students' data and focuses on how different feature selection methods followed by different machine learning procedures can affect the prediction result.

Major requirement of educational data mining lies in the improvement of students' performance. Results obtained from such machine learning procedure help educational institutions to nourish students' educational practice. That is why such experiments require dataset of considerable size to be divided into training and test parts as required. Such database can be constructed from two sources: school reports, based on paper sheets and including few attributes and questionnaires, used to complement the previous information. Data stored in database must go through data cleaning procedures before proceeding toward actual task. Later, insignificant attributes are discovered using feature selection algorithms. Then different results obtained from different feature selection procedures are considered and passed through machine learning step. At last, prediction results are compared in terms of accuracy.

For this work, some well-known feature selection methods and machine learning procedures have been chosen. Among all features, very few plays important role in result estimation. So, feature selection algorithms are required to obtain a subset of significant features to avoid over-fitting problem. Four filtering-type feature selection methods used here include chi-square test, correlation-based feature selection method, ROC score-based method, and mean squared error (MSE)-based method. Lasso and Ridge regression procedures embody embedded feature selection method. Recursive feature selection using random forest method represents wrapper feature selection methods. In the second and most crucial part of this work, selected attributes obtained from feature selection steps were used to build models using chosen machine learning algorithms. As classification methods, logistic regression, support vector machine (SVM), K-nearest neighbor (KNN) method, naive Bayes classification, and random forest classification were applied. Linear regression was another method used on the chosen attribute set. These machine learning procedures were applied separately on the results obtained from each individual feature selection method. Finally, accuracy score was used to compare the performance of feature selection methods and machine learning procedures.

4 Result and Discussion

Proposed procedures as described in the previous section have been implemented with a system having AMD Ryzen 3 2200G processor in Windows 10 Operating System. The software tool used to implement the algorithms was python 3.7.4 in Jupyter notebook platform.

For this comparative study, the student prediction dataset from the UCI machine learning repository [7] has been used. This data was collected from the Alentejo region of Portugal from two public schools during the 2005–2006 session. The dataset has been split into the two parts one comprising the core subject of Mathematics and the other Portuguese. During a year, the student's marks evaluation is divided into three phases G1, G2, and G3. G3 corresponds to the final grade which is our target attribute to be predicted. The data attributes include student grades, demographic, social, and school-related features, and it was collected by using school reports and questionnaires. The math dataset contains 365 examples, and the Portuguese dataset has 649 examples (Fig. 2).

In the above-mentioned dataset with 32 attributes, feature selection and machine learning procedures have been applied. Results obtained are shown in the next section

Attribute	Description
school	student's school (binary: "GP"-Gabriel Pereira or "MS"- Mousinho da Silveira
sex	student's sex (binary: "F" – female or "M" – male)
age	student's age (numeric: from 15 to 22)
address	student's home address type (binary: "U" – urban or "R" – rural)
famsize	family size (binary: "LE3"- less or equal to 3 or "GT3" – greater than 3)
Pstatus	parent's cohabitation status (binary: "T"- living together or "A"- apart)
Medu	mother's education (numeric: 0 – none, 1 – primary education (4th grade), 2- 5th to 9th grade, 3- secondary education or 4- higher education)
Fedu	father's education (numeric: 0 – none, 1 – primary education (4th grade), 2- 5th to 9th grade, 3- secondary education or 4- higher education)
Mjob	mother's job (nominal: 'teacher', 'healthcare' related, 'civil services' (e.g., administrative or police, 'at-home' or 'other')
Fjob	father's job (nominal: 'teacher', 'healthcare' related, 'civil services' (e.g., administrative or police), 'at-home' or 'other')
reason	reason to choose this school (nominal: 'close to home', 'school reputation', 'course preference' or 'other'
guardian	student's guardian (nominal: 'mother', 'father' or 'other')
traveltime	home to school travel time (numeric: 1- <15 min, 2- 15 to 30min, 3- 30min to 1 hour or 4- >1 hour)
studytime	weekly studytime (numeric: 1- <2hours, 2- 2 to 5 hours, 3- 5 to 10 hours or 4- >10 hours)
failures	Number of past class failures (numeric: n if 1<=n<3, else 4)
schoolsup	extra educational support (binary: yes or no)
famsup	family educational support (binary: yes or no)
paid	extra paid classes within the course subject (Math or Portuguese) (binary: yes or no)
activities	extra-curricular activities (binary: yes or no)
nursery	attended nursery school (binary: yes or no)
higher	wants to take higher education (binary: yes or no)
internet	internet access at home (binary: yes or no)
romantic	with a romantic relationship (binary: yes or no)
famrel	quality of family relationship (numeric: from 1- very bad to 5- excellent)
freetime	free time after school (numeric: from 1- very low to 5- very high)
goout	go out with friends (numeric: from 1- very low to 5- very high)
Dalc	workday alcohol consumption (numeric: from 1- very low to 5- very high)
Walc	weekend alcohol consumption (numeric: from 1- very low to 5- very high)
health	current health status (numeric: from 1- very bad to 5- very good)
absences	number of school absences (numeric: from 0 to 93)
G1	first period grade (numeric: from 0 to 20)
G2	second period grade (numeric: from 0 to 20)
G3	final period grade (numeric: from 0 to 20)

Fig. 2 Attribute description of dataset [7]

4.1 Feature Selection Results

4.1.1 Filter Methods

In any prediction procedure, the filter method acts as a preprocessing step before going into actual result generation. It does not depend on any machine learning algorithm.

Chi-Square Ranking of the Chosen Set of Features The chi-square method is believed to be the right choice to get the significance value of each feature. It has calculated the significant value of each feature toward the target. It determines if the association between two categorical variables of the sample would reflect their real association in the population (Table 1).

Chi-square score is given by:

$$x^2 = \frac{(\text{Observed frequency} - \text{Expected frequency})^2}{\text{Expected Frequency}} \tag{1}$$

Table 1 Chi-square test result

Attribute	Ranking
*G*2	665.656818
Absences	627.431411
*G*1	462.985299
Failures	272.095228
Reason	41.985012
School	35.849805
Walc	35.742843
Schoolsup	32.65607
Mjob	31.270835
Dalc	26.196743

Result shows the attribute 'absences' has got more importance than attribute '*G*1'.

Correlation Result with Target Variable To find out the correlation between any two variables, *X* and *Y* (say), a well-known method is to use linear correlation coefficient 'r' given by:

$$r = \frac{\sum(Xi - \overline{Xi})\sum(Yi - \overline{Yi})}{\sqrt{\sum(Xi - \overline{Xi})^2}\sqrt{\sum(Yi - \overline{Yi})^2}} \quad (2)$$

Correlation is similarity measures between two features. If two features are linearly dependent, then their correlation coefficient is ± 1. If the features are uncorrelated, the correlation coefficient is 0. Result shows that the previous two grades (*G*2 and *G*1) are the most important ones (Table 2).

Table 2 Correlation with target variable (*G*3)

Attribute	Correlation coefficient
*G*2	0.910743
*G*1	0.809142
Failures	0.383145
Higher	0.236578
Medu	0.201472
Studytime	0.161629
Fedu	0.159796
Dalc	0.129642
School	0.127114
Age	0.125282

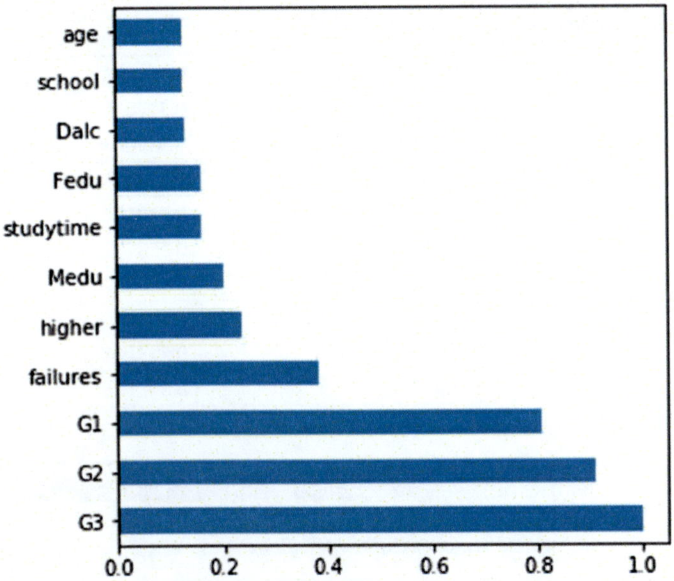

Fig. 3 Graphical representation of correlation result

4.1.2 Graphical Representation

See Fig. 3.

ROC Scores for Our Dataset *Receiver operating characteristic* (ROC) curve tells us about how good the model can distinguish between two things (*e.g., if a patient has a disease or no*). Better models can accurately distinguish between the two, whereas a poor model will have difficulties in distinguishing between the two.

4.1.3 How It Works:

- First, it builds one decision tree per feature, to predict the target.
- Second, it makes predictions using a decision tree and the considered feature.
- Third, it ranks the features according to the machine learning metric (ROC).
- It selects the highest-ranked features (Table 3).

4.1.4 Graphical Representation of the ROC Result

From this result, we can observe that, except G2, all other attributes have a score more than or near 0.5. So, from this observation, the ROC-AUC score concludes that G2 influences prediction model far more than other attributes does (Fig. 4).

Table 3 ROC score result

Attribute	ROC score
*G*2	0.700498
*G*1	0.577021
Absences	0.518068
Medu	0.514094
Age	0.510107
Health	0.508772
Reason	0.50803
Traveltime	0.503155
Nursery	0.503144
School	0.502107

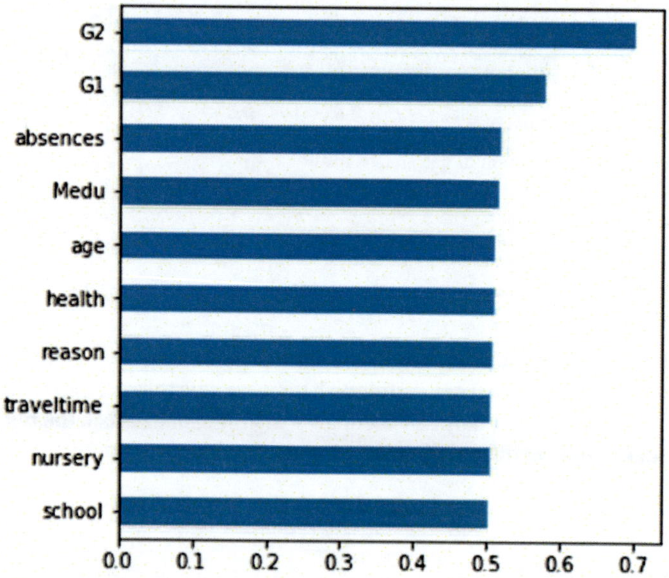

Fig. 4 Graphical representation of ROC score result

MSE (Mean Squared Error) Values for Our Feature Set See (Table 4).

$$\mathrm{MSE} = \frac{1}{n}\sum_{1}^{n}\left(Yi - \widehat{Yi}\right), \tag{3}$$

where *Yi* is the value of dependent attribute.

Table 4 MSE value result

Attribute	MSE
*G*2	2.113615
*G*1	4.170515
Failures	9.640718
Higher	10.308597
Medu	10.52278
Fedu	10.857638
Studytime	11.004261
Dalc	11.00577
Age	11.072356
Goout	11.137201

Using MSE, errors that present in the prediction of the target attribute are computed with respect to each feature. Depending on a particular threshold value for error, features having more error value can be removed from the feature set.

4.1.5 Graphical Representation

See Fig. 5.

4.1.6 Embedded Methods

Embedded methods combine properties of both filter and wrapper methods. It does not separate learning from feature selection. An intrinsic model building metric is used during the learning process (Tables 5 and 6).

Lasso regression is applied on set of features which performs both variable selection and regularization to enhance prediction quality. Here *G*2, *G*1, Absence, Famrel, paid these attributes are obtained in the order of decreasing coefficient values irrespective of their sign.

Ridge regression does not attempt to select features at all, it instead uses a penalty applied to the sum of the squares of all regression coefficients. So, here we are getting two top attributes Dalc and romantic different from the ridge regression result.

4.1.7 Wrapper Methods

Wrapper methods find out the relevance of each feature in predicting the target attribute but suffers from over-fitting issues.

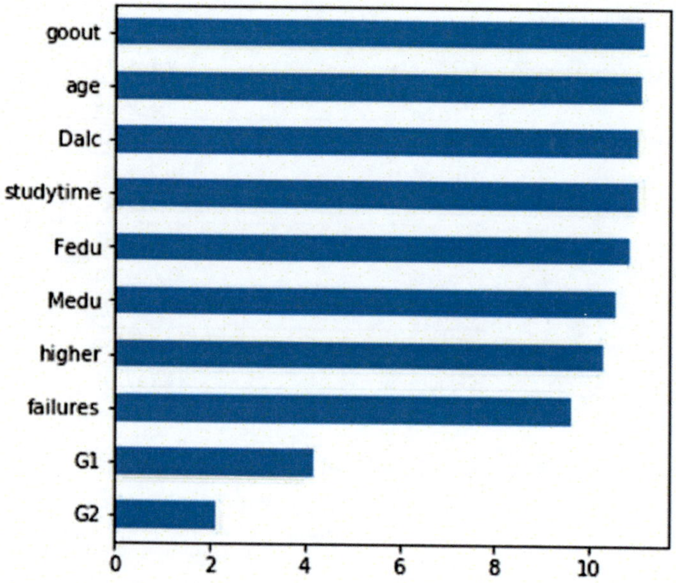

Fig. 5 Graphical representation of MSE value result

Table 5 Lasso regression result

Selected attributes
Failures
Paid
Absences
G1
G2

Table 6 Ridge regression result

Selected attributes
Fedu
Traveltime
Failures
Paid
Higher
Romantic
Dalc
Absences
G1
G2

Table 7 Recursive feature selection using random forest result

Attribute	Feature importance
*G*2	0.207977
*G*1	0.133048
Absences	0.081647
Age	0.055740
Goout	0.054908
Health	0.052090
Mjob	0.050624
Walc	0.049776
Freetime	0.049764
Fedu	0.048255

Recursive Feature Selection Using Random Forest Recursive feature elimination selects predictors in a backward selection approach. In this technique, at first a model is built on the whole set of predictors and then it computes importance score for each one. Predicto *r* (*s*) having least score are removed, and procedure continues by rebuilding the model and so on. The analyst must decide on number of predictor subsets to evaluate and each subset's size. This result has also given importance to 'age' and 'goout' attribute apart from '*G*2' and '*G*1' (Table 7).

So, results obtained from different feature selection algorithms provide various subset of features as they differ in methodology. Filter methods discussed show statistically that *G*1, *G*2, Absence, School, Reason, these five attributes are the most relevant ones in case of discrete target and dataset. But in case of continuous values, *G*2, *G*1, Failure, Higher education, Medu, Fedu, these are more relevant.

Among all the results obtained, *G*1 and *G*2 come out to be the most significant feature for prediction but correlation feature selection shows that *G*1 and *G*2 are highly correlated, and hence for getting good prediction result in the final work, it will be beneficial to remove *G*1 from the feature set.

4.2 *Performance Prediction Results (G3 Being Target Attribute)*

After feature selection procedure, major focus was on the selection of machine learning method for prediction. Classification methods have been used greatly. There lie certain reasons for which classification methods are better candidate for this dataset. Here *G*3 being target attribute contains discrete values (0 to 20). To predict future values of *G*3 attribute obviously classification procedures will be more beneficial to use as it works on labeled category (Figs. 6 and 7).

Only linear regression is used here. In case of regression methods, straight line is drawn having *G*3 in X-axis. Due to the general characteristics of regression method,

Machine Learning Algorithms / Feature Selection Result Used	Logistic Regression (accuracy score)	SVM (accuracy score)	KNN (accuracy score)	Naïve Bayes (accuracy score)	Random Forest (accuracy score)	Linear Regression (MSE)
Correlation	71	74	68	69	70	0.277
ROC	74	75	63	75	74	81.83
MSE	72	75	63	51	67	0.277
Recursive Feature Selection	73	73	58	73	70	0.2816
Lasso Regression	35	36	26	20	42	1.13
Ridge Regression	74	76	65	72	74	0.2779

Fig. 6 Prediction results

Feature Selection Algorithm	Attributes Used							
Chi-square	G2	absences	G1	failures	reason	school	Walc	schoolsup	
Correlation with target variable	G2	G1	failures	higher	Medu	studytime	Fedu	Fedu	
ROC	G2	G1	absences	Medu	age	health	reason	Traveltime	
MSE	G2	G1	failures	higher	Medu	Fedu	studytime	Dalc	
Recursive Feature Selection	G2	G1	absences	age	goout	health	Mjob	Walc	
Lasso Regression	failures	paid	absences	G1	G2				
Ridge Regression	Fedu	traveltime	failures	paid	higher	romantic	Dalc	Dalc	G1

Fig. 7 Attributes are taken into consideration during prediction

it may happen that resulting prediction value will go below zero and above 20, but it is not possible as $G3$ value cannot be negative and cannot be greater than 20. That is why we have restricted our work within linear regression.

As given in result part, accuracy score is deciding factor for prediction models. For all the feature selection methods, accuracy scores obtained by using K-nearest neighbor (KNN) classifier are very low. So KNN is not at all suitable classifier to be used for this dataset. Influences of attribute combinations provided by Chi-square and Lasso regression method is very low on the prediction. Other results are almost satisfactory. However, the work can be extended to use more machine learning models to get better prediction performance.

5 Conclusion and Future Scope

The focus of the paper was to examine how the results of various feature selection tests influence different prediction model performance. Such experiments also find out the accuracy of choice of attributes during data collection. In case of any educational institution, attributes influencing their students' performance changes as per its regional culture, habits, socioeconomic scenario. In the future, same experiment can be performed on different regional data. Depending on region (generally country), attribute choice is very important. Online data collection procedures can be chosen. As stated earlier in previous part. This work only includes the use of linear regression method. In the future, further progress in this experiment may include knowledge about other types of regression methods so that how other regression methods work on this experiment can be checked.

References

1. Romero C, Ventura S (2017) Educational data mining: a Survey from 1995 to 2005. Exp Syst Appl 136
2. Asif R, Hina S, Haque S (2017) Predicting student academic performance using data mining methods. In: Semantic scholar, vol 17, no 5. N.E.D University of Engineering & Technology /Department of Computer Science & Software Engineering
3. Cortez P, Silva A (2008) Using data mining to predict secondary school student performance. In: 5th annual future business technology conference. Porto, pp 5–12
4. Yassein NA, Helali RGM, Mohomad SB (2017) Predicting student academic performance in KSA using data mining techniques. J Inform Tech Softw Eng 7–9
5. Hasan R, Palaniappan S, Raziff ARA, Mahmood S, Sarker KU (2018) Student academic performance prediction by using decision tree algorithm. In: 4th international conference on computer and information sciences (ICCOINS). Kuala Lumpur, pp 1–5
6. Rahman MH, Islam MR (2017) Predict student's academic performance and evaluate the impact of different attributes on the performance using data mining techniques. In: 2nd international conference on electrical & electronic engineering (ICEEE). Rajshahi, pp 1–4
7. Cortez P, University of Minho, Student Performance Dataset. https://archive.ics.uci.edu/ml/datasets/Student+Performance

Diabetes Disease Prediction Using Classification Algorithms

Taiba Sangien, Tabinda Bhat, and Misbah Shafiq Khan

Abstract Diabetes mellitus is a clinical syndrome characterized by hyperglycemia due to absolute or relative deficiency of insulin. Hyperglycemia is the increase in blood glucose concentration. To aid the practice of diabetes detection, we present a method of categorizing patients, which will predict the likelihood of diabetes in them with maximum accuracy. In this paper, three most common predictive algorithms, namely support vector machine (SVM), logistic regression, and random forest (RF), are used to detect whether a person is diabetic or not. A survey is performed on individuals of Pima Indians heritage. Pima Indians Diabetes Dataset (PIDD) is sourced from the UCI machine learning repository. The three machine learning algorithms are evaluated based on performances on various measures like accuracy, precision, F-Measure, and recall. Results show that SVM surmounts with the highest accuracy of 80% as compared to the other two algorithms. Ultimately, the results are efficiently shown using receiver operating characteristic area under the curve.

Keywords Machine learning · Diabetes mellitus · Support vector machine · Logistic regression · Random forest · Classification

1 Introduction

Diabetes is a chronic, metabolic disease specified by raising levels of blood glucose (hyperglycemia) [1]. Diabetes mellitus is caused due to deficiency of insulin in the body. The three main types of diabetes are categorized as: gestational diabetes, type 1 diabetes, and type 2 diabetes. Gestational diabetes is a type of diabetes that develops with the onset of pregnancy. In type 1 diabetes, the immune system inaccurately attacks the pancreas with antibodies causing permanent damage to it due to which it is unable to produce enough insulin. The type 2 diabetes starts with insulin resistance where the pancreas produces insulin, but the body is not able to use insulin effectually.

Diabetes is worldwide in distribution. A study by the International Diabetes Federation revealed that the number of people with diabetes is constantly rising. The cases

T. Sangien (✉) · T. Bhat · M. S. Khan
SSM College of Engineering, Srinagar, India

K. Dahal et al. (eds.), *Internet of Things and Its Applications*, Lecture Notes in Electrical Engineering 825, https://doi.org/10.1007/978-981-16-7637-6_17

of diabetes patients were 382 million in 2013, and by 2035, the cases are likely to see a sharp rise and are expected to be 595 million [2]. The global pandemic mostly involves type 2 diabetes and is analogous with some of the contributory factors, including increased longevity, obesity, unsatisfactory diet, sedentary lifestyle, and increasing urbanization. 10% cases of diabetes are mainly due to type 1 and gestational diabetes while the remaining 90% account for type 2 diabetes.

Health issues associated with diabetes are chronic as well as acute. Diabetes increases the risk of cardiovascular disease (the most common diabetes complication) nerve damage (neuropathy), kidney damage (nephropathy), eye damage (retinopathy), Alzheimer's disease, and depression. Pregnant women who have diabetes (gestational diabetes) have special health concerns. High blood sugar levels can cause birth defects. They also can increase the risks of miscarriage and other complications.

A major step towards keeping people suffering from diabetes mellitus in good health is the early diagnosis and treatment of the disease. An initial assessment can be made using machine learning about diabetes mellitus by examining their daily physical data, and it can work as a reference for medical practitioners [3]. These raw data are huge and are generated from different sources and formats. It can have irrelevant features as well as some data might be missing. The process of extorting hidden information from large volumes of raw data is called data mining. The study of large datasets to select valid features and to extract hidden and formerly unknown patterns, relationships, and knowledge are hard to detect with conventional statistical techniques [4]. Many data mining techniques like clustering, classification algorithms like naïve Bayes, decision tree, and neural networks are used for medical diagnoses of various diseases. In this study, we used SVM, random forest, and logistic regression as classifiers to predict whether a person is diabetic or not. A nonlinear classification can be efficiently performed by support vector machines. Decision tree has an appreciative classification power and is one of the most popular machine learning methods in the field of medical care. Random forest generates many decision trees [3]. Logistic regression is a supervised classification algorithm. With the help of these techniques, we validated the accuracies of diabetes mellitus.

The rest of the paper is organized as follows: The next section provides an overview of the related work; the following section illustrates the methodology; the Sect. 4 presents the data modeling techniques; the Sect. 5 includes the result analysis, and finally, the conclusion and future scope are given in Sect. 6.

2 Related Work

A number of exceptional writers have put forth their ideas about the use of various machine learning techniques in diabetes disease classification and some of them are explained below.

Pima Indians Diabetes Dataset (PIDD) was used to detect diabetes using support vector algorithms [5]. Diabetes disease prediction is one of the major machine learning classification problems, and hence, diagnosis and analysis of diabetes are a topic worth studying [3]. Substantial studies concerning diabetes prognosis have been undergone for several years. Recently, some reports have been distinguished between different learning methods. Such comparisons are generally a few and conducted on Pima Indians Diabetes Database with a limited number of datasets [6]. Data mining models and various machine learning techniques are progressively employed in research related to diabetes mellitus from electronic health record data [7].

In study [8], a number of machine learning algorithms, including decision tree, logistic regression, and naive Bayes, are used in order to apply them to real dataset and make use of disease classifiers. In a study, the independent features like maternal age, body mass index, racial origin, previous history of gestational diabetes mellitus, and macrosomic neonate also known as big baby syndrome are some prominent independent predictors for the classification of diabetes [9]. A diabetes prediction system is developed using different machine learning and data mining techniques; however, the artificial neural network (ANN) turns out to give better accuracy than any other models that included expectation maximization (EM), Gaussian mixture modeling (GMM), logistic regression, and SVM [10].

Research shows that the processing of data plays a vital role in model efficiency, for example, in Bhat et al. [11] the accuracy of the model is 82.33%, while that for an unprocessed neural network prediction model comes out to be 75.81% accurate. The experiments performed on Pima Indians Diabetes (PID) dataset in Sisodia and Sisodia [12] give an accuracy of 76.30% using naïve Bayes classification algorithm. K-nearest neighbor (KNN) provided a good, accurate result of 76.96% in Hashi et al. [13]. An intelligent-based classification and prediction model in Sreelakshmi and Preetha [14] shows that the accuracy for FFB-NN feedforward backpropagation neural network without feature selection is 76.2%, whereas with feature selection goes up to 84.8%; however, for decision tree (C4.5) the percentage varies from 57 to 80%, respectively.

In Xu et al. [15], RF was experimented and came out with better accuracy than other data mining algorithms used in the study. Author [16] has efficiently given the best-predicted outcome for diabetic nephropathy using decision tree, RF, SVM, and naïve Bayes. A better diagnosis of diabetes by predictive analysis of selected features in J48 with a good performance is eminent in this research [17]. ID3 decision tree is used to build the predictive model with 80% accuracy [18].

In Mahboob Alam [19], ANN, RF, and K-means clustering algorithms were executed for the prognosis of diabetes mellitus and ANN outperformed by delivering an accuracy of 75.7%. Comparative analysis between ANN, SVM, naïve Bayes, logistic regression, and C5.0 decision tree in Varma and Panda [20] shows that C5.0 and logistic regression are equivalently accurate followed by naïve Bayes with the second-highest accuracy, and the least accuracy is shown in case of SVM. In Radja and Emanuel [21], SVM aids the prediction of diabetes disease by showing an accuracy value of 77.3%. The results of the analysis in MR [22] show that logistic regression gives an accuracy of 77.78% on the PID dataset. Logistic regression generated an accuracy score of 77.9% while as coarse Gaussian SVM approach produced an accuracy score of 65.5% in Al-Zebari and Sengur [23].

Research has been done before on large datasets, and a lot of literature work is published on predictive models with the demand for an accurate prediction. There can be many complications for the classification of diabetes using comparative analysis of the algorithms. Our model with the view of increased accuracy is based on the comparative study of the combination of three machine learning algorithms with similar parameters.

The main objective of this paper is to demonstrate the best predictive model, using a combination of classification algorithms to detect diabetes considering all models to have the same preprocessing and split architecture. The Pima Indians Diabetes Dataset is trained in this system using SVM, logistic regression, and random forest. The sample data is tested which predicts the patient's being diabetic or non-diabetic.

3 Methodology

Preprocessing of data accompanied by data modeling techniques can be performed to improve the performance of the developed prediction model. A conceptual diagram is shown in Fig. 1 that provides a summarized information in a concise manner. The initial phase incorporates the gathering of data which after preprocessing and applying machine learning algorithms are comparatively analyzed to test the accuracy.

3.1 Dataset

The dataset on which the training phase is done is the Pima Indian Diabetes Dataset [PIDD] obtained from the University of California, Irvine (UCI), machine repository standard dataset. This dataset is a collection of indicative remedial reports from

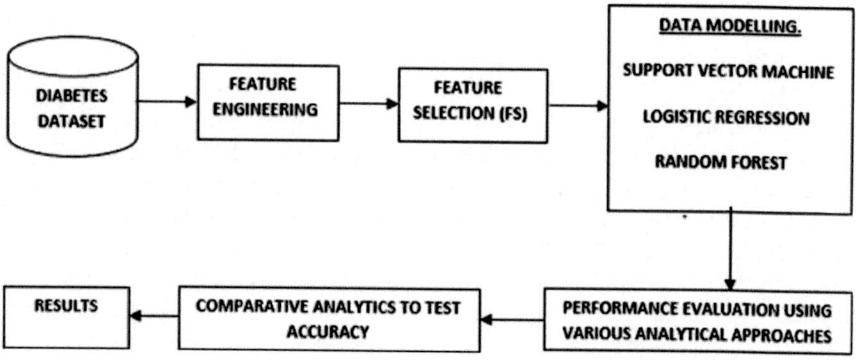

Fig. 1 Proposed method

768 instances of patients. In particular, all patients are female and at least 21 years old of Pima India legacy, a populace living in the Gila River Indian Community in southern Arizona, USA [24]. There are 9 attributes, namely Number of times pregnant, Plasma Glucose concentration 2 hours in an oral glucose tolerance test, Diastolic blood pressure (mm Hg), Triceps skinfold thickness (mm), 2-Hour serum insulin (mu U/ml), body mass index(weight in kg/(height in m)^2) Hereditary factor-Pedigree Function, Age (years), and class variable (0 or 1). The class (target) variable takes a binary (0 or 1) value, while 0 implies the patient has tested negative for diabetes and 1 implies the patient has tested positive for diabetes. There are 268 cases in class 1 and 500 cases in class 0. It indicates that about 34% of patients are diabetic and the remaining 65% of patients are non-diabetic.

3.2 *Feature Engineering*

There were no missing values reported in the Pima Indian Dataset. However, there were missing values disguised as zeros. 28 cases had a diastolic blood pressure of 0, five patients had a sugar of 0, 11 more had a mass body record of 0, 140 others had serum insulin levels of 0, and 192 others had a skinfold thickness readings of 0. Deleting quality features results in the possible loss of important data and also replacing the missing values with zeros prompts poor classification [25]. The desired accuracy might not be acquired in that case. By applying statistical analysis and based on the distribution of data, zeros are replaced with the median value of the corresponding column in the global dataset.

The data was demonstrated visually using box plot which was used to gather the information regarding the variability of statistical data and detect the outlier according to which insulin and skin thickness had 3 and 1 outliers, respectively, which after removing reduced the dataset size from 768 to 764 rows. There were outliers present in other variables as well, but the removal of all the outliers would result in a reduced dataset which would not have been appropriate for classification.

3.3 *Feature Selection*

Feature selection is a vital part of the data modeling workflow. Preprocessing of data, i.e., removal of redundant data or the transformation of unstructured data, is important before data mining [26]. Feature subset selection is significant to amplify the performance without changing the original structure of data mining methods [26]. Selecting proper feature set to include in the model increases the computing process as well. There are many feature selection methods. In this study, we take into account all the 8 independent feature variables in order to produce higher classification accuracy.

3.4 Data Modeling

The heart of model building is the implementation of algorithms. Several widely used classification models such as k-nearest neighbors (KNN), naïve Bayes (NB), decision tree (J48), random forest (RF), support vector machine (SVM), and logistic regression (LR) are used to model patterns of cases and controls based on the extracted features and then use the models to test the ability of our extracted features on the identification of type 2 diabetes mellitus subjects [7]. In this study, we make use of three classification algorithms, i.e., SVM, logistic regression, and random forest to build our model. The input in the form of training and testing data is given to the algorithms to execute and provide an output for further evaluation.

4 Data Modeling Techniques

4.1 Support Vector Machine

Support vector machine first proposed by Vipnik is a set of supervised machine learning method used in medical diagnosis for classification and regression. The objective of the support vector algorithm is to find a hyperplane in N-dimensional space (N-number of features) that separates the positive and negative data points with maximum class distance [27]. SVM converts low-dimensional data into higher dimensions, for higher margin there is a better control over generalization error else the model falls the victim to overfitting. In statistics, it is called structured risk minimization. SVM uses something called kernel function, polynomial function, linear function, etc., to systematically find support vector classifier in higher dimension. The training dataset contains a lot of overlapping and one way to deal with overlapping of data is to use SVM with radial basis function (RBF). The kernel trick allows SVM to construct the classifier without explicitly knowing the feature space. Recently, SVM has attracted a high degree of interest in the machine learning, research community [28]. Radial kernel finds SVC in infinite dimension. It determines how much influence each observation in training dataset has on classifying new observation. The RBF kernel function can be defined as $K(a, b) = \exp(-\gamma \|a - b\|)$ where γ is the kernel parameter and a, b are the training vectors. Kernel functions only calculate the relationships between every pair of points as if they are in higher dimensions. This trick of calculating relationships in higher dimension without actually transforming the data to higher dimensions is called as kernel trick which reduces the amount of computation required for SVM by avoiding the math that transforms the data from higher dimensions to lower dimension. Table 1 shows the confusion matrix of SVM.

Table 1 Confusion matrix for support vector machine

Predicted			
Actual		Diabetic	Non-diabetic
	Diabetic	41	33
	Non-diabetic	13	143

Table 2 Confusion matrix for logistic regression

Predicted			
Actual		Diabetic	Non-diabetic
	Diabetic	40	34
	Non-diabetic	14	142

4.2 Logistic Regression

Logistic regression is a specific type of generalized linear model used for classifying records of a dataset based on the values of the input fields. It is a classification algorithm for categorical or discrete target field instead of a numeric one. In logistic regression, we predict a variable that is binary all of which can be coded as 0 or 1 like whether something is true or false, pass or fail, win or lose, alive or dead, default or no-default, healthy or sick where the dependent variable is dichotomous. Logistic regression can be used for both binary classification and multi-class classification. The curve in logistic regression returns a probability score between 0 and 1for a given sample of data. The main objective of training a model is to change the parameters of the model so as to be the best estimation of the labels of the samples in the dataset.

Logistic regression is widely applicable to a number of fields that include medical science, social sciences, etc. For example, the Trauma and Injury Severity Score (TRISS), which is widely used to predict mortality in injured patients, was originally developed using logistic regression [29]. It is also used in marketing applications such as prediction of a customer's propensity to purchase a product or halt a subscription. In economics, it can be used to predict the likelihood of a person's choosing to be in the labor force, and a business application is about to predict the likelihood of a homeowner defaulting on a mortgage [10].

In this paper, we used logistic regression to predict whether a patient is diabetic or not based on the specific observed characteristics with an accuracy of 0.79. Table 2 shows the confusion matrix of logistic regression.

4.3 Random Forest

Random forest also known as the random decision forest is an ensemble learning technique which is used both as a classification and regression method. Random forest is built from decision trees. Decision trees are easy to build, easy to use, and easy to

Table 3 Confusion matrix for random forest

Predicted			
Actual		Diabetic	Non-diabetic
	Diabetic	42	32
	Non-diabetic	17	139

interpret, but they are not flexible when it comes to classifying new samples. Random forest combines the understandability of decision trees with flexibility resulting in a vast improvement in accuracy. This algorithm uses a Bootstrap dataset while considering a random subset of variables to extract numerous version of the sample sets from the original training datasets, resulting in a wide variety of decision trees [15]. The variety of trees makes random forest more productive than individual decision trees. The classification is done by combining all the results of decision trees, and the sample is classified by confirmed voting procedures [15]. Bootstrapping the data plus using the aggregate to make a decision is called "bagging." About one-third of the cases are left out while sampling the data and this data is called out-of-bag (OOB) data. Accuracy of the random forest classifier is determined by the proportion of the OOB samples that were correctly classified by the classifier. The proportion of OOB samples that are misclassified account for OOB error. For a better accuracy, the number of variables per each step is altered by using the square of the number of variables at the start and by trying few settings above and below that value in the process. The acquired accuracy in the study comes out to be 0.78. Table 3 shows the confusion matrix of random forest.

5 Result Analysis

In order to evaluate the results of the classification, analysis of performance metrics has been adopted. In performance matrices, there are four classes:

True positive (TP): The number of features that are correctly classified to a particular class. It is when the actual result gives diabetic as output and the predicted output comes out to be the same.

True negative (TN): The number of features correctly rejected from a particular class. It is when the actual result gives non-diabetic as output and the predicted output comes out to be the same.

False positive (FP): The number of features incorrectly rejected from a particular class. It is when the actual result gives non-diabetic as output and the predicted output comes out to be diabetic, also known as type I error.

False negative (FN): The number of features incorrectly classified to a particular class. It is when the actual result gives diabetic as output and the predicted output comes out to be non-diabetic, also known as a type II error.

The experiments for the classification of diabetes performed using different classification algorithms give different results. Table 4 shows the number of data points

Table 4 Classification accuracy

Algorithm	Samples	Training data	Testing data	Classes	Accuracy (%)
Support vector machine	764	534	230	2	80
Logistic regression	764	534	230	2	79
Random forest	764	534	230	2	78

used for testing and training, number of classes, and the respective accuracies by different models. A comparative analysis of support vector machine, logistic regression, and random forest by implementing the same random split 70–30 with 70% (534 records) as training data and 30% (230 records) as testing data shows that the prediction accuracy for support vector machine is higher as compared to random forest and logistic regression. However, there is a slight difference in the accuracies of the models but comparatively SVM proves to be better classifier.

The level of efficacy of a classification model is determined by the number of correctly classified data points given by true positives and true negatives. In any classification model, the main aim should be to reduce type I and type II errors. The following equation is used to calculate the accuracy of the classifiers.

$$\text{Accuracy} = \frac{\text{TP} + \text{TN}}{\text{TP} + \text{FP} + \text{TN} + \text{FN}} \tag{1}$$

The receiver operating characteristic (ROC) curves were generated based on the predicted outputs. The ROC curve is a tradeoff between true positive rate (TPR) and false positive rate (FPR). The FPR is plotted on the x-axis and the TPR on y-axis. The true positive rate is also known as sensitivity or recall for true values, and true negative rate or recall for false values is known as specificity. For true values, precision is the positive prediction value; i.e., it gives the percentage of actual positive results out of total positive predicted results, whereas for false values precision is the percentage of actual negative results out of total negative results.

For positive results:

$$\text{Sensitivity} = \text{TPR} = \text{Recall} = \frac{\text{TP}}{\text{TP} + \text{FN}} \tag{2}$$

$$\text{Positive Precision Value} = \text{Precision} = \frac{\text{TP}}{\text{TP} + \text{FN}} \tag{3}$$

For negative results:

$$\text{Specificity} = \text{TNR} = \text{Recall} = \frac{\text{TN}}{\text{TN} + \text{FP}} \tag{4}$$

$$\text{Precision} = \frac{\text{TP}}{\text{TP} + \text{FN}} \tag{5}$$

The following equations are used to calculate the false positive rate (FPR) and false negative rate (FNR) also known as fall-out and miss-rate, respectively.

$$\text{Fall} - \text{Out} = \text{FPR} = \frac{\text{FP}}{\text{FP} + \text{TN}} \tag{6}$$

$$\text{Miss} - \text{Out} = \text{FNR}\frac{\text{FN}}{\text{FN} + \text{TP}} \tag{7}$$

Precision and recall together combined with new measure is known as F1-Score and is given as:

$$\text{F1 - Score} = \frac{2 * \text{Precision} * \text{Recall}}{\text{Precision} + \text{Recall}} \tag{8}$$

The results in Fig. 2 show the positive that is diabetic and negative that is non-diabetic outcomes for different classifiers. The sensitivity for SVM and random forest comes out to be higher than logistic regression, and hence, SVM and random forest are slightly better at correctly identifying positives which in this case are patients that are diabetic. Similarly, the specificity for SVM and random forest is higher than logistic regression; hence, SVM and random forest are better at correctly identifying negatives, i.e., patients that are not diabetic.

By analyzing Fig. 2, the values of sensitivity for SVM are higher than logistic regression and random forest that indicates SVM is slightly better at correctly identifying positives which in this case are patients that are diabetic. Similarly, the specificity for SVM is higher than logistic regression and random forest implying SVM to be better at correctly identifying negatives, i.e., patients that are not diabetic. In other words, there is an improvement in the classification by support vector machine classifier.

Classifiers	Precision.		Recall		F1-Score		ROC-Curve Area
	Non-diabetic	Diabetic	Non-diabetic	Diabetic	Non-diabetic	Diabetic	Diabetic and Non-diabetic.
Support Vector Machines using RBF.	0.81	0.76	0.92	0.55	0.86	0.64	0.671
Logistic Regression	0.81	0.74	0.91	0.54	0.86	0.62	0.871
Random Forest	0.81	0.71	0.89	0.54	0.85	0.63	0.843

Fig. 2 Performance of SVM, logistic regression, and random forest classifiers

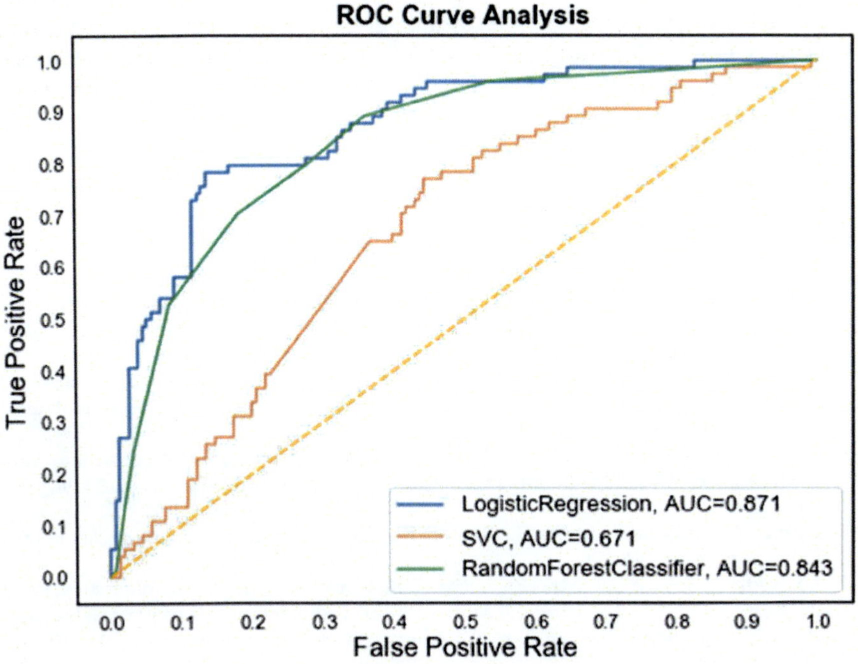

Fig. 3 ROC curve for SVM, logistic regression, and random forest

The receiver operating characteristic curves for diabetes data using SVM, logistic regression, and random forest are shown in Fig. 3.

The accuracy also increases when desirable hyperparameters are set. Hyperparameters determine the structure of the model and are arbitrarily set before training the model. Tuning of hyperparameters, i.e., the right combination of their values, gives a higher accuracy of the function.

6 Conclusion and Future Scope

Diabetes is a major public health concern and diagnosis of the disease at an early stage is a real-world complication in medical field. In this study, various classification methods are designed in order to bracket diabetic patients from non-diabetic ones. This work includes the study of three machine learning algorithms, namely support vector machine (SVM), logistic regression, and random forest. These classification algorithms were assessed on different measures in search of discovering the best result in terms of accuracy. Experiments were conducted on Pima Indians Diabetes Database and the experimental results gave a desired accuracy of 80% in case of the support vector classifier.

In the future, many other life-threatening diseases can be predicted in advance before they prove lethal using the designed system with the used machine learning algorithms. Moreover, there are many other machine learning classification algorithms that can be used to improve the prediction of diabetes mellitus.

References

1. Barakat N, Bradley AP, Barakat MNH (2010) Intelligible support vector machines for diagnosis of diabetes mellitus. IEEE Trans Inf Technol Biomed 14(4):1114–1120. https://doi.org/10.1109/TITB.2009.2039485
2. Mercaldo F, Nardone V, Santone A (2017) Diabetes mellitus affected patients classification and diagnosis through machine learning techniques. Procedia Comput Sci 112:2519–2528. https://doi.org/10.1016/j.procs.2017.08.193
3. Zou Q, Qu K, Luo Y, Yin D, Ju Y, Tang H (2018) Predicting diabetes mellitus with machine learning techniques. Front Genet 9(November):1–10. https://doi.org/10.3389/fgene.2018.00515
4. Aparna K, Reddy NCS, Prabha IS, Srinivas KV (2014) Disease prediction in data mining techniques 1 8491:246–249
5. Karatsiolis S, Schizas CN (2012) Region based support vector machine algorithm for medical diagnosis on pima Indian diabetes dataset. In: IEEE 12th international conference bioinformatics bioengineering BIBE 2012, vol November, pp 139–144. https://doi.org/10.1109/BIBE.2012.6399663
6. Perveen S, Shahbaz M, Guergachi A, Keshavjee K (2016) Performance analysis of data mining classification techniques to predict diabetes. Procedia Comput Sci 82(March):115–121. https://doi.org/10.1016/j.procs.2016.04.016
7. Zheng T et al (2017) A machine learning-based framework to identify type 2 diabetes through electronic health records. Int J Med Inform 97:120–127. https://doi.org/10.1016/j.ijmedinf.2016.09.014
8. Nai-Arun N, Moungmai R (2015) Comparison of classifiers for the risk of diabetes prediction. Procedia Comput Sci 69:132–142. https://doi.org/10.1016/j.procs.2015.10.014
9. Kleijer WJ, van der Sterre MLT, Garritsen VH, Raams A, Jaspers NGJ (2011) Evolution of prenatal detection of neural tube defects in the pregnant population of the city of Barcelona from 1992 to 2006. Prenat Diagn 31(10):1184–1188. https://doi.org/10.1002/pd
10. Komi M, Li J, Zhai Y, Xianguo Z (2017) Application of data mining methods in diabetes prediction. In: 2017 2nd international conference image, visual computer ICIVC 2017, vol S Ix, pp 1006–1010. https://doi.org/10.1109/ICIVC.2017.7984706
11. Bhat VH, Rao PG, Shenoy PD, Venugopal KR, Patnaik LM (2009) An efficient prediction model for diabetic database using soft computing techniques. In: Lecture notes computer science (including Subser. Lect. Notes Artif. Intell. Lect. Notes Bioinformatics), vol 5908. LNAI, p. 328–335. https://doi.org/10.1007/978-3-642-10646-0_40
12. Sisodia D, Sisodia DS (2018) Science direct prediction of diabetes using classification algorithms. Procedia Comput Sci 132(Iccids):1578–1585. https://doi.org/10.1016/j.procs.2018.05.122
13. Hashi EK, Uz Zaman MS, Hasan MR (2017) An expert clinical decision support system to predict disease using classification techniques. In: ECCE 2017—International conference electric computer communication engineering, pp 396–400. https://doi.org/10.1109/ECACE.2017.7912937
14. Sreelakshmi S, Preetha KG (2016) Innovations in bio-inspired computing and applications. Adv Intell Syst Comput 424(Ibica):139–149. https://doi.org/10.1007/978-3-319-28031-8

15. Xu W, Zhang J, Zhang Q, Wei X (2017) Risk prediction of type II diabetes based on random forest model. In: Proceeding 3rd IEEE international conference advance electric electronic information, communication bio-informatics, AEEICB 2017, pp 382–386. https://doi.org/10.1109/AEEICB.2017.7972337
16. Huang GM, Huang KY, Lee TY, Weng JTY (2015) An interpretable rule-based diagnostic classification of diabetic nephropathy among type 2 diabetes patients. BMC Bioinf 16(1):1–10. https://doi.org/10.1186/1471-2105-16-S1-S5
17. Pradeep KR, Naveen NC (2016) Predictive analysis of diabetes using J48 algorithm of classification techniques. In: Proceeding 2016 2nd international conference contemporary computer informatics, IC3I 2016, vol 6, pp 347–352. https://doi.org/10.1109/IC3I.2016.7917987
18. Han J, Rodriguez JC, Beheshti M (2009) Discovering decision tree based diabetes prediction model. Commun Comput Inf Sci 30:99–109. https://doi.org/10.1007/978-3-642-10242-4_9
19. Mahboob Alam T, et al. (2019) A model for early prediction of diabetes. Inf Med Unlocked, 16(January):100204. https://doi.org/10.1016/j.imu.2019.100204
20. Varma KM, Panda BS (2019) Comparative analysis of predicting diabetes using machine learning techniques. Jetir 6(6):522–530. Available: www.jetir.org
21. Radja M, Emanuel AWR (2019) Performance evaluation of supervised machine learning algorithms using different data set sizes for diabetes prediction. In Proceeding—2019 5th international conference science information technology embrace industry 4.0 toward innovation cyber physics system ICSITech 2019, pp 252–258. https://doi.org/10.1109/ICSITech46713.2019.8987479
22. MR (2017) Prediction of diabetes disease using classification data mining techniques. Int J Eng Technol 9(5):3610–3614. https://doi.org/10.21817/ijet/2017/v9i5/170905319
23. Al-Zebari A, Sengur A (2019) Performance comparison of machine learning techniques on diabetes disease detection. In: 1st international informatics software engineering conference innovation technology digital transformation IISEC 2019—Proceedings, pp 2–5. https://doi.org/10.1109/UBMYK48245.2019.8965542
24. Chitra K (2018) Classification of diabetes disease using support vector machine 3(2):1797–1801. Available: https://www.researchgate.net/publication/320395340
25. Vijayan VV, Anjali C (2016) Prediction and diagnosis of diabetes mellitus—A machine learning approach. IEEE Recent Adv Intell Comput Syst RAICS 2015(December):122–127. https://doi.org/10.1109/RAICS.2015.7488400
26. Hamza A, Moetque H (2017) Diabetes disease diagnosis method based on feature extraction using K-SVM. Int J Adv Comput Sci Appl 8(1):236–244. https://doi.org/10.14569/ijacsa.2017.080130
27. Odedra SD, Samanta S, Vidyarthi AS (2010) Computational intelligence in early diabetes diagnosis: a review. Rev Diabet Stud 7(4):252–261. https://doi.org/10.1900/RDS.2010.7.252
28. Auria L, Moro RA (2011) Support vector machines (SVM) as a technique for solvency analysis. SSRN Electron J August. https://doi.org/10.2139/ssrn.1424949
29. Voyd CR, Tolson MA, Copes W (1987) Evaluating trauma care: the TRISS method. Trauma score and the injury severity score. J Trauma 27(4):370–378

Dissected Scene Character Recognition Using HOG Descriptors

Payel Sengupta and Ayatullah Faruk Mollah

Abstract Automatic scene text recognition is an interesting problem in computer vision and Internet of things. It may facilitate intelligent interaction between machines and mankind in today's cloud-enabled civilization. In this paper, we present a method for dissected scene character recognition. At first, color images are converted into grayscale and then some noise removal and pre-processing operations are applied. Next, we normalize them to bring them to a uniform dimension and compute features for training and prediction. Experimenting on scene characters at three different levels of complexities i.e. relatively good images, relatively bad images, and combined images with multiple classifiers such as naïve Bayes, KNN, MLP, random forest and SVM, detail results are reported. Highest accuracies i.e. 74.48% for good images only, 59.13% on bad images only and 71.52% for overall images, are obtained with the SVM classifier. Comparison with similar state-of-the-art methods is also included and our method is found to outperform others.

Keywords Scene character recognition · Scene character pre-processing · HOG descriptors · Pattern classification

1 Introduction

In the era of Internet of things, the development of automatic scene text recognition system is an important necessity. It is a complex problem in retrieval of meaningful information from scene images. Unlike conventional character recognition problems, character or word recognition from scene text components is very challenging due to many factors including varying and/or poor camera resolution, illumination, noise, background, etc. Sample images reflecting different challenges are shown in Fig. 1. Some methods or approaches have already been reported in the literature to address the issues related to scene character recognition. Two types of recognition methods

P. Sengupta (✉) · A. F. Mollah
Department of Computer Science and Engineering, Aliah University, IIA/27 New Town, Kolkata 700160, India

K. Dahal et al. (eds.), *Internet of Things and Its Applications*, Lecture Notes in Electrical Engineering 825, https://doi.org/10.1007/978-981-16-7637-6_18

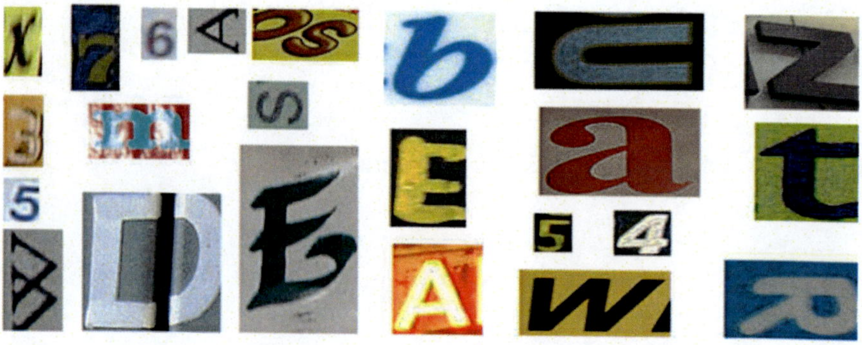

(a) (b)

Fig. 1 Sample dissected scene character images from Chars74k dataset [9]. **a** Bad images dataset. **b** Good images dataset

have been widely applied [1, 2], viz. traditional classifiers driven by handcrafted features and CNN or other deep learning approaches [3, 4]. Though deep learning approaches usually produce relatively better results, they require massive datasets. At present, available scene character and word datasets [5–9] are small in size, specifically in view of requirements for deep learning. Chars74k [9] is a popular benchmark dataset containing 12,505 dissected scene character images.

As the current work focuses on dissected scene character recognition, only scene character recognition works are reported here. Neumann et al. [10] proposed a text localization and recognition method on Chars74k (with only 7705 scene character images) and ICDAR 2003 datasets. In this work, Maximally Stable Extremal Region (MSER) is used for character proposal generation followed by character vs. non-character classification with support vector machine (SVM). Chekol et al. [11] also proposed a character recognition method using curvature-based features. The curvature values of each point are calculated by boundary values of the dissected scene character images and used as feature descriptors. Bai et al. [12] proposed a character recognition method using strokelets, which can automatically detect the annotations of characters. As strokelets are very efficient in characterizing objects, this method is applicable on different languages also. Local binary pattern network called LBPNet is proposed by Lin et al. [13] for character recognition by incorporating two operations, namely LBP and channel fusion. A random projection is also used to recognize characters with memory efficiency in their work. Sundin et al. [14] applied CNN on the synthetic images (scene and font characters) of the Chars74k dataset. A character recognition method using KNN classifier has been proposed by Barnouti et al. [15]. Another CNN-based recognition method is proposed by Abdali et al. [16] on EMNIST and Chars74k datasets where images have been augmented to train the CNN model. Jaderberg et al. [17] reported another method based on CNN with conditional random field (CRF) to detect and recognize characters.

In short, it may be stated that studies carried out so far either rely on augmentation or go for designing composite techniques for the recognition of complex scene character images. In both cases, recognition accuracies are far away from practical applicability. Moreover, the performance of such methods is limited to good or well-acquired scene character images and they often fail to perform equally well for complex scene character images. In this paper, we present a method for the recognition of complex characters, and detail results with multiple classifiers are separately reported on good, bad, and combined sets of the Chars74k dataset. It is found that our method outperforms state-of-the-art methods on the same dataset in all the three combinations of experiments.

2 Applied Approach

In this work, we consider complex color scene character images obtained from Chars74k dataset. First of all, these color scene characters are converted into grayscale images [18] that are preprocessed for noise removal and quality improvement as discussed in Sect. 2.1. Then, feature descriptors are extracted and passed to classifiers as outlined in Sects. 2.2 and 2.3, respectively.

2.1 Pre-processing

Grayscale character images often contain different kinds of noises, degradation and deformation. Hence, pre-processing may greatly help the performance of subsequent steps. Here, we apply median filter [19] to remove impulse noises. It may be noted that median filtering is very effective to remove small noises from images without changing the original size and shape of the images. We have experimentally chosen kernels of size 5×5 pixels. Effect of median filtering on a gray character image may be observed from Fig. 2. After that, all character images are normalized to 32 $\times$ 32 pixels. It is also noticed that character images usually have different contrast and brightness. Consequently, histogram equalization is applied [20, 21] to improve the contrast quality of an image. Illustration of histogram equalization on a sample filtered and normalized image is made in Fig. 3.

2.2 Feature Extraction

In our experiments, histogram of oriented gradients (HOG) [22, 23] feature descriptor is used to calculate the feature vector. We have chosen 324 HOG features to classify 12,505 scene character images (only scene characters are taken from Chars74k dataset) with 62 classes [9]. First, we break scene character images obtained after

Fig. 2 Demonstration of pre-processing on grayscale character image. **a** Grayscale character image before median filtering. **b** Grayscale character image after noise removal operation using median filtering

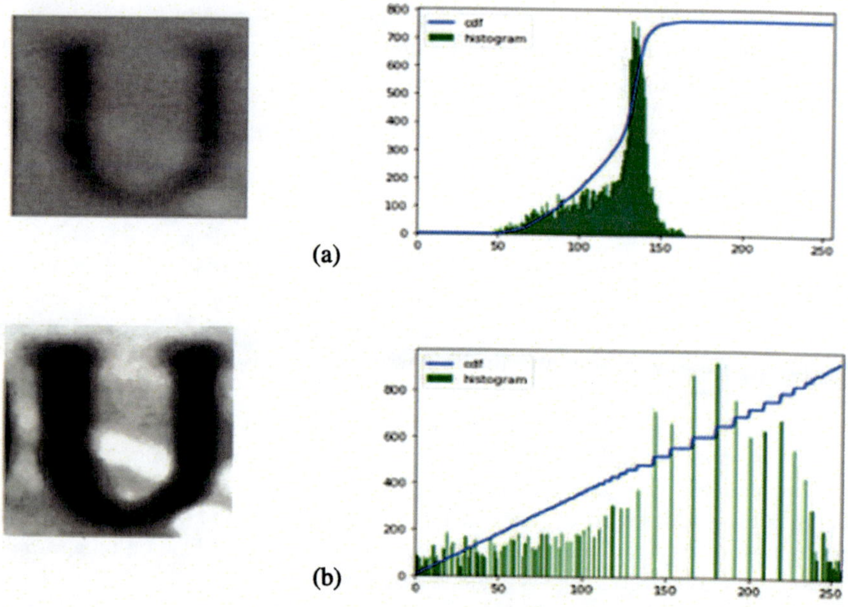

Fig. 3 Illustration of histogram equalization on a sample filtered and normalized image. **a** Grayscale character image with pre-processing before histogram equalization. **b** Grayscale character image with pre-processing after histogram equalization

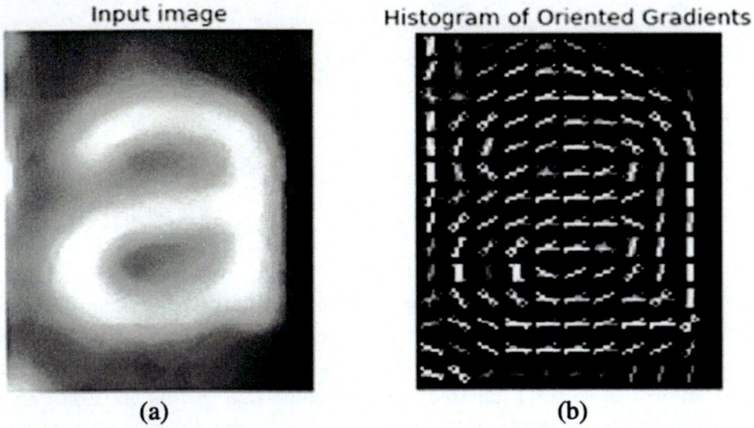

Fig. 4 Overview of HOG descriptors for a sample pre-processed scene character image. **a** Original gray image. **b** Its HOG descriptors

applying operations discussed in Sect. 2.1 into cells (a tiny connected region). Then, we compute a histogram of gradient directions from the pixels within the generated cells. After that, we consider a set of blocks which are formed by group of adjacent cells. In our work, we have experimentally chosen 9 orientations, 8 × 8 pixels per cell, 2 × 2 cells per block for HOG descriptors. In Fig. 4, a pictorial representation of these descriptors obtained for a sample scene character image is shown.

2.3 Scene Character Classification

After feature extraction from obtained grayscale character images, feature vectors are applied to multiple traditional classifiers like naive Bayes, K-nearest neighbors (KNN), multilayer perceptron (MLP), random forest, and support vector machine (SVM). As massive scene character datasets are not available, we have chosen conventional yet high-performing classifiers [24] for our experiments. It is worth mentioning that the size of the dataset we are working with i.e. Chars74k contains only 12,505 images of 62 categories. Like any standard supervised approach, we split the samples into training and test sets that are respectively used for training and prediction.

3 Experimental Results and Discussion

Separate experiments have been carried out on good, bad as well as combined sets of dissected color scene character images of Chars74k dataset. Altogether, it has 12,505 images of 62 character classes of English language, viz. digits (0–9), uppercase alphabets (A-Z) and lowercase alphabets (a-z). In all the three cases, 80% samples are randomly taken for training and the rest 20% samples are used for quantitative assessment of prediction performance.

3.1 Results on Good and Bad Sets

As shown in Fig. 1, image samples of the good set are relatively less complex than that of the bad set of Chars74k dataset. The good set contains a total of 7110 characters and the bad set contains 4506 characters. Classification performance on these two subsets is shown in Tables 1 and 2. It may be noted that results are reported with a number of classifiers in terms of standard metrics. Normalized confusion matrices are also shown in Fig. 5. It may be noted that overall performance on bad set is comparatively less than that of the good set, which is due to increased complexity in the samples of the bad set. For instance, SVM yields 74.48 and 59.13% classification accuracy on good and bad sets, respectively.

Table 1 Recognition performance on the subset of good scene character images of Chars74k dataset (with multiple classifiers and multiple standard evaluation metrics)

Classifier	Precision	Recall	F-measure	Accuracy
KNN	67.19	57.81	62.14	74.78
MLP	59.02	60.59	59.79	70.13
Naive Bayes	53.67	58.16	55.82	63.39
Random Forest	67.78	48.44	56.50	69.79
SVM	70.96	64.64	67.65	74.48

Table 2 Performance on the subset of bad/complex scene character images of Chars74k dataset (with multiple classifiers)

Classifier	Precision	Recall	F-measure	Accuracy
KNN	48.19	40.05	43.74	40.36
MLP	36.52	36.66	36.58	52.15
Naive Bayes	34.42	37.67	35.97	44.51
Random Forest	36.12	26.69	30.69	50.09
SVM	48.90	45.44	47.10	59.13

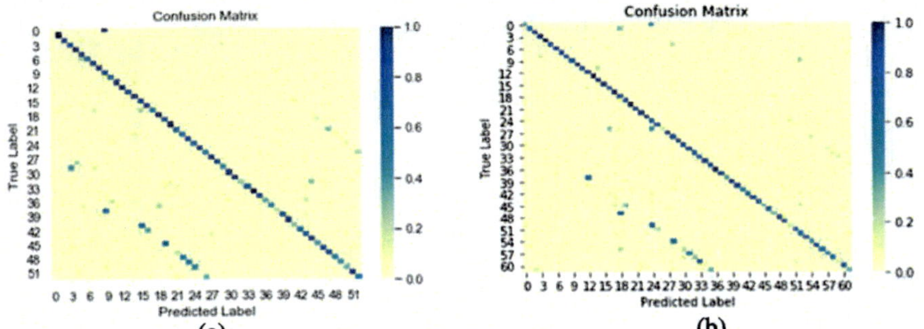

(a) (b)

Fig. 5 Confusion matrices obtained for SVM classifier with polynomial kernel to classify 62 classes (10 for digits, 26 for uppercase alphabets, and 26 for lowercase alphabets). **a** Only good images and **b** Only bad images

3.2 Performance on Combined Set

Considering the combined set of good as well as bad subsets of character samples and splitting in 4:1 ratio, we obtain 10,004 character images for training and 2501 character images for testing. Table 3 shows the obtained results for popular classifiers. It may be observed that classification performance with this combined set is greater than that of the bad set, however lesser than that of the good set, which is quite understandable. A pictorial view of classification and misclassification is included in Fig. 6.

Parameters of the classifiers are empirically tuned. For MLP, considered parameters are three hidden layers of 120 neurons each, 'ReLU' activation function, 4000 maximum iteration and back-propagation learning algorithm. In KNN, number of neighbors is taken as 5. In case of random forest classifier, 400 trees are constructed and final class is determined on the basis of majority of voting. Initially, we experimented with three types of SVM kernels, i.e., linear, radial basis function (RBF) and polynomial, and we found that polynomial kernel with gamma = 2 generates the highest classification accuracy of 71.52% for overall (bad and good images), 59.13% for only bad images and 74.48% for only good images.

Table 3 Summary of recognition results on combined Chars74k (bad and good sets) scene character images

Classifier	Precision	Recall	F-measure	Accuracy
KNN	60.05	52.48	56.01	67.03
MLP	54.49	53.34	53.90	64.41
Naive Bayes	48.49	51.56	49.97	58.58
Random Forest	61.69	40.33	48.77	62.10
SVM	65.70	59.75	62.58	71.52

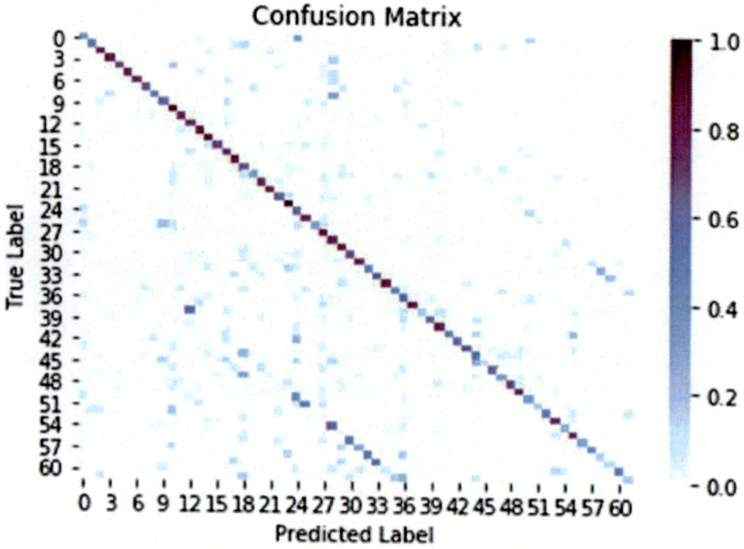

Fig. 6 SVM classifier with polynomial kernel to classify 62 classes (10 for digits, 26 for uppercase alphabets and 26 for lowercase alphabets) using confusion matrix

3.3 Comparative Study and Discussion

Comparison with similar works of the current state-of-the-art is often desirable. Hence, we have shown such comparison in Table 4. As reflected in Table 4, our method outperforms other methods on the same dataset in all the three cases of experiments, i.e., good set, bad set, and combined sets. It is worth mentioning that deep learning-based works are not considered for comparison.

Taking the complexity associated with dissected scene characters into account, obtained results may be considered encouraging. However, insights into the problem reveal that major misclassifications are due to complex background around character strokes, and small number of samples for few classes [25] like in digits '5', '7', uppercase alphabets, '*Q*', '*X*', '*Z*', and lowercase characters '*b*', '*j*', '*q*', '*z*'. Another major problem is that, identical looking characters only differ in size, e.g., {'*C*', '*c*'}, {'*O*', '*O*'}, {'*X*', '*x*'}, {'*Z*', '*z*'} that are quite difficult to discriminate after normalization. Many character images are also orientated at some angles. Thus, deeper insights are required and more issues should be taken into account to address efficient scene character recognition.

Table 4 Performance comparison with similar methods of the current state-of-the-art and the proposed method on Chars74k dissected scene character dataset

Method	Year	Recognition accuracy			
		Bad images	Good images	Overall images	Remarks
Chekol et al. [11]	2019	–	–	65.30	1068 randomly chosen images
Campos et al. [9]	2009	–	54.30	–	7705 natural images
Neumann et al. [10]	2010	–	71.60	–	7705 natural images
Lin et al. [13]	2020	58.31	–	59.46	4798 bad natural images, bad and good 12,505 natural images
Bai et al. [12]	2016	–	–	62.00	12,505 bad and good natural images
Proposed	2020	59.13	74.48	71.52	Only 12,505 natural images

4 Conclusion

In this paper, we propose a scene character recognition method using different pre-processing operations and HOG feature descriptor. Experiments have been carried out on dissected scene character images at three different levels of complexities, relatively good set, bad/complex set, and their combination. With a moderately trained SVM classifier, 74.48, 59.13, and 71.52% accuracies are obtained for only good images, only bad images, and overall images (good and bad), respectively. Comparison with state-of-the-art reveals that in all these three cases, our method has surpassed the results of similar methods. However, there is scope of further improvement by addressing the undertraining issue due to small number of samples for few classes, applying certain post-processing techniques to avoid misclassification between similar-shaped alphabets, and designing efficient features to deal with background complexities.

Acknowledgements The authors are thankful to the Department of Computer Science and Engineering of Aliah University, Kolkata, India, for providing every kind of support for carrying out this research work. P. Sengupta is further grateful to Dept. of MA & ME, Govt. of West Bengal for providing Swami Vivekananda Merit cum Means Fellowship.

References

1. Lin H, Yang P, Zhang F (2020) Review of scene text detection and recognition. Arch Comput Meth Eng 27(2):433–454
2. Zhu Y, Yao C, Bai X (2016) Scene text detection and recognition: recent advances and future trends. Front Comput Sci 10(1):19–36
3. Sengupta P, Mollah AF (2020) Journey of scene text components recognition: progress and open issues. Multimedia Tools Appl. Springer, 80(4):6079–6104
4. Liu X, Meng G, Pan C (2019) Scene text detection and recognition with advances in deep learning: a survey. Int J Doc Anal Recogn 22(2):143–162
5. Karatzas D, Shafait F, Uchida S, Iwamura M (2013) ICDAR 2013 robust reading competition. In: Proceedings of 12th international conference on document analysis and recognition. IEEE, pp 1484–1493
6. Iwamura M (2018) Adv Scene Text Datasets: arXiv preprint arXiv:1812.05219
7. Lucas S, Panaretos M, Sosa AL, Tang A Wong, S and Young R (2003) ICDAR 2003 robust reading competitions. In: Proceedings of seventh international conference on document analysis and recognition. IEEE, pp 682–687
8. Karatzas D, Gomez-Bigorda L, Nicolaou A, Ghosh S, Bagdanov A, Iwamura M, Matas J, Neumann L, Chandrasekhar VR, Lu S, Shafait F (2015) ICDAR 2015 competition on robust reading. In: Proceedings of 13th international conference on document analysis and recognition. IEEE, pp 1156–1160
9. De Campos TE, Babu BR, Varma M (2009) Character recognition in natural images. In: Proceedings of international conference on computer vision theory and application, pp 273–280
10. Neumann L, Matas J (2010) A method for text localization and recognition in real-world images. In: Proceedings of Asian conference on computer vision. Springer, pp 770–783
11. Chekol B, Celebi N, Taşci T (2019) Segmented character recognition using curvature based global image feature. Turkish J Electric Eng Comput Sci 27(5):3804–3814
12. Bai X, Yao C, Liu W (2016) Strokelets: a learned multi-scale mid-level representation for scene text recognition. IEEE Trans Image Proc 25(6):2789–2802
13. Lin JH, Lazarow J, Yang A, Hong D, Gupta R, Tu Z (2020) Local binary pattern networks. In: Proceedings of IEEE winter conference on applications of computer vision, pp 825–834
14. Sundin H, Josefsson J (2020) Evaluating synthetic training data for character recognition in natural images. Degree Project of KTH Royal Institute of Technology, Sweden
15. Abdali AR, Ghani RF (2019) Robust character recognition for optical and natural images using deep learning. In: Proceedings of IEEE student conference on research and development, pp 152–156
16. Barnouti NH, Abomaali M, Al- MHN (2018) An efficient character recognition technique using K-nearest neighbour classifier. Int J Eng Technol 7(4):3148–3153
17. Jaderberg M, Simonyan K, Vedaldi A, Zisserman A (2014) Deep structured output learning for unconstrained text recognition. In: Proceedings of international conference on learning representations, pp 1–10
18. Mollah AF, Basu S, Nasipuri M (2012) Computationally efficient implementation of convolution-based locally adaptive binarization techniques. In: Proceedings of international conference on information processing. Springer, pp 159–168
19. Bapu J, Florinabel DJ (2020) Real-time image processing method to implement object detection and classification for remote sensing images. In: Proceedings of international conference on earth science informatics. Springer, pp 1–13
20. Goyal V, Shukla A (2020) An enhancement of underwater images based on contrast restricted adaptive histogram equalization for image enhancement. In: Smart innovations in communication and computational sciences. Springer, pp 275–285
21. Mollah AF, Basu S, Nasipuri M, Basu DK (2013) Handheld mobile device based text region extraction and binarization of image embedded text documents. J Intell Syst 22(1):25–47
22. Gogna A, Majumdar A (2019) Discriminative autoencoder for feature extraction: application to character recognition. Neural Proc Lett 49(3):1723–1735

23. Sengupta P, Mollah AF (2020) Scene character recognition with morphological filtering and HOG features. In: Proceedings of an international conference on computing & communication. Springer, pp 1–9
24. Bakas J, Mahalat MH, Mollah AF (2016) A comparative study of various classifiers for character recognition on multi-script databases. Int J Comput Appl 155(3):1–5
25. Mollah AF, Basu S, Nasipuri M (2018) An automatic annotation scheme for scene text archival applications. In: Proceedings of international conference on advances in computing and data sciences. Springer, pp 66–76

A Comparative Analysis of Network Intrusion Detection System for IoT Using Machine Learning

Bhaskar Mondal and Sunil Kumar Singh

Abstract The size of Internet of Things (IoT) networks is increasing exponentially, and parallelly, the security threats are also escalating. As many of the IoT devices run on limited resources, any intrusion attack based on DoS, packet flooding, man-in-middle and probing is effective to disturb, distract and defunct the networks. A intrusion detection is always a challenging task for the network administrator or any automated software system. A machine learning network-based intrusion detection system (IDS) works efficiently and detects such attacks in any type of networks. It analyzes the packets transmitting over the networks without bothering the IoT devices. Hence, IDS systems are highly crucial and important for IoT network security. This paper proposes a machine learning network-based IDS for securing IoT networks. The proposed technique uses classification techniques to classify a network packet as normal or some kind of malicious attacks. The model was trained with a dataset which is a network logs collected from a network transmitting data from NodeMCU with ESP8266 wi-fi module to a server. The data was captured from the ultrasonic sensor with Arduino and NodeMCU used to monitor a network. For choosing the best detection model out of eight classification based models were studied. The decision tree and random forest are most accurate models as compared to other classification techniques. The comparative analysis of these models is analyzed and discussed in the result section.

Keywords IoT · Network security · Intrusion detection system · Machine learning · Attacks

B. Mondal (✉)
Department of Computer Science and Engineering, National Institute of Technology Patna, Patna, India
e-mail: bhaskar.cs@nitp.ac.in

S. K. Singh
School of Computer Science and Engineering, VIT-AP University, Near Vijayawada, Andhra Pradesh, India

K. Dahal et al. (eds.), *Internet of Things and Its Applications*, Lecture Notes in Electrical Engineering 825, https://doi.org/10.1007/978-981-16-7637-6_19

1 Introduction

Internet of Things (IoT) has grown up very faster compared with any other technology around the world. IoT is a smart communication technology where different devices are connected to each other, and the end devices, known as sensor nodes, sense their surroundings and send alert messages to their peer devices [1, 14, 15, 20]. Here, two similar as well as different devices can communicate, and finally, data was transferred to cloud or Internet [17]. Now, in COVID-19 pandemic, many countries are using IoT in their health services as well as other areas to maintain the social distance. It also reduces the direct contact of human beings in their working environment [19]. Along with this, online attacks are also increased drastically. According to Akamai report, the distributed denial-of-service (DDoS) attacks increase a 20% in daily between February and March. As per the McAfee threat Labs report of 2019–2020 (Q1) [12], it observed 375 threats per minutes and new IoT Malware increased by 58%. According to the report, the major threats areas are financial services, health care, education and government sectors. Due the advancement of technologies, the attackers are also using new and powerful attacking techniques.

An IDS monitors network packets for illegal and suspicious activity and issues alerts if such type of events encounters. When any anomalies behavior or breach detected, it mostly informed to an administrator. In few cases, it stored centrally using a security information and event management (SIEM) system. Based on the implementation, IDS systems can be network-based, host-based, distributed or virtual machine-based IDS [10]. Though signature-based approaches are capable to detect the malicious traffic, but they cannot detect any unknown attacks efficiently. IoT is a huge network and content large amount of data (bigdata), and therefore, conventional IDS techniques and network administrators are not sufficient to detect any unusual or malicious data. There are two kinds of IDS systems available depending on the placement of sensor nodes in IoT network: host-based IDS (HIDS) and network-based IDS (NIDS) [9]. In HIDS, sensors commonly called agents installed to each system, and each agent collects the data about malicious activity. Whereas, NIDS works differently as it scans networks at host/router level and examines the network traffic and makes a logs report or generates an alert if any malicious nodes encounters. In IoT, NIDS is popularly used due to its huge size and generation of large amount of data. Figure 1 shows the complete process of NIDS system. Network-based IDS performs better on IoT devices as they run on network devices instead running on the IoT device itself. NIDS used three basic types of techniques to detect the malicious traffic: signature-based, anomaly-based and stateful protocol analysis. Signature-based are old techniques inspired by anti-virus working mechanism where it checks for a specific patterns like byte sequence in network traffic or known unusual activity sequence used by malware. It has its databases to match these traffic and take the decision. Whereas, anomaly-based techniques observe the network behavior and make an alert when any anomaly detects. Since sensor nodes have limited resources, this needs to be considered when we design any new IDS system for IOT. Therefore, NIDS system is better as it works at host level or router level. Additionally, IoT

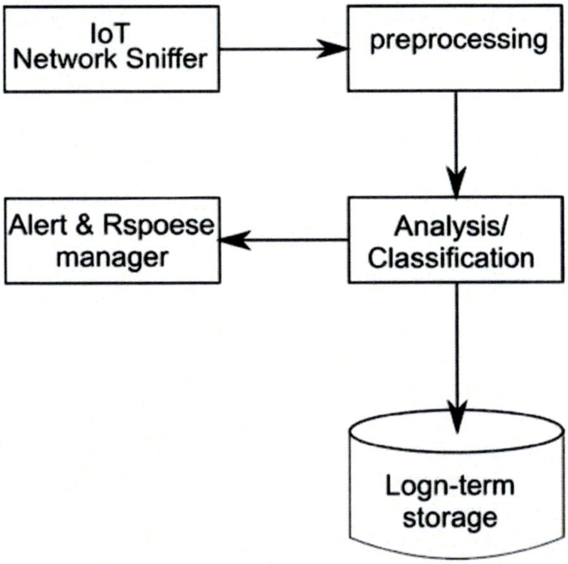

Fig. 1 Network-based intrusion detection system

systems are mostly wireless in nature which is more vulnerable for various security attacks. Nowadays, ML-based automated IDS techniques are more popular, and so many schemes are already proposed using different ML methods [2, 4–6, 16, 18]. In this paper, we propose an intrusion detection system based on machine learning (IDS-ML) which clearly identifies the novel as well as conventional attacks. We have used seven ML-based classifier models to analyze their performance such as (i) Logistic regression (LR), (ii) K-nearest neighbors classifier (KNN), (iii) Decision tree (DT), (iv) Random forest (RF), (v) Gaussian Naive Bayes classifier (GNB), (vi) AdaBoost classifier (ABC), (vii) Gradient boosting (GB) and (viii) Linear discriminant analysis (LDA). The experiment is conducted on a benchmark intrusion dataset, and the accuracy of the gradient boosting algorithm has reached 100%, which is higher than other comparison algorithms. This shows that the proposed algorithm provides an improvement solution for the performance of intrusion detection.

The rest of the paper is arranged as follows: Some latest related work is discussed in Sect. 2. In Sect. 3, the proposed IDS model and all classification models performance are discussed, Sect. 4 deals with the result and analysis of various DL classifiers using the dataset, and finally, Sect. 5 concludes this work with future scope.

2 Related Work

In this section, some important and recent papers related to IDS and ML-IDS are discussed. An IDS based on SVM classifier is proposed by Deng et al. for distributed and hierarchical IDS [7]. It achieved accuracy rate more than 90% by using biasing in feature selection. Mitrokotsa et al. [13] have done a performance analysis of five popular supervised classification algorithms (the Gaussian mixture model,the Naïve Bayes model, multilayer perceptron, (SVM) model and the linear model) to detect the malicious node for mobile ad-hoc networks. The multilayer perceptron classifier has shown best performance, while the Naïve Bayes classifier performs poorly. Azmoodeh and Choo [3] proposed a malware detection for "Internet Of (Battlefield) Things Devices" using the deep eigenspace learning techniques. They achieved 99.68%, 98.59%, 98.37% and 98.48% of accuracy, precision, recall and F-measure, respectively. Doshi et al. [8] have applied five machine learning methods to classify normal IoT data from denial-of-service (DoS) attack packets. The five models are based on KNN "KDTree" algorithm, decision tree using Gini impurity scores, SVM with linear kernel (LSVM), random forest using Gini impurity scores and artificial neural network (ANN). Based on the accuracy, recall, precision and F1-score, the random forest performed well among all the models.

Zwane et al. [22] have analzed seven machine learning classifiers, namely multilayer perceptron, Bayesian network, support vector machine (SMO), Adaboost, random forest, bootstrap aggregation and decision tree (J48). They have used WEKA tool to implement and performance evaluation of the classifiers. According to their results, ensemble-based learning methods performed better as compared to single learning methods on accuracy metrics, but it is little bit slower in terms of build time and model test time. Kumar et al. [11] proposed a unified IDS (UIDS) for IoT environment to prevent the network from four different types of attacks: exploit, DoS, probe and generic. They have used a benchmark dataset UNSW-NB15 to detect the malicious traffic in IoT. Their results claim that this scheme detects more malicious packets as compared to existing methods designed on same dataset. Verma et al. [21] have investigated ML-based classifiers for security of IoT against DoS attacks. They have done a comprehensive study on classifiers for the advancement of anomaly IDS. They have used popular benchmark datasets like CIDDS-001, UNSWNB15 and NSL-KDD. In their results, it concludes that classification and regression trees and extreme gradient boosting methods are best choice to make a balance between metrics and response time for building IoT specific anomaly-based IDS.

3 The Data Collection and Preparation

An open dataset is chosen for the experiment purpose. The dataset is a network logs collected from a network transmitting data from NodeMCU with ESP8266 wi-fi module to a server. The data was captured from the ultrasonic sensor with Arduino and NodeMCU used to monitor a network. The dataset consists of 477,426 odd packets captured with 13 features as described in Table 1.

The dataset is already labeled with six different classes as given in Table 2.

A packet normal means that the packet is not malicious or is not intended to initiate an attack. Wrong setup denotes that the packet carries some erroneous information, DDoS means that packet has potential to make a distributed denial-of-service attack

Table 1 Feature details

Sl. No.	Feature	Data type
1	Frame number	int64
2	Frame time	int64
3	Frame length	int64
4	eth source	int64
5	eth destination	int64
6	Source IP	int64
7	Destination IP	int64
8	IP Protocol	float64
9	IP length	float64
10	TCP length	float64
11	TCP source port	float64
12	TCP destination port	float64
13	Value	float64
14	Normality	int64

Table 2 Class labels details

Sl. No.	Class label
1	Normal
2	Wrong setup
3	ddos
4	Data type probing
5	Scan attack
6	Man in the middle

which means it can keep the resources busy by different kinds of packet flooding or keep sending the same packet. Scan attacks are for intention to scan the network for vulnerabilities and open and free ports and IPs. Man-in-middle attacks involve attacker who is eavesdropping, capturing and resending the packets pretended that they are the legitimate user.

The features are being selected based on the correlation among them. A heat map representation of the correlation among the features is presented in Fig. 2. The features are not highly correlated to each other, and hence, those features are selected for the training of the model.

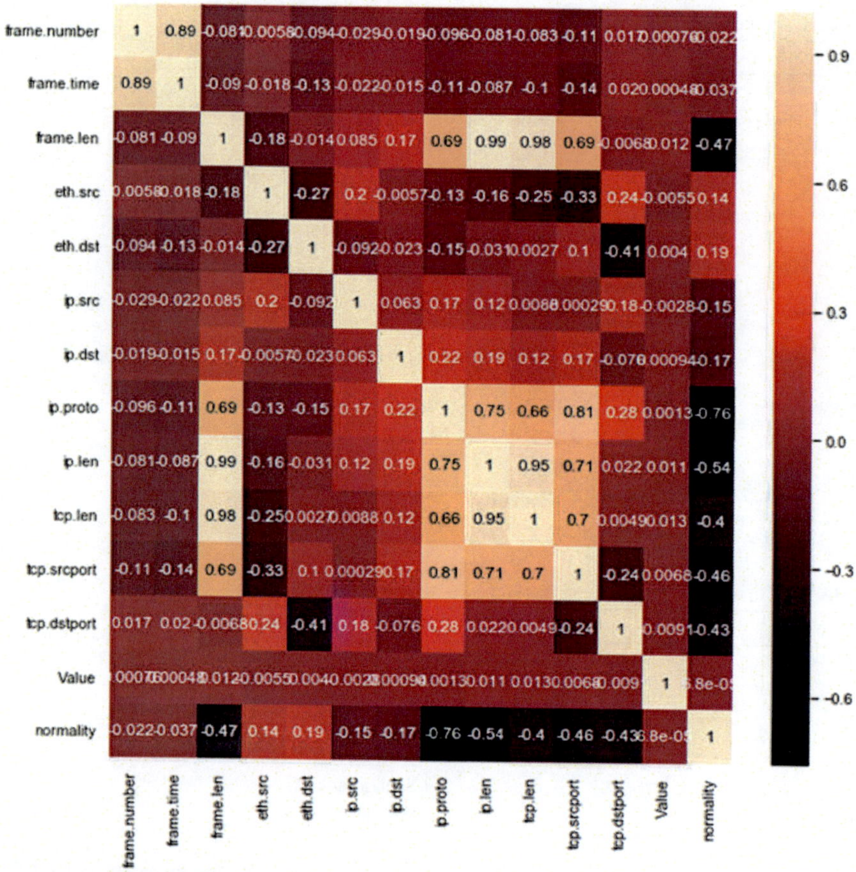

Fig. 2 Heat map of the features

4 Results and Analysis

The results and performance comparison of the models are presented in this section. The overall process and experimental setup are visualized in Fig. 3. The whole process may be separated in two major parts; first data collection and preprocessing, second build the model for intrusion detection. the first part is already discussed in the previous section.

For building the models, the dataset is divided into training and test set. Where 75% data is considered as training data, and 25% data is used for testing purpose. Each classification model is trained and tested with the test set using Python 3 on Windows 10 platform.

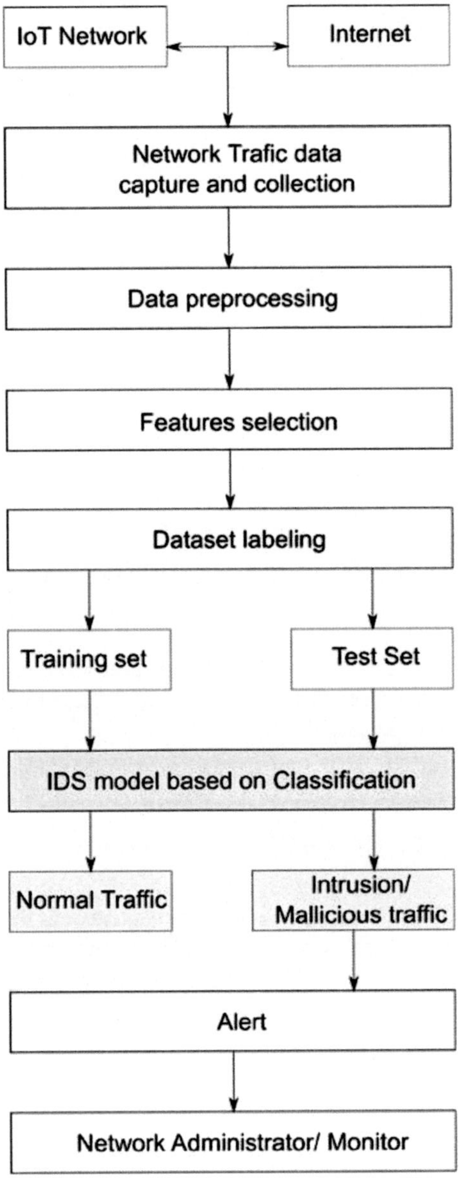

Fig. 3 Working process of the IDS

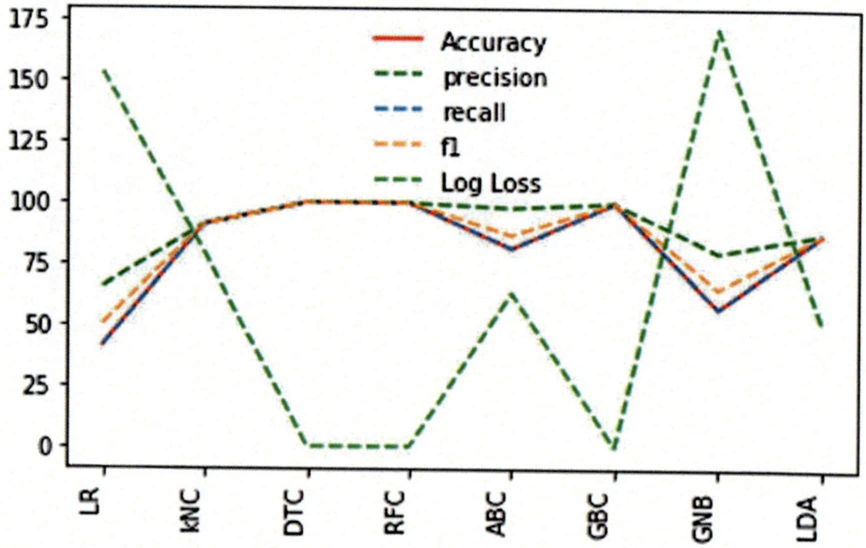

Fig. 4 Plot of accuracy, precision, recall, F1 and log loss of all eight models

For each model, accuracy, precision, recall, F1-score and log loss are calculated. We can see the performance of different models based on precision, recall, F1-score and log loss in Fig. 4.

4.1 Accuracy and Correctness

The accuracy represents the closeness of the prediction to the correct class. Accuracy a is the ratio of sum of true negatives and true positives to total number of tests which is given by Eq. 1. The comparative results are presented in Table 3.

$$a = \frac{1}{n_{\text{samples}}} \sum_{i=0}^{n_{\text{samples}}-1} 1(\hat{y}_i = y_i) \tag{1}$$

4.2 Precision and Recall

The accuracy Precision is the ratio of true positives to the sum of true negatives and true positives. On the other hand, recall is the ratio of true positive to the sum of true positive and false negative. The comparative results are presented in Table 3.

Table 3 Accuracy, precision, recall, F1 and log loss

Classifier	Accuracy	Precision	Recall	F1-score	Log loss
Logistic regression	41.215	65.323	41.215	49.964	153.406
KNN	90.814	90.840	90.814	90.820	79.211
Decision tree	100.000	100.000	100.000	100.000	0.000
Random forest	99.999	99.999	99.999	99.999	0.003
AdaBoost	81.272	98.092	81.272	86.794	63.246
Gradient boosting	100.000	100.000	100.000	100.000	0.000
Gaussian NB	56.922	79.583	56.922	64.898	371.774
LDA	86.875	87.399	86.875	86.916	50.980

4.3 F1-Score

In IDS systems, false positive and false negatives are equally important, and therefore, $F1$ score is also calculated. The $F1$ score is given by harmonic mean of precision (P) and recall (R). The comparative results are presented in Table 3.

$$f1 = 2 \times \frac{P \times R}{P + R} \tag{2}$$

4.4 Log Loss

Log loss gives the probability of prediction of a model. A lower value of log loss shows a higher accuracy of the model. It is given by Eq. 3. The comparative results are presented in Table 3.

$$L_{\log}(Y, P) = -\log \Pr(Y|P) = -\frac{1}{N}\sum_{i=0}^{N-1}\sum_{k=0}^{K-1} y_{i,k} \log p_{i,k} \tag{3}$$

4.5 Other Parameters

There are some other parameters in IoT networks on which the IDS performance depended which are resource available on the IDS device, deployment and network bandwidth, etc.

5 Conclusion

This paper proposes a IDS detection model for IoT network using ML classifiers. For our model, we have used eight ML classifiers to evaluate the performance using Python tools. The result analysis demonstrates that gradient boosting and decision tree show the best performance. These two classifiers recorded 100% of accuracy, precision, recall and F1 score. The random forest achieves 99.99% detection rate, while in AdaBoost, LDA and KNN, the detection rate percentage are 81.23%, 86.88% and 90.81%, respectively. The logistic regression and Gaussian NB perform poorly, and their accuracy rate is less than 60%. The performance analysis concludes that gradient boosting and decision tree are the best choice for the ML-IDS in IoT networks. The network datasets are collected using Arduino and NodeMCU with ESP8266 wi-fi module. This dataset is new and related to IoT network, but they cannot be used for harsh environment or any mobile networks. In future scope, we will try to design a new IDS system for these issues with new relevant datasets.

References

1. Al-Fuqaha A, Guizani M, Mohammadi M, Aledhari M, Ayyash M (2015) Internet of things: a survey on enabling technologies, protocols, and applications. IEEE Commun Surv Tutorials 17(4):2347–2376
2. Asharf J, Moustafa N, Khurshid H, Debie E, Haider W, Wahab A (2020) A review of intrusion detection systems using machine and deep learning in internet of things: challenges, solutions and future directions. Electronics 9(7):1177
3. Azmoodeh A, Dehghantanha A, Choo KKR (2018) Robust malware detection for internet of (battlefield) things devices using deep eigenspace learning. IEEE Trans Sustain Comput 4(1):88–95
4. Bhaskar M, Patra O, Satapathy A, Behera, Soumya R (2020) A comparative study on financial market forecasting using ai: a case study on nifty. In: International conference on emerging technologies in data mining and information security, vol 3, pp 1–12
5. Buczak AL, Guven E (2016) A survey of data mining and machine learning methods for cyber security intrusion detection. IEEE Commun Surv Tutorials 18(2):1153–1176
6. Chaabouni N, Mosbah M, Zemmari A, Sauvignac C, Faruki P (2019) Network intrusion detection for iot security based on learning techniques. IEEE Commun Surv Tutorials 21(3):2671–2701
7. Deng H, Zeng QA, Agrawal DP (2003) Svm-based intrusion detection system for wireless ad hoc networks. In: 2003 IEEE 58th Vehicular Technology Conference. VTC 2003-Fall (IEEE Cat. No. 03CH37484), vol 3. IEEE, pp 2147–2151
8. Doshi R, Apthorpe N, Feamster N (2018) Machine learning ddos detection for consumer internet of things devices. In: 2018 IEEE Security and Privacy Workshops (SPW). IEEE, pp 29–35
9. Elrawy MF, Awad AI, Hamed HF (2018) Intrusion detection systems for iot-based smart environments: a survey. J Cloud Comput 7(1):21
10. Khraisat A, Gondal I, Vamplew P, Kamruzzaman J (2019) Survey of intrusion detection systems: techniques, datasets and challenges. Cybersecurity 2(1):20
11. Kumar V, Das AK, Sinha D (2019) Uids: a unified intrusion detection system for iot environment. In: Evolutionary intelligence, pp 1–13
12. Labs M (2020) Mcafee labs 2020 threats predictions report

13. Mitrokotsa A, Dimitrakakis C (2013) Intrusion detection in manet using classification algorithms: the effects of cost and model selection. Ad Hoc Netw 11(1):226–237
14. Mondal B, Behera PK, Gangopadhyay S (2020) A secure image encryption scheme based on a novel 2d sine–cosine cross-chaotic (sc3) map. J Real-Time Image Process
15. Mondal B, Mandal T (2020) A secure image encryption scheme based on genetic operations and a new hybrid pseudo random number generator. Multimedia Tools Appl 79:17497–17520
16. Mondal B, Patra O, Mishra S, Patra P (2020) A course recommendation system based on grades. In: 2020 International Conference on Computer Science, Engineering and Applications (ICCSEA). IEEE, pp 1–5
17. Mondal B, Sinha D, Gupta NK, Kumar N, Goyal P (2012) An optimal (n, n) secret image sharing scheme. UACEE Int J Comput Sci Appl 2(3):61–66
18. Shukla P (2017) Ml-ids: a machine learning approach to detect wormhole attacks in internet of things. In: 2017 Intelligent Systems Conference (IntelliSys), pp 234–240
19. Singh RP, Javaid M, Haleem A, Suman R (2020) Internet of things (iot) applications to fight against covid-19 pandemic. Diabetes Metabolic Syndrome: Clin Res Rev 14(4):521–524
20. Stoyanova M, Nikoloudakis Y, Panagiotakis S, Pallis E, Markakis EK (2020) A survey on the internet of things (iot) forensics: challenges, approaches, and open issues. IEEE Commun Surv Tutorials 22(2):1191–1221
21. Verma A, Ranga V (2020) Machine learning based intrusion detection systems for iot applications. Wireless Personal Commun 111(4):2287–2310
22. Zwane S, Tarwireyi P, Adigun M (2018) Performance analysis of machine learning classifiers for intrusion detection. In: 2018 International Conference on Intelligent and Innovative Computing Applications (ICONIC). IEEE, pp 1–5

Feature Extraction and Classification of ECG Signals Through Dimension Reduction

Sumanta Kuila, Namrata Dhanda, and Subhankar Joardar

Abstract Arrhythmia is one of the major heart diseases which causes the death of a large number of people every year. So for the prediction and treatment of cardiac failure the arrhythmia detection is important. The main goal of this research is to work with the arrhythmic electrocardiogram (ECG) data, where feature selection, dimension reduction, and feature extraction worked with the ECG data. The classification model of arrhythmia was designed with the process which can be divided into three significant stages. In the initial stage, the temporal and statistical features of the heartbeat selection were done. In the second stage, feature size reduction was done by genetic algorithms, principal component analysis, and in the final stage support vector machine, K-nearest neighbor, neural network used for the classification of the ECG data. Here the proposed scheme used nine different types of ECG beats for classification. Accuracy, specificity, sensitivity are the parameters shown in the experimental results and in the performance metrics. The experimental results show that the proposed approach has the capacity to classify different types of ECG arrhythmias which helps the physicians and cardiologists to give proper cardiac treatment.

Keywords Arrhythmia · Classification · Feature selection · Genetic algorithm · Support vector machine (SVM) · Principal component analysis (PCA) · K-nearest neighbor (K-NN)

1 Introduction

The heart abnormalities and the change of electrical activity of the heart cause the heart arrhythmia which blocks the normal behavior of the heart. The computer-aided cardiac diagnosis makes automatic division of ECG heartbeats into various subcategories which decreases time for the physicians to analyze these heartbeats for the

S. Kuila (✉) · S. Joardar
Haldia Institute of Technology, Haldia, India

N. Dhanda
Amity University Uttar Pradesh, Noida, India

K. Dahal et al. (eds.), *Internet of Things and Its Applications*, Lecture Notes in Electrical Engineering 825, https://doi.org/10.1007/978-981-16-7637-6_20

treatment of arrhythmia. For ECG beat classification, feature selection was done by dimensionality reduction, beat calculation was done by feature extraction, and construct classifier was used for classification. The adequate feature extraction and their dimensionality and size reduction were the key activities for optimal arrhythmia classification. The wavelet transform, Fourier transform, statistical methods, spectral correlation, Lyapunov exponent, and vector quantization are the various feature extraction methods. For dimensionality reduction ICA, GA, PCA, self-optimizing map (SOM) methods are used. The feature extraction was done by discrete wavelet transform using the sub-band decomposition [1, 2]. The QRS complex of the ECG signal is the main component for this decomposition. Here in this work, the time interval of the two pulses of heartbeat is considered as features. For feature extraction, PCA and spectral correlation were used and SVM was used for classification. For feature extraction, the total heartbeat signals were used where the QRS complex, P wave, U and T waves have specific impact on arrhythmia classification. Here each individual beat signal was divided into nine identical sections; also, the features were analyzed and the information were extracted [3, 4].

2 Methods of Selection and Classification

Figure 1 describes the feature selection and the classification of the heartbeat. In the preprocessing step, the fluctuations and noise were removed.

Using beat-parsing steps the signal was parsed where the time interval and statistical features of the ECG signal were analyzed there. The raw ECG data is preprocessed for feature extraction and dimension reduction. For this, genetic algorithm works with the methodology of selection, crossover, and mutation of the selected chromosome. Here three important algorithms related to dimension reduction are used. The ICA and PCA were used to reduce the feature vector size. The main activity of genetic algorithm (GA) was to select the best features among the various available features. In GA to get better results, calculation of the parameters was done by optimizing the parameters [5, 6]. The classification was done by different

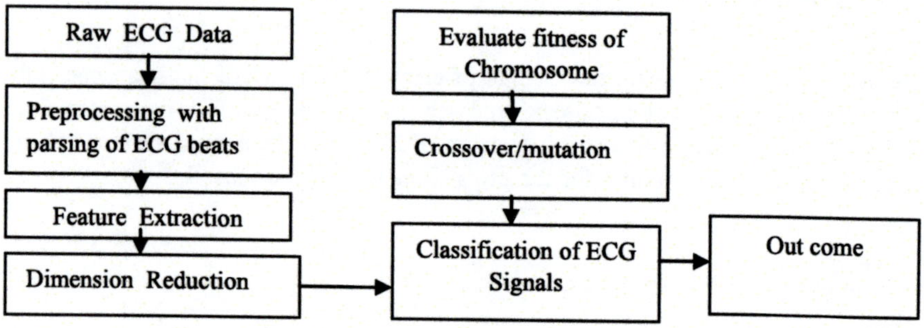

Fig. 1 Block diagram of feature selection and classification schema

algorithms like SVM, NN algorithms, K-NN, and DT. By comparing the different algorithms, the best result was achieved with the selected ECG dataset [7].

3 ECG Database Used

The experiment worked with MIT-BIH arrhythmia database with the dataset containing 48 signals. Each of the signal contains two significant leads (Lead I and Lead II), and the duration of the signal is 30 min. For the experiment, 28 files were selected randomly to identify the ventricular and supraventricular arrhythmias. The MIT-BIH arrhythmia database is clean, and it is annotated with timing and beat level information [8]. In the signal files, position of the beats is determined by labels and the random selection procedure was applied in this work. In this work, nine various categories of ECG data beats were taken where the total number of 6298 samples are used.

4 Preprocessing of Data

The ECG signal oscillation creates an unwanted effect on the related features where the performance of the signal processing depends on specific type of ECG signals recorded from the people of different age category. The noise in the signal depends on the patients daily natural activities such as coughing, bending, and breathing [10]. The setting of the ECG signal starts from 0, where the zero-mean signal is implemented as $Y(t)$, where $t = 1, 2, 3, \ldots, p$ and the calculated equation is

$$Y(t) = x(t) - \overline{x} \tag{1}$$

where $\overline{x}$ is the corresponding arithmetic mean of the raw ECG signal of $x(t)$, $t = 1, 2, 3, \ldots, p$ and p is the length of the signal. Now by using cascade filter, the frequency components which lie below 0.5 and 2 Hz are totally removed.

Now by using cascade filter, the frequency components which lie below 0.5 and 2 Hz were totally removed. Hence, from the signal the baseline wonder was totally removed, and in the situation of stress test, the components work with the higher frequency [11]. The issues worked here was the raw ECG signal, the filter signal with certain time frame, and the results of the cascading filter were important. Figure 2 describes the various stages of the outcome of the raw ECG signal. Initially, the output from cascading filter was shown and then the final filtered signal outcome was displayed.

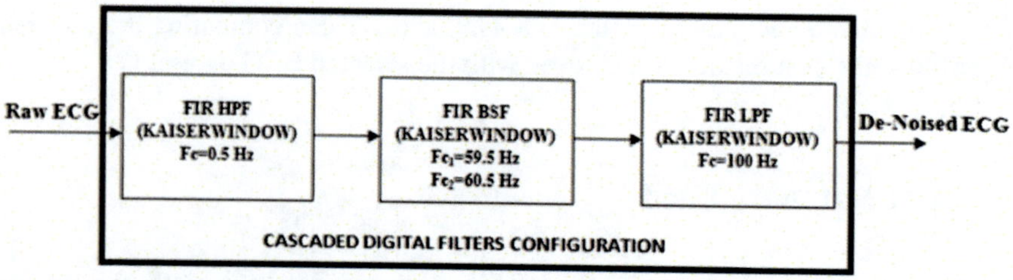

Fig. 2 ECG signal with cascading filter configuration [11]

4.1 Parsing of Beat

Before feature extraction, each and every ECG data beat would be separated and rated in a single beat format. The QRS complex of the ECG signal was read first, and it extracts the maximum window length from each beat and these particular points resides right and left sides of the R point. The annotation files worked with the MIT-BIH database where the R point location activates the connected beat [12].

4.2 Feature Extraction

For arrhythmia classification feature based on statistics, morphological features and domain features based on subspace were used. In the experiment, 200 sample points were taken for the amplitude values where for the calculation of statistical features beat-parsing steps are used. The distribution range, kurtosis, skewness, standard deviation, inter-quartile range, and statistical features were taken for calculation. To calculate the features, the entire beat was broken into four regions of same width [13].

Figure 3 describes the features of the ECG signal where P, Q, R, S, T points analyzed, which are the key components for this feature extraction process. The border of the area is indicated by dashed lines at the top of the p wave and the R point indicates the border of the second area where T wave indicates the third area border. The main components to calculate the features are skewness and kurtosis where skewness is the measure of symmetry which looks same to right and left center points. Kurtosis measures a normal distribution whether the ECG data is lightly tailed or heavily tailed [14]. The datasets containing low kurtosis have low outliner or have light tails whereas high kurtosis has good outliner or heavy tails.

The difference between the smallest and largest values is calculated which finds the distribution range. The difference between the upper quartile and lower quartiles is statistically measured the distribution, and it is equal to this difference. The quartiles work, and it splits a ranked order dataset with four equal segments. Using these values, the fragments are parted and these are called three quartiles which are represented

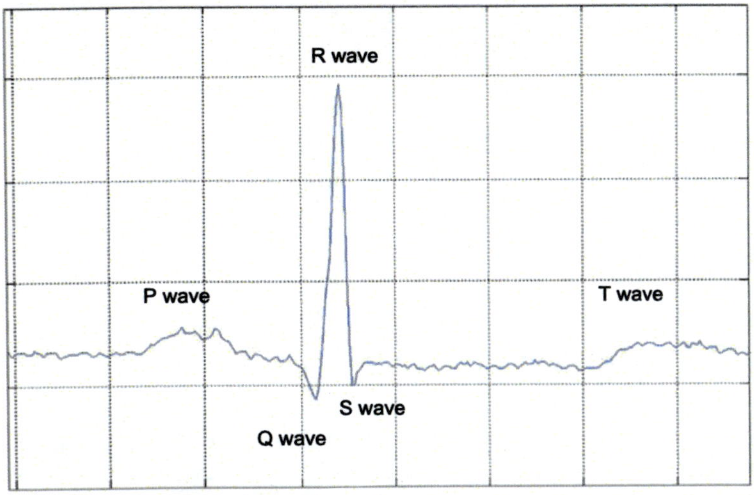

Fig. 3 Components to calculate features of a single beat ECG signal [13]

Table 1 Different statistical features

Feature	Kurtosis	Skewness	Range	Interquartile range	Standard deviation	Mean
Whole Beat	KUR0	SKE0	RAN0	IR0	SD0	MN0
Segment 1	KUR1	SKE1	RAN1	IR1	SD1	MN1
Segment 2	KUR2	SKE2	RAN2	IR2	SD2	MN2
Segment 3	KUR3	SKE3	RAN3	IR3	SD3	MN3
Segment 4	KUR4	SKE4	RAN4	IR4	SD4	MN4

by Q_1, Q_2, and Q_3. The statistical distribution is measured by standard deviation of the data which is equal to the square root of the data variance [15].

Table 1 shows the different statistical features. For the active beat, the time interval between next and previous beat was treated as features.

4.3 *Measurement of the Performances*

The accuracy, specificity, and sensitivity were the performance metrics used to implement and compare classification results. In the multiclass classification, the accuracy metric was a single value, whereas specificity and sensitivity were estimated separately by each class.

$$\text{Sensitivity} = \text{TP}/\text{TP} + \sum \text{FN} \tag{2}$$

$$\text{Specificity} = \text{TP}/\text{TP} + \sum \text{FP} \tag{3}$$

$$\text{Accuracy} = \sum_{i=1}^{9} \text{TP}/\text{Total Number of beats} \tag{4}$$

True positive was symbolized by TP, false positive was symbolized by FP, false negative was symbolized by FN for each of the class [16].

5 Methodology Used

The heartbeat classification performance depends on several issues of the classification methods such as the quality of the signal of ECG data, features that will represent the beats, the quality of test data which was used to train the classification algorithms. Here MIT-BIH arrhythmia database is used to read the ECG signals [17]. The performance of the classification of the ECG beat that classifies the system validated by 6234 number of ECG data beats where in the dataset several (nine) common ECG beat types are available. The ECG beat contains R samples, V samples, P samples, L samples, and N samplers which contain 1000 data beats each whereas F samples contain 260 beats, a samples have 150 beats, E samples have 106 beats, and F samples have 802 beats. All the testing and experiments were done in Python 3.7 with Anaconda Jupyter notebook environment. In the study, tenfold cross-validation method was used for testing and training the classifier for classification. The overall performance was also calculated by tenfold cross-validations where the ECG dataset was parted randomly into subsets of 10 mutually exclusive parts of equal size. Each of the part of the dataset was trained and tested for 10 times. For the simulation in every iteration 1 subset was tested and 9 subsets were trained by the classifier. The accuracy estimation was done by number of overall accurate samples of the classifier in the working dataset. Thus, in the training process the negative and positive effects of the data samples produce a standard result. The experiments work with several classification algorithms to test performance, and the algorithms are neural network (NN), k-nearest neighbors (K-NN), decision tree (DT), and support vector machine (SVM). The values of average specificity, accuracy, and average sensitivity were calculated to estimate the classification algorithm performance, and tenfold cross-validation is used for this purpose [18, 19]. The classification algorithm initially worked with time-domain and statistical features of 32 beats. Then the dimension reduction algorithms, GA, PCA were implemented to the feature set by which results were distributed to the classifier [20].

Table 2 Confusion matrix generated for the distribution of the ECG features

Types of heartbeat	Kurtosis	Skewness	Range	Interquartile range	Standard deviation	Mean
Normal beat	94.32	94.16	94.26	98.23	98.27	95.12
Fusion ventricular	98.12	97.43	96.13	97.58	95.37	94.36
Left bundle block	97.23	96.28	95.19	96.28	96.18	93.16
Right bundle block	96.17	95.17	94.27	97.28	97.36	97.16
Unclassifiable beat	95.12	94.15	97.28	96.15	96.36	96.26

6 Measurement of the Performances

Different fiducial points are available in the heartbeat waves, and they are P, T, U, and QRS complex. To recognize arrhythmia, the shapes of these wave have great importance. The features were analyzed, and more information regarding the ECG wave were calculated to determine accurate features. The measure of peak end is described as kurtosis where probability distribution worked on random variables or the similar datasets. R-R interval is described as time-based feature which describes time between two R points [21, 22]. The measure of symmetry of probability distribution for random variables of the dataset is skewness, and it works either side of its mean of the dataset.

Table 2 describes the confusion matrix generated for the distribution of kurtosis and skewness features where the allocation of R-R and kurtosis features works separately with the whole ECG beats. Between the classes, these features create a significant clustering. The different heartbeat types are taken such as normal beat, fusion ventricular, left bundle block, right bundle block, and unclassifiable beat to generate the confusion matrix.

7 Results for Time-Domain and Statistical Features

The performance evaluation was done by 32 features, and four classifiers were used to evaluate that. For the K-NN classifier, after testing the accuracy rate achieved was 97.65% where the K parameters which were basically the data found from the number of the nearest neighbors to classify every sample when it is required. Here training and testing were done, and after that, the best classification performance was produced. Initially, it was taken $K = 1$, and then in the classification method, the Euclidean distance function was implemented in K-NN classification method. SVM, DT, NN were also applied with this experiments to compare the results which were

Table 3 Classification results of temporal and statistical features

	Sensitivity	Specificity	Accuracy
Decision tree	0.952	0.963	0.972
Support vector machine	0.992	0.987	0.989
K-nearest neighbors	0.983	0.964	0.972
Neural network	0.979	0.962	0.969

obtained by the tests. One hidden layer is found in NN whose size was set as 24 with grid search. Apart from that, another grid search method was applied to resolve the parameters of SVM classifiers where the polynomial kernel function worked whose gamma was set as 2, C set as 2 and degree set as 4. In the experiment, when DT classification works the gain ratio was obtained by criterion parameter and here it was set as 0.2.

The results of the classification are demonstrated in Table 3 which uses 32 temporal and statistical features. Analyzing the results, it was clear that the highest success is achieved by SVM classifier, the K-NN, and NN achieves almost same accuracy of the classification, the performance accuracy of the DT classifier is the lowest among all these four classifiers.

7.1 Results from Dimension Reduction Methods

For size reduction, feature selection and feature extraction are the important steps, wherein classification schema additional features were used which would raise the cost of calculation. So using the minimum features, the system should be developed which will reduce the cost of development, and hence, the importance of dimension reduction comes. Two important techniques were used for dimension reduction first was principal component analysis (PCA) and the second was genetic algorithm (GA). These two methods were used to move data so that the new features can adopt the data for logical representation [23, 24]. The another technique to work with the available features and select the best representation is genetic algorithm. In the PCA, grid search method was used. The main objective here was to calculate the count of the principal component (PC) where the evaluation of each PC count was done in the search process of the K-NN classifier. Table 4 displays the values of the performance metrics of the input features using principal component analysis.

Table 4 Performance metrics of the principal components as the input feature

	Sensitivity	Specificity	Accuracy
Decision tree	0.942	0.953	0.932
Support vector machine	0.989	0.982	0.987
k-nearest neighbors	0.962	0.954	0.963
Neural network	0.971	0.958	0.961

Table 5 Result of the classification when genetic algorithm was taken as an input vector

	Sensitivity	Specificity	Accuracy
Decision tree	0.951	0.966	0.947
Support vector machine	0.972	0.974	0.965
k-nearest neighbors	0.987	0.989	0.992
Neural network	0.974	0.962	0.957

In the PCA, grid search method was used. The main objective here was to calculate the count of the principal component (PC) where the evaluation of each PC count was done in the search process of the K-NN classifier [25, 26].

The genetic algorithm (GA) was applied to reduce the volume of the input feature vector which contains 32 features as a standard selection procedure. Initially, GA selected 17 features and then gradually the features increase by standard dimension reduction procedure. So selection of GA as a standard input vector, the experiment achieved the maximum classification accuracy. With using the K-NN classifier, the accuracy reached at 99.2% with the selected GA features.

Table 5 shows the classification performance of by the selected features using GA. The classifier architecture and input vector size control the classification times for the input vectors when the tenfold cross-validation is used. Analyzing the computational load of all the classifiers, it was found that the slowest classifier was NN for all feature sets.

8 Conclusion

This study worked with the classification schema with the ECG dataset where 9 different types of arrhythmia were detected and classified by using MIT-BIH arrhythmia database. Temporal and statistical features were selected from the time series of single ECG data beat. The single beat ECG signal was partitioned with 4 identical parts where the features were calculated using NN, K-NN, DT, and SVM algorithms. For the classification, 32 feature vectors were identified in the feature set to achieve the arrhythmia classification successfully. Using dimensionality reduction methods, the total count of the features were decreased and the comparison between the test results were analyzed. The better classification results were achieved by genetic algorithm with the selected features. Using the feature selection and classification schema, the computer-aided automation of arrhythmia classification can be achieved.

References

1. de Chazal P, O'Dwyer M, Reilly RB (2004) Automatic classification of heartbeats using ECG morphology and heart beat interval features. IEEE Trans Biomed Eng 1196–1206
2. Inan T, Giovangrandi L, Kovacs JTA (2006) Robust neural network based classification of premature ventricular contractions using wavelet transform and timing interval features. IEEE Trans Biomed Eng 53(12):2507–2515
3. Raut R, Dudul SV (2008) Arrhythmias classification with MLP neural network and statistical analysis. In: First international conference on emerging trends in engineering and technology. IEEE
4. Jadhav SM, Nalbalwar SL, Ghatol AA (2012) Artificial neural network models based cardiac arrhythmia disease diagnosis from ECG signal data. Int J Comput Appl 8–13
5. Joachims T (1997) Text categorization with support vector machines: learning with many relevant features. Technical report, University of Dortmund
6. Azemi A, Sabzevari VR, Khademi M, Gholizadeh H, Kiani A (2006) Intelligent arrhythmia detection and classification using ICA. In: EMBS annual international conference. IEEE, New York City, USA
7. Margarita RGA, Christian GQM (2013) Using genetic algorithm feature selection in neural classification systems for image pattern recognition. Int J Modern Phys 52–58
8. Kulkarni SP (2015) DWT and ANN based heart arrhythmia disease diagnosis from MIT-BIH ECG signal data. Int J Rec Innov Trends Comput Commun 276–279
9. Gharaviri A, Dehghan F, Teshnelab M (2008) Comparision of neural network Anfis, and SVM classification for PVC arrhythmia detection. In: Seventh international conference on machine learning and cybernetics, Kunming, July 2008, pp 24–29
10. Pasolli E, Melgani F (2015) Genetic algorithm-based method for mitigating label noise issue in ECG signal classification. Biomed Sig Process Control 19:130–136
11. Patro KK, Kumar PR (2015) De-noising of ECG raw signal by cascaded window based digital filters configuration. In: Communication and information technology conference (PCITC), Siksha 'O' Anusandhan University, Bhubaneswar. IEEE Power
12. Aslantas G, Gurgen F, Salah AA (2014) GA-NN approach for ECG feature selection in rule based arrhythmia classification. Neural Netw World 24:267–283
13. Rodríguez PR, Mexicano A, Bila J, Cervantes S, Ponce R (2015) Feature extraction of electrocardiogram signals by applying adaptive threshold and principal component analysis. J Appl Res Technol 261–269
14. Oh IS, Lee JS, Moon BR (2004) Hybrid genetic algorithms for feature selection. IEEE Trans Pattern Anal Mach Intell 26(11):1424–1437
15. Handels H, Ross T, Kreusch J, Wolff HH (1999) Feature selection for optimized skin tumor recognition using genetic algorithms. Artif Intell Med 16:283–297
16. Ghumbre S, Patil C, Ghatol A (2011) Heart disease diagnosis using support vector machine. In: International conference on computer science and information technology, Thailand, Pattaya
17. Mark R, Moody G MIT-BIH arrhythmia database 1997. Available http://ecg.mit.edu/dbinfo.html
18. Song MH, Lee J, Cho SP, Lee KJ (2005) Support vector machine based arrhythmia classification using reduced features. Int J Control Autom Syst 3(4):571–579
19. Punch WF, Goodman ED, Pei M, Chia-shun L, Hovland L (1993) Further research on feature selection and classification using genetic algorithms. In: Fifth international conference on genetic algorithms and their applications (ICGA), p 557
20. Pei M, Goodman ED, Punch WF, Ding Y (1995) Genetic algorithms for classification and feature extraction, classification society of North America. In: Annual meeting, June 1995
21. Laguna P, Jane R, Caminal P (1994) Automatic detection of wave boundaries in multilead ECG signals: validation with the CSE database. Comput Biomed Res 27:45–60
22. Hu YH, Tompkins WJ, Urristi JL, Valtino XA (1993) Application of artificial neural networks for ECG signal detection and classification. J Electrocardiol 26:66–73

23. Wu T, Hung K, Liu J, Liu T (2013) Wavelet-based ECG data compression optimization with genetic algorithm. Biomed Sci Eng 746–753
24. Ubeyli ED (2007) ECG beats classification using multiclass support vector machines with error correcting output codes. Dig Sig Process 17:675–684
25. Takiguchi T, Ariki Y (2006) Robust feature extraction using kernel PCA. In: International conference on acoustics, speech and signal processing, pp 509–512
26. Tantawi MM, Revett K, Salem AB, Tolba MF (2015) A wavelet feature extraction method for electrocardiogram (ECG)-based biometric recognition, pp 1271–1280

Human Identification Using Low-Resolution Thermal and High-Resolution RGB Images

Pavan Talluri and Mohit Dua

Abstract Object detection and identification with great accuracy and speed in real time are a major issue in present growing applications, deep learning models, such as deep neural networks, convolutional neural networks, and modified models like YOLO, SSD are achieving sufficient performance in object identification with great computational power and better processing environment, and consequently this is the great drawback to run this deep learning (DL) models in low-end devices like embedded applications and portable applications. In this paper, we have used Tiny Yolo v3, K-means clustering algorithm of unsupervised learning and IR images has been used to address the issue of high-computational power and low-resolution human detection. We have used IR images instead of RGB images, and thermal images are great solution to the issue of low-illumination object identification. K-means clustering algorithm retrieves the object and classifies the object by segmenting the given image. The deep learning model Tiny Yolo extracts the object features and identifies the proper bounding box. Proposed model describes real-time low-resolution human identification in two steps, identifying anchor boxes by using K-means clustering algorithm of unsupervised learning model and predicting proper bounding boxes based on the anchor boxes by using Tiny Yolo of deep learning model.

Keywords Thermal images · RGB images · Tiny Yolo v3 · Human identification · K-means clustering · Low resolution

1 Introduction

Image processing mechanism involves many applications like object detection, image encryption, image fragmentation, image to data conversion, and many more. Currently we are focusing on object detection from an image. Usually in computer

P. Talluri (✉) · M. Dua
National Institute of Technology, Kurukshetra, India

M. Dua
e-mail: er.mohitdua@nitkkr.ac.in

K. Dahal et al. (eds.), *Internet of Things and Its Applications*, Lecture Notes in Electrical Engineering 825, https://doi.org/10.1007/978-981-16-7637-6_21

vision human detection or object detection involves three steps, human pedestrian detection, feature extraction, and classification of extracted features.

Accuracy of an object detection mode mostly depends on what type of an image we are applying present techniques that can perform better for RGB Images but those methods are not suitable for less illumination and black and white images. We have many types of images available in the market; they are thermal images and near-infrared (NIR) images taken by a CCD camera. Thermal cameras are capable of capturing infrared radiation and collecting the infrared radiation emitted by all objects in the particular movement. Pixel values of a thermal image specify heat or temperature. Thermal cameras do not depend on light as other visible cameras, and they are able to see objects in the less illumination. Presently the cost of thermal cameras increases more and more due to rising necessity in the market for thermal imaging applications.

Currently conventional neural networks (CNN) best paradigm for object identification in RGB images. In the year 2012, AlexNet successfully developed an ImageNet Large Scale Visual Recognition [1] to detect objects, and subsequently convolutional neural networks started getting successes for object detection with modified models, such as single-shot detector (SSD), convolutional neural networks (CNN), R-FCN [2], mask regional convolutional neural network (RCNN) [3], and YOLO [4].

An event of identifying humans is very usual but very destructive. Currently, the most successful paradigm for object detection in RGB images is based on convolutional neural networks (CNNs). Development started with the great success of AlexNet in the ImageNet Large Scale Visual Recognition Challenge in 2012 [5] for the image recognition task. Since then, several successful CNN architectures have been developed for the object detection task, such as RCNN [6], SSD [7], Mask R-CNN [3], R-FCN [8], and YOLO [9] detection. Unexpectedly, a man or an animal came in the middle of the roadway; it had to be identified in quick time. Existed [10] available image detection models are taking a high amount of time to process the images and detect a human object from an image which has low brightness or low resolution. In the presented work, we have explained a solution to real-time problems like the detection of intruders in night time (for security), pedestrian detection for self-driving autonomous cars (or) unnamed vehicles to avoid accidents and obstacles. Object identification system based on K-means algorithm of unsupervised learning method followed by a deep learning approach on thermal images [9].

We have seen in many different popular fields that object detection is a mandatory task. A lot of approaches are available in object detection on the bases of deep learning [11], SSPNet [12], RFCN [2], FPN [13], YOLO giving better performance than classical models. Now Yolo is the best-known method for good performance and better accuracy, and Yolo has different flavor's YoloV1, Yolo-V2, Yolo-V3. The size of Yolo-V1 is expanded and it is up to 1 GB in size, due to the model having twenty-four (24) convolutional layers and two (2) fully connected-layers and best configuration platform needed for execution of this model [4]. When it comes to Yolo-V2 model size has reduced a bit and accuracy has improvised a lot and it became more robust than previous model (Yolo-V1), due to an introduction of new concept called anchor boxes for bounding box prediction and elimination of complex fully connected layers [14].

The final flavor of YoloV3 [15] started revolution in accuracy by introduction of residual structure and the powerful platform of GPU, and eventually Yolo model acquires breakthrough in accuracy limitation of YOLO model, not suitable for constrained domain with real-time performance due to large amount of increased size of the algorithm. In present work, we have used a Tiny Yolo v3 advanced version of YOLO, and it is suitable for medium size environments. In this work, we trained our proposed model with FLIR dataset and considered only four (4) classes such as van, person, parking meter, and car. The previous version of Tiny Yolo [16] detection process starts with anchor boxes and with help of this anchor boxes it identifies the bounding boxes and this can identify different classes and different labels. In some cases, one object can be considered as many different class types with different scores. If the present image car is identified twice, the model filters the bounding boxes of the car to get a unique identified car.

2 Existing Solution

Artificial neural networks (ANNs) are the base for deep learning (DL) and machine learning (ML) [17]. From last few decades, ANN becomes a major wing in study region, to detect any object convolutional neural networks are mandatory, CNN is constructed with prime element of neurons; depending on the detection requirement, we construct the neural networks; if detection computations are complex, increase in the number of layers solves the problem.

There are many object detection methods are available for specific task, such as face detection [18], pedestrian detection [19], vehicle detection [20], and text detection [21]. Additionally, there are some techniques for generic object detection and object class detection [22]. There are different object detection frameworks available, in that two major frameworks perform better detection results, they are region-based framework (two-stage) and unified-based framework (one stage).

Two-stage detector performs better compared with single-stage detector when high computational power is allowed to calculate detection results, and two-stage frameworks like RCNN, RFCN, and Mask RCNN are more flexible to deal with high computational challenges, their structures are more flexible to face the complex challenges [3]. Single-stage detection accuracy is less compared with region-based two-stage detectors due to the backbone network quality [23]. YOLO [4] and SSD [24] measure under single-stage detectors, and they perform usually faster than multi-stage region-based detectors, due to introducing backbone networks and elimination of preprocessing layers. In any stage, detectors are suitable for real-time object detection except the time consumption on feature extraction stage (backbone layers) [24].

3 Proposed Solution

In the present method, image is passing through many modified convolutional layers while training the data; when image is passed through a first convolutional layer, it scans entire image once and divide into small grids (size is 19 $\times$ 19). Each grid has to predict a maximum of five (5) bounding-boxes if more than one object presents in that particular grid; if we take a normal resolution image, we will get more than 1600 bounding-boxes per image.

Tiny Yolo algorithm is applying a different kind of approach for model training and image processing. Classical models follow regional approaches but Tiny Yolo follows a regression methodology, which involves classification of images based on the objects and predicting the bounding boxes. Bounding box has parameters such as height, width, value corresponding to class, center of the bounding box, and probability of bounding box.

After bounding boxes are predicted for each grid, more than one bounding boxes are possible for each grid, and some bounding boxes do not contain any object or image, and to eliminate this duplication of bounding boxes or redundancy, we will calculate the predicted score for each bounding boxes by considering the four parameters of the bounding boxes. Whichever bounding boxes achieves the highest predicted score that bounding boxes is the final boxes, this press is known as non-max suppression (Identifying max-predicted bounding boxes).

Proposed model describes real-time low-resolution human identification in two steps, identifying anchor boxes by using K-means clustering algorithm of unsupervised learning model and predicting proper bounding boxes based on the anchor boxes by using Tiny Yolo of deep learning model. Object detection from IR images is the challenging task in real-time object detection due to the blank shapes of objects of IR images than RGB image objects. The reason of performance and size of the model attracts the researchers to deploy the model in small embedded applications like mobiles and surveillance cameras. This proposed model is perfect fit for real-time object detection in portable applications, such as embedded applications and mobiles; so, it can achieve a better rate of processing speed when compared to classical object detection approaches [25].

3.1 *Proposed Architecture*

Tiny Yolo v3 has a smaller number of convolutional layers and fully connected layers, and it makes the model small and occupies a small amount of space; Tiny Yolo is a compressed version of YOLO model, and it has reduced the computational complexity for real-time single shot object detection algorithms. Tiny Yolo has achieved accuracy 58% above so far with a smaller number of hidden layers (sixteen).

In the present architecture (Fig. 1), we have explained four steps namely data merging step, clustering, training, and human object identification. In architecture, we have mainly focused on training and detection steps, training the model takes inputs are anchorsdata .txt, classes file, dataset for training and pretrained weighted file (pretrained_data.weights) for default detection configurations. After training the completed model ready to detect the input images, the prediction step takes the input as an image and produces the predicted reliable output.

Based on proposed architecture, human identification is improved. We have used thermal dataset to train the model, and we have used the K-means clustering algorithm to identify the good priors; to improve the object detection and feature extraction, we have used the Tiny Yolo v3 model. This model contains a very small number of convolutional and fully connected layers, and the size of convolutional layers is 1×1 and fully connected layers are 3×3. 1×1 convolutional layers are helpful for reducing the features calculation overhead and simplify the human identification in real time.

The present method image is passing through many modified convolutional layers while training the data, and when image is passed through a first convolutional layer, it scans entire image once and divides into small grids (size is 19 X 19). Each grid has to predict a maximum of five (5) bounding boxes, if more than one object are

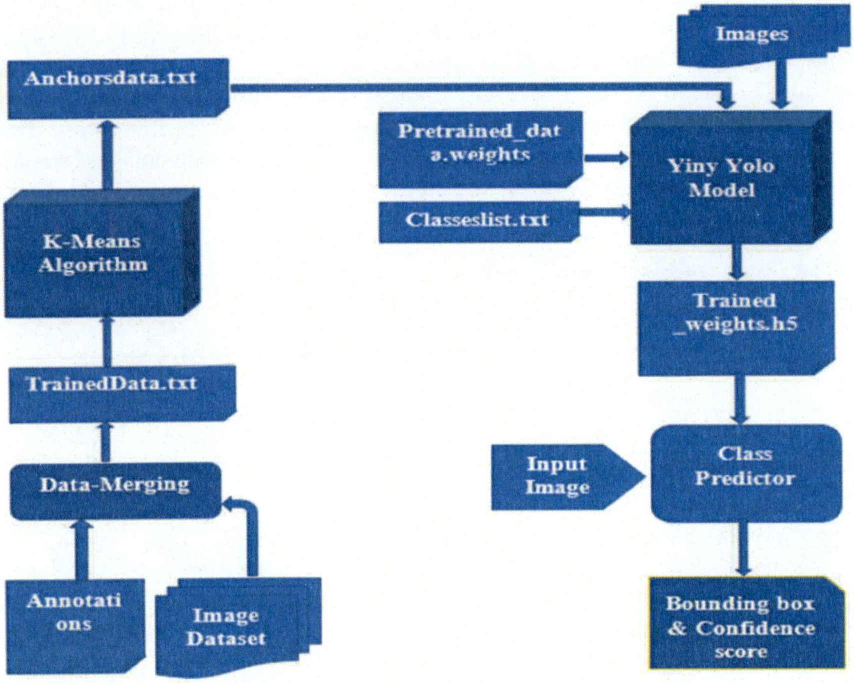

Fig. 1 Complete architecture of proposed model

present in that particular grid, and if we take a normal resolution image, we will get more than 1600 bounding-boxes per image.

4 Results and Analysis

The best performance of the combinations is highlighted in Table 1 and represents the different proposed models of true positive, false positive and precision, when model trained on IR images and tested on IR images model producing average precision is 0.7210, model trained on IR images and tested on RGB images model producing average precision 0.6534 and tested on IR images and trained on RGB images model getting better average precision and better accuracy compared with other variations above defined in table. We have achieved better performance for the reason of training model with RGB image dataset.

After completion of training, we have tested our trained weights with a tested dataset. Table 1 represents the results. For training we have used 2291 thermal images in which model finds 1652 under true positive remaining as false positive, finally we got precision 0.7210. The classical model HAAR cascade gave around 0.6554 and HOG has precision 0.6485 as compared with this model our model performs better. Tiny Yolo is also an efficient base network for object identification system. We are showing results of Tiny Yolo trained for object identification on FLIR dataset depending on the latest work [27]. Table 1 Tiny Yolo is compared with HAAR cascade [28] and a conventional HOG framework [29], and in our experiment, we have used images of 350 and 650 input resolution images. For our framework, Tiny Yolo achieves comparatively best results to other frameworks with only less amount of computational complexity, a smaller number of layers and size of model.

Table 1 Result analysis of proposed work

Approach used	Techniques used	Trained data	Tested data	TP	FP	Precision
De Oliveira [1, 14]	HAAR Cascade	–	–	59	31	0.6554
Rujikietgumjorn [15, 26]	Conventional HOG: Atrium-range dataset	–	–	2615	1415	0.6485
Talluri et al. [25]	Tiny Yolo v3 + K means Clustering	FLIR	FLIR	1652	639	0.7210
Proposed approach	Tiny Yolo v3 + K means Clustering	RGB	FLIR	1005	275	0.78515

$$\text{Precision}(p) = \frac{\text{TP}}{\text{TP} + \text{FP}} \tag{1}$$

$$\text{Recall}(r) = \frac{\text{TP}}{\text{TP} + \text{FN}} \tag{2}$$

$$\text{AP} = \int_{0}^{1} p(r)\text{d}r \tag{3}$$

where TP = true positives, FP = false positives, TN = true negatives, and FN = false negatives [30].

5 Resultant Images from Proposed Model

We have used FLIR Thermal dataset available in [2, 31] to identify human objects in low-resolution images, out of which we have used 2304 thermal images for training purposes and 1378 thermal images for testing.

We have used following configurations for testing and training the model as follows: Param Shavak (high performance computing) (or) Google Colab GPU for training the model, operating systems Ubuntu or Windows or Unix, installed RAM (primary memory) capacity minimum of 8 GB or use Google Colab GPU, if we use integrated design environment go to sublime-text or Visual-Studio, Programming-Language Python_3 or above and packages are Kera's, OpenCV, NumPy, TensorFlow. Once model training is completed successfully, we took a snapshot of trained weights for future purposes and tested input images and achieved satisfactory results. This snapshot was further useful for deploying on any other devices like embedded systems and small ended handheld applications like mobiles, PDA, and camera applications for real-time human detection. The obtained results are tabulated with related existing frameworks to verify and compare the performance and speed (Figs. 2 and 3).

Figures show the expected curve, precision is high and recall is low at threshold and precision begins to drop at very high recall. Above two classifiers represent the similar performance, but there is a slight variation is there (Fig. 4).

6 Conclusion

In this paper, we have presented an explanation for the problem of "low-resolution human identification in thermal imagery" with help of thermal images, and thermal images solve the issue of detection of low illumination background images and dark background images. We trained model with thermal images and RGB images, and

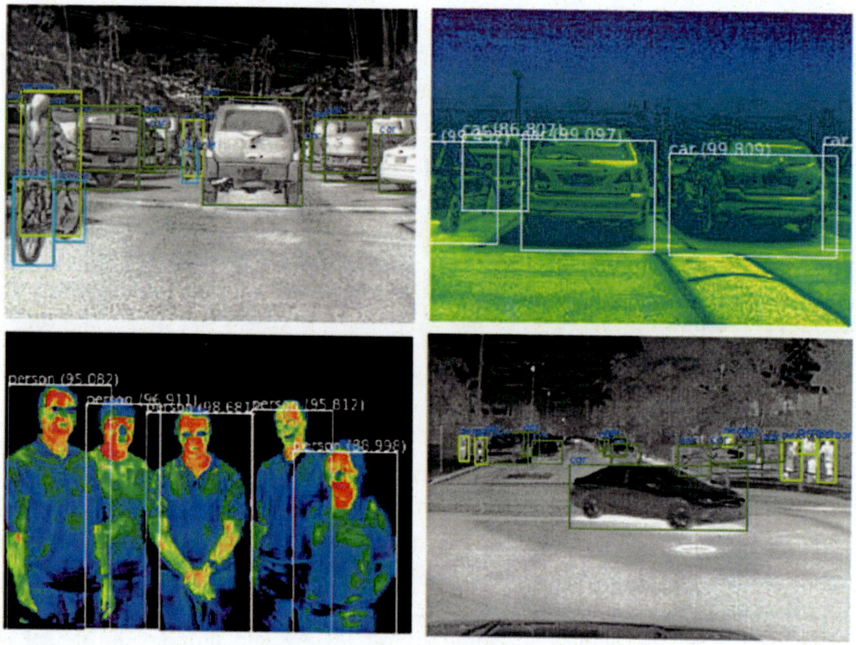

Fig. 2 Resultant object detection images, trained on RGB images and tested on FLIR dataset

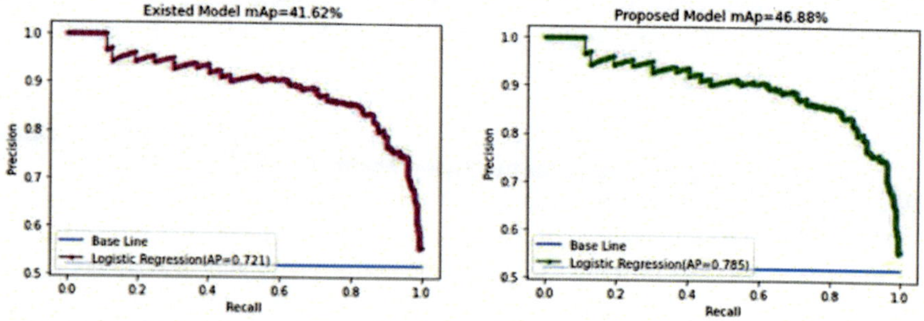

Fig. 3 Precision–recall curves of existed (left) and proposed (right) model

we considered only four types of classes of thermal image objects while training and testing. To address the above issue, we have taken Tiny Yolo v3 model, and another variation of YOLO v3 model, YOLO model needs high-performance computational power; to perform object detection in real time, to reach the model top performance, it needs a GPU support. The major advantage of tiny model can be capable of run in any portable or embedded applications, and the size of the model is very small <50 Mb (less than fifty Mb), by reason of elimination of 3 × 3 convolutional with 1 × 1 convolutional layers. The Tiny Yolo v3 model was trained on over 1600 thermal

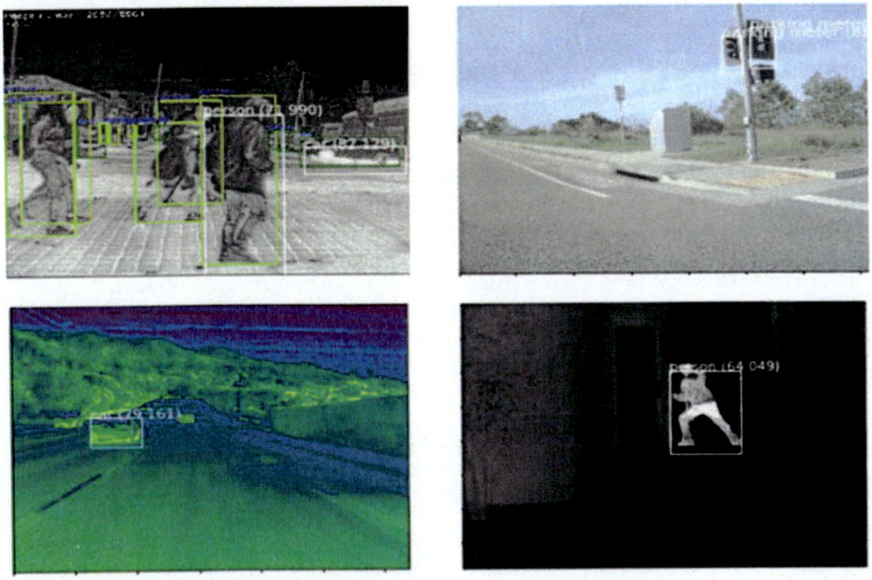

Fig. 4 Resultant object detection images, trained and tested on FLIR dataset

images and tested on more than 290 thermal images, and the performance of the model (trained and tested on IR images) is 41.62%. With variation Tiny model tested on the normal RGB images (trained on thermal images and tested on RGB images), we have achieved performance of 39.86%. In another variation, Tiny Yolo model is trained with 1005 RGB images which we have collected from COCO dataset and tested with 275 FLIR thermal dataset, and we have reached the performance of model (trained on RGB images and tested in IR images) over 46.88%. Finally, concluding that, when a model trained on RGB images and tested on IR images, the performance of the model differs from the model trained and tested on IR images. Tiny model scans and tests the entire image in single scan, and so this model can be used in real-time object identification like advanced automated driving autonomous cars, surveillance camera systems to detect intruders, and pedestrian detection in nighttimes, and the performance of the model attracts the researchers to deploy in portable applications.

References

1. Krizhevsky AS (2012) Imagenet classification with deep convolutional neural networks. In: Advances in neural information processing systems, pp 1097–1105
2. Dai J, Li Y (2016) R-FCN: object detection via region-based fully convolutional networks. In: Proceedings of the neural information processing systems, pp 379–387
3. He KG (2017) Mask R-CNN. In: 2017 IEEE international conference on computer vision

(ICCV), pp 2980–2988
4. Redmon J, Divvala S (2016) You only look once: unified, real time object detection. In: Proceedings of the IEEE conference on computer vision and pattern recognition (CVPR), pp 779–788
5. Gaszczak A, Breckon TP (2011) Real-time people and vehicle detection from UAV, p 7878
6. Kanungo TD (2002) An efficient k-means clustering algorithm: analysis and implementation. IEEE Trans Pattern Anal Mach Intell 24:881–892
7. Yang KP (2020) Unsupervised K-means clustering algorithm. IEEE, pp 80716–80727
8. Fayyad UM, Piatetsky-Shapiro G, Smith P (1996) Advances in knowledge discovery and data mining. AAAI/MIT Press
9. Gersho A, Gray RM (1992) Vector quantization and signal compression. Kluwer, Boston
10. Dua MA (2020) K-means and DNN-based novel approach to human identification in low resolution thermal. In: Advancements in computer vision applications in intelligent systems and multimedia technologies. IGI Global, pp 25–37
11. Fang WL (2019) Tinier-YOLO: a real-time object detection method for constrained environments. IEEE Access 8:1935–1944
12. He K, Zhang X (2014) Spatial pyramid pooling in deep convolutional networks for visual recognition. In: Proceedings of European conference on computer vision. Springer, pp 346–361
13. Lin T-Y, Dollar P (2017) Feature pyramid networks for object detection. In: Proceedings of CVPR, vol 1(2), pp 2117–2125
14. Redmon J, Farhadi A (2017) YOLO9000: better, faster, stronger, vol 8. arXiv: 1612.08242. [Online], Available https://arxiv.org/abs/1612.08242
15. Redmon J, Farhadi A (2018) YOLOV3: an incremental improvement. arXiv: 1804.02767. [Online]. Available https://arxiv.org/abs/1804.02767
16. Girshick R (2015) Fast R-CNN. In: Proceedings of the IEEE international conference on computer vision, pp 1440–1448
17. Hinton GE, Salakhutdinov R (2006) Reducing the dimensionality of data with neural networks. Science 313(5786):504–507
18. Zafeiriou SZ (2015) A survey on face detection in the wild: past, present and future. Comput Vis Image Understanding 138:1–24
19. Enzweiler M (2009) Monocular pedestrian detection: survey and experiments. IEEE TPAMI 2179–2195
20. Sun ZB (2006) On road vehicle detection: a review. IEEE TPAMI 28(5):694–711
21. Ye Q (2015) Text detection and recognition in imagery: a survey. IEEE TPAMI 37(7):1480–1500
22. Zhu J, Jiang Z (2019) Efficient registration of multi-view point sets by k-means clustering. Inform Sci 488:205–218
23. Huang JR (2017) Speed/accuracy trade-offs for modern convolutional object detectors. In: CVPR
24. Liu W (2016) SSD: single shot multi-box detector. In: European conference on computer vision, pp 21–37
25. Talluri P (2020, June) Low-resolution Human Identification in thermal imagery. In: 2020 5th international conference on communication and electronics systems (ICCES). IEEE, pp 1283–1287
26. Donahue JJ (2014) A deep convolutional activation feature for generic visual recognition. ICML 32:647–655
27. Huang J, Rathod V (2016) Speed/accuracy trade-offs for modern convolutional object detectors. arXiv preprint arXiv:1611.10012
28. De Oliveira DC (2016) Towards real-time people recognition on aerial imagery using convolutional neural networks. In: 2016 IEEE 19th international symposium on real-time distributed computing (ISORC)
29. Rujikietgumjorn S (2017) Real-time hog-based pedestrian detection in thermal images for an embedded system. In: 14th IEEE international conference on advanced video and signal based surveillance (AVSS). IEEE, pp 1–6

30. Mann W (1953) Mean value methods in iteration. In: Proceedings of the American mathematical society, pp 506–510
31. Oltean GC (2019) Towards real time vehicle counting using yolo-tiny and fast motion estimation. In: 2019 IEEE 25th international symposium for design and technology in electronic packaging (SIITME), pp 240–243

Supervised Machine Learning Approaches for Attack Detection in the IoT Network

Mir Shahnawaz Ahmad and Shahid Mehraj Shah

Abstract In recent years, the Internet of Things (IoT) has emerged to be one of the vital fields with a vast application domain. IoT network brings heterogeneous devices closer to each other, which eventually helps to solve various real-world problems. Apart from all the applications, the devices in the IoT network are more prone to malicious attacks because of their limited computational power and other resources thereby shrinking the application domain of IoT. Due to the massive size of the IoT network, it becomes very challenging to monitor each and every device for the purpose of attack detection. However, because of the intelligent, self-learning and adaptive nature of various machine learning approaches, they can be used for detecting the malicious traffic in the IoT network. In this regard, we have investigated the applicability of various supervised machine learning classifiers on network security laboratory knowledge discovery in databases (NSL-KDD) dataset and compared their performances. This study shows that random forest classifier and stochastic gradient descent classifier (with log loss function) outperform the other classifiers, studied in this paper, for classifying the IoT network traffic into malicious and non-malicious.

Keywords Internet of Things · Machine learning · Self-learning · Malicious activities · Attack detection

1 Introduction

With the recent advancement in the technologies behind the Internet, a humongous number of devices have started connecting, resulting in the Internet of things (IoT) network. Thus, IoT is a network where even the ordinary heterogeneous devices used by human beings gather data from the surroundings using sensors/actuators and share this data over the Internet for accomplishing a particular task(s). IoT has transformed the classical field into a smart field like smart healthcare, smart transport,

M. S. Ahmad (✉) · S. M. Shah
Communication Control and Learning Lab, Department of Electronics and Communication Engineering, National Institute of Technology Srinagar, Srinagar, J&K, India

K. Dahal et al. (eds.), *Internet of Things and Its Applications*, Lecture Notes in Electrical Engineering 825, https://doi.org/10.1007/978-981-16-7637-6_22

smart grid, smart home, smart waste management, and many more [1]. Although the enhancement in scalability of IoT networks has increased its applicability in various fields, it has also resulted in the serious issue of security and privacy. Due to the limited computation power available at IoT devices, many devices can be easily targeted by a malicious attacker, thereby causing severe damage to the network. The IoT network should have additional security procedures which can achieve data privacy, confidentiality, encryption, device naming, management, and safety [2], but the feature of heterogeneity makes it challenging to implement the already available security procedures. Also, for such a vast network like IoT, it becomes difficult to monitor each device for detecting malicious behavior. Thus, an intelligent security procedure is needed for detecting such attacks in the IoT network along with updating its attack database.

Machine learning algorithms have the capability to identify various patterns in the network traffic by using available training data. Various machine learning approaches can be used for detecting various attacks like denial of service (DoS) attacks, malware detection, jamming, and eavesdropping attacks [3]. The machine learning algorithms are classified as supervised and unsupervised. In supervised learning, a labeled dataset is used to train a model, whereas in unsupervised learning, we use an unlabeled set of data. It depends on the availability of labeled/unlabeled data which depicts the use of a particular learning technique. The trained model is then used to identify similar patterns in the input and update the system.

To analyze the performance of various machine learning procedures for detecting various malicious attacks in the IoT network, we have used NSL-KDD [4] which is the updated version of KDD99 dataset. This dataset has been widely used to evaluate the performance of the attack detection system in IoT [5]. The various attacks included in NSL-KDD dataset include: DoS attack, user-to-root attack, remote-to-local attack, and probes.

In DoS attack, the adversary launches an attack with which the legitimate users are not able to access the available resources. The various varieties of DoS attack included in the dataset are apache2: Launched in apache Web server using HTTP header, arppoison: Attacker translates wrong MAC from IP address, back: It is also launched in apache Web server using multiple front-slashes in URL, Crashiis: Attack is launched using modified GET requests, dosnuke: In this attack, the attacker sends message out of band (MSG-OOB) to NetBIOS thereby crashing server, land: It is the attack launched using spoofed-SYN packets, mailbomb: Here, attacker sends many mail messages which lead to system failure, Neptune: Flooding of SYN packets, ping of death (PoD): Sending of oversized packets to crash a particular node, process table: It uses for (k) function to launch an attack, Smurf: It broadcasts Internet control message protocol (ICMP) request packets to overwhelm the target node, teardrop: It tampers IP header, UDPStorm: It causes congestion in the network by bombarding user datagram protocol (UDP) packets.

In user-to-root attack, the attacker gets access to the network using techniques like password sniffing and then increases its privileges from a regular user to root user, the various user-to-root attacks in the dataset are LoadModule: Here, attacker creates a program to create a bug in the kernel to allow an unauthorized users to access root,

Perl: It exploits a bug in the system to access user-IDs and group-IDs, PS: Attack module used for Solaris 2.5, Xterm: It uses buffer overflows in Xaw library to launch an attack.

Remote-to-local attack is an attack where an illegitimate user tries to gain access to a network/system by exploiting some vulnerabilities, which include password-guessing, HTTP-tunnel, FTP-write, IMAP (it is similar to buffer-overflow and uses Internet message access protocol), named PHF (uses vulnerabilities in some CGI script), sendmail, Xlock (attacker takes the access of some legitimate unprotected user), Xsnoop (the attacker can analyze the traffic of a legitimate user with some server to gain access).

Probes are programs that can gather information regarding a particular network, and then the attacker can use this gathered information to launch attacks. The various types of probes are Ipsweep (identifies listening nodes in a network), Mscan (uses domain name server or brute force technique to identify network devices), Nmap (scans complete network using TCP/UDP/ICMP), SAINT (security administrator's integrated network tool), Satan (SAINT's predecessor) [6].

In this paper, we describe the details of various machine learning algorithms used in our study, and apply them to the NSL-KDD dataset for classifying the data into malicious and non-malicious. We also discuss various features in the dataset and describe the data preprocessing steps, which helps in enhancing the performance of various classifiers. Finally, the performance of various machine learning classifiers in detecting the malicious traffic is compared to identify the best method(s) for processing the given dataset.

The rest of the paper is organized as Sect. 2 discusses the NSL-KDD dataset in details and describes its features and various data preprocessing steps; Sect. 3 highlights various supervised machine learning algorithms used in our study and performance parameters to measure their performance in classifying the network traffic into malicious and non- malicious; Finally, Sect. 4 concludes the paper.

2 Data Specifications and Preprocessing

To compare the efficiency and applicability of various machine learning algorithms for detecting malicious behaviors in IoT, we have used a popular intrusion detection dataset—NSL-KDD. The training data are available in "KDDTrain+" (consisting of 125,973 data entries), and the testing data are available in "KDDTest+" (consisting of 22,543 entries), the details are shown in Table 1. Each dataset has 41 features and a

Table 1 Number of records in dataset

Dataset	Record count		
	Total entries	Normal entries	Malicious entries
KDDTrain+	125,973	67,343	58,630
KDDTest+	22,543	9710	12,833

single class label details of which are shown in Table 2. The "protocol type," "service," and "flag" attributes in the dataset contained string type data, for computational

Table 2 Details of features in NSL-KDD dataset

Features	Description	Possible feature values
duration (1)	Time during which the devices were connected	0–to–54451
protocol type (2)	Type of protocol used at transport layer: ICMP/TCP/UDP	1, 2, 3
service (3)	Service used by the destination network (70 different Web services)	1–to–70
flag (4)	This feature depicts the connection status (details of various flags given in Fig. 1)	1–to–11
src_bytes (5)	Data (in bytes) transferred from source device to destination device in single connection	0–to–1379963888
dst_bytes (6)	Data (in bytes) transferred from destination device to source device in single connection	0–to–309937401
land (7)	It depicts whether the source and destination devices are having same port numbers and IP addresses. (for same, land = 1, and for different, land = 0)	0, 1
wrong_fragment (8)	Count of wrong fragments sent in a single connection	0, 1, 3
urgent (9)	Count of urgent packets sent in a single connection	0–to–3
hot (10)	Count of critical data in the information shared between source and destination devices like program creation, program execution, etc	0–to–101
num_failed_logins (11)	Number of failed login attempts	0–to–4
logged_in (12)	Login status (for successfully login = 1, otherwise = 0)	0, 1
num_compromised (13)	Count of "compromised" conditions	0–to–7479
root_shell (14)	Depicts if the root shell is obtained or not (obtained = 1, otherwise = 0)	0, 1
su_attempted (15)	Depicts if the "su root" command was attempted or not	0, 1

(continued)

Table 2 (continued)

Features	Description	Possible feature values
num_root (16)	Count of operations performed as root	0–to–7468
num_file_creations (17)	Count of file creation operations performed in a single connection	0–to–100
num_shells (18)	Count of shell prompts	0–2
num_access_files (19)	Count of access operations performed on files	0–9
num_outbound_cmds (20)	Number of outbound commands in an ftp session	0
is_hot_logins (21)	Depicts whether the device is root/admin	0, 1
is_guest_login (22)	Depicts whether the device is logged in as guest	0, 1
count (23)	Count of connections with the same destination device within past 2 s	0–511
srv_count (24)	Count of connections with the same port number of destination device within past 2 s	0–511
serror_rate (25)	Connection percentage out of connections in "count" that have set flags: s0, s1, s2, or s3, on	0–to–1
srv_serror_rate (26)	Connection percentage out of connections in "srv_count" that have set flags: s0, s1, s2, or s3, on	0–to–1
rerror_rate (27)	Connection percentage out of connections in "count" that have set flags: REJ, on	0–to–1
srv_rerror_rate (28)	Connection percentage out of connections in "srv_count" that have set flags: REJ, on	0–to–1
samc_srv_rate (29)	Connection percentage out of connections in "count" that belong to same service	0–to–1
diff_srv_rate (30)	Connection percentage out of connections in "count" that belong to different services	0–to–1
srv_diff_host_rate (31)	Connection percentage out of connections in "srv_count" that belong to different destination	0–to–1
dst_host_count (32)	Count of connections with same destination host IP address	0–to–255
dst_host_srv_count (33)	Count of connections with same port number	0–to– 255

(continued)

Table 2 (continued)

Features	Description	Possible feature values
dst_host_same_srv_rate (34)	Connection percentage out of connections in "dst_host_count" that belong to same service	0–to–1
dst_host_diff_srv_rate (35)	Connection percentage out of connections in "dst_host_count" that belong to different services	0–to–1
dst_host_same_src_port_rate (36)	Connection percentage out of connections in "dst_host_srv_count" that use same source port	0–to–1
dst_host_srv_diff_host_rate (37)	Connection percentage out of connections in "dst_host_srv_count" that were connected to different destination devices	0–to–1
dst_host_serror_rate (38)	Connection percentage out of connections in "dst_host_count" that have set flags: s_0, s_1, s_2, or s_3, on	0–to–1
dst_host_srv_serror_rate(39)	Connection percentage out of connections in "dst_host_srv_count" that have set flags: s_0, s_1, s_2 or s_3, on	0–to–1
dst_host_rerror_rate (40)	Connection percentage out of connections in "dst_host_count" that have set flags: REJ, on	0–to–1
dst_host_srv_rerror_rate (41)	Connection percentage out of connections in "dst_host_srv_count" that have set flags: REJ, on	0–to–1
class_label	Classification of the traffic input as normal (1) and attack (0)	0, 1

purposes, we have converted them into numeric data. In "protocol type," we have used one, two, and three for representing ICMP, TCP, and UDP protocols, respectively. The "service" attribute contained 70 different Web services used by the devices in the destination network, and the services are represented numerically (ranging from 1 to 70) in our study. Also, the various values taken by "flag" attributes are shown in Fig. 1. For training and testing, all the entries in the respective datasets were used.

The data available is assumed to be sampled from normal distribution and we normalize the data using standard scalar.

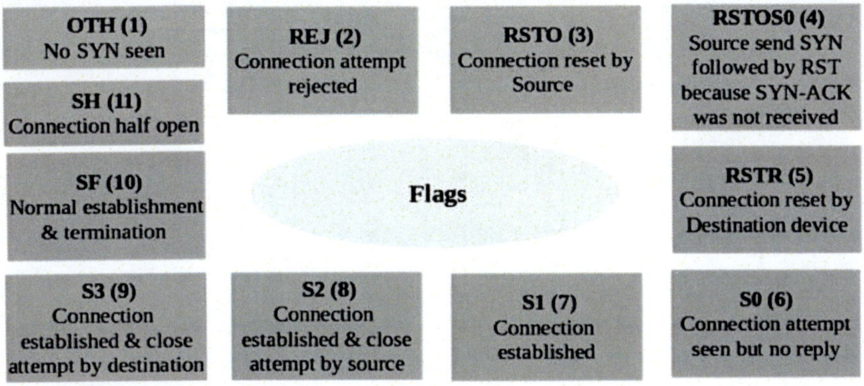

Fig. 1 Flag feature in dataset and its representation

3 Numerical Results

3.1 Supervised Machine Learning Algorithms Used

In this paper, we have used the following machine learning algorithms for the attack detection in IoT network: logistic regression [7], K-nearest neighbor [8], linear support vector machine (SVM), SVM with RBF kernel, SVM with polynomial kernel [9], SVM and logistic regression with stochastic gradient descent (hinge loss, log loss, Huber loss, and modified huber loss function) [10, 11], Gaussian Naive Bayes, Bernoulli Naive Bayes [12], decision tree [13], random forest [14], and multilayer perceptron (MLP) classifier [15].

Logistic regression: While implementing logistic regression on the dataset, we have set maximum iteration of 100 and used L2 norm for regularization and also we have used Broyden-Fletcher-Goldfarb-Shanno (BFGS) optimization algorithm [16].

KNN: We have implemented the KNN classifier for various values of K on the given dataset. To choose the optimal value of *K*, we observed that the accuracy of the KNN algorithm increases with *K* but for $K > 10$, the accuracy saturates, hence, choose the value of *K* as 10 (as shown in Fig. 2). We have used Minkowski distance to find K-nearest neighbors for a new data point.

SVM: For SVM, we have used L2 for regularization in all kernel types. For linear SVM, maximum iteration was set to 10,000 and kernel coefficient was $1/N$, where N is the total features in the dataset. For polynomial kernel type, the degree was chosen to be 3.

SGD: We also implemented logistic regression and linear SVM with SGD using various loss functions like: hinge, log, Huber, and modified Huber. During simulation, maximum number of iterations was set to 10,000 with $\alpha = 0.1$, where α is a constant which is multiplied to regularization term to control the degree of regularization. Here, again, we have used L2 for regularization.

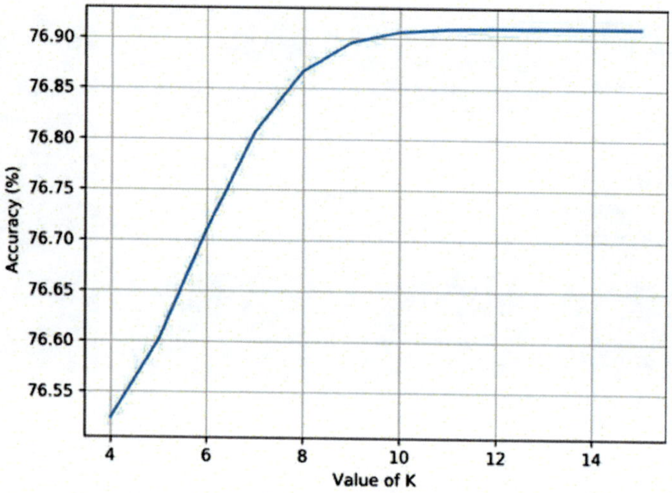

Fig. 2 Attack detection accuracy for various values of K using KNN classifier on KDDTest+ dataset

Decision tree classifier: In this classifier, we choose the best split at each node in a decision tree using the Gini index with a minimum allowed splits at each node as two. The formula for calculating Gini index is given as

$$\text{Gini Index} = i - \sum_{i=1}^{n} (p_i)^2$$

where p_i is the probability with which an element is classified to a particular class.

Random forest classifier: Here, again, we choose the best split at each node in a decision tree using the Gini index with a minimum allowed splits at each node as two. The classifier was iteratively run on the training data to find the optimal number of trees in the forest and maximum training samples to be used to train each base estimator. We observed that if we increase the number of trees in the forest beyond ten and maximum training samples to be used for base estimator training beyond 1000, then there was no change in classifier's accuracy (as shown in Figs. 3 and 4). So, we choose the number of trees in the forest as ten and maximum training samples to be used for base estimator training as 1000.

Artificial neural network (ANN) classifier: For ANN-based classification, we have used multilayer perceptron (MLP) classifier. For implementing MLP, we have used a single hidden layer of neurons (with number of neuron = 90 and the activation function as a rectified linear unit function) and a single neuron at the output layer to classify the input data into malicious (abnormal) or non-malicious (normal). We train the neural network using SGD algorithm with a learning rate = 0.001 and L2 norm for regularization.

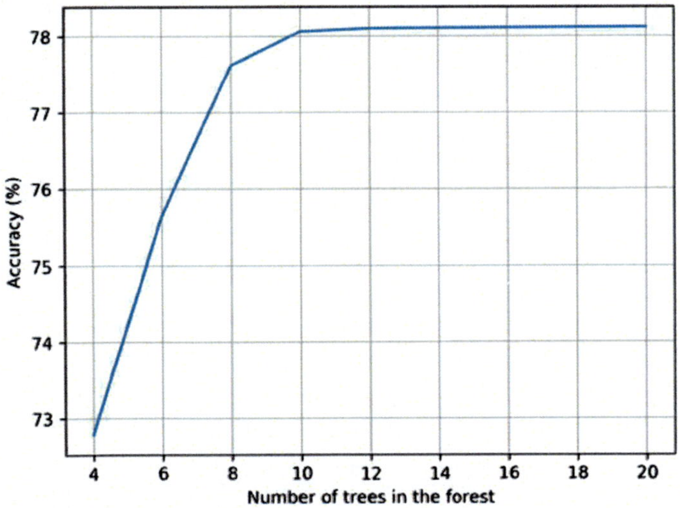

Fig. 3 Attack detection accuracy for varying number of trees in random forest classifier on KDDTest+ dataset

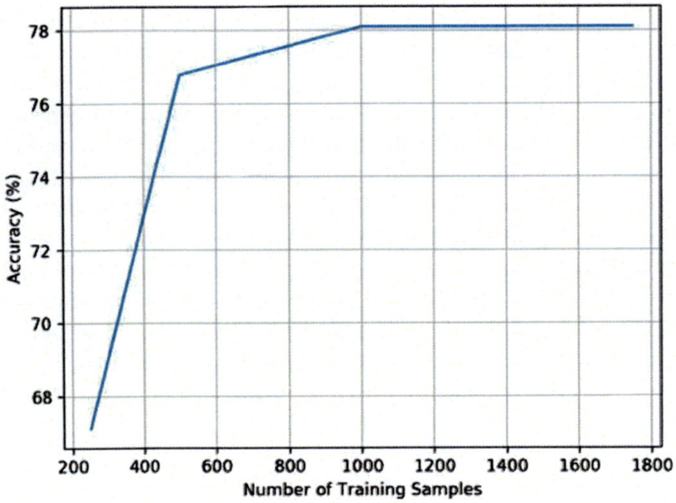

Fig. 4 Attack detection accuracy for varying number of training samples used to train the base estimator in random forest classifier on KDDTest+ dataset

The simulation was carried out on Intel core i3-5005U CPU @ 2.00 GHz * 4, RAM: 4 GB and operating system: Ubuntu 14.04 LTS (64bit), using Scikit-learn and pandas package.

3.2 Performance Evaluation

The experimental evaluation of various supervised machine learning algorithms was carried out on "KDDTrain+" dataset, and the efficiency of each trained model was tested on "KDDTest+" dataset.

We use confusion matrix to evaluate the performance of different machine learning models as shown in Table 3. Using the values obtained from the confusion matrix for each classifier, the following performance parameters were used [17]:

$$\text{Accuracy} = (\text{TP} + \text{TN})/(\text{TP} + \text{FN} + \text{FP} + \text{TN})$$

$$\text{Precision} = \text{TP}/(\text{TP} + \text{FP})$$

$$\text{Recall} = \text{TP}/(\text{TP} + \text{FN})$$

$$\text{F1 score} = 2\text{TP}/(2\text{TP} + \text{FP} + \text{FN})$$

The accuracy gives the percentage of samples that are correctly classified out of all the tested samples, precision depicts the classifier's ability to detect only relevant data, recall gives the efficiency of a classifier to detect all interested data points in the dataset (i.e., total positives detected by the system to that of actual positives throughout the system), and F1 score is the harmonic mean of both precision and recall. In Table 4, we summarize the performance of detecting attacks by all the classifiers used in our study on KDDTest + dataset, in terms of various performance parameters shown in Eqs. 1 to 4. We have also used area under the curve (AUC) of a receiver operating characteristic (ROC) curve as a performance matrix, which gives the overall performance of each classifier used in classifying the malicious and non-malicious traffic in KDDTest+ dataset (Figs. 5 and 6). The value of AUC lies between [0, 1] and the performance of a particular classifier is directly proportional to the value of AUC obtained, i.e., a classification model is considered efficient if its AUC value lies close to 1 and inefficient if its value lies close to 0. Besides these performance parameters, we have also noted the time, a particular classifier takes to train on KDDTrain+ dataset (training time) and the time it takes to predict a label for data in KDDTest+ dataset (test time). Evidently, more the training time of a particular classifier, more computation resources are needed for achieving classification, and lesser the test time indicates that the classifier will have quick response.

Table 3 Confusion matrix

Actual value	Predicted value	
	Attack	Normal
Attack	True positive (TP)	False negative (FN)
Normal	False positive (FP)	True negative (TN)

Table 4 Performance analysis of various machine learning classifiers on KDDTEST+ dataset

ML Classifiers	TP	FP	TN	FN	ACC (%)	Precision	Recall	F1 score	Training time (s)	Test time (s)	AUC
C1	9071	4953	7880	639	75.19	0.81	0.75	0.75	410.2	0.002	0.87
C2	9508	4699	8134	202	78.25	0.84	0.78	0.78	117.9	9.68	0.94
C3	9500	5092	7741	210	76.48	0.83	0.76	0.76	100.8	7.85	0.89
C4	9482	4977	7856	228	76.91	0.84	0.77	0.77	74.39	184.5	0.85
C5	9070	4899	7934	640	75.42	0.81	0.75	0.75	11.38	0.002	0.87
C6	9515	4931	7902	195	77.26	0.84	0.77	0.77	0.36	0.002	0.90
C7	9342	4781	8052	368	77.16	0.83	0.77	0.77	0.50	0.002	0.92
C8	9550	4915	7918	160	77.49	0.84	0.77	0.77	0.71	0.002	0.90
C9	9084	5064	7769	626	74.76	0.80	0.75	0.74	0.40	0.002	0.88
C10	8883	4326	8507	827	77.14	0.81	0.77	0.77	0.12	0.019	0.86
C11	9430	5141	7692	280	75.95	0.83	0.76	0.76	0.12	0.021	0.93
C12	9421	4694	8139	289	77.89	0.84	0.78	0.78	1.24	0.005	0.81
C13	9430	4655	8178	280	78.10	0.84	0.78	0.78	14.05	0.32	0.96
C14	9466	4581	8252	244	78.59	0.84	0.79	0.78	406.9	0.14	0.96

*C1: Linear SVM; C2: SVM with RBF kernel; C3: SVM with polynomial kernel; C4: KNN; C5: Logistic regression; C6: Stochastic gradient descent with hinge loss function; C7: Stochastic gradient descent with log loss function; C8: Stochastic gradient descent with Huber loss function; C9: Stochastic gradient descent with modified Huber loss function; C10: Gaussian Naive Bayes; C11: Bernoulli Naive Bayes; C12: Decision tree; C13: Random forest; C14: Multilayer perceptron (MLP) classifier

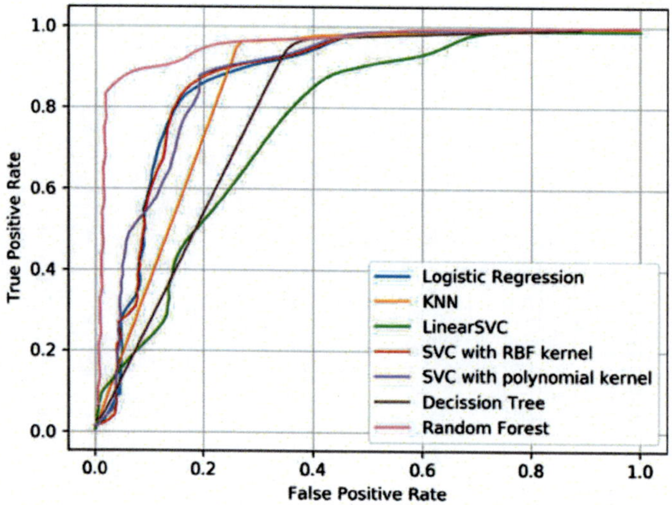

Fig. 5 ROC curves for various classifiers after classifying IoT network traffic in KDDTest+ dataset (Part 1)

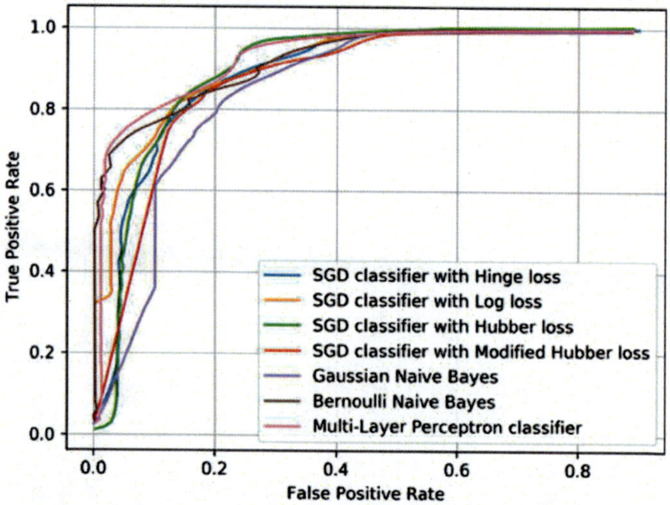

Fig. 6 ROC curves for various classifiers after classifying IoT network traffic in KDDTest+ dataset (Part 2)

From Table 4 and Figs. 5 and 6, it is evident that SVM with RBF kernel, random forest classifier, and multilayer perceptron have the highest accuracy (78.25%, 78.10%, and 78.59%, respectively) of detecting malicious traffic with an AUC values of 0.94, 0.96, and 0.96, respectively, but among them multilayer perceptron and SVM with RBF kernel has higher training time which restricts their use in IoT network.

Also, among all the classifiers used in our study, stochastic gradient descent with log loss function has very less training time with AUC value of 0.92 (which is highest among other loss functions used with SGD classifier) and an accuracy of 77.16%. Thus, we conclude that random forest classifier and SGD with log loss function perform better than the other discussed classifiers for detecting malicious traffic and are more suitable for IoT environments due to their low training and test time. Also, the random forest classifier can prove efficient in IoT for attack detection when fog computing paradigm is used, because in fog computing, the attack detection module can be deployed on fog devices (which have higher computational ability than edge devices) where each fog device can monitor a particular group of IoT devices and due to this limited number of devices to monitor the performance of attack detection can further increase. Whereas, the SGD classifier with log loss function will be more suitable for attack detection in cloud-based IoT network, since the cloud device has to monitor a huge number of IoT devices for attack detection, so a classifier with least training and test time is preferred.

4 Conclusion

In this paper, we studied the applicability of various supervised machine learning algorithms: SVM, logistic regression, KNN, stochastic gradient descent, Naive Bayes, decision tree, random forest, and multilayer perceptron, on NSL-KDD dataset for the detection of malicious traffic. After the thorough analysis of various performance parameters, the study concluded that random forest classifiers and stochastic gradient descent with log loss function are more accurate and better suited for detecting the attacks in IoT than the other discussed classifiers. Even though the above-mentioned classifiers outperform the other methods, the detection accuracy can further be enhanced in future by applying new and updated optimization algorithms.

References

1. Pawar AB, Ghumbre S (2016) A survey on IoT applications, security challenges and counter measures. In: 2016 international conference on computing, analytics and security trends (CAST). IEEE, pp 294–299
2. Jaiswal S, Gupta D (2017) Security requirements for internet of things (IoT). In: Proceedings of international conference on communication and networks. Springer, pp. 419–427
3. Xiao L, Wan X, Lu X, Zhang Y, Wu D (2018) IoT security techniques based on machine learning: how do IoT devices use AI to enhance security? IEEE Sig Process Mag 35(5):41–49
4. Nsl-kdd dataset. https://www.unb.ca/cic/datasets/nsl.html. Accessed: 02.03.2020
5. Gunupudi RK, Nimmala M, Gugulothu N, Gali SR (2017) Clapp: a self constructing feature clustering approach for anomaly detection. Futur Gener Comput Syst 74:417–429

6. Tavallaee M, Bagheri W, Lu W, Ghorbani AA (2009) A detailed analysis of the kdd cup 99 data set. In: 2009 IEEE symposium on computational intelligence for security and defense applications. IEEE, pp. 1–6
7. Hosmer Jr DW, Lemeshow S, Sturdivant RX (2013) Applied logistic regression, vol 398. Wiley
8. Cover T, Hart P (1967) Nearest neighbor pattern classification. IEEE Trans Inf Theor 13(1):21–27
9. Cortes C, Vapnik V (1995) Support-vector networks. Mach Learn 20(3):273–297
10. Mehrkanoon S, Huang X, Suykens JA (2014) Non-parallel support vector classifiers with different loss functions. Neurocomputing 143:294–301
11. Xu G, Cao Z, Hu B-G, Principe JC (2017) Robust support vector machines based on the rescaled hinge loss function. Pattern Recogn 63:139–148
12. John GH, Langley P (2013) Estimating continuous distributions in Bayesian classifiers. arXiv preprint arXiv:1302.4964
13. Safavian SR, Landgrebe D (1991) A survey of decision tree classifier methodology. IEEE Trans Syst Man Cybern 21(3):660–674
14. Dietterich TG (2000) An experimental comparison of three methods for constructing ensembles of decision trees: bagging, boosting, and randomization. Mach Learn 40(2):139–157
15. Guezzaz A, Asimi A, Mourade A, Tbatou Z, Asimi Y (2019) A multilayer perceptron classifier for monitoring network traffic. In: International conference on big data and networks technologies. Springer, pp 262–270
16. Battiti R, Masulli F (1990) Bfgs optimization for faster and automated supervised learning. In: International neural network conference. Springer, pp 757–760
17. Hasan M, Islam MM, Zarif MII, Hashem M (2019) Attack and anomaly detection in IoT sensors in IoT sites using machine learning approaches. Internet Things 7:100059

Brain Tumor Detection: A Review of Early Stage Tumor Detection Techniques

Rohit Mohanty, Santosh Kumar Mahto, and Rashmi Sinha

Abstract A multitude of methods are being deployed today in the field of medical diagnosis, and of those, the field that is growing most rapidly is of biomedical imaging. A plethora of these techniques have allowed us to identify abnormalities, ranging from trivial to the deadliest of forms, in the human body. As a result of this, the demand for better and improved methods of biomedical image processing has reached its zenith in today's era. Image processing helps us to remove the excess unwanted information from the images and provide substantial meaningful information to help in accurate diagnosis. Of the various medical imaging methods available to us like MRI, CT scan, X-rays, and so on, MRI seems to be the safest and most reliable one as it does not need the patient to be exposed to harmful radiations. The image processing techniques of preprocessing, segmentation, optimization, and feature extraction can then be used to detect and classify tumors from MRI images. In this survey, we will be looking into the steps involved in medical image processing and will review a variety of research done in this field.

Keywords Image processing · Segmentation · Filtering techniques · Tumor detection

R. Mohanty
Department of Electrical and Electronics Engineering, International Institute of Information Technology, Bhubaneswar, India

S. K. Mahto
Department of Electronics and Communication Engineering, Indian Institute of Information Technology, Ranchi, India
e-mail: skumar@iiitranchi.ac.in

R. Sinha (✉)
Department of Electronics and Communication Engineering, National Institute of Technology, Jamshedpur, India
e-mail: rsinha.ece@nitjsr.ac.in

K. Dahal et al. (eds.), *Internet of Things and Its Applications*, Lecture Notes in Electrical Engineering 825, https://doi.org/10.1007/978-981-16-7637-6_23

1 Introduction

An early detection, diagnosis and classification of brain tumors can be of paramount importance in the treatment and care of the patients affected with this disease. This registers a pressing need for advancements in the field of image processing which can help in the accurate diagnosis of these tumors. Brain tumor is caused by an abnormal growth of cells in the brain. They can be of two types: malignant or cancerous tumors and benign or non-cancerous tumors. Benign tumors have a slow growth rate and can be identified easily. On the other hand, malignant (cancerous) tumors are very aggressive and life threatening. Symptoms of brain tumor are headaches, nausea, and in some cases vision loss (Fig. 1).

Magnetic resonance imaging (MRI) is considered most appropriate for brain tumor detection as it gives the neurosurgeons sufficient information without exposing the patient to harmful radiations. An accurate processing and segmentation of MRI images can help the doctor to pin-point the exact location of the tumor in the brain, thus helping in precise treatment. But there are multiple challenges in the segmentation of MRI images. One of the major challenges is the extreme variation in the size, position, shapes, appearances, and properties of tumors across various patients. The overlap of intensities of tumors with normal brain tissues is another major challenge. In many cases, the expansion of tumor creates a deformity in other nearby parts of the brain which leads to an abnormal geometry even in the healthy tissues.

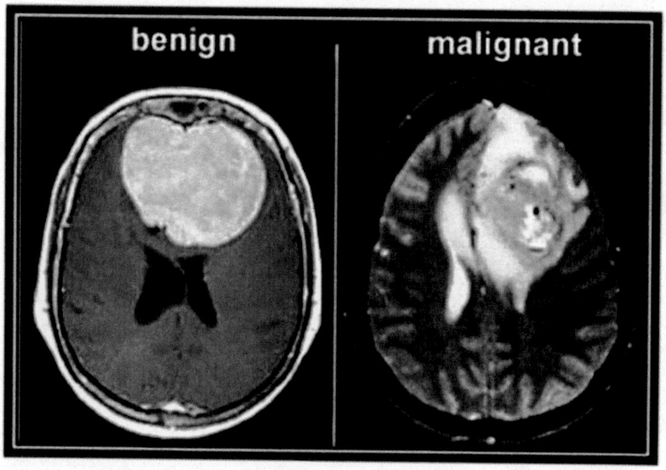

Fig. 1 Benign tumor (left) versus malignant tumor (right) [11]

2 Steps in Brain Tumor Detection

In this section, we will present the steps that are needed to analyze the MRI and detect abnormalities in the brain and will discuss various that have been developed for brain tumor detection. The flowchart shown below gives a bird's eye view of the process (Fig. 2).

2.1 Preprocessing

In the preprocessing step, we eliminate the unwanted or obstructive information from the MRI images. This would typically involve processes like converting the image to grayscale, salt and pepper, noise reduction and removal, and image enhancement. Another important step in the preprocessing of brain MRI images is skull stripping. In this process, we remove the non-brain tissues like skin, fat, skull from the brain tissues.

Before proceeding with other preprocessing steps, the first thing that we need to do is convert the image into grayscale in which every pixel represents the intensity value at that pixel. Converting the image into grayscale is really helpful for the process of image segmentation.

After converting the image into grayscale, we need to apply filters to reduce noise from the image. A filter can be used to blur the noise leaving behind a smoother image but losing the finer details of the image in the process. On the other hand, a filter can also be used to sharpen the image which enhances the finer details but increases the noise in the process, which then needs to be removed before further usage of the image in order to achieve a high level of accuracy. The two most commonly preferred filters are

- **Median Filter:**

  It works by changing the intensity of each pixel to the median of the intensity values of its adjacent pixels. One of its best attributes is that it maintains the edges

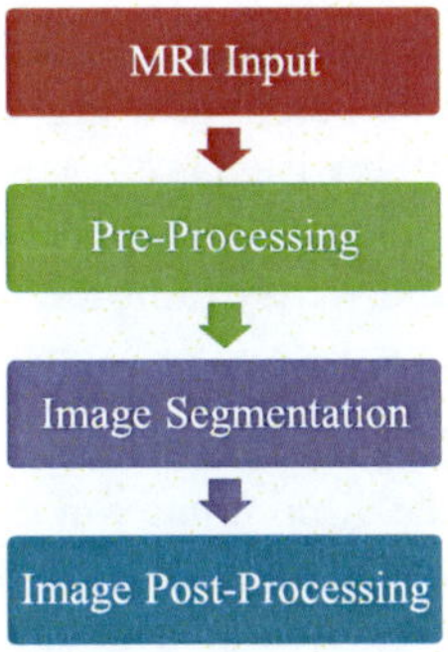

Fig. 2 Process of brain tumor detection

but removes the noise. It is most suitable for Poisson's noise and salt and pepper noise. This filter was used for removing noise by Lakshmi [1], Bhalchandra [2] and Karnan [3] in their experiments.

- **High Pass Filter:**

  It is used to detect the edges of the image and increase the sharpness which can later be removed by other filtering techniques. It is normally used in highlighting the image and the abnormalities which cannot be seen by the naked eye.

The research by Dubey [4] uses a Gaussian filter which removes noise from MRI. After applying a Gaussian filter, we should expect to get a smooth image which is much similar to viewing the image through a translucent screen. A Gaussian filter gives results similar to Weierstrass transform which involves convolving using a Gaussian function. Since the noise is usually present in the high-frequency regions of the image, hence, a Gaussian filter is a low pass filter.

Murthy [5], on the other hand, used the Sobel filter in her research. "The Sobel–Feldman operator is based on convolving the image with a small, separable, and integer-valued filter in the horizontal and vertical directions and is, therefore, relatively inexpensive in terms of computations" [6]. In scenarios where the output image has well defined edges, it is used in conjunction with various edge detection algorithms. It uses a derivative mask that works on the difference in pixel intensities.

The pseudo color translation method proposed by Wu [7] converts a grayscale image to a pseudo color image by assigning a color value to each intensity using a predefined function. It is preferable for use only when a single channel of data is available.

2.2 Image Segmentation

Segmentation helps us to make the image more expressive and easier to examine. In this process, we divide and partition the image into multiple segments. These segments contain sets of pixels that share similar characteristics. Over the years, a multitude of methods have been deployed for the segmentation. Below are the most common methods used by today's researchers:

Thresholding Method: In this method, a threshold value is set and if the intensity of a pixel in the grayscale image is less than the threshold, the pixel is turned black otherwise it is converted to white. Otsu thresholding is an extension of this method in which the image is classified to have two classes, the background and the foreground, and we try to fit the pixels into one of the classes. Murthy [5] and Patil [2] applied this method of segmentation in their research work.

Region Growing: This method strives to achieve image partitioning by exploiting the discontinuity of colored and grayscale regions. In this method, we need to select a seed point. We use this seed to examine its neighboring pixels and decide whether they should be included in the image. This method has an advantage that it allows us to choose multiple seed points at a time. Also, it distinctly highlights the edges of the images. But it also has the disadvantage of being a costly and local method. It is also quite sensitive to noise. Seed-based region growing (SBRG) is a very common image segmentation method in which the seed is often selected manually but an automated version of this method has also been discussed by various researchers.

Watershed Transform: It is a region-based segmentation technique that strives to find the similarities in regions and pixels, i.e., it finds the region which a pixel belongs to. In this process, a watershed transform function is first defined using elementary morphological operations which help to create an intensity gradient of the input image. Finally, we perform a contour search on the gradient image using the watershed transform function. It helps us to differentiate the overlap of the margins of different organs. But sometimes it can result in an over-segmented image and to avoid that from happening, we use marker-controlled watershed Segmentation in which certain elements of the image are predetermined. This method has been used by Badran [8], Vasikarla [4], and Bhalchandra [2] in their research. Watershed algorithms are based on two classes, namely flood-based watershed algorithm and rainfall-based watershed algorithm.

Edge-based Technique: Since edge detection is based on continuity of images, we use this method to find discontinuity in the images. There are a variety of edge detection methods that are being used these days, namely Krish edge detection, Prewitt edge detection, Sobel edge detection, Mare-Hilderth edge detection, log edge detection, Canny edge detection, Roberts edge detection, and Robinson edge detection. Bilateral symmetry is another simple and effective method that makes use of various edge detection algorithms.

Level Set Method: This method makes use of partial differential equations to continuously calculate the difference between pixels after which the segmentation of image becomes easy using well-defined mathematical techniques and methods. This method was proposed by Dubey [4] and Cho [9] in their research.

Genetic Algorithm: The basis of this algorithms is natural selection and evolution. These algorithms belong to a subclass of an evolutionary algorithm as they use evolutionary techniques and natural selection to solve optimization problems. A heuristic and an iterative model are available for implementing this algorithm. It is particularly efficient when the search space is significant. By evaluating the fitness function of every pixel, it helps us to find the optimal path for brain tumor segmentation in MRIs. This method of image segmentation was used by Khare [10] in her research.

Clustering Algorithm: Image segmentation makes use of a multitude of clustering techniques. These techniques group a set of pixels into a cluster based on some similarities. There are two types of clustering techniques, namely hard clustering and soft clustering. In hard clustering, an object can be classified to belong to a single cluster only. This gives us a very definitive segmentation of the image but it becomes very difficult in scenarios where the image has low resolution and contrast. K-means clustering is a good example of hard clustering algorithm. In this algorithm, we try to group the points based on their proximity to a mean point and this creates partitions in the image. Lakshmi [1] and Wu [7] used this technique in their research. Although these algorithms are easy to implement, they have the disadvantages of having sensitive results. On the other hand, in soft clustering techniques like fuzzy c-means algorithm, we do not assign a single cluster to a point. Contrary to it, we find the probability of each point belonging to each cluster hence assuming that an object can belong to more than one cluster. This method is used by Karnan [3] and Lakshmi [1] in their research. Derikvand [12] presented a patch and pixel-based brain tumor segmentation using convolutional neural networks. It was found to improve segmentation results especially in the tumor core region. The research by Yin [13] introduced us to a novel brain tumor segmentation method called partial view segmentation. It not only achieves a good accuracy with artificial neural networks but also requires less time for training. Though it shows some issue of over-fitting in the training process, but it can be offset by a balanced training and testing data set.

2.3 Post-processing

After the segmentation of the image, we need to apply post-processing activities to the image so as to assess its size and classify its type. It involves various optimization techniques that further enhance the accuracy of the results.

A common post-processing technique that was used by Lakshmi [1] in their research is the Canny edge detection. This method exploits the sharp change in the contrast at the boundaries of the objects present in the image. Due to this sharp change in the brightness, a Gaussian filter can be used to remove noise and hysteresis can be used to detect the edges.

Another common tool that is used in image post-processing is the use of morphological operations. In this method, we use a mathematical structuring element to define the size of the tumor and extract meaningful information from the image. Harris-Laplace or LOG-Lindeberg algorithm was used by Badran [8] to optimize the segmentation results. On the other hand, Sadashivappa [5] proposed morphological operations like binary dilation and binary erosion for further optimizations.

Post-processing by convolutional neural network was also proposed by Badran [8] in his research. It is an intriguing deep learning model that can further enhance the accuracy of the results.

Another substantial post-processing method is the use of particle swarm optimization as demonstrated by Karnan [3] in his research. This meta-heuristic technique creates a sample population and searches iteratively for the optimal solution by upgrading generations. Unlike genetic algorithm, it does not require an evolutionary operator.

Histogram equalization is another method that uses contrast adjustment. It is useful in scenarios where background and foreground are both dark such as X-rays where it enhances the visible bone structure in the image. It was used by Murthy [5] and Lin [7] in their research.

Support vector machines are machine learning algorithms that can be used for regression and classification. It learns from predefined labeled data and has shown to have a higher accuracy than other classification methods. It was used by Khare [10] in conjunction with curve fitting, by Hussain [14] in conjunction with GLCM and by Gurbina [15] in conjunction with wavelet transforms in their research for further classification.

Another simple but powerful machine learning classification algorithm that can be used for brain tumor detection is Naive Bayes classifier. It was used by Divyamary [16] in their research and has produced efficient results in classifying malignant and benign tumors.

3 A Comparative Analysis of Methods of Tumor Detection

When it comes to filtering, median filter is most commonly used because of its efficiency in removing salt and pepper noise and its simplicity of use. It uses convolution for filtering, and since it is a nonlinear filter, it is effective in preserving the edges of the image. Contrary to this, since a Gaussian filter is a low pass filter, hence, it is unable to preserve the information from the edges. But the Gaussian filter is very effective in smoothening Gaussian noise and is less complex and cheap to implement than the median filter. But the best filter to preserve the edges would be the Sobel filter.

Thresholding is the easiest and most widely used segmentation method. When the contrast between the background and foreground image is relatively high, thresholding can be used very effectively. Thresholding might not very useful in extracting information from MRIs because it completely depends on the contrast in the image.

As opposed to thresholding technique, watershed technique is much slower and requires a lot of calculations. Watershed is also affected by the presence of noise as opposed to thresholding which is quite resistant. The selection of seed region is vital in determining the success of watershed technique but the segmented image might have holes due to noise. Clustering algorithms like fuzzy c-means and fuzzy k-means clustering both fuzzy logic-based unsupervised techniques that are widely used for image segmentation. But compared to watershed or thresholding, these algorithms are more CPU and memory intensive. Since these techniques are pixel based, their susceptibility to noise is quite high and hence they require a good amount of preprocessing. Table 1 shown below gives a list of segmentation techniques along with their advantages and disadvantages.

Based on the level of human intervention needed, segmentation methods can be classified into three classes, namely manual, semi-automated, and fully automated segmentation. But the manual segmentation is never used with digital images as it is prone to errors.

Table 1 Comparison of methods of segmentation

S. No	Methods of segmentation	Advantages	Disadvantages
1	Watershed	It has a high range of capture It has mathematical morphology	It can cause over segmentation of the images
2	Threshold method	It can be used when pre-filtering is not done because it has low susceptibility to noise It is good for quick analysis of image Is good for scenarios when contrast is high	It cannot be used in images with poor contrast It is not accurate in images with lots of background and foreground objects It does not provide enough information from the MRI
3	Fuzzy c-means, k-means, and level set methods	These are unsupervised methods and require least human intervention Level set technique can be used for computer aided vision These are useful for large images with poor contrast	These are highly susceptible to noise The process might be long and laborious These methods are more CPU and memory intensive.

4 Conclusion

In this paper, we conducted a survey of the various steps involved in medical image processing that helps us in the detection and diagnosis of brain tumors through MRI inputs and analyzed various techniques that are being used in each step. A brief discussion of the current research in each step of the process is provided along with a description of the techniques used. At the end, a comparison of these techniques has been provided to give a clear view of the strengths and weaknesses of the current techniques in the steps involved in processing of the brain images. Since segmentation is the most important aspect of image processing, a more detailed discussion is presented on the current research in the methods of image segmentation. The ultimate goal of this survey is to find gaps in the current research and pave way for future work in the field of brain tumor detection so that it can benefit the entire mankind.

References

1. Selvakumar J, Lakshmi A, Arivoli T (2012) Brain tumor segmentation and its area calculation in brain MR images using K-mean clustering and fuzzy C-mean algorithm. In: 2012 IEEE-international conference on advances in engineering, science and management (ICAESM-2012), March 30, 31 2012
2. Patil RC, Bhalchandra AS (2012) Brain Tumour extraction from MRI images using MATLAB. Int J Electron Commun Soft Comput Sci Eng 2(1). ISSN 2277-9477

3. Gopal N, Karnan M (2010) Diagnose brain tumor through MRI using image processing clustering algorithms such as fuzzy C Means along with intelligent optimization techniques. In: 2010 IEEE international conference on computational intelligence and computing research
4. Dubey R, Hanmandlu M, Vasikarla S (2011) Evaluation of three methods for MRI brain tumor segmentation. In: 2011 eighth international conference on information technology: new generations
5. Murthy TSD, Sadashivappa G (2014) Brain tumor segmentation using thresholding, morphological operations and extraction of features of tumor. In: 2014 international conference on advances in electronics computers and communications
6. Wikipedia contributors. Sobel operator. Wikipedia, The Free Encyclopedia. https://en.wikipedia.org/w/index.php?title=Sobel_operator&oldid=750123462
7. Wu M-N, Lin C-C, Chang C-C (2007) Brain tumor detection using color-based K-means clustering segmentation. In: 2007 third international conference on intelligent information hiding and multimedia signal processing (IIH-MSP 2007)
8. Badran EF, Mahmoud EG, Hamdy N (2010) An algorithm for detecting brain tumors in MRI images. In: 2010 international conference on computer engineering and systems
9. Taheri S, Ong SH, Chong VFH (2010) Level-set segmentation of brain tumors using a threshold-based speed function. Image Vison Comput 28(1):26–27
10. Khare S, Gupta N, Srivastava V (2014) Optimization technique, curve fitting and machine learning used to detect Brain Tumor in MRI. In: 2014 proceedings of IEEE international conference on computer communication and systems ICCCS14
11. https://www.researchgate.net/figure/Benign-Tumor-left-and-Malignant-Tumor-Right-5_fig1_334947653. Last accessed 2020/10/10
12. Derikvand F, Khotanlou H (2019) Patch and pixel based brain tumor segmentation in MRI images using convolutional neural networks. In: 2019 IEEE 2019 5th Iranian conference on signal processing and intelligent systems (ICSPIS), Shahrood, Iran (2019.12.18-2019.12.19)
13. Yin Y (2020) Partial view segmentation: a novel approach to the brain tumor segmentation. In: 2020 IEEE 2020 3rd international conference on computer and communication engineering technology (CCET), Beijing, China (2020.8.14-2020.8.16)
14. Hussain A, Khunteta A (2020) Semantic segmentation of brain tumor from MRI images and SVM classification using GLCM features. In: 2020 IEEE 2020 second international conference on inventive research in computing applications (ICIRCA), Coimbatore, India (2020.7.15-2020.7.17)
15. Gurbina M, Lascu M, Lascu D (2019) Tumor detection and classification of MRI brain image using different wavelet transforms and support vector machines. In: 2019 IEEE 42nd international conference on telecommunications and signal processing (TSP), Budapest, Hungary (2019.7.1-2019.7.3)
16. Divyamary D, Gopika S, Pradeeba S, Bhuvaneswari M (2020) Brain tumor detection from MRI images using Naive classifier. In: 2020 IEEE 6th international conference on advanced computing and communication systems (ICACCS), Coimbatore, India (2020.3.6-2020.3.7)

A Novel Stroke Measurement Operator for Visual Objects

Tauseef Khan and Ayatullah Faruk Mollah

Abstract Stroke is an important intrinsic property of objects which may be effectively used for object characterization. However, stroke measures reported so far lack accurate measurement of stroke width of general objects, and varying degree of stroke uniformity, which is very crucial for characterization, is not taken into consideration. In this paper, a novel stroke measurement operator is proposed using distance transform and medial skeleton map of visual objects, which effectively measures stroke width and degree of stroke uniformity for any kind of object or component. Possible utilities of this operator are also outlined in the context of object detection and classification. Finally, in order to validate and demonstrate the applicability of this operator, computed stroke features are applied in text non-text classification on a benchmark dataset and 94.5% accuracy is achieved, which establishes its robustness and effectiveness.

Keywords Stroke width · Stroke uniformity · Object characterization · Distance transformation · Medial skeleton map

1 Introduction

Automated object identification from scene images and video frames is a pioneer research area owing to huge advancement of computer vision and machine intelligence. Detection of objects in practical scenario is still challenging due to complex background, arbitrary shape, and other confusions. Extraction of intrinsic and discriminative properties from target objects may lead to accurate object detection. In general, classification of objects presents within an image that can be attempted from its shape information. In this context, stroke width of foreground objects may

T. Khan (✉) · A. F. Mollah
Department of Computer Science and Engineering, Aliah University, New Town Campus, Kolkata, West Bengal 700160, India

T. Khan
Department of Information Technology, Haldia Institute of Technology, Haldia, West Bengal 721657, India

K. Dahal et al. (eds.), *Internet of Things and Its Applications*, Lecture Notes in Electrical Engineering 825, https://doi.org/10.1007/978-981-16-7637-6_24

significantly represent its geometric shape, and thus, it may be effective for classification. Identification of objects with uniform stroke has a wide scope of applicability in text localization, traffic sign detection, etc. Adequate number of works are reported on text detection and object identification using stroke-based features so far [1–5, 13]. However, most of these methods have applied stroke-width transform (SWT) map to segment text region from complex background. Epstein et al. [1] were the first to exploit the stroke feature of text components and generate SWT map of corresponding image where every pixel holds stroke-width value. Huang et al. [3] have designed extended version of SWT by incorporating color information of text pixels. Few works have been reported on structural object detection using stroke-width feature [5, 10]. Nevertheless, apart from texts, any work is hardly found on general object detection and classification using stroke feature descriptors to the best of our knowledge.

In this paper, a novel stroke measurement operator (SMO) is introduced for general objects that effectively computes the stroke width and degree of stroke uniformity across a component. Stroke width is measured using distance value of medial skeleton points of a component image, whereas amount of dispersion of stroke-values signifies the uniformity of stroke width across the object. In order to validate and demonstrate its applicability, extracted features from the designed operator are evaluated in a component-level classification problem on a public dataset comprised of text and non-text components. Contribution of this paper may thus be stated as: (i) a novel stroke measurement operator is introduced for general objects; (ii) besides measuring stroke width, it also computes degree of stroke uniformity across a component; (iii) outcomes of the designed operator may be treated as feature descriptors for component-level classification; and (iv) possible scope of applications and demonstrated results followed by insightful analysis are also presented.

2 Related Works

Majority of the works have been found on text detection and recognition using stroke-based feature descriptors. Mosleh et al. [2] have proposed a text detection framework for scene images using set of feature descriptors obtained from stroke-width transform map which is generated from connected component images. Subsequently, feature vector is fed into pattern classifier to distinguish text and non-text components. Su et al. [6] have designed an enhanced version of SWT for scene text detection by identifying seed point within stroke and gradually grew from it to find more stroke segments. Recently, Saha et al. [7] have designed an end-to-end text detection and script identification framework using shallow stroke-based features and deep neural network. It generates coarse-level candidate text proposals using MSER and stroke-width transform map. Similar kind of work is reported in [8], where candidate text regions are localized using stroke-width transform map, and subsequently, false positives are removed using one-class classifier driven by some shallow texture-based features.

Apart from text detection, few works have been reported on medical image analysis as well. Cheng et al. [9] have proposed a vessel segmentation approach from retinal image using a set of hybrid feature descriptors. Vessels with constant stroke width and limited lengths are detected using SWT, and extracted features are used for discriminative classification. In another work [11], skeleton map is generated from retinal vessels in order to compute width and length of arbitrary-oriented vessels. Zhou et al. [12] have first proposed a road detection method by exploiting the width of different roads generated using SWT from road image and achieved high detection accuracy.

Recently, stroke-based features are effectively introduced for component-level text non-text classification which is a non-trivial task especially under scene environment. Khan et al. [14] have introduced a text non-text classification framework using stroke-value distribution profile of stroke-potential pixels which is generated from distance transform map of component images. A medial skeleton-based stroke feature descriptor for text non-text classification is reported by Khan et al. [14], where distance values of stroke pixels are extracted from both distance transform and medial skeleton map. Finally, a distribution profile is generated upon frequency of distinct distance values of both text and non-text components. Later, Khan et al. [16] have designed an area occupancy profile of farthest equidistant stroke pixels that effectively distinguish text and non-text components from complex document and scene images. In the light of the above discussion, it is realized that most of the related works are found on document and scene-level text images, rather than general scenario of objects or components. Moreover, studies on stroke uniformity in regard to stroke width are also left. Therefore, an effective stroke measurement operator reflecting intrinsic stroke information might bring a lot of utilities in object characterization.

3 Proposed Methodology

In order to design the stroke measurement operator for object components, binarization is a prerequisite. Although appropriate binarization is crucial to overcome the contextual intrusion problem, it may not play decisive part in the proposed work. Given object component is first binarized using global Otsu's method [17], and then two feature maps, viz. distance transform and medial skeleton, are generated in parallel from the binary image. Finally, both the maps are fused together to design the proposed SMO for the given object component. It is worth mentioning that, part of the proposed work is inspired by [15]. Complete pipeline of the proposed work is pictorially shown in Fig. 1. A step-wise description is presented in the following subsections.

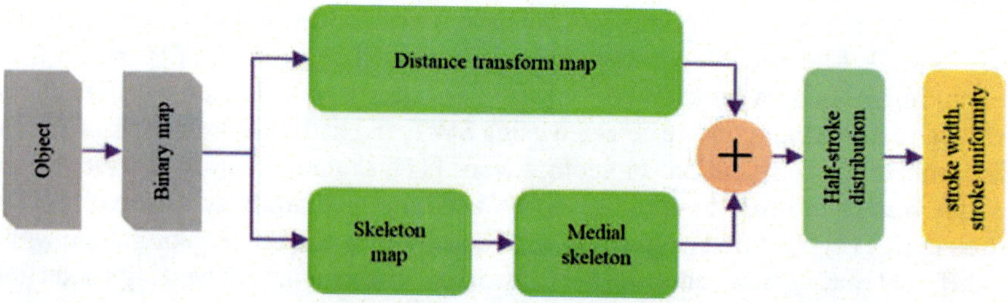

Fig. 1 Pipeline of the developed operator. Two feature maps (distance transform map and medical skeleton map) are generated from binarized object component image and these maps are fused together to identify distance values of stroke-potential pixels. Finally, stroke width and stroke uniformity are computed from generated distribution profile of distance values of stroke-potential pixels

3.1 Generation of Distance Transform Map

Distance transformation [18] carries distance information of each object pixels computed with respect to its nearest background pixels using Euclidean distance metric. To provide a clear understanding, Fig. 2 shows the generated distance transform map from few synthetic object components along with some scene images. It is observed that, pixels positioned across center of stroke carry the highest distance values and slowly decrease toward the contour of the object.

Fig. 2 Step-wise generation of distance transform maps from synthetic and scene images are visually represented (top-down). **a, b, c** Row-wise input synthetic geometric objects and text component, corresponding binary image and generated distance transform map. **d** Scene-level text component, corresponding binary image, and final distance transform map

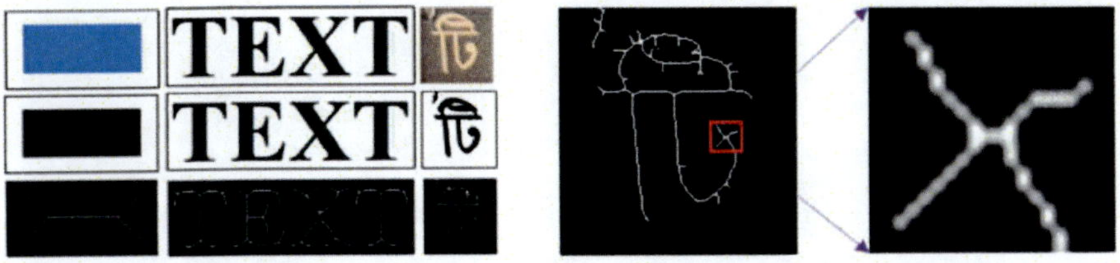

(a) (b)

Fig. 3 Visualization of skeleton map of synthetic and scene object components. **a** Object components and corresponding skeleton images (top-down). **b** A magnified view of bifurcated branches of skeleton map generated (marked in color for better visibility)

3.2 *Generation of Medial Skeleton Map*

Single pixel wide thin skeleton map is generated from binary image of object components using skeletonization technique. Skeleton map gradually removes the object pixels stating from border regions and keeps only medial axis pixels. Figure 3 shows some skeleton maps from given binary images. It is observed that skeleton map preserves the bifurcated branches toward the end of strokes. A magnified view of bifurcated end-points of skeleton map is clearly visualized in Fig. 3b. Bifurcated spurious branches mislead the stroke measurement and yield erroneous output. Hence, these spurious edges are unwanted for current state of work, and therefore, a branch removal algorithm reported in [15] has been applied to remove such bifurcated end-points of stroke.

Normally, pixels at the end of stroke have exactly one neighbor object pixels called *leaf node*; on the other side, pixels in bifurcated branches have more than two object neighbors called *junction*. In between them, all pixels have exactly two object neighbors in both directions. Current algorithm gradually removes pixels starting from leaf node and stop when it reaches to *junction*. The strategy behind the removal of spurious branches is demonstrated in Fig. 4 where the generated medial skeleton map removes bifurcated end-points of skeleton image.

3.3 *Stroke Measurement Operator*

In order to measure the stroke width of object components, both distance transform and medial skeleton maps are considered. Let, distance transform map that contains real-value distance weights may be denoted as $[D(m,n)]_{P\times Q} \in \mathbb{R}^{+}$ where P and Q denote the number of rows and columns, respectively. Similarly, medial skeleton map may be denoted as $[\text{MS}(m,n)]_{P\times Q}$, where $\text{MS}(m,n) \in [0, 1]$. Then, pixel-wise multiplication of both maps is performed and a new feature map $[F(m,n)]_{P\times Q}$ is obtained. The generated feature map contains distance values (s) of object pixels of medial skeleton points only. Figure 5 shows the distribution profile of s of medial

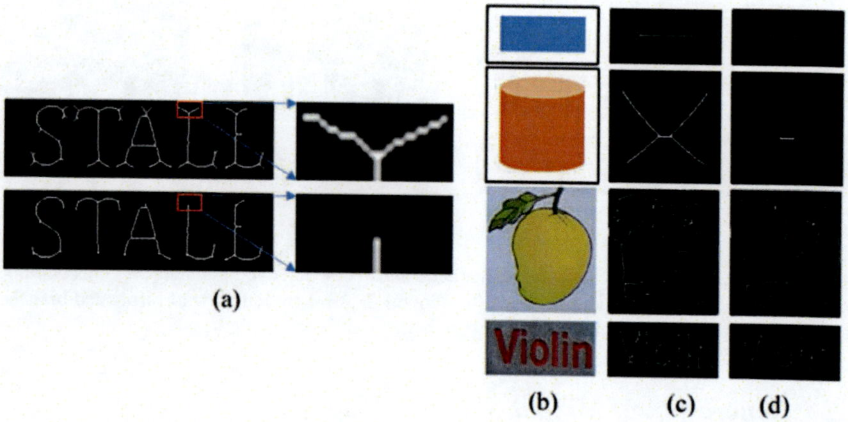

Fig. 4 Generation of medial skeleton map from skeleton image for object components. **a** Skeleton map with magnified view of bifurcated branches, corresponding medial skeleton map with no spurious bifurcated branches (magnified view of same slice may be observed), **b** Few object components, **c** corresponding skeleton maps of (b), **d** generated medial skeleton maps from (c)

skeleton pixels for object components. It has been observed that for texts, dispersion of s is very low and concentrated in a narrow range due to its near-uniform stroke width, whereas for non-texts dispersion is very high and distributed in a wide range due to non-uniformity of stroke width. Moreover, it is noticed that frequency of s denoted as $f(s)$ is highest for a specific distance value and comparatively low for adjacent distance values. Distribution of distance values for nearly unistroked objects formed normal distribution which is distinct in nature from objects with non-uniform strokes. It is worth mentioning that, object with uniform stroke will have only one peak value in distribution profile which is observed in Fig. 5 (first row).

Stroke-width measurement. Distance values of medial skeleton pixels are half of the actual width of the object stroke. Very few distance values are extracted from near-uniform strokes with one carry significantly high occurrence which is unlikely for non-uniform strokes. Thus, mean of all distance values may compute the actual width of strokes for any objects which is denoted by λ. Stroke width of any object is mathematically expressed using Eq. 1.

$$\lambda = 2 \int_{-\infty}^{+\infty} s f(s) \mathrm{d}s \quad (1)$$

Stroke confidence score. Degree of stroke uniformity across the object is measured using stroke confidence score denoted as SCS. Stroke regularity is measured by amount of dispersion of distance values of stroke pixels. Mathematical expression for SCS is defined in Eq. 2.

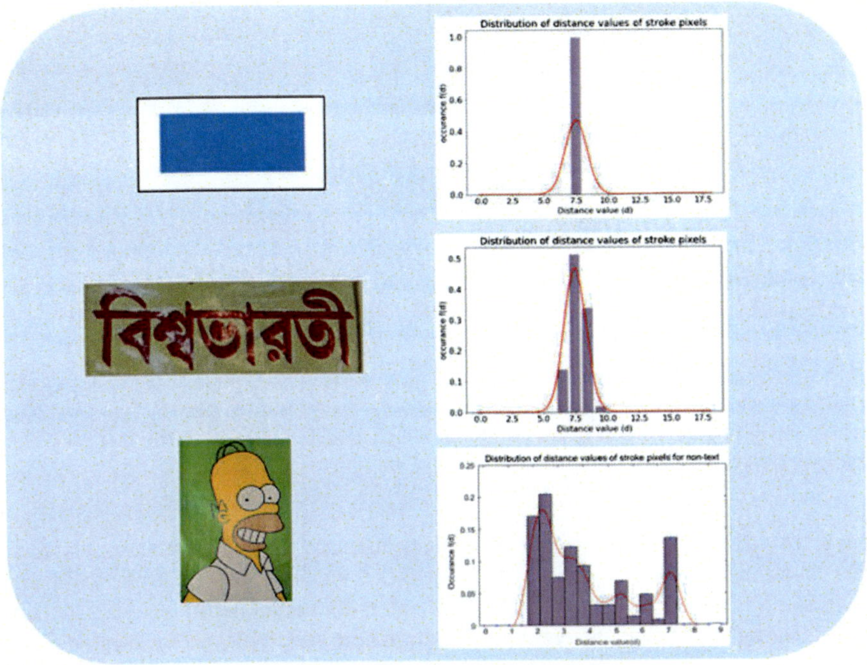

(a) (b)

Fig. 5 Distribution profile of distance values of stroke pixels for different object components. **a** Object components, **b** corresponding distribution of distance values of stroke-potential pixels generated from medial skeleton map. Distribution profile for unistroked objects formed normal distribution that significantly distinguish it from other non-uniform stroke objects

$$\text{SCS} = \frac{1}{1 + \sigma^s} \tag{2}$$

Here, σ is the standard deviation of all distance values of medial skeleton pixels. In ideal situation, SCS yields exactly 1 for unistroked objects when dispersion of distance values is zero, whereas confidence score decreases with higher amount of dispersion. Figure 6 shows the application of SMO on both synthetic and real object components that effectively measures stroke width and degree of stroke-uniformity.

4 Evaluation and Analysis

In order to validate the developed operator, a number of experiments have been carried out on publicly available Aliah University Text Non-text (AUTNT) dataset [19] which is comprised of component-level text and non-text images. To the best of

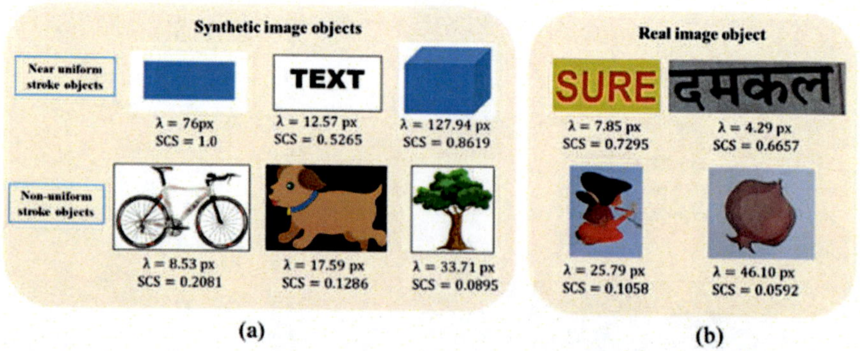

Fig. 6 Measurement of stroke-based parameters using SMO for object components. (a) Stroke-width value (λ) and confidence score (SCS) for synthetic objects. **b** Same measurement is reported for real image objects. 1st row shows the objects with near-uniform stroke widths that carry high confidence score compared to objects with non-uniform stroke width (2nd row)

our knowledge AUTNT is the only available benchmark dataset reported for text non-text classification in component-level. In the following experiments, stroke width and confidence score are treated as feature descriptors for component-level text non-text classification. Moreover, different pattern classifiers are employed for evaluation.

4.1 Dataset Description

AUTNT comprises of total 10,771 component images (7890 text and 2881 non-text). Entire dataset is divided into 5:1 ratio for training and test purpose for model evaluation. Images of AUTNT are captured using handheld mobile camera from complex document and indoor–outdoor scene images under unconstrained environment. AUTNT is a multilingual dataset containing English, Devanagari, and Bangla scripts. Figure 7 shows some sample images of this dataset.

Fig. 7 Sample text and non-text images from AUTNT dataset

Table 1 Performance of the proposed stroke-based feature descriptors in text non-text classification using different classification models

Classifier	*P*	*R*	FM	Accuracy
NB	0.884	0.887	0.885	0.886
MLP	0.949	0.945	0.947	**0.945**
SVM	0.936	0.935	0.935	0.934
RF	0.940	0.939	0.939	0.938
AdaBoost	0.942	0.939	0.940	0.937

4.2 Performance Obtained

A set of experiments are steered out for text non-text classification using two stroke-based feature descriptors. To train the model, text and non-text components are labeled as 0 and 1, respectively. Table 1 reports the obtained results on AUTNT dataset for different classification models. The proposed feature descriptors are evaluated using standard statistical metrics such as accuracy, precision (*P*), recall (*R*), and f-measure (FM). It is worth mentioning that, both *P* and *R* are separately calculated for the two output classes, and at the end, weighted average is taken. However, FM is simply the harmonic mean of *P* and *R*. Five different classifiers are adopted, viz. naïve Bayes (NB), multilayer perceptron (MLP), support vector machine (SVM), random forest (RF), and AdaBoost.

It may be observed from Table 1 that the highest classification accuracy; i.e., 94.5% is achieved with MLP and performance of other classifiers are also reasonably high (over 90% in most cases), which unquestionably establish the prowess and robustness of the proposed features. Usually, text strokes are near-uniform irrespective of image source which consequently yields high degree of stroke uniformity, whereas non-text strokes mostly have arbitrary stroke width with low confidence score.

4.3 Discussion

Usage of the proposed operator in object classification is demonstrated in Sect. 4.2. However, it has multifold utilities that may be effectively applied in various object-level analysis which are mentioned below:

Stroke-width estimation. Given an image object, SMO computes its width of stroke.

Degree of stroke uniformity. The SCS signifies the degree of stroke uniformity.

Object characterization. The outcomes of this operator, viz. stroke width (λ) and confidence score (SCS). may be treated as a pair of feature descriptors for object tracking, clustering, classification, and other analysis.

Identification of unistroked objects. Objects with uniform or near-uniform stroke width may be effectively identified by its degree of stroke uniformity. Objects with near-uniform stroke yield high confidence scores and vice-versa.

5 Conclusion

A novel stroke measurement operator for general object components is proposed in this paper. The proposed operator effectively measures the stroke width and regularity of stroke across object components which may be useful for object characterization, clustering, classification, recognition, or other analysis. Moreover, to demonstrate the strength and effectiveness of the operator, the pair of features is applied in text non-text classification problem on a benchmark dataset and pretty high classification accuracy with multiple classifiers (i.e., over 93%) is obtained. It may also be stated that the operator has manifold utilities in characterization and analysis of objects in general. The operator and the feature pair may be applied on other object-level classification problems which can be considered for potential future work.

References

1. Epshtein B, Ofek E, Wexler Y (2010) Detecting text in natural scenes with stroke width transform. In: Proceedings of IEEE computer society conference on computer vision and pattern recognition, pp 2963–2970
2. Mosleh A, Bouguila N, Hamza AB (2012) Image text detection using a Bandlet-based edge detector and stroke width transform. In: Proceedings of British machine vision conference, pp 1–12
3. Huang W, Lin Z, Yang J, Wang J (2013) Text localization in natural images using stroke feature transform and text covariance descriptors. In: Proceedings of IEEE international conference on computer vision, pp 1241–1248
4. Bosamiya JH, Agrawal P, Roy PP, Balasubramanian R (2015) Script independent scene text segmentation using fast stroke width transform and GrabCut. In: Proceedings on 3rd IAPR Asian conference on pattern recognition, pp 151–155
5. Xu L, Kong M, Pan B (2017) Building extraction by stroke width transform from satellite imagery. In: Proceedings on Chinese conference on computer vision, pp 340–351
6. Su F, Xu H (2015) Robust seed-based stroke width transform for text detection in natural images. In: Proceedings of 13th international conference on document analysis and recognition, pp 916–920
7. Saha S, Chakraborty N, Kundu S, Paul S, Mollah AF, Basu S, Sarkar R (2020) Multi-lingual scene text detection and language identification. Pattern Recogn Lett 138:16–22
8. Mukhopadhyay A, Kumar S, Chowdhury SR, Chakraborty N, Mollah AF, Basu S, Sarkar R (2019) Multi-lingual scene text detection using one-class classifier. Int J Comput Vis Image Process 9(2):48–65
9. Cheng E, Du L, Wu Y, Zhu YJ, Megalooikonomou V, Ling H (2014) Discriminative vessel segmentation in retinal images by fusing context-aware hybrid features. Mach Vis Appl 25(7):1779–1792
10. Cho H, Yoon HJ, Jung JY (2018) Image-based crack detection using crack width transform (CWT) algorithm. IEEE Access 6:60100–60114

11. Mayrhofer-Reinhartshuber M, Cornforth DJ, Ahammer H, Jelinek HF (2015) Multiscale analysis of tortuosity in retinal images using wavelets and fractal methods. Pattern Recogn Lett 68:132–138
12. Zhou H, Kong H, Wei L, Creighton D, Nahavandi S (2016) On detecting road regions in a single UAV image. IEEE Trans Intell Transp Syst 18(7):1713–1722
13. Khan T, Mollah AF (2018) A novel text localization scheme for camera captured document images. In: Proceedings of 2nd international conference on computer vision and image processing, pp 253–264
14. Khan T, Mollah AF (2019) Distance transform-based stroke feature descriptor for text non-text classification. Rec Dev Mach Learn Data Anal 189–200
15. Khan T, Mollah AF (2019) AUTNT—a component level dataset for text non-text classification and benchmarking with novel script invariant feature descriptors and D-CNN. Multimedia Tools Appl 78(22):32159–32186
16. Khan T, Mollah AF (2020) Text non-text classification based on area occupancy of equidistant pixels. Procedia Comput Sci 167:1889–1900
17. Otsu N (1979) A threshold selection method from gray-level histograms. IEEE Trans Syst Man Cybern 9(1):62–66
18. Maurer CR, Qi R, Raghavan V (2003) A linear time algorithm for computing exact Euclidean distance transforms of binary images in arbitrary dimensions. IEEE Trans Pattern Anal Mach Intell 25(2):265–270
19. Aliah University Text Non-text dataset. https://github.com/iilabau/AUTNTdataset

Malaria Detection Using VGG19 and Deep Convolutional Neural Network

Gaurav Prasad, Angana Chakraborty, and Ananya Banerjee

Abstract Malaria is caused when a person is bitten by a female Anopheles mosquito. It is a fatal disease. Lots of work has been done in the detection of this disease all over the world. Every year millions of people all over the world are affected by malaria. It is a contagious disease. Traditionally, malaria is diagnosed by distinguishing healthy red blood cells from the infected ones. It is also very time consuming and it may produce inaccurate reports due to human error. Compared with the traditional methods used to extract the engineering feature of the hand, the proposed method uses an in-depth study of end-to-end design that makes both the extraction features and direct separation from the dots separated by red blood bands. The data, used in this study, were taken from the National Institute of Health called the NIH malaria dataset. A pre-trained model and a custom model have been compared here to analyze the better model. To maximize performance, standard preprocessing methods have been used to extract various features of the data. Besides, more complex structures have been used and tested to select the most efficient model. Machine learning and deep learning have played a great role in the detection and prediction of numerous diseases all over the world. Machine learning algorithms also help people to analyze large and complex medical databases and analyze them with clinical understanding. This paper uses these advanced techniques for prediction of malaria. In this paper, convolutional neural network (CNN) has been used to build a model and to train the model to detect the parasitized from non-parasitized samples. The dataset, used here, contains stained red blood cell images. The accuracy of the custom model is 97.50%.

Keywords Malaria parasite detection · Convolutional neural network (CNN) · VGG19

G. Prasad (✉) · A. Banerjee
Department of Computer Science and Engineering, Narula Institute of Technology, Kolkata, India

A. Banerjee
e-mail: ananya.banerjee@nit.ac.in

A. Chakraborty
Department of Computer Science and Engineering, Haldia Institute of Technology, Haldia, India

K. Dahal et al. (eds.), *Internet of Things and Its Applications*, Lecture Notes in Electrical Engineering 825, https://doi.org/10.1007/978-981-16-7637-6_25

1 Introduction

Plasmodium bacteria causes malaria. It is a life-threatening disease which puts an estimated 3.2 million people are at high risk, worldwide. The African region has the highest rates of malaria in the world (90%), followed by the Southeast Asia Region (7%) and the Eastern Mediterranean Region (2%) [1]. Visual detection and Plasmodium recognition in RBC are possible chemically process [2]. The staining process creates the colors of the RBCs but highlights Plasmodium, WBCs, and platelets. The discovery of Plasmodium requires the discovery of stains. However, we need to analyze the colored objects continuously to find out if parasites or not to prevent false diagnoses. Several ways are available for the detection of malaria. Malaria is caused by one cell microorganism belonging to the genus Plasmodium. There are a total of five species. These include P. vivax, P. ovale, P. knowlesi, and P. malariae. P. falciparum is that the deadliest among them [3]. The symptoms of this disease include headaches, fever, fatigue, and vomiting. This virus spread by RBCs in human blood which are carried throughout the body. If not given proper treatment, malaria is dangerous enough to be fatal.

In this paper, we have developed a model to predict malaria parasite from red blood cells using deep convolutional network technique, where we have used a VGG19 model for prediction and have compared it with a custom model. We have used several data augmentation techniques to achieve better result and high performance. The dataset has been divided into test and train sets which have been used rigorously to train the models. The convolution layers, pooling layers, and dropout layers have been used in the custom model for training and prediction.

2 Dataset

This data have been taken from a dataset by Natural Institute of Health. There are a total of 27,558 images in this dataset. There are equal number of parasitized and non-parasitized images, which is 13,799. This dataset is a collection of research images containing slide images of RBCs. From Chittagong Medical College Hospital, Bangladesh, the photographs of Giemsa smear slides with spots from 150 P. falciparum-infected and 50 healthy patients were collected. In this paper, we have taken those images and converted them to equal size for compliance with the model. The functions performed for optimizing these images have been mentioned in the Sect. 4 of this paper named data augmentation (Figs. 1 and 2).

3 System Requirements

See Table 1.

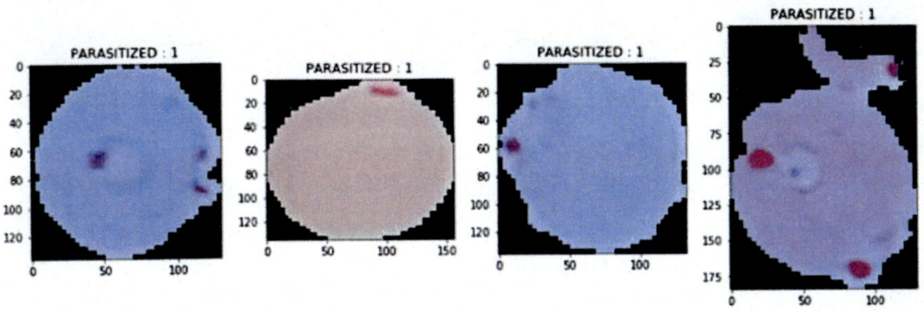

Fig. 1 Parasitized image sample

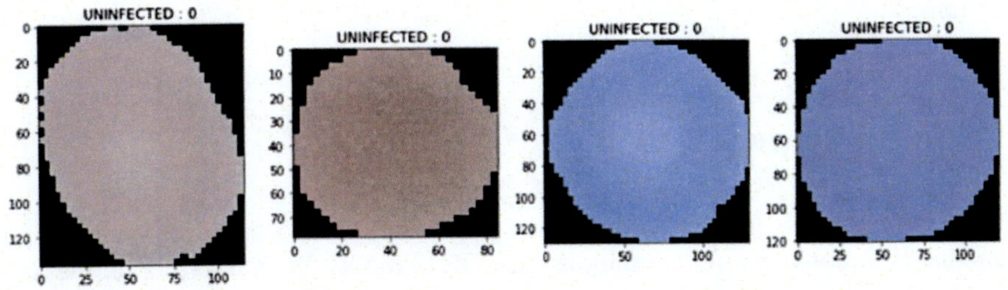

Fig. 2 Uninfected image sample

Table 1 System requirements

System	Requirements
OS	Windows 10 (64 bit)
Processor	Intel® Core(TM) i7-7700 CPU @3.60 GHz processor
Graphics card	Nvidia® GTX 1060 6GB
Python version	3.8.6
Keras version	2.2.5
Tensorflow version	2.2.0

4 Data Augmentation

Data augmentation is the process used for enhancing the data [4]. It contributes greatly in enhancing the performance of the model. The NIH dataset has both parasitized and non-parasitized images. The most basic problem we face while training a dataset is the problem of overfitting. Data augmentation helps us overcome this problem. It develops a variety in images protecting the basic features in it. Various techniques are used to get images closer to a real-world scenario. The images on the training set play a great role in training a model. Data augmentation provides a broader

Table 2 Data augmentation

Augmentation type	Parameters
Rescale	1/255
Shear range	0.2
Zoom range	0.3
Fill mode	Nearest
Rotational range	30
Horizontal flip	True

base to these training features [5, 6]. The data augmentation applied in here includes rescaling, shearing, zooming, and horizontally flipping the images. All are mentioned in Table 2.

5 Model Network Design

We have used two models to depict the prediction. One is a custom model and another is through transfer learning using VGG19 model. We are testing for making the machine learning with better accuracy.

5.1 VGG Model

Here, as a form of transfer learning, the VGG19 model has been pre-trained over ImageNet [7]. ImageNet is a huge database of images. There are 19 convolution layers on which this model has been trained (refer to Fig. 3). With each convolution 2D layer and max-pooling layer, the number filters and parameters both keep on increasing. The activation function used here is rectified linear units (ReLU). The ReLU activation function uses these convolution blocks as input. In the final output layer, Softmax has been used as an activation function as it is a classification problem. We end up with a total of 20, 07, 562 parameters with 50, 178 being trainable parameters. The output has a dense layer 50, 178 output neurons. Refer to Fig. 3.

The weights and biases of this model have been optimized by using "Adam" optimizer. It helps in efficiently updating and optimizing the weights using the stochastic gradient descent approach. Error handling is done by cross-entropy. Using mini batches at every epoch,

$$\text{Cross-Entropy} = (y \log(p) + (1 - y) \log(1 - p))$$

The last four layers have been trained to optimize the functioning of this VGG19 model. The other layers are kept frozen. The results are shown on training last four

Model: "model_1"

Layer (type)	Output Shape	Param #
input_1 (InputLayer)	[(None, 224, 224, 3)]	0
block1_conv1 (Conv2D)	(None, 224, 224, 64)	1792
block1_conv2 (Conv2D)	(None, 224, 224, 64)	36928
block1_pool (MaxPooling2D)	(None, 112, 112, 64)	0
block2_conv1 (Conv2D)	(None, 112, 112, 128)	73856
block2_conv2 (Conv2D)	(None, 112, 112, 128)	147584
block2_pool (MaxPooling2D)	(None, 56, 56, 128)	0
block3_conv1 (Conv2D)	(None, 56, 56, 256)	295168
block3_conv2 (Conv2D)	(None, 56, 56, 256)	590080
block3_conv3 (Conv2D)	(None, 56, 56, 256)	590080
block3_conv4 (Conv2D)	(None, 56, 56, 256)	590080
block3_pool (MaxPooling2D)	(None, 28, 28, 256)	0
block4_conv1 (Conv2D)	(None, 28, 28, 512)	1180160
block4_conv2 (Conv2D)	(None, 28, 28, 512)	2359808
block4_conv3 (Conv2D)	(None, 28, 28, 512)	2359808
block4_conv4 (Conv2D)	(None, 28, 28, 512)	2359808
block4_pool (MaxPooling2D)	(None, 14, 14, 512)	0
block5_conv1 (Conv2D)	(None, 14, 14, 512)	2359808
block5_conv2 (Conv2D)	(None, 14, 14, 512)	2359808
block5_conv3 (Conv2D)	(None, 14, 14, 512)	2359808
block5_conv4 (Conv2D)	(None, 14, 14, 512)	2359808
block5_pool (MaxPooling2D)	(None, 7, 7, 512)	0
flatten_2 (Flatten)	(None, 25088)	0
dense_3 (Dense)	(None, 2)	50178

Total params: 20,074,562
Trainable params: 50,178
Non-trainable params: 20,024,384

Fig. 3 VGG19 architecture

Table 3 List of parameters used for training VGG19 model

Parameter name	Type/value
Epochs	50
Batch size	32
Optimizer, learning rate	Adam
Error function	Categorial cross entropy
Input size	224 X 224
Pooling	Max
Pre-trained weights	ImageNet

layers that clearly outperform the performance of frozen layers. This method saves time and memory and makes our model more efficient (Table 3).

5.2 *Custom Model*

In the custom model, the input format of image is RBG (three filters) of 50 × 50 pixels. This architecture consists of three layers. Each layer consists of four components, a convolution 2D layer, a max-pooling layer, and a layer for batch normalization along with a dropout layer in the end. The convolution 2D layer has a dimension of 3 × 3 with 32 filters. It is followed by a max-pooling layer of dimensions 2 × 2. Next to that is a layer for batch normalization [8]. To be precise what a batch normalization, it is a technique for standardizing the inputs to a layer far each of the mini batches. The axis used for this purpose is chandim. In the end, it is a 20% dropout layer [9]. The three layers of these four components are followed by a flattening layer after which we have a dense layer with rectified linear unit (ReLU) activation function [10–12]. Next is a 50% dropout layer, and finally, a dense layer resulting ultimately to a total of 285,506 parameters (refer to Fig. 4).

This model based is based on supervised learning method. This approach is same to that of a classification problem. Categorical cross-entropy is the loss function that is used here. This loss function calculates the difference between the original data and our predicted output. We have used "Adam" optimizer here as an adaptive learning rate. Through backpropagation, it optimizes the weights and biases during training. Softmax is the activation function used here for the output layer. This balancing of weights and biases helps in making the model more efficient. 50 epochs each with a batch size of 32 have been used to train the model. All the parameters are mentioned in Table 4.

Layer (type)	Output Shape	Param #
conv2d_1 (Conv2D)	(None, 48, 48, 32)	896
max_pooling2d_1 (MaxPooling2	(None, 24, 24, 32)	0
batch_normalization_1 (Batch	(None, 24, 24, 32)	128
dropout_1 (Dropout)	(None, 24, 24, 32)	0
conv2d_2 (Conv2D)	(None, 22, 22, 32)	9248
max_pooling2d_2 (MaxPooling2	(None, 11, 11, 32)	0
batch_normalization_2 (Batch	(None, 11, 11, 32)	128
dropout_2 (Dropout)	(None, 11, 11, 32)	0
conv2d_3 (Conv2D)	(None, 9, 9, 32)	9248
max_pooling2d_3 (MaxPooling2	(None, 4, 4, 32)	0
batch_normalization_3 (Batch	(None, 4, 4, 32)	128
dropout_3 (Dropout)	(None, 4, 4, 32)	0
flatten_1 (Flatten)	(None, 512)	0
dense_1 (Dense)	(None, 512)	262656
batch_normalization_4 (Batch	(None, 512)	2048
dropout_4 (Dropout)	(None, 512)	0
dense_2 (Dense)	(None, 2)	1026

Total params: 285,506
Trainable params: 284,290
Non-trainable params: 1,216

Fig. 4 Custom model architecture

Table 4 List of parameters for training custom model

Parameter name	Type/value
Epochs	50
Batch size	32
Optimizer, learning rate	Adam
Error function	Categorical crossentropy
Input size	50 × 50
Pooling	Max
Dropout layers	4

6 Results

We have performed the entire prediction based on two models. Various preprocessing techniques have been used to enhance the performance and accuracy of the models. The VGG19 model performs prediction with an accuracy of 95.5% with 50 epochs in the process. On the other hand, the custom model predicts the data with an accuracy of 97.5% with 50 epochs in the process (Fig. 5; Table 5).

The above graph represents the training and validation logs for the custom model after 50 epochs. The blue line indicates the training accuracy, while the orange line indicates the training loss. From the graph, we can see the model converging at above

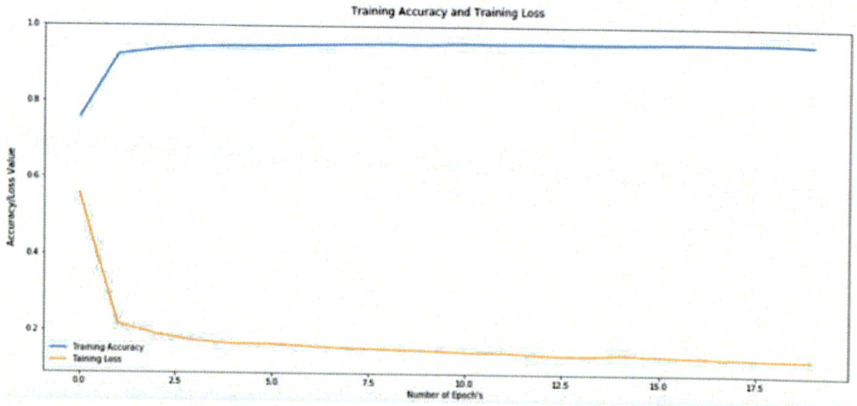

Fig. 5 Training accuracy and training loss

Table 5 Testing accuracy of the models

Model	Testing accuracy (%)
VGG19	95.50
Custom	97.50

95.77% accuracy for training data. The loss graph shows the fall in the loss as the model's accuracy increases.

7 Conclusion

In this paper, we have developed models to predict malaria through image classification using deep learning techniques. We have preprocessed the data and the data augmentation techniques have helped in converting images to a form that has greatly optimized the task. We have used a VGG19 model trained on ImageNet and a custom model to compare the accuracies, we achieve by modifying models according to our needs and provided dataset. Our custom model towers over the VGG19 model with an accuracy of 95.77% for training data and 97.50% for testing data. Thus, we can conclude that data augmentation can help a great way in optimizing the data and enhancing the models and a combination of data augmentation along with the correct use of normalization and dropout features CN outperform VGG19 models in image processing.

References

1. WHO (2016) Fact sheet: world Malaria report. In: World health organization. World Health Organization
2. Girshick RB, Donahue J, Darrell T, Malik J (2014) Rich feature hierarchies for accurate object detection and semantic segmentation. In: Proceedings of the IEEE conference on computer vision and pattern recognition (CVPR), pp 580–587
3. Gopakumar GP, Swetha M, Siva GS, Subrahmanyam GRKS (2018) Convolutional neural network-based malaria diagnosis from focus stack of blood smear images acquired using custom-built slide scanner. Int J Biophotonics 11(3)
4. Liang Z et al (2016) CNN-based image analysis for malaria diagnosis. In: IEEE international conference on bioinformatics and biomedicine (BIBM), pp 493–496
5. Bloland PB (2001) Drug resistance in Malaria, WHO/CDS/CSR/DRS/2001.4. World Health Organization, Switzerland
6. Caraballo H, King K (2014) Emergency department management of mosquito-borne illness: malaria, dengue, and West Nile virus. Emerg Med Prac 16(5):1–23
7. Liang Z, Andrew P, Ilker E, Mahdieh P, Kamolrat S, Kannappan P, Peng G (2016) CNN-based image analysis for malaria diagnosis. In: 2016 IEEE international conference on bioinformatics and biomedicine (BIBM). IEEE, pp 493–496
8. Szegedy C, Wei L, Yangqing J, Pierre S, Scott R, Dragomir A, Dumitru E, Vincent V, Andrew R (2015) Going deeper with convolutions. In: Proceedings of the IEEE conference on computer vision and pattern recognition, pp 1–9
9. Hung J, Anne C (2017) Applying faster R-CNN for object detection on malaria images. In: Proceedings of the IEEE conference on computer vision and pattern recognition workshops, pp 56–61
10. Deng J, Wei D, Richard S, Li L-J, Li K, Fei-Fei L (2009) Imagenet: a large-scale hierarchical image database. In: IEEE conference on computer vision and pattern recognition, pp 248–255
11. Ioffe S, Szegedy C (2015) Batch normalization: accelerating deep network training by reducing internal covariate shift. ICML (cited 946 times, HIC: 56 , CV: 0)

12. Hinton GE, Krizhevsky A, Srivastava N, Sutskever I, Salakhutdinov (2014) Dropout: a simple way to prevent neural networks from overfitting. J Mach Learn Res 15:1929–1958 (cited 2084 times, HIC: 142 , CV: 536), R

PANDIT: An AI Twin-Based Radiography Image-Assisted nCOVID-19 Identification and Isolation

Swarnava Biswas, Debajit Sen, and Moumita Mukherjee

Abstract The unprecedented rise and spread of the pandemic in form of nCOVID-19 has really raised high concerns in the socioeconomic front. The usual diagnosis is made by an RT-PCR test, which is highly specific can incorrectly identify some nCOVID-19 individuals to cause a serious compromise in overall accuracy. Since the drug application in its full swing is still some months away, hence, the need of the hour is to find a more accurate technique which can be used by health care centers having basic point of care facilities. The increase in the number of cases in India and lack of test kits in some of the less known diagnostic centers has added more concerns to the increasing problems. Additionally, the test kits incur a significant cost making it less affordable to some of the diagnostic centers. Hence, this research group in this article has proposed an algorithm centered around the concept of Internet of Things, a dual deep learning based algorithm, and collating the decision by a strong decision fusion technique. The objective of the algorithm is to detect and isolate the nCOVID-19 subjects in a cost-effective way to keep a check on the spread. This pandemic detection and isolation technique (PANDIT) is based on two different radiography image technology and uses a state-of-the-art deep learning algorithm for the purpose. The radiography technique has long been the most acceptable technique for cases related to pneumonia. The group has developed the algorithm based on X-ray and CT scan as its training data. The novelty of this paper is best described by a multi-fold methodology. Firstly, the significance of radiography imaging for detecting and identification of COVID-19 subjects. A simple connected value chain driven by Internet of Things (IoT) would enable the isolation process in an efficient and accelerated manner.

S. Biswas
The Neotia University, Sarisha, West Bengal 743368, India

D. Sen
ELMAX Systems and Solutions, Kolkata, West Bengal 700045, India

M. Mukherjee (✉)
Adamas University, Kolkata, West Bengal 700126, India
e-mail: moumita.mukherjee@adamasuniversity.ac.in

K. Dahal et al. (eds.), *Internet of Things and Its Applications*, Lecture Notes in Electrical Engineering 825, https://doi.org/10.1007/978-981-16-7637-6_26

Keywords Internet of Things · Deep learning · Machine learning · COVID-19 detection and isolation · X-ray · CT scan · Decision fusion algorithm · Raspberry Pi · Intel® Movidius™ Neural compute stick

1 Introduction

With inception of data and analytics enabling direct and continuous connect with users, and other machine learning-based image processing analyzing vast amounts of people health data, the focus has clearly shifted to prevention ahead of cure [1, 2]. The predicted persistence of COVID in the future will necessitate diagnostic testing in remote location limited by point of care facilities. IoT in healthcare is challenged due to low levels of Internet coverage, especially in countries like India. As a holistic and preventive system, IoT in healthcare provides a shift from cost-based to value-based healthcare thereby strengthening existing medical centers limited by facilities. Its socioeconomic benefits to people, end users, and governments through advanced and improved healthcare delivery is unmatched. The reason the research group thought of proposing a solution for an IoT enabled healthcare is to make preventative and control the spread as part of the modern health informatics system [3, 4]. All things considered, radiography assessment can be directed quicker and have more prominent accessibility given the commonness of chest radiology imaging frameworks in present-day human services frameworks and the accessibility of compact units, making them a decent supplement to RT-PCR testing especially since CXR imaging is regularly proceeded. In any case, perhaps, the greatest bottleneck confronted is the requirement for master radiologists to decipher the radiography pictures, since the visual markers can be inconspicuous. All things considered, and PC helped demonstrative frameworks that can help radiologists to even more quickly and precisely decipher radiography pictures to recognize COVID-19 cases which is exceptionally wanted [5–7].

2 Materials and Methods

Inspired by the requirement for quicker understanding of radiography pictures, various artificial intelligence (AI) frameworks dependent on deep learning have been proposed, and results have demonstrated to be very encouraging as far as precision in recognizing patients tainted with COVID-19 by means of radiography imaging, with the attention principally on CT imaging [8–10]. This AI-enabled method is powered by a powerful multi-layer convolutional neural network, also known as deep neural network in short. Many pretrained networks are available, but we have chosen the one which has been successfully used in the field of medical imaging and related works [11–15]. The deep network used by us is RESNET-50. Moreover, keeping the objective of the paper in mind, we have restricted our classification problem to a binary class problem only viz-a-viz COVID and normal [16–19]. The next important step is to tune the network to achieve the optimum-most performance. In order to obtain the right model, we have framed our design methodology in the following manner:

(i) For the first type, we have divided the model in training and hold-out sets in the split percentage of 80–20. We have tuned the hyperparameters as well to make the most out of this method.
(ii) In the second type, we have used the K-fold cross-validation concept, to reduce our misclassification error and increase the model accuracy in return. The K-fold validation shuffles the training set randomly and iteratively and deploys K different algorithms to automatically return five models which can individually be evaluated on some performance criteria to obtain the best out of the lot. We have chosen the value of K as '5' in our case.

3 Decision Fusion Algorithm

Any decision fusion algorithm plays a crucial role in multi-hypothesis situation, and hence, universally applicable in AI-enabled clinical diagnostic procedures as well. The concept of multi-hypothesis will act like a valuable second opinion to doctors and hence will be an enabler for any diagnostic-based approach. In our case, we have proposed series of screening phases for the correct detection and isolation of the COVID subjects.

Our algorithm finally scans through the sets of inferences drawn from the different AI engines and finally returns the output of the class which has been inferred the most. The overall computational flow of our architecture is given below (Fig. 1).

4 IoT-Enabled Deployability

To make this composite system available to the diagnostic facilities, the authors have used the concept of Internet of Things (IoT), a technology extremely popular in telemediation. Internet of Things (IoT) has been a trusted friend of various medical facilities ever since its inception. The concept of telemedicine or remote patient monitoring system has evolved in many ways with use of IoT-based devices. The authors have used the concept of IoT to complement our AI-based patient diagnostic system to increase the reliability and performance. The Intel® Movidius™ Neural Compute Stick (NCS) is a new piece of hardware used for enhancing the inference process of computer vision models on low-powered edge devices.

The philosophy of the method used for deployment of our algorithms is given below (Fig. 2).

A flow diagram of the entire value chain of our intelligent framework is shared below (Fig. 3).

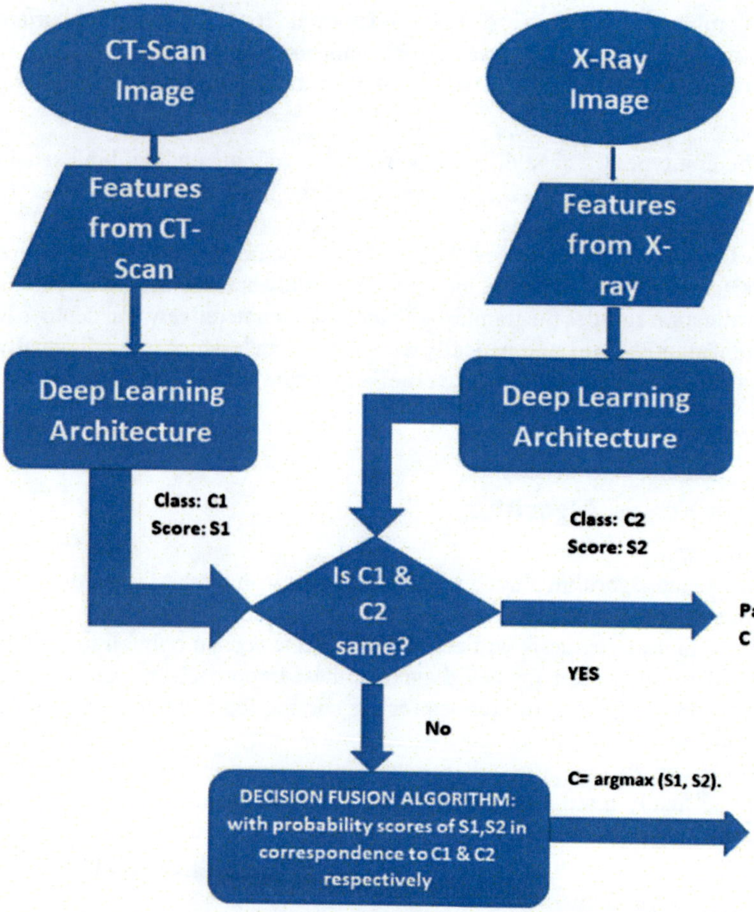

Fig. 1 Flowchart for decision fusion algorithm

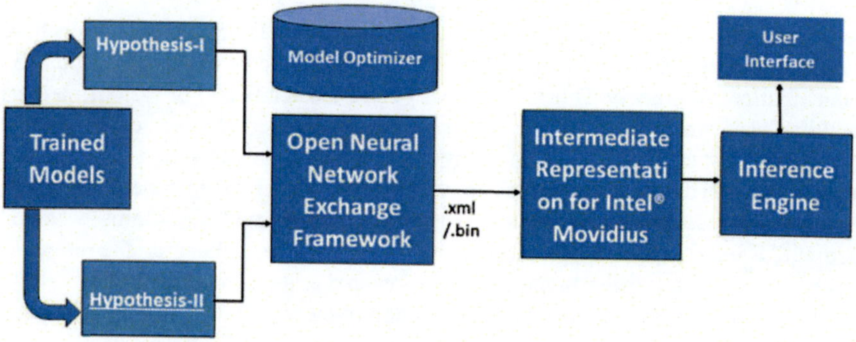

Fig. 2 Philosophy of our AI software architecture to be used for deployment

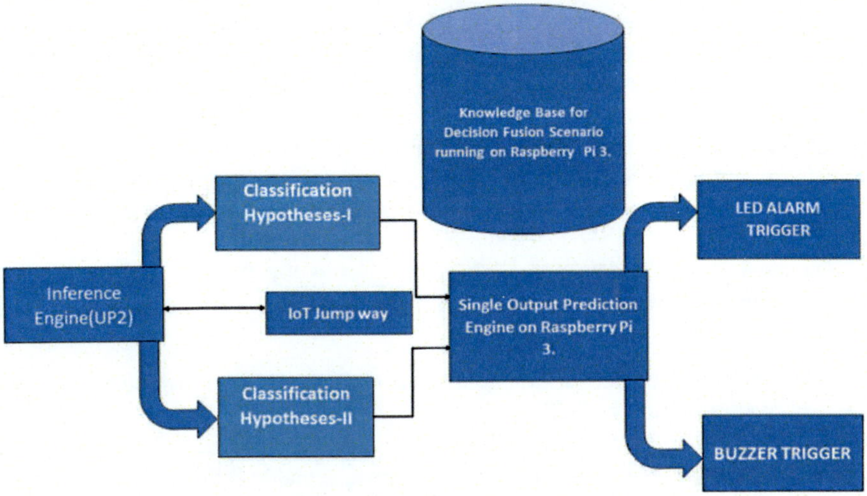

Fig. 3 Flow diagram of the entire value chain of our intelligent framework

The novelty of our detection procedure is pivoted around an already established IoT connectivity concept. The device we prototyped is an IoT-enabled alarm system built on a Raspberry Pi device. The Raspberry Pi is virtually connected with the server, so that for any COVID-19 positive subject, the server can trigger specific actions on the IoT network for communicating with the Raspberry Pi device. The results are captured from the classification of images in the server and are sent to the Raspberry Pi device where the actions like turning on a red LED and a triggering a buzzer when COVID-19 is detected and turning on a blue LED when a normal subject is detected have been programmed and activated. This is an amazingly simple proof of concept (POC) which shows a possibility of a powerful application that can save time for medical staff and could help save lives through early and accurate detection and isolation.

5 Results and Discussions

To choose the right methodology, we have directly compared the hold-out and one of the cross-validation models and its performance on a new test set which was not a part of our training or validation samples. The test data was also previously labeled to validate the accuracy of our model.

The two main factors that play a pivotal role in selecting the correct model are sensitivity and specificity. In simple terms and as an explanation, we find that COVID (target class) detected as COVID (output class) define the sensitivity. Hence, in our, case, we find that the deep network model for K-fold validation exhibits higher accuracy in terms of detecting COVID for the test set. The other part analogous to it is specificity being defined by normal (target class) detected as normal (output class). We find that the radiography images are overly sensitive in nature and hence can be the ideal tool for detecting such kind of cases. The other two important factors which help us to know the right model are false negatives and false positives. COVID detected as normal accounts for the false negative part. Normal detected as COVID accounts for the false positive part. The K-fold validation model again stands out for this case as well.

The results for the above methodology are given below (Fig. 4).

The indicative inference that we may draw from the above comparison chart is as follows:

(1) RESNET-1 deep network model has outperformed the other models in terms of overall accuracy (75.3%).
(2) The sensitivity rate (98%) is also higher in the case of RESNET-1
(3) The false negative rate (2%) is also within acceptable limits.
(4) The only drawback of the model lies in the number of false positives, and hence, the specificity as well. A little compromise on this point is still acceptable since the true objective of the model is met. In line with this, the precision rate is hence considered to be under reasonable limits around 74.6%.

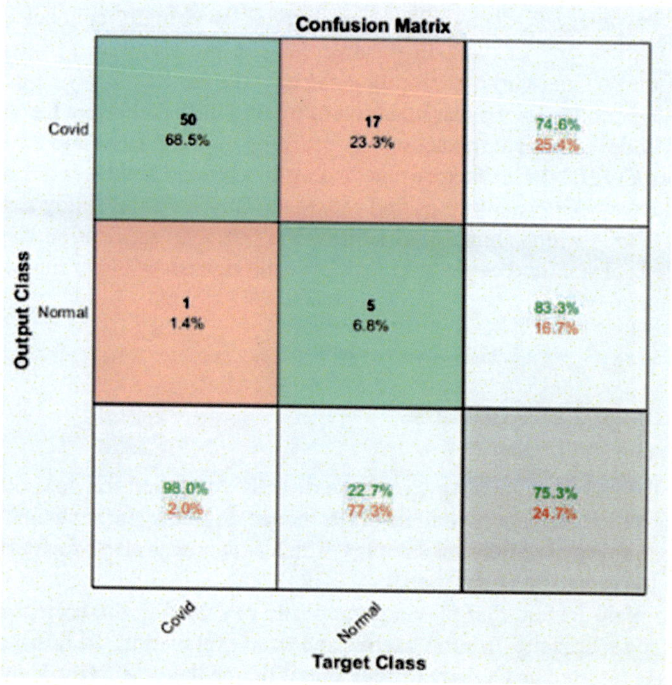

Fig. 4 K-fold cross-validation performance on new test data

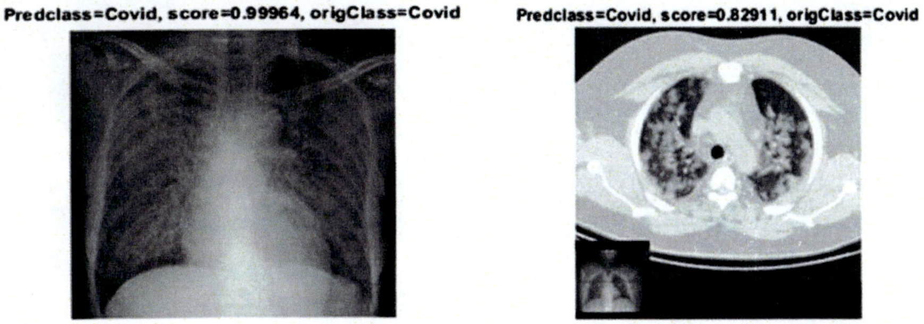

Fig. 5 Classification results for the two different sets of radiography images

The classification results for the two different sets of radiography images for the above model are also given below (Fig. 5).

The classification or detection results using decision fusion for the above sample can be concluded straightway to be COVID. For decision conflicting scenarios, it takes a maximum likelihood estimate as the flowchart has explained in one of the previous sections. The isolation for COVID subjects can be taken care by the IoT-enabled alarm system for taking precautionary measures at the point of care facilities. The authors have thus proposed a network architecture for making the medical facility an IoT-enabled unit. This is proposed for faster detection and isolation of COVID-19 subjects. The same has been discussed vividly in one of the previous sections.

The performance of the model was further validated by another indicative visualization known as the class activation mapping (CAM). The mathematics behind any deep learning network is overly complex in nature. Hence, in most cases, the network is treated like a black-box. This method can help us to understand the reason behind a classification in terms of a gradient map or score. In our case, the red colored zones enabled the model to classify the image under any class.

Some sample observations/classifications on images along with the CAM zones is given in the following figures (Fig. 6).

6 Conclusions

The novelty of our detection procedure is pivoted around the IoT connectivity concept. The device we prototyped is an IoT-enabled alarm system built on a Raspberry Pi device. The Raspberry Pi is virtually connected with the server, so that for any COVID-19 positive subject, the server can trigger specific actions on the IoT network for communicating with the Raspberry Pi device. The results that are captured from the classification of images in the server are sent to the Raspberry Pi device where the actions like turning on a red LED and a buzzer when COVID-19 is detected and turning on a blue LED when the classification results when normal being detected have been programmed and activated. This is an amazingly simple

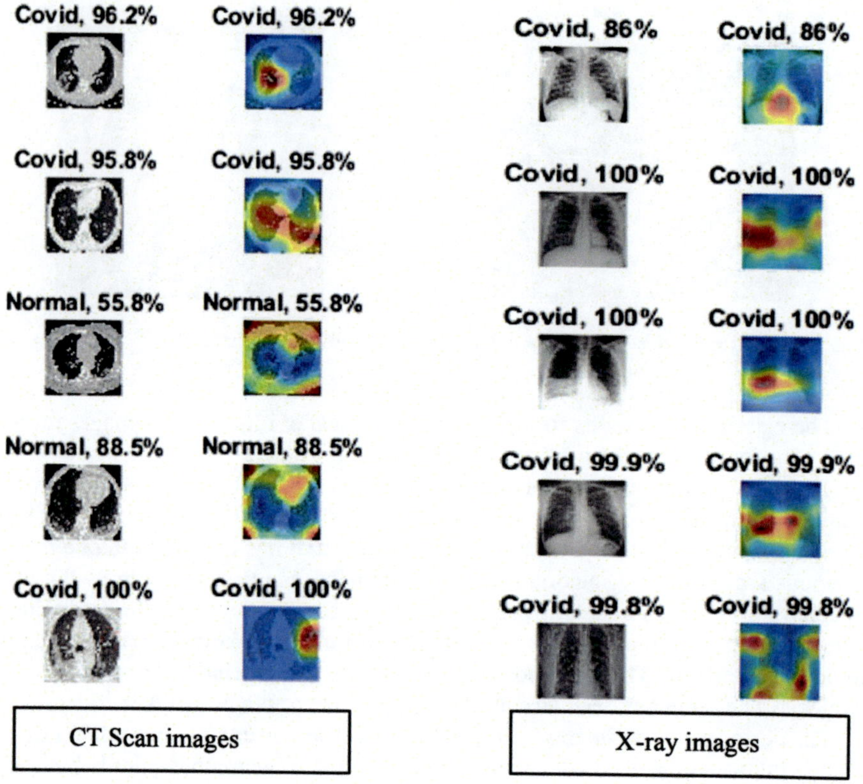

Fig. 6 Classification activation mapping (CAM): annotated results to show affected lung for X-ray images and CT scan images

proof of concept POC) which shows a possibility of its powerful applications that can save time for medical staff and could help save lives through early and accurate detection. The proposed architecture as discussed is a more composite and robust one since it includes a two-step radiography imaging-based screening process. Our algorithm has been developed into a deployable form and hence would act as an aid to the health practitioners and other medical staff and would eventually provide a valuable second opinion in the existing diagnostic process. Our algorithm is not intended to replace the more acceptable RT-PCR process but would act like an aid to the same. In addition to being an aid, it is also a highly cost-effective solution for the prediction and isolation of COVID-19 cases.

The aim of our work lies in the effective and efficient combination of IoT-based technology with state-of-the art artificial intelligent frameworks for the prediction and isolation purpose.

References

1. Cascella M, Rajnik M, Cuomo A, Dulebohn SC, Di Napoli R (2020) Features, evaluation and treatment coronavirus (COVID-19). In Statpearls [internet]: StatPearls Publishing
2. Adhikari SP, Meng S, Wu YJ, Mao YP, Ye RX, Wang QZ, Sun C, Sylvia S, Rozelle S, Raat H, Zhou H (2020) Epidemiology, causes, clinical manifestation and diagnosis, prevention and control of coronavirus disease (COVID-19) during the early outbreak period: a scoping review. Infect Dis Poverty 9(1):1–12
3. Lai CC, Shih TP, Ko WC, Tang HJ, Hsueh PR (2020) Severe acute respiratory syndrome coronavirus 2 (SARS-CoV-2) and corona virus disease-2019 (COVID-19): the epidemic and the challenges. Int J Antimicrobial Agents 105924
4. Novel CPERE (2020) The epidemiological characteristics of an outbreak of 2019 novel coronavirus diseases (COVID-19) in China. Zhonghua liuxingbingxue zazhi 41(2):145
5. Monllor P, Su Z, Gabrieli L, Montoro A, Taltavull de La Paz MDLP (2020) COVID-19 infection process in Italy and Spain: are the data talking?
6. http://www.cdc.gov/coronavirus/2019-ncov/cases-updates/cases-in-us.html. Accessed 15 October 2020
7. https://www.ecdc.europa.eu/en/geographical-distribution-2019-ncov-cases. Accessed 15 October 2020
8. Mahalle P, Kalamkar AB, Dey N, Chaki J, Shinde GR (2020) Forecasting models for coronavirus (COVID-19): a survey of the state-of-the-art
9. Petropoulos F, Makridakis S (2020) Forecasting the novel coronavirus COVID-19. PloS One 15(3):e0231236
10. Liu Z, Guo W (2020) Government responses matter: predicting COVID-19 cases in US under an empirical Bayesian time series framework. medRxiv
11. Al-Qaness MA, Ewees AA, Fan H, Abd El Aziz M (2020) Optimization method for forecasting confirmed cases of COVID-19 in China. J Clin Med 9(3):674
12. Perone G (2020) An ARIMA model to forecast the spread of COVID-2019 epidemic in Italy. arXiv preprint arXiv:2004.00382
13. Fanelli D, Piazza F (2020) Analysis and forecast of COVID-19 spreading in China, Italy and France. Chaos Solitons Fractals 134:109761
14. Gupta R, Pal SK (2020) Trend analysis and forecasting of COVID-19 outbreak in India. medRxiv
15. Sugishita Y, Kurita J, Sugawara T, Ohkusa Y (2020) Forecast of the COVID-19 outbreak, collapse of medical facilities, and lockdown effects in Tokyo, Japan. medRxiv
16. Kumar P, Kalita H, Patairiya S, Sharma YD, Nanda C, Rani M, Rahmani J, Bhagavathula AS (2020) Forecasting the dynamics of COVID-19 Pandemic in Top 15 countries in April 2020: ARIMA model with machine learning approach. medRxiv
17. https://people.duke.edu/~rnau/411arim.htm. Accessed 15 October 2020
18. https://www.mohfw.gov.in/. Accessed 15 October 2020
19. http://covidindiaupdates.in/. Accessed 15 October 2020

Analysis of Various Noise Reduction Techniques for Breast Ultrasound Image Enhancement

Gaurav Makwana, Ram Narayan Yadav, and Lalita Gupta

Abstract Ultrasound imaging is an essential diagnostic tool for analyzing the body's internal structure to find disease or abnormal tissues. The presence of noise in ultrasound images affects edges and fine details, making diagnosis more difficult by reducing contrast and resolution. It is vital to preserving the feature of the image by reducing the noise level. In this paper, we have explained the types of noises present in the ultrasound image and the filters used to remove the noises. The noise introduced in the breast ultrasound image is impulse noise, Gaussian noise, and speckle noise. Several denoising algorithms have been proposed for image denoising: filter, median filter, Gaussian filter, Butterworth filter, wavelet-based filter, and CNN for efficient noise reduction. A quantitative measure of comparison is provided by the parameters like mean square error (MSE), peak signal-to-noise ratio (PSNR), signal-to-noise ratio (SNR), mean, variance, standard deviation, kurtosis, and skewness of the image.

Keywords Ultrasound images · Gaussian noise · Impulse noise · Speckle noise · Average filter · Median filter · Gaussian Filter · Butterworth filter · Wavelet · CNN

1 Introduction

Breast cancer is the most common, life-threatening cancer in women all around the world. Breast disease often does not cause any agony or uneasiness until it has spread to close-by tissue. Early diagnosis can minimize cancer-related death. A breast ultrasound finds an abnormality in the breast using high-frequency sound waves that penetrate the breast tissues and converts them into images. Ultrasound can distinguish between solid and cystic breast masses, which are very difficult to identify in the mammogram. Still, it also has some limitations, like low resolution and low contrast. Also, speckle noise and blurry edges also affect the diagnosis accuracy to distinguish between various organs, so a lot of expertise is required for an exact diagnosis.

G. Makwana (✉) · R. N. Yadav · L. Gupta
Department of Electronics and Communication Engineering, Maulana Azad National Institute of Technology, Bhopal, India

K. Dahal et al. (eds.), *Internet of Things and Its Applications*, Lecture Notes in Electrical Engineering 825, https://doi.org/10.1007/978-981-16-7637-6_27

Image enhancement and noise removal can reduce this fatigue up to some extent. The enhancement method's selection should be such that the intensity difference among the object and pixel must be large with no image distortion. Otherwise, it may cause some regions to be under enhanced or some over enhance, in both condition information will be lost. Image enhancement techniques are of two types: spatial domain methods, and other is frequency domain methods. The spatial domain method directly operates on image pixels. Pixel values are changed to get better enhancement, whereas the frequency domain method rate of change of pixel value is identified, i.e., frequency distribution of the image. This transformed image is then processed; then, it is converted back into an image in the spatial domain by inverse transform operation. One of the most useful methods for enhancing low contrast images is histogram equalization (HE). This method uniformly distributed the pixel intensities in the new image [2]. Other types of HE methods are local area histogram equalization (LHE) and histogram stretching. LHE defines the histogram of a local region of an image. Still, LHE causes computational complexity and background distortion because this transformation is not monotonic, changing the gray level of the original image. So it is required to employ image enhancement techniques that can improve the contrast of breast ultrasound images [3, 4] that can represent breast tissue structure more effectively to make a better diagnosis.

Another cause that makes ultrasound images more difficult for diagnosis is image noise. There are many reasons for image noise like imperfect instruments, wrong image acquisition steps, power line interference, patient discomfort, image transmission, compression, etc. It is required to denoise the image before image data analysis. In ultrasound imaging, signal backscattered from the background soft tissue causes speckle noise that affects image quality, making diagnosis more difficult, so it becomes imperative to remove the noise to improve tissue structures and image quality. Another noise that affects the image quality is Gaussian noise. The cause of Gaussian noise is low sensor illumination, temperature, and electronic interference during transmission due to poor sensor connection. Impulse noise (salt and pepper) is another form of noise that affects an image's diagnosis accuracy. It occurs due to the sharp disturbance in an image because of a defect of the camera sensor, system software, or hardware failure during image acquisition and data transmission.

2 Methodology

In this paper, several filters have been proposed for reducing noises like mean filter [5], median filter [6–8], Butterworth filter, Gaussian filter, wavelet filter [9, 10], and convolutional neural network model [11, 12] for image denoising. Adam et al. [13] combine averaging and nonlinear Gaussian filtering to reduce speckle and additive random noise. Chen et al. [14] suggested median filtering to eliminate impulsive noise. Adaptive median filters [15] can also be used for noise reduction, which replaces the pixel value with the local neighborhood's weighted median. Still, it has some demerits; image smoothing also removes some sufficient detail in the image [16].

2.1 Mean (Averaging) Filter

It replaces each pixel in an image with the average value of all its neighbors. Mean filtering is like a convolution filter based around a kernel. This kernel controls the shape and size of the neighborhood that represents the amount of filtering.

2.2 Gaussian Filter

It is another linear smoothing filter. It is defined as [2, 17]

$$h(x, y) = \frac{1}{\sqrt{2\pi}\sigma} e^{-\frac{x^2+y^2}{2\sigma^2}} \tag{1}$$

The above function shows that the Gaussian filter is separable. The amount of smoothing depends upon the standard deviation. Mathematically, a Gaussian filter modifies the image by convolving with a Gaussian function.

2.3 Butterworth Filtering

Butterworth filter of order n and cutoff frequency D_0 is defined as [2, 17]

$$H(u, v) = \left[\frac{1}{1 + [D(u, v)/D_0]^{2n}}\right] \tag{2}$$

By increasing the order of the filter, we can control the sharpening of an image.

2.4 Discrete Wavelet Transform

DWT decomposes the image into four frequency components. Low–low (LL) frequency components give average information of an image, and other frequency components give directional information like low–high (LH), high-low (HL), and high-high (HH) which gives horizontal, vertical, and diagonal coefficients, respectively [9, 10]. Figure 1 shows the image denoising process using a wavelet transform.

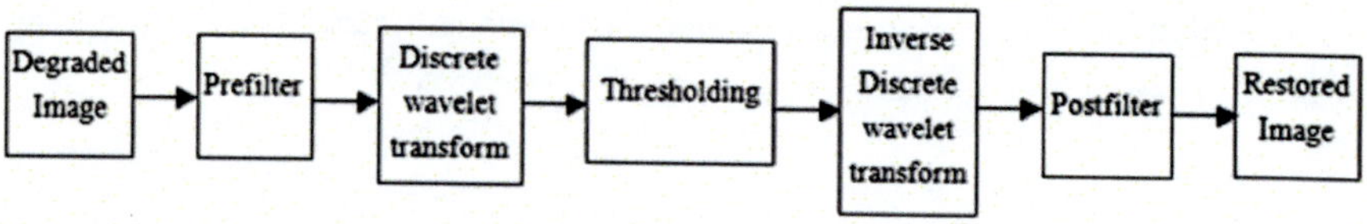

Fig. 1 Wavelet-based image denoising

Thresholding is a fundamental process in image denoising. Proper selection of thresholding can denoise the image because small value thresholding cannot remove noise, and a larger value may cause loss of image detail. Thresholding is of two types hard and soft thresholding, as defined below.

Hard thresholding is defined as

$$g\prime(m,n) = \begin{cases} g(m,n), & \text{if } |g(m,n)| > t \\ 0, & \text{otherwise} \end{cases} \tag{3}$$

And soft theresholding is defined as

$$g\prime(m,n) = \begin{cases} g(m,n) - t, & \text{if } |g(m,n)| > t \\ g(m,n) + t, & \text{if } |g(m,n)| < -t \\ 0 & \text{if } |g(m,n)| \leq t \end{cases} \tag{4}$$

Generally, soft thresholding is better option to minimize the MSE.

2.5 Convolutional Neural Network

Convolution neural network (CNN) is a deep learning tunable architecture that is feed-forward neural network consists of multiple trainable nonlinear transformation stages. It extracts the feature vector in every step, which represents the input and output vector, respectively. CNN consists of two networks that extract a feature of the input image, and the second classifies the extracted feature [18]. CNN has three types of network layers that are convolution, pooling, and fully connected layers. Convolution kernel is also known as convolution filter pass over the image and feature map value are calculated as follows

$$\begin{aligned} G(m,n) &= x(m,n) * h(m,n) \\ &= \sum_j \sum_k h(i,j)x(m-j,n-k) \end{aligned} \tag{5}$$

The convolution kernel is a feature identifier that filters the image's vital feature and produces a convolution map about the distribution of these features. The neurons figure out how to assemble the data to gain a higher-order feature of the image in the next convolutional layers [19]. Since convolution operation shrinks the image, when the kernel moves through the image, the impact of the outskirts pixels on the feature map is very small than the center pixels. It might decrease the image's spatial resolution. CNN uses image padding to minimize this problem. Another method to reduce the feature map size is stride. It functions just like the sliding window, and it controls the amount of movement of the convolution kernel to the input image.

The convolutional layer typically includes a nonlinear activation function. Nonlinearity is important to model the complex function. Rectified linear unit (ReLU) activation function is generally used because it shows a better convergence rate as compared to the other activation function. It gives zero output for the negative input value, which means it does not activate neurons with a negative value and for the positive input values it gives output exactly the same as the input.

$$\psi(x) = \max(0, x) \tag{6}$$

The last hidden layer of CNN is known as fully connected layers [20]. The output of this layer is a column vector and each row of the column vector is corresponding to a different class. Classification of these different classes based on the probability estimation of each class. Practically input to the fully connected layer must be as small as possible, pooling and stride reduce the size of the data matrix of fully connected layers. During the training process, we calculate the MSE between the restored image and degraded image to optimize the output. Optimizer optimizes the parameter to decrease the MSE [21]. In this training model, stochastic gradient descent (SGD) is used for parameter optimization. This can be represented as

$$w_n^{(i+1)} = w^{(i)} - \beta.\frac{\delta\xi}{dw} \quad \forall i = 1, 2, 3, \ldots \tag{7}$$

where $w_n^{(i)}$, β, and ξ is the iteration parameters, learning rate, and loss function, respectively. Now if g_{mn} is the pixel at (m, n) position and y_{mn} is the expected output, then the loss function is defined by Eq. (8)

$$\xi(w, m, n) = h(N(f(w_g, g_{mn})), y_{mn}) \tag{8}$$

where $f(.)$ is output of CNN network. MSE will be minimum for the $\delta\xi/\delta w = 0$.

3 Results

Ultrasound imaging is an incontestable noninvasive imperative instrument for diagnosis internal structure of the body to recognize inevitably sicknesses or abnormality in tissues. To remove noises from the ultrasound images, different filtering techniques have been proposed in this paper. We have used some performance measure such as MSE, PSNR, SNR, mean, variance, standard deviation, skewness, and Kurtosis to validate the efficiency of our method.

Figure 2 is a breast ultrasound image. Figure 3a shows the image with speckle noise. Figure 3b, c is the result obtained by the averaging filtering method with 3

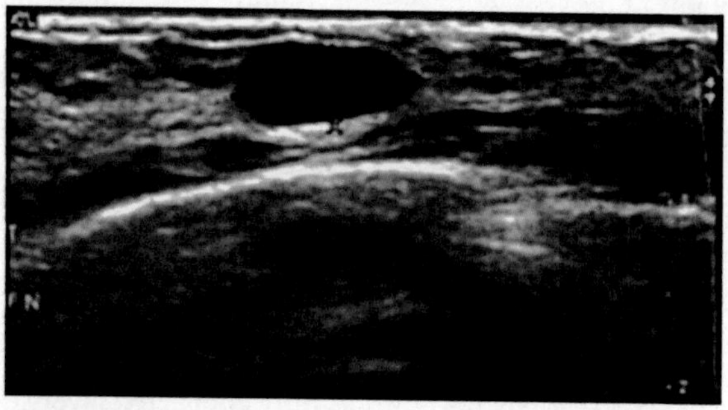

Fig. 2 Breast ultrasound image

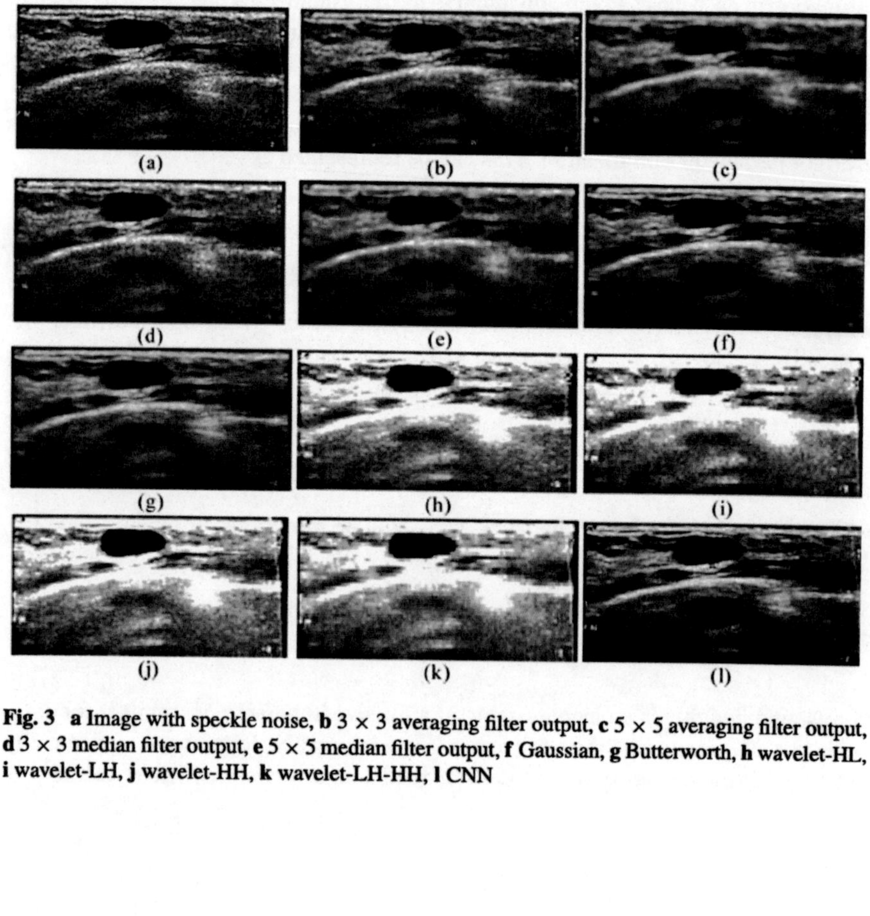

Fig. 3 **a** Image with speckle noise, **b** 3 × 3 averaging filter output, **c** 5 × 5 averaging filter output, **d** 3 × 3 median filter output, **e** 5 × 5 median filter output, **f** Gaussian, **g** Butterworth, **h** wavelet-HL, **i** wavelet-LH, **j** wavelet-HH, **k** wavelet-LH-HH, **l** CNN

Table 1 Performance evaluation for speckle noise

		MSE	SNR	PSNR	Mean	STD	Kurtosis	Skewness	Variance
Mean filter	3 × 3	8.58e+03	0.9469	8.7975	79.127	8.9457	3.5631	0.9249	2.29e+03
	5 × 5	8.31e+03	1.0879	8.937	79.091	8.9857	3.237	0.8128	2.03e+03
Median	3 × 3	0.0073	61.6123	69.4705	0.3046	0.0376	4.1546	1.0672	0.0372
	5 × 5	0.0097	60.4064	68.2585	0.2991	0.0369	3.6585	0.8927	0.0323
Gaussian		8.48e+03	0.9742	8.8451	79.029	9.0034	3.5937	0.9528	2.23e+03
Butterworth		8.78e+03	0.8502	8.6969	79.162	8.9387	3.9058	1.0551	2.49e+03
Wavelet	HL	0.1364	48.8217	56.782	0.0025	0.25	8.7881	0.5297	0.0159
	LH	0.1358	48.8411	56.8014	0.003	0.0244	7.5862	0.2511	0.0171
	HH	0.1376	48.7844	56.7447	0.0015	0.0237	8.7642	0.5308	0.0146
	LH-HH	0.1352	48.8607	56.821	0.0033	0.0195	7.479	0.28	0.0138
CNN		0.0018	67.7065	75.6817	0.3102	0.0346	3.831	1.0393	0.0375

× 3 and 5 × 5 filter masks, respectively. Figure 3d, e are the results obtained after applying the median filter of 3 × 3 and 5 × 5 filter mask, respectively. Figure 3f is the result obtained after applying the convolutional neural network algorithm for noise reduction. Figure 3g–j is the result obtained after applying second-level decomposition of the wavelet transform. To evaluate the performance of the methods, MSE, SNR, PSNR, mean, standard deviation, kurtosis, skewness, and variance are calculated. From Table 1, we can easily evaluate that the median filter and CNN give better performance for speckle noise reduction. But the median filter adds image smoothing, i.e., it also remove some high-frequency pixel (edge) while filtering, but CNN gives better subjective analysis and statistical data shows that CNN has minimum MSE, high SNR, and PSNR. This shows that CNN is a better method for speckle noise reduction.

Figure 4a shows the image with Gaussian noise; Fig. 4b, c is the result obtained by the averaging filtering method with 3 × 3 and 5 × 5 filter mask, respectively. Figure 4d, e are the results obtained after applying the median filter of 3 × 3 and 5 × 5 filter mask, respectively. Figure 4f, g are the result of Gaussian and Butterworth filter, respectively. Figure 4h–k is the result obtained after applying second-level decomposition of the wavelet transform. Figure 4l is the result obtained after applying the convolutional neural network algorithm for noise reduction. Statistical analysis for Gaussian noise reduction is given in Table 2. Median filter, wavelet decomposition, and CNN show better performance for Gaussian noise reduction. As we can subjectively analyze, CNN enhances the ultrasound image better as compared to all other filtering techniques and statistical data also verify the result. CNN method is having minimum MSE and high SNR and PSNR.

Figure 5a shows the image with impulse noise. Figure 5b, c is the result obtained by the averaging filtering method with 3 × 3 and 5 × 5 filter masks, respectively. Figure 5d, e are the results obtained after applying the median filter of 3 × 3 and 5 × 5 filter mask, respectively. Figure 5f, g are the result of Gaussian and Butterworth filter, respectively. Figure 5h–k is the result obtained after applying second-level

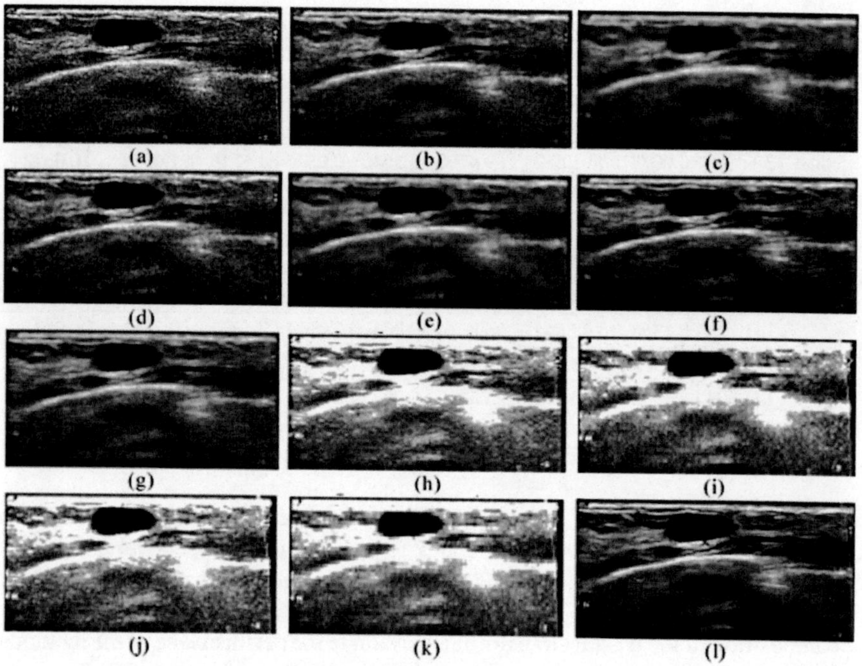

Fig. 4 **a** Image with Gaussian noise, **b** 3 × 3 averaging filter output, **c** 5 × 5 averaging filter output, **d** 3 × 3 median filter output, **e** 5 × 5 median filter output, **f** Gaussian, **g** Butterworth, **h** wavelet-HL, **i** wavelet-LH, **j** wavelet-HH, **k** wavelet-LH-HH, **l** CNN

Table 2 Performance evaluation for Gaussian noise

		MSE	SNR	PSNR	Mean	STD	Kurtosis	Skewness	Variance
Mean Filter	3 × 3	8.81e+03	1.0174	8.6833	80.4269	8.6133	3.762	1.0301	2.32e+03
	5 × 5	8.43e+03	1.1693	8.8701	79.96	8.4847	3.346	0.9348	2.03e+03
Median	3 × 3	0.0099	60.4796	68.1668	0.3114	0.0362	3.8686	1.0095	0.0386
	5 × 5	0.0125	59.469	67.1605	0.3081	0.0368	3.6564	0.9193	0.0347
Gaussian		8.71e+03	1.0401	8.7306	80.2519	8.9812	3.8125	1.0806	2.27e+03
Butterworth		9.02e+03	0.907	8.5784	80.5094	8.6259	3.8638	1.0704	2.53e+03
Wavelet	HL	0.1407	48.8089	56.6467	0.0035	0.232	6.7832	0.381	0.178
	LH	0.14	48.8302	56.668	0.0042	0.0241	5.8475	0.1668	0.0188
	HH	0.0142	48.7687	56.6064	0.0026	0.0238	7.0446	0.362	0.0167
	LH-HH	0.1394	48.8517	56.6895	0.0042	0.0192	6.2295	0.2067	0.0151
CNN		0.0033	65.86	72.9252	0.3147	0.0353	4.1967	1.2207	0.0374

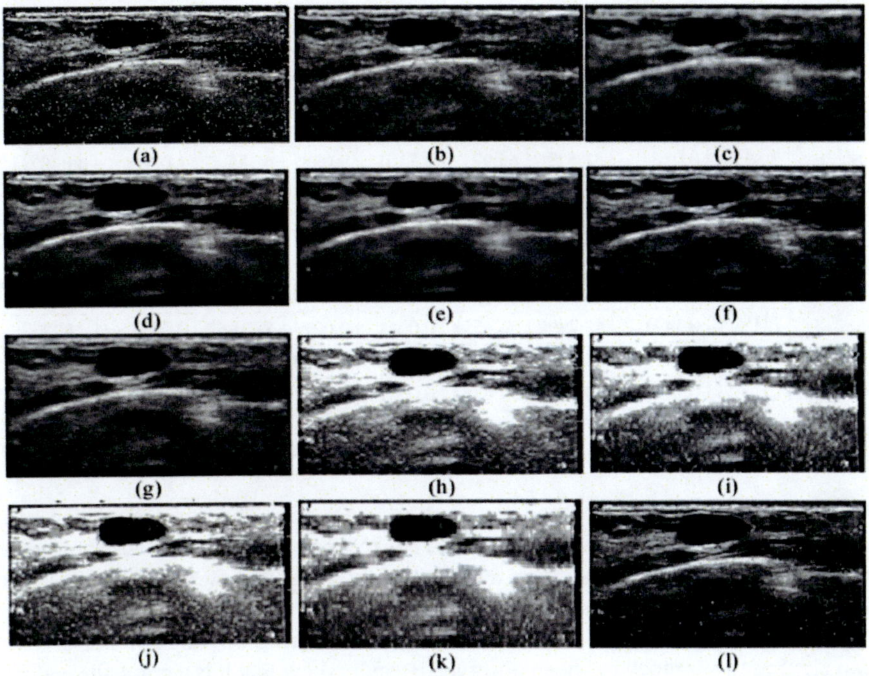

Fig. 5 **a** Image with impulse noise, **b** 3 × 3 averaging filter output, **c** 5 × 5 averaging filter output, **d** 3 × 3 median filter output, **e** 5 × 5 median filter output, **f** Gaussian, **g** Butterworth, **h** wavelet-HL, **i** wavelet-LH, **j** wavelet-HH, **k** wavelet-LH-HH, **l** CNN

decomposition of the wavelet transform. Figure 5l is the result obtained after applying the convolutional neural network algorithm for noise reduction. Statistical analysis for impulse noise reduction is given in Table 3. Subjective analysis of different filtering techniques in Fig. 5 shows that median filtering is the best choice for impulse noise reduction.

4 Conclusions

This paper discussed the recent developments in image denoising methods and shown their merits and demerits. A breast ultrasound image enhancement by removal of Gaussian, salt pepper, and speckle noise based on spatial domain filtering, frequency domain filtering, and CNN based approach is developed. The proposed method is effective in image noise reduction and gives better enhancement for diagnosis. Subjective analysis as well as statistical data shows that ultrasound image is better enhanced and preserved the image detail. It is also shown that the performance of different filters is varied and depends upon the type of noise. Medical ultrasound

Table 3 Performance evaluation for impulse (salt and pepper) noise

		MSE	SNR	PSNR	Mean	STD	Kurtosis	Skewness	Variance
Mean Filter	3 × 3	8.95e+03	1.1695	8.6108	81.6964	8.5741	3.741	0.9387	2.26e+03
	5 × 5	8.71e+03	1.3428	8.7304	82.0656	8.7464	3.399	0.8325	1.96e+03
Median	3 × 3	0.0164	58.5296	65.97	0.3101	0.0381	4.125	1.0869	0.0387
	5 × 5	0.0187	57.9773	65.4157	0.3065	0.038	3.8977	0.9843	0.0351
Gaussian		8.88e+03	10,214	8.6459	81.8089	8.5666	3.8195	0.9907	2.18e+03
Butterworth		8.78e+03	1.2872	8.6935	82.0226	8.4829	3.7314	0.9742	2.05e+03
Wavelet	HL	0.1478	48.77	56.4355	0.0032	0.0245	6.2649	0.4682	0.0206
	LH	0.1471	48.7898	56.4553	0.0041	0.0274	5.7196	0.2527	0.022
	HH	0.1491	48.7314	56.3969	0.0022	0.0233	6.25	0.5107	0.0185
	LH-HH	0.146	48.8208	56.4863	0.0041	0.0198	6.1332	0.3205	0.0162
CNN		0.0021	67.1691	74.841	0.3221	0.0334	3.9674	1.0656	0.0381

imaging is an important method for medical diagnosis. The experimental result shows that the CNNs method can effectively reduce speckle and Gaussian noise while preserving image details like mass, microcalcification, internal echo, etc. The CNN based filtering approach has been proved to be the best when the image is corrupted with speckle and Gaussian and median filtering for impulse noise. The wavelet-based approach has been observed. Results show that it is a good alternative for denoising images corrupted with Gaussian and impulsive noise.

References

1. Galatsanos NP, Segall CA, Katsaggelos AK (2003) Digital image enhancement. In: Drigger R (ed) Encyclopedia of optical engineering. Taylor & Francis, pp 388–402
2. Gonzales R, Woods R (2008) Digital image processing, 3 ed. Pearson Education Prentice Hall
3. Dale-Jones R, Tjahjadi T (1993) A study and modification of the local histogram equalization algorithm. Pattern Recogn 26(9):1373–1381
4. Zhang XY, Ge L, Wang TF (2008) Entropy-based local histogram equalization for medical ultrasound image enhancement. In: International conference on bioinformatics and biomedical engineering, Shanghai, pp 2427–2429
5. Lin P, Chen B, Cheng F, Huang S (2016) A morphological mean filter for impulse noise removal. J Display Technol 12(4):344–350
6. Tang J, Wang Y, Cao W, Yang J (2019) Improved adaptive median filtering for structured light image denoising. In: 2019 7th international conference on information, communication and networks (ICICN), Macao, pp 146–149
7. Erkan U, Enginoglu S, Thanh DNH, Hieu LM (2020) Adaptive frequency median filter for the salt and pepper denoising problem. IET Image Proc 14(7):1291–1302
8. Monajati M, Kabir E (2020) A modified inexact arithmetic median filter for removing salt-and-pepper noise from gray-level images. IEEE Trans Circuits Syst II Exp Briefs 67(4):750–754
9. Mohideen SK, Perumal SA, Sathik MM (2008) Image de-noising using discrete wavelet transform. IJCSNS Int J Comput Sci Netw Sec 8(1):213–216

10. Kimlyk M, Umnyashkin S (2018) Image denoising using discrete wavelet transform and edge information. In: 2018 IEEE conference of Russian young researchers in electrical and electronic engineering (EIConRus), Moscow, pp 1823–1825
11. Liu Z, Yan WQ, Yang ML (2018) Image denoising based on a CNN model. In: 2018 4th international conference on control, automation and robotics (ICCAR), Auckland, pp 389–393
12. Tian C, Xu Y, Zuo W (2020) Image denoising using deep CNN with batch renormalization. Neural Netw 121:461–473
13. Adam D, Beilin-Nissan S, Friedman Z, Behar V. (2006) The combined effect of spatial compounding and nonlinear filtering on the speckle reduction in ultrasound images. Ultrasonics 44(2):166–181
14. Chen Y, Yin RM, Flynn R, Broschat S (2003) Aggressive region growing for speckle reduction in ultrasound images. Pattern Recogn Lett 24(4):677–769
15. Djurovic I (2016) BM3D filter in salt-and-pepper noise removal. EURASIP J Image Video Process 1–11
16. Michailovich OV, Tannenbaum A (2006) Despeckling of medical ultrasound images. IEEE Trans Ultrason Ferroelectr Freq Control 53(1):64–78
17. Jayaraman S, Veerakumar T, Esakkirajan S (2010) Digital image processing, 3 ed. Tata McGraw-Hill Education
18. Thakur RS, Yadav RN, Gupta L (2019) State-of-art analysis of image denoising methods using convolutional neural networks. IET Image Proc 13:2367–2380
19. Ilea DE, Whelan PF (2006) Color image segmentation using a spatial k-means clustering algorithm. In: International conference on machine vision and image processing, Dublin, pp 1–8, Aug–Sept 2006
20. Cheng HD, Shan J, Ju W, Guo Y, Zhang L (2010) Automated breast cancer detection and classification using ultrasound images: a survey. Pattern Recogn 43(1):299–317
21. Liu Z, Yan WQ, Yang ML (2018) Image denoising based on a CNN model. In: International conference on control, automation and robotics (ICCAR), Auckland, pp 389–393

Wireless Network

Novel Range-Free Localization for Wireless Sensor Networks Using Fuzzy Logic

Arindam Giri, Subrata Dutta, and Sarmistha Neogy

Abstract Location-aware applications are of no use without knowing the locations of sensors. Decision making is done once the location of data is identified. For example, irrigation is initiated at the crop area where from collected moisture data falls under specified limit of moisture in a smart agriculture. Range-free locations become popular due to cost effectiveness. All range-free localization methods based on DV-Hop algorithm consider only hop distance and assume that an anchor closer to an unknown node can help localization better than anchors far apart. However, selecting an anchor in localization depends on factors like residual energy and density of anchors in addition to hop distance to unknown nodes. In a view to prefer an anchor over others, they are assigned weights based on the above-mentioned three factors in this paper. In a densely deployed wireless sensor network, assigning weights of anchors based on the three parameters is associated with uncertainties arising out of hop distance ambiguity. In this paper, the problem of assigning weights of anchors is performed by a fuzzy logic model. Simulation results reveal that our approach is superior to others in terms of localization accuracy.

Keywords Sensor network · Range-free localization · Hop distance · Residual energy · Node density · Fuzzy logic

1 Introduction

Due to the availability of low-cost tiny sensors, they are being used in every sector of society. Generally, sensors are used to collect information parameters like temperature form an unattended environment like military surveillance area. In order to gather information over a wide area, large numbers of sensors are deployed over the

---

A. Giri (✉)
Haldia Institute of Technology, Haldia, West Bengal 721657, India

S. Dutta
National Institute of Technology, Jamshedpur, Jharkhand 831014, India

S. Neogy
Jadavpur University, Kolkata, West Bengal 700032, India

K. Dahal et al. (eds.), *Internet of Things and Its Applications*, Lecture Notes in Electrical Engineering 825, https://doi.org/10.1007/978-981-16-7637-6_28

application area. Allowing them to communicate each other so as to forward sensed data via other sensors to a central location forms a wireless sensor network (WSN). WSNs are successfully applied in many applications like smart agriculture, ambient assisted living (AAL), and smart city [1]. The authors in [2] have proposed a smart agricultural model using WSN to collect different soil parameters like moisture so as to assist farmers in taking timely decision. Location-dependant application like smart agriculture initiates irrigation to the dried up crop area after knowing the location of moisture data. Without knowing locations of sensors water may be wasted, thereby increasing cost of cultivation. Such an application saves natural resources like water. The method of estimating locations of sensors in WSN is called localization [3]. Location-aware applications come to know locations of data with the help of localization. WSN used in military surveillance must have accurate localization capability to monitor foes. Localization methods must be efficient in terms of computational complexity as sensors have limited computational resources like energy.

Attaching global positioning system (GPS) for the purpose of locating sensors is not possible as they incur cost. Generally, a few sensors are attached with GPS making them location-aware, called anchors. Rest of the sensors in WSN adopts techniques to find their locations using locations of anchors. The techniques are known as localization. Existing localization algorithms are of two types: range-based and range-free. The former algorithms use range measurements like distance and angle to find unknown locations [4, 5]. In spite of higher localization accuracy, range-based techniques do not become popular due to higher cost of range measurement devices. On the contrary, range-free algorithms do not use absolute range measurements [6, 7]. So, there is no need of additional hardware which makes them suitable to be used in large-scale WSN applications. Range-free localization algorithms were inspired by popular DV-Hop algorithm [8]. This algorithm relies on the hop count which is calculated as the minimum number of hops among the paths between an anchor and an unknown node. Due to its simplicity, DV-Hop becomes popular. However, it prefers all anchors equally in calculation irrespective of the distance from the unknown sensor. In fact, the closer the anchor is to a node the more is the accuracy of localization. Weights are assigned to anchors based on hop count in order to improve accuracy of localization in weighted DV-Hop.

In a densely deployed WSN, hop count values are not always accurate as one anchor counts another nearby anchor in a dense WSN. Fuzzy logic has been used to resolve ambiguity in hop count values. As the nodes in WSN are randomly deployed, a hop count does not always reflect the actual physical distance. Inaccurate hop count may result in inaccurate weigh of anchor. In this circumstance, fuzzy logic may be used to calculate anchor weights dealing with uncertainties in hop count. In FWDV-Hop [9], authors map erroneous hop count to anchor weighs using fuzzy logic. This algorithm only considers hop count in weight calculation. However, selecting an anchor in localization not only depends on hop count but also on other factors like residual energy and density of anchors. An anchor with more residual energy and node density may provide accurate location information to an unknown node. In other words, two anchors with same hop count to an unknown node with different residual energy and node density may have different weights. The anchor with more residual energy and node density may be preferred over the other in unknown location estimation. Motivated by the approach, we develop a fuzzy model named NFWDV-Hop for selecting weights based on the above-mentioned three factors in this paper.

The remainder of this paper is organized as follows. In Sect. 2, existing literature is discussed. The proposed fuzzy model for calculating anchor weights is given in Sect. 3. Simulation results along with analysis can be found in Sect. 4, and conclusion is given in Sect. 5.

2 Literature Review

2.1 DV-Hop Algorithm

Range-free localization algorithms are inspired by DV-Hop algorithm. The beauty of this algorithm is it uses only few anchor nodes. It calculates hop distance by estimating hop count between nodes. Initially, all nodes share hop count vector to neighbors. Then average hop count is calculated by Eq. (1). The actual hop distance is calculated as the product of hop count and average hop distance as in Eq. (2).

The average hop size of ith anchor is calculated as:

$$\text{Hopsize}_i = \frac{\sum_{j\neq i}\sqrt{(x_i - x_j)^2 + (y_i - y_j)^2}}{\sum_{j\neq i} h_{ij}}. \tag{1}$$

where (x_i, y_i), (x_j, y_j) are the coordinates of anchors i and j and h_{ij} is the minimum hop count between them. An anchor node broadcasts average hop size value, and a node calculates the distance to it using Eq. (2).

$$d_{ij} = \text{Hopsize}_i. \tag{2}$$

Finally, unknown node P finds its location, $X = (x, y)$ as follows:

$$\begin{cases} (x - x_1)^2 + (y - y_1)^2 = d_1^2 \\ (x - x_2)^2 + (y - y_2)^2 = d_2^2 \\ \quad\quad\quad\quad . \\ \quad\quad\quad\quad . \\ (x - x_n)^2 + (y - y_n)^2 = d_n^2 \end{cases}. \tag{3}$$

Now, Eq. (3) can be rewritten as:

$$\begin{cases} 2(x_1 - x_n)x + 2(y_1 - y_n)y = x_1^2 - x_n^2 + y_1^2 - y_n^2 - d_1^2 + d_n^2 \\ 2(x_2 - x_n)x + 2(y_2 - y_n)y = x_2^2 - x_n^2 + y_2^2 - y_n^2 - d_2^2 + d_n^2 \\ \vdots \\ 2(x_{n-1} - x_n)x + 2(y_{n-1} - y_n)y = x_{n-1}^2 - x_n^2 + y_{n-1}^2 - y_n^2 - d_{n-1}^2 + d_n^2 \end{cases} \tag{4}$$

Now, Eq. (4) can be expressed using matrices as $AX = B$, where

$$A = -2 \times \begin{bmatrix} x_1 - x_n y_1 - y_n \\ x_2 - x_n y_2 - y_n \\ . \\ . \\ x_{n-1} - x_n y_{n-1} - y_n \end{bmatrix},$$

$$B = \begin{bmatrix} d_1^2 - d_n^2 - x_1^2 + x_n^2 - y_1^2 + y_n^2 \\ d_2^2 - d_n^2 - x_2^2 + x_n^2 - y_2^2 + y_n^2 \\ . \\ . \\ d_{n-1}^2 - d_n^2 - x_{n-1}^2 + x_n^2 - y_{n-1}^2 + y_n^2 \end{bmatrix}, \text{ and}$$

$$X = \begin{bmatrix} x \\ y \end{bmatrix}$$

So, X can be calculated as:

$$X = \left(A^T A\right)^{-1} A^T B. \tag{5}$$

Though DV-Hop is simple and good enough to localize sensor nodes, it only considers anchors closer to unknown nodes. It believes that closer anchors can provide better location estimation than other anchors. However, this fact is not always happens in a randomly deployed WSN. In a densely populated network, hop count is not accurate. In such situation, an anchor at more distance may give better hop count estimation. So, through DV-Hop location estimation is not accurate. As a solution, anchors are assigned weights on order to reflect importance.

2.2 Additions to DV-Hop

Guadane et al. [10] suggest weight of an anchor to be calculated as:

$$w_i = \frac{1}{h_{ij}}. \tag{6}$$

Here, h_{ij} is the hop count between node i and j. w_i's are inaccurate if hop counts are inaccurate. Weighted DV-Hop algorithm follows least square method using calculated weights in Eq. (7):

$$f(x, y) = \min \sum_{i=1}^{n} w_i^2 \left(\sqrt{(x_i - x_j)^2 + (y_i - y_j)^2} - d_i \right)^2. \quad (7)$$

where n is number of nodes.

Equation (7) may be formed as $A'X = B'$, where

$$A' = -2 \times \begin{cases} w_1^2 w_n^2 (x_1 - x_n) & w_1^2 w_n^2 (y_1 - y_n)_1 \\ w_2^2 w_n^2 (x_2 - x_n) & w_2^2 w_n^2 (y_2 - y_n) \\ \vdots & \vdots \\ w_{n-1}^2 w_n^2 (x_{n-1} - x_n) & w_{n-1}^2 w_n^2 (y_{n-1} - y_n) \end{cases}, \text{ and}$$

$$B' = \begin{bmatrix} w_1^2 w_n^2 (d_1^2 - d_n^2 - x_1^2 + x_n^2 - y_1^2 + y_n^2) \\ w_2^2 w_n^2 (d_2^2 - d_n^2 - x_2^2 + x_n^2 - y_2^2 + y_n^2) \\ . \\ . \\ w_{n-1}^2 w_n^2 (d_{n-1}^2 - d_n^2 - x_{n-1}^2 + x_n^2 - y_{n-1}^2 + y_n^2) \end{bmatrix}$$

So, location of P is calculated below:

$$X = \left(A'^T A'\right)^{-1} A'^T B'. \quad (8)$$

In [10], weights are estimated using Eq. (9). This algorithm not only prefers a closer anchor but also an anchor with correct hop count.

$$w_i = \frac{\frac{1}{h_{ij}}}{\sum_{k=1}^{n} \frac{1}{h_{kj}}}, \quad k \neq j. \quad (9)$$

Localization error for the position of node i follows:

$$e = \frac{1}{n-1} \sum_{i=1}^{n} \sqrt{(x - x_i)^2 + (y - y_i)^2}. \quad (10)$$

where (x, y) and (x_i, y_i) are the actual and estimated coordinates of i, respectively.

In HWDV-hop [11], a node finds hop size using one-hop distances to all anchors. This algorithm minimizes localization error.

In FWDV-Hop [9], weights are allocated based on only hop count. Uncertainty in hop count is managed by fuzzy logic. The fuzzy logic model takes a single input hop count and generates weight as output. Triangular memberships are used to map inputs to outputs. The algorithm provides accurate localization. However, it ignores other influencing parameters like residual energy and node density of anchor node in assigning weight.

3 Overview of Proposed Algorithm

In this section, we outline the overview of our fuzzy model. The fuzzy inputs hop count, residual energy, and node density are applied to fuzzy system and the system returns weight as output as shown in Fig. 1. Here, hop count represents the number of hops in the shortest path from unknown node to anchor, residual energy is the energy of a sensor node at the moment, and the node density signifies the number of nodes surrounding an anchor. For simplicity, all fuzzy input and output variables are represented by the same triangular membership function given in Fig. 2. The fuzzy variables possess five membership values: very low (VL), low (L), moderate (M), high (H), and very high (VH). Anchors with lower hop count, higher residual energy as well as higher node density provide more accurate location information to an unknown node. Accordingly, input and output variables are related as follows.

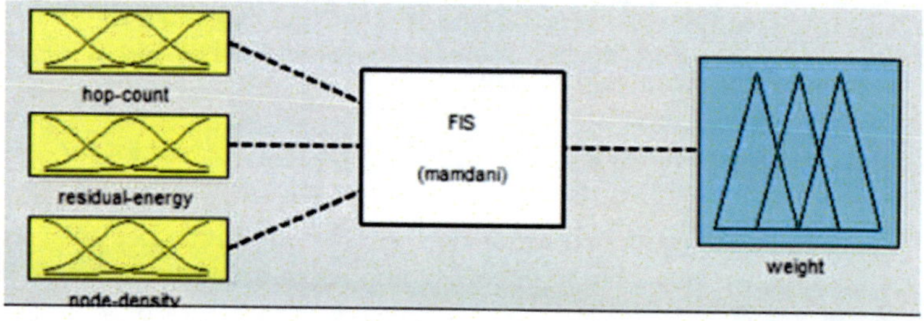

Fig. 1 Fuzzy logic model built with inputs to calculate weight

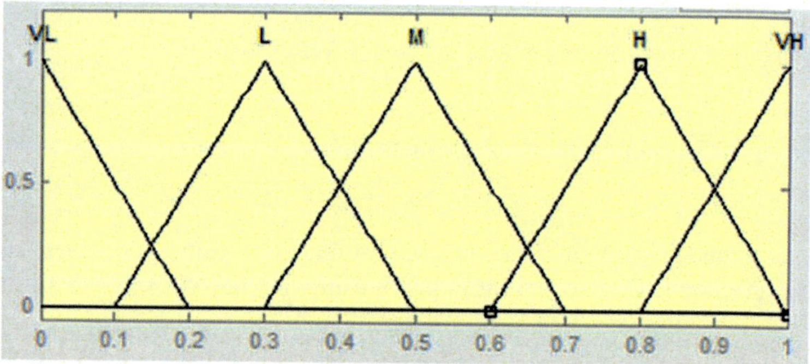

Fig. 2 Membership functions for input/output variable

An anchor with higher hop count is assigned lower weight. The higher the residual energy the more is weight. Similarly, the more is the node density around an anchor the more is the weight.

3.1 Introduction to Fuzzy Logic

Fuzzy logic [12] was introduced by Zadeh to deal with uncertainty in data. Fuzzy logic is used in situation consisting of entities with imprecise data. Fuzziness in data is represented by fuzzy set along with membership to the set. Unlike classical sets, elements in fuzzy sets may have partial membership values. Membership values are calculated by membership functions like trapezoidal function. Mapping of fuzzy inputs to outputs is executed by fuzzy inference system (FIS) as given in Fig. 3. The FIS first converts crisp inputs to fuzzy values by fuzzifier. Now the outputs are calculated by an inference engine. The decision making is performed inside inference engine. Generally, it uses IF-ELSE rules to map inputs to outputs.

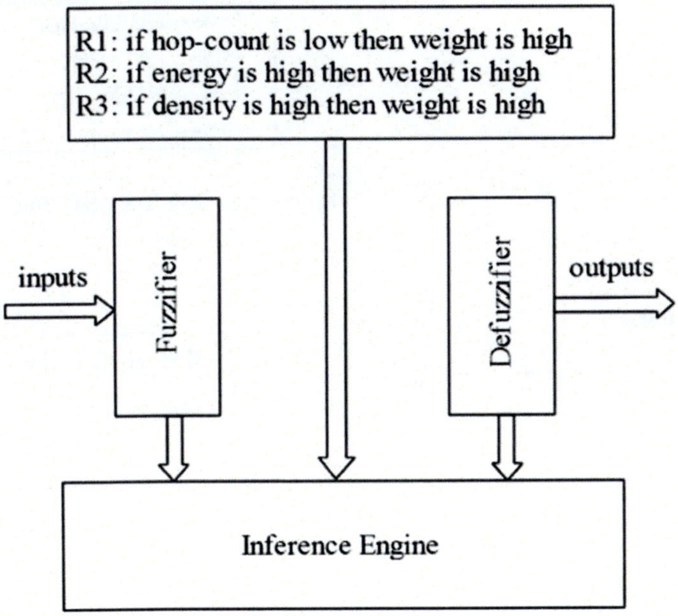

Fig. 3 Components of a fuzzy inference system

4 Simulation Results and Analysis

We use MATLAB for simulation of proposed algorithm, NFWDV-Hop. We consider a WSN with randomly deployed sensors in an area of 50 m^2 × 50 m^2. All nodes are set with an initial energy of 2 J and communication radius of 15 m. Localization error in terms of RMSE is estimated with nodes in the range 100 to 500 and communication radius of nodes within 10–20 m. Localization error is plotted after changing anchor-to-sensor node ratio from 5% to 40%. Other algorithms are also executed in the same simulation platform and parameters. The proposed algorithm is compared with other algorithms in Figs. 4, 5 and 6.

Fig. 4 Normalized localization error vs. number of sensor nodes, with 10% of anchor nodes

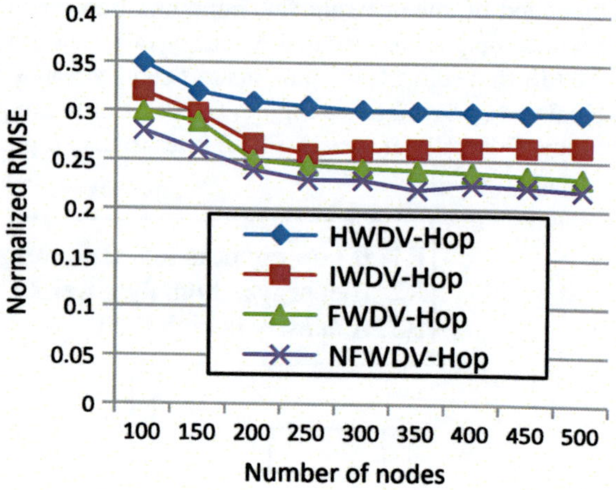

Fig. 5 Normalized localization error vs. anchor nodes ratio (in percentage of anchor nodes) with 100 sensor nodes

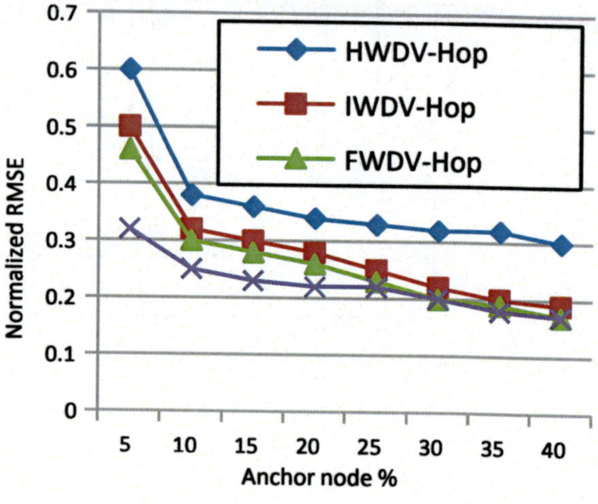

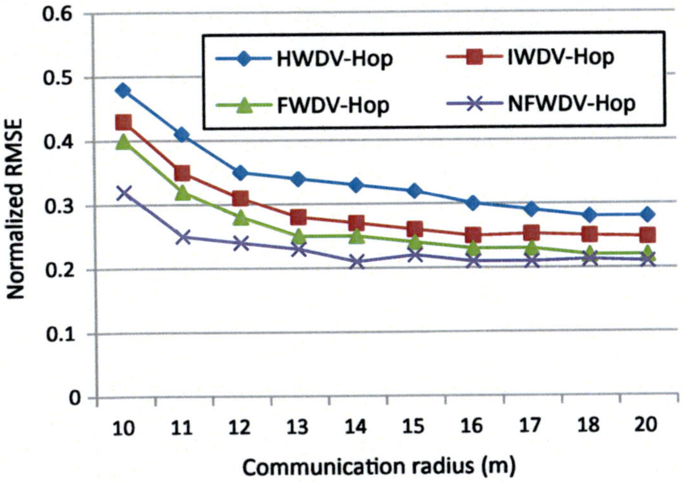

Fig. 6 Normalized localization error vs. radio range of sensors, with 100 sensors and 10% of anchor nodes

In Fig. 4, localization error is shown with varying number of sensor nodes of which 10% nodes are anchors. The more the nodes the less is the localization error. There is no significant change in error as soon as number of nodes reaches to 250. With 10% of anchors, proposed algorithm provides accurate localization than other algorithms because of ability to deal with hop count ambiguity in randomly deployed WSN. The error of NFWDV-Hop reaches to 20% with 500 nodes as weights are calculated using fuzzy logic.

Number of anchors deployed in WSN influences localization error. Localization error is illustrated in Fig. 5 with varying anchor nodes from 5 to 40%. Localization error is decreased with increase of anchor ratio. Fuzzy logic can represent imprecise hop count and estimates weight. That is why our approach provides more localization accuracy compared to others.

With increased communication radius sensors can communicate more neighbors. The effect of communication radius on localization error is given in Fig. 6. It reveals that error is decreased with increased radius. In simulation, 100 nodes of which 20% are anchor nodes are considered for experiment. The RMSE value approaches to nearly 20% with a communication radius of 20 m. As anchor weights are calculated using fuzzy logic, our approach outperforms others in terms of RMSE.

5 Conclusion

Location-aware applications like crop monitoring heavily depend on localization. Decision making in such applications becomes efficient once locations of nodes are known. Range-free localization helps in developing low-cost application as no extra hardware is needed. Choosing anchors in localization with inaccurate hop count is done by proposed approach. Fuzzy logic is used to overcome uncertainty in hop count estimation. Weights of anchors are estimated through inference engine based on hop count, residual energy, and node density. Performance of proposed algorithm is measured in terms of localization error. Comparing with similar algorithms proves its superiority over other.

References

1. Akyildiz IF, Su W, Sankarasubramaniam Y, Cayirci E (2002) Wireless sensor networks: a survey. Comput Netw 38:393–422
2. Giri A, Dutta S, Neogy S, Dahal K, Pervez Z (2017) Internet of Things (IoT): a survey on architecture, enabling technologies, applications and challenges. In: Proceedings of the 1st international conference on internet of things and machine learning. ACM, New York, USA, pp 7:1–7:12
3. Patwari N, Ash JN, Kyperountas S, Hero AO, Moses RL, Correal NS (2005) Locating the nodes: cooperative localization in wireless sensor networks. IEEE Sig Process Mag 22:54–69
4. Shanshan W, Jianping Y, Zhiping C, Guomin Z (2008) A RSSI-based self-localization algorithm for wireless sensor networks. J Comput Res Dev 45:385–388
5. Niculescu D, Nath B (2003) Ad hoc positioning system (APS) using AOA. In: IEEE INFOCOM 2003. twenty-second annual joint conference of the IEEE computer and communications. IEEE societies. IEEE, pp 1734–1743
6. Shang Y, Ruml W (2004) Improved MDS-based localization. In: IEEE INFOCOM. IEEE, pp 2640–2651
7. He T, Huang C, Blum BM, Stankovic JA, Abdelzaher T (2003) Range-free localization schemes for large scale sensor networks. In: Proceedings of the 9th annual international conference on mobile computing and networking. ACM, pp 81–95
8. Niculescu D, Nath B (2001) Ad hoc positioning system (APS). In: IEEE global telecommunications conference, GLOBECOM'01, pp 2926–2931
9. Giri A, Dutta S, Neogy S (2020) Fuzzy logic-based range-free localization for wireless sensor networks in agriculture. In: Advanced computing and systems for security. Springer, pp 3–12
10. Guadane M, Bchimi W, Samet A, Affes S (2017) Enhanced range-free localization in wireless sensor networks using a new weighted hop-size estimation technique. In: IEEE 28th annual international symposium on personal, indoor, and mobile radio communications (PIMRC). IEEE, pp 1–5
11. Hadir A, Zine-Dine K, Bakhouya M, El Kafi J (2014) An optimized DV-hop localization algorithm using average hop weighted mean in WSNs. In: IEEE 5th workshop on codes, cryptography and communication systems (WCCCS). IEEE, pp 25–29
12. Zadeh LA (1996) Fuzzy logic = computing with words. IEEE Trans Fuzzy Syst 4:103–111

Design of Wideband Dual-Polarized Antenna for LTE Application

Santimoy Mandal and Chandan Kumar Ghosh

Abstract In this article, we have proposed a new technique for dual-polarized microstrip antenna for MIMO system. The patch is consists of circularly shaped and fed by a gap among the patch edge and microstrip open end. Dual-polarization can be achieved by feeding a circular patch which is orthogonal to the gap. Measured result shows that the projected antenna has the wide bandwidth of 5.25 to 6.37 GHz for $S_{dd11} \leq -10$ dB with a port isolation of 39 dB along with a stable gain of 13.05 dBi.

Keywords Dual-polarization · Wideband · MIMO · Circular microstrip antenna

1 Introduction

The wireless communication system has been increased its demand for the wide-band applications with the improvement of multi-band technology. To achieve the required goal, dual-band microstrip patch antennas are widely used in the fields of wireless communication system because of its compact size, adaptability, low cost, and high performance characteristics. Owing to the advantages in dual-band antenna, it gives flexibilities of connecting different communication devices for transmitting and receiving signals. In [1] dual-polarized antenna, the diversity gain is maximized if both the input ports receive radiation in an orthogonal manner. Numerous methods of dual-polarized antennas have been proposed in an open research, which includes crossed dipole antennas in [2, 3], patch antennas in [4, 5], and slot antennas in [6, 7]. In [8], a relative bandwidth of 45% (1.71–2.69 GHz) is obtained for VSWR < 1.5 by loading long and short dipoles. In addition, the proposed antenna has high port isolation, a stable gain, and a stable radiation pattern across the whole operating bandwidth. In [9], combining magnetic and electric dipoles, a new type of dipole antenna is developed. Even though, it has a broad impedance bandwidth of 65.9%

S. Mandal (✉)
RVS College of Engineering and Technology, EdalberaBhilai Pahari, Jamshedpur 831012, India

C. K. Ghosh
Dr. B.C. Roy Engineering College, Jemua Road, Fuljhore, Durgapur 713206, India

K. Dahal et al. (eds.), *Internet of Things and Its Applications*, Lecture Notes in Electrical Engineering 825, https://doi.org/10.1007/978-981-16-7637-6_29

(1.7185–3.409 GHz) for VSWR < 2 and a high port isolation of 36 dB, because of its all-metal structure, the weight of the antenna is increased.

In this article, a dual-band dual-polarized patch antenna is proposed that consists of four straight probes for orthogonal polarization. The antenna is designed based on a distinctive dual-band patch antenna employed by two circular radiating patches. In contrast with the previous designs, the projected antenna achieved dual bands, dual-polarizations, high isolation, simultaneously, and finds its application in WiMAX, WLAN/IoT. The proposed technique has been simulated through the process of numerical simulation using the method of moment (MOM)-based IE3D electromagnetic simulator, and the electrical characteristics of the same have been studied. The schematic diagram of the proposed antenna design is shown in Fig. 1c.

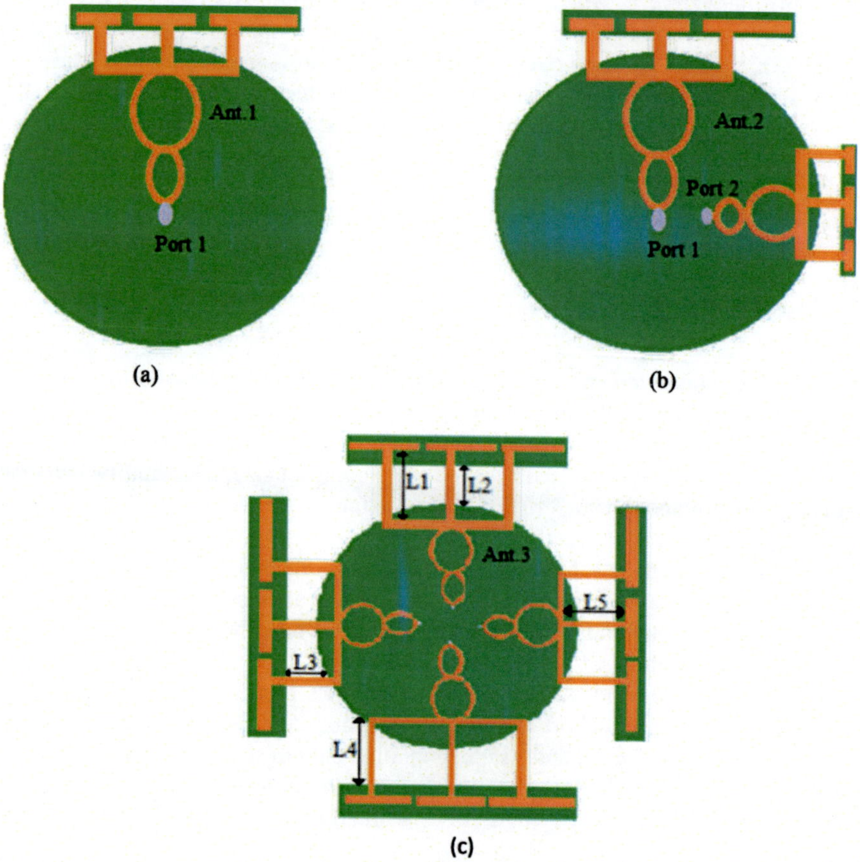

Fig. 1 **a** Schamatic diagram of antenna.1, **b** schematic diagram of antenna.2, **c** schamatic diagram of proposed antenna

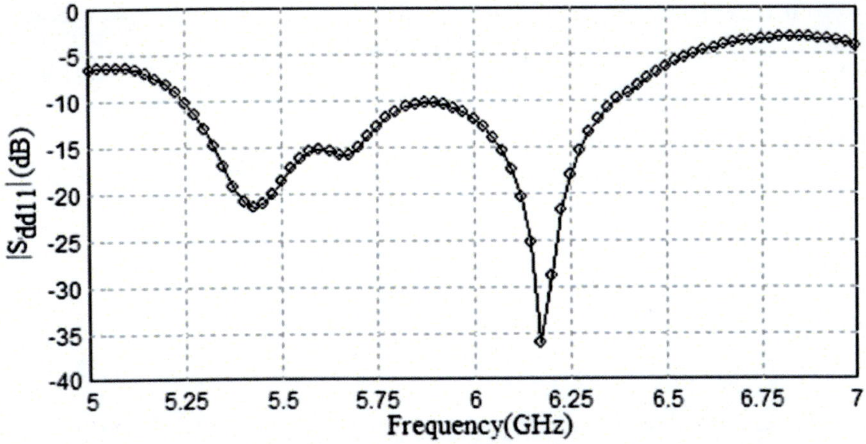

Fig. 2 Simulated S_{dd11} of the simplified reference ant. 3 (proposed antenna)

2 Antenna Design and Analysis

A FR4 substrate with permittivity (ε_r) 4.4, thickness 1.6 mm, and loss tangent 0.02 is used for designing this antenna. Hence, the S-parameters of the differentially driven antenna can be obtained by using the single-ended four-port S-parameters shown in Fig. 2. We have optimized the dimension of dual-polarized circularly shaped antenna is shown in Fig. 1 a–c) by the process of simulation. Optimized dimension of the proposed antenna is as follows: L1 = 22 mm, L2 = 12 mm, L3 = 10 mm. From the given Fig. 3, it shows the multiresonance behavior of ant.1 and ant.2. After optimizing, the dimension of our projected dual-polarized differentially feed circular shaped patch antenna in ant.3 shows the wide operating bandwidth in Fig. 2.

Due to the symmetry property of the antenna, very high isolation is measured within the whole operation band which is higher than 39 dB as shown in Fig. 3.

3 Parametric Study

A parametric study has been carried out to optimize the dimension of the projected antenna. When the length of the parameter, L1 and L2 become longer or shorter from the optimized dimension in Figs. 4 and 5. It can be observed that the resonance at the lower frequency moves to the higher frequencies, whereas the other resonances are unaltered. For L3 grows longer or shorter from the optimized value, the center frequency also moves to the higher frequency end in Figs. 6 and 7.

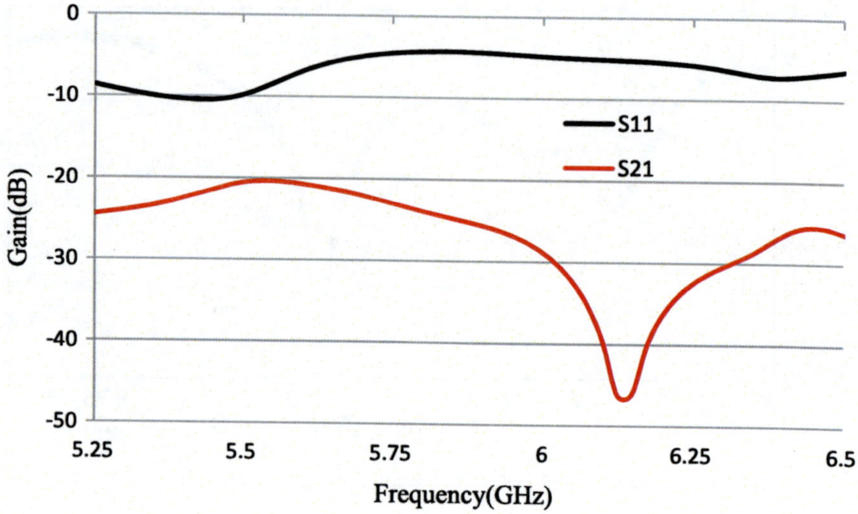

Fig. 3 Reflection and isolation performance characteristic of ant. 3

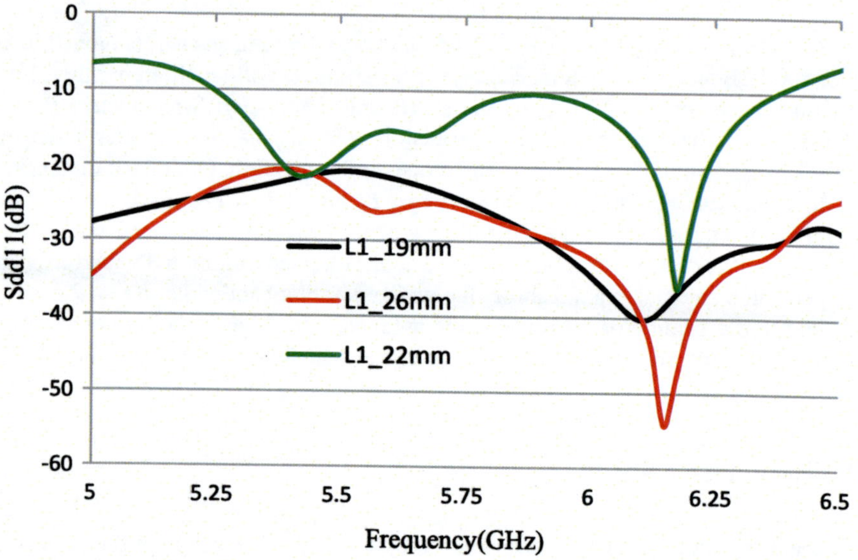

Fig. 4 Parametric study by varying the length (L1) of the patch

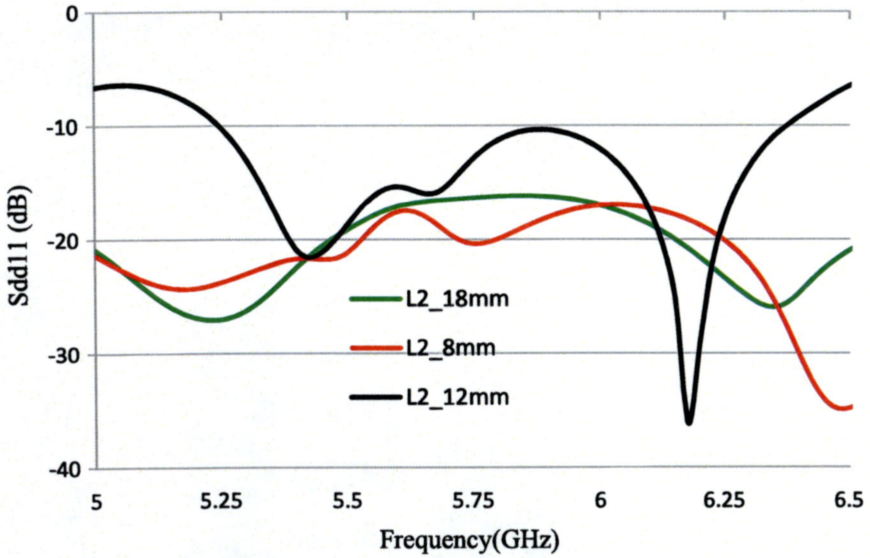

Fig. 5 Parametric study by varying the length (L2) of the patch

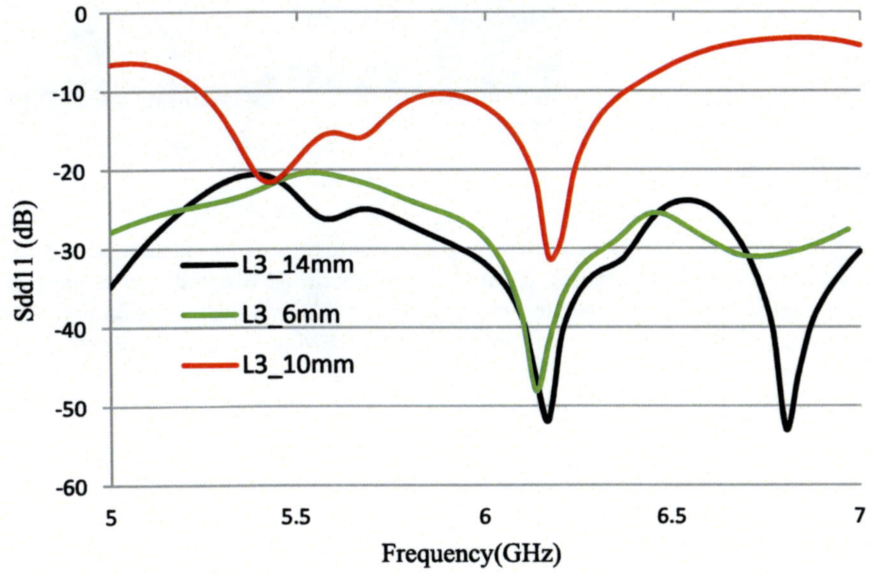

Fig. 6 Parametric study by varying the length (L3) of the patch

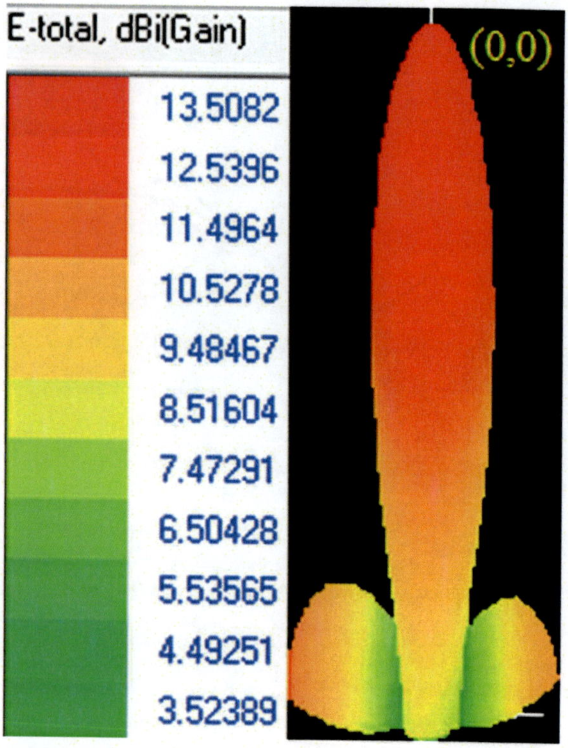

Fig. 7 Measured 3D radiation pattern of the proposed antenna (ant.3)

4 Conclusion

In this article, we proposed a dual-polarized microstrip patch antenna with dual-polarized radiation by feeding a circular patch between two orthogonal gaps. It shows good impedance matching characteristics with high port isolation and also we have optimized the antenna dimension after varying the length and width of the patch. From the investigational result shown above, it can be concluded that the proposed antenna element is used in LTE base station applications as well as IoT-based applications in WLAN and WiMax.

References

1. Wong KL (2002) Compact and broadband microstrip antennas. Wiley, Hoboken, NJ, USA
2. Zuo S, Liu Q-Q, Zhang Z-Y (2014) Wideband dual-polarized crossed-dipole antenna with parasitical crossed-strip for base station applications. Progress In Electromagnet Res C 48:159–166
3. Liu C, Liu J-L, Huang Y-H, Zhou L-Y (2012) A novel dual-polarized antenna with high isolation and low cross polarization for wireless communication. Progress In Electromagnet Res Lett

32:129–136
4. Deng JY, Guo LX, Yin YZ, Qiu J, Wu ZS (2013) Broadband patch antennas fed by novel tuned loop. IEEE Trans Antennas Propag 61(4):2290–2293
5. Gou YS, Yang SW, Zhu QJ, Nie ZP (2013) A compact dual-polarized double-shaped patch antenna with high isolation. IEEE Trans Antennas Propag 61(8):4349–4353
6. Jiang XL, Zhang ZJ, Tian ZJ, Li Y, Feng ZH (2013) A low-cost dual-polarized array antenna etched on a single substrate. IEEE Antennas Wireless Propag Lett 12:265–268
7. Lian RN, Wang ZD, Yin YZ, Wu JJ, Song XY (2016) Design of a low-profile dualpolarized stepped slot antenna array for base station. IEEE Antennas Wireless Propag Lett 15:362–365
8. Luo Y, Chu QX (2015) Oriental crown-shaped differentially fed dual-polarized multidipole antenna. IEEE Trans Antennas Propag 63(11):4678–4685
9. Wu BQ, Luk KM (2009) A broadband dual-polarized magneto-electric dipole antenna with simple feeds. IEEE Antennas Wireless Propag Lett 8:60–63

Prolonging Lifetime of Wireless Sensor Networks Using Modified N Policy Queueing Model

Veena Goswami and G. B. Mund

Abstract This paper considers a queue-based technique for the improvement of the lifetime of wireless sensor networks (WSNs). We investigate the impact of queueing scheme on the power consumption of nodes in the networks and consider three possible states for each node: sleep, idle, and busy. We model the queue at each node in the network into an M/M/1 queueing system employing modified N threshold lifetime improvement method, where the service starts after the number of packets in the system reaches the threshold N. When the threshold attains N, the system serves the packets in a single batch. This policy reduces the waiting time of the initial packets in the network. Arrivals during the busy state receive single service. The node switches to sleep mode when the queue becomes empty. We estimate the power consumption, which provides a measure of the efficiency that can improve the lifetime of WSNs. This paper also explores the dependence of power consumption on various system parameters.

Keywords Wireless sensor networks · Modified N policy · Prolonging lifetime · Power consumption · Queue

1 Introduction

Wireless sensor networks (WSN) have been useful in many industrial applications, and they are vital components in the advancing Internet of things (IoT). These networks contain several hundred self-powered nodes, and each node linked with one or more sensors. Every single sensor node has four essential components: sensing, transceiver, power, and processing units [1–3]. Although these sensor networks have pioneered infinite opportunities, they set some intimidating challenges, the primary being that of power scarceness. To maintain these nodes, we need non-renewable

V. Goswami (✉) · G. B. Mund
School of Computer Applications, Kalinga Institute of Industrial Technology, Bhubaneswar, India
e-mail: veena@kiit.ac.in

G. B. Mund
e-mail: mund@kiit.ac.in

K. Dahal et al. (eds.), *Internet of Things and Its Applications*, Lecture Notes in Electrical Engineering 825, https://doi.org/10.1007/978-981-16-7637-6_30

energy. Due to workload fluctuations and heterogeneous hardware, the nodes in a network inclined toward the dissolution of unequal energy. Usually, sensor networks experience many-to-one traffic networks, which lead to a gain in the diversification of node power absorption.

Various analysis neglects the energy consumption in all states except busy state, stating that most of the energy consumption is during packet transmission. However, this assumption has some shortcomings [4, 5]. In the idle mode, a substantial amount of power systematically gets consumed during signal processing and detection. Even sleep mode has a corresponding amount of power detriment. The dissimilarity in the energy levels of sleep–wake-up scheduling is discussed in [6]. Yuan et al. [7] have presented a dynamic sleep scheduling algorithm. Their method tries to balance the energy used up by a sensor node. Lin et al. [4] discussed on dynamic power management in wireless sensor networks. Their method tries to balance the power used up by a sensor node. Research interests on energy efficiency have been increasing day by day [8–11].

The variations of the N threshold queueing system have been studied in [12–15]. It is an efficient method to preserve energy, and the benefit of this policy is that N data packets are required for a server to switch to busy state. By this, the probability of an unplanned transition to an active state decreases. Also, the threshold policy adds simplicity and tractability. The utility of servers may be altered by varying the threshold, and this may be applied in a particular process [16–19]. Mann et al. [20] analyzed an optimal resource replication for wireless sensor networks using a queueing model. Nidhi and Goswami [21] have discussed a finite buffer queueing method that uses randomized start-up state in order to prolong lifetime of WSNs.

In this paper, we propose an M/M/1 queueing model that follows a modified N-policy to reduce the energy consumed by nodes. The state change is not randomized, and we consider the system to have an infinite buffer. In the beginning, when the system does not have any data packets, the server is in a sleep state. Arrival of a packet results in a transition to the idle state. Here, the server waits for N packets to be accumulated. Once the threshold is reached, the N packets are served in a single batch. The data packets are initially served in batch to cut down the waiting time of the packets in the system, and hence to speed up the service. Each arrival during busy state receives single service. The node switches to sleep mode when the queue becomes empty.

The rest of the paper is organized as follows: Sect. 2 describes the mentioned model, while Sect. 3 deals with computation of various performance indices. Section 4 analyzes the numerical results. Section 5 concludes the paper.

2 System Description

We assume half-duplex communication and consider the functioning of a single server, that is, a node. This mode of communication exhausts less power and permits for both up and down data streams. The suggested model depicts by the flowchart in Fig. 1.

We consider a modified N policy M/M/1 queueing model. The N-policy queueing model discussed in this paper handles the control of service. Initially, the system has no data packets, and is thus considered to be in the sleep state. There is a transition to idle state on arrival of data packets. The state of the server stays idle when the number of data packets ranges from 1 to N-1. When the system reaches N-1 number of data packets, the threshold value attains with the arrival of the next packet. When N data packets are accumulated, there is a single batch service. Each of the arrivals after initiation of batch service while the node is in busy state receives single service.

Figure 2 describes the different states that the server can attain, along with the corresponding transitions. We consider an infinite buffer space queueing system with single server in which the arrival of customers is a Poisson process with rate λ.

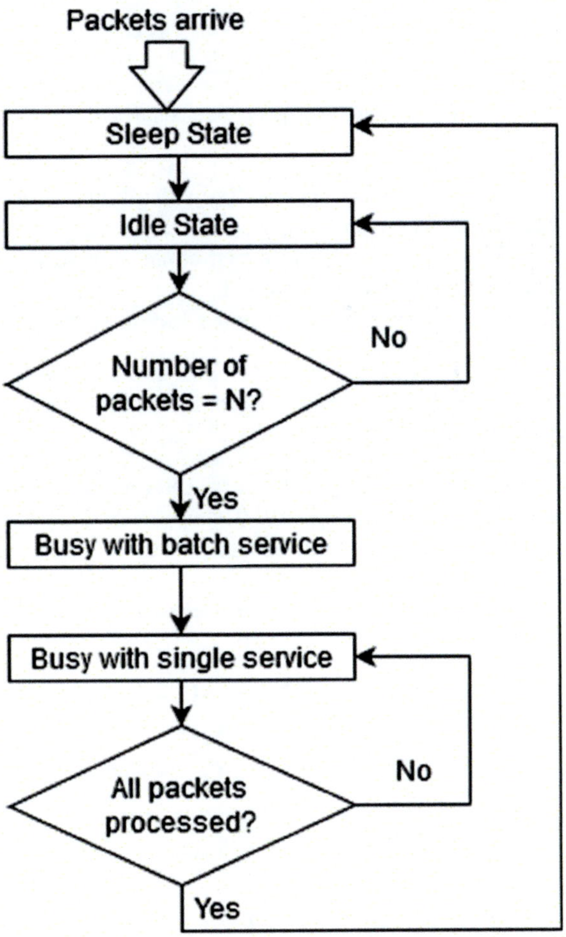

Fig. 1 Suggested system model

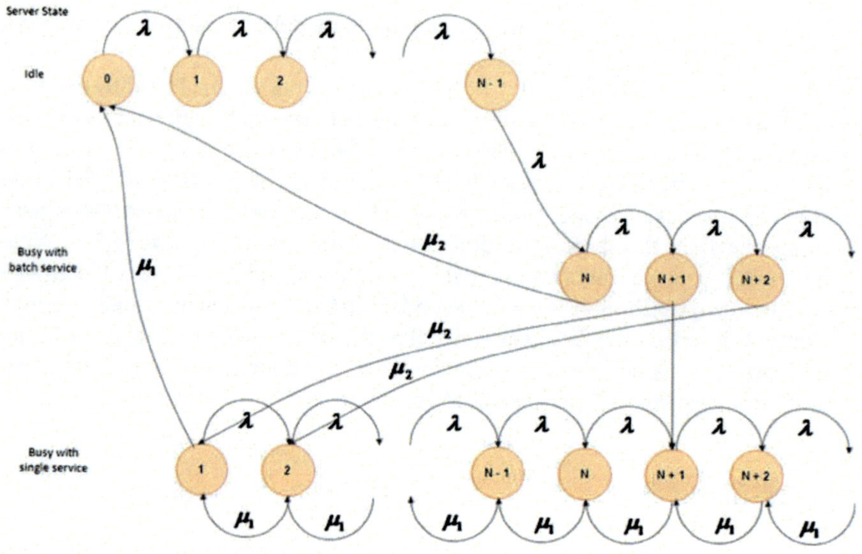

Fig. 2 State transition rate diagram for modified N policy model

The service times are exponentially distributed with rates μ_1 and μ_2, depending on whether the service is single or batch. We assume that $\mu_2 < \mu_1$. We also consider that the server executes in a first-come, first-served basis. It is also assumed that traffic intensity is $\rho = \lambda/\mu_1 < 1$.

The state of the system at time t is defined by following random variables:

$-N_s(t) =$ number of data packets in the sensor node,
$-\varphi(t) = 0$ $\{1\}$ [2], if the server is on idle mode {busy with single service} [busy with batch service].

The process $\{N_s(t), \varphi(t) : t \geq 0\}$ is a continuous time Markov chain (CTMC) with state space $\Omega = \{(n, 0) : 0 \leq n \leq N-1\} \cup \{(n, 1) : n \geq 1\} \cup \{(n, 2) : n \geq N\}$. Using the one-step transition analysis, the steady-state equations can be written as

$$\lambda P_{0,0} = \mu_1 P_{1,1} + \mu_2 P_{N,2}, \tag{1}$$

$$\lambda P_{n,0} = \lambda P_{n-1,0}, \ 1 \leq n \leq N-1, \tag{2}$$

$$(\lambda + \mu_1) P_{1,1} = \mu_1 P_{2,1} + \mu_2 P_{N+1,2}, \tag{3}$$

$$(\lambda + \mu_1) P_{n,1} = \lambda P_{n-1,0} + \mu_1 P_{n+1,1} + \mu_2 P_{N+n,2}, \ n \geq 2, \tag{4}$$

$$(\lambda + \mu_2) P_{N,2} = \lambda P_{N-1,0}, \tag{5}$$

$$(\lambda + \mu_2)P_{n,2} = \lambda P_{n-1,2}, n \geq N + 1. \tag{6}$$

From (1, 2, 5), and (6), we obtain

$$P_{n,0} = P_{0,0}, 1 \leq n \leq N - 1, \tag{7}$$

$$P_{n,2} = \left(\frac{\lambda}{\lambda + \mu_2}\right)^{n-N+1} P_{0,0}, n \geq 0, \tag{8}$$

$$P_{1,1} = \frac{\lambda^2}{\mu_1(\lambda + \mu_2)} P_{0,0}. \tag{9}$$

Using (8) in (4) and applying the displacement operator in the resultant equation, we obtain

$$P_{n,1} = \frac{\lambda}{\lambda - \mu_1 + \mu_2}\left[\left(\frac{\lambda}{\mu_1}\right)^n - \left(\frac{\lambda}{\lambda + \mu_2}\right)^n\right] P_{0,0}, n \geq 1. \tag{10}$$

Substituting (7–10) in the normalization condition $\sum_{n=0}^{N-1} P_{n,0} + \sum_{n=1}^{\infty} P_{n,1} + \sum_{n=N}^{\infty} P_{n,2} = 1$, the only unknown $P_{0,0}$ can be computed as

$$P_{0,0} = \frac{\mu_2(\mu_1 - \lambda)}{\lambda\mu_1 + N\mu_2(\mu_1 - \lambda)}. \tag{11}$$

3 Performance Indices

Average queue length when server is busy: The average queue length when the server is busy may be given as

$$\begin{aligned} L_q &= \sum_{n=1}^{\infty}(n-1)P_{n,1} + \sum_{n=N}^{\infty}(n-N)P_{n,2} \\ &= \frac{\lambda^2[\lambda\mu_2 + \mu_1(\mu_1 - \lambda)]}{\mu_2(\mu_1 - \lambda)[\lambda\mu_1 + N\mu_2(\mu_1 - \lambda)]}. \end{aligned}$$

Expected length of busy period, idle period, and busy cycle: The expected duration of a busy period for an ordinary $M/M/1$ queue is $\frac{1}{\mu_1 - \lambda}$. So, the expected duration of a busy period is

$$E(B) = \frac{1}{\mu_2} + \sum_{j=0}^{\infty} \frac{j}{\mu_1 - \lambda} \frac{e^{-\lambda/\mu_2}(\lambda/\mu_2)^j}{j!} = \frac{\mu_1}{\mu_2(\mu_1 - \lambda)}.$$

Thus, the expected length of idle period $E(I)$ and busy cycle $E(C)$ are given by

$$E(I) = \frac{N}{\lambda},\ E(C) = \frac{N}{\lambda} + \frac{\mu_1}{\mu_2(\mu_1 - \lambda)}.$$

The probability of server in different states: Let P_I, P_B, and P_S denote the probability that the server is in idle, busy with batch service or busy with single service states. They are given by

$$P_I = \sum_{n=0}^{N-1} P_{n,0},\ P_B = \sum_{n=N}^{\infty} P_{n,1},\ P_S = \sum_{n=1}^{\infty} P_{n,2}.$$

Total energy consumption rate: We define the associated cost factors as follows: C_1 represents the waiting time per data packet per unit time when server is busy. C_2 and C_4 denote the service cost per unit time, associated with batch and single service, respectively. C_3 is a fixed cost associated with the commencement of each busy period, and C_5 is the cost towards waiting per unit time until service starts. For each cycle, we formulate the power consumption using the derived performance measures.

$$\begin{aligned} F(N) &= C_1 L_q + (C_2 + C_3)\frac{1}{E[C]} + C_3 L_1 + C_4 \frac{N(N-1)}{2\lambda} \\ &= C_1 \frac{\lambda^2[\lambda\mu_2 + \mu_1(\mu_1 - \lambda)]}{\mu_2(\mu_1 - \lambda)[\lambda\mu_1 + N\mu_2(\mu_1 - \lambda)]} \\ &+ (C_2 + C_3)\frac{\lambda\mu_2(\mu_1 - \lambda)}{\lambda\mu_1 + N\mu_2(\mu_1 - \lambda)} \\ &+ C_4 \frac{\lambda^2[\mu_1\mu_2 + \lambda(\mu_1 - \lambda)]}{\mu_2(\mu_1 - \lambda)[\lambda\mu_1 + N\mu_2(\mu_1 - \lambda)]} \\ &+ C_5 \frac{N(N-1)}{2\lambda}. \end{aligned}$$

4 Numerical Results

In this section, we evaluate our analytical model through simulation and illustrate numerical results in the form of tables and graphs. This provides a managerial insight on optimal decisions that can be taken. Also, they help examine the qualitative views of the analytical model under consideration. For the numerical analysis of this model, we introduce a parameter α that defines the relationship between μ_1 and μ_2 as $\mu_2 = \mu_1/(\alpha N)$ where $0.5 < \alpha < 1$. Thus, the time needed for N single services is assuredly greater than the expected service time for a batch of N data packets.

Table 1 explores the impact of the threshold value N on the state of the system, when adopting the modified N policy model. When N increases, the expected number of data packets in the system, represented by L, increases. This results in an increase in the expected waiting time in system (W). Also, there is an increase in the expected length of the queues in busy state and in the busy cycle, i.e., $E(B)$ and $E(C)$, respectively. The average queue size when server is busy, L_b also increases. As a result, the power consumption per cycle $F(N)$ decreases.

The variation of L_q with respect to α and N is studied in Fig. 3. It is observed that L_q increases with an increase in the value of the parameter α. Similarly, as the threshold increases, so does the value of L_q. Figure 4 investigates the relationship between F and λ for different values of the parameter α. An increase in arrival rate

Table 1 Sensitivity analysis for various N when $\mu = 5.0$, $\lambda = 3.0$, $\alpha = 0.7$

N	L	W	F	$E(B)$	$E(C)$	L_b
2	1.85220	0.617398	167.251	0.70	1.36667	0.89122
3	2.51610	0.838699	164.431	1.05	2.05000	1.10634
4	3.10080	1.060000	193.238	1.40	2.73333	1.32146
5	3.84390	1.281300	242.694	1.75	3.41667	1.53659
6	4.50780	1.502600	309.143	2.10	4.10000	1.75170
7	5.17166	1.723890	391.014	2.45	4.78333	1.96679
8	5.83539	1.945130	487.523	2.80	5.46667	2.18176
9	6.49870	2.166230	598.228	3.15	6.15000	2.39636
10	7.16103	2.387010	722.856	3.50	6.83333	2.61014

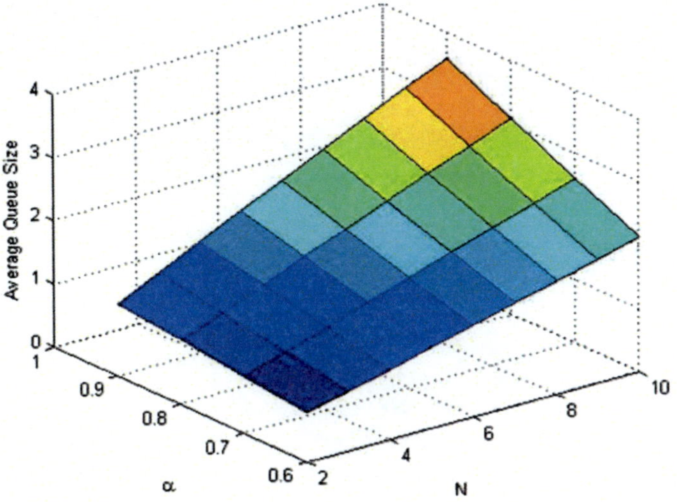

Fig. 3 L_q versus α and N

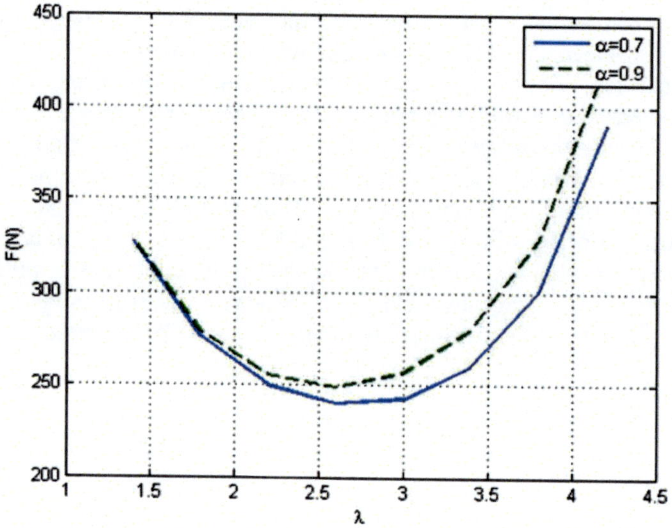

Fig. 4 F versus λ for different α

initially results in a decrease in F. Further increase, however, results in an increase in power consumption per cycle. Greater the value of α chosen, greater is the value of F. Thus, the parameters α and λ have to be carefully chosen in order to reduce power consumption per cycle.

The dependence of expected waiting time of packets (W) on threshold N and parameter α is studied in Fig. 5. For a fixed N, an increase in α results in an increased

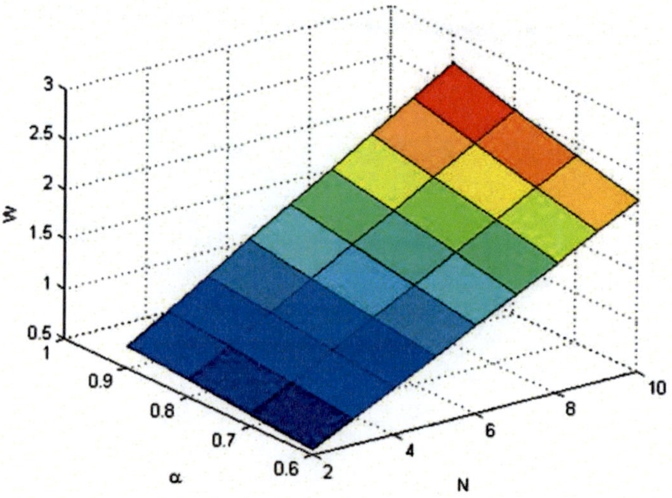

Fig. 5 W versus α and N

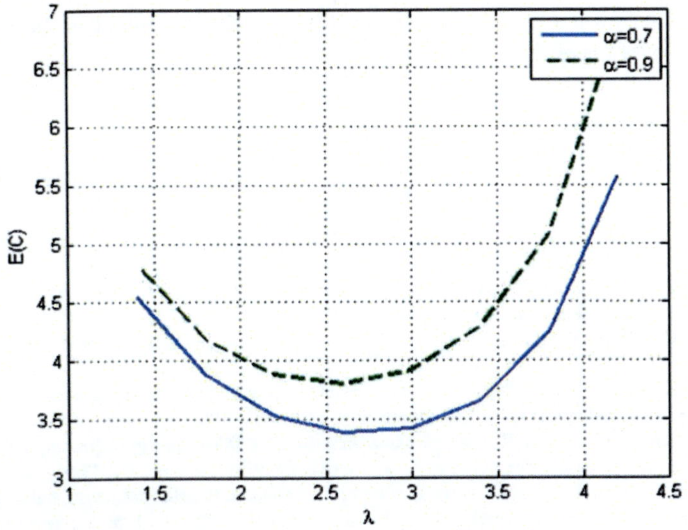

Fig. 6 E(C) versus λ for different α

value of W. Similarly, if the value of N is increased, keeping α fixed, W increases. Figure 6 analyzes the relationship between $E(C)$ and λ for different values of α. As arrival rate increases, there is initially a decrease in $E(C)$, after which it shows an increasing trend. When an increased value of α is considered, the trend remains the same, but is shifted upward.

5 Conclusion

We proposed an M/M/1 queueing model with modified N policy in order to prolong the lifetime of wireless sensor networks. Each node in the network was modeled as a server that followed the proposed model. We presented the variation of performance measures with respect to defined system parameters. We also compared the performance of both models in terms of expected number of data packets in system and expected waiting time in system. The result obtained was that modified N showed a significantly lower value for both the parameters used for comparison. However, the associated cost in case of modified N policy model was higher as the hardware required is more complex and requires greater installation and initialization cost. We calculated an estimate of the total energy consumed per cycle and investigated its dependence on other parameters. Based on the results obtained in this paper, a system manager can choose the required model and the appropriate system parameters for any given system. The parameters required for least power consumption are discussed in the numerical analysis. We proposed an efficient method to reduce the energy consumed per cycle, which in turn leads to improvement in the lifetime of wireless sensor networks.

References

1. Akyildiz IF, Su W, Sankarasubramaniam Y, Cayirci E (2002) Wireless sensor networks: a survey. Comput Netw 38(4):393–422
2. Chen P, Zhou W, Zhou Y (2015) Equilibrium customer strategies in the queue with threshold policy and setup times. Math Problems Eng
3. Culler DE, Hong W (2004) Special issue: wireless sensor networks—introduction. Commun ACM 47(6):30–33
4. Lin C, Xiong N, Park JH, Kim TH (2009) Dynamic power management in new architecture of wireless sensor networks. Int J Commun Syst 22(6):671–693
5. Rajendran V, Obraczka K, Garcia-Luna-Aceves JJ (2006) Energy-efficient, collision-free medium access control for wireless sensor networks. Wireless Netw 12(1):63–78
6. Yang X, Vaidya NH (2004) A wakeup scheme for sensor networks: achieving balance between energy saving and end-to-end delay. In: Proceedings of the 10th IEEE real-time and embedded technology and applications symposium, pp 19–26
7. Yuan Z, Wang L, Shu L, Hara T, Qin Z (2011) A balanced energy consumption sleep scheduling algorithm in wireless sensor networks. In: International conference in wireless communications and mobile computing, July 2011, pp 831–835
8. Miller MJ, Vaidya NH (2005) A MAC protocol to reduce sensor network energy consumption using a wakeup radio. IEEE Trans Mob Comput 4(3):228–242
9. Wang L, Xiao Y (2006) A survey of energy-efficient scheduling mechanisms in sensor networks. Mobile Netw Appl 11(5):723–740
10. Jones CE, Sivalingam KM, Agrawal P, Chen JC (2001) A survey of energy efficient network protocols for wireless networks. Wireless Netw 7(4):343–358
11. Liu M, Cao J, Zheng Y, Gong H, Wang X (2008) An energy-efficient protocol for data gathering and aggregation in wireless sensor networks. J Supercomput 43(2):107–125
12. Deepak TG (2001) Analysis of some queueing models related to N-policy (Doctoral dissertation, Cochin University of Science and Technology)
13. Yang DY, Wang KH (2013) Interrelationship between randomized F-policy and randomized N-policy queues. J Ind Prod Eng 30(1):30–43
14. Ke JC (2003) The operating characteristic analysis on a general input queue with N policy and a startup time. Math Methods Oper Res 57(2):235–254
15. Kuo CC, Wang KH, Pearn WL (2011) The interrelationship between N-policy M/G/1/K and F-policy G/M/1/K queues with startup time. Int J Quality Technol Quantit Manage 8:237–251
16. Huang DC, Tseng HC, Deng DJ, Chao HC (2012) A queue-based prolong lifetime methods for wireless sensor node. Comput Commun 35(9):1098–1106
17. Huang DC, Lee JH (2013) A dynamic N threshold prolong lifetime method for wireless sensor nodes. Math Comput Model 57(11):2731–2741
18. Jiang FC, Huang DC, Yang CT, Leu FY (2012) Lifetime elongation for wireless sensor network using queue-based approaches. J Supercomput 59(3):1312–1335
19. Jiang FC, Wu HW, Huang DC, Lin CH (2010) Lifetime security improvement in wireless sensor network using queue-based techniques. In: International conference on broadband, wireless computing, communication and applications, pp 469–474
20. Mann CR, Baldwin RO, Kharoufeh JP, Mullins BE (2008) A queueing approach to optimal resource replication in wireless sensor networks. Perform Eval 65(10):689–700
21. Nidhi M, Goswami V (2016) A randomized N-policy queueing method to prolong lifetime of wireless sensor networks. In: Proceedings of 3rd international conference on advanced computing, networking and informatics, pp 347–35

Analysis of Energy Harvesting Techniques for Wireless Sensor Networks Deployment Scenarios

Abhay Joshi, Sai Deepika Machavaram, and Hara Gopal Mani Pakala

Abstract The exponential scaling of wireless sensor networks (WSNs) has a far-reaching impact on operational energy requirements. Naturally occurring or ambient environmental energy (e.g., light, radio frequency, vibration, wind, or thermal energy) can be utilized to improve the lifetime of WSNs. This paper attempts to provide an overview of energy harvesting sources available in a target operating environment and demonstrates power consumption system models of an application low-power embedded system-on-chip (SoC) platform capable of harvesting ambient energy and communicating wirelessly. The said node is responsible for harvesting several sources of energy to electricity and contains a power management module for processing and distributing the harvested power across the system and a low-power communications radio. The idea is to understand the end deployment environment through experimentation and study the energy harvesting methodology.

Keywords Energy harvesting · Wireless sensor networks · Node deployment

1 Introduction

The wireless sensor networks (WSNs), with their large-scale and dense deployment, form an interesting paradigm over the existing systems. The nodes often use batteries as their source of energy but with a very limited capacity to sustain as nodes. Their replacement is time-consuming, tedious, and expensive in many applications especially where node deployment is in not so easily accessible areas such as furnace rooms, factories with high roofing, and hazardous areas such as chemical factories and battery charging stations [1, 2]. Energy harvesting (EH) is a unifying term for methods and techniques that utilize energy sources that are normally insufficient and of very low quality for powering WSN nodes directly or through their onboard batteries. Due to extremely strict energy constraints, energy conservation becomes a major design consideration of WSNs [2]. Although a plethora of energy-conserving

A. Joshi (✉) · S. D. Machavaram · H. G. M. Pakala
Institution of Electrical and Electronics Engineers (IEEE), Hyderabad, India
e-mail: abhay.joshi.in@ieee.org

K. Dahal et al. (eds.), *Internet of Things and Its Applications*, Lecture Notes in Electrical Engineering 825, https://doi.org/10.1007/978-981-16-7637-6_31

techniques have been developed for WSNs, e.g., energy-efficient medium access control (MAC) and energy-efficient routing protocols, the energy consumed by a node remains one of great significance in the development of WSNs.

The motivation behind this paper is to understand the end deployment environment and thus choose the adaptability of the energy harvesting methodology as it is the key to ensuring an optimal deployment. A significant portion of the circuitry can be minimized if proper matching of the application, the end node, and harvesting techniques is performed.

2 Related Work

Heliomote [3] is built on the Mica2 platform with NiMH storage-based energy harvesting system for harvesting ambient solar energy. The analysis resulting from the work suggests that if the battery size is enough to accommodate the variability in extracted energy and the rate of consumption of power is less than the rate of sourced power, then the perpetual operation can be achieved. Prometheus [4], based on TelosB, is also a solar energy harvesting-based system that has two energy buffers, viz primary energy buffer (supercapacitor) and secondary energy buffer (Li-ion battery). The primary buffer forms the primary energy source for the node and gets replenished as and when the solar energy is available. When excess energy is incident, the primary buffer is responsible for charging the secondary buffer, which is utilized by the node when the primary buffer energy is below a set threshold. The buffer management is handled by the software. AmbiMax [5] is built on the EcoNode platform and is capable of harvesting both solar and wind energy. The energy buffers are formed by an array of supercapacitors of 10F and a Li-Po battery of 70mAh. The buffer management is handled by the hardware.

The paper focuses on the hardware models of energy harvesting sensor nodes and the analysis of harvesting power from three sources, viz radio frequency (RF), solar energy, and piezoelectric energy. It details the WSN node prototype for three applications: cold storage unit, precision agriculture, and pothole detection and discusses the key results obtained from the prototype test and analysis.

3 Energy Harvesting System

A sensor node typically contains several components: one or several sensors, a transceiver to communicate with other nodes, energy source, external and/or on-chip memories, and a microcontroller unit (MCU) to interface the above components. The energy harvesting system consists of the following building blocks (Fig. 1):

- Input: An energy generator, usually solar, piezo, RF;

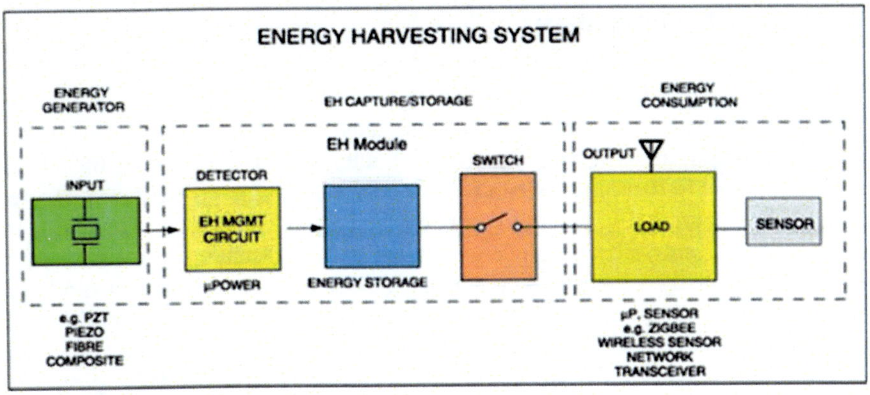

Fig. 1 Block diagram of energy harvesting system

Table 1 Power density and cost summary of energy harvesting sources

Energy source	Power density	Cost [9]	Refs.
Solar (Photovoltaic cell)	15 mW/cm^2 (direct sun) 10 μW/cm^2 (indoor)	\$0.5–\$10	[6]
Thermal (Thermoelectric generator)	15 μW/cm^3 20–60 μW/cm^2 40 μW/cm^3	\$1.0–\$30	[6] [7] [8]
Radio frequency (Antenna)	12 nW/cm^2 0.2 mW/cm^2 –1 μW/cm^2	\$0.5–\$25	[6] [7]
Wind (Micro-turbine)	28.5 mW/cm^2 3.8 × 10^{-4} W/cm^3	\$2.5–\$50	[6] [8]

- Power electronics and management circuit: Detector responsible for transducing the ambient energy, conditioning, and managing the flow of energy to the load;
- Energy storage: Device responsible for the storage of the energy harvested, usually a capacitor (or) rechargeable battery;
- Switch: To engage and disengage from the load based on the architecture described further in the paper;
- Load: Consumer of the energy stored in the system, usually a sensor node.

Table 1 summarizes the energy sources, their power density, and indicative costs. These energy sources are unique in their availability and variability in a given environment.

4 System Architectures

The dynamical variation of energy harvesting makes it difficult to predict and control the power profiles of sensor nodes. Since the availability of harvesting energy varies over time, some systems may have queues for holding data and processing them later.

As shown in Fig. 2a, some systems consider that energy storage is either connected to harvesting sources for recharging or connected to loads for discharging. Sensors start recharging only after their energy storage is fully discharged by loads. After being fully recharged, sensors wait for being selected and activated to discharge energy. After running out of energy, sensor nodes switch to recharging mode again. The system can be modeled as a three-state process: a "passive" state for recharging, a "ready" state waiting for activation, and an "active" state in discharging. The system structure in Fig. 2a has low efficiency in terms of energy harvesting, but the prediction of available energy is straightforward. The system architecture shown in Fig. 2b considers that energy harvesting and energy consumption occur at the same time. The harvested energy is first stored into energy storage which then serves energy supply to loads if necessary. The system structure in Fig. 2b is more efficient than Fig. 2a. However, the former is more difficult to model and analyze than the latter one. If the maximum harvesting power and the maximum consuming power are given, then a power management scheme is provided to ensure that energy harvesting sensor nodes never run out of energy. Furthermore, the power management schemes utilize data queues to store unprocessed packets and maximize the data rate of a single sensor node by jointly controlling the data queue and the energy buffer.

To consume harvesting power most efficiently, harvesting sources are directly connected to loads as illustrated in Fig. 2c. With this connection structure, harvested

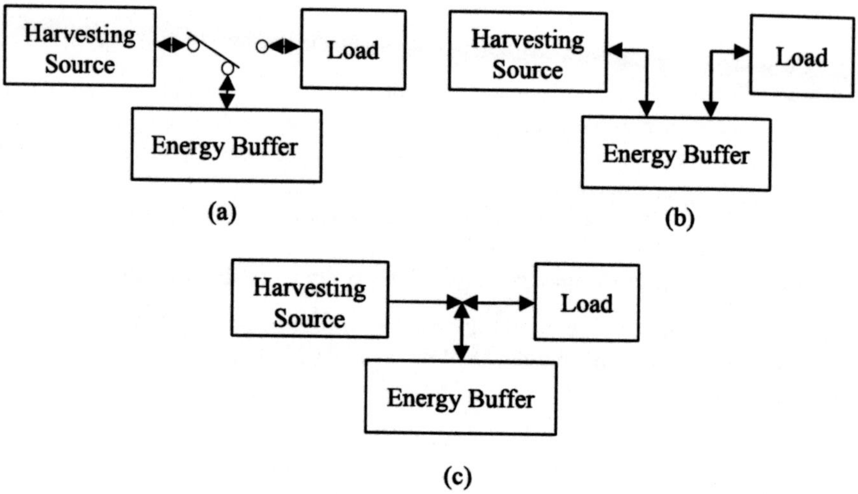

Fig. 2 Energy harvesting system architectures

energy first supplies loads for energy consumption, then the remaining energy is forwarded to energy storage. However, the system structure in Fig. 2c is too complex to model. Figure 2c is the most efficient structure among the three configuration structures in Fig. 2, due to its direct consumption of harvesting power. If the leakage current is negligible, then the efficiency of the system structure in Fig. 2b is almost the same as the structure in Fig. 2c.

5 WSN Node Prototype

The current analysis utilizes the Cypress Energy Harvesting Power Management Integrated Circuit (PMIC) S6AE101A, Bluetooth Low Energy (BLE) connectivity with an ARM Cortex M0 CPU Core. As shown in Fig. 3, the node is based on Cypress (Infineon) CYBLE-022001–00 which is a 32-bit, 48-MHz ARM Cortex-M0 based node equipped with BLE 4.1 radio. The rationale behind choosing this platform is its low consumption current of 250 nA and low startup power of 1.2 μW which are essential for a node to operate on harvested energy.

The system architecture of the S6AE101A is on the lines of Fig. 2a. As described in Fig. 4, it integrates a power gating switch (SW1), a power storage switch (SW2), and a harvested energy/battery changeover switch (SW4). Switches SW7 and SW9 are used to charge a capacitor (CVINT) that drives the internal circuit. SW7 turns ON when operating with the power source of the harvested energy. SW9 turns ON when operating with the power source of the battery. Power provided by harvested energy is charged in CVSTORE1 via SW2. Then, SW1 is turned on when the VSTORE1

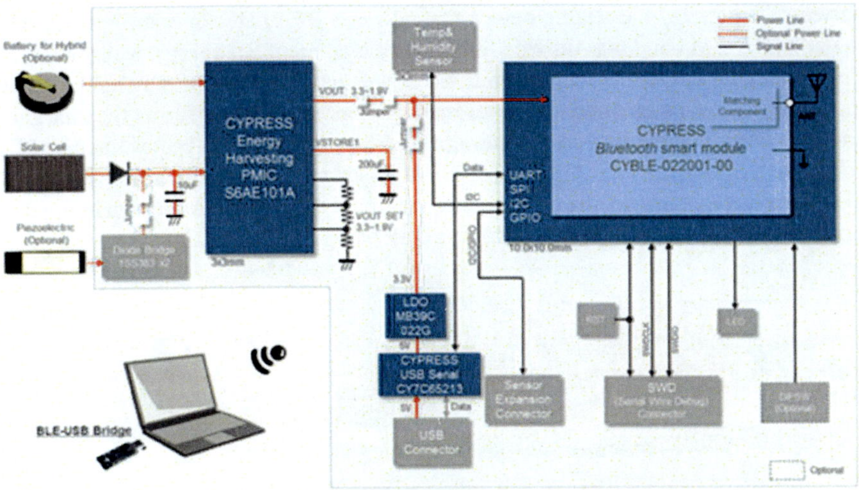

Fig. 3 Wireless sensor node architecture based On CYBLE-022001–00 and S6AE101A

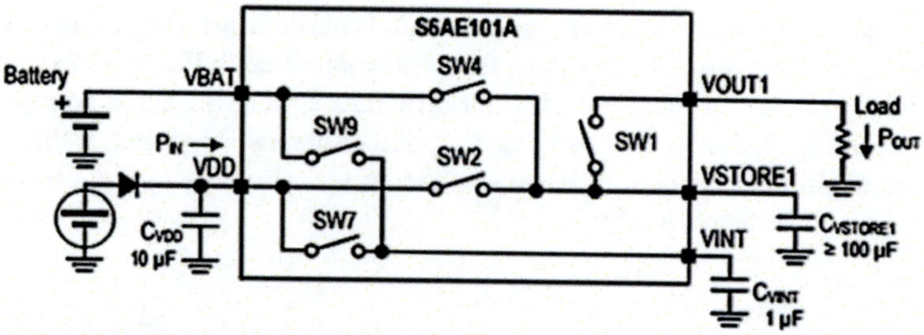

Fig. 4 Node power management system architecture

voltage is within the range of the VOUT maximum voltage (VVOUTH) to VOUT minimum voltage (VVOUTL), and the power is applied to a load. When power is not generated by harvested energy, power from the connected battery is supplied to CVSTORE1 via SW4 [10].

5.1 Cold Storage Unit: RF Energy Harvesting

A cold storage unit requires constant monitoring to maintain the quality of products. Inadequate management of the unit can reduce the acceptable quality limit of the temperature/humidity-dependent products resulting in approximately one-third of these products being thrown away [11]. The measuring system cannot make use of wires since each sensor should be able to move independently of the other. Hence, an RF energy-powered sensor node can be beneficial. The Powercast P1110B Powerharvester, an energy harvesting device that converts RF to DC, operating in the 902 to 928 MHz ISM band was utilized to harvest the ambient RF energy. Murata Ultracapacitors EDLC Double Cell 470mF 4.2 V was used as the storage element. The module also houses an onboard power management unit and performs the charging and discharging according to the storage threshold and system requirements. TI's HDC1010 for humidity and OPT3002 light to digital sensor for temperature measurements were used to model the power consumption of the system. As shown in Fig. 5a, a prototype PCB (32.00 × 17.00 mm) was built using FR4 material with 35 μm thick

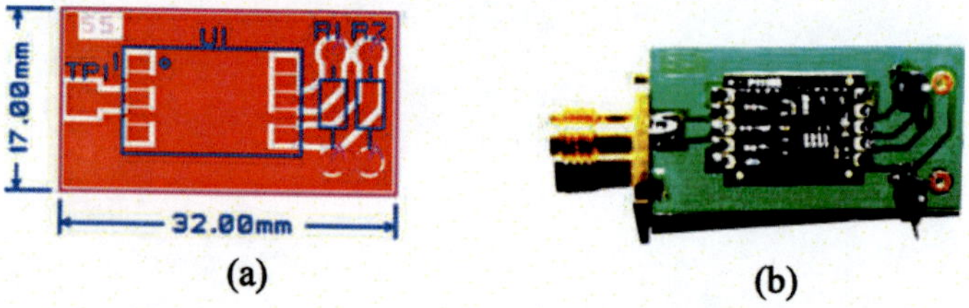

Fig. 5 Prototype and fabricated RF energy harvesting P1110B PCB with SMA connector

copper. The dielectric constant of the laminate was 4.4. As the module is already balanced to 50 Ω, no external RF matching circuitry was utilized. A 50 Ω transmission to TP1, with an RF feedline connected to an SMA connector, was used as the input to the module. The bottom layer of the prototype PCB acts as a ground plane, with the GND pins on the top layer connected to the ground by plated through holes. The fabricated device is shown in Fig. 5b which has the P1110B SMD and SMA connector soldered onto it along with probe points for power measurement.

5.2 *Precision Agriculture: Solar Energy Harvesting*

A typical green-house equips the farmer with the ability to provide the desired environmental conditions at an affordable cost and efficiency. The environmental parameters, for e.g., temperature, humidity, etc., need to be constantly monitored using sensors. For minimal precision monitoring, at least 40 to 50 wireless sensor nodes are required [12]. The nodes can be placed in monitoring zones equipped with solar energy harvesting. The sensors placed at regular intervals monitor the temperature and humidity with the help of power from the solar modules. The Texas Instruments eZ430-RF2500-SHE-01 solar energy harvesting module was used to harvest the ambient light energy. It has high-efficiency solar (2.25 in × 2.25 in) panel which is optimized for indoor operations, wherein the intensity of light is usually low. As described in Sect. 4, the ambient light must be transduced and conditioned which is handled by the EnerChip EH CBC5300 mounted on the solar energy harvesting module (SEH-01). The power management block prevents the EnerChip from a deep discharge when deployed in a low-light environment or when the load tries to draw a very high current. The block also provides a smooth transition during power on. The measurements from the onboard Si7020 I2C humidity and temperature sensor on the Cypress (Infineon) node were communicated to the host bridge using BLE communications for this particular use case.

5.3 *Pothole Detection: Piezoelectric Energy Harvesting*

The pothole detection system aims to detect uneven aberrations and potholes on the road. The typical system architecture consists of an edge node and an access point. The edge node consists of an accelerometer and a GPS module. It is responsible for translating the X–Y plane changes and thus recording the information pertaining (using GPS) to the road condition. The piezoelectric energy harvester converts the vibrations induced due to the potholes and the vehicle into electrical output. The information stored is dumped into the access point and can be effectively utilized to monitor road conditions and improve the conditions in case of potholes [13]. A piezoelectric crystal from manufacturer Midé Technology Corporation, PPA-1011 based on PZT-5H, was used to harness the vibrations. The total thickness of PPA-1011 is

0.71 mm. Its resonant frequency is 135 Hz aligns well with the low vibrations encountered under the pothole monitoring deployment scenario and provides a capacitance of 100 nF. Murata Ultracapacitors EDLC Double Cell 470mF 4.2 V was used as the storage element. The mCube M3635 accelerometer and maestro A2135-H GPS receiver were used to model the power consumption of the system. The LTC3588-1 was used as an ultralow quiescent current power supply. By interfacing it directly to the PPA-1011, its function was to rectify the voltage waveform, store harvested energy onto the capacitor, bleed off any excess power via an internal shunt regulator, and maintain a regulated output voltage by means of a nanopower high-efficiency synchronous buck regulator. The typical acceleration rate for the described use case ranges in the order of ~0.25–0.5 g.

6 Results

To verify the operational functionality of the sensor node utilizing various energy sources, the measurements from the onboard Si7020 I2C humidity and temperature sensor were communicated to the host bridge node using BLE advertising data (30 bytes). In the Beacon protocol packet, temperature and humidity sensor information was stored in the minor region (2 bytes). Humidity data was stored in the upper byte, and temperature data was stored in the lower byte. For RF and piezoelectric energy harvesting scenarios, power consumption system models were created which puts forth various operational phases of the node as a whole, viz sense, aggregate, transmit, receive, and sleep and their associated power consumption values.

6.1 Cold Storage Unit: RF Energy Harvesting

The power consumption system model of the RF energy harvesting based node for cold store unit monitoring is described in Table 4. Operating at 3.6 V, communicating temperature and humidity measurements using HDC1010 and OPT3002, respectively, around 11.73 A of current is required for a day's operation, taking into account various leakages as well for the prototype custom SoC. Hence, that translates to a power requirement is around 42.25 W. The commercially available Keysight's N5173B EXG X-Series Microwave Analog Signal Generator, 9 kHz to 40 GHz, was used as the source RF energy for the system. The RF OUT on board the signal generator was directly coupled to the end-launched SMA female connector on the fabricated RF-DC Converter module. The results obtained by performing the load test, at various frequencies, by connecting a resistive load of 66.6 Ω are given in Table 2.

The module was directly connected to the RF signal generator without any cables or links which add to the RF link budget losses. Since the module was directly coupled, the practical air measurements were not conducted which are expected to

Table 2 RF energy harvester power output

Input	915 MHz		923 MHz		915 MHz		923 MHz	
(dBm)	(mV)	(dBm)	(mV)	(dBm)	(mV)	(dBm)	(mV)	(dBm)
−4					3.8	21.59	3.7	21.36
−2					11.7	31.36	11.3	31.06
0			58.0	45.26	58.1	45.28	55.8	44.93
2	89.6	49.04	90.2	49.10	88.9	48.97	86.1	48.70
4	133.8	52.52	133.8	52.52	131.2	52.35	129.4	52.23
6	192.4	55.68	192.0	55.66	187.6	55.46	186.6	55.41
8	268.5	58.57	267.6	58.54	261.1	58.33	261.9	58.36
10	362.6	61.18	360.8	61.14	352.3	60.93	355.3	61.01
12	472.0	63.47	471.0	63.46	459.0	63.23	465.0	63.34
14	622.0	65.87	619.0	65.83	604.0	65.62	614.0	65.76
16	791.0	67.96	788.0	67.93	773.0	67.76	784.0	67.88
18	891.0	68.99	892.0	69.00	883.0	68.91	888.0	68.96

drop exponentially as per the Friis transmission formula. Based on the measurements obtained, it is observed that high incident RF energy, around 4–6 dBm, is required for the node to operate, which is difficult to achieve using ambient sources. Hence, the RF energy harvesting technique is a suitable option only with a dedicated transmitter block for reliable applications. It also requires careful know-how of the environment and is feasible only with a dedicated transmitter for a scaled node deployment.

6.2 *Precision Agriculture: Solar Energy Harvesting*

A system model for the solar energy harvesting based node for precision agriculture was not created as it is a well-known technique. As shown in Fig. 6, the DC output across the J5 connector of SEH-01 was connected to the sensor node.

To measure the resulting power from the optimized panel, the available light was varied, and the energy output of the solar module was calculated for ambient operation, using a 10Ω resistor (as the modeled load) and a multimeter to measure the voltage and current. The following measurements were performed to obtain the maximum power point (MPP) energy:

Under constant incident ambient light

$$E_a = (0.8 \times V_a) \times (0.8 \times I_a) = (0.8 \times 5.82\,V) \times (0.8 \times 41.4\,\mu\text{A}) = 154.20\,\mu\text{W} \quad (1)$$

Upon covering harvester panel:

Table 3 Power consumption system model for RF energy harvesting powered nodes for humidity and temperature measurement

Operating Voltage @ 3.6 V I Current @ mA I Sensors: Humidity & Temperature HDC1010, Light to Digital OPT3002										
State	MCU	HDC1010	OPT3002	Tx	Rx	Misc.	Duration (s)		Frequency (/d)	Total
Sense	0.04	0.0013	0.003			0.02	0.0012	0.03	720	43.2659232
Aggregate	4					0.02	0.00385	0.803	72	289.44
Tx	12			60		0.02	0.1		144	1037.088
Rx	12				60	0.02	0.5		144	5185.44
Sleep	0.04	0.0001	0.0004			0.02	0	0	85680	5183.64
Total									(mA)	11738.87392

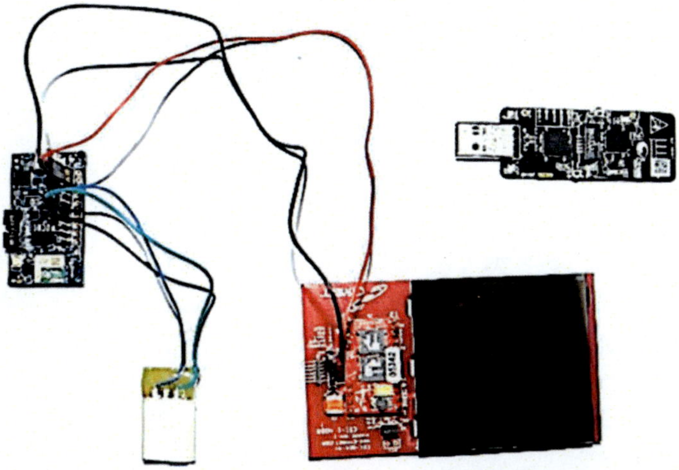

Fig. 6 Solar energy harvested system using TI eZ430-RF2500-SEH-01

$$E_c = (0.8 \times V_c) \times (0.8 \times I_c) = (0.8 \times 5.74\,V) \times (0.8 \times 26.0\,\mu\text{A}) = 95.51\,\mu\text{W} \quad (2)$$

6.3 *Pothole Detection: Piezoelectric Energy Harvesting*

The power consumption system model of the piezoelectric energy harvesting based node for road condition monitoring is described in Table 3. Operating at 3.6 V for 14 h/day, using the mCube M3635 accelerometer and maestro A2135-H GPS receiver to capture displacement and location data, respectively, around 8.06 A of current is required for a day's operation, taking into account various leakages as well for a prototype custom SoC, hence, that translates to a power requirement is around 29.03 W. We would need an electromagnetic shaker in a closed-loop configuration, a potentiometer, and multimeters to calculate the overall power from the crystal at varying acceleration rates. As shown in Fig. 7, PPA-1011 was connected to the LTC3588-1, whose rectified DC output was fed to the sensor node. The Murata EDLC 470mF supercapacitor was used as the storage element. The material was excited by the means of human excitation which was successfully able to generate enough power to start BLE advertising packets transmission from the node. An impulse power measurement met the requirements for the node power as detailed in the system model. The key drawback of the analyzed system is the narrow peak of the resonance frequency. Since the vibrations are largely restricted to one direction, careful tuning of the crystal has to be performed after a detailed vibration analysis of the prospective asset on which it is about to be mounted, in order to ensure that

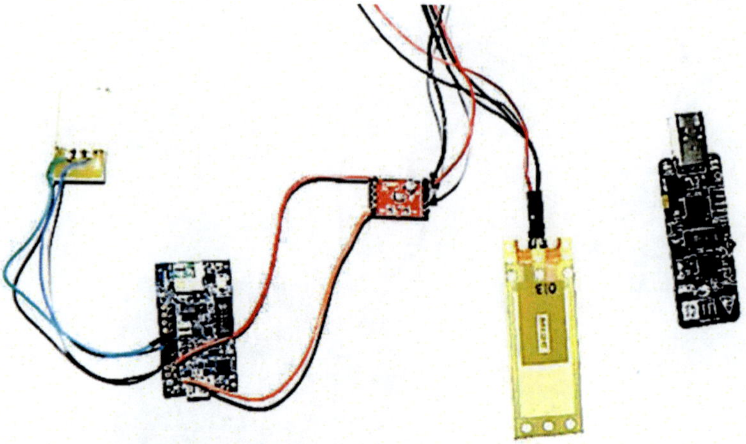

Fig. 7 Piezoelectric energy harvested system using LTC3588-1

maximum vibrations can be harvested. The trim weights can be configured to further tune the crystal leading to a higher peak.

7 Conclusion

The end sensor node power requirements will also largely influence the harvesting technique. The analysis of three sources of energy, viz solar, piezo, and RF, showed that solar energy harvesting provides the power best density and trumps any other source of energy. Piezo energy harvesting requires careful analysis of source vibration frequency and is a viable option for impulse transmission nodes. RF energy harvesting using ambient sources is found to be the most ineffective technique due to its extremely low-power density. The knowledge of the energy content and behavior of the energy source is important for the optimization and choice of the energy harvesting technique. This leads us to pursue an opportunity, wherein several of these techniques can be combined together, wherever they may be available, to create a unified sensor node capable of harvesting possible ambient sources of energy. This is also possible because the techniques act independently of each other with no contention.

Table 4 Power consumption system model for piezo energy harvesting powered nodes for road condition monitoring

Operating Voltage @ 3.6 V | Current @ mA | Sensors: mCube M3635 Accelerometer, maestro A2135-H GPS Receiver

State	MCU	M3635	A2135-H	Tx	Rx	Misc.	Duration (s)		Frequency (/d)		Total
Sense	0.04	0.01	32			0.02	0.08	40	28800	720	2004.48
Aggregate	4					0.02	0.1		2618.181818		1052.509091
Tx	4			60		0.02	0.8192		1.00		52.445184
Rx	4				60	0.02	5		1.00		320.1
Sleep	0.04	0.001	0.1			0.02					4636.8
Total									(mA)		8066.334275

Acknowledgements We would like to thank Dr. Aanandh Balasubramanian, DMTS at Ineda Systems Pvt. Ltd. (now Intel Corporation), Hyderabad, for co-supervising this research.

References

1. Adu-Manu KS, Nadir A, Cristiano T, Hoda A, Wendi H (2018) Energy-harvesting wireless sensor networks (EH-WSNs) a review. ACM Trans Sensor Netw (TOSN) 14(2):1–50
2. Babayo AB, Mohammad HA, Ihsan A (2017) A review on energy management schemes in energy harvesting wireless sensor networks. Renew Sustain Energy Rev 76:1176–1184
3. Aman K, Jason H, Sadaf Z, Srivastava MB (2007) Power management in energy harvesting sensor networks. Trans Embedded Syst 6(4):32
4. Jiang X, Polastre J, Culler D (2005) Perpetual environmentally powered sensor networks. In: Fourth international symposium on information processing in sensor networks. ACM, Los Angeles, pp 463–468
5. Chulsung P, Chou PH (2006) AmbiMax: autonomous energy harvesting platform for multi-supply wireless sensor nodes. In: 3rd annual IEEE communications society on sensor and Ad Hoc communications and networks, vol 1. IEEE, Reston, pp 168–177
6. Habibzadeh M, Hassanalieragh M, Ishikawa A, Soyata T, Sharma G (2017) Hybrid solar-wind energy harvesting for embedded applications: supercapacitor-based system architectures and design tradeoffs. IEEE Circuits Syst Mag 17:29–63
7. Kim S, Vyas R, Bito J, Niotaki K, Collado A, Georgiadis A, Tentzeris M (2014) Ambient RF energy-harvesting technologies for self-sustainable standalone wireless sensor platforms. Proceedings of IEEE 102:1649–1666
8. Raghunathan V, Kansal A, Hsu J, Friedman J, Srivastava M (2005) Design considerations for solar energy harvesting wireless embedded systems. In: Proceedings of the 2005 4th international symposium on information processing in sensor networks. vol 2005. IEEE, Boise, ID, USA, pp 457–462
9. PSMA (2014) Energy harvesting market requirements, economics and technology drivers. In: APEC industry session—breakthrough technologies driving successful energy harvesting-powered products. PSMA-Power Sources Manufacturers Association
10. Cypress Corporation (Infineon) AN213948—Basic Concepts for Energy Delivery with S6AE101A, S6AE102A, and S6AE103A. https://www.cypress.com/documentation/application-notes/an213948-basic-concepts-energy-delivery-s6ae101a-s6ae102a-and. Last Accessed 12 Nov 2020
11. Badia-Melis R, Ruiz-Garcia L, Garcia-Hierro J, Villalba JI (2015) Refrigerated fruit storage monitoring combining two different wireless sensing technologies: RFID and WSN 15(3):4781–4795. Sensors, Basel
12. Lufeng M, Yuan H, Yunhao L, Jizhong Z, Shao-Jie T, Xiang-Yang L, Guojun D (2009) Canopy closure estimates with GreenOrbs: sustainable sensing in the forest. In: Proceedings of the 7th ACM conference on embedded networked sensor systems (SenSys'09), Association for Computing Machinery, New York, NY, USA, pp 99–112
13. Zorlu O, Topal ET, K ¨ulah H (2011) A vibration-based electromagnetic energy harvester using mechanical frequency up-conversion method. IEEE Sensors J 11(2):481–488

Performance Analysis of Consensus-Based Time Synchronization Algorithms for Wireless Sensor Network

Suresh Kumar Jha, Anil Gupta, and Niranjan Panigrahi

Abstract Time synchronization is an indispensable requirement for wireless sensor network (WSN) applications. But, the traditional, centralized time synchronization algorithms are not suitable for WSN. Nowadays, consensus-based time synchronization (CTS) algorithms are becoming popular due to its distributed nature. In this context, various consensus-based time synchronization algorithms have been proposed, but their performance vary depending on various factors. This paper presents a thorough performance analysis of some state-of-the-art CTS algorithms in terms of random network topology, synchronization period, convergence speed, and synchronization error. This analysis will give an insight into the behavior of different CTS algorithms which will help the research community to understand the intrinsic property of this class of synchronization algorithms.

Keywords Wireless sensor network (WSN) · Consensus-based time synchronization (CTS) · Clock skew · Clock offset

1 Introduction

The clock synchronization is a critical requirement for distributed systems due to unavailability of a centralized clock. The WSN being a distributed system faces the same issue. Nowadays, WSN has taken more attention due to its increasing demand in many applications like healthcare monitoring, area monitoring, industrial domain, threat detection, etc. A WSN generally consists of a huge number of densely deployed, low-cost, battery-operated, small devices. Time synchronization

S. K. Jha (✉) · A. Gupta
JNVU, Jodhpur, Rajasthan, India

A. Gupta
e-mail: anilgupta@jnvu.edu.in

N. Panigrahi
PMEC, Berhampur, Odisha, India
e-mail: niranjan.cse@pmec.ac.in

K. Dahal et al. (eds.), *Internet of Things and Its Applications*, Lecture Notes in Electrical Engineering 825, https://doi.org/10.1007/978-981-16-7637-6_32

between these small devices is required for different applications such as multi-agent networks, data fusion, moving objects trajectory estimations, and for different monitoring systems [1].

Broadly there are two types of approaches followed to achieve time synchronization algorithms in WSN; centralized and distributed approach. There are many centralized algorithms that have been proposed in literature like RBS, FTSP, and TPSN which were based on hierarchical organization of the network as a rooted tree, and gain synchronization with reference (root) node's time. This type of approach has many shortcomings like root node failure problem, and large overhead to divide the network into hierarchical clusters [2].

The second approach is distributed consensus-based approach whose basic principle is to agree upon a common value. In this approach, there is no leader or root node; despite all the nodes run a common algorithm to achieve consensus on a common value. It is very robust in terms of node failure or new node addition in the network [3]. Though CTS algorithms are well-suited for WSN due to its distributed nature, the performance of distributed CTS algorithms is mainly affected by different factors like network topology, averaging principle, and network connectivity. In this context, a thorough insight is necessary about CTS algorithms' behavior.

The main work of this paper are given below:

i. Some recent state-of-the-art CTS algorithms are briefly presented, and their basic averaging method is described.
ii. A thorough simulation-based performance analysis is carried out using some standard performance criteria like convergence speed, global synchronization error, and local synchronization error using Pymote, a Python-based discrete-event simulator for WSN.

2 Related Work

In the literature, there are so many time synchronization algorithms that have been proposed for WSN. It can be divided in two types: distributed and centralized time synchronization. Centralized time synchronization algorithms in WSNs include TPSN, RBS, and FTSP [4] for sensor networks, which required referenced node to be synchronized. While distributed time synchronization algorithm includes time diffusion protocol (TDP), a more popular protocol average time-sync (ATS) was proposed which made by two averaging consensus algorithms. So far, this protocol has required a huge data interchange and the converging speed might also steady when the network size growing exponential. The idea behind ATS is to averaging the local information again and again till all nodes reached a common clock value, but the drawback is its slower convergence rate. Later on, MTS algorithm was developed to increase the convergence speed [5].

Maximum time synchronization (MTS), average time synchronization (ATS), consensus clock synchronization (CCS), and selective averaging consensus algorithm (SATS) [6], in these protocols, there are not required any reference node, all nodes synchronize to their neighbor's node. So the distributed approaches are more robust, scalable, and paying more attention in recent technologies.

Leblanc proposed a resilient consensus protocol by which synchronization can achievable in the presence of malicious node by using the graph robustness property [7]. Zhu proposed sensor fusion with asymmetric links over networks [8]. Dhuli explained the convergence analysis of consensus algorithm used in WSN [9]. Duan has proposed accurate synchronization for WSNs with bounded noise using maximum consensus-based approach [10].

3 System Models

This part briefly explains the different models adopted for performance analysis of CTS algorithms.

3.1 *Clock Model*

Each sensor node is equipped with its own clock in a sensor network. Ideally, the sensor node's clock should be equipped as, where *t* represents the ideal or reference time. Generally, the functioning of the sensors clock for the ith node can be express as

$$C_i(t) = \theta + \text{s.t} \tag{1}$$

A pictorial representation can be express in Fig. 1.

From Eq. 1, the relation among two nodes, node x and node y, can be represent as

$$C_x(t) = \theta^{xy} + s^{xy}.C_x(t) \tag{2}$$

where θ^{xy} and s^{xy} represent relative clock offset and skew between node x, y. If two clock are absolutely synchronized, then $\theta^{xy} = 0$ and s^{xy} should always be 1.

3.2 *Network Model*

Graph data structure gives us a natural abstraction that how wireless sensors shared information among networks. It contains a high-level abstraction of network topology where objects referred to as vertices and edges. The graph can be defined as the set that has a finite number of elements, we denote it as (V, E), where v represents the set of vertices $\{v1, v2, v3\ldots\}$ and denote as a set of edges $\{e1, e2, e3\ldots\}$. Now consider

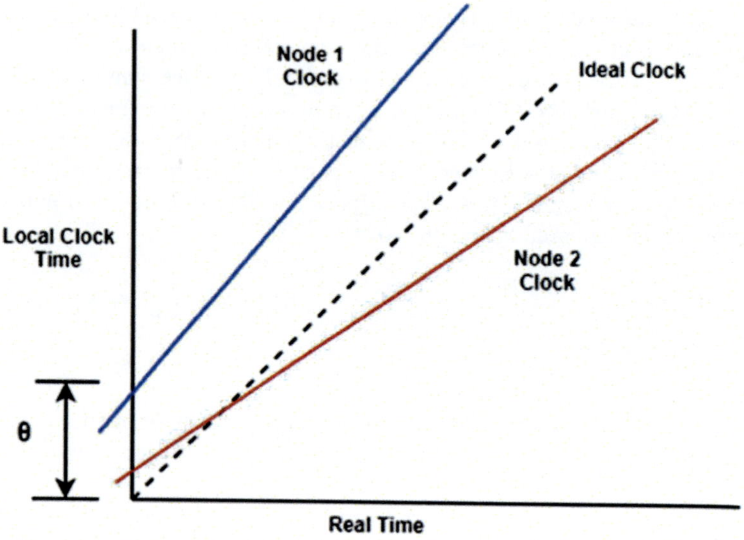

Fig. 1 Clock model of sensor node

the set of two elements, which consist of elements $\{vi, vj\}$ such that $i, j = 1,2,3\ldots.n$ and $i \neq j$. The graph G called connected if every vertex in $V(G)$, there is a path that has them as its end vertices [11].

3.3 Consensus Time Synchronization Model

Every cycle of the CTS calculation, each node starts the synchronization process by sending an initial message. After accepting the messages by its neighbors, it calculates the arrival time of the neighbor's messages. At that point, every node updates its local clock time by utilizing pair-wise or weighted averaging strategies, until converging to the average of the initial clock among all nodes. The mathematical model of the clock updates rule which can be rewritten as.

$$C_i(k+1) = x_i(k) + \varepsilon \sum_{j \in N_i}^{n} (x'_j(k) - x_i(k))$$
$$i = 1, 2, 3\ldots n \tag{3}$$

where $x_i(k)$ is the local time at the node '*i*' during each iteration '*k*' and 'ε' is the constant step size for each iteration and N_i is the set of neighbor nodes by which node *i* can communicate.

4 State-of-the-Art CTS Algorithms

The following consensus-based time synchronization algorithms are considered for analysis.

4.1 Average Time Synch (ATS) [4]

This protocol averages the local information to achieve the global agreement.

$$\left.\eta_{ij}\left(t^{+}\right)=\rho_{n}\eta_{ij}(t)+\left(1-\rho_{\eta}\right)\frac{T_{j}^{\text{new}}-T_{j}^{\text{old}}}{T_{ij}^{\text{new}}-T_{ij}^{\text{old}}}\right\},\ t=t_{e}^{j}$$
$$\left(T_{ij}^{\text{old}},T_{j}^{\text{old}}\right)=\left(T_{ij}^{\text{new}},T_{j}^{\text{new}}\right)$$
$$\eta_{ij}(t)=\eta_{ij}\left(t^{+}\right),\ t\in\left(t^{+},t_{e+1}^{j}\right. \tag{4}$$

4.2 Average Time Synchronization Pair-Wise [11]

Average time synchronization method can be explore mathematically by

$$P_{\text{avg}(t)}=wt+\text{avg}(P(0)) \tag{5}$$

for every nodes, where in Eq. (6)

$$\text{avg}(\text{p}(0))=1/n\sum_{i=1}^{n}\text{p}_{\text{i}}(0) \tag{6}$$

Two connected nodes can exchange the local clock by

$$p_{i}(t-1)\text{ and }p_{j}(t-1) \tag{7}$$

and update their local clocks to

$$p_{i}(t)=p_{j}(t)=(p_{i}(t-1)+p_{j}(t-1))/2 \tag{8}$$

4.3 Consensus Clock Synchronization (CCS) [12]

There are three methods given for evaluation of the convergence in consensus algorithm, but we are using confidence weighted averaging method.

Confidence weighted average (CWA):-A confidence parameter is assigned to more weightage to the clock readings for the node which is nearest to the consensus value.

$$C_i(t_j) = \frac{\lambda_i C_i(t_j) + \lambda_j C_j(t_j)}{\lambda_i + \lambda_j} \tag{9}$$

where λ is confidence parameter.

4.4 Selective Averaging Time Synchronization (SATS) [13]

Let the node's which have maximum relative offset among all the neighbors node is μ_k, then μ_i updates their local clock as

$$\frac{X_1^k + X_k^i + \mathrm{d}_1 + \mathrm{d}_2^k}{2} = (X_2^k + d_1) - \delta_{ik}/2 \tag{10}$$

where X_1^n is receiving time stamp, X_2^n is sending timestamp, and d is delay.

5 Simulation

5.1 Simulation Setup

To analyze the CTS algorithms, simulation has been implemented on Pymote simulator which provides event-based simulation and special builds for analysis of distributed algorithms. It is designed to rapidly testing of new algorithms and their analysis. Basically, Pymote relies on the graph data structure to explore the communication among the nodes. It is mainly developed on top of NetworkX and also supports different scientific packages like Numpy, matplotlib, scipy, etc. Apart from that it also provides a GUI on which we can configure the different scenarios of the network. Its GUI work on PySide and the console provides IPython [14]. To clarify the connected topology, the deployment of sensors done by the radius(r) using the formula [15].

$$r = L * \sqrt{2\log n/n} \tag{11}$$

5.2 *Simulation Results*

This section shows the behavior of different CTS algorithms w.r.t various performance metrics.

5.2.1 Convergence Speed

See Fig. 2.

5.2.2 Global Synchronization Error (GSE)

Figure 2e showing the comparison of the different algorithms in terms of global synchronization error. The SATS algorithm is showing minimum error.

5.2.3 Local Synchronization Error (LSE)

Figure 2f is showing the comparison of the different algorithms in terms of the local synchronization error of each node. The nodes in the SATS algorithm are showing minimum error.

5.3 *Observations*

To study the different algorithms by their behavior and correctness, implemented in Python using Pymote simulator, the simulation has been performed by using the criteria mention in Table 1.The analysis is done for 50 iterations and found the result mentioned following.

- In Fig. 2a, the average TimeSynch (ATS) protocol, which uses the averaging local information method, achieves convergence between 20 and 30 iterations.
- In Fig. 2b, the consensus clock synchronization (CCS) uses confidence weighted averaging method which converges near about 20 iterations.
- In Fig. 2c, the average time synch pairwising (ATSP) uses random pair-wise averaging method which converges near 10 iterations.
- In Fig. 2d, the selective averaging time synchronization (SATS) uses maximum difference-based selective pair-wise averaging method converges near five iterations.

6 Conclusion and Future Work

In this paper, different consensus-based time synchronization protocols are studied and analyzed under different performance metrics such as convergence speed, global synchronization error, and local synchronization error. From the simulation, it is concluded that the SATS algorithm is showing optimum convergence speed under minimum error. Apart from these factors, other factors may be considered like robustness, scalability, and energy efficiency when implemented in real scenarios. In future, different CTS algorithms with security aspects will be studied and analyzed.

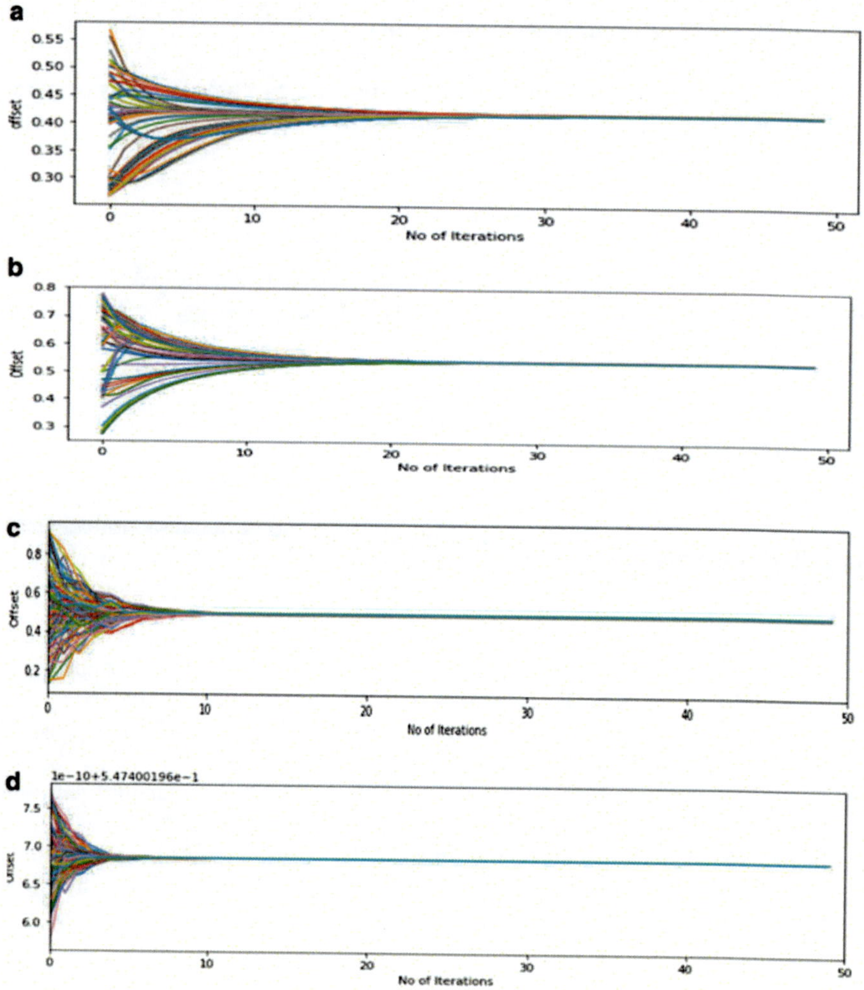

Fig. 2 **a** ATS **b** CCS **c** ATSP **d** SATS. **e** Global synchronization error. **f** Local error of individual nodes

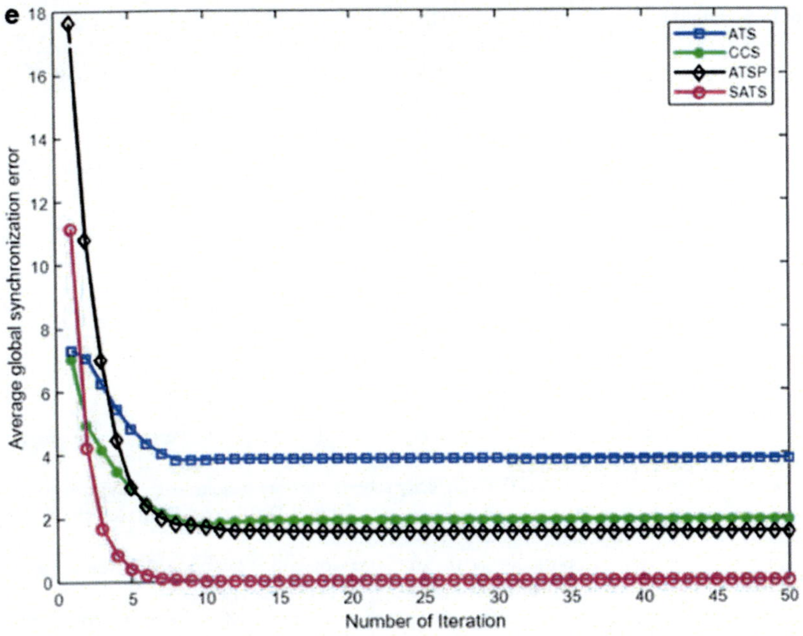

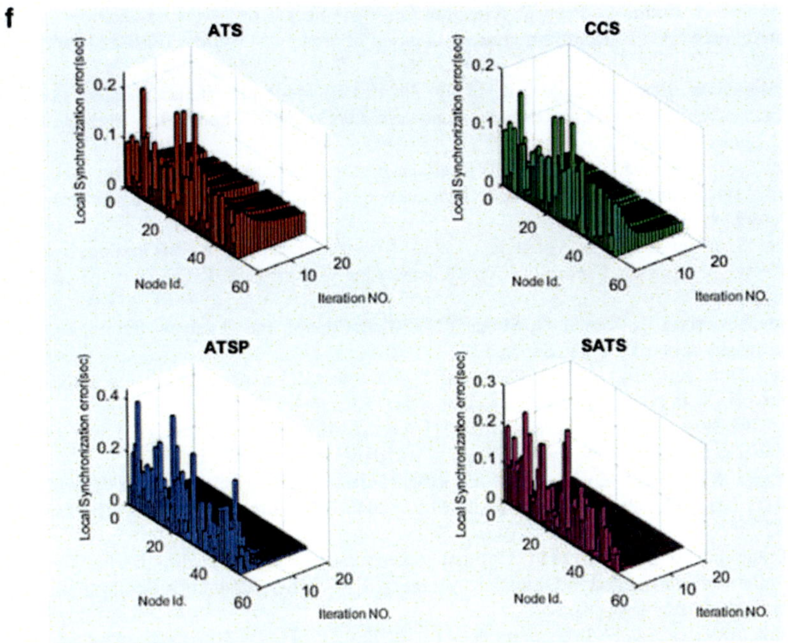

Fig. 2 (continued)

Table 1 Simulation parameters

S. no	Parameter	Values
1	Deployment area	10 × 10 square unit
2	Topology	Random
3	No. of nodes	100
4	Interval of iteration	10 s
5	Connectivity radius (r)	2 unit
6	Skew	Uniform (−5, 5)
7	Offset	Uniform (0, 1)

References

1. Phan L-A, Taejoon K, Taehong K, JaeSeang L, Jae-Hyun H (2019) Performance analysis of time synchronization protocols in wireless sensor networks, MDPI, Sensor Netw
2. Jha SK, Panigrahi N, Gupta A (2020) Security threats for time synchronization protocols in the internet of things. In book: Principles of internet of things (IoT) ecosystem: insight paradigm. pp 495–517. https://doi.org/10.1007/978-3-030-33596-0_20
3. Dong W, Liu X (2015) Robust and secure time synchronization against Sybil attacks for sensor networks. IEEE Trans Ind Inform.https://doi.org/10.1109/tii.2015.2495147
4. Schenato L, Fiorentin F (2012) Average TimeSynch: a consensus-based protocol for time synchronization in wireless sensor networks. Elsevier
5. Jianping H, Jiming C, Peng C, Xianghui C (2014) Secure time synchronization in wireless sensor networks: a maximum consensus-based approach. https://doi.org/10.1109/TPDS.2013.150
6. Zhaowei W, Peng Z, Linghe K, Dong, Jin (2018) Node-Identification-based secure time synchronization in industrial wireless sensor networks. MDPI Sens 18:2718. https://doi.org/10.3390/s18082718
7. LeBlanc HJ, Koutsoukos X (2018) Resilient first-order consensus and weakly stable, higher order synchronization of continuous time networked multiagent systems. IEEE Trans Control Netw Syst 5(3):1219–1231
8. Zhu S, Chen C, Xu J, Guan X, Xie L, Johansson KH (2018) Mitigating quantization effects on distributed sensor fusion: A least squares approach. IEEE Trans Signal Process 66(13):34593474
9. Sateeshkrishna D, Kumar G, Singh YN (2014) Convergence analysis for regular wireless consensus networks. IEEE Sensors J
10. He J, Duan X, Cheng P, Shi L, Cai L (2017) Accurate clock synchronization in wireless sensor networks with bounded noise. Automatica 81:350–358
11. Jianshe W, Licheng J, Ranran D (2011) Average time synchronization in wireless sensor networks by pairwise messages. Elsevier. 10.1016/ j.comcom. 2011.09.007
12. Maggs KT (2012) Consensus clock synchronization for wireless sensor networks. June 2012. https://doi.org/10.1109/JSEN.2011.2182045 DOI: https://doi.org/10.1109/JSEN.2015.2420952 ·
13. Niranjan P, Pabitra MK (2014) Optimal consensus-based clock synchronization algorithm in wireless sensor network by selective averaging. 1st October 2014. https://doi.org/10.1049/iet-wss.2013.0102
14. https://pymote.readthedocs.io/en/latest/
15. Xiong G, Shalinee K (2009) Analysis of distributed consensus time synchronization with Gaussian delay over wireless sensor networks. EURASIP J Wireless Commun Netw. https://doi.org/10.1155/2009/528161

Internet of UAV Mounted RFID for Various Applications Using LoRa Technology: A Comprehensive Survey

Priti Mandal, Lakshi Prosad Roy, and Santos Kumar Das

Abstract Dominance of IoT in today's market has expedited unmanned aerial vehicle (UAV) in various applications. UAV is borne with different sensors to detect and localize the concerned target in various environmental conditions. According to the government survey analysis, it is predicted that RFID tags will see a growth of 7.7% CAGR by 2023. So, both RFID and UAV are the revolutionary technology which can be used to loom toward adverse situation applications. In this paper, a brief review is done about how data collected by UAV mounted RFID can be used accessed by the users from a far distance using LoRa technology (LPWAN technology) and discussed briefly about the real-time applications and the upcoming challenges in the concerned research field.

Keywords Unmanned aerial vehicle (UAV) · RFID sensor · Internet of Things (IoT) · Wireless sensor network (WSN) · LoRa

1 Introduction

Popularity of unmanned aerial vehicles (UAVs) increased drastically in a blink of eyes. UAVs omnipresent nature makes it place in both military as well as civilian segment. In military sector, it can be used for security [1], patrolling in the border areas [2], etc. where as in civilian sector it is used in agriculture [3], disaster relief operation [4, 5], traffic monitoring [6], accident and crowd monitoring [7], constructional sites [8], Internet delivery [9], etc. Due to the accessibility in the deserted area where it is not appropriate for the human to reach, drones is apt for IoT applications, where end users can access the information from anywhere.

In general, IoT is the interconnection of different things (such as sensor and actuator) may be placed on any platform according to the requisite of the application through internet for collecting data, transmitting it to the cloud, and finally accessed by the user without the involvement of human [10]. So, Io(UAV) denoted that UAV

P. Mandal (✉) · L. P. Roy · S. K. Das
Department of Electronics and Communication Engineering, National Institute of Technology Rourkela, Rourkela, Odisha, India

K. Dahal et al. (eds.), *Internet of Things and Its Applications*, Lecture Notes in Electrical Engineering 825, https://doi.org/10.1007/978-981-16-7637-6_33

borne sensors will be used for getting the data and uploading it in cloud by various wireless technologies.

In this paper, a comprehensive survey on various applications of Io(UAV) with LoRa technology is made. The remainder of the paper is arranged as follows. Section 2 discuss about the growth of WSN. In Sect. 3, basics function of WSN is discussed. Section 4 describes various applications of Io(UAV) with LoRa. In Sect. 5, challenges in future research are pointed out.

2 History of Wireless Sensor Network (WSN)

Alike other booming technologies, wireless sensor network flourished for fulfilling the necessity of military as well as the growing demand of consumers, i.e., for industrial applications. Wireless sensor network first came into picture in 1950s as *Sound Surveillance System* (SOSUS), which is similar to the contemporary wireless sensor network. SOSUS was used for the detection and the tracking of submarines during the cold war by the US Military. In 1969, US DARPA used a network of radar as sensors for air defense engaged by Aerostats. Meanwhile, hardware test bench set up was made in late 1970s. Then, distributed WSN makes it impact in 1980 by collaborating with educational institutes (MIT, Carnegie Mellon University). After this, various initiatives were taken to promote the significance of WSN like UCLA Wireless integrated Network sensor in 1993, different awareness programs were conducted by MIT in 2000, NASA in 2001 and many more which are represented in Fig. 1. It makes a leap when a vast mass marketing of IoT. Up to last decade, wireless sensor network was not that prominent because of the expenses, network prototype, power consumption, and low range. After the widespread of IoT, WSN [12] finds its new ray of optimism.

In the next section, inherent functions of wireless sensor network is discussed.

3 Function of Wireless Sensor Network (WSN)

The elementary function of any sensor node is to sense in the required field for knowing the parameters of interest. So the basis step which involves in the process is the collection of data using appropriate communication technology and finally delivers it to the end users.

3.1 Data Collection

The aim of any application is to collect the data from the environment. Depending upon the application for collecting data, UAV boarded with many IoT devices such

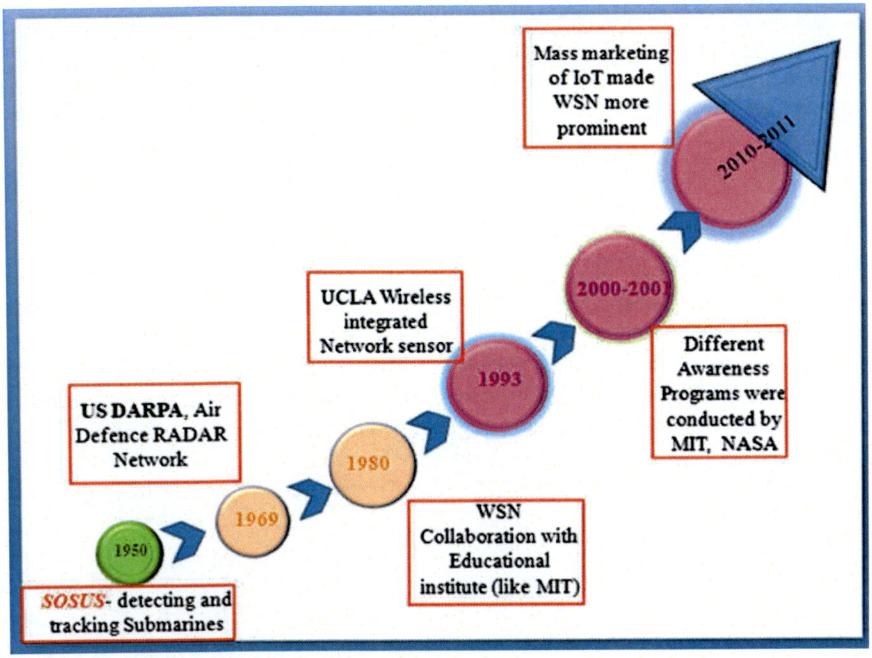

Fig. 1 Growth of wireless sensor network in few decades

as sensors, actuators, or any other intelligent devices [13]. After collecting the data, it is transmitted to the base station. Most fundamental IoT device for collection of data is sensor, which is used to determine different parameters [14]. Though RFID was originated in 1999 [15], it still have a revolutionary impact in many upcoming technologies. It is used for sensing, detecting and tracking various elements or physical attributes [16]. RFID is feasible for managing the supply of construction work [17]. In the construction sites, there are many activities like supervision of material, machinery; workers were still done by man power, which is not possible in every situation. So, UAV borne RFID was used to have an eye on them. Wu et al. used machine learning algorithm, k-nearest neighbor for localization. Not only in the outdoor localization, it can be used for indoor applications as well [18]. RFID placed on drone can get the information even in harsh condition like fire break out, gas leakage, [19] etc.

3.2 Communication Technology

On the basis of the network access, communication technology can be classified as short-range technology and long-range technology. Short-range technology includes

Bluetooth, Zigbee, and Wi-Fi; and long-range technology includes NB-IoT, LoRa, and SigFox. Comparison between both the technologies is done in Table 1 [21].

Zhou et al. [22] designed a UAV for patrolling against the traditional method. In this whole system, UAV is connected to the base station through Zigbee network, though Bluetooth is used for localization where GPS signal does not work efficiently. This system provides the image data of surveillance along with less probability of error caused by Bluetooth.

Zhou et al. [23] used Zigbee technology to estimate the location of UAV, but it was not that reliable, and Kalman filter is to be used to get the exact location. For indoor localization [24] of UAV which is done using Zigbee and inertial navigation system (INS), accurate data is analyzed by using extended Kalman filter.

UAV is used to localize the target in the IoT network. Accuracy of the localization is improved in a LoRa network for the first time. Bi-directional information exchange by an autonomous system was used to get the coordinates of UAV in a network [25]. In this direction, research challenges are very fascinating.

3.3 *Data Delivery*

After collecting data by the active sensors, next step is to deliver the data. Depending upon the scenario or the demand of situation whether it requires quick response in emergency situation or it can tolerate a certain amount of delay. There are two methods of transferring data. One is direct method of sending it to base station. This case is prominent when only one UAV is used. But when multi-UAV comes into picture, then there will be congestion in the network. If the application can tolerate delay, then drone can wait when the network is free or good, then it can transmit but in the case of power consumption will be more. In order to reduce power consumption, optimization of network congestion and efficient communication technology is to be used [20].

Few real-time applications are discussed in the next section in which UAV can use LoRa technology for transmission of data to the end users, uploaded in the cloud from a far distance where accessing network is difficult and consumes less energy.

4 Applications

Each UAV U is located in the 3D space. In 3D space, UAV U is located within R set of circular region whose radius r_u is in three-dimensional spaces. Let $r = (x_u + y_u + h_u) \in R$, x_u, y_u is the position of UAV in two-dimension, h_u is the height or altitude. Position of the target (construction material, wildlife, gas leakage point, etc.) is estimated at each discrete time $t \in [0, T_m]$, T_m is the total time duration for monitoring. The position of the target present on ground is given in coordinates (x_n^t, y_n^t). The angle of visibility is considered as θ of UAV U [26].

Table 1 Comparison of short-range technology and long-range technology [21]

Technology	Wireless technology (short range)				LPWAN technology (long range)			
	Zigbee	Wi-Fi	BLE	Z Wave	LoRa	SigFox	Ingenu	NB-IoT
Range	75–100 m	70–250 m	>100 m	~100 m	2–5 km (urban) 15 km (rural)	3–10 km (urban) 30-50 km (rural)	15 km urban	10–15 km
Bandwidth	2 MHz	22 MHz	2.4 GHz ISM band each channel has 1 MHz	908.42 MHz	<500 kHz (channel bandwidth of 125 kHz)	100 Hz	1 MHz	180 kHz (200 kHz carrier bandwidth)
Data rate	250 kbps	600 Mbps	1 Mbps	40–100 kbps	50 kbps with FSK	Less than 100 bps	624 kbps DL 156 kbps UL	200 kbps
Latency	15 ms	<26 ms	~3 < 10 ms	1 s (wake up)	1–10 ms	High 1–30 ms	~10 s	<10 s
Lifetime of battery	For years	–	–	10 years life in a coined size battery	>10 years	10 years (depends upon the number of messages transmitted)	10–20 years	>10 years
Cost	Low	Moderate	Less	Low	Low	Low	Low	Low
Mobility	Yes but low	Yes	Limited	Yes	Yes	Yes	No	Nomadic
Advantage	Highly reliable	Easy to deploy	Provide more privacy	Low-cost less vulnerable signal remain strong up to 100 feet	Highly immune to interference, adaptive data rate longer battery life	Highly reliable, complexity is low	Extreme coverage, transmitted power control to extend battery lifetime	Use PSM obsolete the requirement of gateway, better range and coverage

(continued)

Table 1 (continued)

Technology	Wireless technology (short range)				LPWAN technology (long range)			
	Zigbee	Wi-Fi	BLE	Z Wave	LoRa	SigFox	Ingenu	NB-IoT
Limitation	Short range, security is vulnerable	Limited range, vulnerable, speed is much slower	Non-flexible	Over long distance is does not able to propagate	Low data rate, no acknowledgment provided, lower latency	Low security	Analytical study is not present	Interference, no support of handoff

So the area of coverage r_u^h by UAV U is represented as,

$$r_u^h = h_u \tan\left(\frac{\theta}{2}\right) \quad (1)$$

where h_u is the height at UAV hovering and θ is the angle of visibility.

The distance between the UAV and target within the coverage area is given by Euclidean distance which is represented as,

$$d(u, n) = \sqrt{\left(x_u - x_n^t\right)^2 + \left(y_u - y_n^t\right)^2} \leq r_u^h \quad (2)$$

where x_u, y_u is the two-dimensional position of the UAV and x_n^t, y_n^t is the position of the target.

If there are multiple numbers of UAVs, then it is given as,

$$D_{u,v} = \sqrt{(x_u - x_v)^2 + (y_u - y_v)^2 + (h_u - h_v)^2} \leq R_u \quad (3)$$

where (x_u, y_u, h_u) is the three-dimensional position of one UAV and (x_v, y_v, h_v) is the three-dimensional position of another UAV.

Based on the above mathematical formulation, various applications are briefly discussed below.

4.1 *Monitoring Agriculture Parameters*

For growing five types of crop, farmer needs to deal with lot of hardship. They need to monitor each and every factor such as humidity, temperature, pests, and soil-moisture time to time. All this consumes a lot of time as well as it requires more man source for monitoring the parameters. So, to ease the adversity of farmers, swarm of UAVs mounted RFID sensor can be used to monitor all the factors. The data can be collected by the sensors, and the data can be transmitted through the LoRa technology to a long range of distance where the gateway will be placed, and finally, it is uploaded to the cloud through the Internet. Then, easily farmer could get all the information through any Web application (Fig. 2).

4.2 *Monitoring Wildlife*

In National Park, Wildlife Sanctuary, and Zoo, wildlife is placed in vast area. It is very difficult to keep an eye on each species of wildlife; few may be endemic or

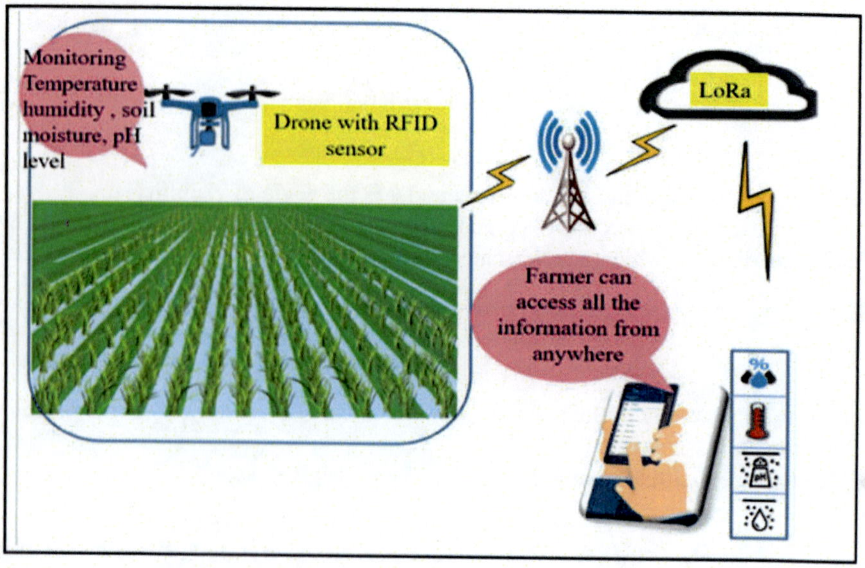

Fig. 2 Monitoring agriculture parameters using Io(UAV) and LoRa technology

endangered. Even to go in close proximity to the wildlife is very risky. So to avoid the risk of human life, UAVs can be used to supervise the area. The animals might be moving in a huge number, so to have a total count of them, we can use the RFID sensors. But as they move from place to place, it is not possible to keep the tags placed everywhere, so we can take the benefit of UAV for this process. This reduces the tedious job of the keepers along with their own safety. They can easily access it through cloud. As the monitoring unit will be place far away from the whole vicinity of the wildlife in order not to disturb them, so LoRa technology would be appropriate for this scenario (Fig. 3).

4.3 Chemical Leakage

Chemical leakage in the factory has always been dangerous for all the people in the surrounding area. If accidentally leakage turns out, immediately it is required to impede. In that situation, it is very difficult to analyze from which pipeline it is leaked. So in that case, UAVs borne RFID sensor can be used to detect and identify which gas is leaking and the exact location of the leakage. Then, the workers can easily turn off the valve of the required pipe. By staying far from the factory, workers can know all the information. This is safeguarding the life of worker also as inhaling harmful gases may lead to death or any other chronic disease which may later destroy their life (Fig. 4).

Fig. 3 Monitoring wildlife in zoo using Io(UAV) and LoRa technology

Fig. 4 Leakage localization using Io(UAV) and LoRa technology

5 Open Challenges

There are many challenges which call for the solution. The following challenges need to address in future research.

- Most challenging issue in Io(UAV) is the connectivity wirelessly. For certain applications like road accidents and terrorist attack where the time plays a significant role, transmission of information with low latency is required. But if more than one UAV is placed, then packets might collide at the receiver due to network congestion. So in order to provide a high-quality data, UAV needs to wait until the network is free.
- In multi-UAV, RFID will be placed. Then, after sensing the data while transmission, there might be collision at the base station (reader). So an anti-collision algorithm is to be established, so that collision might affect the quality of service.
- The major concern in today's world is security. When the data acquired by the sensors is to be sent to the base station, there is a possibility that any intercept drone may try to hack the data and transmit incorrect data to the base station. For this problem, proper authentication of the sensor data is to be done.
- Data rate of LoRa technology is 27 kbps which is very low and insufficient to stream any image or video. To send the image, there is an experiment performed in [27], where the image captured by the sensor is converted into hexadecimal data, and this data is transmitted in 84-channel packets, but it may be used only for the scenario where delay is allowed.

6 Conclusion

In this paper, journey of wireless sensor network is discussed briefly. Then, how the Io(UAV) collects and delivers the data using RFID sensors to the users along with the communication technology used so far is reviewed. Few applications where Io(UAV) can utilize emerging LoRa technology is recommended. Then, finally, the possible future research work is pointed out.

References

1. Shakhatreh H, Sawalmeh AH, Al-Fuqaha A, Dou Z, Almaita E, Khalil I, Othman NS, Khreishah A, Guizani M (2019) Unmanned aerial vehicles (UAVs): a survey on civil applications and key research challenges. IEEE Access 7:48572–48634
2. Bein D, Bein W, Karki A, Madan B (2019) Optimizing border patrol operations using unmanned aerial vehicles. In: International conference on information technology—new generations
3. Tokekar P, Hook J, Mulla D, Isler V (2016) Sensor planning for a symbiotic UAV and UGV system for precision agriculture. IEEE Trans Rob 32
4. Panda K, Das S, Sen D, Arif W (2019) Design and deployment of UAV-aided post-disaster emergency network. IEEE Access 7

5. Zhao N, Lu W, Sheng M, Chen Y, Tang J, Richard F, Wong K (2019) UAV-assisted emergency networks in disasters. IEEE Wirel Commun 26
6. Nejjari F, Benhlima L, Bah S (2016) Event traffic detection using heterogenous wireless sensors network. In: International conference of computer systems and applications (AICCSA)
7. Moraes R, Freitas E (2020) Multi-UAV based crowd monitoring system. IEEE Trans Aerospace Electron Syst 56
8. Pi H, Li X, Yuan C, Yang Z, Wei L, Lian Z (2020) Application of multi-rotor UAV patrol system in safety and quality management of power grid construction projects. In: International conference on artificial intelligence and electromechanical automation (AIEA)
9. Ortiz S, Calafate CT, Cano J, Manzoni PT, Toh CK (2019) A UAV-based content delivery architecture for rural areas and future smart cities. IEEE Internet Comput 23
10. Zhang Q, Jiang M, Feng Z, Li W, Zhang W, Pan M (2019) IoT enabled UAV: network architecture and routing algorithm. IEEE Internet Things J 6
11. Buurman B, Kamruzzaman J, Karmakar G, Islam S (2020) Low-power wide-area networks: design goals, architecture, suitability to use cases and research challenges. IEEE Access 8
12. Khan I, Belqasmi F, Glitho R, Crespi N, Morrow M, Polakos P (2016) Wireless sensor network virtualization: a survey. IEEE Commun Surv Tutorials 18
13. Connected living: mobilising the Internet of Things. Connected Living, June 2014
14. Irwin JD (1997) The industrial electronics handbook. CRC Press, Boca Raton, FL, USA
15. Landt J (2005) The history of RFID. IEEE Potentials 4:8–11
16. Jo M, Yong H, Cha SH, Choo H (2009) Mobile RFID tag detection influence factors and prediction of tag detectability. IEEE Sens J 9
17. Hubbard B, Wang H, Leasure M (2015) Feasibility study of UAV use for RFID material tracking on construction sites. In: Proceedings of 51st ASC annual international conference, College Station, TX, USA
18. Choi JS, Son B, Kang H, Lee D (2012) Indoor localization of unmanned aerial vehicle based on passive UHF RFID systems. In: Proceedings of 9th international conference on ubiquitous robots ambient intelligence (URAI), pp 188–189
19. Greco G, Lucianaz C, Bertoldo S, Allegretti M (2015) A solution for monitoring operations in harsh environment: a RFID reader for small UAV. In: Proceedings of international conference on electromagnetics in advanced applications (ICEAA), Turin, Italy, pp 859–862
20. Motlagh N, Taleb T, Arouk O (2016) Low-altitude unmanned aerial vehicles-based Internet of Things Services: comprehensive survey and future perspectives. IEEE Internet Things J 3
21. Popli S, Jha RK, Jain S (2018) A survey on energy efficient Narrowband Internet of Things (NBIoT): architecture, application and challenges. IEEE Access
22. Zhou M, Lin J, Liang S, Du W, Cheng L (2017) A UAV patrol system based on Bluetooth localization. In: Asia-Pacific conference on intelligent robot systems (ACIRS)
23. Zhou Q, Wang L, Yu P, Huang T, Zhou M (2009) Unmanned patrol system based on Kalman filter and Zigbee positioning technology. J Physics Conf 1168
24. Delafontaine V, Schiano F, Cocco G, Rusu A, Floreano D (2020) Drone-aided localization in LoRa IoT network. In: IEEE international conference on robotics and automation (ICRA)
25. Choi C, Baccelli F, Veciana G (2020) Modeling and analysis of data harvesting architecture based on unmanned aerial vehicles. IEEE Trans Wirel Commun 19
26. Caillout C, Giroire F, Razafindralanbo T (2019) Efficient data collection and tracking with flying drones. AdHoc Netw 89:35–46
27. Jebril A, Sali A, Ismail A, Rasid M (2018) Overcoming limitations of LoRa physical layer in image transmission. Sensor
28. Chen E, Chen J, Mohamed A, Wang B, Wang Z, Chen Y (2020) Swarm intelligence application to UAV aided IoT data acquisition deployment optimization. IEEE Access 8
29. Arafat Y, Moh SM (2019) Localization and clustering based on swarm intelligence in UAV networks for emergency communications. IEEE Internet Things J 6

30. Tuna, Nefzi B, Conte G (2014) Unmanned aerial vehicle-aided communications system for disaster recovery. J Netw Comput Appl 41:27–36
31. Hayat S, Yanmaz E, Muzaffar R (2016) Survey on unmanned aerial vehicle networks for civil applications: a communications viewpoint. IEEE Commun Surv Tutorials 18(4):2624–2661

A Compact ZOR Antenna with Defective Ground for Wireless Data Transmission and Short-Range Radar Applications

Komal Roy and Rashmi Sinha

Abstract A compact ZOR antenna with omni-directional radiation pattern is sketched and implemented in this paper. The design consists of S-shaped patch with defective ground. The patch is fed with direct contact feed line. The −10 dB bandwidth achieved is 210 MHz ranging from 7.20 to 7.41 GHz operating at central frequency of 7.3 GHz. The overall dimension of the antenna is 5 mm × 5 mm × 1.6 mm ($0.125\lambda_0 \times 0.125\lambda_0 \times 0.04\lambda_0$). The proposed antenna is fabricated on FR-4 substrate, having pattern which is dipole in XZ-plane and omni-directional in YZ-plane with return loss of −23 dB. The current distribution is uniform throughout which confirms the ZOR phenomenon. The antenna is measured and tested with the help of vector network analyzer (VNA), and results are almost similar to each other. At last antenna is compared with already reported ZOR antennas. The proposed antenna has excellent radiation characteristics, ease of fabrication, and find applications for wireless data transmission and short-range radar applications.

Keywords Omni-directional · Radiation pattern · Bandwidth

1 Introduction

The demands of wireless communication are increasing very rapidly from the last few decades. As the demand are increasing in the same promotional, the size of the devices is also becoming smaller and smaller [1]. To reduce the size of a device, the components associated with it must be reduce in such a way so that their performance should be maintained. The antennas associated till date with wireless devices are not much compact, and as the size of antenna is reduced, their performance gets affected [2].

To overcome the above problem, many concepts are provided by the researchers, and concept based on metamaterial is one of the best solutions which attracted

K. Roy (✉) · R. Sinha
National Institute of Technology, Jamshedpur, Jharkhand, India

R. Sinha
e-mail: rsinha.ece@nitjsr.ac.in

K. Dahal et al. (eds.), *Internet of Things and Its Applications*, Lecture Notes in Electrical Engineering 825, https://doi.org/10.1007/978-981-16-7637-6_34

researchers [3]. Metamaterial is a mannered designed structure which have simultaneously negative or zero permittivity and permeability for a particular or a range of frequency [4]. Due to these unique properties, metamaterial helps backward waves [5], inverse Snell's law [6], negative index of refraction [7], zero propagation [8], anti-parallel group and phase velocities [9], etc. Considering the above artificial properties, metamaterial devices are gaining their impact over conventional devices such as cloaking [10], superlense [11], resonators [12], sensors [13], absorbers [14], cross-polariser [15], and antennas [16].

Different antennas such as ultra-wideband metamaterial antennas [17], fractal metamaterial antennas [18], metamaterial patch antennas [19], and ZOR antennas [20] are proposed in the past to enhance the performance.

Caloz et al. in 2002 use the concept of metamaterial in transmission line [21]. Metamaterial which supports backwards waves are also termed as also termed as left-handed material (LHM) but they are lossy in nature and antenna design using LHM possesses narrow bandwidth and hard to implement. The LH-TL supports the backward wave but physical existence of LH-TL is not possible due to the presence of practise elements. To overcome this, caloz et al. combine LH and RH material in such a way that LH-TL operates in low frequency and RH-TL operates in high frequency and they termed this TL CRLH-TL [21]. The structure which uses the concept of CRLH-TL resonant in omni-directional radiation pattern and bandwidth achieved is larger as compared to LH-TL structure.

The concept of CRLH-TL can be used for different applications such as ZOR filters [22], ZOR antennas [23]. ZOR stands for zeroth-order resonator; this term is used because at one particular frequency the permittivity and permeability simultaneously become equal to zero.

This paper presents a compact ZOR antenna with omni-directional radiation pattern, which consists of S-shaped patch with defective ground. The patch is feeded with direct contact feed line. The proposed antenna uses the properties of CRLH-TL to achieve the omni-directional radiation pattern. The −10 dB bandwidth achieved is 210 MHz ranging from 7.20 to 7.41 GHz operating at central frequency of 7.3 GHz. The overall dimension of the antenna is 5 mm × 5 mm × 1.6 mm ($0.125\lambda_0 \times 0.125\lambda_0 \times 0.04\lambda_0$). The proposed antenna is fabricated on FR-4 substrate, having pattern which is dipole in XZ-plane and omni-directional in YZ-plane with return loss of −23 dB. The current distribution is uniform throughout which confirms the ZOR phenomenon. The simulation result is carried out using ANSYS HFSS 19.1. The proposed antenna has excellent radiation characteristics, ease of fabrication, and find applications for wireless data transmission and short-range radar applications.

2 Omni-Directional Antenna

The omni-directional antenna is also termed as ZOR antenna, and it is one of the applications of CRLH-TL which supports an infinite wavelength at a finite microwave frequency resulting in uniform distribution of current or voltage in the entire TL structure [24]. CRLH-TL is a combination of series impedance and parallel admittance. Occurrence of impedance is due to the effect of L_R and C_L. Similarly, occurrence of admittance is due to the effect of L_L and C_R [25]. Figure 1 depicts the equivalent circuit of CRLH-TL.

Series impedance and parallel admittance can be calculated as [26]

$$\begin{aligned} Z &= j(L_R - 1/C_L) \\ Y &= j(C_R - 1/L_L) \end{aligned} \tag{1}$$

From Eq. (1), propagation constant is defined as [26]

$$\gamma = \alpha + j\beta = \sqrt{ZY} \tag{2}$$

where α represents attenuation constant and β represents phase constant. β (phase constant) can be one of the two, purely real or purely imaginary. When β is purely real, wave propagation is acting as a pass band ($\gamma = j\beta$) and when β is purely imaginary, wave propagation is acting as a stop band ($\gamma = \alpha$) [27]. The presence of stop band is a differentiating characteristic of CRLH-TL, and it is neither present in RH-TL nor in LH-TL [27].

In general, two resonance conditions occur in CRLH-TL, i.e., series resonance (ω_{se}) and parallel resonance (ω_{sh}). When series resonance and parallel resonance of the CRLH-TL are not equal, then it is referred as unbalanced CRLH-TL ($\omega_{se} \neq \omega_{sh}$) whereas when both are equal it referred as balanced CRLH-TL ($\omega_{se} \neq \omega_{sh}$) [28]. The series and parallel resonance can be defined as [29]

$$\omega_{se} = \frac{1}{\sqrt{L_R C_L}} \text{and } \omega_{sh} = \frac{1}{\sqrt{L_L C_R}} \tag{3}$$

By applying Bloch-Floquet theory to the equivalent circuit, dispersion equation for CRLH-TL can be obtained as [30]:

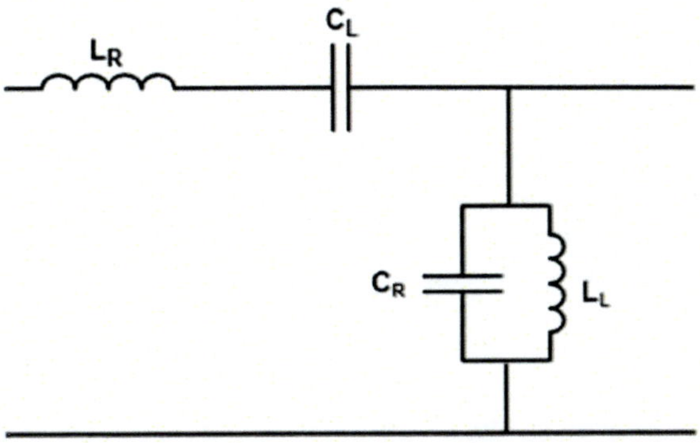

Fig. 1 CRLH-TL equivalent circuit

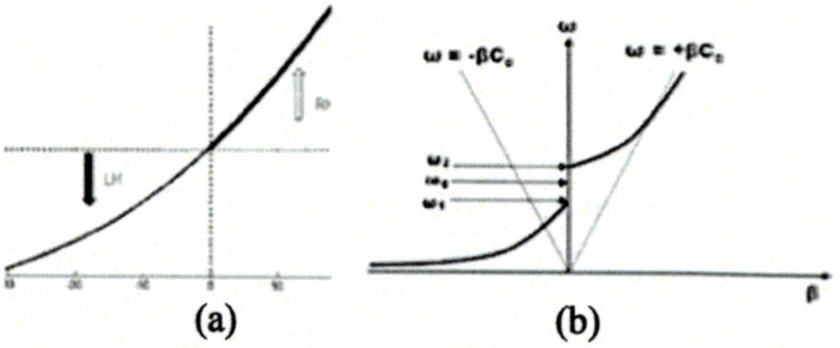

Fig. 2 **a** Balanced CRLH-TL and **b** unbalanced CRLH-TL

$$\cos\beta l = 1 - \frac{1}{2}\left(\frac{\omega_L^2}{\omega^2} + \frac{\omega^2}{\omega_R^2} - \left|\frac{\omega_{\text{sh}}^2}{\omega_R^2} + \frac{\omega_{\text{se}}^2}{\omega_R^2}\right|\right) \tag{4}$$

where $\omega_R = 1/\sqrt{C_R, L_R}$, $\omega_L = 1/\sqrt{C_L, L_L}$, $\omega_{\text{se}} = 1/\sqrt{C_L, L_R}$, $\omega_{\text{sh}} = 1/\sqrt{C_R, L_L}$, d = length of unit cell.

The dispersion relation calculated from Eq. (4) is portrayed in Fig. 2a, b. Figure 2a shows the balanced CRLH-TL which includes cut-off region, LH region, and RH region, and Fig. 2b explains unbalanced CRLH-TL which includes cut-off region, LH region, RH region, and band stop region [31].

3 The Proposed Omni-Directional Antenna Design

The proposed S-shaped ZOR antenna is designed on FR-4 substrate with defective ground, having a thickness of 1.6 mm which is sandwiched between the conducting materials, i.e., copper having thickness of 0.035 mm. The top is patch and bottom is a ground plane. The overall dimension of the antenna is 5 mm × 5 mm, which is very compact. The proposed S-shaped ZOR antenna is fed by direct feeded line. The top view and bottom view of antenna are portrayed in Fig. 3a, b, respectively.

4 Result and Analysis

The proposed S-shaped antenna is fabricated on FR4 material with overall dimension of 5 mm × 5 mm × 1.6 mm (0.125 λ_0 × 0.125 λ_0 × 0.04 λ_0) and is simulated at operating frequency of 7.3 GHz. The simulated return loss is portrayed in Fig. 4. It can also be noticed that the proposed antenna has 210 MHz bandwidth range from 7.20 to 7.41 GHz.

The voltage standing wave ratio (VSWR) comes to be 1.43 as evident from portrayed Fig. 5, which is less than 2 in the operating frequency. This indicates that more than 90% of power is reflected from the antenna.

Figure 6 shows the HFSS simulated dispersion curve for the proposed S-shaped ZOR antenna. The dispersion relation of the proposed structure was calculated by the

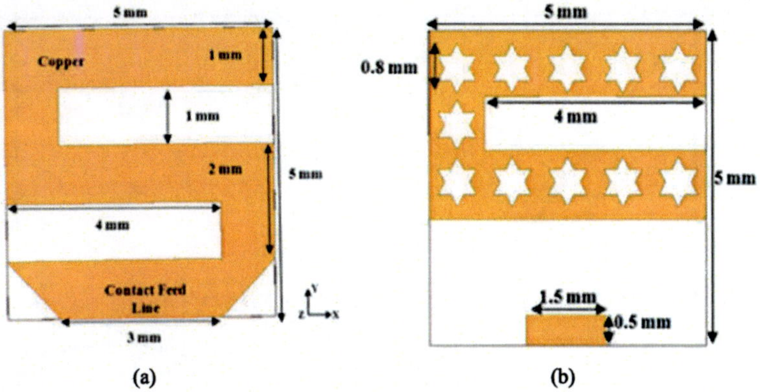

Fig. 3 **a** Top view and **b** bottom view

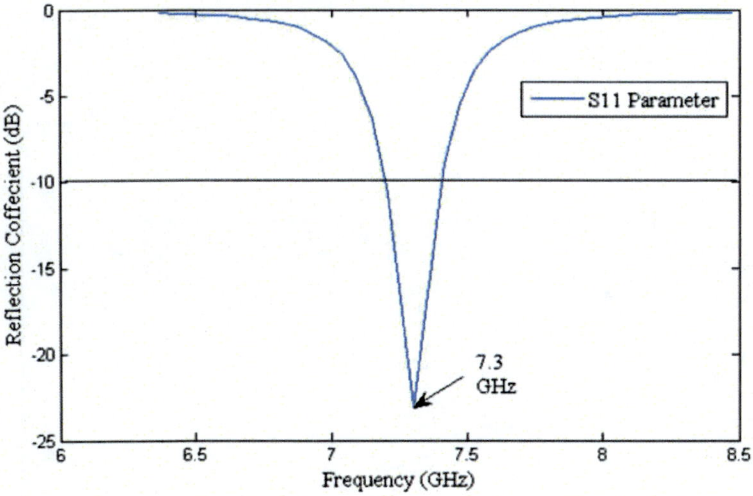

Fig. 4 Reflection coefficient of proposed antenna

simulated S-parameters in HFSS where the periodic boundary condition was applied on both the sides of proposed structure.

Simulated radiation patterns of electric plane and the magnetic plane of the proposed ZOR antenna at 7.3 GHz are portrayed in Fig. 7a, b, respectively. Results indicate that it offers dipole in XZ-plane and omni-directional in YZ-plane with return loss of −23 dB.

Figure 8 shows a 3D plot of the power pattern expressed in decibels. Careful observation of the figures reveals that power is radiated equally in all directions confirming the omni-direction radiation pattern of the ZOR antenna.

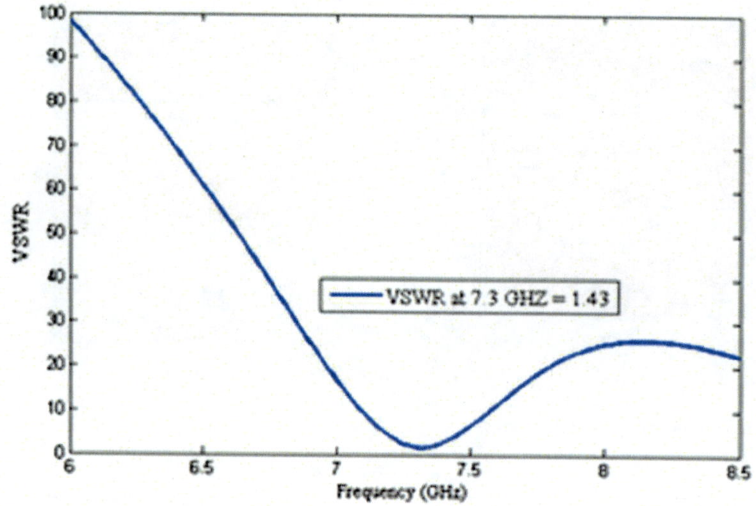

Fig. 5 VSWR of proposed antenna

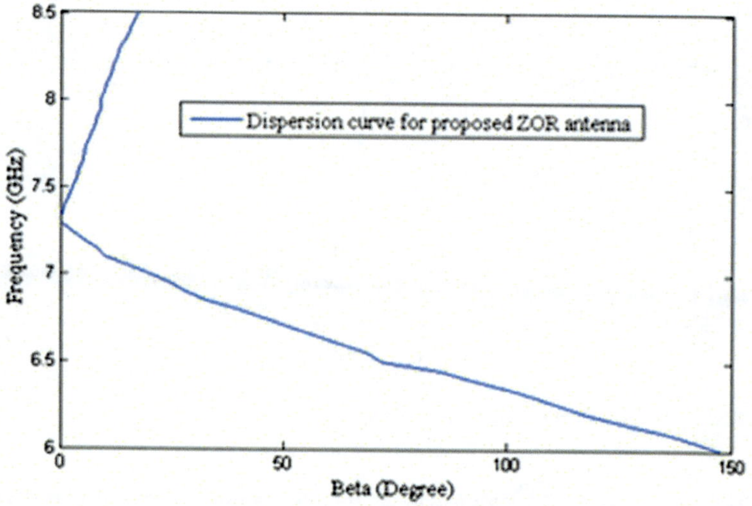

Fig. 6 Dispersion curve for proposed ZOR antenna

Figure 9 represents the distributions of current field at ZOR frequency. It can be concluded that all field lines are uniform and in phase throughout the structure, which indicates that $\beta = 0$ at ZOR frequency.

The major advantage of this antenna is that it provides a reasonable gain and peak gain as it offers 1.99 dB gain and 2.32 dB peak gain at 7.3 GHz in the bore sight direction. In general, metamaterial structures suffer from the low gain or negative

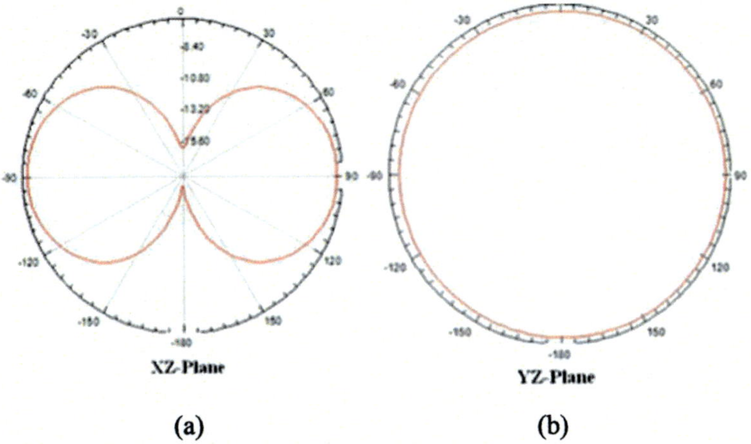

Fig. 7 **a** Dipole pattern and **b** omni-directional pattern

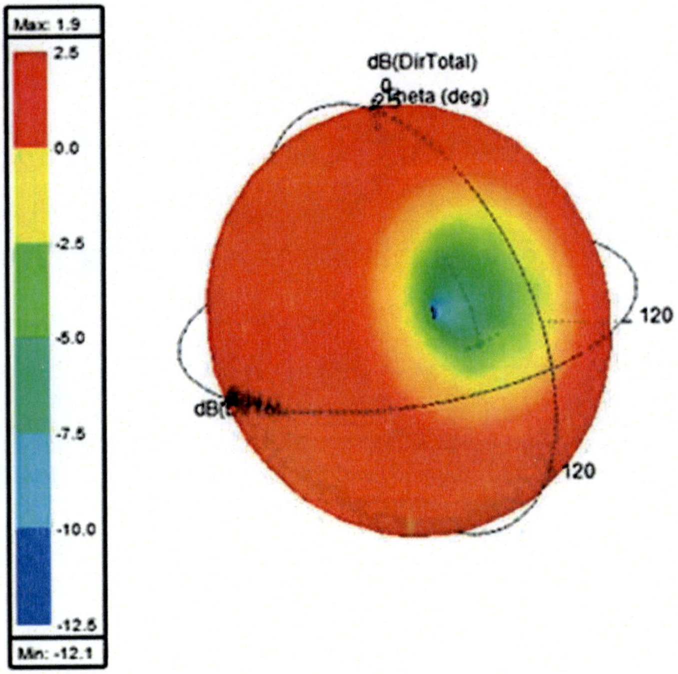

Fig. 8 3D polar plot with omni-directional pattern

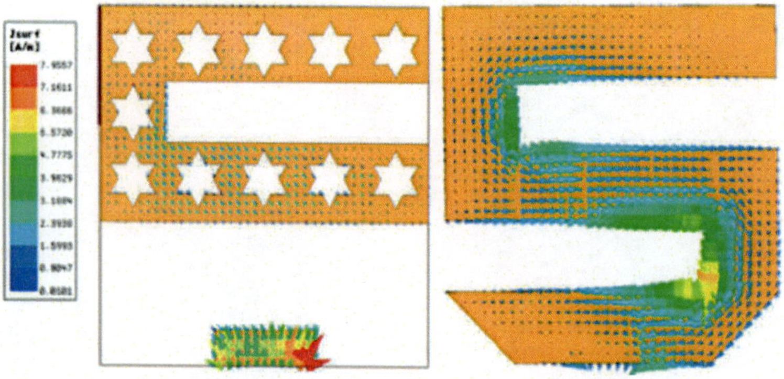

(a) (b)

Fig. 9 Current distribution **a** bottom view, **b** top view

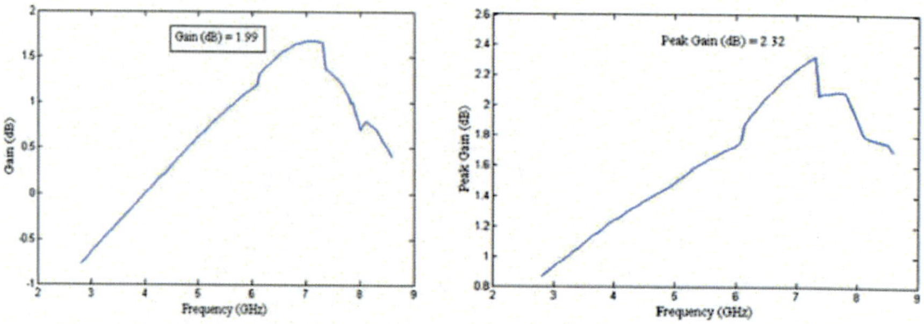

(a) (b)

Fig. 10 **a** Gain (dB), **b** peak gain (dB)

gain problem. It is evident that at ZOR frequencies fields are tightly bounded on the surface of the structure due to high level of miniaturization. Figure 10a, b shows the gain and peak gain results of the proposed antenna, respectively.

5 Experimental Setup

Reflection coefficient is measured with the help of vector network analyzer (VNA). Figure 11 shows the experimental setup for measurement of antenna.

The simulated and the measured result of S_{11} parameter is shown in Fig. 12. It is observed that an experimental and simulated result approximately matches within the tolerance limit. Comparison between the measured and simulated result is done in Table 1.

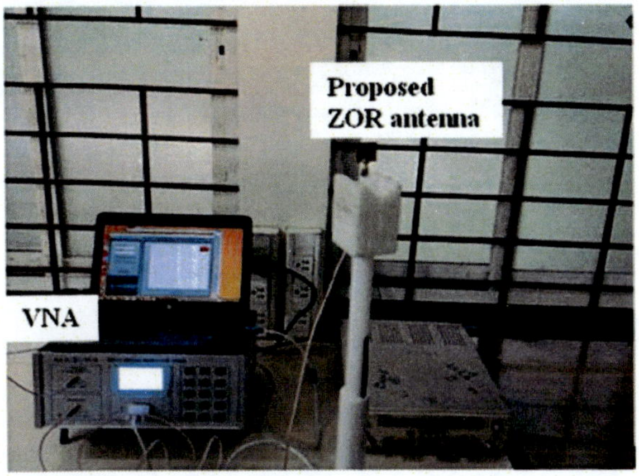

Fig. 11 Experimental setup

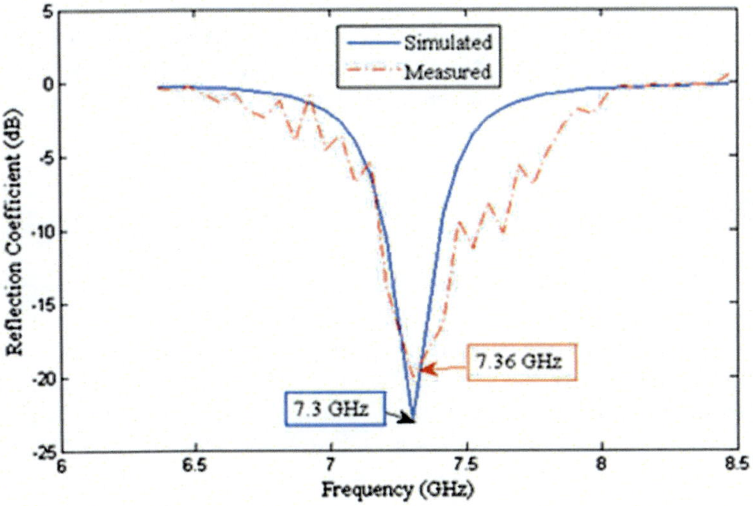

Fig. 12 Simulated and measured result S11 parameter

Table 1 Measured and simulated result of proposed ZOR antenna

Measured			Simulated		
f_0 (GHz)	S_{11} (dB)	−10 dB bandwidth (MHz)	f_0 (GHz)	S_{11} (dB)	−10 dB bandwidth (MHz)
7.3	−23	210	7.36	−20	330

Table 2 Comparison from different papers

Ref. No	f_0 (GHz)	Total size (mm) (λ_0)	Bandwidth (MHz)	Peak gain (dBi)	Thickness (λ_0)
[32]	9.8	3.5 × 7 (0.11 × 0.23)	N/A	2.11	1.57 (0.05)
[33]	5	16.5 × 16.5 (0.28 × 0.28)	150	0.08	1.27 (0.02)
[34]	12	11.8 × 11.8 (0.47 × 0.47)	160	1.21	0.508 (0.020)
[35]	10.4	25 × 15 (0.89 × 0.53)	80	1.82	1.6 (0.057)
[36]	5.78	40 × 40 (0.70 ×0.70)	45	2.29	1.96 (0.038)
Proposed work	7.3	5 × 5 (0.12 × 0.12)	210	2.32	1.6 (0.04)

The results obtained by proposed ZOR antenna are analyzed with antennas as reported in Table 2. It is observed that 210 MHz bandwidth is achieved in the proposed antenna which is much higher as compared to other published results.

6 Conclusion

A compact ZOR antenna with omni-directional radiation pattern is designed and developed in this paper. The −10 dB bandwidth achieved is 210 MHz ranging from 7.20 to 7.41 GHz operating at central frequency of 7.3 GHz. The design consists of S-shaped patch with defective ground. The patch is feed with direct contact feed line. The overall dimension of the antenna is 5 mm × 5 mm × 1.6 mm. The proposed antenna is fabricated on FR-4 substrate, having dipole like pattern in xz-plane and omni-directional pattern in yz-plane with cross-polarization level of −23 dB. The current distribution is uniform throughout which confirms the ZOR phenomenon. The antenna is measured and tested with the help of vector network analyzer (VNA). The measured and simulated results are almost similar to each other. The bandwidth achieved is much higher than the reported results till now. The proposed antenna has excellent radiation characteristics, ease of fabrication, and find applications for wireless data transmission and short-range radar applications.

References

1. Yang H-L et al (2016) A dual-band low-profile metasurface-enabled wearable antenna for WLAN devices. Prog Electromagn Res 61:115–125

2. Khanjari, Poorgholam S, Jarchi S, Mohammad-Taheri M (2019) Compact and wideband planar loop antenna with microstrip to parallel strip balun feed using metamaterials. AEU Int J Electron Commun 111:152883
3. Rajesh kumar V, Raghavan S (2015) A compact metamaterial inspired triple band antenna for reconfigurable WLAN/WiMAX applications. AEU Int J Electron Commun 69(1):274–280
4. Lapine M, Tretyakov S (2007) Contemporary notes on metamaterials. IET Microwaves Antennas Propag 1(1):3–11
5. Barde C, Choubey A, Sinha R (2019) Wide band metamaterial absorber for Ku and K band applications. J Appl Phys 126(17):175104
6. Xi S et al (2009) Experimental verification of reversed Cherenkov radiation in left handed metamaterial. Phys Rev Lett 103(19):194801
7. Shalaev VM et al (2005) Negative index of refraction in optical metamaterials. Miniaturized Hexagonal Shaped Omni-directional ZOR Antenna 19. Opt Lett 30(24):3356–3358
8. Capolino F (2009) Theory and phenomena of metamaterials. CRC press
9. Barde C et al (2020) A compact wideband metamaterial absorber for Ku band applications. J Mater Sci Mater Electron 31(19):16898–16906
10. Zhang J, Mortensen NA (2011) Ultrathin cylindrical cloak. Prog Electromagn Res 121:381–389
11. Fang N, Zhang X (2002) Imaging properties of a metamaterial superlens. In: Proceedings of the 2nd IEEE conference on nanotechnology. IEEE
12. Bonache J et al (2006) On the electrical characteristics of complementary metamaterial resonators. IEEE Microwave Wirel Components Lett 16(10):543–545
13. Wang W et al (2017) Ultrasensitive terahertz metamaterial sensor based on vertical split ring resonators. Photonics Res 5(6):571–577
14. Barde C, Choubey A, Sinha R (2019) A set square design metamaterial absorber for X-band applications. J Electromagn Waves Appl 1–14
15. Chin JY, Lu M, Cui TJ (2008) Metamaterial polarizers by electric field coupled resonators. Appl Phys Lett 93(25):251903
16. Barde C et al (2019) A low profile pentagonal shape zeroth order resonator antenna for K_a band applications. In: 2019 photonics and electromagnetics research symposium-spring (PIERS-Spring). IEEE
17. Heydari S et al (2017) Dual band monopole antenna based on metamaterial structure with narrowband and UWB resonances with reconfigurable quality. AEU Int J Electron Commun 81:92–98
18. Mansouri Z et al (2016) Dual notch UWB fork monopole antenna with CRLH metamaterial load. Prog Electromagn Res 65:111–119
19. Dong Y, Toyao H, Itoh T (2011) Compact circularly-polarized patch antenna loaded with metamaterial structures. IEEE Trans Antennas Propag 59(11):4329–4333
20. Mishra N, Chaudhary RK (2016) A miniaturized ZOR antenna with enhanced bandwidth for WiMAX applications. Microwave Opt Technol Lett 58(1):71–75
21. Lai A, Itoh T, Caloz C (2004) Composite right/left-handed transmission line metamaterials. IEEE Microwave Mag 5(3):34–50
22. Jang G, Kahng S (2011) Design of a metamaterial band-pass filter using the zor of a modified circular mushroom structure. Microwave J 54(5)
23. Barde C et al (2019) A novel ZOR-inspired patch antenna for vehicle mounting application. In: Ambient communications and computer systems. Springer, Singapore, pp 47–53
24. Lee J-G, Lee J-H (2007) Zeroth order resonance loop antenna. IEEE Trans Antennas Propag 55(3):994–997
25. Sanada A et al (2004) A planar zeroth-order resonator antenna using a left-handed transmission line. In: 34th European microwave conference, vol 3. IEEE
26. Chiu S-C, Lai C-P, Chen S-Y (2012) Compact CRLH CPW antennas using novel termination circuits for dual-band operation at zeroth-order series and shunt resonances. IEEE Trans Antennas Propag 61(3):1071–1080
27. Nuthakki, Rajasekhar V, Dhamodharan SK (2018) Bandwidth enhancement of ZOR antenna by loading novel via-Less CRLH-TL unit cells. AEU Int J Electron Commun 83:501–511

28. Lee H, Woo D-J, Nam S (2015) Compact and bandwidth-enhanced asymmetric coplanar waveguide (ACPW) antenna using CRLH-TL and modified ground plane. IEEE Antennas Wirel Propag Lett 15:810–813
29. Kang H, Lim S (2013) Electric and magnetic mode-switchable dual antenna for null compensation. IEEE Antennas and Wirel Propag Lett 12. Miniaturized Hexagonal Shaped Omni-directional ZOR Antenna 21:300–303
30. Jang T, Choi J, Lim S (2010) Compact coplanar waveguide (CPW)-fed zeroth-order resonant antennas with extended bandwidth and high efficiency on vialess single layer. IEEE Trans Antennas Propag 59(2):363–372
31. Wang W, Xing M, Guo X (2016) A novel design of quad-band combination of circularly polarized microstrip antenna. In: 2016 progress in electromagnetic research symposium (PIERS). IEEE
32. Park J-H et al (2007) Epsilon negative zeroth-order resonator antenna. IEEE Trans Antennas Propag 55(12):3710–3712
33. Pyo S et al (2009) A slot-loaded composite right/left-handed transmission line for a zeroth-order resonant antenna with improved efficiency. IEEE Trans Microwave Theor Tech 57(11):2775–2782
34. Giuppi F et al (2012) A compact, single-layer substrate integrated waveguide (SIW) cavity-backed active antenna oscillator. IEEE Antennas Wirel Propag Lett 11:431–433
35. Rajasekhar NV, Kumar DS (2018) Metamaterial-based compact UWB planar monopole antennas. Microwave Opt Technol Lett 60(6):1332–1338
36. Boukarkar A et al (2016) Miniaturized single-feed multiband patch antennas. IEEE Trans Antennas Propag 65(2):850–854

Information Security

Intrusion Detection System Performance Comparison Using Dimensionality Reduction Techniques

Angana Chakraborty and Subhankar Joardar

Abstract Ineffective and huge network data traffic hampers the efficiency of the intrusion detection system (IDS). Removing the irrelevant features from the high-dimensional dataset is very challenging due to its intrinsic dimension. So, we have implemented two feature reduction techniques, i.e., principal component analysis (PCA), linear discriminant analysis (LDA). Feature reduction techniques are also called dimensionality reduction techniques where dimensions of input features are reduced. Dimensionality reduction techniques increase the accuracy and speed of machine learning classifiers. We have compared the performance of PCA and LDA concerning accuracy, training time, and testing time by using Naïve Bayes and support vector machine (SVM) classifiers. LDA outperforms PCA for binary, multi-class classification of intrusion detection system for both the classifiers individually. SVM provides better classification accuracy than Naïve Bayes for PCA and LDA both.

Keywords Intrusion detection system (IDS) · Dimensionality reduction technique · Principal component analysis (PCA) · Linear discriminant analysis (LDA) · Naïve Bayes · SVM

1 Introduction

Cybersecurity falls under the trending research areas in the domain of information and communication technology. Intrusion detection system (IDS) belongs to one of the demanding research tools for providing cybersecurity. The main challenge of IDS is to obtain high accuracy rate, low false alarm rate, and reduced detection time. Various machine learning algorithms (i.e., decision tree (DT), random forest (RF), Naïve Bayes (NB), support vector machine (SVM)) are developed to improve the performance of intrusion detection system. Naïve Bayes and SVM are one of the powerful supervised machine learning approaches. These approaches cannot ensure high accuracy and reduced detection time due to a large number of irrelevant input

A. Chakraborty (✉) · S. Joardar
Department of Computer Science and Engineering, Haldia Institute of Technology, Haldia, India

K. Dahal et al. (eds.), *Internet of Things and Its Applications*, Lecture Notes in Electrical Engineering 825, https://doi.org/10.1007/978-981-16-7637-6_35

features in a dataset. Redundant and unnecessary features degrade the performance of IDS as well as it makes the dataset very complex. So, extracting important and relevant input features is an important part of IDS. Feature extraction is also termed as dimensionality reduction technique where the dimension of input features is reduced. In this technique, input features small subset is selected to focus on target labels. So, dimensionality reduction technique deducts the cost of counting unselected features [1]. Execution speed of machine learning algorithms becomes fast and memory requirement is also reduced by feature extraction technique [2]. There are several popular dimensionality reduction techniques used by researchers. These techniques are classified as linear technique and nonlinear technique. Linear dimensionality reduction techniques are principal component analysis (PCA), linear discriminant analysis (LDA), etc. In our research, we have compared the performance of PCA and LDA using Naïve Bayes and SVM classifiers. We have implemented binary and multi-class classification of Kyoto 2006+ network traffic dataset by applying PCA and LDA individually. Performance analysis of two classifiers (Naïve Bayes, SVM) is also done in this paper.

The paper is organized as follows: Sect. 2 represents literature survey of different dimensionality reduction techniques. Proposed methodology is explained in Sect. 3. Section 4 gives an experimental result analysis. Work is concluded in Sect. 5.

2 Literature Survey

We have surveyed different research papers on dimensionality reduction techniques. The literature survey is presented as follows:

Aburomman et al. [1] have proposed a combination of binary SVM classifiers on KDD99 dataset using PCA, LDA feature extraction techniques. Their proposed combined PCA-LDA-SVM method provides the highest accuracy rate of 0.92162 than PCA-SVM (0.8902) and LDA-SVM (0.8993). Authors have implemented KDD99 dataset which cannot handle the current network situation properly.

Singh et al. [3] have built an efficient feature reduction technique generalized discriminant analysis (GDA) for multi-class classification of KDDCup99 dataset. They have performance comparison (detection rate, false alarm rate, training, and testing time) of GDA technique with PCA technique using self-organizing map (SOM) and C4.5 decision tree classifiers. GDA outperforms PCA concerning detection rate, and GDA gives lower false alarm rate, reduced training, and testing time than PCA for both the classifiers. Researchers have also observed that C4.5 classifier provides better detection rate and reduced training and testing time than SOM for all the classes (Normal, DOS, R2L, U2R, Prob). Authors have used KDDCup99 dataset which is an old dataset and cannot handle the recent network situation properly.

Datti et al. [4] have compared the performance of two feature reduction techniques (PCA, LDA) for multi-class classification of the NSL-KDD dataset. They have trained the feed-forward backpropagation neural network algorithm. Researchers have chosen one small dataset and one large dataset to investigate the efficiency of

PCA and LDA. They have observed that PCA outperforms LDA concerning accuracy and PCA gives reduced training time (45609 ms) than LDA for small dataset. But, LDA gives higher accuracy and reduced training time (72094 ms) than PCA for large dataset. Authors have not considered one single dataset to analyze the performance of PCA, LDA.

Murugesan et al. [5] have made a performance analysis of three feature reduction techniques like PCA, LDA, and independent component analysis (ICA) using SVM for binary classification of a simulated dataset. PCA outperforms LDA and ICA in terms of accuracy of 94%. Authors have not implemented multi-class classification of the simulated dataset, and they have not used any real-time traffic data for simulation.

Ikram et al. [6] have proposed a hybrid intrusion detection model by combining PCA, RBF kernel-based SVM. They have used NSL-KDD dataset and Gurekddcup dataset for experiment. Authors have obtained an accuracy of 99%, correlation coefficient of 0.99 after reducing the dimension of input features using both two datasets. Researchers have not tested other kernel parameters of SVM.

Shen et al. [7] have developed an efficient intrusion detection method by integrating PCA, LDA, and Naïve Bayesian classification (BC) algorithm for binary classification of CICIDS2017 dataset. They have compared the accuracy of their proposed classifier (PCA-LDA-BC) with other classifiers (logistic regression, SVM, SVM (kernel function)). Proposed classifier gives highest accuracy of 96%. Researchers have not considered multi-class classification of the dataset, and they have not used any method to handle more complex multilateral network environment.

Ammar [8] has measured the classification performance (accuracy, false positive rate, and false negative rate), execution time of a heuristic for variable selection (HVS) feature selection technique, and traditional PCA for binomial classification of Kyoto 2006+ dataset under 13 different testing scenarios. Author has trained the dataset using feed-forward neural network algorithm. Researcher has obtained higher accuracy and higher false positive rate for all testing scenarios. But, scenario 1b, 1c, 5b give reduced execution time than other scenarios. Author has not implemented multi-class classification of Kyoto 2006 + dataset to investigate the performance of HVS and PCA.

3 Proposed Methodology

In our research, we have proposed a method for intrusion detection using Kyoto 2006+ dataset. We have implemented binary, multi-class classification of the dataset using proposed methodology. The proposed intrusion detection framework is represented below in Fig. 1.

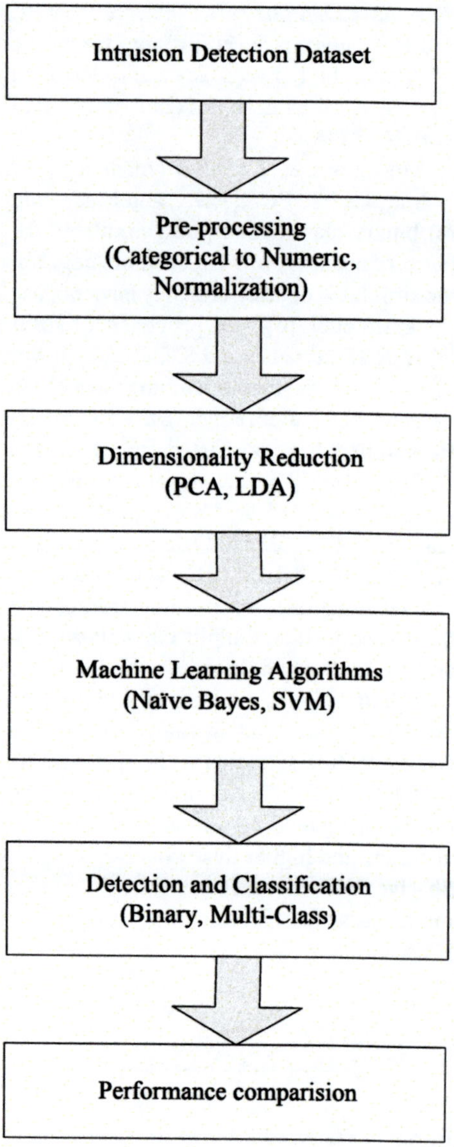

Fig. 1 A proposed Intrusion Detection framework

3.1 Dataset

We have simulated a real-time efficient dataset called Kyoto 2006+ dataset [9] which was based on real data traffic from November 2006 to August 2009. Among the 24 features of Kyoto 2006 + dataset, we have chosen all the14 features of KDD Cup 99 dataset [10]. We have also selected feature no 10 ('Protocol') [11] among

Table 1 Features selected for binary classification of Kyoto 2006+ dataset

Feature no	Name of the feature
1	Duration
2	Service
3	Source bytes
4	Destination bytes
5	Count
6	Same_srv_rate
7	Serror_rate
8	Srv_serror_rate
9	Dst_host_count
10	Dst_host_srv_count
11	Dst_host_same_src_port_rate
12	Dst_host_serror_rate
13	Dst_host_srv_serror_rate
14	Flag
Additional Feature (4)	Label
Additional Feature (10)	Protocol

the additional features of Kyoto 2006+ dataset for binary classification. Additional feature no 4 'Label' is selected to specify the types of intrusions (i) -1 for 'Attack' (ii) 1 for 'Normal'. These features are written below in Table 1.

In this paper, we have also implemented multi-class classification of Kyoto 2006 + dataset (Nobember, 2015). Out of the 24 features of Kyoto 2006 + dataset, we have investigated that additional feature no 1 'IDS_detection' represents IDS attack, additional feature no 2 'Malware_detection' represents malware attack, additional feature no 3 'Ashula_detection' represent shellcode andexploit codes attacks. We have classified total 5 intrusion types based on additional feature no 1, 2, 3. The intrusion types are categorized based on the following rules:

if (IDS_detection = '0' and Malware_detection = '0' and Ashula_detection = '0') then Intrusion_type = 'Normal'.
else if (IDS_detection != '0' and Malware_detection = '0' and Ashula_detection = '0') then Intrusion_type = 'IDS'.
else if (IDS_detection = '0' and Malware_detection != '0' and Ashula_detection = '0') then Intrusion_type = 'Malware'.
else if (IDS_detection = '0' and Malware_detection = '0' and Ashula_detection != '0') then Intrusion_type = 'Shellcode'.
else Intrusion_type = 'Unknown'.

3.2 *Preprocessing*

Data pre-processing converts raw dataset to clean dataset. It deals with missing values, removes duplicate values, transforms categorical data to numeric data and performs normalization of data.

- **Dealing with missing, duplicate values:** Dealing with missing values is an important part of pre-processing. We have checked that there are no missing values in our dataset and we have removed duplicate records from the dataset.
- **Transformation of categorical data into numeric data:** Among the selected features, we have checked that feature no 2('Service'), 14 ('Flag'), additional feature no 10 ('Protocol') contain categorical data. So, we have transformed those categorical data into numeric data. After that the dataset is split into two parts: (i) 80% for training (ii) 20% for testing. 247,254 numbers of training records and 61,814 numbers of testing records have been selected for binary classification. 207,725 numbers of training records and 51,932 numbers of testing records have been selected for multi-class classification. Binary, multi-class classification of the training, testing dataset have been shown in Tables 2 and 3, respectively.
- **Normalization:** Normalization or data scaling method is used to organize records or features in proper format. We have used min–max scaling for data normalization. In min–max scaling, all the features for training will be scaled within a fixed range from 0 to 1.

Min–max scaling is illustrated in the equation below:

$$X_{\text{norm}} = \frac{X - X_{\min}}{X_{\max} - X_{\min}} \tag{1}$$

Table 2 Binary classification of the training and testing dataset

Intrusion types	Number of records of the training dataset	Number of records of the testing dataset
Attack	228,815	57,250
Normal	18,439	4564

Table 3 Multi-class classification of the training and testing dataset

Intrusion types	Number of records of the training dataset	Number of records of the testing dataset
Normal	201,086	50,279
IDS	6304	1563
Malware	102	25
Shellcode	10	2
Unknown	223	63

In this equation, X_{norm} is the normalization result, X is the initial value before normalization, X_{max} denotes the maximum values of every feature; X_{min} denotes the minimum values of every feature.

3.3 *Dimensionality Reduction*

We have implemented principal component analysis (PCA), linear discriminant analysis (LDA) for dimensionality reduction we have analyzed their performances.

3.3.1 Principal Component Analysis (PCA)

PCA is also called as Karhunen–Loeve transform which falls under the most efficient dimensionality reduction techniques for statistical data analysis. It is used to find out similarities and differences between data [12]. PCA transforms large set of input features to a smaller set of input features (called principal component (PC)) while preserving the maximum information in the large set. For most of the datasets, first several PCs with large variances are chosen and the remaining PCs are omitted with a minimum variance loss for dimensionality reduction [2].

Suppose, there are K features in every datum and there are N data which is denoted by $y_{11}, y_{12}, y_{13} \ldots y_{1k}; y_{21}, y_{22}, y_{23} \ldots y_{2k}$. Similarly, the final datum can be represented by $y_{N1}, y_{N2}, y_{N3} \ldots y_{NK}$.

The matrix $A = [\Phi_1, \Phi_2 \ldots \Phi_K](N*K$ matrix).

Consider y_j, where $j = 1, 2 \ldots K$ are randomly selected data records. Mean $\overline{y}$ is represented by the following equation:

$$\overline{y} = \frac{1}{k}\sum_{j=1}^{k} y_j \quad (2)$$

The derivation from the mean is defined by the following equation:

$$\Phi_j = y_j - \overline{y} \quad (3)$$

The covariance matrix C of the dataset is represented as:

$$C = \sum_{j=1}^{K} \Phi_j \Phi_j^T \quad (4)$$

Eigenvalues ($\lambda_1, \lambda_2, \ldots, \lambda_N$), eigenvectors ($u_1, u_2, \ldots, u_N$) of the covariance matrix C are calculated. The L eigenvectors which have the highest eigenvalues are chosen. Subspace L dimensionality can be calculated by the equation below:

$$\frac{\sum_{j=1}^{L}\lambda_j}{\sum_{j=1}^{N}\lambda_j} > \text{Threshold value (e.g., 0.9 or 0.95)} \tag{5}$$

3.3.2 Linear Discriminant Analysis (LDA)

Linear discriminant analysis is also called as Fisher mapping which is one of the traditional and efficient dimensionality reduction techniques for statistical data analysis. Group of biased information is sorted out to the maximum by LDA at the time of dimensionality reduction and it identifies the directions [13]. It is a well-known strategy of recent times to detect network intrusions. LDA detects new, unknown, and pattern of intrusions [14].

The between-class scatter matrix S_b, the within-class scatter matrix S_w are represented by the equations below for all records of all intrusion types.

$$S_b = \sum_{m=1}^{C}(\mu_m - \mu)(\mu_m - \mu)^T \tag{6}$$

$$S_w = \sum_{m=1}^{C}\sum_{n=1}^{P_m}(x_n - \mu_m)(x_n - \mu_m)^T \tag{7}$$

P_m represents number of training records for each class m, C refers to the distinct classes numbers, μ_m represents mean vector of records that falls under class m, x_n denotes the records set that belongs to class m with x_n being the nth data of that particular class. S_w denotes the features scatter on the mean of each class, and S_b denotes the features scatter on the overall mean for all classes. The objective is to maximize S_b and to minimize S_w. So, we can say that the ratio $\frac{\det|S_b|}{\det|S_w|}$ is maximum when the projection matrix column vectors are the eigenvectors of $S_w^{-1}S_b$.

3.4 Classification

We have used machine learning algorithms like Naïve Bayes (NB), support vector machine (SVM) for binary and multi-class classification of the Kyoto 2006 + dataset after reducing the features by PCA, LDA individually. These algorithms are illustrated below:

3.4.1 Naïve Bayes (NB)

Naïve Bayes falls under the intelligent supervised machine learning algorithms that depend on the number of network connection records. Naive Bayes algorithm is divided into two categories: training and detection. The n-dimensional vectors $(A_1, A_2, ..., A_n)$ represent network connection records, here n denotes the n characteristic features of the network connection record [7]. Naïve Bayes depends on Bayes theorem where conditional probabilities are utilized. Data samples' probabilities are calculated by multiplication of all attribute's conditional probabilities. Prediction is evaluated by considering probabilities of every class sample and by choosing the class value with highest probability [15].

$$P(M|N) = \frac{P(N|M) \times P(M)}{P(N)} \tag{8}$$

3.4.2 Support Vector Machine (SVM)

State vector machine is the maximum publicly utilized supervised machine learning algorithm which was first discovered in the mid-1990 [16, 17]. The algorithm uses labeled training data, and it deals with hyperplane-based vectors that separate output data into classes to increase the margin between all attack classes [16]. SVM is a binary classifier but cascade manner based multi-class classification is also supported by it. This algorithm depends on the kernel types and parameters used [18]. In our experiment, we have used RBF kernel method of SVM.

The RBF kernel $K(x, x_j) = \exp(-\gamma||x - x_j||^2)$ is one of the most utilized kernel functions.

$$f(x) = \sum_{j=1}^{m} \alpha_j \exp(-\gamma||x - x_j||^2) + b \tag{9}$$

α_j is the language multiplier,$||x - x_j||^2$ represents the squared Euclidean distance measure between two vectors and b is the bias. σ is an unrestricted parameter which is defined as.

$$\gamma = \frac{1}{2\sigma^2} \tag{10}$$

SVM is specifically designed for binary classification, but, multi-class classification is not supported by SVM. So, we have implemented one-versus-rest strategy on RBF kernel-based SVM for multi-class classification. One-vs-rest strategy divides multi-class classification into one binary classification problem per class. Class labels numbers of the dataset should match with the number of generated binary classifiers.

4 Experimental Result Analysis

This section provides a comparative analysis of PCA and LDA for binary, multi-class classification of Kyoto 2006 + dataset through some experiments based on machine learning approaches (i.e., Naïve Bayes (NB), support vector machine (SVM)). The experiments are performed on Anaconda Python 3.7.1 Individual Edition (provides free and open-source Python distribution for scientific computing) using Jupyter Notebook Editor (consists of many python packages).

We have compared the accuracy, training time, and testing time taken by PCA and LDA for binary and multi-class classification individually.

Figure 2 depicts accuracy comparison between PCA and LDA for binary classification of Kyoto 2006+ dataset.

From Fig. 2, we have observed that LDA gives higher accuracy than PCA for Naïve Bayes and SVM algorithms both.

PCA and LDA training time and testing time comparison for binary classification of Kyoto 2006+ dataset are presented in Tables 4, 5 respectively.

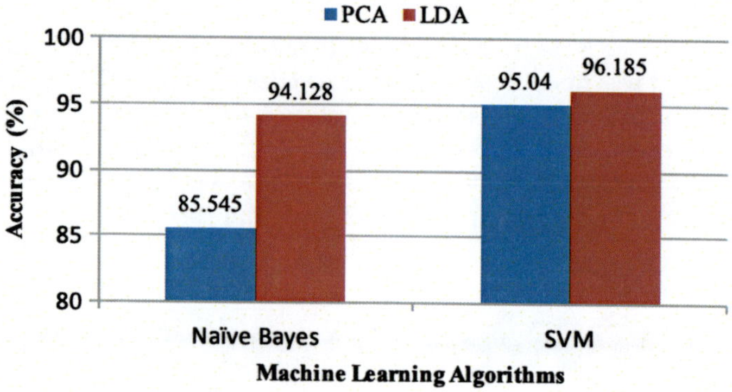

Fig. 2 Comparison of accuracy between PCA and LDA for binary classification of Kyoto 2006+ dataset

Table 4 PCA and LDA training time comparison for binary classification of Kyoto 2006 + dataset

Machine learning algorithms	Training time(s)(PCA)	Training time(s)(LDA)
Naïve Bayes	1.342	*1.201*
SVM	2738.003	*853.252*

Table 5 PCA and LDA testing time comparison for binary classification of Kyoto 2006 + dataset

Machine learning algorithms	Testing time(s)(PCA)	Testing Time(s)(LDA)
Naïve Bayes	0.094	*0.016*
SVM	73.585	*45.536*

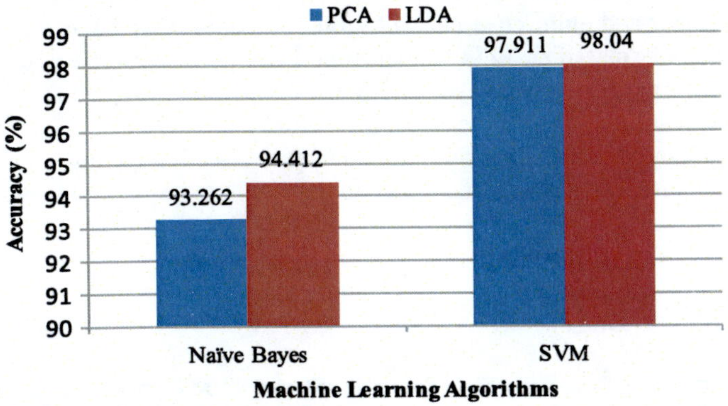

Fig. 3 Comparison of accuracy between PCA and LDA for Multi-Class Classification of Kyoto 2006 + dataset

Table 6 PCA and LDA training time comparison for multi-class classification of Kyoto 2006+ dataset

Machine learning algorithms	Training time(s)(PCA)	Training time(s)(LDA)
Naïve Bayes	0.749	*0.705*
SVM	1277.419	*804.8*

Table 7 PCA and LDA comparison of testing time for multi-class classification of Kyoto 2006 + dataset

Machine learning algorithms	Testing time(s)(PCA)	Testing time(s)(LDA)
Naïve Bayes	0.203	*0.062*
SVM	19.781	*18.533*

By analyzing Tables 4 and 5, we have checked that LDA has taken reduced training time and testing time than PCA for Naïve Bayes and SVM both.

Figure 3 depicts accuracy comparison between PCA and LDA for multi-class classification of Kyoto 2006 + dataset.

Comparison of training time and testing time taken by PCA and LDA for multi-class classification of Kyoto 2006 + dataset are presented in Tables 6, 7, respectively.

By analyzing Tables 6 and 7, we have checked that LDA has taken reduced training time and testing time than PCA for Naïve Bayes and SVM both.

5 Conclusion

This paper represents an efficient intrusion detection framework using two dimensionality reduction methods, i.e., principal component analysis (PCA), and linear discriminant analysis (LDA). We have analyzed the performance of PCA and LDA based on binary and multi-class classification of Kyoto 2006+ dataset using Naïve Bayes and SVM classifier. LDA outperforms PCA concerning accuracy, training time, testing time for Naïve Bayes and SVM classifiers individually. SVM gives better classification accuracy than Naïve Bayes for PCA and LDA both. In the future, we will implement genetic algorithm based feature selection technique to optimize the accuracy of the intrusion detection system.

References

1. Aburomman AA, Reaz MBI (2016) Ensemble of binary SVM classifiers based on PCA and LDA feature extraction for intrusion detection. In: IEEE Advanced Information Management, Communicates, Electronic and Automation Control Conference
2. Neethu B (2012) Classification of intrusion detection dataset using machine learning approaches. Int J Electron Comput Sci Eng ISSN. 2277–1956
3. Singh S, Silakari S, Patel R (2011) An efficient feature reduction technique for intrusion detection system. In: International conference on machine learning and computing, Vol.3
4. Datti R, Lakhina S (2012) Performance comparison of features reduction techniques for intrusion detection system. In: Int J Comput Sci Technol 3(1), ISSN. 2229-4333
5. Murugesan U, Padmavathi G (2013) An accurate method for detection of cyber attacks. Aust J Basic Appl Sci 940–944, ISSN- 1991–8178
6. Ikram ST, Cherukuri AK (2016) Improving accuracy of intrusion detection model using PCA and optimized SVM. J Comput Inform Technol 24(2):133–148
7. Shen Z, Zhang Y, Chen W (2019) A Bayesian classification intrusion detection method based on the fusion of PCA and LDA. Secur Commun Networks
8. Ammar A (2015) Comparison of feature reduction techniques for the binominal classification of network traffic. J Data Anal Inform Process 11–19
9. http://www.takakura.com/Kyotodata
10. KDD Cup (1999) http://kdd.ics.uci.edu/databases/kddcup99/kddcup99.html
11. Jabbar MA, Aluvalu R, Reddy S, Satyanarayana S (2017) RFAODE: A novel ensemble intrusion detection system. In: 7th International Conference on Advances in Computing & Communications, pp 226–234
12. Smith LI (2002) A tutorial on principal components analysis. In: Computer Science Technical Report
13. Noel GE, Gustafson SC, Gunsch GH (2001) Network-based anomaly detection using discriminant analysis. J Inform Warfare 1(2):12–22
14. Hyvärinen A, Oja E (2000) Independent component analysis: algorithms and applications. In: Neural Networks Research Centre, Helsinki University of Technology, Neural Networks, pp 411–430
15. Murphy KP (2012) Machine learning: a probabilistic perspective. The MIT Press
16. Saranyaa T, Sridevib S, Deisyc C, Chungd TD, Khane MA (2020) Performance analysis of machine learning algorithms in intrusion detection system: a review. In: Third international conference in computing and network communications, procedia computer science, pp 1251–1260
17. Boser EB, Guyon MI, Vapnik NV (1992) A training algorithm for optimal margin classiers. In: Proceedings of the 5th annual ACM workshop on computational, pp 144–152
18. Mehmood T, Rais HBM (2016) Machine learning algorithms in context of intrusion detection. In: IEEE third international conference on computer and information sciences, pp 369–373

SMLHADC: Security Model for Load Harmonization and Anomaly Detection in Cloud

Mahima Sandeep Bakshi, Drashti Banker, Vivek Prasad, and Madhuri Bhavsar

Abstract Cloud computing (CC) emerged as one of the important utility such as water and electricity bills, where the user has to pay for whatever quantity they have utilized for their tasks. As cloud users are increasing; so the crucial challenges for the cloud service provider (CSP) is to balance the load among the resources shared by the end-users and curb increased security risks, misuse, or malicious attacks in cloud computing. The functionality of load balancing is divided into two functions, first, there will be allocation of resources and the second is the provisioning of resources along with task scheduling in the distributed system such as CC. Although many state-of-the-art approaches are available in the literature which provides load balancing and better resource utilization. In this paper, we build a dynamic prediction approach for cloud resource usage, discuss a load balancing technique, and identify sudden spikes and failures that are causes of the anomaly using a reactive and proactive approach. The final result demonstrates a comparative analysis of deep learning models for the prediction of cloud resource utilization, insight into dynamic load balancing technique and anomaly detection.

Keywords Anomaly detection · Cloud computing · Cloud service provider · Deep learning · Load balancing · Outliers

M. S. Bakshi · D. Banker · V. Prasad (✉) · M. Bhavsar
Department of Computer Science and Engineering, Institute of Technology, Nirma University, Ahmedabad, India
e-mail: vivek.prasad@nirmauni.ac.in

M. S. Bakshi
e-mail: 16bit057@nirmauni.ac.in

D. Banker
e-mail: 16bce012@nirmauni.ac.in

M. Bhavsar
e-mail: madhuri.bhavsar@nirmauni.ac.in

K. Dahal et al. (eds.), *Internet of Things and Its Applications*, Lecture Notes in Electrical Engineering 825, https://doi.org/10.1007/978-981-16-7637-6_36

1 Introduction

Cloud computing which is termed a computing system that offers Internet-based services on demand in a parallel and distributed environment. Cloud providers are facing rapidly increasing traffic loads [1]. As a result, the cloud service provider is using proper strategies for their large-scale data centers. A concept termed as load balancing plays an efficient role in making proper usages of resources. Load balancing is a technique that distributes the input workload over the available computing resources for faster and reliable output/services [1, 2]. So each resource does approximately equal amount of task. Starting with static load balancing, it is used when the input has a fixed lower and upper threshold for traffic. And the load over physical machines (PM) oscillates between the lower and upper threshold values. Also, the traffic will be divided equivalently among all servers of a particular cloud [3]. Another technique which is dynamic load balancing is used when the operating environment is very much complex and which cannot be handled properly by the previous technique. It uses the present state of the system during the decision-making process of load balancing, which is more suitable for dynamic applications such as in a heterogeneous cloud environment [3]. This technique is used particularly when the states and workload over the virtual machines (VMs) and PMs do not have fixed boundaries [4, 5]. To make cloud resource management more efficient in a dynamic environment, simulation of cloud resource usage based on prediction patterns will be taken into account [6].

Additionally, in order to detect failures in terms of resource utilization early, we need a robust mechanism to handle and alert the cloud service provider [7].

1.1 Problem Statement

Cloud platforms are ubiquitous in today's world. Reliability, security, flexibility, scalability, and reduced costs are some of the advantages because of which users switch to cloud services to handle their data [6–8]. But sometimes, these are obscured by the disadvantages that are prevalent for instance like the job arrival pattern which is not predictable and there is a difference in the capacity of each node for managing load balancing and cyber-attacks that lead to the leaking of crucial data to the third party or over exhaustion in terms of consumption of resources [9]. The problem statement for load balancing is, therefore, how to properly manage the load balancing scheme that distributes Session Initiation Protocol (SIP) requests to a collection of servers to effectively utilize the resources at those servers and build a robust predictive model with the desired accuracy and security [10].

1.2 Proposed Solution

The solution to above problem is threefold. First, we need to propose a technique that will identify and analyze the number of VM's required and to schedule those VM's among cloud users according to their needs. Second step is to implement a cloud resource prediction model that determines future resource usage and utilization. Thirdly, to prevent misuse of crucial data of cloud users to third party, we will focus on drastic changes in resource usage during an attack by tracing and alerting the system when anomalous points are encountered.

1.3 Architecture

In this module, architecture has been designed for secure dynamic load balancing that will monitor and predict the cloud environment. In Fig. 1, a machine learning algorithm can be taken into consideration to properly estimate the incoming traffic size of the task, look up for the available free server instances at the data center to schedule the tasks, and to identify the path. Basically, we monitor the resources using the particle swam intelligence algorithm to effectively satisfy customer needs and requirements. A prediction model using deep learning algorithm is used to estimate incoming resource usage demand in an advanced time frame and also serves as a powerful method for enabling automated secure resource scaling management. All of these factors will contribute in building a productive reliable dynamic load balancing model.

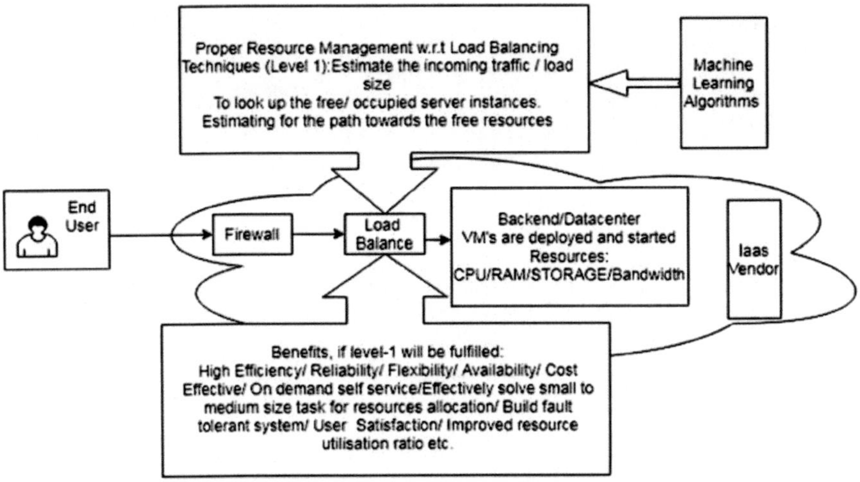

Fig. 1 System architecture

2 Methodology

2.1 Monitoring of the Cloud Resources Using Particle Swarm Optimization (PSO)

Cloud users generally submit computationally intensive applications to resource brokers or resource managers with their application choices and required resources (such as software, hardware, network, and QoS) [11–13]. The proposed research study was performed to investigate a swarm intelligence method for building a profile for each request submitted to the big data monitoring system and cloud resource choice which are optimally represented in Fig. 2.

The particle swarm optimization was applied in the test analysis and is bio-inspired algorithm [11, 14, 15]. The PSO-based application profiling based on similarity match retrieves a desired resource list which is created; otherwise, it would be stored in the database as a new application profile for further analysis. The research framework implements a PSO-based resource selection mechanism that optimally selects resources for the provision of virtual instances and the running of large data applications. The fitness function used to select an ideal resource from the desired list of resources is described. It takes into account both processing time and the cost of processing.

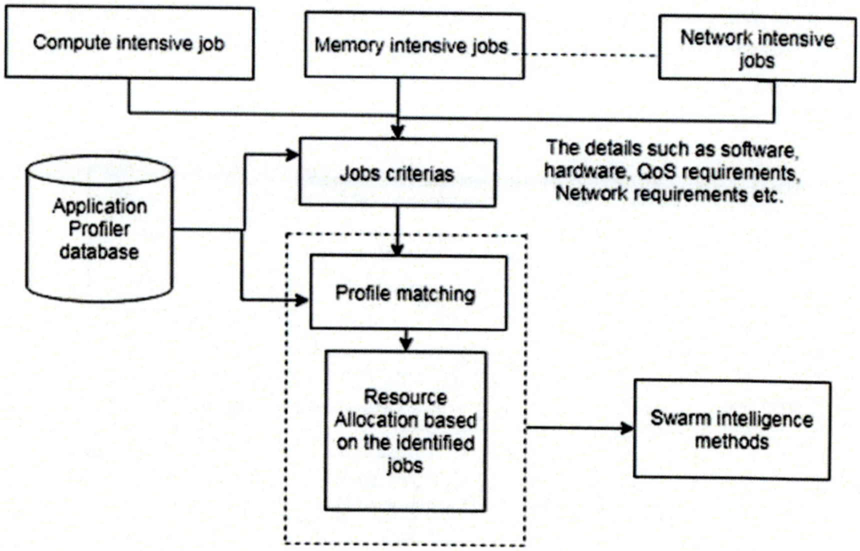

Fig. 2 Swarm intelligence and cloud resource monitoring

2.2 *Prediction of Resource Usage*

The dataset used for the purpose of forecasting cloud resources contains a timestamp (ms) and resource parameter values. The parameters used are extracted from the real-time cloud service. For instance, to practically demonstrate the prediction method, we use the following parameters—CPU utilization (MHZ), memory usage (MB), disk read throughput (KB/s), disk write throughput (KB/s), network transmitted throughput (KB/s), and network received throughput (KB/s) [10]. In order to adjust hyperparameters according to the dataset, grid search is used. Table 1 shows parameters used in deep learning models.

Table 2 depicts a summary of one of the models of deep learning that is bidirectional LSTM. In this implementation, mini batching is used to divide training and test into batches of size 64. The size of input data is adjusted by specifying the size of batches, historical timestamp data (10), and the actual output value (1). The input layer is fetched to the neural network containing BLSTM layer (10 units) sandwiched between the LSTM layer (10 units) [16].

Finally, we get the predicted value by output layer. Table 3 illustrates a summary of GRU model, in which input layer is fetched to neural network containing two layers of GRU each containing ten units and an output layer of one unit. Similarly, Table 4 depicts a summary of LSTM model, in which the input layer is fetched to multiple LSTM layers arranged in a sequential manner an output layer that predicts future resource usage value.

Table 1 Grid search parameters for memory usage

Models	Grid search parameters
LSTM	Batch size:64, epochs:120, cv:3, n jobs:−1
BLSTM	Batch size:64, epochs:120, cv:3, n jobs:−1
GRU	Batch size':64, epochs:100, cv:3, n jobs:−1

Table 2 BLSTM model summary

Layer (type)	Output shape	Param
input 1(Input layer)	(64, 10, 1)	0
lstm 1(LSTM)	(64, 10, 10)	480
bidirectional 1(Bidirection)	(64, 10, 20)	1680
lstm 3(LSTM)	(64, 10, 10)	1240
dense 1(Dense)	(64, 10, 1)	11

Total params: 3411
Trainable params: 3411
Non-trainable params: 0

Table 3 GRU model summary

Layer(type)	Output shape	Param
input 1(Input layer)	(64, 10, 1)	0
gru 1(GRU)	(64, 10, 10)	360
gru 2(GRU)	(64, 10, 10)	630
dense 1(Dense)	(64, 10, 1)	11

Total params: 1001
Trainable params: 1001
Non-trainable params: 0

Table 4 LSTM model summary

Layer(type)	Output shape	Param
input 1(Input layer)	(64, 10, 1)	0
lstm 1(LSTM)	(64, 10, 10)	480
lstm 2(LSTM)	(64, 10, 20)	2480
lstm 3(LSTM)	(64, 10, 30)	6120
lstm 4(LSTM)	(64, 10, 40)	11,360
dense 1(Dense)	(64, 10, 1)	41

Total params: 20,481
Trainable params: 20,481
Non-trainable params: 0

2.3 Anomaly Detection

We take into account an uncommon CPU utilization dataset with regard to the time stamp for a practical demonstration of this approach in this detection process. So, we are taking two separate approaches here.

- Reactive Method: Using the isolation forest and interquartile range system, this is applied. Next, the dataset is split into training and test sets with a 7:3 ratio. Forest-based isolation scores and anomalous data points based on this label are marked as 1, and regular data points are marked as 0. We will measure the interquartile range of the dataset to further improve our results. Any data points lying beyond the lower and upper bound range are considered anomalous on the basis of this knowledge.
- Proactive Method: The threshold-based LSTM model implements this. First, dataset segregation and processing are conducted in the same way as in a reactive approach. Second, in order to train and predict values, the LSTM model is described. The loss function is measured and plotted for both sets in relation to the actual value, based on the model's forecast. We manually determine the threshold value of the loss to be taken into account according to the results, after plotting the loss function and examining the trend. We mark as an exception the data points that are above the threshold and below them as regular data points.

3 Experimentation and Results

The dataset used for the purpose of prediction takes into account real-world cloud watch AWS data, and it is filtered on the basis above-mentioned parameters considered for resource usage patterns.

3.1 Prediction of Resource Usage

In this module, we will be demonstrating the prediction result graphs of memory usage using deep learning algorithms. Figures 3, 4, 5, 6, 7, and 8, respectively, display the graph of actual and predicted memory usage for long short-term memory networks (LSTM), bidirectional long short-term memory networks (BLSTM), and

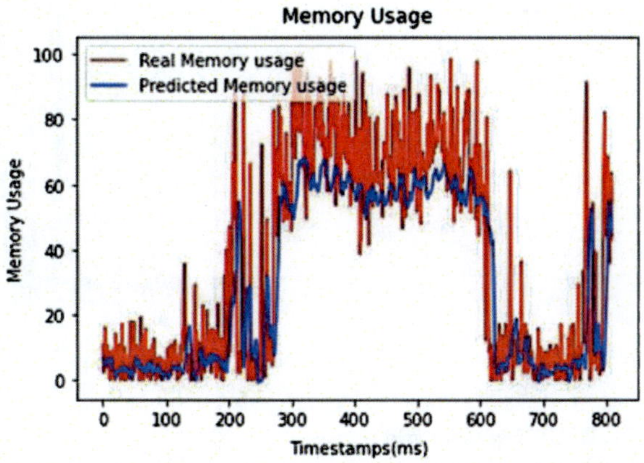

Fig. 3 LSTM without grid search

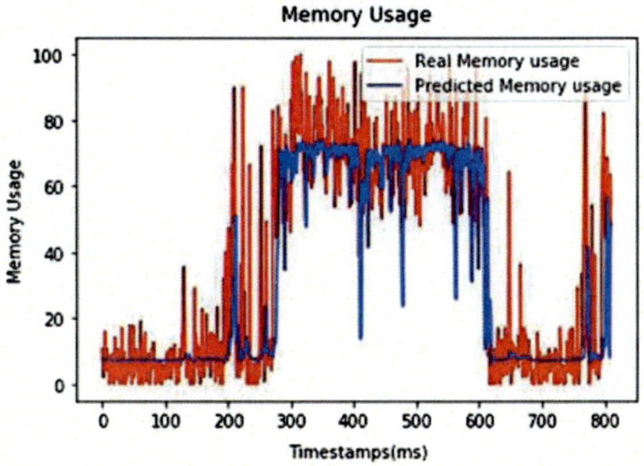

Fig. 4 LSTM with grid search

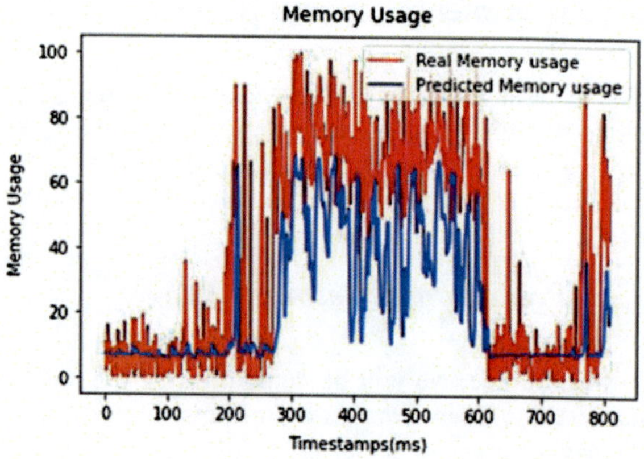

Fig. 5 BLSTM without grid search

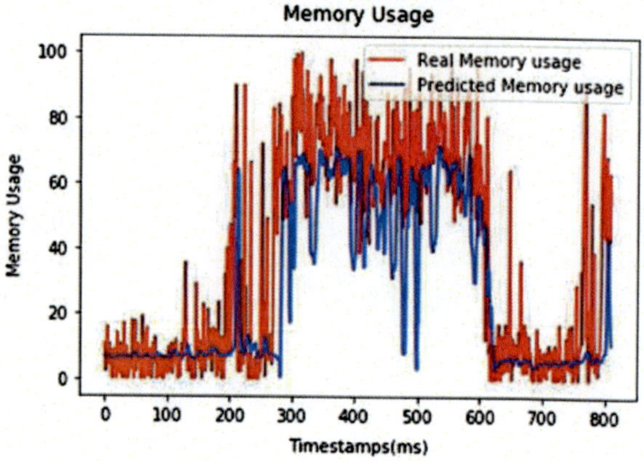

Fig. 6 BLSTM with grid search

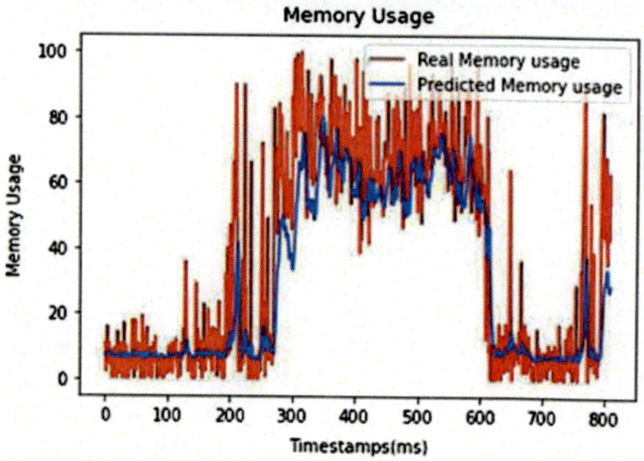

Fig. 7 GRU without grid search

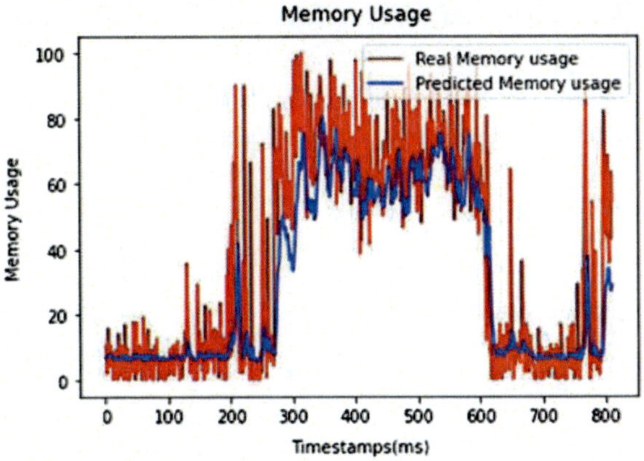

Fig. 8 GRU with grid search

Table 5 RMSE and MAE values for prediction of memory usage parameter

Prediction model	Memory usage parameters			
	Without grid		With grid	
	RMSE	*MAE*	*RMSE*	*MAE*
LSTM	15.9796	10.0377	9.9427	8.9338
BLSTM	20.6983	13.2350	3.2840	2.7399
GRU	13.8488	8.6183	2.8640	1.9459

*RMSE—Root mean square error
*MAE—Mean absolute error

gated recurrent units (GRU). Similarly, we have implemented methods for other resources as mentioned above. For the purpose of comparing the accuracy of models for each cloud resource, we will take into consideration root mean square error (RMSE) and mean absolute error (MAE) values [8]. Table 5 depicts the performance of deep learning models on memory usage in terms of RMSE and MAE values. If we consider *r* as actual resource usage value and ˆ*r* as predicted resource usage value and N is the total number of values, then RMSE is given as follows:

$$\mathrm{RMSE} = \sqrt{\left(\frac{1}{N}\right)\sum_{i=1}^{N}\left(r_i - \hat{r}_i\right)^2} \tag{1}$$

MAE is given as follows:

$$\mathrm{MAE} = \left(\frac{1}{N}\right)\sum_{i=1}^{N}\left(r_i - \hat{r}_i\right) \tag{2}$$

3.2 Anomaly Detection

We will consider an anomalous dataset of CPU usage and implement both proactive and reactive methods to illustrate anomaly detection.

Figure 9 represents anomaly detection using reactive approach, wherein red data points represent anomaly. In this method, isolation forest segregates anomalies at sample points instead of profiling regular data points. To detect anomalies, contamination factor of 0.02 is used. To further make our observations more accurate, interquartile range method is used. Here, lower boundary has value of −0.0872, and higher boundary has value of 0.4891. Any value lying outside this range is termed as an anomaly.

Figure 10 represents anomaly detection using proactive approach, wherein red data points represent anomaly. In this method, we take threshold value of 0.6 by taking into consideration mean absolute error value of test dataset. If test set loss is greater than threshold, then it is termed as anomaly; otherwise, it is considered as normal data.

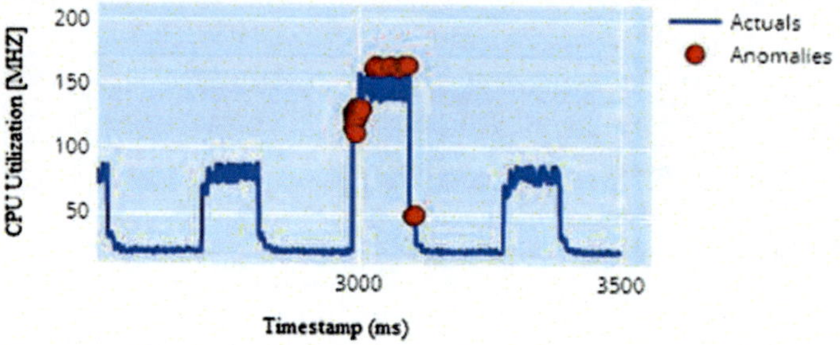

Fig. 9 Anomaly detection using isolation forest

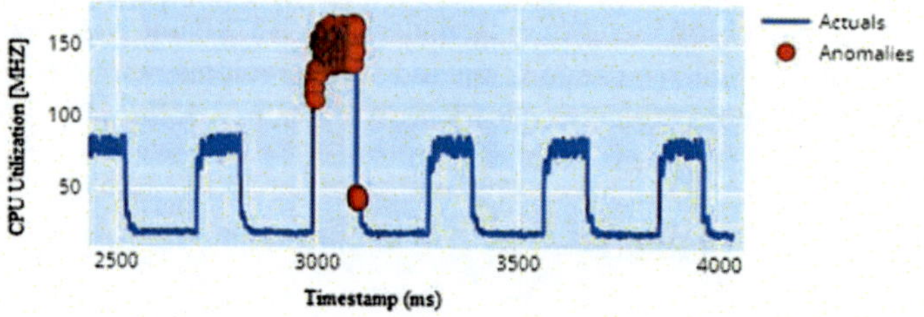

Fig. 10 Anomaly detection using LSTM autoencoder

4 Conclusion and Future Work

In this paper, we present an approach to monitor cloud resources using swarm intelligence and discuss security methods to build reliability and trust in our system. Also, to estimate cloud resource consumption and utility in future, we propose an efficient prediction model. In the implementation of prediction model, we conclude that grid search provides the best parameters for training deep learning models, so that loss is minimum. In memory usage dataset, the performance of GRU model is better as compared to the other two in terms of RMSE and MAE values. Using proactive and reactive approaches, we proposed an anomaly detection-based approach. Proactive models are more effective from these two models as it warns the device before a possible attack. In the future, we will analyze different load balancing algorithms and compare their performance. We will plan to incorporate models like a hierarchical reinforcement learning and supervised prediction classification approach in the prediction part. Also, explore root cause of observed irregularities will be better established as this will be helpful in future cloud computing resource forecasting.

References

1. Al Nuaimi K, Mohamed N, Al Nuaimi M, Al-Jaroodi J (2012) A survey of load balancing in cloud computing: Challenges and algorithms. In: 2012 second symposium on network cloud computing and applications, IEEE, pp 137–142
2. Deepa T, Cheelu D (2017) A comparative study of static and dynamic load balancing algorithms in cloud computing. In: 2017 International conference on energy, communication, data analytics and soft computing (ICECDS), IEEE, pp 3375–3378
3. Prasad VK, Bhavsar M (2017) Exhausting autonomic techniques for meticulous consumption of resources at an IaaS layer of cloud computing. In: International conference on future internet technologies and trends, Springer, pp 37–46
4. Dashti SE, Rahmani AM (2016) Dynamic vms placement for energy efficiency by psoin cloud computing. J Exp Theor Artif Intell 28(1–2):97–112
5. Prasad VK, Bhavsar MD (2020) Monitoring IaaS cloud for healthcare systems: healthcare information management and cloud resources utilization. Int J E-Health Med Commun (IJEHMC) 11(3):54–70
6. Ramachandra G, Iftikhar M, Khan FA (2017) A comprehensive survey on security in cloud computing. Procedia Comput Sci 110:465–472
7. Prasad VK, Bhavsar M (2017) Efficient resource monitoring and prediction techniques in an IaaS level of cloud computing: survey. In: International conference on future internet technologies and trends, Springer, pp 47–55
8. Prasad VK, Bhavsar MD (2021) SLAMMP framework for cloud resource management and its impact on healthcare computational techniques. Int J E-Health Med Commun (IJEHMC) 12(2):1–31
9. Wang P, Lei Y, Agbedanu PR, Zhang Z (2020) Makespan-driven workflow scheduling in clouds using immune-based PSO algorithm. IEEE Access 8:29281–29290
10. Chen Z, Hu J, Min G, Zomaya AY, El-Ghazawi T (2019) Towards accurate prediction for high-dimensional and highly-variable cloud workloads with deep learning. IEEE Trans Parallel Distrib Syst 31(4):923–934

11. Luong NC, Wang P, Niyato D, Wen Y, Han Z (2017) Resource management in cloud networking using economic analysis and pricing models: A survey. IEEE Commun Surv Tutorials 19(2):954–1001
12. Prasad VK, Bhavsar M (2019) Preserving SLA parameters for trusted IaaS cloud: an intelligent monitoring approach. Recent Patents Eng 13(1)
13. Prasad VK, Bhavsar MD (2020) Monitoring and prediction of SLA for IOT based cloud. Scalable Comput: Pract Experience 21(3):349–358
14. Al-Maamari A, Omara FA (2015) Task scheduling using pso algorithm in cloud computing environments. Int J Grid Distrib Comput 8(5):245–256
15. Shekhar S, Gokhale A (2017) Dynamic resource management across cloud-edge resources for performance-sensitive applications. In: 2017 17th IEEE/ACM international symposium on cluster, cloud and grid computing (CCGRID), IEEE, pp 707–710
16. Kumar J, Goomer R, Singh AK (2018) Long short term memory recurrent neural network (lstm-rnn) based workload forecasting model for cloud data centers. Procedia Comput Sci 125:676–682

FPGA-Based Low Delay Adjacent Triple-Bit Error Correcting Codec

Raj Kumar Maity, Jagannath Samanta, and Jaydeb Bhaumik

Abstract Error correcting codes (ECCs) are employed in most of the modern communication systems and semiconductor memories to detect and correct errors introduced by noise and radiation effects, respectively. In these systems, multiple errors are more common nowadays due to the rapid advancement in semiconductor technology. Various types of multi-bit ECCs like Bose-Chaudhuri-Hocquenghem (BCH) and Reed-Solomon (RS) codes are capable of handling these multiple errors at the cost of complex decoding method. Alternatively, multi-bit adjacent error correcting codes such as single error correction, double error detection, and double adjacent error correction (SEC-DED-DAEC) and single error correction, double error detection, double adjacent error correction, and triple adjacent error correction (SEC-DED-DAEC-TAEC) codes have comparatively simpler decoder structure for handling multi-bit adjacent errors. But the main drawbacks of these codes are the growing codec design constrains as correction capability increases. In this paper, a new SEC-DED-DAEC-TAEC code has been proposed for the word lengths of 16, 32, and 64 bits. The proposed codecs for three different message lengths have been simulated and synthesized in FPGA platform. The proposed codecs have lower delay compared to the existing-related works. Also miscorrection rate of the proposed codes is lower compared to the related works.

Keywords ECCs · SEC code · SEC-DED-DAEC code · SEC-DED-DAEC-TAEC code · FPGA · Miscorrection rate

1 Introduction

The protection of semiconductor devices like memory against soft errors is generally achieved by employing suitable error correcting codes (ECCs) [1]. The simplest form

R. K. Maity (✉) · J. Samanta
Department of ECE, Haldia Institute of Technology, Haldia, W.B., India

J. Bhaumik
Department of ETCE, Jadavpur University, Kolkata, W.B., India
e-mail: jaydeb.bhaumik@jadavpuruniversity.in

K. Dahal et al. (eds.), *Internet of Things and Its Applications*, Lecture Notes in Electrical Engineering 825, https://doi.org/10.1007/978-981-16-7637-6_37

of ECCs, which are applied in this regard, is known as single error correction (SEC) and single error correction-double error detection (SEC-DED) codes [2]. But the possibilities of more than single bit errors have been increased a lot due to rapid upgradation of very large scale integration (VLSI) technology [3]. So ECC schemes with higher correction capabilities [4] are required to handle these multiple bit errors. The Bose-Chaudhuri-Hocquenghem (BCH) code [5] is capable of correcting multi-bit errors. But the decoder for this code is very complex in nature.

As an alternative, the adjacent error correcting codes are capable of correcting single and multi-bit adjacent errors with simpler decoder circuitry. The most common adjacent error correcting code is the single error correction, double error detection, and double adjacent error correcting (SEC-DED-DAEC) codes [6–14] which are capable of correcting single and adjacent double errors and detecting any type of double errors. Another category of adjacent error correcting codes is the single error correction, double error detection, double adjacent error correcting, and triple adjacent error correcting (SEC-DED-DAEC-TAEC) codes [15, 16] which have comparatively complex decoding circuitry compared to SEC-DED-DAEC decoders. So the main drawback of adjacent error correcting code is their decoder's overheads (area, delay, and power) which increase with the increasing of error correction capability. Another issue of adjacent error correcting codes is the miscorrection rate which must be lower while correcting adjacent errors. Several proposals for reducing the design overheads of adjacent ECCs have been found in the literature. Two techniques, presented in [11], have been adopted for decoding of SEC-DED-DAEC codes with lower area and delay, respectively. Different overheads reduction schemes of adjacent ECCs have been found in [9, 10, 12, 13, 17]. The main advantages of triple adjacent and burst ECCs are lesser number of parity bits and moderate design overheads [16]. Adjacent ECCs with lower miscorrection rate have been explored in [7, 8, 14, 15].

In this paper, a new SEC-DED-DAEC-TAEC code has been proposed for the word lengths of 16, 32, and 64 bits. The proposed codecs for all the word lengths have been simulated and synthesized in FPGA platform. The proposed codecs have lower delay compared to the related works. Also miscorrection rate of the proposed codes is lower compared to those related works.

The rest of this paper is organized as follows. Section 2 presents the basics of SEC-DED-DAEC-TAEC code. The overview of existing adjacent ECCs is provided in Sect. 3. The proposed SEC-DED-DAEC-TAEC codes and their miscorrection rates are discussed in Sect. 4. Section 5 provides the proposed codec design. The FPGA-based synthesis results are summarized in Sect. 6, and finally, the paper is concluded in Sect. 7.

2 Basics of SEC-DED-DAEC-TAEC Code

The structure of parity check matrix (H) of an (n, k) SEC-DED-DAEC-TAEC code in systematic form is provided in Eq. (1).

$$H = \left[DI_{n-k}\right] \quad (1)$$

where D and I_{n-k} are two sub-matrices which represent data matrix and identity matrix, respectively. Also n and k represent codeword and message lengths, respectively. The D-matrix consists of k number of data columns, and these columns are selected in the manner such that the functionality of SEC-DED-DAEC-TAEC code is satisfied. The column selection rules of H-matrix for SEC-DED-DAEC-TAEC code and its corresponding functionalities have been listed in Table 1. The H-matrix of any SEC-DED-DAEC-TAEC code is constructed by incorporating these basic rules. The SEC-DED-DAEC-TAEC encoder generates $(n-k)$ parity bits by XOR-ing the data bits based on the H-matrix. The combined form of data and parity bits is termed as the codeword. These codewords are transmitted or stored to protect the valuable information in communication systems or memories, respectively. The SEC-DED-DAEC-TAEC decoder accepts the received codeword (r) and produces corrected version of the data by employing the following functionality:

- Syndrome generation: Syndrome bits are computed by multiplying transpose of H-matrix (H^T) with the received codeword r. The XOR gates are used to construct the syndrome circuit. All zero syndromes imply no error in the received codeword; whereas, non-zero syndrome indicates errors in received codewords. So error detection is performed based on syndrome values.
- Error location determination: The precise locations of detected errors are also determined by syndrome values. In case of SEC-DED-DAEC-TAEC decoder, the error location determination block for a particular data bit compares at most the syndromes for one single error, two double adjacent errors, and three triple adjacent errors. In error location detection block of the decoder, all these syndromes are represented by employing AND and NOT logic gates and then they are compared by employing OR logic gates. So this part of the TAEC decoder is more complex than that of syndrome generation and error correction part.

Table 1 Column selection rules for different types of coding functionality

Coding functionality	Column selection rules
SEC	All the columns in H-matrix must be distinct and non-zero
DED	The minimum distance ($d$min) of the H-matrix must be greater than equals to four for DED functionality
DAEC	The XOR sum of any two adjacent column in the H-matrix must be different from: (i) each other and (ii) any column of H-matrix
TAEC	The XOR sum of any three adjacent column in the H-matrix must be different from: (i) each other (ii) any column of H-matrix, and (iii) XOR sum of any two adjacent column of H-matrix

- Error correction: Error correction block of the decoder corrects the errors which are located by error location determination block. For this purpose, XOR logic gates are employed which invert the bits in error.

3 Overview of Existing Adjacent ECCs

Several adjacent ECCs are available in the literature for protecting memories from soft errors. Two of them which are related to the present work have been summarized here.

3.1 SEC-DED-DAEC and SEC-DED-DAEC-TAEC Code by Neale et al. [15]

Neale et al. [14] have introduced a class of SEC-DED-DAEC code which is capable of detecting scalable adjacent errors. Identity matrices in the lower portion and variable weight columns in the upper portion are the main structural differences of the Neale et al. *H*-matrices from others. The SEC-DED-DAEC code matrix with various sizes of identity matrices has been presented in this work. The SEC-DED-DAEC codes presented in [14] have been further upgraded to include triple adjacent correction capability in [15]. The DAEC and TAEC codes presented in [14, 15] have the conventional decoding technique. But the encoding of these codes differs from the conventional encoding in the sense that some of the parity bits not only depends on the data bits but some other parity bits. The main features of Neale et al. DAEC and TAEC codes are scalable adjacent error detection and lower miscorrection rate with moderate design overheads.

3.2 SEC-DAEC-TAEC Code by Adalid et al. [16]

A class of SEC-DAEC-TAEC and three-bit burst ECCs have been familiarized by Adalid et al. [16]. These codes have designed with the minimum number of parity bits. Also minimization of (i) total number of one's in the *H*-matrix and (ii) weight of its heaviest row is the two techniques which have been employed for the construction of the *H*-matrices. The SEC-DAEC-TAEC codes by Adalid et al. have the features of low redundancy and better error coverage.

4 Proposed SEC-DED-DAEC-TAEC Code

In this section, details of proposed SEC-DED-DAEC-TAEC code are presented. The H-matrices of proposed SEC-DED-DAEC-TAEC codes have been constructed by considering the following rules:

1. All the columns are different from each other, non-zero and have a constant weight equals to three.
2. All the XOR sums of any two adjacent columns which include at least one data column are different from each other and have a constant weight equals to four.
3. All the XOR sums of any three adjacent columns which include at least one data column are different from each other and from any column of H-matrix and have a constant weight equals to three.

The first condition provides SEC functionality of the proposed codes. The second and third conditions are for the double and triple adjacent errors correction capabilities, respectively. Also the DED feature of the proposed codes has been confirmed by the first and second conditions jointly as the DED syndrome is even weighted in proposed scheme. The required number of parity bits for the feasibility of the proposed (n, k) SEC-DED-DAEC-TAEC code in systematic form is bounded by the following equation:

$$\binom{n-k}{3} \geq (k+n-2) \text{ and } \binom{n-k}{4} \geq k \tag{2}$$

where $(n-k)$ is the number of parity bits. In Eq. (2), k and $(n-2)$ numbers of different weight three syndromes which are required for SEC and TAEC functionality, respectively. So the total number of combinations of weight three in $(n-k)$ bits must be greater than equals to required number of weight three syndromes. Similarly, total number of combinations of weight four syndromes in $(n-k)$ bits must be greater than or equals to the required number of weight four syndromes.

The required numbers of parity bits for the proposed SEC-DED-DAEC-TAEC codes with different word lengths have been listed in Table 2 based on Eq. (2). The proposed H-matrices with word lengths 16, 32, and 64 bits have been presented in Figs. 1, 2, and 3, respectively.

Table 2 Required number of parity bits for proposed codes

k	$n-k$	Number of weight three syndromes		Number of weight four syndromes	
		Available	Required	Available	Required
16	8	56	38	70	16
32	9	84	71	136	32
64	11	165	137	330	64

Fig. 1 Proposed *H*-matrix for (24, 16) SEC-DED-DAEC-TAEC code

$$\begin{bmatrix}
010110011010101000000001 \\
100100000101010100000000 \\
011010101000000010000000 \\
110000110110000001000010 \\
000011010000110000100000 \\
001001000001100000010011 \\
000000001100010000001010 \\
101101100011000000000101
\end{bmatrix}$$

$$\begin{bmatrix}
00100000010100010110000000100110100000000000 \\
11000000110000001010101100000001010000000001 \\
10110001101101011000110000000011001000000000 \\
01011000000110000011000101100000001000000 \\
00001101000001100000000011011010000100000 \\
00000110100000000101010110000000000010001 \\
10000011011000110000011000101000000001000 \\
00010100001011000001100000010100000000101 \\
01101010000010101100001010110000000000010
\end{bmatrix}$$

Fig. 2 Proposed *H*-matrix for (41, 32) SEC-DED-DAEC-TAEC code

$$\begin{bmatrix}
0000000000000000000100010000110000000000010110001010001000001010000000000000001 \\
00001001011011010100000000010010101100000000000101101100000000010100000000000 \\
0011010110110110110110000000000001100001101100000000000110110000000100000000000 \\
1010000000001000000100000110110000001010110000000001100000000101000100000000000 \\
1101101100000000110000101100000000110110000000000110000001011000000010000000010 \\
000000000000010110000000000110101010101100000000000001010110000000000010000000100 \\
0100011000000010000001010101101100000000000001011000000000101100000000100000000 \\
00010000010110000110110000000000001000011000010101101100000000000000001000000 \\
000001011000000000000011010000000001100000010110000000000001011000000000100101 \\
011011000010000010110110000000000000101001100000100000010110000000000000010110 \\
10000000100000010000000000000100110000000101011000000101100000000000000001011
\end{bmatrix}$$

Fig. 3 Proposed *H*-matrix for (75, 64) SEC-DED-DAEC-TAEC code

4.1 Miscorrection Rate of Proposed Codes

The miscorrection rate of an adjacent ECC is the rate of misinterpreting nonadjacent random errors as adjacent errors. So an adjacent ECC with lower miscorrection rate is the preferred choice for any application. The miscorrection rate is generally computed against two-random and three-random errors. The mathematical expression of miscorrection rate (*MR*) of an (n, k) adjacent ECC has been presented in [15] which is reproduced here in Eqs. (3) and (4) for two-random and three-random errors, respectively.

$$MR_{2\text{-random errors}} = \frac{SS_2}{\binom{n}{2} - (n-1)} \quad (3)$$

where SS_2 is the number of combinations of two-random columns in H-matrix which falsely produce either an error correcting syndrome or an all zero syndrome and the denominator represent total number of combination of two in n bits excluding the number of double adjacent error correcting syndromes.

$$MR_{3\text{-random errors}} = \frac{SS_2}{\binom{n}{3} - (n-2)} \quad (4)$$

where SS_3 is the number of combinations of three-random columns in H-matrix which falsely produce either an error correcting syndrome or an all zero syndrome and the denominator represent total number of combination of three in n bits excluding the number of triple adjacent error correcting syndromes.

The miscorrection rates of proposed and existing codes in [15, 16] have been computed based on Eqs. (3) and (4) for two-random and three-random errors, respectively, and summarized in Table 3. The miscorrection rates against two-random errors of proposed codes are slightly lower compared to Neale et al. codes for all the word lengths except 64-bit. But in the case of three-random errors, the proposed and Neale et al. [15] codes have nearly equal miscorrection rates for all the word lengths. The codes by Adalid et al. have been observed to have very higher miscorrection rates against both types of random errors for all the word lengths compared to our proposed and Neale et al. codes.

Table 3 Comparison of miscorrection rate of proposed codec

Codec	Miscorrection rate (%)	
	Two-random errors	Three-random errors
Proposed (24, 16) SEC-DED-DAEC-TAEC	13.83	38.71
Neale (24, 16) I5 SEC-DED-DAEC-TAEC [15]	16.20	39.70
Adalid (22, 16) SEC-DAEC-TAEC [16]	98.57	98.22
Proposed (41, 32) SEC-DED-DAEC-TAEC	16.03	36.92
Neale (41, 32) I5 SEC-DED-DAEC-TAEC [15]	21.80	43.70
Adalid (39, 32) SEC-DAEC-TAEC [16]	89.76	89.65
Proposed (75, 64) SEC-DED-DAEC-TAEC	12.74	24.36
Neale (75, 64 I5 SEC-DED-DAEC-TAEC [15]	11.40	25.30
Adalid (72, 64) SEC-DAEC-TAEC [16]	84.71	83.79

5 Proposed Codec Design

In proposed codec, the parity bits are computed using encoder circuit. The expressions of the parity bits for proposed (24, 16) SEC-DED-DAEC-TAEC code are shown in Eq. (5).

$$\begin{aligned}
c_1 &= d_2 + d_4 + d_5 + d_8 + d_9 + d_{11} + d_{13} + d_{16} \\
c_2 &= d_1 + d_4 + d_{10} + d_{12} + d_{14} \\
c_3 &= d_2 + d_3 + d_5 + d_7 + d_9 \\
c_4 &= d_1 + d_2 + +d_7 + d_8 + d_{10} + d_{11} + d_{15} \\
c_5 &= d_5 + d_6 + d_8 + d_{13} + d_{14} \\
c_6 &= d_3 + d_6 + d_{12} + d_{13} + d_{15} + d_{16} \\
c_7 &= d_9 + d_{10} + d_{14} + d_{15} \\
c_8 &= d_1 + d_3 + d_4 + d_6 + d_7 + d_{11} + d_{12} + d_{16}
\end{aligned} \tag{5}$$

where c_i for $i = 1, 2, \ldots, 7, 8$, and d_j for $j = 1, 2, \ldots, 15, 16$ represent parity bits and data bits respectively. The expressions of proposed encoder circuit are presented in Eq. (5). These equations are implemented by employing XOR logic gates. On the other hand, the proposed decoder accepts the received codeword (r) as input and produces the corrected version of data as output. The flow diagram of proposed decoding technique has been shown in Fig. 4. The first task of proposed decoder is to compute the syndrome bits (S). The expressions of eight syndrome bits for proposed (24, 16) SEC-DED-DAEC-TAEC code have been shown in Eq. (6).

$$\begin{aligned}
S_1 &= r_2 + r_4 + r_5 + r_8 + r_9 + r_{11} + r_{13} + r_{15} + r_{24} \\
S_2 &= r_1 + r_4 + r_{10} + r_{12} + r_{14} + r_{16} \\
S_3 &= r_2 + r_3 + r_5 + r_7 + r_9 + r_{17} \\
S_4 &= r_1 + r_2 + r_7 + r_8 + r_{10} + r_{11} + r_{18} + r_{23} \\
S_5 &= r_5 + r_6 + r_8 + r_{13} + r_{14} + r_{19} \\
S_7 &= r_9 + r_{10} + r_{14} + r_{21} + r_{23} \\
S_8 &= r_1 + r_3 + r_4 + r_6 + r_7 + r_{11} + r_{12} + r_{22} + r_{24}
\end{aligned} \tag{6}$$

The second task of proposed decoder is to locate the errors in case of non-zero syndrome bits. For this purpose, two complementary signals, namely notDED and DED are employed based on odd and even number of errors, respectively, in the proposed decoder. As the proposed H- matrices have odd weighted columns, so weight of syndrome is also odd in case of odd number of errors. Similarly, even weighted syndrome implies even number of errors. So, non-zero syndrome value implies that either notDED or DED will be satisfied. In the proposed decoder single and triple adjacent errors are corrected, when notDED signal is satisfied and this signal is expressed as the XOR sum of all the syndrome bits. Also the DED signal

Fig. 4 Flowchart of proposed decoding technique

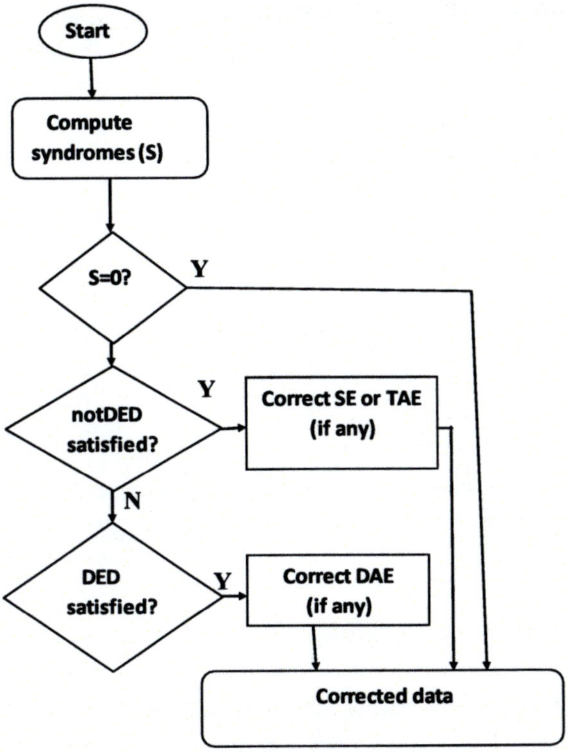

which is the complement of notDED signal is used for correcting double adjacent errors. Also errors are located by employing the only the one's in the syndrome values instead of all the syndrome bits. The error location determination (ELD) logic for the first data bit of proposed (24, 16) decoder has been shown in Eq. (7).

$$ELD_1 = ((S_2\&S_4\&S_8)|(S_1\&S_2\&S_6)\&nD)|(((S_1\&S_2\&S_3\&S_8))\&D \quad (7)$$

where "&" and "|" indicate AND and OR operations, respectively, nD and D are the notDED and DED signals, respectively, which have been presented in Eqs. (8) and (9).

$$nD = S_1 + S_2 + S_3 \cdots + S_7 + S_8 \quad (8)$$

$$\text{and } D = (nD)' \quad (9)$$

So, the ELD expressions of proposed decoder are more simplified compared to traditional SEC-DED-DAEC-TAEC decoder. This is due to the requirement of lesser number of AND logic gates and only one NOT logic gates in the proposed decoder.

Table 4 Comparisons of FPGA-based synthesis results

Codec	Area (LUTs)	Delay (ns)	Improvement (%)	
			Area	Delay
Proposed (24, 16) SEC-DED-DAEC-TAEC	83	2.82	10.84	25.17
Neale (24, 16) I5 SEC-DED-DAEC-TAEC [15]	94	3.53	−31.33	18.79
Adalid (22, 16) SEC-DAEC-TAEC [16]	57	3.35	–	–
Proposed (41, 32) SEC-DED-DAEC-TAEC	163	3.99	0	5.26
Neale (41, 32) I5 SEC-DED-DAEC-TAEC [15]	163	4.20	1.22	10.53
Adalid (39, 32) SEC-DAEC-TAEC [16]	165	4.41	–	–
Proposed (75, 64) SEC-DED-DAEC-TAEC	338	4.01	7.70	18.20
Neale (75, 64) I5 SEC-DED-DAEC-TAEC [15]	364	4.74	−2.66	14.71
Adalid (72, 64) SEC-DAEC-TAEC [16]	329	4.60	10.84	25.17

The third and final task of the proposed decoder are to correct located errors, and this is achieved by employing XOR logic gates.

6 FPGA-Based Synthesis Results

The proposed codecs have been simulated and synthesized on field programmable gate array (FPGA)-based vertex 7 (xc7vx330t-3ffg1157) device families. This synthesis results of proposed and existing-related works in [15, 16] have been summarized in Table 4. As shown in Table 4, the SEC-DAEC-TAEC codec by Adalid et al. exhibits minimum number of lookup tables (LUTs) requirement for 16-bit word length. But the proposed 16-bit codec has lesser number of LUTs requirement compared to Neale et al. 16-bit codec. In case of 32-bit word length, the LUTs requirement of the proposed and existing codecs is almost equal. The LUTs requirement of proposed and Adalid et al. 64-bit codecs is nearly equal and slightly lower compared to Neale et al. 64-bit codec. The proposed codecs experience lower delay compared to all the existing codecs for all three word lengths as indicated in Table 4. But maximum improvement in delay has been achieved against Neale et al. 16-bit codec.

7 Conclusion

In this paper, a new SEC-DED-DAEC-TAEC code has been proposed for the word lengths 16, 32, and 64 bits, respectively. The proposed codecs have lower delay compared to the existing-related works. Also miscorrection rate of the proposed codes has been computed for two-random and three-random errors. It has been observed that the proposed codes have lower miscorrection rates compared to those related works. So the proposed scheme can be applied to the embedded SRAM memory where high speed and lower miscorrection rate are desired for correcting single, double, and triple adjacent error.

References

1. Chen CL, Hsiao MY (1984) Error-correcting codes for semiconductor memory applications: a state-of-the-art review. IBM J Res Develop 28(2):124–134
2. Hsiao MY (1970) A class of optimal minimum odd-weight-column SEC-DED codes. IBM J Res Develop 14(4):301–395
3. Ibe E, Taniguchi H, Yahagi Y, Shimbo K, Toba T (2010) Impact of scaling on neutron-induced soft error in SRAMs from a 250 nm to a 22 nm design rule. IEEE Trans Electron Devices 57(7):1527–1538
4. Samanta J, Bhaumik J, Barman S (2017) Compact CA-based single byte error correcting codec. IEEE Trans Comput 67(2):291–298
5. Reviriego P, Argyrides C, Maestro JA (2012) Efficient error detection in double error correction BCH codes for memory applications. Microelectron Reliab 52(7):1528–1530
6. Dutta A, Touba NA (2007) Multiple bit upset tolerant memory using a selective cycle avoidance based SEC-DED-DAEC code. In: Proc. 25th IEEE VLSI Test Symp., pp 349–354
7. Ming Z, Yi XL, Wei LH (2011) New SEC-DED-DAEC codes for multiple bit upsets mitigation in memory. In: Proceeding IEEE/IFIP 20th International Conference VLSI System Chip, pp 254–259
8. Dutta A (2012) Low cost adjacent double error correcting code with complete elimination of miscorrection within a dispersion window for multiple bit upset tolerant memory. In: Proceeding IEEE/IFIP 20th International Conference VLSI System Chip, pp 287–290
9. Maity RK, Samanta J, Bhaumik J (2019) An area and power efficient double adjacent error correcting parallel decoder based on (24, 12) extended golay code. In: 2019 IEEE international conference on electrical, computer and communication technologies (ICECCT), IEEE, pp 1–6
10. Maity RK, Samanta J, Bhaumik J (2020) New compact SEC-DEDDAEC code for memory applications. In: Proceedings of the 2nd international conference on communication, devices and computing, Springer, Singapore, pp 321–329
11. Reviriego P, Martinez J, Pontarelli S, Maestro JA (2014) A method to design SEC-DED-DAEC codes with optimized decoding. IEEE Trans Device Mater Reliab 14(3):884–889
12. Reviriego P, Liu S, Xiao L, Maestro JA (2016) An efficient single and double-adjacent error correcting parallel decoder for the (24,12) extended golay code. IEEE Trans Very Large Scale Integr (VLSI) Syst 24(4):1603–1606
13. Li J, Reviriego P, Xiao L, Liu Z, Li L, Ullah A (2019) Low delay single error correction and double adjacent error correction (SEC-DAEC) codes. Microelectron Reliab 97:31–37
14. Neale A, Sachdev M (2013) A new SEC-DED error correction code subclass for adjacent MBU tolerance in embedded memory. IEEE Trans Device Mater Rel 13(1):223–230
15. Neale A, Jonkman M, Sachdev M (2015) Adjacent-MBU-tolerant SEC-DED-TAEC-yAED codes for embedded SRAMs, IEEE Trans Circuits and Systems-II: Express Briefs, vol. 62, no. 4, (2015) 387–391.
16. Saiz-Adalid L-J, Reviriego P, Gil P, Pontarelli S, Maestro JA (2015) MCU Tolerance in SRAMs through low-redundancy triple adjacent error correction. IEEE Trans Very Large Scale Integr (VLSI) Syst 23(10):2332–2336
17. Maity RK, Tripathi S, Samanta J, Bhaumik J (2020) Lower complexity error location detection block of adjacent error correcting decoder for SRAMs. IET Comput Digit Tech

Security of Cloud Computing Using Quantum Zero-Knowledge Proof System

Surya Bhushan Kumar, Ranjan Kumar Mandal, Kuntal Mukherjee, and Rajiv Kumar Dwivedi

Abstract Security and privacy play a significant role while using cloud computing. Users can use cloud services, if and only if they find it trust worthy. Users can outsource their sensitive data to cloud environment provided they fill that their data is secured at cloud environment. The popularity of cloud computing depends upon how it can provide the security and privacy of users' valuable data. Thus, a secured cloud computing is the basic agenda prior to adoption of cloud computing. It is doubtless that the security of cloud computing can only be achieved provided by users of it is the authentic one. In this context, this endeavor has proposed the unique mechanisms to verify the authenticity of the users of the cloud. Apart from this trivial concept of verifying the authenticity, the Zero-Knowledge Proof concept has been introduced here. Furthermore, the concept of quantum mechanics has been incorporated here to provide Zero-Knowledge Proof Engine (ZKE) more effective. Thus, a quantum-based ZKE framework along with algorithm is the main concept of this manuscript. An experimental result has been shown to lay bare the effectiveness of the proposed algorithm.

Keywords Zero-Knowledge Proof System · Secured cloud services · Polarization of photons · Quantum Zero-Knowledge Proof · Quantum security · Cloud security from DDoS attack

1 Introduction

Cloud computing has gained huge goodwill and support by scaling on-demand Virtual Machine (VM) infrastructure with the help of accessibility, flexibility, and

S. B. Kumar · R. K. Dwivedi
Department of Computer Applications, VBU University, Hazaribag, Jharkhand, India

R. K. Mandal
Department of Computer Science and Engineering, Ranchi University, Ranchi, Jharkhand, India

K. Mukherjee (✉)
Department of Computer Science and Engineering, BIT Mesra, Ranchi, Jharkhand, India
e-mail: kmukherjee@bitmesra.ac.in

K. Dahal et al. (eds.), *Internet of Things and Its Applications*, Lecture Notes in Electrical Engineering 825, https://doi.org/10.1007/978-981-16-7637-6_38

computing of assets when user requests. VMs as cloud infrastructure will never endure insufficient of assets due to enough amount of as and when require assets in the cloud. This characteristic of "flexibility" or "auto-scaling" comes out into economic losses-based DDOS attacks which is called as economic denial of sustainability (EDoS) attack or Fraudulent Resource Consumption attack (FRC). Nowadays, DDoS becomes EDoS, when the target service is offered by the cloud for a victim or competitor. The Attackers use Bots and Trojans to disrupt the competitor's services on cloud for blackmailing, personal benefits, economical jealously, and political enmity among countries in order to create cyber battles [1]. As found in the research work in [2], DDoS attacks now called as economic denial of sustainability (EDoS) due to pay per use payment method in cloud environment. The main objective of EDoS attack is to make cloud costing model unsustainable. The benefit of cloud infrastructure is to only use hardware and no need to worry about such as room size, electricity bill, and air cooling. The cloud servers which are mostly paid on an hourly basis become EDoS then pricing becomes more important. Like DDoS attacks, EDoS attacks are also based upon three types, i.e., connection layer exhaustion attacks, bandwidth consuming attacks, and specific target attacks. To mitigate attacks and resources distribution, VMs need additional demanding resources allocation in general situation in cloud [1]. In addition to this, EDoS and DDoS attacks are very high yielding for attackers to gain benefits over cloud functionalities like multi-tenant, auto-scaling or pay-as-you-billing. This permits a cloud user to use services without buying them. This also allows more benefits to work on a physical server notwithstanding VM from separate VM owners. The problem occurs when any VM is affected with venomous application then DDoS attack on a physical host is initiated by that VM which causes disrupt the services [3]. Again, DDoS is observed in the research work of [3] like collateral damage. Amazon EC2 cloud servers confront a huge loss due to DDoS attacks by the attackers. This results in heavy monetary losses and many brief and everlasting effects on their business and the clients too. Furthermore, according to Verisign iDefense Security Intelligence Services, the most attacked spot of DDoS is cloud and software as a service (SaaS) in end region. It is also noticed that mostly prevention techniques were used only for cloud in order to rescue from huge economic losses. Due to DDoS attack, approximately, 444 K USD total financial loss has been recorded and the total financial loss in research of DDoS attack holds out to 66 K per hours [3].

2 State of the Art

Virtualization of cloud infrastructure as a service provides server flexibility and requirement by their users. The more VMs resources and services are used by the cloud customers and become inoperative when do not utilize. This kind of scenario in medical terms is called organ donation. Nowadays, cloud server infrastructure is on very high demand and that is why following organ (infrastructure) donation trend [4]. An interesting approach is stated in [5] is a Parallel and Multi-Stage Security Mechanism (PMSSM). This mechanism is based upon encryption approach in order to achieve intrusion identification and authentication. This multi-staging defense mechanism is used to ensure cloud data security and of course intrusion detection. Furthermore, the authors proposed an Artificial Immune System in [6] which is designed to mitigate DDoS attacks. This intelligent system behaves as an impersonator to find out DDoS attacks. This artificial system is plotted on the human biological resistance scheme. A KDD cup 99 dataset is brought into experimental assessment to design the system. This system accuracy for DDoS attack was 96.56%. The third-party auditors authorize the data reliability authentication process by entailing conditional proxy re-encryption. In this, in order to deal the issue with remote data storage, the auditor uses metadata. This proposed approach reduces the additional cost of metadata generation [7]. A novel approach has been proposed in order to identify the DDOS attacks using a machine learning optimization technique. This system learns from usual and unusual situation in cloud server to mitigate DDoS and EDoS attacks and also distinguish these attacks. This approach combines a Cuckoo Search (CS) approach with artificial neural network (ANN) approach is called CS—ANN. The CS optimizes the features of cloud user and attacker and then put forward to ANN structure. The CS—ANN experiment results confirm true and false positive rates and detection accuracy of 0.99%, 0.0105%, and 0.9865%, respectively [8]. Precisely, the security plays an important role to become accustomed the use of cloud servers such that most of the users avoid cloud services offered by the providers. The main security issue is identity theft due to password authentication in cloud servers. Some decades ago, the physical protection of user id and password was stored in PC and these data were difficult to access by illegitimate users. Woefully, this type of physical protection is not allowed by cloud computing [9]. Mostly, the online users set their easy passwords such as family member names, favorite stuff, games, and mobile numbers. By the help of dictionary attack, anyone can get password easily. In addition to this, man-in-the-middle (MITM) attack, offline attack, and others online attacks are used to gather the passwords. In MITM, the attackers eavesdrop between the two users and gain access to their passwords to login into the cloud server and kick out the authentic user and hack his or her account. But, in traditional approach, offline attackers are hard to get any vital information regarding the user to gain access on his or her account [10, 11]. The research work in [12], the two-way authentication mechanism is proposed. In this smart card-based mechanism, the card contains users' credentials and safeguards by password. The lack of this mechanism is it is a technical device and the smart card has extra cost. Additionally, a middleware application is used to initiate the smart card communication process. The digital identity is recognized by biometric aspects such as iris, speech to verify the user on behalf of the biometric data stored. It is limited and time-consuming process when it additionally undergoes from reply attack such that it needs additional devices to gather the information and to process it. The serious drawback of this system is the failure to accept logins of a large number of users at concurrent time [13].

3 Proposed Approach

3.1 Proposed Framework

In this Proposed Framework (Fig. 1), the user of cloud computing has to prove himself/herself that he/she is authorized user of it (Prover). For the whole process, the Zero-Knowledge Proof Engine (Verifier) would check the credibility of Prover. The Zero-Knowledge Proof Engine (ZKE) would neither ask login ID of Prover nor would his/her password rather only check the credentiality of his/her by putting a series of question before him/her. If Prover provides the satisfactory answer to those questions then only he/she would be allowed to access the cloud services. The main challenges are to design such series of questions for ZKE.

The success of ZKE would depend on the design of efficient algorithms for the whole verification process. The next subsection of this manuscript has proposed algorithm for the same.

There are three types of cloud computing services, namely "Infrastructure As A Service (IAAS)", "Platform As A Service (PAAS)", and "Software As A Service (SAAS)".

IAAS offers on-demand storage servers, computing resources, etc.

PAAS provides platform for developers to develop their own application software and models.

Furthermore, SAAS provides different applications on-demand based on "pay-as-you-use" model.

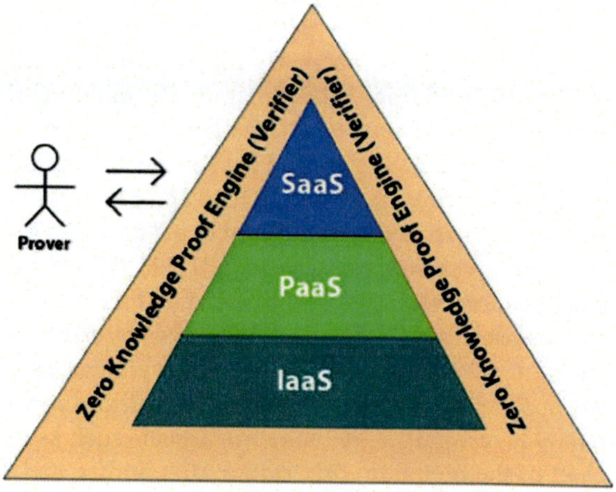

Fig. 1 Proposed framework of secured cloud computing

3.2 Proposed Algorithm

The proposed algorithm is given below:

Step 1: The user willing to get cloud services first applies for registration.

Step 2: Cloud Service Provider (CSP) would allow the user to do the whole registration through public channel.

Step 3: After a complete interaction between the CSP and the user through public channel, a full documentation as follows is finally stored at the CSP server.

Step 4: After Step 3, the CSP would provide the user his/her userid and password through private channel as:

Step (a) The user would find the numerical equivalent of his fundamental attributes as shown in Table 1.

Step (b) The user has to find out the co-prime of his/her fundamental entities (shown in Table 1) with the random number sent to him/her during online services. If two fundamental entities ties up then the entity with higher priority would be considered.

Step (c) Now, the user would consider random number sent to him/her by the CSP as "x" whereas co-prime to x from his/her fundamental attributes (described in Table 1) would be considered as "y".

Step (d) Now, the user would find out new number "z" such that (x, y, z) is primitive.

Step (e) Now, "z" is taken as the basis of photon for the secured communication between cloud services and the users.

Now, when the user interacts with the CSP for services through public channel using his/her user id the ZKE engine will generate a random number, say, x and would send it to user for verification. If the user is an authorized user then he/she would find "z" as per above-mentioned protocol. Now ZKE would generate a "welcome" signal to user with "z" as basis of photon and awaits response from user at same basis "z". If both matches then ZKE would declare user as authorized user and starts to provide cloud services to the users for same basis "z" is shown in Fig. 2.

Table 1 User details

Name: Ray
Date of Birth: 01/03/1996
Place of Birth: Ranchi
Time of Birth: 10:09:00
Security Question: Cafe

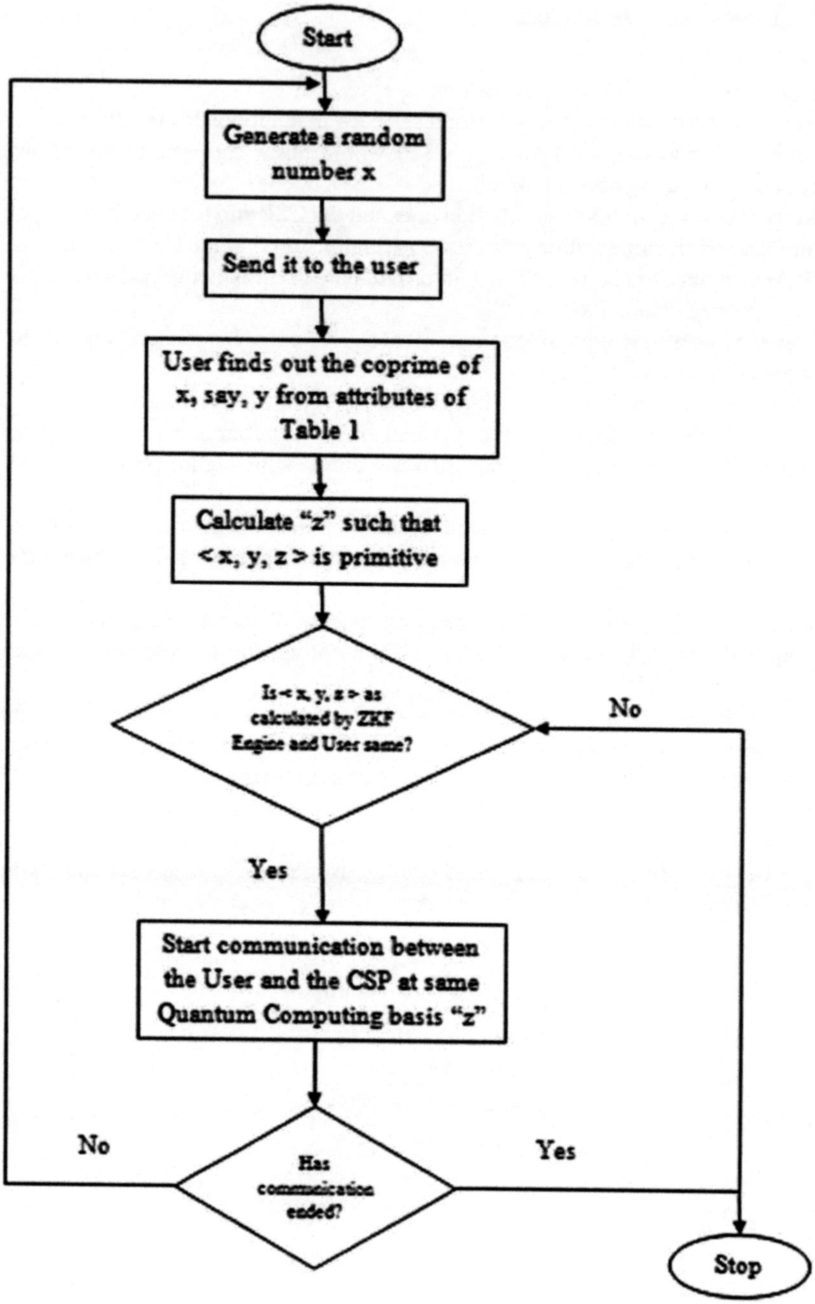

Fig. 2 Proposed algorithm

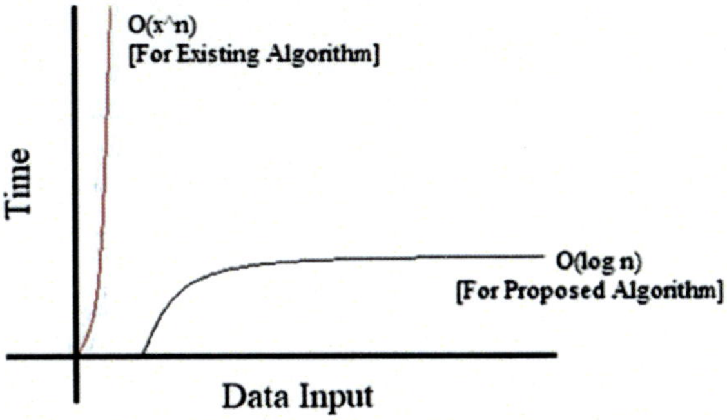

Fig. 3 Comparative study between proposed algorithm and existing algorithm

3.3 Analysis of Proposed Algorithm

By inspection, it is observed that the time complexity of proposed algorithm is $O(\log(n))$. The comparative study of the proposed algorithm and the existing algorithm is shown in Fig. 3.

3.4 Explanation of Proposed Algorithm

For sake of simplicity, let us assume the following example –

Name: Ray.

Date of Birth: 14/08/1996.

Place of Birth: Ranchi.

Time of Birth: 10:09:00.

Security Question: Cafe.

The whole detail is shown in Table 2. The Unicode value of name at Table 2 is 192, the numeric value of Date of Birth at Table 2 is 14081996, the Unicode value of Place of Birth at Table 2 is 313 + 6E, the numeric value of Time of Birth at Table 2 is 100900, and the Unicode value of Security Question at Table 2 is 235.

Let the random number generated by ZKE, i.e., "x" be 7 and ZKE sends this number, i.e., 7 (i.e., x) to the users to check for verification of his authorization. Now

Table 2 Unicode values of user details

Name: Ray	Unicode value is 192
Date of Birth: 01/03/1996	Num value is 01031996
Place of Birth: Ranchi	Unicode value is 313 + 6E
Time of Birth: 10:09:00	Num value is 100900
Security Question: Cafe	Unicode value is 235

users would find co-prime of 7 (i.e., x) from the data available at Table 2. If ties then that number from Table 2 would be considered whose rank is high.

The co-prime of "x" (i.e., 7) is 192 from above Table 2. Hence, here the value for "y" is "192". Now, let us calculate "z", such that (x, y, z) is primitive, here "z" is "192.13". Now ZKE would send "Welcome message" to user using photon of basis z i.e., "192.13".

If the user is an authorized user, then the user would also be able to count "z" as "192.13" and would response to ZKE at same polarize angle "192.13". Once ZKE finds the same polarize angle "192.13", would declare user as authorized user and then after the cloud services by the CSP would be provided to the users at basis "z" of photon.

4 Experimental Setup and Results

The simulation of the proposed algorithm has been done using Python programming language. The users will interface with the system through apps. The overall process is shown in Fig. 4. After insertions of the required data by users through the interface apps, the calculation for value of "z" is done. The overall working process is shown

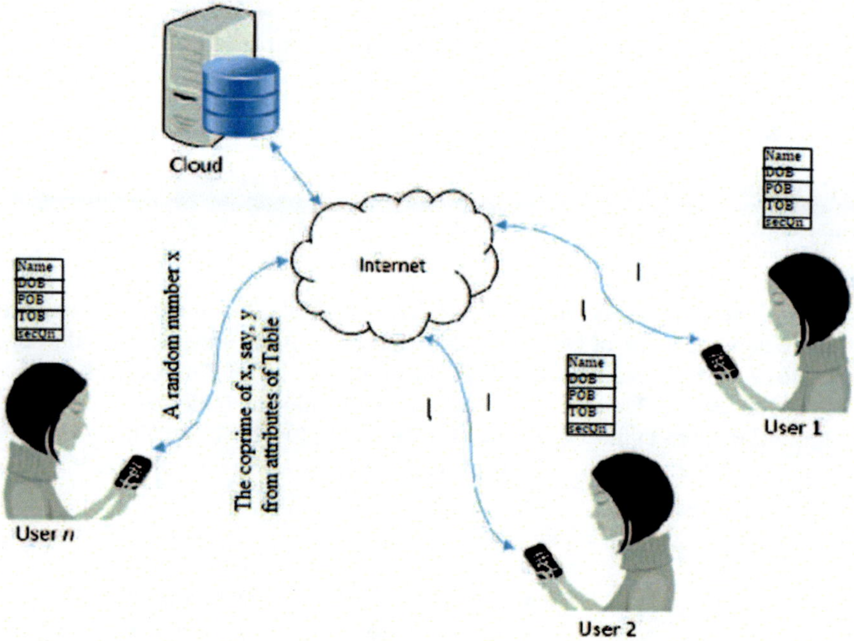

Fig. 4 User interface

in Fig. 5. Furthermore, the polarization of the photon is done based upon the above calculations. The final simulation result is shown in Fig. 6.

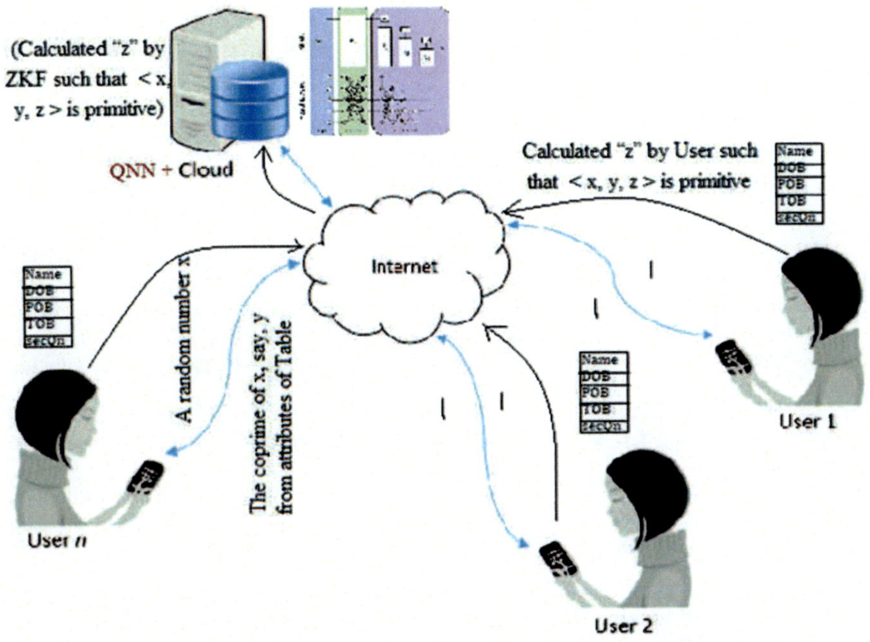

Fig. 5 Process for calculation of z

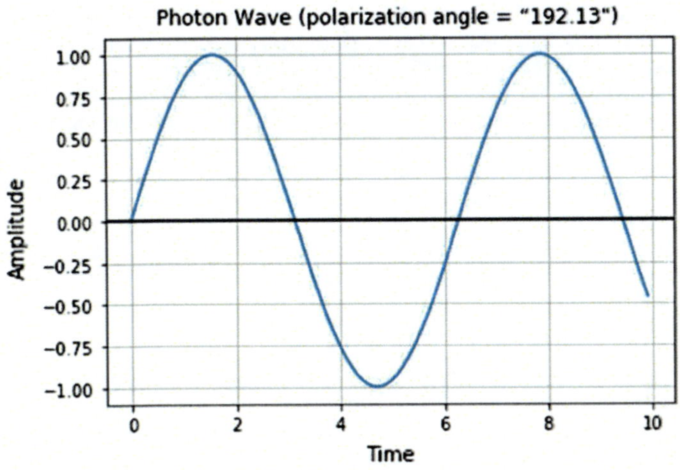

Fig. 6 Simulated result

5 Conclusions

In this manuscript, photon is polarized at a specific angle for verifying the user of the cloud by ZKE which acts as a verifier. When ZKE finds that the users are the authorized user without asking the password of the user, then only the cloud services are provided to the user.

References

1. Arvindhan M, Ande BP (2020) Data mining approach and security over Ddos attacks. ICTACT J Soft Comput 10(2):2061–2065. https://doi.org/10.21917/ijsc.2020.0292
2. Bunkar K, Singh UK, Pandya B, Bunkar R (2012) Data mining: prediction for performance improvement of graduate students using classification. In: 2012 ninth international conference on wireless and optical communications networks (WOCN), Indore, pp 1–5. https://doi.org/10.1109/WOCN.2012.6335530
3. Somani G, Gaur MS, Sanghi D, Conti M (2016) DDoS attacks in cloud computing: collateral damage to non-targets. Comput Netw 109(2):157–171. https://doi.org/10.1016/j.comnet.2016.03.022
4. Badr A, William A (2017) Proactive approach for the prevention of DDoS attacks in cloud computing environments. In: Lee R (ed) Applied computing and information technology. Studies in Computational Intelligence, vol 695, Springer. https://doi.org/10.1007/978-3-319-51472-7_9
5. Goyal R, Manoov R, Sevugan P, Swarnalatha P (2020) Securing the data in cloud environment using parallel and multistage security mechanism. In: Soft computing for problem solving, vol 1057, pp 941–949, Springer, Singapore. https://doi.org/10.1007/978-981-15-0184-5_80
6. Prathyusha DJ, Kannayaram G (2020) A cognitive mechanism for mitigating DDoS attacks using the artificial immune system in a cloud environment. Evol Intel. https://doi.org/10.1007/s12065-019-00340-4
7. Salim A, Tiwari RK, Tripathi S (2020) An efficient public auditing scheme for cloud storage with secure access control and resistance against DOS attack by iniquitous TPA. Wireless Pers Commun pp 1–26. https://doi.org/10.1007/s11277-020-07079-7
8. Alzahrani AS (2020) An optimized approach-based machine learning to mitigate DDoS attack in cloud computing. Int J Eng Res Technol 13(6):1441–1447. https://doi.org/10.37624/IJERT/13.6.2020.1441-1447
9. Ramachandra AC, Bhattacharya S (2020) Literature survey on log-based anomaly detection framework in cloud. In: Computational intelligence in pattern recognition, vol 1120, pp 143–153, Springer, Singapore. https://doi.org/10.1007/978-981-15-2449-3_12
10. Mondal HS, Hasan MT, Hossain MB, Rahaman ME, Hasan R (2017) Enhancing secure cloud computing environment by Detecting DDoS attack using fuzzy logic. In: 2017 3rd international conference on electrical information and communication technology (EICT), pp 1–4, IEEE Dec 2017. https://doi.org/10.1109/EICT.2017.8275211
11. Biswas R, Wu J (2018) Filter assignment policy against distributed denial-of-service attack. In: 2018 IEEE 24th international conference on parallel and distributed systems (ICPADS), pp 537–544, IEEE Dec 2018. https://doi.org/10.1109/PADSW.2018.8644584
12. SecureList, Kaspersky Lab Report, Distribution of DDoS attacks by type, Q1 2019, Available Online at https://securelist.com/ddos-report-q1-2019/90792
13. Verma P, Tapaswi S, Godfrey WW (2020) An adaptive threshold-based attribute selection to classify requests under DDoS attack in cloud-based systems. Arab J Sci Eng 45:2813–2834. https://doi.org/10.1007/s13369-019-04178-x

Survey on Botnet Detection Techniques

Rahul Mishra and Sudhanshu Kumar Jha

Abstract Due to unpleasant market competition, IT companies are releasing the software or applications without much considering the unintended security breaches presented inside it. The malware programs moving around the Internet is looking for such kind of the security breaches to attain the malicious intention. Botnet is a kind of malware program(s) looking for the vulnerable system and has now become the worldwide epidemic due to mostly applied for a malicious purpose. Many malware detection techniques have been discussed so far for botnet detection in the literature, however typically considering it on host device or on the traffic of a network. IoT devices are much vulnerable to botnet as the manufacturer has the main concern to releases with new feature in order to compete the market without much attention to weak point(s) inside it. This paper discusses the basic botnet detection techniques. For this purpose, the paper covers both static and dynamic detection techniques along with its advantages and shortcomings. On the basis of characteristics for botnet detection, this paper also deliberates the basic procedures to create botnet detectors, defining some parameters for botnet detection and categorizing the detection methodologies. Further, the paper reveals an implementable position in the system with advantages and drawbacks on the detection performance.

Keywords Botnet · Botnet detection technique · Anomaly-based detection · Signature-based detection · Specification-based detection · IoT Vulnerability

1 Introduction

Due to the present coronavirus COVID-19 pandemic situation, a huge demand is looking for various ICT tools to meet the immediate business need without much concern to the security breaches inside it. Self-propagating botnet programs are moving around the Internet in search of vulnerable system. As per report by World Health Organization (WHO), more than 450 e-mails addresses and passwords of WHO's were compromised during April 2020 [1] and a forecasted report by P&S

R. Mishra (✉) · S. K. Jha
Department of Electronics and Communication, University of Allahabad, Prayagraj 211002, India

K. Dahal et al. (eds.), *Internet of Things and Its Applications*, Lecture Notes in Electrical Engineering 825, https://doi.org/10.1007/978-981-16-7637-6_39

intelligence, Market Research Future (MRFR) shows a compound annual growth rate (CAGR) of 12.6% with total $119.9 billion financial loss by the end of 2030 due to cyber-attacks [2, 3], and thus, this area needs immediate attention among researcher.

The journey of malwares was first appeared in 1988 and continues their impact till date. McGraw and Morrisett [4] define malicious code or malware as "any code added, changed, or removed from a software system in order to intentionally cause harm or subvert the intended function of the system." Initially, Internet users were attacked with malwares, propagating through e-mails, freeware, lucrative software, and games or with some other medium [5]. However, by the time, this kind of approaches for infecting devices become not much effective, as the infection processes were easily detectable due to the actions initiated by the malware during infection or changes caused on the infected devices. With the advancement in anti-malware protection and elimination programs, the malware presence is being detected and immediately removed from the infected devices, and thus, the infected devices are no longer remain asset for the attacker. With the advancement in technology, the intention of trend for infection was changed, and now, attacker wish to hold the control of infected device as long as possible. Therefore, the presence of malware in the infected device needs to be shield. Various kinds of malwares are being used; botnet is one of them.

Botnets are utilizing the feature of self-propagation to target more and more devices by exploiting the vulnerabilities present in the devices such as open ports and default credentials. IoT devices are most suitable for performing such activities as the devices are implemented with least security mechanisms, continuous network connectivity, and limited computing resources in order to retain it simple to use. IoT devices are dynamic, heterogeneous, and interoperable, and due to these features, a uniform solution to prevent or mitigate the botnet in IoT devices is not feasible [6]. Most of the IoT devices are connected to Internet without firewall and available round the clock and are mostly configured with factory-enabled default username and password with open ports for various protocol, customer-care service, and thus, to contaminating an IoT device becomes quite easy. The device infected with malicious code called bot and group of infected devices together with Command and Control Centre (C&C) termed botnet [7]. In a botnet, infected devices (bots), take command(s) from the C&C to perform predefined action(s) on the basis of the received commands. These commands are given by the attacker who controls the botnet. Typically, the commands are being used for performing DDoS attack, sending spam mails, click fraud, or stealing financial and sensitive information from the infected device [8]. Many researchers believe that more than 25% of the IoT devices which are connected to the Internet without any proxy server are member of botnets [8–10].

Primarily botnets are configured for testing various features of network and however, later on, intruders started implementing bots with intention to perform malicious activities such as purloin financial information, security credential, sending spam mails, or performing DdoS attack to slow down or sometimes stop the services of the targeted system [11]. These botnets are capable of utilizing exact vulnerabilities available on specific devices from a manufacturer, in order to the keep the device easy to use, companies provide details on their website such as default credentials, open telnet ports for remote access of device. This publicly available information is utilized by the attackers to target any IoT device or a model of a manufacturer. Mostly, the botnet exploits the codes of exiting botnets or append new features to exploit the new vulnerabilities of the IoT devices [12]. The malware detector is a kind of program which basically looks for the description and the basic functionalities of malicious program. Unlike an anti-virus software, a malware detector is not necessarily supposed to reside on the device under observation (DUO) and senses the presence of malware on the basis of set of rules of the detection techniques. The performance of a malware detectors depends on the set of detection techniques it uses [7]. Many researches have been carried out to predict or mitigate the botnet attack. These detection techniques utilize various parameters of botnet or network such as botnet signature, network traffic, very-long connection time between client–server and so on [13–15].

The motivation and main contribution of this paper are as follows: Sect. 2 of the paper discusses the taxonomy of a botnet with common IoT vulnerabilities. Section 3 deliberates the basic detection techniques and a discussion about categories of botnet, based on common botnet attack along with a detailed review and comparison analysis of botnet detection techniques and tools followed by conclusion in Sect. 4.

2 Botnet Taxonomy

As described, a botnet is a network of connected bots that spread over network to perform various malicious activities such as spam mail generation, distributed denial of service (DDoS) attack, stealing sensitive financial (credit/debit card data) and security information, and tricking personal information for identity theft [13]. An autonomous program performing the above actions without taking instructions from any intruder is called bot, whereas the network of bots connected to Command and Control Server (C&C) taking commands to perform actions based on the commands are called botnet [8–10]. Botnets utilize the vulnerabilities present in the devices to infect them, and once a device is infected, it starts working as a bot and further search for the new device on the network for further infection. List of common vulnerabilities in Table 1 [13–15]. These botnets generally utilize default credentials of IoT devices, open ports or sometimes the vulnerabilities present in the software. Mirai botnet is the famous botnet that performs dictionary-based default credential attack on the devices.

The connection between bots and C&C defines the architecture of the botnet. The architecture of the botnet can be categorized as centralized, peer to peer (P2P), and hybrid [7]. The centralized architecture of botnet is easy to implement, generate quick response to bots' requests, quick and direct update to bots but the dependability on a single source make is less reliable.

Whereas the P2P architecture does not directly communicate to the bots, rather command is sent via another bot in the network so the detection of the C&C becomes

Table 1 Common IoT vulnerabilities

Vulnerability	Description
Default credentials	Many IoT devices have no mechanism to change the default credentials. The credentials of the devices are available on the manufacturer's website for open access
Un-authorized access	IoT devices do not ensure who can access the data available with them
Insecure software/firmware	These devices also lack the update mechanism or if the patch is available, then it is not verified
Physical security	Sometimes these devices are installed on open places or their physical security is not considered and open to disassemble or can be accessed via USB or some removal storage medium
Open ports and network services	These devices are vulnerable to DDoS attack, and network ports are open for remote login purposes

next to impossible. The hybrid architecture utilizes the features of both architectures to control the bots. Figure 1 shows the architecture of the botnet.

Authors [11, 12] classified the IoT attacks on the basis of how the attacker utilizes these devices after successful infection. Table 2 shows the categories of attacks.

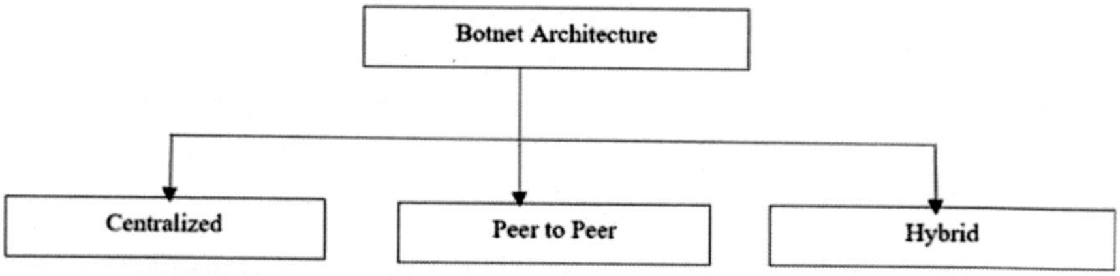

Fig. 1 Botnet command and control topology

Table 2 Common botnet attack categories

Category	Description
Ignoring the functionality	In this type of attack, the intended works of the IoT devices are ignored and the IoT devices are considered only as the computing device that is connected to the Internet
Reducing the functionality	These attacks are designed to limit the functionalities of the IoT devices. But these kinds of attacks cost human life if the target is medical equipment
Misusing the functionality	Misusing the functionality of devices may sometime cause reverse impact of the intended functionality or doing something that is not expected from the device
Extending the functionality	These attacks are designed to extend the functionalities of the infected IoT devices; i.e., infected devices are performing works for which they were not designed

These attacks were categorized on the basis of the impact caused by the malicious code after they have successfully infected the device or network.

3 Botnet Detection Techniques

Botnet detection techniques are the most discussed topic nowadays. Many works have been done to address this issue. Here in this paper, we are trying to find out the most relevant works and categories them how they are addressing detection of botnet. The botnet detector takes two inputs [12]. The first input is the knowledge of the malicious behavior of the botnet. Second input is the program that needs to be observed. Once the botnet detector has the knowledge of what is considered malicious behavior and the program under inspection, then it employs its detection techniques to decide that the program is malicious or benign. Sometimes IDS and malware detectors are used interchangeably but a malware detector is usually only a component of a complete IDS. Techniques used for detecting malware can be categorized into three categories: anomaly-based detection, specification-based detection, and signature-based detection [16]. Figure 2 provides the information about various botnet detection approaches. All the three categories have three subcategories, namely static, dynamic, and hybrid [13, 17–19].

An anomaly-based botnet detection technique uses its gathered information during the run-time of a program to constitute normal behavior of the program to decide the maliciousness of a program under observation. A sub-category of anomaly-based detection is referenced as specification-based detection.

Specification-based botnet detection techniques control some specification or set of rules of what is valid behavior in order to decide the maliciousness of a program under observation. Programs violating the set of rules of specification are considered anomalous and usually malicious [20].

Signature-based botnet detection technique uses its predefined set of rules to what is known to be malicious for the host under observation to decide the maliciousness of a program under inspection. It is clear that characterization of properties or signature

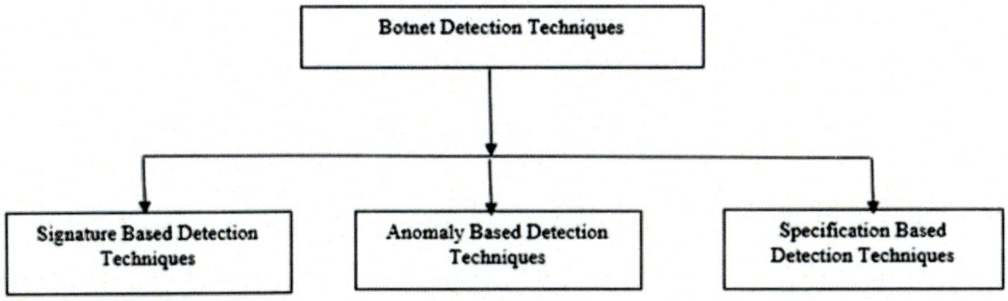

Fig. 2 Botnet detection techniques

of the malicious behavior is the key to a signature-based botnet detection method's effectiveness.

Table 3 presents a detailed analysis of variances among static and dynamic approach. Static analysis uses syntax or structural properties of the program under observation to predict its maliciousness. Whereas the dynamic approach works on various features on the host or the network under observation such as connection time between client–server, DNS and starts when the code starts executing [21]. The specific approach of an anomaly-based or signature-based technique is determined by how the techniques gather information to detect malware.

Table 3 Review of botnet detection techniques and tools

Techniques	Description	Tools
Dynamic anomaly-based detection technique	The information is gathered from the program under observation (POU) when starts executing on the host. This technique also	PAYL[22], Computer Forensic Method for Privacy invasive Software [23]
Static anomaly-based detection technique	Characteristics of file's structure of the PUO are used to detect. The malware can be detected even it is not being executed	Fileprint Analysis [24]
Hybrid anomaly-based detection technique	Uses the features of both dynamic anomaly-based detection and static-based detection	Strider GhostBuster [25]
Dynamic specification-based detection techniques	Tries to categories specification-based behavior at run-time to detect the malicious code	Monitoring Security-Critical Programs [26], Using Dynamic Information Flow to Protect Application [27]
Static specification-based detection techniques	Focuses on structural properties of files of PUO	Static Detection of Malicious Code in Executables [28], Detecting malware in Firmware [29]
Hybrid specification-based detection techniques	Specification-based behavior at run-time to detect the malicious code and structural properties of PUO	DOME [30]
Dynamic signature-based detection techniques	Uses the information gathered during execution of PUO	Rule-Based IDS Approach [31], Behavioral Approach to Worm Detection [32]
Static signature-based detection techniques	This technique uses the sequence of the code of PUO	Generic Virus Scanner [33]
Hybrid signature-based detection techniques	Uses the properties of static and dynamic detection techniques to detect the malwares	Analyzing and Detecting Malicious Mobile Code [30]

These botnet detection techniques can be further classified as on host and on network-based detection techniques [9, 11]. On host-based detection technique, the detection of malware is done locally on the host itself. These techniques are local to the machines under observation. While in network-based detection techniques, the network traffic is monitored for the detection of malware. Network-based monitoring technique can be further divided into active monitoring techniques and passive monitoring techniques.

In active monitoring techniques especially, crafted packets are injected into the network traffic and their responses are apprehended for presence of malware in the network. Many legitimate DNS that expires are used by intruders for malicious activities. Such domains are specially used for sending the patches for many software, and when these patches are installed on machine, it starts performing many malicious activities such as keystroke recording, stealing valuable information from the device and sending it to the attacker's device. Passive detection technique monitors such DNS queries for malicious activities that can be a part of any botnet.

3.1 Limitations in Botnet Detection

Botnet detection has many limitations [20, 25, 26] for designing single uniform solution such as heterogeneity, functionalities, and management policies for IoT devices. Sometimes the governing policies, goals of the Internet or network may also limit the applicability of botnet detection mechanisms. Lack of information related to connected devices to a network, sometimes it tough to decide that a particular device belongs to a network. The Command and Control channels also cause problem in detection of botnet many C&C channels use push methodology, whereas some C&C channels use pull methodology for communication. The protocols such as HTTP and IRC are used for communication between client and server. These protocols are also one of the limiting factors for botnet detection.

4 Conclusion

Due to the wide range of applications and easy deployment, IoT devices become a popular choice among people; however, due to unpleasant market demand, manufacture is not much concern about the security breaches in their products. Malware(s) is (are) one of the malicious programs always looking for such kind of vulnerable devices, and thus, IoT devices become one of the great choices. Due to the availability of open source-code of many malwares on various online forums and available for free download, intruders are using their knowledge to add new features to utilize the new vulnerabilities available in the system. In this paper, various methods have been compared based on the static analysis of malware detection. A detailed common IoT vulnerabilities have been presented based on common botnet attack categories.

A brief review of botnet detection techniques and tools also has been discussed so far. Static botnet detection techniques have fixed set of rules to detect the botnet that makes them quick and easy to implement but this also becomes their limitation when it comes to detect any new botnet as their rules are not known to the static-based detection techniques. Whereas dynamic detection techniques detect malware by utilizing the features of botnet when they are executing and trying to detect the malicious action being performed by any device. If so then the alarm is triggered for botnet. This detection technique is complex to implement but has its own advantages as it works on the attributes of the traffic and behavior of the PUO. Discussion on detailed common IoT vulnerabilities and review of botnet detection techniques and tools are main contributions of this paper.

References

1. WHO's report, April 2020. WHO reports fivefold increase in cyber-attacks, urges vigilance. https://www.who.int/news/item/23-04-2020-who-reports-fivefold-increase-in-cyber-attacks-urges-vigilance. Accessed on 22 Nov 2020
2. MRFR report on (Sept 2020). Botnet Detection Market Research Report, by Vertical (Government & Defense, IT & Telecommunications), by Organization Size (Large Enterprise, SMEs), by Application (Mobile-based, Web-based) by Deployment (On-Cloud, On-premise) — Global Forecast till 2023, https://www.marketresearchfuture.com/reports/botnet-detection-market-6477. Accessed on 22 Nov 2020
3. P&S Intelligence press journals. July, 2020. Cyber Security Market Research Report: By Component (Solutions, Services), Security Type (Application, Network, Endpoint, Cloud, Enterprise), Deployment (On-Premises, Cloud), Enterprise (Large Enterprises, SME), Use Case (Security Monitoring, Network Traffic Analysis, Threat Hunting, Incident Response, Data Exfiltration), Industry (Aerospace & Defense, Government, BFSI, Healthcare, Retail, IT & Telecom, Manufacturing) - Global Industry Analysis and Growth Forecast to 2030. https://www.psmarketresearch.com/market-analysis/cyber-security-market. Accessed on 22 Nov 2020
4. McGraw G, Morrisett G (2000) Attacking malicious code: a report to the InfoSec research council. IEEE Softw 17(5):33–41
5. Biozid B, Mohiuddin A (2020) Deep learning meets malware detection: an investigation. In: Combating security challenges in the age of big data, Springer, pp 137–155
6. Yerima SY, Alzaylaee MK (2020) Mobile Botnet detection: a deep learning approach using convolutional neural networks. In: 2020 International Conference Cyber Situational Awareness, Data Analysis Assessment, Cyber SA 2020
7. Karim A, Bin Salleh R, Shiraz M, Shah SAA, Awan I, Anuar NB (2014) Botnet detection techniques: review, future trends, and issues. J Zhejiang Univ Sci C 15(11):943–983
8. Alieyan K, Almomani A, Manasrah A, Kadhum MM (2017) A survey of botnet detection based on DNS. Neural Comput Appl 28(7):1541–1558
9. Tyagi AK, Aghila G (2011) A wide scale survey on botnet. Int J Comput Appl 34(9):9–22
10. Feily M, Shahrestani A, Ramadass S (2009) A survey of botnet and botnet detection. In: Proceeding—2009 3rd International conference emerging security information, system technology security 2009, pp 268–273
11. Daya AA, Salahuddin MA, Limam N, Boutaba R (2019) A graph-based machine learning approach for bot detection. In: 2019 IFIP/IEEE Symposium on Integrated Network and Service Management IM 2019, pp 144–152 April 2019

12. Li H, Hu G, Yuan J, Lai H (2012) P2P botnet detection based on irregular phased similarity. In: International Conference on Instrumentation, Measurement, Computer, Communication and Control IMCCC 2012, pp 79–82
13. Meidan Y, Sachidananda V, Peng H, Sagron R, Elovici Y, Shabtai A (2020) A novel approach for detecting vulnerable IoT devices connected behind a home NAT. Comput Secur 97(101968)
14. Rahal BM, Santos A, Nogueira M (2020) A distributed architecture for DDoS prediction and bot detection. IEEE Access 8:159756–159772
15. Mohd Aman AH, Yadegaridehkordi E, Attarbashi ZS, Hassan R, Park YJ (2020) A survey on trend and classification of internet of things reviews. IEEE Access 8:111763–111782
16. Dong X, Hu J, Cui Y (2018) Overview of botnet detection based on machine learning. In: Proceeding—2018 3rd International Conference Mechanical Control Computer Engineering ICMCCE 2018, pp 476–479
17. Soe YN, Feng Y, Santosa PI, Hartanto R, Sakurai K (2020) Machine learning-based IoT-botnet attack detection with sequential architecture. Sensors (Switzerland) 20(16):1–15
18. Blaise et al. A (2020) Detection of zero-day attacks : an unsupervised port-based approach to cite this version : HAL Id : hal-02889708. Comput Netw, Elsevier 180:107391
19. Damodaran A, Di Troia F, Visaggio CA, Austin TH, Stamp M (2017) A comparison of static, dynamic, and hybrid analysis for malware detection. J Comput Virol Hacking Tech 13(1)
20. Ngo QD, Nguyen HT, Le VH, Nguyen DH (2020) A survey of IoT malware and detection methods based on static features. ICT Express, no. xxxx, 2020 (in press)
21. Thorat SA, Khandelwal AK, Bruhadeshwar B, Kishore K (2008) Payload content based network anomaly detection. In: 1st International Conference on the Applications of Digital Information and Web Technologies ICADIWT, pp 127–132
22. Oliveira LB, Pereira FMQ, Misoczki R, Aranha DF, Borges F, Nogueira M, Wangham M, Wu M, Liu J (2018) The computer for the 21st century: present security & privacy challenges. J Internet Serv Appl 9(1):24
23. Cui B, Jin H, Carullo G, Liu Z (2015) Service-oriented mobile malware detection system based on mining strategies. Pervasive Mob Comput 24:101–116
24. Richer TJ, Neale G, Osborne G (2015) On the effectiveness of virtualisation assisted view comparison for rootkit detection. In: Proceedings of the 13th Australasian information security conference (AISC 2015), vol 27, pp 30 Jan 2015
25. Chandramohan M (2018) Scalable analysis for malware and vulnerability detection in binaries (Doctoral dissertation)
26. Muppidi SR, Kodeswaran PA, Mukherjea S, Nandakumar V, Kapoor S International Business Machines Corp, 2016. Policy-based dynamic information flow control on mobile devices. U.S. Patent 9,253,209
27. Cui Z, Du L, Wang P, Cai X, Zhang W (2019) Malicious code detection based on CNNs and multi-objective algorithm. J Parallel Distrib Comput 129:50–58
28. Sallam AS, McAfee LLC (2016) System and method for firmware based anti-malware security. U.S. Patent 9,317,690
29. Khan WZ, Khan MK, Muhaya FTB, Aalsalem MY, Chao HC (2015) A comprehensive study of email spam botnet detection. IEEE Commun Surv Tutorials 17(4):2271–2295
30. Kumar V, Sinha D, Das AK, Pandey SC, Goswami RT (2020) An integrated rule-based intrusion detection system: analysis on UNSW-NB15 data set and the real time online dataset. Clust Comput 23(2):1397–1418
31. Norouzi M, Souri A, Samad Zamini M (2016) A data mining classification approach for behavioral malware detection. J Comput Netw Commun
32. Wressnegger C, Freeman K, Yamaguchi F, Rieck K (2017) Automatically inferring malware signatures for anti-virus assisted attacks. In: Proceedings of the 2017 ACM on Asia conference on computer and communications security, pp 587–598, Apr 2017
33. Alme C, McAfee LLC (2016) System and method for detecting malicious mobile program code. U.S. Patent 9,246,938

A Comparative Study of Data Encryption Techniques for Data Security in the IoT Device

Rameez Raja Kureshi and Bhupesh Kumar Mishra

Abstract The Internet of things (IoT) has been widely used around the world and the use of IoT applications has been increasing. Because of its accessibility with the existing communication and computing capacity, the IoT application domain has been widening. But, as it has been the issue of security and privacy in other communication-based application, the IoT device-based applications have also been under the threat of attacks. There have been risks of both physical damaging on the devices and digital cyber attacks. To apply the security on the digital IoT system, lightweight cryptography techniques have been used as an alternative to minimize any potential threats and hence make the system more secure. Over the years, different encryption algorithms such as Advanced Encryption Standard (AES), Data Encryption Standard (DES), Triple Data Encryption Standard (3DES), Blowfish, and many others have been implemented in different IoT applications. However, selecting the right algorithm has always been challenging. In this paper, a comparative analysis in terms of the encryption process and throughput has been analyzed for AES, DES, 3DES, and Blowfish algorithms. As an IoT application, Raspberry Pi 3B+ based system for sensing humidity and temperature has been designed for the experimental comparison among these algorithms. In the application, the sensor data has been transmitted to the server using different encryption algorithms having varying key lengths. The comparative analysis of the encryption process time and throughput has been analysed to evaluate the strength and weaknesses of each encryption algorithm in our IoT application. The experimental results have shown that DES and 3DES algorithm has the minimum time requirement for encryption whereas Blowfish has the highest throughput.

Keywords IoT · Sensor · Encryption · Decryption · Lightweight cryptography · Raspberry Pi · Throughput

---

R. R. Kureshi (✉) · B. K. Mishra
Faculty of Engineering and Informatics, University of Bradford, Bradford, UK
e-mail: krameezr@bradford.ac.uk

B. K. Mishra
e-mail: b.mishra@bradford.ac.uk

K. Dahal et al. (eds.), *Internet of Things and Its Applications*, Lecture Notes in Electrical Engineering 825, https://doi.org/10.1007/978-981-16-7637-6_40

1 Introduction

IoT applications are being used in all aspect of day-to-day life such as health, safety, transportation, agriculture, industry to name a few. The IoT has been widely used technology which implies Internet communication protocols for communication. Cost in terms of memory, time, and energy requirements is among the key aspects while using the IoT devices in any applications. However, security is a major concern of the IoT applications since any unauthorized access to the application data and confidential information is in threats [1]. Protected communication is required to achieve higher security in IoT application which measures whether the communication is secure or not in terms of data confidentiality and authentication [2].

In general, any IoT devices have mainly four major functional units: (i). sensing the environment, (ii). process the data coming from the environment, iii. communicate the data over the network, and iv. apply security and privacy measures. Among these four, the security and privacy measures have been challenging as it requires careful concern on the issues such as confidentiality, integrity, and authentication (CIA) of data been used in the IoT applications [3]. Also, the IoT is the modification of the traditional system which uses Internet, sensors, and mobile host networks to communicate with each other (wireless sensor network—WSN). With the modification of the system, the new security and privacy safeguarding challenges have emerged. To do so, cryptography is being used to secure the data while transmitting over the network. Different algorithms are used where keys have been applied for encryption and decryption [4]. The encryption algorithms perform either using bit by bit on plain text or the whole block of plain text [5]. Encryption of data has been one of the approaches to mitigate any attack or threats in the data security of IoT.

There has been a trade-off among performance measures while improving the security in the IoT system. The trade-off provides a sense of the efficiency of any alternatives being used to improve the IoT security [6]. In most of the IoT applications, the time measures and security aspects are usually been addressed lightly. In this paper, we have compared different encryption algorithms in two different aspects: (i). encryption process time requirement and (ii). throughput of different encryption algorithms with varying key length. The main objective of the work is to explore the efficient encryption algorithm to apply in IoT system. To validate this, the comparison analysis has been applied in terms of the time required by different encryption algorithms. The experiment shows how different encryption algorithms took different time ranges with a different key length and hence affecting the system performance. The comparative analysis has been performed among AES, DES, 3DES, and Blowfish to evaluate the encryption time and throughput. The result showed that AES has the highest time requirement for encryption process whereas the DES and 3DES have a nearly similar but lower time requirement. The Blowfish algorithm has comparatively highest throughput among these algorithms.

In the rest of the paper, Sect. 2 represents a review of related works. Section 3 decribes encryption algorithms. Section 4 shows the experimental setup of the IoT application. This section also describes the results obtained and the comparative analysis of different algorithms. In Sect. 5 conclusion and future work has been presented.

2 Related Work

Literature has highlighted the advancement of research actions in IoT frameworks. It is a glimpse of a general survey on security issues and treated in a generic manner [7]. Some of the research has already been done in various aspects of IoT data transmissions in different categories such as authentication, identification, data integrity, privacy and access control. Also, some of the challenges and future direction are given in these works of literature [7, 8]. This literature argued that data security and privacy measures must be applied in any IoT application.

Over the years, there have been events where the IoT devices have been targeted by the attackers. The security of IoT devices has been broken and it will remain a threat in future as the number of IoT devices is increasing and also there have been heterogeneous protocols for IoT devices [9]. The IoT network architecture is different than traditional network architecture [10]. The IoT network is characterized by a weak wireless signal with the low power supply, limited computing, and storage capacity which has a direct impact on safe data transmission. Also, the network topology used for IoT devices is dynamic. The IoT nodes are often open which makes it susceptible to any potential attack [11]. Besides, more often, the firmware used in IoT devices is not regularly updated which also contribute to the vulnerability of IoT devices. Considering these, encryption plays a vital role in security measures and privacy protection. However, there have been challenges which enhance the scope of more work in the area of IoT security. The traditional encryption algorithms implementation over the IoT devices has been challenging as those traditional algorithms often work with large key lengths and consume more resources. But, the IoT devices have limited memory and work with low power supply. Despite this, there is some lightweight encryption techniques are available for IoT devices [12]. The lightweight encryption algorithms enhance security measures.

In literature, comparative study of AES has been applied in a low response embedded system as a feasible encryption algorithm in the IoT platform [13]. This study showed that the AES algorithm required several encryptions and decryption operations, and required more resources in terms of energy and memory. DES encryption algorithm has been used with FPGA with the minimization as an objective in the execution time and maximization in throughput [14]. However, at the higher frequency, the FPGA temperature has been increased which caused the system failure. 3DES algorithm was applied in a system where the software control mechanism was designed for data acquisition and control switching but the system had consumed a large number of resources [15]. Blowfish is another type of encryption algorithm applied in a different type of IoT applications, for example, biometrics authentication systems [16], cloud security [17], and iris-based authentication [18]. This work has proven that the Blowfish algorithm has improved accuracy and performance. In general, the literature has shown that there is a need for the security mechanism to protect digital data. There are different encryption algorithms that have been implemented in several IoT applications with its strength.

3 Encryption Algorithms

To enhance the security in IoT applications, different approaches such as privacy control and authentication have been applied. But, these approaches always have threats with the attack. To increase the security perspective, encryption and decryption algorithms have met the security challenge in a large aspect of IoT applications. Some of the commonly implemented encryption algorithms are AES, DES, 3DES, and Blowfish. AES is a round-based block cipher algorithm which supports the key sizes 128 bits, 192 bits, and 256 bits. These keys are rounded with 10 rounds, 12 rounds, and 14 rounds, respectively. The AES encryption has been standardized under the FIP-197 in 2001 by the National Institute of Standard and Technology (NIST) and added in ISO/IEC 18033–3 [19]. In any communication application, this algorithm is applied in the background where a block cipher is executed with modes such as Electronic Codebook (ECB), Cipher Feedback (CFB), and Counter (CTR). However, the literature showed that CFB is better in execution as it requires only one-side encryption/decryption [20]. DES has been another encryption standard to be recommended by NIST. The DES use a symmetric-key block cipher for the encryption based on Lucifer algorithm proposed by IBM [21]. 3DES is a modification of DES which is also based on symmetric-key block cipher but this encryption technique applies the DES three times on each data block. Because of these time implementations of the encryption, 3DES increase the security level [21]. Blowfish encryption applies a symmetric block cipher with variable key lengths from 32 to 448 bits.

4 System Design and Experimentation

This section includes the hardware design with a brief description of the major hardware components and software been used. Experimental results of all the techniques.

4.1 Software/Hardware Tools

1. **Raspberry Pi 3B+**

Raspberry Pi 3B+ is a small electronic device with the power of a quad-core processor running at 1.4 GHz and support dual-band wireless LAN with 2.4 GHz, faster Ethernet, and Bluetooth 4.2/BLE communication.

2. **Raspbian OS**

Raspbian is Debian GNU/Linux-based free operating system to support a Raspberry Pi hardware connectivity, communication and logical processing. Raspbian OS comes with a large number of packages (over 35,000) and proper formatted pre-compiled software that helps in easy installation of the system on a Raspberry Pi.

3. **DHT22 Sensor**

In our experiment, the DHT22 sensor is measuring humidity and temperature. The sensor generates digital signals which are directly accessed by Raspberry Pi.

4.2 Experimental Setup

For electronic components connections are shown in Fig. 1 where the Pin 1 of DHT22 is connected to a 3.3 V source, Pin 4 to ground (GND), and Pin 7 to the General-Purpose Input–Output (GPIO) pin on the Raspberry Pi. Also, the circuit shows that a 10 kΩ resistor is used between Pin 1 and Pin 7.

For the network communication, another Raspberry Pi has been used which acts as a server/access point in our experiment as shown in Fig. 2. For the server process functioning, Apache and PhpMyAdmin have been installed on the Raspberry Pi.

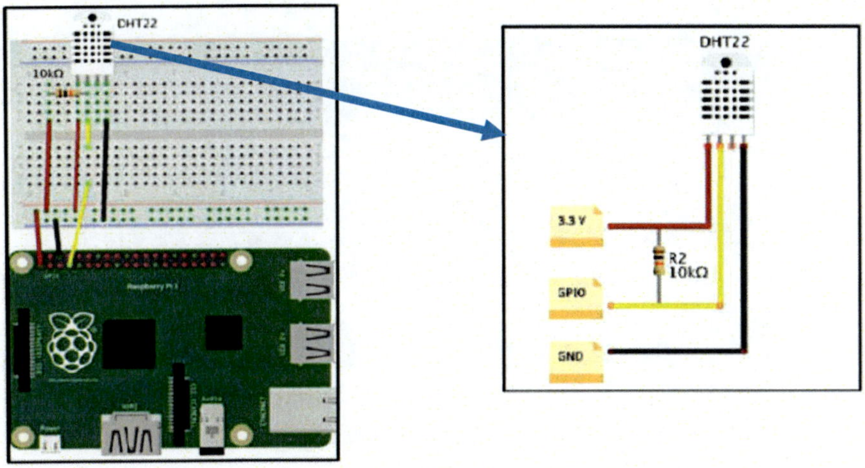

Fig. 1 Circuit connection diagram

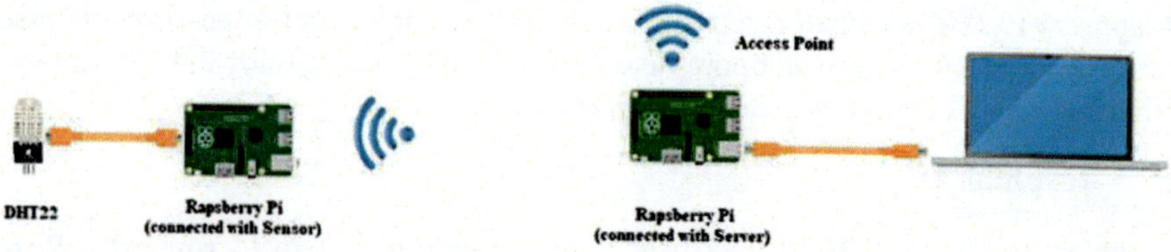

Fig. 2 Network communication diagram

Besides this, the Raspberry Pi worked as wireless access point (WAP) in the network for the small range communication. Python script has been used for at the server side to handle multiple TCP/IP connections.

4.3 *Experimental Results*

Experiments have been performed for different encryption algorithms. First execution has been ignored for all the encryption algorithms to avoid any potential setup error or initial bias. For the AES algorithm, the experiment has been performed with the block size of 320-bits (40-bytes) data and with three different key lengths of 128 bits, 192 bits, and 256 bits. Processing time for each key lengths has been evaluated which included key generation time, encryption time at sensor side and decryption time at server side on Raspberry Pi. Average time has been calculated by executing the experiments 10 times as shown in Fig. 3. From Fig. 3 of the AES execution time plot, it is observed that the key generation required nearly 6 times more than the encryption time. In the whole sequence of secure communication, the decryption time has the minimum execution time.

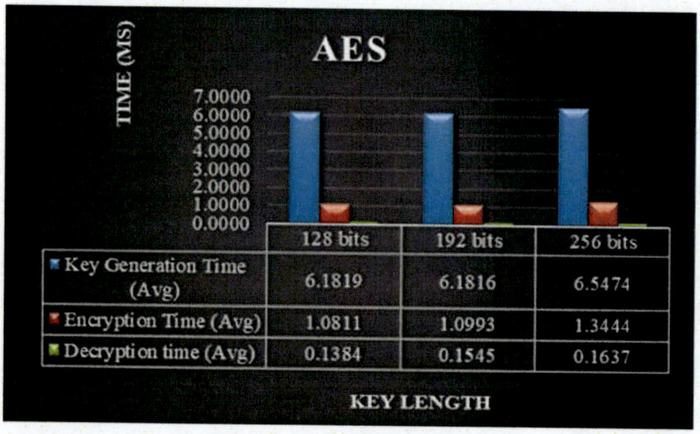

Fig. 3 AES average execution time plot for 10 experiments

For the DES, we have implemented the same as AES approach. Besides, we have also measured DES performance with the key length of 64-bit as shown in Fig. 4. From Fig. 4, it is observed that the key generation time is higher in comparison with encryption time and decryption time but there has not been a large variation as in AES.

In the case of 3DES, the experimental results of 128-bits and 192-bits key length are presented in Fig. 5 with the same block data size as of AES and DES algorithms. Here, among the three execution times, the key generation time is higher. Overall, the process of complete encryption is faster than AES.

For the Blowfish algorithm, a similar experiment has been performed with of 320-bits (40-bytes) the block size data and 128 bits, 192 bits, and 256 bits key length as shown in Fig. 6. From Fig. 6, we can observe that key generation time is much higher when 256 key length has been used. The Blowfish has a similar time requirement

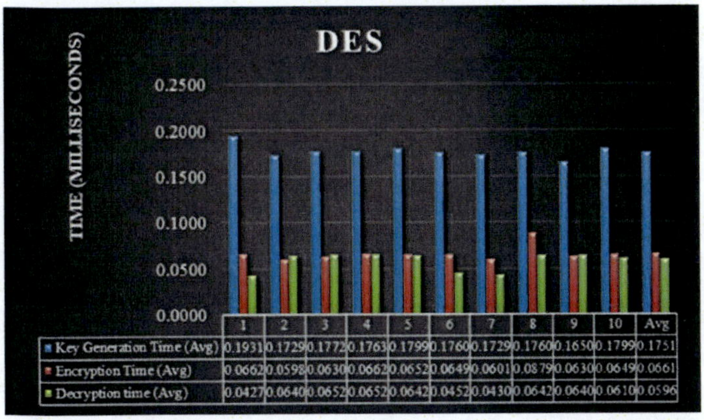

Fig. 4 DES average execution time plot for 10 experiments

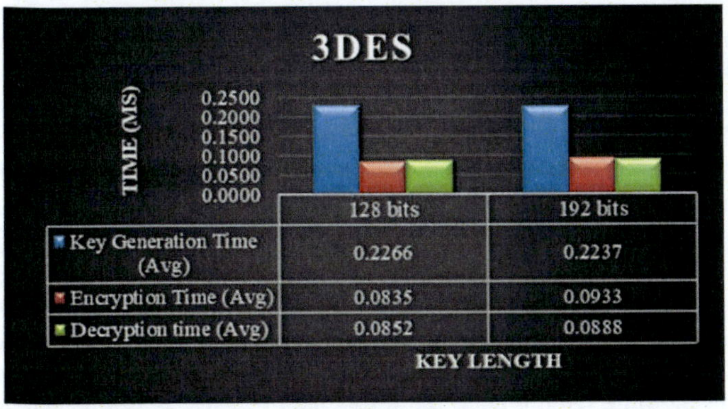

Fig. 5 3DES average execution time plot for 10 experiments

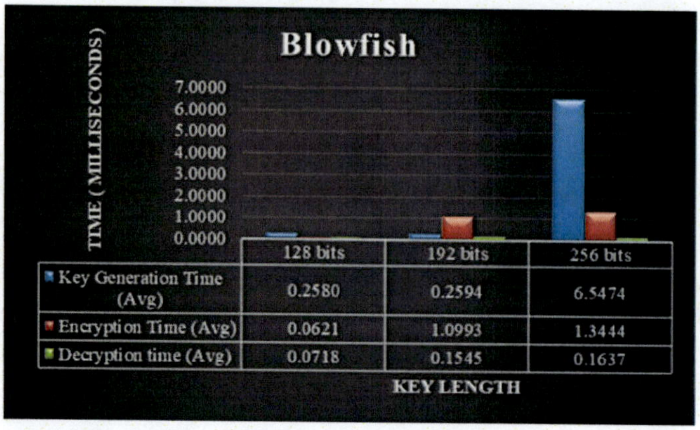

Fig. 6 Blowfish average execution time plot for 10 experiments

as of AES for key generation; however, for other key lengths it showed lower time requirement. From Fig. 6, it can be observed that our approach has been faster than Blowfish when 256 key lengths have been used and faster than AES for all key lengths.

4.4 Comparative Analysis

For the comparative analysis between the encryption algorithms, we have used "Key Generation Time," "Encryption Time," and "Decryption Time" as the performance measures as shown in Figs. 3, 4, 5 and 6. Average time requirements after 10 times of execution for each algorithm show that the AES algorithm has the maximum time requirements whereas the DES and 3DES have lower time requirements for the overall encryption process. The Blowfish algorithm has lower time requirement when lower-key lengths (128 bits, 192 bits) have been used, but when 256 bits key length has been used, the time requirement has been observed very high. In our experiment, the Blowfish algorithm is not optimum when 256 key length has been applied.

The DES algorithm showed that the total time required for the complete encryption process is lower than AES and Blowfish. The time requirement for 3DES is as comparable as DES. In all these algorithms, when the individual time requirement is analyzed, it has been observed that the "Key Generation Time" has been the highest time requirement as compared with "Encryption Time" and "Decryption Time." Among the three, the "Decryption Time" requirement has been the lowest. The experiment also highlighted that there has been an increasing trend in the time required for the whole encryption process when the key length has been increased. Apart from the time requirement, the throughput of each algorithm has also been analyzed. The throughput has been calculated using the relation: Throughput =

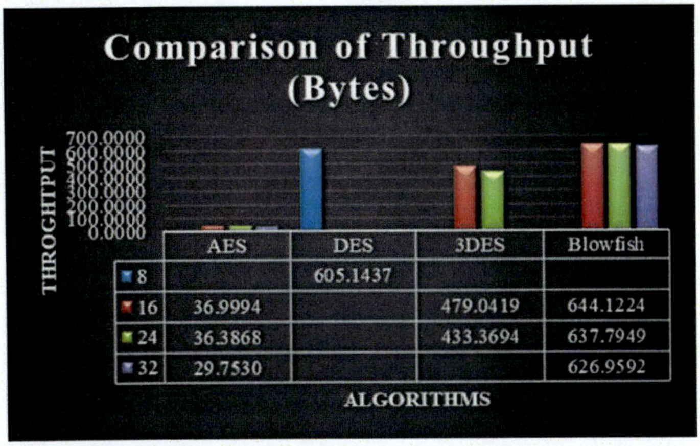

Fig. 7 Comparison plot of different encryption algorithms for throughput

(Total Plaintext Bits/Encryption Time). For the calculation, the ciphertext has been generated from the plain text using the encryption algorithms. Throughput of different algorithms has been plotted as shown in Fig. 7. It has been observed that the AES has the lowest throughput whereas the Blowfish algorithm has the highest throughput.

5 Conclusion and Future Work

The IoT system still requires a mechanism to protect data from any potential threats. Different encryption algorithms have been applied to improve security measures. However, choosing the right algorithms is always challenging as there has been a trade-off between different system performance measures. So there has been essentially a need for balancing between system performance measures. In this study, a comparative analysis with the system performance measures: encryption process time requirement and throughput have been analyzed among AES, DES, 3DES, and Blowfish encryption algorithm. The analysis has shown that the trade-off among these algorithms where in some aspects DES is better whereas in some aspects Blowfish is a better alternative. In future work, we try to do a more critical analysis of these algorithms in terms of power, memory usage with different data size and key lengths.

References

1. Punia A, Gupta D, Jaiswal S (2017) A perspective on available security techniques in IoT. In: 2017 2nd IEEE international conference on recent trends in electronics, information & communication technology (RTEICT), IEEE, pp 1553–1559

2. Goyal P, Batra S, Singh A (2010) A literature review of security attack in mobile ad-hoc networks. Int J Comput Appl 9:11–15
3. Suo H, Wan J, Zou C, Liu J (2012) Security in the internet of things: a review. pp 648–651
4. Surendran S, Nassef A, Beheshti BD (2018) A survey of cryptographic algorithms for IoT devices. In: 2018 IEEE Long Island systems, applications and technology conference (LISAT), pp 1–8
5. Meneghello F, Calore M, Zucchetto D, Polese M, Zanella A (2019) IoT: internet of threats? A survey of practical security vulnerabilities in real IoT devices. IEEE Internet Things J 6:8182–8201
6. Suárez-Albela M, Fernández-Caramés TM, Fraga-Lamas P, Castedo L (2018) A practical performance comparison of ECC and RSA for resource-constrained IoT devices. In: 2018 global internet of things summit (GIoTS), IEEE, pp 1–6
7. Sfar AR, Natalizio E, Challal Y, Chtourou Z (2018) A roadmap for security challenges in the internet of things. Digital Commun Netw 4:118–137
8. Mendez Mena D, Papapanagiotou I, Yang B (2018) Internet of things: survey on security. Inform Secur J: Global Perspect 27:162–182
9. Mohamad Noor Mb, Hassan WH (2019) Current research on internet of things (IoT) security: a survey. Comput Netw 148:283–294
10. Wen Q, Dong X, Zhang R (2012) Application of dynamic variable cipher security certificate in internet of things. In: 2012 IEEE 2nd international conference on cloud computing and intelligence systems, IEEE, pp. 1062–1066
11. Thangavel R, Palanisamy B (2011) Efficient approach towards an agent-based dynamic web service discovery framework with QoS support. In: International symposium on computing, communication, and control (ISCCC), Citeseer, pp 74–78
12. Singh S, Sharma PK, Moon SY, Park JH (2017) Advanced lightweight encryption algorithms for IoT devices: survey, challenges and solutions. J Ambient Intell Humanized Comput 1–18
13. Maitra S, Richards D, Abdelgawad A, Yelamarthi K (2019) Performance evaluation of IoT encryption algorithms: memory, timing, and energy. In: 2019 IEEE sensors applications symposium (SAS), pp 1–6
14. Zeebaree SR (2020) DES encryption and decryption algorithm implementation based on FPGA. Indones J Electr Eng Comput Sci 18:774–781
15. Wang Z, Yao Y, Tong X, Luo Q, Chen X (2019) Dynamically reconfigurable encryption and decryption system design for the internet of things information security. Sensors 19:143
16. Joshy A, Jalaja M (2017) Design and implementation of an IoT based secure biometric authentication system. In: 2017 IEEE international conference on signal processing, informatics, communication and energy systems (SPICES), IEEE, pp 1–13
17. Hussaini S (2020) Cyber security in cloud using blowfish encryption. Int J Inform Technol (IJIT) 6
18. Mohammed AF, Qyser AAM (2019) A hybrid approach for secure iris-based authentication in IoT. In: International conference on intelligent computing and communication technologies, Springer, pp 159–167
19. Bui D-H, Puschini D, Bacles-Min S, Beigné E, Tran X-T (2016) Ultra low-power and low-energy 32-bit datapath AES architecture for IoT applications. In: 2016 International conference on IC design and technology (ICICDT), IEEE, pp 1–4
20. Fahd S, Afzal M, Abbas H, Iqbal W, Waheed S (2018) Correlation power analysis of modes of encryption in AES and its countermeasures. Futur Gener Comput Syst 83:496–509
21. Srinivasarao D (2011) Analyzing the superlative symmetric cryptographic encryption algorithm. J Global Res Comput Sci 2:101–105

Secure Outsourcing of Image Editing Based on Homomorphic Encryption

Aniket Das, Somdatta Mukherjee, Akarsh Srivastava, Kanishka Gupta, and Sujoy Datta

Abstract As cloud computing has become more popular and easier to access, many online applications now provide image storage and editing services to their users. However, an image may contain personal information which the user does not want disclosed. In this paper, we focus on secure outsourcing of image storage and editing to an untrusted cloud environment using homomorphic encryption. We use Paillier's Homomorphic Cryptosystem to encrypt pixels of an image and then convert an existing image editing method into one that can operate over the encrypted pixels. The homomorphic property is showcased by implementing a brightness transform function that can operate over the encrypted image without revealing the original image data.

Keywords Homomorphic encryption · Paillier's cryptosystem · Image editing · Cloud computing

1 Introduction

Cloud computing service delivers computing support to resource constrained users to outsource large-scale computational tasks to the cloud with massive computational power. It reduces the overhead by delivering the right IT resources such as computing power, storage, bandwidth, etc. This technology offers several advantages like storage capacity, reliability, agility, scalability and dynamic data availability. Therefore, it is increasingly fast, and it is inevitably integrated into everyday life. However, outsourcing services to the third party public cloud restrains the naive consumers of cloud services and denies them direct control over the virtual servers that captures and generates their data throughout the computation, which renders reduced visibility and control thus imposing a risk on the security front of the end user's personal data.

A. Das (✉) · S. Mukherjee · A. Srivastava · K. Gupta · S. Datta
KIIT Deemed to be University, Bhubaneswar, India
e-mail: daniket@protonmail.com

S. Datta
e-mail: sdattafcs@kiit.ac.in

K. Dahal et al. (eds.), *Internet of Things and Its Applications*, Lecture Notes in Electrical Engineering 825, https://doi.org/10.1007/978-981-16-7637-6_41

With the emergence of social media platforms, digital images that were considered personal data for users are widespread on the Internet. Consequently, several image editing apps have surfaced online to cater to the users' needs. Firstly, there are two types of photo editing applications: Device-based application and server-based/ cloud-based application. The cloud-based photo applications require users to upload their photos, which is then sent to the server/cloud computers, which sends back the resultant image after editing. When these photo editing applications are moved to the cloud, the traditional controls such as firewalls, proxies are no longer effective enough for protection, as these applications are running on untrusted network.

A homomorphic encryption(HE) scheme allows construction of programs which feeds encrypted data to the system and obtains an encrypted result. Since the computations are performed on the encrypted data, the raw data is never revealed thus ensuring the confidentiality of processed data. The ciphertext retains the same structure as the plaintext, thus equivalent results are obtained. Likewise if a user encrypts the pixels of an image using an HE scheme, then the services can yield results by operating on the encrypted pixels. In view of this, the encrypted image can be edited without decryption, ensuring privacy from curious application servers [1]. Fully homomorphic cryptosystems supports each and every addition and multiplication functions thus conserving the ring structure [2]. This means we can consider a ring (R,+,), where R represents the bits on which one can perform (R,+) and (R,)operations. With both addition and multiplication over bits, NAND gates can be created, using which every other Boolean gate can be derived thus making computation on the encrypted data feasible. Gentry's [3] homomorphic scheme shows us that full homomorphism is possible.

Partially homomorphic encryption schemes are much simpler and offer only one kind of computation. They are practical, offer an improved performance over FHE schemes and find applications in private information retrieval, electronic voting, multiparty computation, collision-resistant hash functions, secure computing, etc. on the cloud. There are several partially homomorphic cryptosystems like Goldwasser and Micali [4], ElGamal [5], Melchor [6] and Paillier [7]. General-purpose homomorphic encryption schemes like Gentry's FHE [3] are highly complex. Instead of using modular arithmetic like most other cryptosystems, Gentry's framework is based on Euclidean lattices. The cryptosystem presented by Gentry requires a very large ciphertext compared to the corresponding plaintext. The fact that it relies on a complex mesh of lattices also makes the implementation very had and operations run very slowly on the ciphertext [2]. Yao proposed a solution with his garbled circuits approach [8] which was documented and proved to be possible by Goldreich [9, 10]. Applying circuit level encryption seems even more complicated in practice when constructing encrypted circuits for all algorithms. We observe that instead of constructing generic homomorphic schemes, finding solutions to specialised cases has provided us with more efficient solutions to those problems [11–13]. In this paper, we explore homomorphic computation for outsourcing of image storage and encryption.

1.1 Problem Statement

We consider a cloud architecture which involves interaction between two objects: the cloud user, who has an image to upload to the cloud and the cloud server, which has significant computational resources to store images and also provides image editing services. We consider a brightness adjustment function to be translated to the encrypted domain. Implementing a brightness adjustment function proves that any image editing operation requiring an addition operation can be performed over an image encrypted using our scheme.

The cloud user has an image I of width n and height m which needs to be stored on the cloud and also needs to have a brightness transform applied on it. I can either be grayscale or RGB.

Considering the cloud server wants to store and process the image, the security threats faced by the user come from two main sources

(i) the malicious behaviour of the cloud server which operates under the semi-honest model [14] and,
(ii) data access by unauthorised users.

The straightforward solution to protect the confidentiality of images is to encrypt the images using traditional image encryption schemes based on chaotic maps [15, 16] before outsourcing. The problem we face when using traditional image encryptions is that no operation can be done on the encrypted data without first decrypting it. Furthermore, the cloud or other unauthorised users can derive sensitive information through methods described by Islam et al. [17] and Williams et al. [18] or even through frequency analysis-based attacks [19].

1.1.1 Design Goals

To construct a solution for the aforementioned model, we need to satisfy the following security and functional requirements.

- Preserve the confidentiality of I
- Be robust against frequency analysis attacks
- Be editable while encrypted.

1.2 Our Proposed Solution

We propose an image encryption scheme based on Paillier's homomorphic encryption. A homomorphic cryptosystem like Pailler's enables secure computation in untrusted environments like a public cloud server. The idea of homomorphism in computation is to perform operations on an encrypted ciphertext and obtaining the same result as applying the operation on the plaintext.

We consider that the user will encrypt the image I before outsourcing. We will be referring to this encrypted image object as I_ε.

In our implementation, we extend Paillier's homomorphic cryptosystem [7] to operate over images. We design our algorithms to operate on the image per pixel, as Paillier's scheme supports encryption of integers $\in Z^*_{N^2}$ where N is the product of two large primes, this makes image pixels the perfect candidates for encryption as each pixel colour is an unsigned integer of 8 bits which can have a maximum value of 255. We can achieve significant security without using very large keys resulting in faster encryption and decryption times.

Our implementation of image cryptography consists of these three algorithms which have been summarised below and will be formally defined later in Sect. 3.

- $\mathbf{ImgEncrypt(pk, I)} \rightarrow \mathbf{I_\varepsilon}$ This algorithm encrypts each pixel of the image I with public key pk and returns the encrypted image I_ε.
- $\mathbf{secureBrightness(pk, I_\varepsilon, \delta)} \rightarrow \mathbf{I'_\varepsilon}$ This algorithm adjusts the brightness of I_ε by using homomorphic addition to add a factor δ to each pixel, and returns the encryption of brightness adjusted I as I'_ε.
- $\mathbf{ImgDecrypt(pk, sk, I_\varepsilon)} \rightarrow \mathbf{I}$ This algorithm decrypts each pixel value in I_ε and then rounds the values to it's nearest integer $\in [0, 255]$ in case the brightness function returns value outside of an 8-bit integer.

We use key generation function from Paillier's cryptosystem for generating the keys, which will be defined in Sect. 2

2 Background and Prerequisites

In this section, we present an overview of the methods we are depending on for our security model in this paper.

2.1 *Scientific Computing Using Python and SciPy*

We will be using the Python programming language [20] for implementation of our algorithms, and the NumPy [21] package from the SciPy [22] library. SciPy is an open-source Python library for scientific and engineering computation. The NumPy package offers fast and powerful N-dimensional arrays and various mathematical tools like random number generators, function vectorisation, *etc.*. This package will be used to load an image into an array for encryption and processing. Let A be an N-dimensional NumPy array. The following array attributes and functions will be relevant to our algorithms.

(i) **A.shape** The shape attribute of an array object returns a tuple of array dimensions. This property is used to get the current array shape

(ii) **A.flatten()** This function returns a copy of the array collapsed into one dimension
(iii) **A.reshape(shape)** The reshape function takes a tuple *shape* as the parameter and returns an array containing the same data as A with a new shape.

2.2 *The Paillier Cryptosystem*

The Paillier Cryptosystem [7] is additive homomorphic as as it only supports addition of encrypted integers or multiplication of an encrypted integer by an unencrypted multiplier by repeated addition. We consider E_{pk} to be the encryption function with public key pk and D_{sk} to be the decryption function with secret key sk. We will be referring to the encryption of integer z as z_ε in further sections of the paper *i.e.* $z_\varepsilon \leftarrow E_{pk}(z)$.

The properties of the Paillier Cryptosystem which will be relevant to our scheme are the following:

(i) **Homomorphic Addition** The homomorphic addition can add two ciphertexts and return the encryption of the sum of plaintexts

$$E_{pk}(a + b) \leftarrow E_{pk}(a) * E_{pk}(b) \bmod N^2 \tag{1}$$

We denote this operation as '⊕' in further sections.
(ii) **Homomorphic Multiplication** The homomorphic multiplication algorithm can multiply a ciphertext with a plaintext.

$$E_{pk}(a * b) \leftarrow E_{pk}(a)^b \bmod N^2 \tag{2}$$

We will denote this operation as '⊗' in further sections.
(iii) **Semantic Security** [23] The encryption scheme uses a random factor in the encryption algorithm thus making sure given a set of ciphertexts, no information can be derived about the plaintext.

3 Homomorphic Image Editing Scheme

In this section, we present our image storage and editing outsourcing scheme which will preserve the privacy of images. The architecture is illustrated in Fig. 1. We want an algorithm to work over grayscale as well as RGB images, for this we design algorithms that are shape agnostic.

As we mentioned earlier in Sect. 1.1 the image I is k pixels wide and m pixels tall, for grayscale images, it can be represented in an array of two dimensions $\langle n, m \rangle$. For RGB images however, each pixel contains red, green and blue values to represent

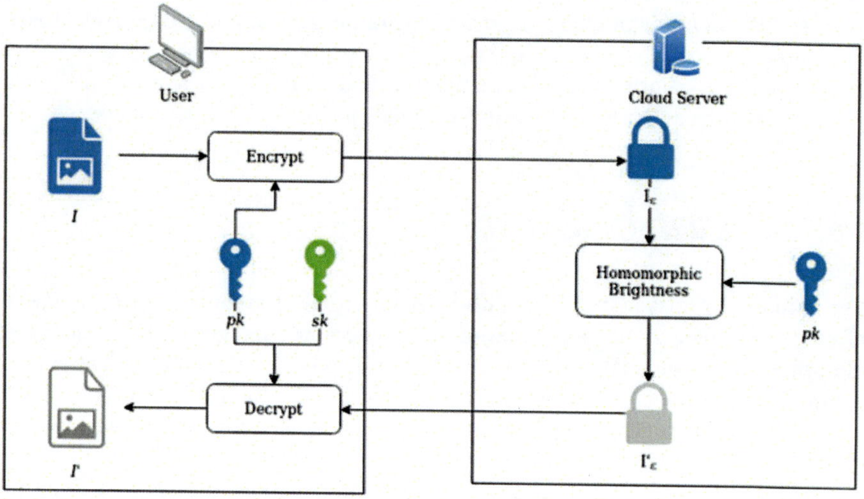

Fig. 1 Homomorphic Image Editing Scheme

which we need an array of three dimensions $\langle n, m, 3\rangle$. For an algorithm to work on both of these, we convert the image into a one dimensional array before processing and then reshape the one dimensional array to the original image shape. For array reshaping operations, we will be using functions from NumPy as mentioned in Sect. 2. We now discuss our proposed solution in detail.

3.1 Image Encryption

For encrypting our image I, we extend Paillier's encryption function E_{pk} to work over each pixel. In the case of RGB images, we need to encrypt each red, green and blue values of each pixel. Since the same operation needs to be applied over each value, the shape of the array does not matter and we can flatten I to one dimension before running encryption operations. The property of semantic security comes into

Algorithm 1 ImgEncrypt(pk, I)$\rightarrow I_\varepsilon$

$\sigma \leftarrow I.shape$
$I \leftarrow I.flatten()$
$I_\varepsilon \leftarrow []$ ▷ [] represents an empty array
for all $i \in I$ **do**
  $I_\varepsilon \stackrel{+}{\leftarrow} E_{pk}(i)$ ▷ $\stackrel{+}{\leftarrow}$ represents append operation
end for
$I_\varepsilon \leftarrow I_\varepsilon.reshape(\sigma)$
return I_ε

play here. As each plaintext can be encrypted to an integer $\in Z_{N^2}$ any information that could be derived from the pattern of pixels is also lost. We will verify this property later in Sect. 3.3.

3.2 Brightness Adjustment Operation over Encrypted Image

This is the algorithm which will run on the server's side. It applies a simple brightness transform over the encrypted Image I_ε, it takes a factor δ as input for the brightness factor, δ takes integer values, positive for increasing the brightness and negative for decrease. This algorithm is truly homomorphic, it does not need to decrypt the array for manipulating the brightness. We achieve this by applying homomorphic addition operation $\oplus$ for every pixel.

3.3 Image Decryption

Decryption is similar to algorithm 1, we take each value for the encrypted array and apply Paillier's decryption function over it. In some instances, the encrypted image I_ε may contain values outside the range of an 8-bit unsigned integer as the

Algorithm 2 secureBrightness(pk, I_ε, δ)$\rightarrow$ I'_ε

$\sigma \leftarrow I_\varepsilon.shape$
$I_\varepsilon \leftarrow I_\varepsilon.flatten()$
$I'_\varepsilon \leftarrow []$
for all $i_\varepsilon \in I_\varepsilon$ **do**
  $I'_\varepsilon \overset{+}{\leftarrow} i_\varepsilon \oplus E_{pk}(\delta)$
end for
$I'_\varepsilon \leftarrow I'_\varepsilon.reshape(\sigma)$
return I'_ε

Algorithm 3 ImgDecrypt(pk, sk, I_ε)$\rightarrow$ I

$\sigma \leftarrow I_\varepsilon.shape$
$I_\varepsilon \leftarrow I_\varepsilon.flatten()$
$I \leftarrow []$
for all $i_\varepsilon \in I_\varepsilon$ **do**
  $i \leftarrow D_{sk}(i_\varepsilon)$
  if $i < 0$ **then** $i \leftarrow 0$
  else if $i > 255$ **then** $i \leftarrow 255$
  end if
  $I \overset{+}{\leftarrow} i$
end for
$I \leftarrow I.reshape(\sigma)$
return I

secureBrightness algorithm does not check for these conditions and just applies an addition operation. To rectify this, we set the closest value of the integer inside the range of an image pixel, *i.e.* an unsigned integer of 8 bits *or* [0, 255].

4 Experimental Results

We will test our implementation of image cryptography on the basis of time taken to encrypt, decrypt and perform a brightness adjustment transform on the encrypted image. Size of the image, type of image (RGB or grayscale) and key length will be variables for our test cases. Theoretically, the time complexity of each algorithm depends only on the shape and type of the image. For any image I of n width and m height, the time complexity would be $n \times m$ for a grayscale image and $3 \times n \times m$ for an RGB image. This is visible in Table 1 for larger image sizes.

All of these tests are being done on a personal computer with the following specifications:

(i) CPU: Intel i7-8750H
(ii) RAM: 16GB
(iii) Python version 3.6.9
(iv) OS: Linux Mint 19.3
(v) Kernel: Linux 5.3.

The algorithm we implemented does not use any parallelism and only runs on one thread. The results in Table 1 are the average of multiple tests done on the machine;

Table 1 Performance analysis over varying images and key lengths

Variables			Time (in seconds)		
Key length (in bits)	Image dimensions	Image type (RGB or grayscale)	Encryption	Decryption	Brightness adjustment
64	128 × 128	Grayscale	1.416	0.862	0.064
64	128 × 128	RGB	1.419	0.884	0.065
64	256 × 256	Grayscale	5.461	3.304	0.262
64	256 × 256	RGB	5.579	3.630	0.269
64	512 × 512	Grayscale	7.455	4.884	0.349
64	512 × 512	RGB	22.6	14.505	1.04
128	128 × 128	Grayscale	4.295	3.273	0.078
128	128 × 128	RGB	4.435	3.308	0.078
128	256 × 256	Grayscale	17.634	13.293	0.321
128	256 × 256	RGB	17.306	13.194	0.321
128	512 × 512	Grayscale	23.087	17.702	0.417
128	512 × 512	RGB	69.59	52.730	1.29

Fig. 2 Original images

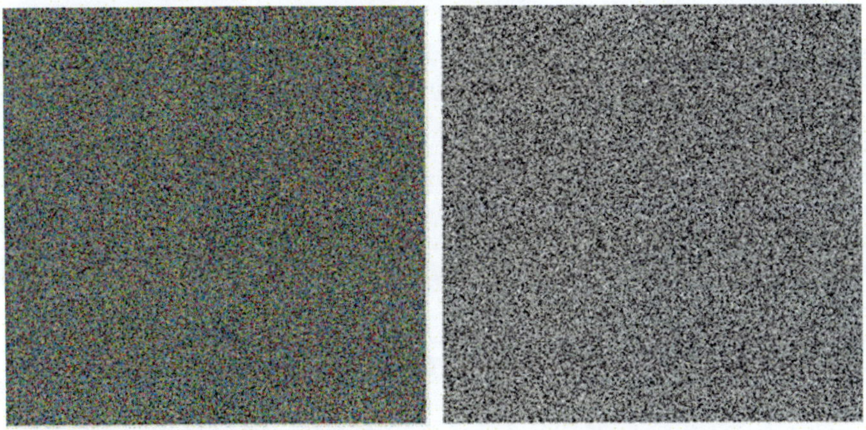

(a) RGB (b) Grayscale

Fig. 3 Visualisation of encrypted objects

each time measurement was done with Python's built-in time module with precision of 15 decimal places, we have averaged and rounded off the readings to 3 decimal places.

For the experiments, we have used the images in Fig. 2 the original size of these images is 512×512, we have resized them to 128×128 and 256×256 to analyse performance on smaller images as well (Fig. 3).

Both of these images were encrypted and then had the *secureBrightness* algorithm run on them with a brightness factor δ of 30. The encrypted objects are then decrypted to get the resulting images in Fig. 4. Looking at the images, we can observe

Fig. 4 Images after brightness adjustment

that the image is visually brighter and therefore conclude that our algorithm works for adjusting the image brightness on both RGB and grayscale images.

For testing the semantic security of our algorithm, we visualise the encrypted object by scaling each encrypted pixel value down to the range of $[0, 255]$ which can then be visualised as it can be represented as an unsigned integer of 8 bits. We construct an image from these values such as the ones we see in Fig. 3. From Fig. 3, we can see that the distribution of values is randomised and not distinguishable from noise; therefore, we conclude that no information can be derived about the plaintext and our scheme is semantically secure.

From Table 1, we observe that the runtime of each function increases with the increase in key length and image size. The encryption and decryption algorithms introduce significant overheads when compared to not encrypting the image at all. In order protect privacy, the compromise in performance can be justified. We would argue that our scheme is suitable for smaller image sizes and can provide significant security even with smaller key lengths.

5 Conclusion

Image storage editing over cloud services has become very common in the current world with many organisations providing unlimited usage for free. Under an outsourced environment, the transactions between the user and the server is always based on trust. In this paper, we have extended the Paillier homomorphic cryptosystem for image encryption and then translated a brightness adjustment algorithm which can work over the encrypted object. This demonstrates that secure image editing is possible in a trustless environment.

Our scheme protects the confidentiality of images, is probabilistic and therefore semantically secure. Further, this proves that any image operation requiring addition, or multiplication by a constant can be translated to work over encrypted images.

This scheme can be further improved by taking advantage of GPU parallelisation for faster encryption and decryption.

References

1. Yi X, Paulet R, Bertino E (2014) Homomorphic encryption and applications, chap. In: Homomorphic encryption. Springer, pp 27–46. https://doi.org/10.1007/978-3-319-12229-8
2. Gentry C (2009) A Fully Homomorphic Encryption Scheme. Ph.D. thesis, Stanford University, Stanford
3. Gentry C (2009) Fully homomorphic encryption using ideal lattices. In: Proceedings of the 41st annual ACM symposium on theory of computing. pp 169–178. STOC '09, Association for computing machinery, New York, NY, USA. https://doi.org/10.1145/1536414.1536440
4. Goldwasser S, Micali S (1982) Probabilistic encryption and how to play mental poker keeping secret all partial information. In: Proceedings of the 14th annual ACM symposium on theory of computing. pp 365–377. STOC '82, Association for Computing Machinery, New York, NY, USA. https://doi.org/10.1145/800070.802212
5. Elgamal T (1985) A public key cryptosystem and a signature scheme based on discrete logarithms. IEEE Trans Inf Theor 31(4):469–472
6. Melchor CA, Castagnos G, Gaborit P (2008) Lattice-based homomorphic encryption of vector spaces. In: 2008 IEEE international symposium on information theory. pp 1858–1862
7. Paillier P (1999) Public-key cryptosystems based on composite degree residuosity classes. In: Stern J (ed) Advances in cryptology—EUROCRYPT '99. Springer, Berlin, pp 223–238
8. Yao ACC (1986) How to generate and exchange secrets. In: 27th annual symposium on foundations of computer science . IEEE (1986). https://doi.org/10.1109/sfcs.1986.25
9. Goldreich O, Micali S, Wigderson A (1987) How to play ANY mental game. In: Proceedings of the 19th annual ACM conference on theory of computing—STOC'87. ACM Press. https://doi.org/10.1145/28395.28420
10. Goldreich O (2003) Cryptography and cryptographic protocols. Distrib Comput 16(2-3):177–199. https://doi.org/10.1007/s00446-002-0077-1
11. Elmehdwi Y, Samanthula BK, Jiang W (2014) Secure k-nearest neighbor query over encrypted data in outsourced environments. In: 2014 IEEE 30th international conference on data engineering. pp 664–675
12. Wang C, Ren K, Wang J (2011) Secure and practical outsourcing of linear programming in cloud computing. In: 2011 proceedings IEEE INFOCOM. pp 820–828
13. Zhou L, Li C (2015) Outsourcing large-scale quadratic programming to a public cloud. IEEE Access 3:2581–2589
14. Yu S, Wang C, Ren K, Lou W (2010) Achieving secure, scalable, and fine-grained data access control in cloud computing. In: 2010 proceedings IEEE INFOCOM, pp 1–9
15. Kumar S, Pradhan C (2020) Color image encryption technique using 4d logistic map. In: Das H, Pattnaik PK, Rautaray SS, Li KC (eds) Progress in computing, analytics and networking. Springer, Singapore, pp 75–82
16. Pareek N, Patidar V, Sud K (2006) Image encryption using chaotic logistic map. Image Vision Comput 24(9):926–934
17. Islam MS, Kuzu M, Kantarcioglu M (2012) Access pattern disclosure on searchable encryption: ramification, attack and mitigation. In: Network and distributed system security symposium (NDSS)

18. Williams P, Sion R, Carbunar B (2008) Building castles out of mud: practical access pattern privacy and correctness on untrusted storage. In: Proceedings of the 15th ACM conference on computer and communications security. CCS '08, Association for Computing Machinery, New York, NY, USA, pp 139-148. https://doi.org/10.1145/1455770.1455790
19. Al-Kadit IA (1992) Origins of cryptology: the arab contributions. Cryptologia 16(2):97–126
20. van Rossum G (1995) Python tutorial. Technical report, CS-R9526, Centrum voor Wiskunde en Informatica (CWI), Amsterdam
21. van der Walt S, Colbert SC, Varoquaux G (2011) The numpy array: a structure for efficient numerical computation. Comput Sci Eng 13(2):22–30
22. Virtanen P, Gommers R, Oliphant TE, Haberland M, Reddy T, Cournapeau D, Burovski E, Peterson P, Weckesser W, Bright J, Van Der Walt SJ, Brett M, Wilson J, Jarrod Millman K, Mayorov N, Nelson ARJ, Jones E, Kern R, Larson E, Carey C, Polat İ, Feng Y, Moore EW, Vand erPlas J, Laxalde D, Perktold J, Cimrman R, Henriksen I, Quintero EA, Harris CR, Archibald AM, Ribeiro AH, Pedregosa F, van Mulbregt P, SciPy 1.0 Contributors (2020) Fundamental algorithms for scientific computing in python. Nat Methods 17:261–272. https://doi.org/10.1038/s41592-019-0686-2
23. Goldwasser S, Micali S, Rackoff C (1989) The knowledge complexity of interactive proof systems. SIAM J Comput 18(1):186–208. https://doi.org/10.1137/0218012